CIVIL ENGINEERING

FE Review Manual

Executive Director of Engineering Education: Brian S. Reitzel, PE

CIVIL ENGINEERING: FE REVIEW MANUAL

Published by Brightwood Engineering Education

2800 E. River Road

Dayton, OH 45439

1-800-420-1432

www.brightwoodengineering.com

Printed in the United States of America.

16 17 18 10 9 8 7 6 5 4 3 2 1

ISBN: 978-1-68338-012-2

CONTENTS

CHAPTER 4

Computational Tools 125

CHAPTER 5

Ethics and Professional Practice 153

CHAPTER 6

Engineering Economics 177

CHAPTER 15

Geotechnical Engineering 649

CHAPTER 16

Transportation Engineering 683

CHAPTER 17

Environmental Engineering 703

CHAPTER 18

Construction 737

CHAPTER 19

Surveying 757

PERMISSIONS

"Rules of Professional Conduct," Chapter 6, reprinted by permission of NCEES
Source: Model Rules, National Council of Examiners for Engineering and Surveying, 2007. www.ncees.org

Figure 10.7 reprinted by permission of ASME.
Source: Moody, L., F, Transactions of the ASME, Volume 66: pp. 671–684. 1944.

The figure on page 732 courtesy of DuPont.
Source: Thermodynamic Properties of HFC-134a, DuPont Company.

Chapter 13 Exhibit 2 and Figures 13.4a-b, 13.5a-b, 13.6a, 13.9, 13.10, 13.18, 13.21, 13.22 and 13.23 reprinted with permission of John Wiley & Sons, Inc.
Source: Callister, William D., Jr. Materials Science and Engineering: An Introduction, 6/e. J. Wiley & Sons. 2003.

Figures 13.4c, 13.5c, and 13.6b reprinted by permission of the estate of William G. Moffatt.
Source: Moffatt, William G. The Structure and Property of Materials, Volume 1. J. Wiley & Sons. 1964.

Tables 13.4 and 13.5 reprinted by permission of McGraw-Hill Companies.
Source: Fontana, M., Corrosion Engineering. McGraw-Hill Companies.

Figure 13.13 used by permission of ASM International.
Source: Mason, Clyde W., Introductory Physical Metallurgy: p. 33. 1947.

Figure 13.26 used by permission of ASM International.
Source: Rinebolt, J.A., and W. J. Harris, Jr., "Effect of Alloying Elements on Notch Toughness of Pearlitic Steels." Transactions of ASM, Volume 43: pp. 1175–1201. 1951

CHAPTER AUTHORS

David R. Arterburn, PhD, New Mexico Institute of Mining and Technology

Gary R. Crossman, PE, Old Dominion University

Fidelis O. Eke, PhD, University of California, Davis

Brian Flinn, PhD, PE, University of Washington

James R. Hutchinson, PhD, University of California, Davis

Ray W. James, PhD, PE, Texas A&M University College Station

Lincoln D. Jones, PE, San Jose State University

Sharad Laxpati, PhD, PE, University of Illinois, Chicago

Robert F. Michel, PE, Old Dominion University

Donald G. Newnan, PhD, PE, San Jose State University

Charles E. Smith, PhD, Oregon State University

CHAPTER 1

Introduction

OUTLINE

HOW TO USE THIS BOOK

Fundamentals of Engineering FE/EIT Exam Preparation is designed to help you prepare for the Fundamentals of Engineering/Engineer-in-Training exam. The book covers the full breadth and depth of topics covered by the new Fundamentals of Engineering exams.

Each chapter of this book covers a major topic on the exam, reviewing important terms, equations, concepts, analysis methods, and typical problems. Solved examples are provided throughout each chapter to help you apply the concepts and to model problems you may see on the exam. After reviewing the topic, you can work the end-of-chapter problems to test your understanding. The problems are typical of what you will see on the exam, and complete solutions are provided so that you can check your work and further refine your solution methodology.

The following sections provide you with additional details on the process of becoming a licensed professional engineer and on what to expect at the exam.

BECOMING A PROFESSIONAL ENGINEER

To achieve registration as a Professional Engineer, there are four distinct steps: (1) education, (2) the Fundamentals of Engineering/Engineer-in-Training (FE/EIT) exam, (3) professional experience, and (4) the professional engineer (PE) exam. These steps are described in the following sections.

Education

Generally, no college degree is required to be eligible to take the FE/EIT exam. The exact rules vary, but all states allow engineering students to take the FE/EIT exam before they graduate, usually in their senior year. Some states, in fact, have no education requirement at all. One merely need apply and pay the application fee. Perhaps the best time to take the exam is immediately following completion of related coursework. For most engineering students, this will be the end of the senior year.

Fundamentals of Engineering/ Engineer-in-Training Examination

This six-hour, multiple-choice examination is known by a variety of names— Fundamentals of Engineering, Engineer-in-Training (EIT), and Intern Engineer— but no matter what it is called, the exam is the same in all states. It is prepared and graded by the National Council of Examiners for Engineering and Surveying (NCEES).

Experience

States that allow engineering seniors to take the FE/EIT exam have no experience requirement. These same states, however, generally will allow other applicants to substitute acceptable experience for coursework. Still other states may allow a candidate to take the FE/EIT exam without any education or experience requirements.

Typically, several years of acceptable experience is required before you can take the Professional Engineer exam—the duration varies by state, and you should check with your state licensing board for details.

Professional Engineer Examination

The second national exam is called Principles and Practice of Engineering by NCEES, but many refer to it as the Professional Engineer exam or PE exam. All states, plus Guam, the District of Columbia, and Puerto Rico, use the same NCEES exam. Review materials for this exam are found in other engineering license review books.

FUNDAMENTALS OF ENGINEERING/ENGINEER-IN-TRAINING EXAMINATION

Laws have been passed that regulate the practice of engineering in order to protect the public from incompetent practitioners. Beginning in 1907 the individual states began passing *title* acts regulating who could call themselves engineers and offer services to the public. As the laws were strengthened, the practice of engineering was limited to those who were registered engineers, or to those working under the supervision of a registered engineer. Originally the laws were limited to civil engineering, but over time they have evolved so that the titles, and sometimes the practice, of most branches of engineering are included.

There is no national licensure law; licensure is based on individual state laws and is administered by boards of registration in each state. You can find a list of contact information for and links to the various state boards of registration at the Brightwood Engineering Web site: *www.brightwoodengineering.com.* This list also shows the exam registration deadline for each state.

Examination Development

Initially, the states wrote their own examinations, but beginning in 1966 NCEES took over the task for some of the states. Now the NCEES exams are used by all states. Thus it is easy for engineers who move from one state to another to achieve licensure in the new state. About 50,000 engineers take the FE/EIT exam annually. This represents about 65% of the engineers graduated in the United States each year.

The development of the FE/EIT exam is the responsibility of the NCEES Committee on Examination for Professional Engineers. The committee is composed of people from industry, consulting, and education, all of whom are subject-matter experts. The test is intended to evaluate an individual's understanding of mathematics, basic sciences, and engineering sciences obtained in an accredited bachelor degree of engineering. Every five years or so, NCEES conducts an engineering task analysis survey. People in education are surveyed periodically to ensure the FE/EIT exam specifications reflect what is being taught. This was last done in 2012-2013, and the survey results drove the format change for the 2014 exam. Previously, a general engineering portion of the exam was given in the morning and a discipline-specific in the afternoon. Now the entire exam is discipline-specific.

The exam questions are prepared by the NCEES committee members, subject matter experts, and other volunteers. All people participating must hold professional licensure. When the questions have been written, they are circulated for review in workshop meetings and by mail. You will see mostly metric units (SI) on the exam. Some problems are posed in U.S. customary units (USCS) because the topics typically are taught that way. All problems are four-way multiple choice.

Examination Structure

The FE/EIT exam will be six hours in length, which includes a tutorial, a break, the exam, and a brief survey at the conclusion of the exam. There are 110 questions total on the exam.

The exam will be divided into two sections with a 25-minute break in the middle. Examinees will be given 5 hours and 20 minutes to complete approximately 55 questions prior to the scheduled break and the remaining questions afterward.

Seven different exams are in the afternoon test booklet, including one for each of the following six branches: civil, mechanical, electrical, chemical, industrial, environmental. An Other Disciplines exam is included for those examinees not covered by the six engineering branches. If you are taking the FE/EIT as a graduation requirement, your school may compel you to take the exam that matches the engineering discipline in which you are obtaining your degree. Otherwise, you can choose the exam you wish to take.

Examination Dates

The FE is administered year-round at NCEES-approved Pearson VUE test centers. Registration will be open year-round. Those wishing to take the exam must apply to their state board several months before the exam date.

Examination Procedure

You will register and schedule your appointment through your My NCEES account on the NCEES Web site. You will first select your exam location, and then you will be presented with a list of available exam dates for your appointment. If you are not happy with the choices, you can browse through the available dates at another NCEES-approved testing center.

The examination is closed book. You may not bring any reference materials with you to the exam. To replace your own materials, NCEES has prepared a *FE Reference Handbook*. The handbook contains engineering, scientific, and mathematical formulas and tables for use in the examination. Examinees will receive the handbook from their state registration board prior to the examination. The *FE Reference Handbook* is also included in the exam materials distributed at the beginning of each exam period.

Examination-Taking Suggestions

Those familiar with the psychology of examinations have several suggestions for examinees:

1. There are really two skills that examinees can develop and sharpen. One is the skill of illustrating one's knowledge. The other is the skill of familiarization with examination structure and procedure. The first can be enhanced by a systematic review of the subject matter. The second, exam-taking skills, can be improved by practice with sample problems—that is, problems that are presented in the exam format with similar content and level of difficulty.
2. Examinees should answer every problem, even if it is necessary to guess. There is no penalty for guessing. The best approach to guessing is to try to eliminate one or two of the four alternatives. If this can be done, the chance of selecting a correct answer obviously improves from 1 in 4 to 1 in 2 or 3.
3. Plan ahead with a strategy and a time allocation. There are 120 morning problems in 12 subject areas. Compute how much time you will allow for each of the 12 subject areas. You might allocate a little less time per problem for the areas in which you are most proficient, leaving a little more time in subjects that are more difficult for you. Your time plan should include a reserve block for especially difficult problems, for checking your scoring sheet, and finally for making last-minute guesses on problems you did not work. Your strategy might also include time allotments for two passes through the exam—the first to work all problems for which answers are obvious to you, the second to return to the more complex, time-consuming problems and the ones at which you might need to guess.
4. Read all four multiple-choice answer options before making a selection. All distractors (wrong answers) are designed to be plausible. Only one option will be the best answer.
5. Do not change an answer unless you are absolutely certain you have made a mistake. Your first reaction is likely to be correct.
6. If time permits, check your work.
7. Do not sit next to a friend, a window, or other potential distraction.

License Review Books

To prepare for the FE/EIT exam you need one or two review books.

1. This book, to provide a review of the discipline-specific examination.
2. *FE Reference Handbook*. At some point this NCEES-prepared book will be provided to applicants by their state registration board. You may want to obtain a copy sooner so you will have ample time to study it before the exam. Pay close attention to the *FE Reference Handbook* and the notation used in it, because it is the only book you will have at the exam.

Textbooks

If you still have your university textbooks, they can be useful in preparing for the exam, unless they are out of date. To a great extent the books will be like old friends with familiar notation. You probably need both textbooks and license review books for efficient study and review.

Examination Day Preparations

The exam day will be a stressful and tiring one. You should take steps to eliminate the possibility of unpleasant surprises. If at all possible, visit the examination site ahead of time to determine the following:

1. How much time should you allow for travel to the exam on that day? Plan to arrive about 15 minutes early. That way you will have ample time, but not too much time. Arriving too early, and mingling with others who are also anxious, can increase your anxiety and nervousness.
2. Where will you park?
3. How does the exam site look? Will you have ample workspace? Will it be overly bright (sunglasses), or cold (sweater), or noisy (earplugs)? Would a cushion make the chair more comfortable?
4. Where are the drinking fountain and lavatory facilities?
5. What about food? Most states do not allow food in the test room (exceptions for ADA). Should you take something along for energy in the exam? A light bag lunch during the break makes sense.

Items to Take to the Examination

Although you may not bring books to the exam, you should bring the following:

- *Calculator*—NCEES has implemented a more stringent policy regarding permitted calculators. For a list of permitted models, see the NCEES Web site *(www.ncees.org)*. You also need to determine whether your state permits pre-programmed calculators. Bring extra batteries for your calculator just in case, and many people feel that bringing a second calculator is also a very good idea.
- *Clock*—You must have a time plan and a clock or wristwatch.

- *Exam Assignment Paperwork*—Take along the letter assigning you to the exam at the specified location to prove that you are the registered person. Also bring something with your name and picture (driver's license or identification card).
- *Items Suggested by Your Advance Visit*—If you visit the exam site, it will probably suggest an item or two that you need to add to your list.
- *Clothes*—Plan to wear comfortable clothes. You probably will do better if you are slightly cool, so it is wise to wear layered clothing.

Special Medical Condition

If you have a medical situation that may require special accommodation, notify the licensing board well in advance of exam day.

Examination Scoring and Results

Examinees will be notified via e-mail when their results are available for viewing in My NCEES. The process is still being finalized, but most examinees should receive their results within 7 to 10 business days.

Errata

The authors and publisher of this book have been careful to avoid errors, employing technical reviewers, copyeditors, and proofreaders to ensure the material is as flawless as possible. Any known errata and corrections are posted on the product page at our Web site, *www.brightwoodengineering.com*. If you believe you have discovered an inaccuracy, please notify Customer Service at *enginfo@brightwood.edu*.

CHAPTER 2

Mathematics

OUTLINE

ALGEBRA

Factorials

Definition. The factorial of a non-negative integer, n, is defined as $n!$ $n! = n(n-1)(n-2)(n-3)\ldots$ and so forth.

For example, $6! = 6(5)(4)(3)(2)(1) = 720$.

Also, $1! = 1$ and $0! = 1$.

Factorials can be written as multiples of other factorials:

$$6! = 6(5!) = 6(5)(4!) = 6(5)(4)(3!) = 6(5)(4)(3)(2!) = 6(5)(4)(3)(2)(1!) = 720$$

Factorials can be multiplied and divided.

$$\frac{n!}{(n-1)!} = \frac{n(n-1)!}{(n-1)!} = n$$

Exponents

Definition. Any number defined as base, b, can be multiplied by itself x number of times, which is denoted as b^x.

For example, $b^4 = b(b)(b)(b)$.

Properties of Exponents

$$b^0 = 1; \quad b^1 = b; \quad b^{-1} = \frac{1}{b}; \quad b^{-x} = \frac{1}{b^x}$$

$$b^{x+y} = \left(b^x\right)\left(b^y\right); \quad b^{x-y} = \frac{b^x}{b^y}; \quad b^{x \times y} = \left(b^x\right)^y$$

$$b^{\frac{1}{2}} = \sqrt{b}; \quad b^{\frac{1}{x}} \sqrt[x]{b}; \quad b^{\frac{x}{y}} = \left(\sqrt[y]{b}\right)^x$$

Logarithms

Definition. If b is a finite positive number, other than 1, and $b^x = N$, then x is the logarithm of N to the base b, or $\log_b N = x$. If $\log_b N = x$, then $b^x = N$.

Properties of Logarithms

$$\log_b b = 1; \ \log_b 1 = 0; \ \log_b 0 = \begin{cases} +\infty, \text{ when } b \text{ lies between 0 and 1} \\ +\infty, \text{ when } b \text{ lies between 1 and } \infty \end{cases}$$

$$\log_b \left(M \cdot N\right) = \log_b M + \log_b N \quad \log_b M/N = \log_b M - \log_b N$$

$$\log_b N^P = p \log_b N \quad \log_b \sqrt[r]{N^P} = \frac{p}{r} \log_b N$$

$$\log_b N = \log_a N / \log_a b; \quad \log_b b^N = N; \quad b^{\log_b N} = N$$

Systems of Logarithms

Common (Briggsian)—base 10.

Natural (Napierian or hyperbolic)—base 2.7183 (designated by e or ε).

The abbreviation of *common logarithm* is log, and the abbreviation of *natural logarithm* is ln.

Example 2.1

(i) Solve for a if $\log_a 10 = 0.25$ and (ii) find $\log\left(\frac{1}{x}\right)$, if $\log x = 0.3332$.

Solution

(i) If $\log_a N = x$, then $N = a^x$ or $a = N^{(1/x)}$

Here, $N = 10$ and $x = 0.25$; then $a = 10^{\frac{1}{0.25}} = 10{,}000$

(ii) Since $\log \frac{M}{N} = \log M - \log N$, $\log\left(\frac{1}{x}\right) = \log 1 - \log x = -0.3332$

Example 2.2

Solve the equation $\log x + \log (x - 3) - \log 4 = 0$.

Solution

Since $(\log M + \log N - \log P) = \log\left(\frac{MN}{P}\right)$, $\log x + \log (x - 3) - \log 4$

$$= \log\left[\frac{x(x-3)}{4}\right] = 0$$

$\left[\frac{x(x-3)}{4}\right] = 10^0 = 1$; simplifying, $x^2 - 3x - 4 = 0$. Finding the roots,

$x = 4$ and -1; then $x = 4$ (-1 is not an answer because the log of a negative number is undefined).

The Solution of Algebraic Equations

Definition. A root, x, is any value such that $f(x) = 0$.

The Quadratic Equation

If $ax^2 + bx + c = 0$, then

$$x = \frac{-b \pm \sqrt{b^2 - 4ac}}{2a}$$

If $b^2 - 4ac > 0$, the two roots are real and unequal; if $b^2 - 4ac = 0$, the two roots are real and equal; if $b^2 - 4ac < 0$, the two roots are imaginary.

Example 2.3

Find the root(s) of the following equations (i) $x + 4 = 0$, (ii) $x^2 + 4x + 3 = 0$, (iii) $x^2 - 4x + 4 = 0$, (iv) $x^2 + 4 = 0$, and (v) $3x^3 + 3x^2 - 18x = 0$.

Solution

For the quadratic equations in (ii), (iii), and (iv), use either the equation or simply a scientific calculator to find the roots. The results yield:

(i) -4, (ii) -1 and -3 (real and distinct roots), (iii) 2 and 2 (real and equal roots),

(iv) $+i2$ and $-i2$ (complex roots; always occur in pairs called *conjugates*)

For (v), factoring, $3x(x^2 + x - 6) = 0$. Roots are 0, -3, and 2.

Example 2.4

Find the equation whose roots are 3 and -2.

Solution

If the roots are x_1, x_2, x_3, etc., the equation is $(x - x_1)(x - x_2)(x - x_3)\ldots = 0$; here, $(x - 3)(x - (-2)) = 0$; $(x - 3)(x + 2) = 0$; $x^2 - x - 6 = 0$.

Progressions

Arithmetic Progression

An arithmetic progression is a, $a + d$, $a + 2d$, $a + 3d$,. . ., where d = common difference.

The nth term is $t_n = a + (n - 1)d$

The sum of n terms is $S_n = \frac{n}{2}[2a+(n-1)d] = \frac{n}{2}(a+t_n)$

Geometric Progression

A geometric progression is a, ar, ar^2, ar^3,..., where r = common ratio.

The nth term is $t_n = ar^{n-1}$

The sum of n terms is $S_n = a\left(\frac{1-r^n}{1-r}\right)$

If $r^2 < 1$, S_n approaches a definite limit as n increases indefinitely, and

$$S_\infty = \frac{a}{1-r}$$

Example 2.5

Consider the arithmetic progression 1, 3, 5, 7, 9, 11, 13, . . . (i) Find the sum of the first seven terms and (ii) the 18th term of the progression.

Solution

(i) First term $a = 1$, number of terms $n = 7$, and the common difference $d = 2$

sum $S_n = \frac{n}{2}[2a+(n-1)d] = \frac{7}{2}[2+6(2)] = 49$

(ii) Number of terms $n = 18$, first term $a = 1$, and the common difference $d = 2$

The last term or the 18th term is $= a + (n - 1)\,d = 1 + (18 - 1)2 = 35$.

Example 2.6

Find the sum of the series 1, 0.5, 0.25, 0.125, 0.0625, . . .

Solution

This geometric series is convergent. First term $a = 1$ and the common ratio $r = 0.5$.

As the number of terms tend to infinity, sum $S = \frac{a}{1-r} = \frac{1}{1-0.5} = 2$

Example 2.7

Consider the geometric progression 2, 4, 8, 16, 32, 64, 128, . . . (i) Find the sum of the first seven terms and (ii) the 20th term of the series.

Solution

(i) First term $a = 2$, common ratio $r = 2$, and the number of terms $n = 7$

sum $S = a\frac{(1-r^n)}{(1-r)} = \frac{2(1-2^7)}{(1-2)} = 256$

(ii) The 20th term of the series is $= ar^{(n-1)} = 2(2)^{(20-1)} = 1{,}048{,}576$.

COMPLEX QUANTITIES

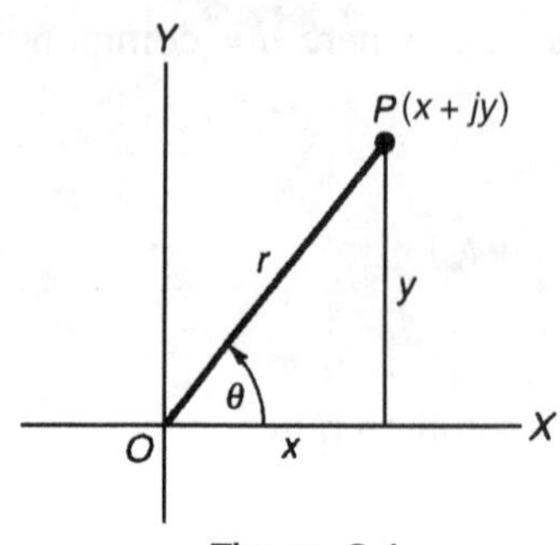

Figure 2.1

Definition and Representation of a Complex Quantity

If $z = x + jy$, where $j = \sqrt{-1}$ and x and y are real, z is called a complex quantity and is completely determined by x and y.

If $P(x, y)$ is a point in the plane (Figure 2.1), then the segment OP in magnitude and direction is said to represent the complex quantity $z = x + jy$.

If θ is the angle from OX to OP and r is the length of OP, then $z = x + jy = r(\cos\theta + j\sin\theta) = re^{j\theta}$, where $\theta = \tan^{-1} y/x$, $r = +\sqrt{x^2 + y^2}$ and e is the base of natural logarithms. The pair $x + jy$ and $x - jy$ are called complex conjugate quantities.

Properties of Complex Quantities

Let z, z_1, and z_2 represent complex quantities; then

Sum or difference: $z_1 \pm z_2 = (x_1 \pm x_2) + j(y_1 \pm y_2)$

Equation: If $z_1 = z_2$, then $x_1 = x_2$ and $y_1 = y_2$

Periodicity: $z = r(\cos\theta + j\sin\theta) = r[\cos(\theta + 2k\pi) + j\sin(\theta + 2k\pi)]$, or $z = re^{j\theta} = re^{j(\theta + 2k\pi)}$ and $e^{j2k\pi} = 1$, where k is any integer.

Exponential-trigonometric relations: $e^{jz} = \cos z + j\sin z$, $e^{-jz} = \cos z - j\sin z$,

$$\cos z = \frac{1}{2}\left(e^{jz} + e^{-jz}\right), \quad \sin z = \frac{1}{2j}\left(e^{jz} - e^{-jz}\right)$$

TRIGONOMETRY

Definition of an Angle

An angle is the amount of rotation (in a fixed plane) by which a straight line may be changed from one direction to any other direction. If the rotation is counterclockwise, the angle is said to be positive; if clockwise, negative.

Measure of an Angle

A degree is $\frac{1}{360}$ of the plane angle about a point, and a radian is the angle subtended at the center of a circle by an arc equal in length to the radius. One complete circle contains 180 degrees or 2π radians; 1 radian = $\pi/180$ degrees.

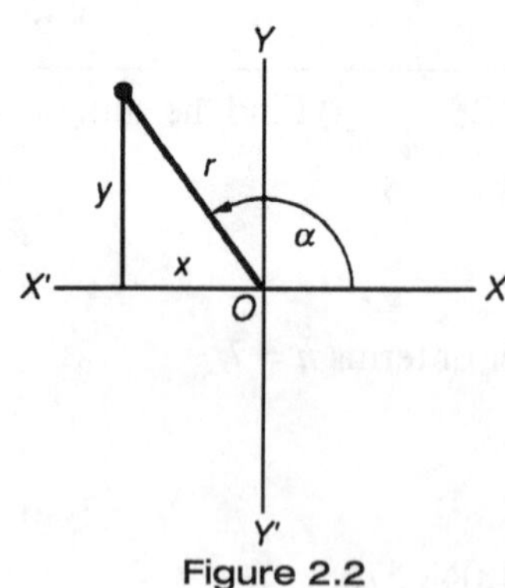

Figure 2.2

Trigonometric Functions of an Angle

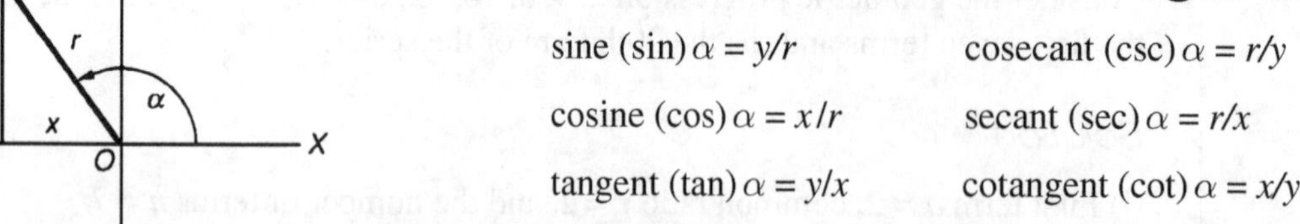

sine (sin) $\alpha = y/r$ cosecant (csc) $\alpha = r/y$

cosine (cos) $\alpha = x/r$ secant (sec) $\alpha = r/x$

tangent (tan) $\alpha = y/x$ cotangent (cot) $\alpha = x/y$

The variable x is positive when measured along OX and negative along OX'. Similarly, y is positive when measured parallel to OY, and negative parallel to OY'.

$$\sin 0° = 0;\ \sin 90° = 1;\ \sin 180° = 0;\ \sin 270° = -1$$

$$\cos 0° = 1;\ \cos 90° = 0;\ \cos 180° = -1;\ \cos 270° = 0$$

Fundamental Relations among the Functions

$$\sin\alpha = \frac{1}{\csc\alpha};\quad \cos\alpha = \frac{1}{\sec\alpha};\quad \tan\alpha = \frac{1}{\cot\alpha} = \frac{\sin\alpha}{\cos\alpha}$$

$$\csc\alpha = \frac{1}{\sin\alpha};\quad \sec\alpha = \frac{1}{\cos\alpha};\quad \cot\alpha = \frac{1}{\tan\alpha} = \frac{\cos\alpha}{\sin\alpha}$$

$$\sin^2\alpha + \cos^2\alpha = 1;\quad \sec^2\alpha - \tan^2\alpha = 1;\quad \csc^2\alpha - \cot^2\alpha = 1$$

Functions of Multiple Angles

$$\sin 2\alpha = 2\sin\alpha\cos\alpha$$

$$\cos 2\alpha = 2\cos^2\alpha - 1 = 1 - 2\sin^2\alpha = \cos^2\alpha - \sin^2\alpha$$

$$\tan 2\alpha = (2\tan\alpha)/(1 - \tan^2\alpha)$$

$$\cot 2\alpha = (\cot^2\alpha - 1)/(2\cot\alpha)$$

Functions of Half Angles

$$\sin\frac{1}{2}\alpha = \sqrt{\frac{1-\cos\alpha}{2}};\quad \cos\frac{1}{2}\alpha = \sqrt{\frac{1+\cos\alpha}{2}}$$

$$\tan\frac{1}{2}\alpha = \frac{1-\cos\alpha}{\sin\alpha} = \frac{\sin\alpha}{1+\cos\alpha} = \sqrt{\frac{1-\cos\alpha}{1+\cos\alpha}}$$

Functions of Sum or Difference of Two Angles

$$\sin(\alpha \pm \beta) = \sin\alpha\cos\beta \pm \cos\alpha\sin\beta$$

$$\cos(\alpha \pm \beta) = \cos\alpha\cos\beta \mp \sin\alpha\sin\beta$$

$$\tan(\alpha \pm \beta) = \frac{\tan\alpha \pm \tan\beta}{1 \mp \tan\alpha\tan\beta}$$

Sums, Differences, and Products of Two Functions

$$\sin\alpha + \sin\beta = 2\sin\frac{1}{2}(\alpha+\beta)\cos\frac{1}{2}(\alpha-\beta)$$

$$\sin\alpha - \sin\beta = 2\cos\frac{1}{2}(\alpha+\beta)\sin\frac{1}{2}(\alpha-\beta)$$

$$\cos\alpha + \cos\beta = 2\cos\frac{1}{2}(\alpha+\beta)\cos\frac{1}{2}(\alpha-\beta)$$

$$\cos\alpha - \cos\beta = 2\sin\frac{1}{2}(\alpha+\beta)\sin\frac{1}{2}(\alpha-\beta)$$

$$\tan\alpha \pm \tan\beta = \frac{\sin(\alpha+\beta)}{\cos\alpha\cos\beta}$$

$$\sin^2\alpha - \sin^2\beta = \sin(\alpha+\beta)\sin(\alpha-\beta)$$

$$\cos^2\alpha - \cos^2\beta = \sin(\alpha+\beta)\sin(\alpha-\beta)$$

$$\cos^2\alpha - \sin^2\beta = \cos(\alpha+\beta)\cos(\alpha-\beta)$$

$$\sin\alpha\sin\beta = \frac{1}{2}\cos(\alpha-\beta) - \frac{1}{2}\cos(\alpha+\beta)$$

$$\cos\alpha\cos\beta = \frac{1}{2}\cos(\alpha-\beta) + \frac{1}{2}\cos(\alpha+\beta)$$

$$\sin\alpha\cos\beta = \frac{1}{2}\sin(\alpha+\beta) + \frac{1}{2}\sin(\alpha-\beta)$$

Example 2.8

Simplify the following expressions to trigonometric functions of angles less than 90 degrees. Note: There is more than one correct answer for each.

i) cos 370°
ii) sin 120°

Solution

Typical strategies for simplifying the functions are shown below:

i) cos 370° = cos (360° + 10°) = (cos 360°)(cos 10°) – (sin 360°)(sin 10°) = (1)(cos 10°) – (0)(sin 10°) = cos 10°

ii) sin 120° = sin (90° + 30°) = (sin 90°)(cos 30°) + (cos 90°)(sin 30°) = (1)(cos 30°) + (0)(sin 30°) = cos 30°

or

sin 120° = sin (180° – 60°) = (sin 180°)(cos 60°) – (cos 180°)(sin 60°) = (0)(cos 60°) – (–1)(sin 60°) = sin 60°

Properties of Plane Triangles

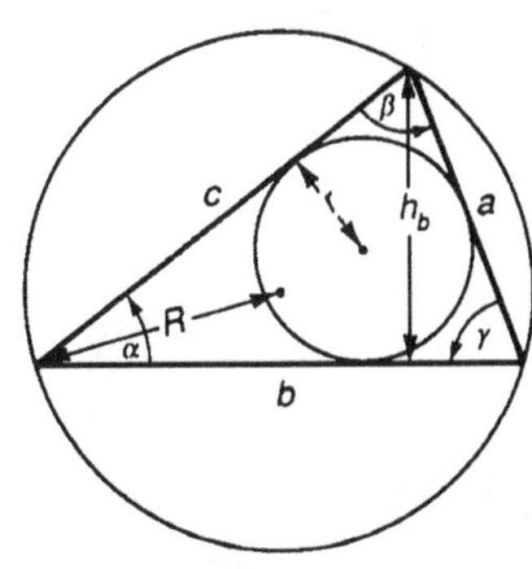

Figure 2.3

Notation. α, β, γ = angles; a, b, c = sides; A = area; h_b = altitude on b; $s = \frac{1}{2}(a+b+c)$; r = radius of inscribed circle; R = radius of circumscribed circle

$$\alpha + \beta + \gamma = 180^\circ = \pi \text{ radians}$$

$$\frac{a}{\sin\alpha} = \frac{b}{\sin\beta} = \frac{c}{\sin\gamma}$$

$$\frac{a+b}{a-b} = \frac{\tan\frac{1}{2}(\alpha+\beta)}{\tan\frac{1}{2}(\alpha-\beta)}$$

$$a^2 = b^2 + c^2 - 2bc\cos\alpha \qquad a = b\cos\gamma + c\cos\beta$$

$$\cos\alpha = \frac{b^2+c^2-a^2}{2bc} \qquad \sin\alpha = \frac{2}{bc}\sqrt{s(s-a)(s-b)(s-c)}$$

$$\sin\frac{\alpha}{2}\sqrt{\frac{(s-b)(s-c)}{bc}} \qquad \cos\frac{\alpha}{2} = \sqrt{\frac{s(s-a)}{bc}}$$

$$\tan\frac{\alpha}{2} = \sqrt{\frac{(s-b)(s-c)}{s(s-a)}} = \frac{r}{s-a}$$

$$h_b = c\sin\alpha = a\sin\gamma = \frac{2}{b}\sqrt{s(s-a)(s-b)(s-c)}$$

$$r = \sqrt{\frac{(s-a)(s-b)(s-c)}{s}} = (s-a)\tan\frac{\alpha}{2}$$

$$R = \frac{a}{2\sin\alpha} = \frac{abc}{4A}$$

$$A = \frac{1}{2}bh_b = \frac{1}{2}ab\sin\gamma = \frac{a^2\sin\beta\sin\gamma}{2\sin\alpha} = \sqrt{s(s-a)(s-b)(s-c)} = rs$$

Example **2.9**

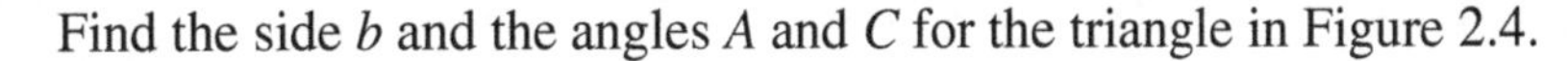

Find the side b and the angles A and C for the triangle in Figure 2.4.

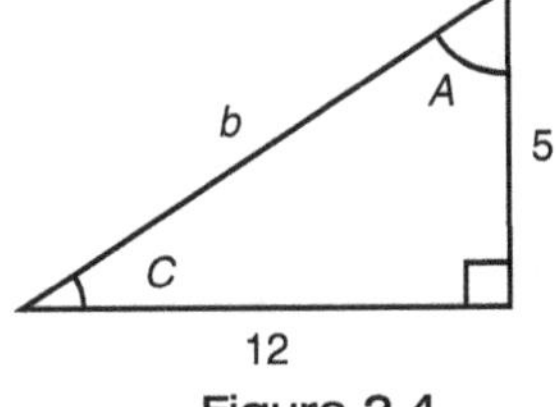

Figure 2.4

Solution

$\tan(A) = 12/5$; then, Angle $A = \tan^{-1}(12/5) = 67.38^\circ$

Since the sum $(A + C + 90^\circ) = 180^\circ$; $C = 90^\circ - A = 22.62^\circ$

Now, $\cos(A) = \frac{5}{b}$; then, $b = \frac{5}{\cos(A)} = 13$

Example 2.10

Find the side c and the angles A and B for the triangle in Figure 2.5.

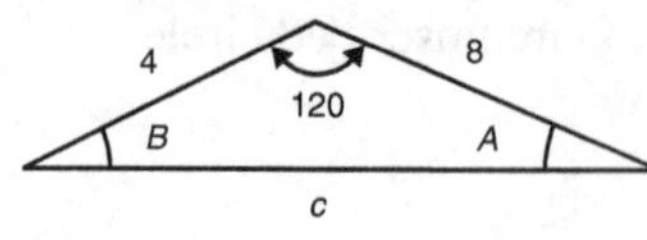

Figure 2.5

Solution

Since two sides and an included angle are given, use the law of cosines.

$c^2 = 4^2 + 8^2 - 2(4)(8)\cos 120$; solving $c = 10.583$

Now use law of sines to find the remaining angles.

$\dfrac{10.583}{\sin 120} = \dfrac{4}{\sin A} = \dfrac{8}{\sin B}$; solving, $A = 19.1°$ and $B = 40.89°$

(Check: sum of the angles = 180°)

Example 2.11

Simplify: (i) $(\sec^2\theta)(\sin^2\theta)$ (ii) $\sin(A + B) + \sin(A - B)$
(iii) $2\sin^2\theta + 1 + \cos^2\theta$

Solution

(i) $(1/\cos^2\theta)\sin^2\theta = \tan^2\theta$ (ii) $2\sin A\cos B$

(iii) $2\sin^2\theta + 1 + (2\cos^2\theta - 1) = 2(\sin^2\theta + \cos^2\theta) = 2$

GEOMETRY AND GEOMETRIC PROPERTIES (MENSURATION)

Notation. a, b, c, d, and s denote lengths, A denotes area, V denotes volume

Right Triangle

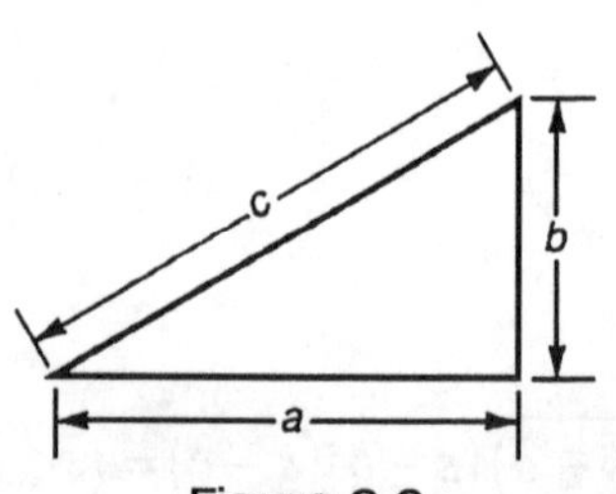

Figure 2.6

$$A = \frac{1}{2}ab$$

$$c = \sqrt{a^2 + b^2},\quad a = \sqrt{c^2 - b^2},\quad b = \sqrt{c^2 - a^2}$$

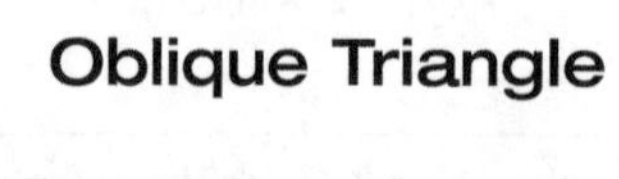

Oblique Triangle

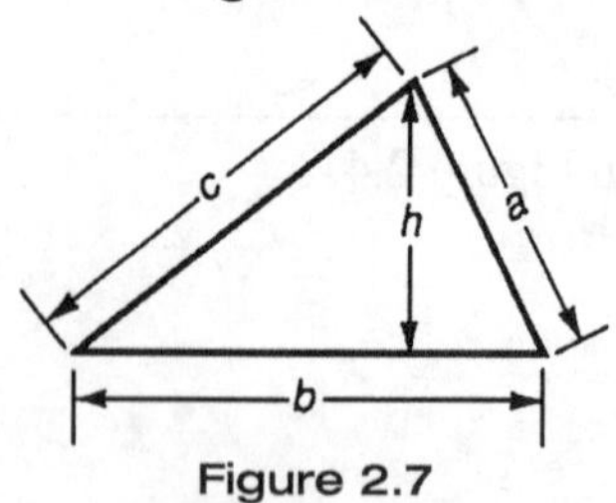

Figure 2.7

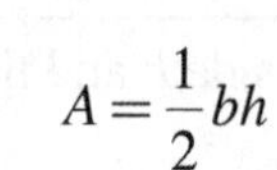

$$A = \frac{1}{2}bh$$

Figure 2.8

Equilateral Triangle

All sides are equal and all angles are 60°.

$$A = \frac{1}{2}ah = \frac{1}{4}a^2\sqrt{3},\qquad h = \frac{1}{2}a\sqrt{3},\qquad r_1 = \frac{a}{2\sqrt{3}},\qquad r_2 = \frac{a}{\sqrt{3}}$$

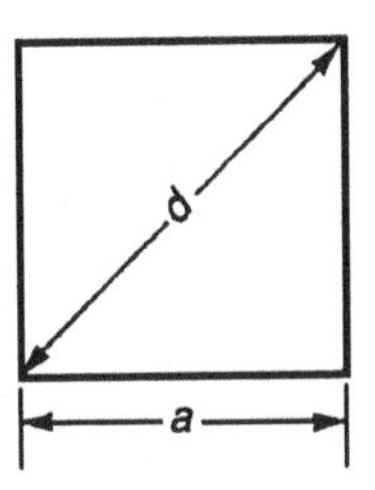

Figure 2.9

Square

All sides are equal, and all angles are 90°.

$$A = a^2, \qquad d = a\sqrt{2}$$

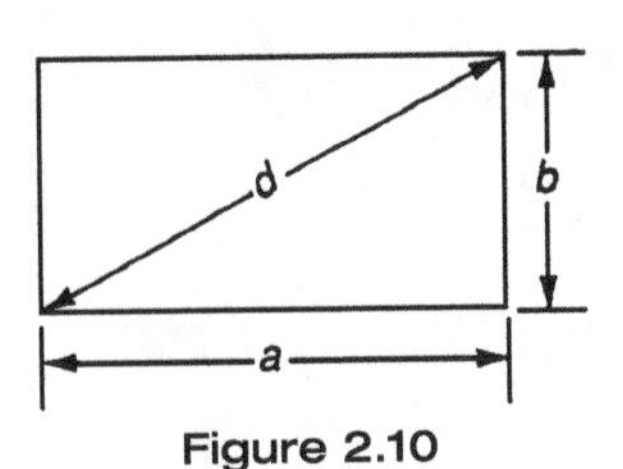

Figure 2.10

Rectangle

Opposite sides are equal and parallel, and all angles are 90°.

$$A = ab, \qquad d = \sqrt{a^2 + b^2}$$

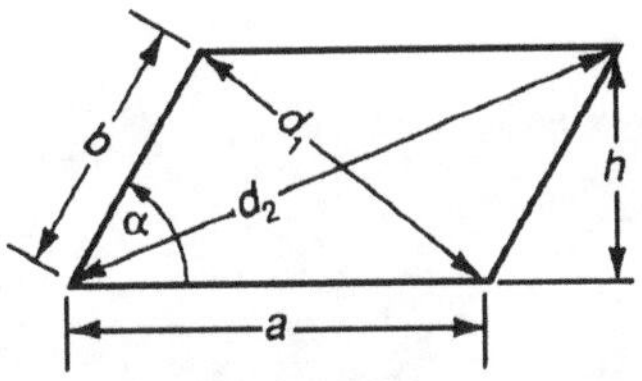

Figure 2.11

Parallelogram

Opposite sides are equal and parallel, and opposite angles are equal.

$$A = ah = ab \sin\alpha, \quad d_1 = \sqrt{a^2 + b^2 - 2ab\cos\alpha}, \quad d_2 = \sqrt{a^2 + b^2 + 2ab\cos\alpha}$$

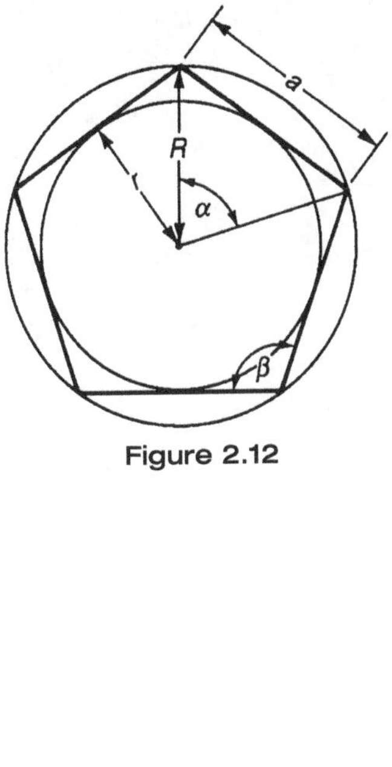

Figure 2.12

Regular Polygon of *n* Sides

All sides and all angles are equal.

$$\beta = \frac{n-2}{n}180° = \frac{n-2}{n}\pi \text{ radians}, \qquad \alpha \frac{360°}{n} = \frac{2\pi}{n} \text{ radians}, \qquad A = \frac{nar}{2}$$

Circle

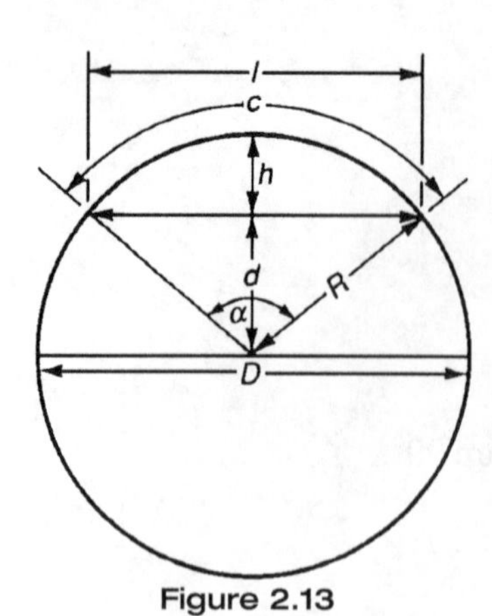

Figure 2.13

Notation. C = circumference, α = central angle in radians

$$C = \pi D = 2\pi R$$

$$c = R\alpha = \frac{1}{2}D\alpha = D\cos^{-1}\frac{d}{R} = D\tan^{-1}\frac{1}{2d}$$

$$l = 2\sqrt{R^2 - d^2} = 2R\sin\frac{\alpha}{2} = 2d\tan\frac{\alpha}{2} = 2d\tan\frac{c}{D}$$

$$d = \frac{1}{2}\sqrt{4R^2 - l^2} = \frac{1}{2}\sqrt{D^2 - l^2} = R\cos\frac{\alpha}{2}$$

$$h = R - d$$

$$\alpha = \frac{c}{R} = \frac{2c}{D} = 2\cos^{-1}\frac{d}{R}$$

$$A_{(circle)} = \pi R^2 = \frac{1}{4}\pi D^2 = \frac{1}{2}RC = \frac{1}{4}DC$$

$$A_{(sector)} = \frac{1}{2}Rc = \frac{1}{2}R^2\alpha = \frac{1}{8}D^2\alpha$$

Ellipse

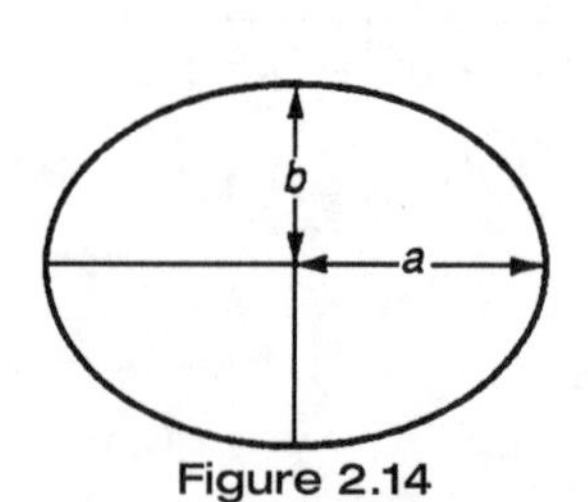

Figure 2.14

$$A = \pi ab$$

$$\text{Perimeter } (s) = \pi(a+b)\left[1 + \frac{1}{4}\left(\frac{a-b}{a+b}\right)^2 + \frac{1}{64}\left(\frac{a-b}{a+b}\right)^4 + \frac{1}{256}\left(\frac{a-b}{a+b}\right)^6 + \cdots\right]$$

$$\text{Perimeter } (s) \approx \pi\frac{a+b}{4}\left[3(1+\lambda) + \frac{1}{1-\lambda}\right], \quad \text{where } \lambda = \left[\frac{a-b}{2(a+b)}\right]^2$$

Parabola

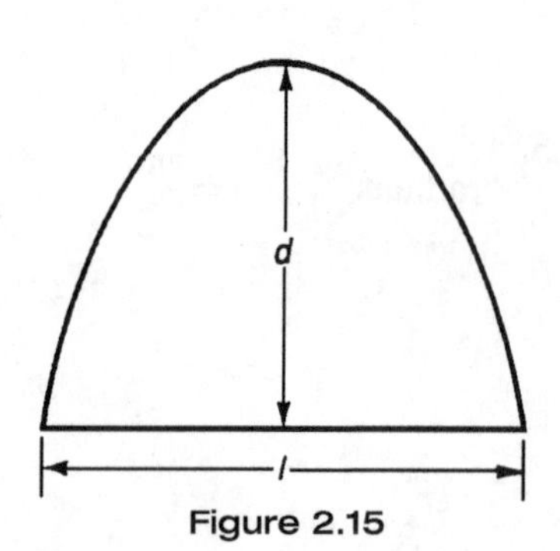

Figure 2.15

$$A = \frac{2}{3}ld$$

Cube

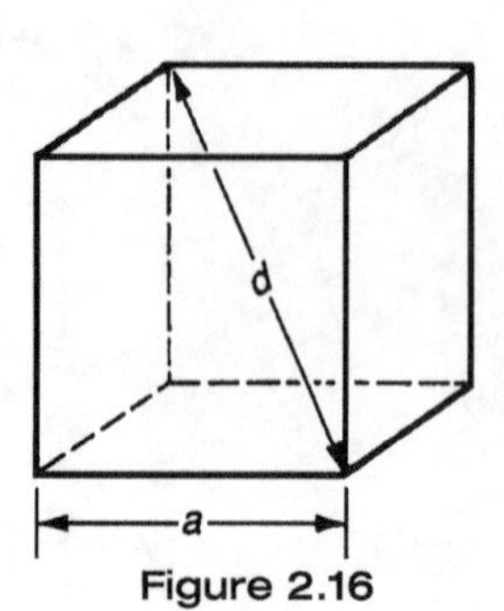

Figure 2.16

$$V = a^3 \qquad d = a\sqrt{3}$$

Total surface area = $6a^2$

Prism or Cylinder

$V =$ (area of base) (altitude, h)

Lateral area = (perimeter of right section)(lateral edge, e)

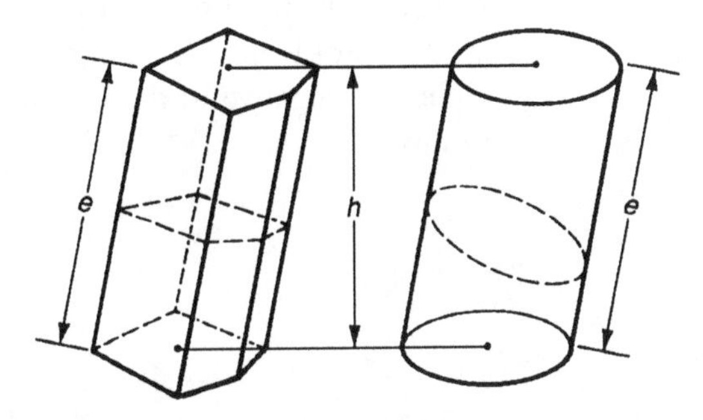

Figure 2.17

Pyramid or Cone

$V = \frac{1}{3}$ (area of base) (altitude, h)

Lateral area of regular figure = $\frac{1}{2}$ (perimeter of base)(slant height, s)

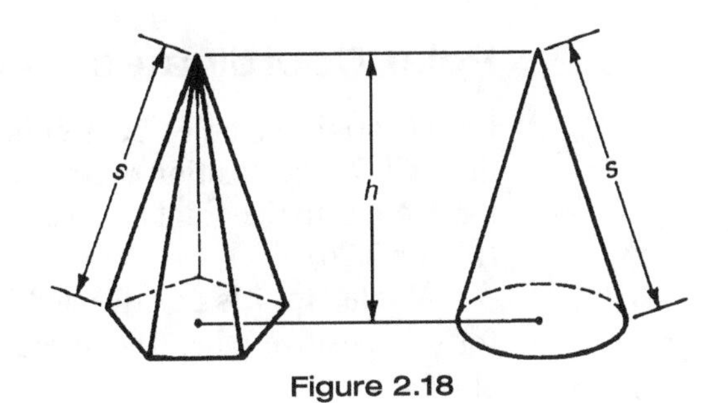

Figure 2.18

Sphere

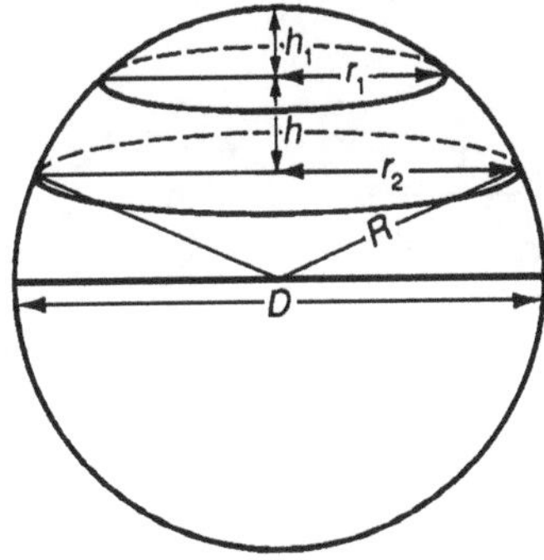

Figure 2.19

$$A_{(sphere)} = 4\pi R^2 = \pi D^2$$

$$A_{(zone)} = 2\pi Rh = \pi Dh$$

$$V_{(sphere)} = \frac{4}{3}\pi R^3 = \frac{1}{6}\pi D^3$$

$$V_{(spherical\ sector)} = \frac{2}{3}\pi R^2 h = \frac{1}{6}\pi D^2 h$$

PLANE ANALYTIC GEOMETRY

Rectangular Coordinates

Let two perpendicular lines, $X'X$ (x-axis) and $Y'Y$ (y-axis) meet at a point O (origin). The position of any point $P(x, y)$ is fixed by the distances x (abscissa) and y (ordinate) from $Y'Y$ and $X'X$, respectively, to P. Values of x are positive to the right and negative to the left of $Y'Y$; values of y are positive above and negative below $X'X$.

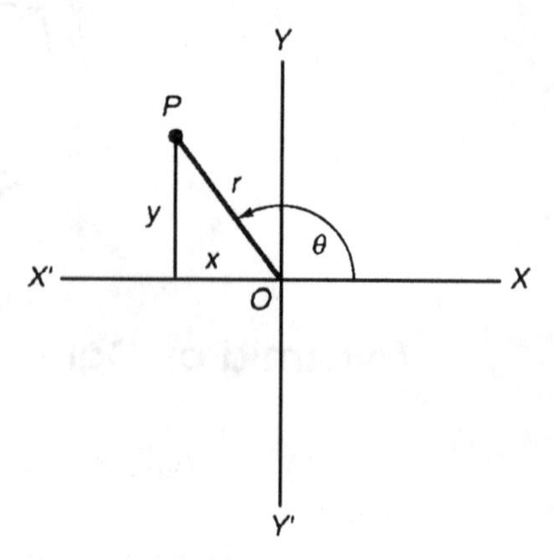

Figure 2.20

Polar Coordinates

Let O (origin or pole) be a point in the plane and OX (initial line) be any line through O. The position of any point $P\ (r, \theta)$ is fixed by the distance r (radius vector) from O to the point and the angle θ (vectorial angle) measured from OX to OP (Figure 2.20).

A value for r is positive and is measured along the terminal side of θ; a value for θ is positive when measured counterclockwise and negative when measured clockwise.

Relations Connecting Rectangular and Polar Coordinates

$$x = r\cos\theta, \quad y = r\sin\theta$$

$$r = \sqrt{x^2 + y^2}, \quad \theta = \tan^{-1}\frac{y}{x}, \quad \sin\theta = \frac{y}{\sqrt{x^2 + y^2}},$$

$$\cos\theta = \frac{x}{\sqrt{x^2 + y^2}}, \quad \tan\theta = \frac{y}{x}$$

Points and Slopes

Let $P_1(x_1, y_1)$ and $P_2(x_2, y_2)$ be any two points, and let α_1 be the angle from the x axis to P_1P_2, measured counterclockwise.

The length P_1P_2 is $d = \sqrt{(x_2 - x_1)^2 + (y_2 - y_1)^2}$.

The midpoint of P_1P_2 is $\left(\frac{x_1 + x_2}{2}, \frac{y_1 + y_2}{2}\right)$.

The point that divides P_1P_2 in the ratio $n_1{:}n_2$ is $\left(\frac{n_1x_2 + n_2x_1}{n_1 + n_2}, \frac{n_1y_2 + n_2y_1}{n_1 + n_2}\right)$.

The slope of P_1P_2 is $\tan \alpha = m = \frac{y_2 - y_1}{x_2 - x_1}$.

The angle between two lines of slopes m_1 and m_2 is $\beta = \tan^{-1} \frac{m_2 - m_1}{1 + m_1m_2}$.

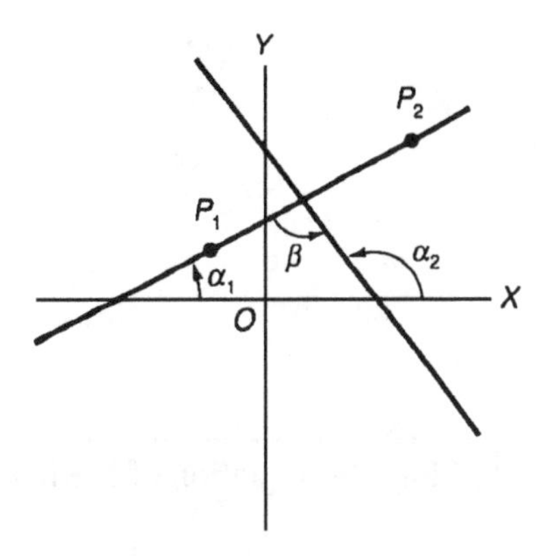

Figure 2.21

Two lines of slopes m_1 and m_2 are perpendicular if $m_2 = -\frac{1}{m_1}$.

Example 2.12

Find the distance between the points (1, –2) and (–4, 2).

Solution

The distance is $d = \sqrt{(x_2 - x_1)^2 + (y_2 - y_1)^2} = \sqrt{[1-(-4)]^2 + [-2-2]^2}$
$= \sqrt{5^2 + 4^2} = 6.4$

Locus and Equation

The collection of all points that satisfy a given condition is called the **locus** of that condition; the condition expressed by means of the variable coordinates of any point on the locus is called the **equation of the locus**.

The locus may be represented by equations of three kinds: (1) a rectangular equation involves the rectangular coordinates (x, y); (2) a polar equation involves the polar coordinates (r, θ); and (3) parametric equations express x and y or r and θ in terms of a third independent variable called a parameter.

The following equations are generally given in the system in which they are most simply expressed; sometimes several forms of the equation in one or more systems are given.

Straight Line

$Ax + By + C = 0$ [$-A/B$ = slope]

$y = mx + b$ [m = slope, b = intercept on OY]

$y - y_1 = m(x - x_1)$ [m = slope, $P_1\,(x_1, y_1)$ is a known point on the line]

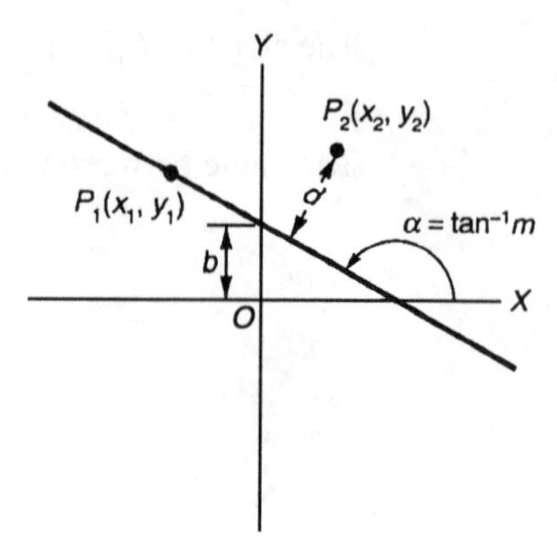

Figure 2.22

Example 2.13

Find the equation of the line passing through the points (2, 1) and (3, –3).

Solution

Slope can be found as $m = \dfrac{y_2 - y_1}{x_2 - x_1} = \dfrac{1-(-3)}{2-3} = -4.$

The point-slope form is $(y - 1) = -4(x - 2)$ or $(y + 3) = -4(x - 3)$.

Simplifying, either equation yields $y = -4x + 9$ or equivalently $4x + y - 9 = 0$.

Example 2.14

Find the equation of the straight line passing through the point (3, 1) and perpendicular to the line passing through the points (3, –2) and (–3, 7).

Solution

Slope of the line passing through (3, –2) and (–3, 7) is

$$m_1 = \frac{y_2 - y_1}{x_2 - x_1} = \frac{7-(-2)}{-3-3} = -\frac{3}{2}$$

Slope of the line passing through (3, 1) is $m_2 = -\dfrac{1}{m_1} = \dfrac{2}{3}$, since the two lines are perpendicular to each other.

Equation is $(y-1) = \dfrac{2}{3}(x - 3)$ or $2x - 3y - 3 = 0$.

Circle

The locus of a point at a constant distance (radius) from a fixed point C (center) is a circle.

$(x-h)^2+(y-k)^2=a^2$	$C(h, k)$, radius $= a$	
$r^2+b^2\pm 2\,br\cos(\theta-\beta)=a^2$	$C(b, \beta)$, radius $= a$	[Figure 2.23(a)]
$x^2+y^2=2ax$	$C(a, 0)$, radius $= a$	
$r=2a\cos\theta$	$C(a, 0)$, radius $= a$	[Figure 2.23(b)]
$x^2+y^2=2ay$	$C(0, a)$, *radius* $= a$	
$r=2a\sin\theta$	$C(0, a)$, *radius* $= a$	[Figure 2.23(c)]
$x^2+y^2=a^2$	$C(0, 0)$, radius $= a$	
$r=a$	$C(0, 0)$, radius $= a$	[Figure 2.23(d)]
$x=a\cos\phi$, $y=a\sin\phi$	$\phi=$ angle from OX to radius	

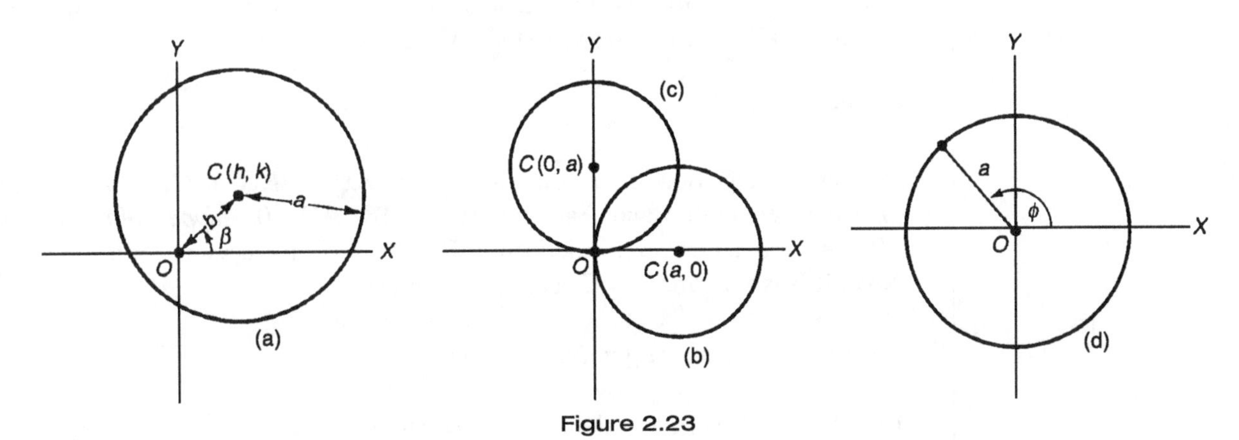

Figure 2.23

Example 2.15

Find the equation of the circle (i) with center at (0, 0) and radius 3, and (ii) with center at (1, 2) and radius 4.

Solution

Equation of a circle with center at (h, k) and radius r is $(x-h)^2+(y-k)^2=r^2$.

i) Here, $h=0$, $k=0$, and $r=3$; equation of the circle is $x^2+y^2=3^2$ or $x^2+y^2-9=0$
ii) Here, $h=1$, $k=2$, and $r=4$; equation of the circle is $(x-1)^2+(y-2)^2=4^2$
Simplifying, $x^2-2x+y^2-4y-11=0$

Conic

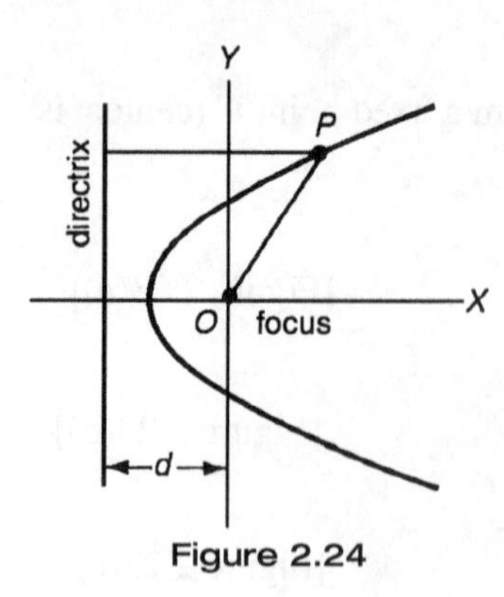

Figure 2.24

A **conic** is the locus of a point whose distance from a fixed point (focus) is in a constant ratio e, called the eccentricity, to its distance from a fixed straight line (directrix).

$$x^2 + y^2 = e^2(d + x)^2 \qquad d = \text{distance from focus to directrix}$$

$$r = \frac{de}{1 - e\cos\theta}$$

The conic is called a parabola when $e = 1$, an ellipse when $e < 1$, and a hyperbola when $e > 1$.

Example 2.16

Which conic section is represented by each of the following equations: (i) $x^2 + 4xy + 4y^2 + 2x = 10$ and (ii) $x^2 + y^2 - 2x - 4y - 11 = 0$.

Solution

The general equation of a conic section is $Ax^2 + 2Bxy + Cy^2 + 2Dx + 2Ey + F = 0$, where both A and C are not zeros. If $B^2 - AC > 0$, a *hyperbola* is defined; if $B^2 - AC = 0$, a *parabola* is defined; if $B^2 - AC < 0$, an *ellipse* is defined. (Note: If B is zero and A = C, a *circle* is defined.)

If A = B = C = 0, a *straight line* is defined.

If B = 0, A = C, a *circle* is defined with equation $x^2 + y^2 + 2ax + 2by + c = 0$.

Center is at (–a, –b) and radius = $\sqrt{a^2 + b^2 - c}$ provided $a^2 + b^2 - c > 0$.

(i) Here, A = 1, B = 2, and C = 4. Then, $B^2 - AC = (2)^2 - (1)(4) = 0$. The equation represents a parabola.

(ii) Here, A = 1, B = 0, C = 1. Because A = C and B = 0, the equation represents a circle.

[Note that the center is at (1, 2), and the radius is $\sqrt{(-1)^2 + (-2)^2 - (-11)} = 4$.]

Parabola

A parabola is a special case of a conic where $e = 1$.

$(y-k)^2 = a(x-h)$ Vertex (h, k), axis OX

$y^2 = ax$ Vertex $(0, 0)$, axis along OX [Figure 2.25(a)]

$(x-h)^2 = a(y-k)$ Vertex (h, k), axis OY

$x^2 = ay$ Vertex $(0, 0)$, axis along OY [Figure 2.25(b)]

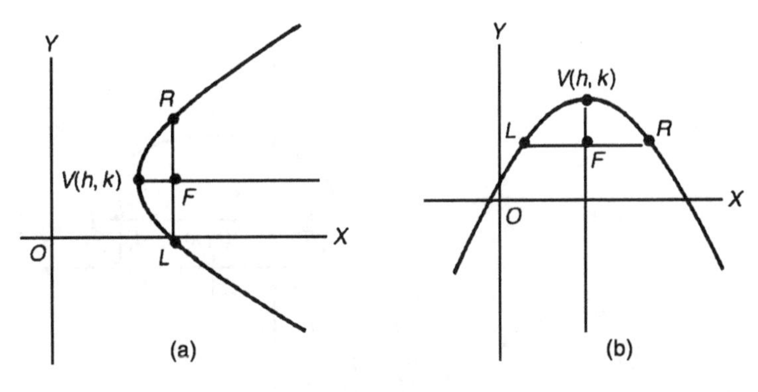

Figure 2.25

Distance from vertex to focus $= VF = \frac{1}{4}a$. Latus rectum $= LR = a$.

Example 2.17

Find the equation of a parabola (i) with center at (0, 0) and focus at (4, 0) and (ii) with center at (4, 2) and focus at (8, 2).

Solution

The equation of a parabola with center at (h, k) and focus at $(h + p/2, k)$ is given as

$(y - k)^2 = 2p(x - h)$

(i) $h = 0$; $k = 0$; $h + p/2 = 4$; then, $p = 8$ and the equation is $y^2 = 16x$.

(ii) $h = 4$; $k = 2$; $h + p/2 = 8$; then, $p = 8$ and the equation is $(y - 2)^2 = 16(x - 4)$.

Ellipse

This is a special case of a conic where $e < 1$.

$$\frac{(x-h)^2}{a^2} + \frac{(y-k)^2}{b^2} = 1 \quad \text{Center } (h, k), \text{ axes} \quad OX, OY$$

$$\frac{x^2}{a^2} + \frac{y^2}{b^2} = 1 \quad \text{Center } (0, 0), \text{ axes along } OX, OY$$

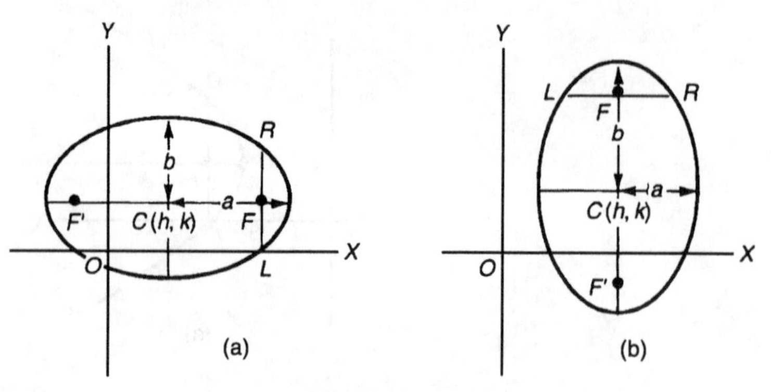

Figure 2.26

	$a > b$, Figure 2.26(a)	$b > a$, Figure 2.26(b)
Major axis	$2a$	$2b$
Minor axis	$2b$	$2a$
Distance from center to either focus	$\sqrt{a^2 - b^2}$	$\sqrt{b^2 - a^2}$
Latus rectum	$\frac{2b^2}{a}$	$\frac{2a^2}{b}$
Eccentricity, e	$\sqrt{\frac{a^2 - b^2}{a}}$	$\sqrt{\frac{b^2 - a^2}{b}}$
Sum of distances of any point P from the foci, $PF' + PF$	$2a$	$2b$

Example 2.18

Find the equation of an ellipse with center at origin, x-axis intercept (4, 0), and y-axis intercept (0, 2).

Solution

Equation of an ellipse with center at origin and x-axis intercept of $(a, 0)$ and y-axis intercept of $(0, b)$ is $\frac{x^2}{a^2} + \frac{y^2}{b^2} = 1$. Here, $a = 4$ and $b = 2$; then, the equation is $\frac{x^2}{4^2} + \frac{y^2}{2^2} = 1$.

Hyperbola

This is a special case of a conic where $e > 1$.

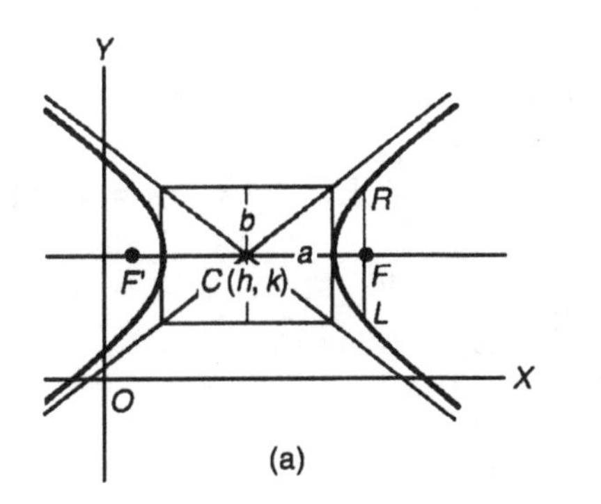

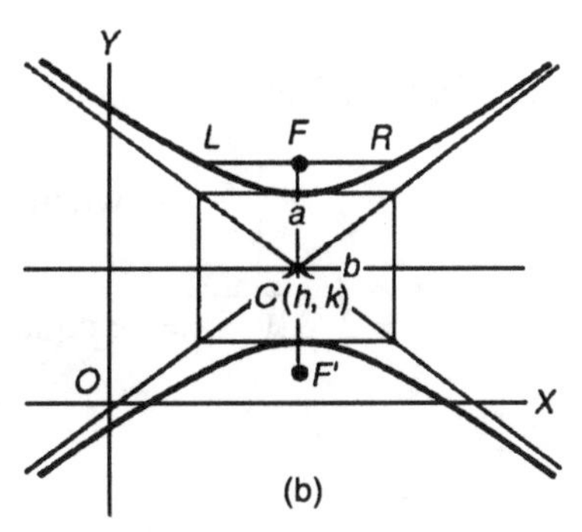

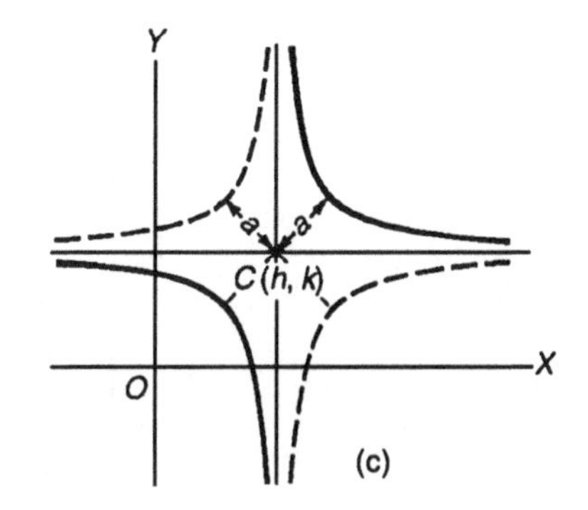

Figure 2.27

$$\frac{(x-h)^2}{a^2}-\frac{(y-k)^2}{b^2}=1 \qquad C(h,k),\ \text{transverse axis} \quad OX$$

$$\frac{x^2}{a^2}-\frac{y^2}{b^2}=1 \qquad C(0,\ 0),\ \text{transverse axis along } OX$$

$$\frac{(y-k)^2}{a^2}-\frac{(x-h)^2}{b^2}=1 \qquad C(h,\ k),\ \text{transverse axis} \quad OY$$

$$\frac{y^2}{a^2}-\frac{x^2}{b^2}=1 \qquad C(0,\ 0),\ \text{transverse along } OY$$

Transverse axis = $2a$; conjugate axis = $2b$

Distance from center to either focus = $\sqrt{a^2+b^2}$

Latus rectum = $\dfrac{2b^2}{a}$

Eccentricity, $e=\dfrac{\sqrt{a^2+b^2}}{a}$

Difference of distances of any point from the foci = $2a$.

The asymptotes are two lines through the center to which the branches of the hyperbola approach arbitrarily closely; their slopes are $\pm b/a$ [Figure 2.27(a)] or $\pm a/b$ [Figure 2.27(b)].

The rectangular (equilateral) hyperbola has $b = a$. The asymptotes are perpendicular to each other.

$$(x-h)(y-k)=\ \pm e=\sqrt{2} \qquad \text{Center } (h,\ k),\ \text{asymptotes} \quad OX, OY$$

$$xy=\ \pm e=\sqrt{2} \qquad \text{Center } (0,\ 0),\ \text{asymptotes along } OX,\ OY$$

The + sign gives the solid curves in Figure 2.27(c); the – sign gives the dotted curves in Figure 2.27(c).

Example **2.19**

What is the equation of the hyperbola with center at origin, passing through (±2, 0), and an eccentricity of $\sqrt{10}$?

Solution

Equation of a hyperbola with center at origin, x-axis intercepts of (± a, 0), and eccentricity e is $\frac{x^2}{a^2}-\frac{y^2}{b^2}=1$, where $b=a\sqrt{e^2-1}$. Here, $a = 2$; $e = \sqrt{10}$; then, $b = a\sqrt{e^2-1} = 2\sqrt{10-1} = 6$; equation is $\frac{x^2}{2^2}-\frac{y^2}{6^2}=1$.

VECTORS

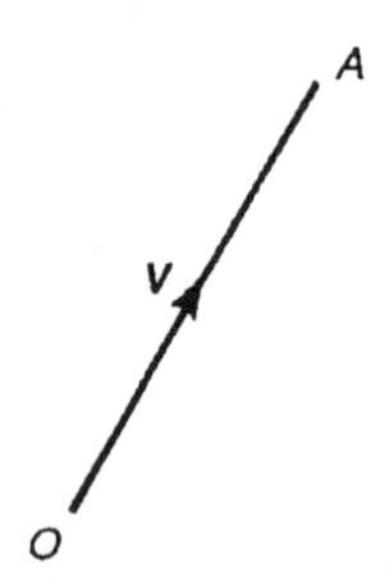

Figure 2.28

Definition and Graphical Representation of a Vector

A vector (**V**) is a quantity that is completely specified by magnitude *and* a direction. A scalar (s) is a quantity that is completely specified by a magnitude *only*.

The vector (**V**) may be represented geometrically by the segment $\overrightarrow{OA}$, the length of OA signifying the magnitude of **V** and the arrow carried by OA signifying the direction of **V**. The segment $\overrightarrow{AO}$ represents the vector –**V**.

Graphical Summation of Vectors

If $\mathbf{V}_1$ and $\mathbf{V}_2$ are two vectors, their graphical sum $\mathbf{V} = \mathbf{V}_1 + \mathbf{V}_2$ is formed by drawing the vector $\mathbf{V}_1 = \overrightarrow{OA}$, from any point O, and the vector $\mathbf{V}_2 = \overrightarrow{AB}$ from the end of $\mathbf{V}_1$ and joining O and B; then $\mathbf{V} = \overrightarrow{OB}$. Also, $\mathbf{V}_1 + \mathbf{V}_2 = \mathbf{V}_2 + \mathbf{V}_1$ and $\mathbf{V}_1 + \mathbf{V}_2 - \mathbf{V} = 0$ (Figure 2.29(a)).

Similarly, if $\mathbf{V}_1, \mathbf{V}_2, \mathbf{V}_3, \ldots, \mathbf{V}_n$ are any number of vectors drawn so that the initial point of one is the end point of the preceding one, then their graphical sum $\mathbf{V} = \mathbf{V}_1 + \mathbf{V}_2 + \ldots + \mathbf{V}_n$ is the vector joining the initial point of $\mathbf{V}_1$ with the end point of $\mathbf{V}_n$ (Figure 2.29(b)).

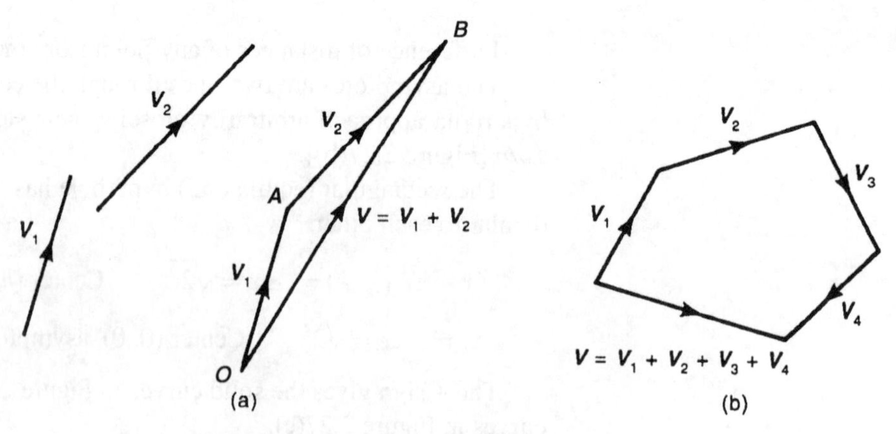

Figure 2.29

Analytic Representation of Vector Components

A vector **V** that is considered as lying in the *x-y* coordinate plane (Figure 2.30(a)) is completely determined by its horizontal and vertical components *x* and *y*. If **i** and **j** represent vectors of unit magnitude along *OX* and *OY*, respectively, and *a* and *b* are the magnitude of *x* and *y*, then **V** may be represented by $\mathbf{V} = a\mathbf{i} + b\mathbf{j}$, its magnitude by $|\mathbf{V}| = +\sqrt{a^2 + b^2}$, and its direction by $\alpha = \tan^{-1} b/a$.

A vector **V** in three-dimensional in space is completely determined by its components *x*, *y*, and *z* along three mutually perpendicular lines *OX*, *OY*, and *OZ*, directed as shown in Figure 2.30(b). If **i**, **j**, and **k** represent vectors of unit magnitude along *OX*, *OY*, *OZ*, respectively, and *a*, *b*, and *c* are the magnitudes of the components *x*, *y*, and *z*, respectively, then **V** may be represented by $\mathbf{V} = a\mathbf{i} + b\mathbf{j} + c\mathbf{k}$, its magnitude by, $|\mathbf{V}| = +\sqrt{a^2 + b^2 + c^2}$, and its direction by $\cos\alpha : \cos\beta : \cos\gamma = a : b : c$.

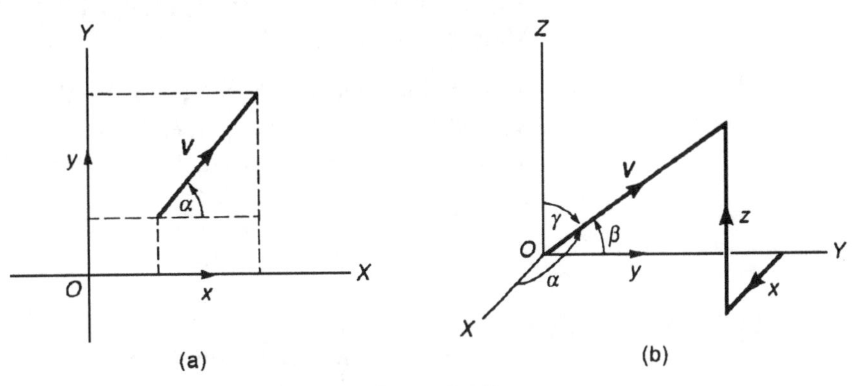

Figure 2.30

Properties of Vectors

$$\mathbf{V} = a\mathbf{i} + b\mathbf{j} \quad \text{or} \quad \mathbf{V} = a\mathbf{i} + b\mathbf{j} + c\mathbf{k}$$

Vector Sum V of any Number of Vectors, $V_1, V_2, V_3, \ldots$

$$\mathbf{V} = \mathbf{V}_1 + \mathbf{V}_2 + \mathbf{V}_3 + \ldots = (a_1 + a_2 + a_3 + \ldots)\,\mathbf{i} + (b_1 + b_2 + b_3 + \ldots)\,\mathbf{j} + (c_1 + c_2 + c_3 + \ldots)\,\mathbf{k}$$

Product of a Vector V and a Scalar *s*

The product $s\mathbf{V}$ has the same direction as **V**, and its magnitude is *s* times the magnitude of **V**.

$$s\mathbf{V} = (sa)\,\mathbf{i} + (sb)\,\mathbf{j} + (sc)\,\mathbf{k}$$

$$(s_1 + s_2)\,\mathbf{V} = s_1\mathbf{V} + s_2\mathbf{V} \qquad (\mathbf{V}_1 + \mathbf{V}_2)s = \mathbf{V}_1 s + \mathbf{V}_2 s$$

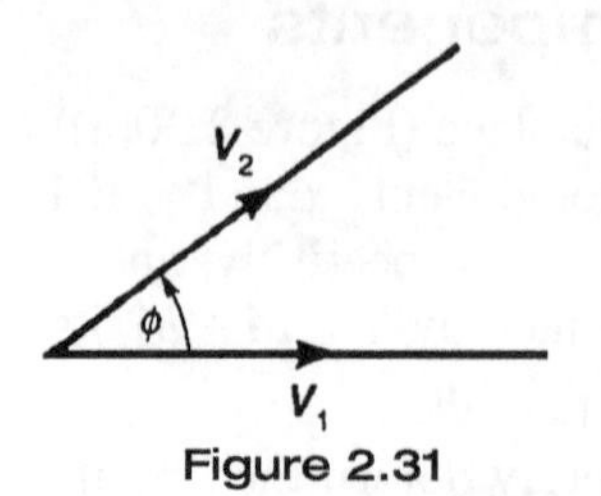

Figure 2.31

Scalar Product or Dot Product of Two Vectors: $V_1 \bullet V_2$

$$\mathbf{V}_1 \bullet \mathbf{V}_2 = |\mathbf{V}_1||\mathbf{V}_2|\cos\phi, \text{ where } \phi \text{ is the angle between } \mathbf{V}_1 \text{ and } \mathbf{V}_2$$
$$\mathbf{V}_1 \bullet \mathbf{V}_2 = \mathbf{V}_2 \bullet \mathbf{V}_1;\quad \mathbf{V}_1 \bullet \mathbf{V}_1 = |\mathbf{V}_1|^2;\quad (\mathbf{V}_1 + \mathbf{V}_2)\bullet \mathbf{V}_3 = \mathbf{V}_1 \bullet \mathbf{V}_3 + \mathbf{V}_2 \bullet \mathbf{V}_3$$
$$(\mathbf{V}_1 + \mathbf{V}_2)\bullet(\mathbf{V}_3 + \mathbf{V}_4) = \mathbf{V}_1 \bullet \mathbf{V}_3 + \mathbf{V}_1 \bullet \mathbf{V}_4 + \mathbf{V}_2 \bullet \mathbf{V}_3 + \mathbf{V}_2 \bullet \mathbf{V}_4$$
$$\mathbf{i}\bullet\mathbf{i} = \mathbf{j}\bullet\mathbf{j} = \mathbf{k}\bullet\mathbf{k} = 1;\quad \mathbf{i}\bullet\mathbf{j} = \mathbf{j}\bullet\mathbf{k} = \mathbf{k}\bullet\mathbf{i} = 0$$

In a plane, $\mathbf{V}_1 \bullet \mathbf{V}_2 = a_1a_2 + b_1b_2$; in space, $\mathbf{V}_1 \bullet \mathbf{V}_2 = a_1a_2 + b_1 b_2 + c_1c_2$.

The scalar product of two vectors $\mathbf{V}_1 \bullet \mathbf{V}_2$ is a scalar quantity and may physically represent the work done by a constant force of magnitude $|\mathbf{V}_1|$ on a particle moving through a distance $|\mathbf{V}_2|$, where ϕ is the angle between the direction of the force and the direction of motion.

Vector Product or Cross Product of Two Vectors: $V_1 \times V_2$

The vector product is $\mathbf{V}_1 \times \mathbf{V}_2 = \mathbf{l}\ |\mathbf{V}_1|\mathbf{V}_2|\sin\phi$, where ϕ is the angle from $\mathbf{V}_1$ to $\mathbf{V}_2$ and $\mathbf{l}$ is a unit vector perpendicular to the plane of the vectors $\mathbf{V}_1$ to $\mathbf{V}_2$ and so directed that a right-handed screw driven in the direction of $\mathbf{l}$ would carry $\mathbf{V}_1$ into $\mathbf{V}_2$.

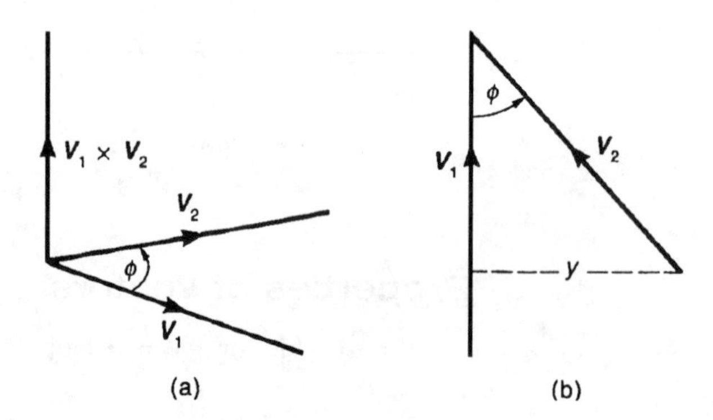

Figure 2.32

$$\mathbf{V}_1 \times \mathbf{V}_2 = -\mathbf{V}_2 \times \mathbf{V}_1;\quad \mathbf{V}_1 \times \mathbf{V}_1 = 0$$
$$(\mathbf{V}_1 + \mathbf{V}_2)\times \mathbf{V}_3 = \mathbf{V}_1 \times \mathbf{V}_3 + \mathbf{V}_2 \times \mathbf{V}_3$$
$$\mathbf{V}_1 \bullet(\mathbf{V}_2 \times \mathbf{V}_3) = \mathbf{V}_2 \bullet(\mathbf{V}_3 \times \mathbf{V}_1) = \mathbf{V}_3 \bullet(\mathbf{V}_1 \times \mathbf{V}_2)$$
$$\mathbf{i}\times\mathbf{i} = \mathbf{j}\times\mathbf{j} = \mathbf{k}\times\mathbf{k} = 0;\quad \mathbf{i}\times\mathbf{j} = \mathbf{k};\quad \mathbf{j}\times\mathbf{k} = \mathbf{i};\quad \mathbf{k}\times\mathbf{i} = \mathbf{j}$$

In the *x-y* plane, $\mathbf{V}_1 \times \mathbf{V}_2 = (a_1b_2 - a_2b_1)\mathbf{k}$.

In space, $\mathbf{V}_1 \times \mathbf{V}_2 = (b_2c_3 - b_3c_2)\mathbf{i} + (c_3a_1 - c_1a_3)\mathbf{j} + (a_1b_2 - a_2b_1)\mathbf{k}$.

The vector product of two vectors is a vector quantity and may physically represent the moment of a force $\mathbf{V}_1$ about a point O placed so that the moment arm is $y = |\mathbf{V}_2| \sin\phi$ (see Figure 2.32(b)).

Example 2.20

Vectors **A** and **B** are defined as: **A** = $i - 2j + 3k$, **B** = $2i + j - 2k$

Find (i) **A** + **B**, (ii) **A** – **B**, (iii) 2**A**, (iv) |**A**|, (v) dot product **A**• **B**, and (vi) the cross product **A** × **B**.

Solution

(i) $\mathbf{A} + \mathbf{B} = 3i - j + k$ (ii) $\mathbf{A} - \mathbf{B} = -i - 3j + 5k$ (iii) $2\mathbf{A} = 2i - 4j + 6k$

(iv) $|\mathbf{A}| = \sqrt{1^2 + (-2)^2 + 3^2} = \sqrt{14}$

(v) Dot product $\mathbf{A} \bullet \mathbf{B} = (2) + (-2) + (-6) = -6$

(vi) Cross product $\mathbf{A} \times \mathbf{B} = \begin{vmatrix} i & j & k \\ 1 & -2 & 3 \\ 2 & 1 & -2 \end{vmatrix} = i + 8j + 5k$

LINEAR ALGEBRA

Matrix Operations

Matrices are rectangular arrays of real or complex numbers. Their great importance arises from the variety of operations that may be performed on them. Using the standard convention, the across-the-page lines are called **rows** and the up-and-down-the-page lines are **columns**. Entries in a matrix are **addressed** with double subscripts (always row first, then column). Thus the matrix

$$\mathbf{A} = \begin{bmatrix} 1 & 2 & 3 \\ 0 & 9 & -3 \end{bmatrix}$$

is 2 × 3, and the "9" is a_{22}. The "2" is a_{12} and the "0" is a_{21}. One also can refer to entries with square brackets, the "9" being $[A]_{22}$ and the "3" $[A]_{13}$.

If two matrices are the same size, they may be added: $[A + B]_{ij} = [A]_{ij} + [B]_{ij}$. Thus,

$$\begin{bmatrix} 1 & 2 & 3 \\ 0 & 9 & -3 \end{bmatrix} + \begin{bmatrix} 1 & 3 \\ 2 & 4 \end{bmatrix}$$

is not defined, but

$$\begin{bmatrix} 1 & 2 & 3 \\ 0 & 9 & -3 \end{bmatrix} + \begin{bmatrix} 1 & 3 & 5 \\ 2 & 4 & 6 \end{bmatrix} = \begin{bmatrix} 2 & 5 & 8 \\ 2 & 13 & 3 \end{bmatrix}$$

is proper.

Any matrix may be multiplied by a **scalar** (a number): $[c\mathbf{A}]_{ij} = c[\mathbf{A}]_{ij}$, so that

$$5 \begin{bmatrix} 1 & 5 \\ 0 & 6 \end{bmatrix} = \begin{bmatrix} 5 & 25 \\ 0 & 30 \end{bmatrix}$$

The most peculiar matrix operation (and the most useful) is matrix multiplication. If **A** is $m \times n$ and **B** is $n \times p$, then **A • B** (or **AB**) is of size $m \times p$, and

$$[AB]_{ij} = \sum_{k=1}^{n} a_{jk} \bullet b_{kj}$$

The **dot product** (scalar product) of the ith row of **A** with the jth column of **B**, as in

$$\begin{bmatrix} 1 & 2 & 3 \\ 0 & 9 & -3 \end{bmatrix} \bullet \begin{bmatrix} 1 & 5 \\ 0 & 6 \end{bmatrix}$$

is not defined (owing to the mismatch of row and column lengths), but

$$\begin{bmatrix} 1 & 2 & 3 \\ 0 & 9 & -3 \end{bmatrix} \bullet \begin{bmatrix} 1 & 5 \\ 0 & 6 \\ 7 & 8 \end{bmatrix} = \begin{bmatrix} 1\,(1)+2\,(0)+3\,(7) & 1\,(5)+2\,(6)+3\,(8) \\ 0\,(1)+9\,(0)-3\,(7) & 0\,(5)+9\,(6)-3\,(8) \end{bmatrix} = \begin{bmatrix} 22 & 41 \\ -21 & 30 \end{bmatrix}$$

is correct.

A matrix with only one row or one column is called a vector, so a matrix times a vector is a vector (if defined). Thus **A** $(m \times n) \bullet$ **X** $(n \times 1) =$ **Y**$(m \times 1)$, so a matrix can be thought of as an **operator** that takes vectors to vectors.

Another useful way of working with matrices is transposition: If **A** is $m \times n$, $\mathbf{A}^t$ is $n \times m$ and is the result of interchanging rows and columns. Hence

$$\begin{bmatrix} 1 & 2 & 3 \\ 0 & 9 & -3 \end{bmatrix}^t = \begin{bmatrix} 1 & 0 \\ 2 & 9 \\ 3 & -3 \end{bmatrix}$$

These various operations interact in the usual pleasant ways (and one decidedly unpleasant way); the standard convention is that all of the following combinations are defined:

$$\begin{gathered}
\mathbf{A}+\mathbf{B}=\mathbf{B}+\mathbf{A} \\
\mathbf{A}+(\mathbf{B}+\mathbf{C}) = (\mathbf{A}+\mathbf{B})+\mathbf{C} \\
c(\mathbf{A}+\mathbf{B})=c\mathbf{A}+c\mathbf{B} \\
(c+d)\,\mathbf{A}=c\mathbf{A}+d\mathbf{A} \\
(-1)\mathbf{A}+\mathbf{A}=(0\mathbf{A}) \\
\mathbf{A}\bullet\mathbf{B}\neq\mathbf{B}\bullet\mathbf{A} \text{ (in general)} \\
\mathbf{A}\bullet(\mathbf{B}\bullet\mathbf{C})=(\mathbf{A}\bullet\mathbf{B})\bullet\mathbf{C} \\
\mathbf{A}\bullet(\mathbf{B}+\mathbf{C})=\mathbf{A}\bullet\mathbf{B}+\mathbf{A}\bullet\mathbf{C} \\
(\mathbf{A}+\mathbf{B})\bullet\mathbf{C}=\mathbf{A}\bullet\mathbf{C}+\mathbf{B}\bullet\mathbf{C} \\
(\mathbf{A}+\mathbf{B})^t=\mathbf{A}^t+\mathbf{B}^t \\
(\mathbf{A}\bullet\mathbf{B})^t=\mathbf{B}^t\bullet\mathbf{A}^t
\end{gathered}$$

In addition, matrices **I**, which are $n \times n$ and whose entries are 1 on the diagonal $i = j$ and 0 elsewhere, are multiplicative identities: **A** • **I** = **A** and **I** • **A** = **A**. Here the two **I** matrices may be different sizes; for example,

$$\begin{bmatrix} 1 & 2 & 3 \\ 0 & 9 & -3 \end{bmatrix} \bullet \begin{bmatrix} 1 & 0 & 0 \\ 0 & 1 & 0 \\ 0 & 0 & 1 \end{bmatrix} = \begin{bmatrix} 1 & 2 & 3 \\ 0 & 9 & -3 \end{bmatrix}$$

but

$$\begin{bmatrix} 1 & 0 \\ 0 & 1 \end{bmatrix} \bullet \begin{bmatrix} 1 & 2 & 3 \\ 0 & 9 & -3 \end{bmatrix} = \begin{bmatrix} 1 & 2 & 3 \\ 0 & 9 & -3 \end{bmatrix}$$

I is called the identity matrix, and the size is understood from context.

Example **2.21**

Verify that the transpose of **A** + **BC** is **C′B′** + **A′** if

$$\mathbf{A} = \begin{bmatrix} 1 & 1 \\ 2 & 3 \end{bmatrix} \quad \mathbf{B} = \begin{bmatrix} 1 & 2 & 3 \\ 0 & 9 & -3 \end{bmatrix} \quad \mathbf{C} = \begin{bmatrix} 1 & 5 \\ 0 & 6 \\ 7 & 8 \end{bmatrix}$$

Solution

$$\mathbf{BC} \begin{bmatrix} 22 & 41 \\ -21 & 30 \end{bmatrix}, \text{ so } \mathbf{A} + \mathbf{BC} = \begin{bmatrix} 23 & 42 \\ -19 & 33 \end{bmatrix} \text{ and } [\mathbf{A} + \mathbf{BC}]' = \begin{bmatrix} 23 & -19 \\ 42 & 33 \end{bmatrix}$$

On the other hand, $\mathbf{C}'\mathbf{B}' = \begin{bmatrix} 1 & 0 & 7 \\ 5 & 6 & 8 \end{bmatrix} \bullet \begin{bmatrix} 1 & 0 \\ 2 & 9 \\ 3 & -3 \end{bmatrix} = \begin{bmatrix} 22 & -21 \\ 41 & 30 \end{bmatrix}$ and

$$\mathbf{A}' = \begin{bmatrix} 1 & 2 \\ 1 & 3 \end{bmatrix}, \text{ so } \mathbf{A}' + \mathbf{C}'\mathbf{B}' = \mathbf{C}'\mathbf{B}' + \mathbf{A}' = \begin{bmatrix} 23 & -19 \\ 42 & 33 \end{bmatrix}$$

Types of Matrices

Matrices are classified according to their appearance or the way they act. If **A** is square and **A′** = **A**, then **A** is called symmetric. If **A′** = –**A**, then it is skew-symmetric.

If **A** has complex entries, **A*** then is called the Hermitian adjoint of **A**. If **A*** = **A′** (complex conjugate), then

$$\begin{bmatrix} 1+i & i \\ 3 & 4-i \end{bmatrix}^* = \begin{bmatrix} 1-i & -i \\ 3 & 4+i \end{bmatrix}' = \begin{bmatrix} 1-i & 3 \\ -i & 4+i \end{bmatrix}$$

If **A** = **A***, then A is called Hermitian. If **A*** = –**A**, the name is skew-Hermitian.

If **A** is square and $a_{ij} = 0$ unless $i = j$, **A** is called diagonal. If **A** is square and zero below the diagonal ($[\mathbf{A}]_{ij} = 0$ if $i > j$), **A** is called upper triangular. The transpose of such a matrix is called lower triangular.

If A is square and there is a matrix $\mathbf{A}^{-1}$ such that $\mathbf{A}^{-1} \bullet \mathbf{A} = \mathbf{A} \bullet \mathbf{A}^{-1} = \mathbf{I}$, **A** is nonsingular. Otherwise, it is singular. If **A** and **B** are both nonsingular $n \times n$ matrices, then **AB** is nonsingular and $(\mathbf{AB})^{-1} = \mathbf{B}^{-1}\mathbf{A}^{-1}$, because $(\mathbf{AB})(\mathbf{B}^{-1}\mathbf{A}^{-1}) = \mathbf{A}(\mathbf{B}\,\mathbf{B}^{-1})\mathbf{A}^{-1} = \mathbf{AIA}^{-1} = \mathbf{AA}^{-1} = \mathbf{I}$, as does $(\mathbf{B}^{-1}\mathbf{A}^{-1}) \bullet (\mathbf{AB})$.

If $\mathbf{A}'\mathbf{A} = \mathbf{AA}' = \mathbf{I}$ and **A** is real, it is called orthogonal (the reason will appear below). If $\mathbf{A}^*\mathbf{A} = \mathbf{AA}^* = \mathbf{I}$ (**A** complex), **A** is called unitary. If **A** commutes with $\mathbf{A}^*$, so that $\mathbf{AA}^* = \mathbf{A}^*\mathbf{A}$, then **A** is called normal.

Elementary Row and Column Operations

The most important tools used in dealing with matrices are the elementary operations: R for row, C for column. If **A** is given matrix, performing $R(i \leftrightarrow j)$ on **A** means interchanging Row i and Row j. $R_i(c)$ means multiplying Row i by the number c (except $c = 0$). $R_j + cR_i$ means multiply Row i by c and add this result into Row j ($i \neq j$). Thus, if

$$\mathbf{A} = \begin{bmatrix} 1 & 2 & 3 \\ 4 & 5 & 6 \\ 7 & 8 & 0 \end{bmatrix}$$

then

$$R(2 \leftrightarrow 3)\,(\mathbf{A}) = \begin{bmatrix} 1 & 2 & 3 \\ 7 & 8 & 0 \\ 4 & 5 & 6 \end{bmatrix} \qquad C_1(2)(\mathbf{A}) = \begin{bmatrix} 2 & 2 & 3 \\ 8 & 5 & 6 \\ 14 & 8 & 0 \end{bmatrix}$$

$$R_1 - R_2(\mathbf{A}) = \begin{bmatrix} -3 & -3 & -3 \\ 4 & 5 & 6 \\ 7 & 8 & 0 \end{bmatrix}$$

These operations are used in reducing matrix problems to simpler ones.

Example 2.22

Solve $\mathbf{AX} = \mathbf{B}$ where

$$\mathbf{A} = \begin{bmatrix} 1 & 2 & 3 \\ 4 & 5 & 6 \\ 7 & 8 & 9 \end{bmatrix}, \quad \mathbf{X} = \begin{bmatrix} x \\ y \\ z \end{bmatrix}, \quad \mathbf{B} = \begin{bmatrix} 1 \\ 1 \\ 1 \end{bmatrix}$$

Solution

Form the "augmented" matrix

$$[\mathbf{A}|\mathbf{B}] = \begin{bmatrix} 1 & 2 & 3 & 1 \\ 4 & 5 & 6 & 1 \\ 7 & 8 & 9 & 1 \end{bmatrix}$$

and perform elementary row operations on this matrix until the solution is apparent:

$$\begin{bmatrix} 1 & 2 & 3 & 1 \\ 4 & 5 & 6 & 1 \\ 7 & 8 & 9 & 1 \end{bmatrix} \begin{matrix} R_2 - 4R_1 \\ R_3 - 7R_1 \end{matrix} \begin{bmatrix} 1 & 2 & 3 & 1 \\ 0 & -3 & -6 & -3 \\ 7 & -6 & -12 & -6 \end{bmatrix} \begin{matrix} R_2\left(-\frac{1}{3}\right) \\ R_3\left(-\frac{1}{6}\right) \end{matrix} \begin{bmatrix} 1 & 2 & 3 & 1 \\ 0 & 1 & 2 & 1 \\ 0 & 1 & 2 & 1 \end{bmatrix}$$

$$R_3 - R_2 \begin{bmatrix} 1 & 2 & 3 & 1 \\ 0 & 1 & 2 & 1 \\ 0 & 0 & 0 & 0 \end{bmatrix}$$

The answer is now apparent: $y + 2z = 1$ and $x + 2y + 3z = 1$, or, z arbitrary, $y = 1 - 2z$, $x = 1 - 2(1 - 2z) - 3z = -1 + z$. This system of equations has an infinite number of solutions.

Example 2.23

Solve the system of equations

$$\begin{aligned} x + y - z &= a \\ 2x - y + 3z &= 2 \\ 3x + 2y + z &= 1 \end{aligned}$$

for x, y, and z in terms of a.

Solution

Strip off the variables x, y, and z:

$$\begin{bmatrix} 1 & 1 & -1 & a \\ 2 & -1 & 3 & 2 \\ 3 & 2 & 1 & 1 \end{bmatrix} \begin{matrix} R_2 - 2R_1 \\ R_3 - 3R_1 \end{matrix} \begin{bmatrix} 1 & 1 & -1 & a \\ 0 & -3 & 5 & 2-2a \\ 0 & -1 & 4 & 1-3a \end{bmatrix}$$

$$\begin{matrix} R_2(2 \leftrightarrow 3) \\ R_2(-1) \end{matrix} \begin{bmatrix} 1 & 1 & -1 & a \\ 0 & 1 & -4 & 3a-1 \\ 0 & -3 & 5 & 2-2a \end{bmatrix} \begin{matrix} R_1 - R_2 \\ R_3 + 3R_2 \end{matrix} \begin{bmatrix} 1 & 0 & 3 & 1-2a \\ 0 & 1 & -4 & 3a-1 \\ 0 & 0 & -7 & 7a-1 \end{bmatrix}$$

The solution is now clear:

$$z = \frac{7a-1}{-7} = -a + \frac{1}{7}$$

$$y = 3a - 1 + 4z = 3a - 1 - 4a + \frac{4}{7} = -a - \frac{3}{7}$$

$$x = 1 - 2a - 3z = 1 - 2a + 3a - \frac{3}{7} = a + \frac{4}{7}$$

Example 2.24

Find $\mathbf{A}^{-1}$ if

$$\mathbf{A} = \begin{bmatrix} 1 & 1 & -1 \\ 1 & 2 & 3 \\ 3 & 2 & 1 \end{bmatrix}$$

Solution

Since this amounts to solving $\mathbf{AX} = \mathbf{B}$ three times, with

$$\mathbf{B} = \begin{bmatrix} 1 \\ 0 \\ 0 \end{bmatrix} \qquad \mathbf{B} = \begin{bmatrix} 0 \\ 1 \\ 0 \end{bmatrix} \qquad \mathbf{B} = \begin{bmatrix} 0 \\ 0 \\ 1 \end{bmatrix}$$

form

$$[\mathbf{A}|\mathbf{I}] = \begin{bmatrix} 1 & 1 & -1 & 1 & 0 & 0 \\ 1 & 2 & 3 & 0 & 1 & 0 \\ 3 & 2 & 1 & 0 & 0 & 1 \end{bmatrix}$$

and perform row operations until a solution emerges.

$$[\mathbf{A}|\mathbf{I}] = \frac{R_2 - R_1}{R_3 - 3R_1} \begin{bmatrix} 1 & 1 & -1 & 1 & 0 & 0 \\ 0 & 1 & 4 & -1 & 1 & 0 \\ 0 & -1 & 4 & -3 & 0 & 1 \end{bmatrix}$$

$$\frac{R_1 - R_2}{R_3 + R_2} z \begin{bmatrix} 1 & 0 & -5 & 2 & -1 & 0 \\ 0 & 1 & 4 & -1 & 1 & 0 \\ 0 & 0 & 8 & -4 & 1 & 1 \end{bmatrix}$$

$$\begin{matrix} R_3\left(\frac{1}{8}\right) \\ \frac{R_2 - 4R_3}{R_1 + 5R_3} \end{matrix} \begin{bmatrix} 1 & 0 & 0 & -\frac{1}{2} & -\frac{3}{8} & \frac{5}{8} \\ 0 & 1 & 0 & 1 & \frac{1}{2} & -\frac{1}{2} \\ 0 & 0 & 1 & -\frac{1}{2} & \frac{1}{8} & \frac{1}{8} \end{bmatrix}$$

Thus,

$$\mathrm{A}^{-1} = \begin{bmatrix} -\frac{1}{2} & -\frac{3}{8} & \frac{5}{8} \\ 1 & \frac{1}{2} & -\frac{1}{2} \\ -\frac{1}{2} & \frac{1}{8} & \frac{1}{8} \end{bmatrix}$$

Example 2.25

Verify that $\mathbf{A}^{-1}\mathbf{A} = \mathbf{I}$ in Example 2.24.

Solution

$$8\mathbf{A}^{-1}\mathbf{A} = \begin{bmatrix} -4 & -3 & 5 \\ 8 & 4 & -4 \\ -4 & 1 & 1 \end{bmatrix} \begin{bmatrix} 1 & 1 & -1 \\ 1 & 2 & 3 \\ 3 & 2 & 1 \end{bmatrix}$$

$$= \begin{bmatrix} -4-3+15 & -4-6+10 & 4-9+5 \\ 8+4-12 & 8+8-8 & -8+12-4 \\ -4+1+3 & -4+2+2 & 4+3+1 \end{bmatrix} = 8 \begin{bmatrix} 1 & 0 & 0 \\ 0 & 1 & 0 \\ 0 & 0 & 1 \end{bmatrix} = 8\mathbf{I}$$

Example 2.26

Describe the set of solutions of $\mathbf{AX} = \mathbf{B}$.

Solution

If $\mathbf{AX}_0 = \mathbf{B}$ is one solution, and $\mathbf{AY} = 0$, then $\mathbf{A}(\mathbf{X}_0 + \mathbf{Y})$ is a solution, so all solutions are of the form $\mathbf{X} = \mathbf{X}_0 + \mathbf{Y}$ where $\mathbf{AY} = 0$. Thus, if $\mathbf{N} = \{\mathbf{Y} : \mathbf{AY} = 0\}$ is the null space of $\mathbf{A}$, the set of solutions to $\mathbf{AX} = \mathbf{B}$ is $\mathbf{X}_0 + \mathbf{N} = \{\mathbf{X}_0 + \mathbf{Y} : \mathbf{Y} \in \mathbf{N}\}$.

Determinants

The determinant of a square matrix is a scalar representing the *volume* of the matrix in some sense. Matrices that are not square do not have determinants.

The determinant is frequently indicated by vertical lines, viz. $|A|$. It is a complicated formula, and one way to find it is by induction. The determinant of a 1×1 matrix is $|a| = a$. The determinant of a 2×2 matrix is

$$\begin{vmatrix} a & b \\ c & d \end{vmatrix} = ad - bc$$

The determinant of an $n \times n$ matrix is given in terms of n determinants, each of size $(n-1) \times (n-1)$. If $\mathbf{A}$ is $n \times n$ and $\mathbf{M}_{ij}$ is the matrix obtained by removing the ith row and the jth column from $\mathbf{A}$, then

$$|A| = \sum_{j=1}^{n} (-1)^{1+j} a_{1j} \, |M_{1j}|$$

Example 2.27

Find the determinant

$$\begin{vmatrix} 1 & 2 & 3 \\ 4 & 0 & 6 \\ 7 & 8 & 9 \end{vmatrix}$$

Solution

$$\begin{aligned} |A| &= (-1)^{1+1} a_{11} |M_{11}| + (-1)^{1+2} a_{12} |M_{12}| + (-1)^{1+3} a_{13} |M_{13}| \\ &= 1 \begin{vmatrix} 0 & 6 \\ 8 & 9 \end{vmatrix} - 2 \begin{vmatrix} 4 & 6 \\ 7 & 9 \end{vmatrix} + 3 \begin{vmatrix} 4 & 0 \\ 7 & 8 \end{vmatrix} \\ &= -48 - 2(36 - 42) + 3(32) = 60 \end{aligned}$$

Example 2.28

Find the determinant

$$\begin{vmatrix} 0 & 0 & 2 & 0 \\ 1 & 2 & 7 & 3 \\ 4 & 0 & 3 & 6 \\ 7 & 8 & -6 & 9 \end{vmatrix}$$

Solution

$$\begin{aligned} |A| &= a_{11}|M_{11}| - a_{12}|M_{12}| + a_{13}|M_{13}| - a_{14}|M_{14}| \\ &= 0|M_{11}| - 0\,|M_{12}| + 2\,|M_{13}| - 0\,|M_{14}| = 2\,(60) = 120 \end{aligned}$$

The last example provides a clue to the evaluation of large determinants, but the use of the first row of **A** in the definition of a determinant was arbitrary. For any row or column (fix i or j),

$$|A| = \sum_{j=1}^{n} (-1)^{1+j} a_{i1j}\, |M_{ij}|$$

The interaction of the determinant with elementary row or column operations is simple: Interchanging two rows changes the sign of the determinant; multiplying a row by a constant multiplies the determinant by that constant.

Example 2.29

Evaluate the determinant

$$\begin{vmatrix} 1 & 2 & 3 & 4 \\ 1 & 1 & 1 & 0 \\ 4 & 0 & 3 & 2 \\ 0 & 3 & 0 & 1 \end{vmatrix}$$

Solution

Choose a row or column with many zeroes and introduce still more:

$$|A| = C_2 - 3C_4\,|A| = \begin{vmatrix} 1 & -10 & 3 & 4 \\ 1 & 1 & 1 & 0 \\ 4 & -6 & 3 & 2 \\ 0 & 0 & 0 & 1 \end{vmatrix} = (-1)^{4+4} a_{44} \begin{vmatrix} 1 & -10 & 3 \\ 1 & 1 & 1 \\ 4 & -6 & 3 \end{vmatrix}$$

$$\overset{R_2 - R_1}{\underset{R_3 - 4R_1}{=}} \begin{vmatrix} 1 & -10 & 3 \\ 0 & 11 & -2 \\ 0 & 34 & -9 \end{vmatrix} = (-1)^{1+1} a_{11} \begin{vmatrix} 11 & -2 \\ 34 & -9 \end{vmatrix} = -99 + 68 = -31$$

Example 2.30

Find which values, if any, of the number c make $\mathbf{A}$ singular if

$$\mathbf{A} = \begin{vmatrix} 1 & 2 & c \\ 4 & 5 & 6 \\ 1 & 1 & 1 \end{vmatrix}$$

Solution

$|\mathbf{A}| = (-1)^2(5-6) + (-1)^3(2)(4-6) + (-1)^4 c(4-5) = -1 + 4 - c = 0.$ Hence $\mathbf{A}$ is singular for only one value of c, $c = 3$.

Cramer's Rule is a consequence of adj(A): If $\mathbf{A}$ is nonsingular, the ith component of the solution of $\mathbf{AX} = \mathbf{B}$ is $x_i = \frac{|A_i|}{|A|}$, where A_i is the result of replacing the ith column of $\mathbf{A}$ by $\mathbf{B}$.

Example 2.31

Find x_2 in $\mathbf{AX} = \mathbf{B}$ by Cramer's Rule if

$$\mathbf{A} = \begin{bmatrix} 1 & 2 & 1 & 1 \\ 3 & 4 & 5 & -2 \\ 6 & 7 & 1 & 5 \\ -1 & 0 & 2 & 0 \end{bmatrix} \quad \text{and} \quad \mathbf{B} = \begin{bmatrix} 1 \\ 2 \\ 3 \\ 4 \end{bmatrix}$$

Solution

First,

$$|A| = \begin{vmatrix} 1 & 2 & 3 & 1 \\ 3 & 4 & 11 & -2 \\ 6 & 7 & 13 & 5 \\ -1 & 0 & 0 & 0 \end{vmatrix} = (-1)^{4+1}(-1) \begin{vmatrix} 2 & 3 & 1 \\ 4 & 11 & -2 \\ 7 & 13 & 5 \end{vmatrix}$$

$$= \begin{vmatrix} 0 & 0 & 1 \\ 8 & 17 & -2 \\ -3 & -2 & 5 \end{vmatrix} = (-1)^{1+3}(1) \begin{vmatrix} 8 & 17 \\ -3 & -2 \end{vmatrix} = -16 + 51 = 35$$

Next, the numerator of x_2 is

$$\begin{vmatrix} 1 & 1 & 1 & 1 \\ 3 & 2 & 5 & -2 \\ 6 & 3 & 1 & 5 \\ -1 & 4 & 2 & 0 \end{vmatrix} = \begin{vmatrix} 1 & 0 & 0 & 0 \\ 3 & -1 & 2 & -5 \\ 6 & -3 & -5 & -1 \\ -1 & 5 & 3 & 1 \end{vmatrix} = \begin{vmatrix} -1 & 2 & -5 \\ -3 & -5 & -1 \\ 5 & 3 & 1 \end{vmatrix} = \begin{vmatrix} -1 & 2 & -5 \\ 0 & -11 & 14 \\ 0 & 13 & -24 \end{vmatrix}$$

$$= -\begin{vmatrix} -11 & 14 \\ 13 & -24 \end{vmatrix} = -\begin{vmatrix} -11 & 14 \\ 2 & -10 \end{vmatrix} = -(110-28) = -82, x_2 = -\frac{82}{35}$$

NUMERICAL METHODS

This portion of numerical methods includes techniques of finding roots of polynomials by the Routh-Hurwitz criterion and Newton methods, Euler's techniques of numerical integration and the trapezoidal methods, and techniques of numerical solutions of differential equations.

Root Extraction

Routh-Hurwitz Method (without Actual Numerical Results)

Root extraction, even for simple roots (i.e., without imaginary parts), can become quite tedious. Before attempting to find roots, one should first ascertain whether they are really needed or whether just knowing the area of location of these roots will suffice. If all that is needed is knowing whether the roots are all in the left half-plane of the variable (such as is in the s-plane when using Laplace transforms—as is frequently the case in determining system stability in control systems), then one may use the Routh-Hurwitz criterion. This method is fast and easy even for higher-ordered equations. As an example, consider the following polynomial:

$$p_n(x) = \prod_{m=1}^{n}(x - x_m) = x^n + a_1x^{n-1} + a_2x^{n-2} + \cdots + a_{n-1} \quad \textbf{(2.1)}$$

Here, finding the roots, x_m, for $n > 3$ can become quite tedious without a computer; however, if one only needs to know if any of the roots have positive real parts, one can use the Routh-Hurwitz method. Here, an array is formed listing the coefficients of every other term starting with the highest power, n, on a line, followed by a line listing the coefficients of the terms left out of the first row. Following rows are constructed using Routh-Hurwitz techniques, and after completion of the array, one merely checks to see if all the signs are the same (unless there is a zero coefficient—then something else needs to be done) in the first column; if none, no roots will exist in the right half-plane. In case of zero coefficient, a simple technique is used; for details, see almost any text dealing with stability of control systems. A short example follows.

$F(s) = s^3 + 3s^2 + 10$ Array: s^3 1 2 Where the s^1 term is formed as

$= (s+?)(s+?)(s+?)$	s^2	3	10	
	s^1	$-\frac{4}{3}$	0	$(3 \times 2 - 10 \times 1)/3 = -\frac{4}{3}$. For details, refer to any text on control systems or numerical methods.
	s^0	10	0	

Here, there are two sign changes: one from 3 to $-\frac{4}{3}$, and one from $-\frac{4}{3}$ to 10. This means there will be two roots in the right half-plane of the s-plane, which yield an unstable system. This technique represents a great savings in time without having to factor the polynomial.

Newton's Method

The use of Newton's method of solving a polynomial and the use of iterative methods can greatly simplify a problem. This method utilizes synthetic division and is based upon the remainder theorem. This synthetic division requires estimating a root at the start, and, of course, the best estimate is the actual root. The root is the correct one when the remainder is zero. (There are several ways of estimating this root, including a slight modification of the Routh-Hurwitz criterion.)

If a $P_n(x)$ polynomial (see Equation (2.1)) is divided by an estimated factor $(x - x_1)$, the result is a reduced polynomial of degree $n - 1$, $Q_{n-1}(x)$, plus a constant remainder of b_{n-1}. Thus, another way of describing Equation (2.1) is

$$P_n(x)/(x - x_1) = Q_{n-1}(x) + b_{n-1}/(x - x_1) \quad \text{or} \quad P_n(x) = (x - x_1)Q_{n-1}(x) + b_{n-1} \tag{2.2}$$

If one lets $x = x_1$, Equation (2.2) becomes

$$P_n(x = x_1) = (0)Q_{n-1}(x) + b_{n-1} = b_{n-1} \tag{2.3}$$

Equation 2.3 leads directly to the remainder theorem: "The remainder on division by $(x - x_1)$ is the value of the polynomial at $x = x_1$, $P_n(x_1)$."[1]

Newton's method (actually, the Newton-Raphson method) for finding the roots for an nth-order polynomial is an iterative process involving obtaining an estimated value of a root (leading to a simple computer program). The key to the process is getting the first estimate of a possible root. Without getting too involved, recall that the coefficient of x^{n-1} represents the sum of all of the roots and the last term represents the product of all n roots; then the first estimate can be "guessed" within a reasonable magnitude. After a first root is chosen, find the rate of change of the polynomial at the chosen value of the root to get the next, closer value of the root x_{n+1}. Thus the new root estimate is based on the last value chosen:

$$\begin{aligned} x_{n+1} &= x_n - P_n(x_n)/P_n'(x_n), \\ \text{where } P_n'(x_n) &= dP_n(x)/dx \text{ evaluated at } x = x_n \end{aligned} \tag{2.4}$$

NUMERICAL INTEGRATION

Numerical integration routines are extremely useful in almost all simulation-type programs, design of digital filters, theory of z-transforms, and almost any problem solution involving differential equations. And because digital computers have essentially replaced analog computers (which were almost true integration

[1] Gerald & Wheatley, *Applied Numerical Analysis*, 3rd ed., Addison-Wesley, 1985.

devices), the techniques of approximating integration are well developed. Several of the techniques are briefly reviewed below.

Euler's Method

For a simple first-order differential equation, say $dx/dt + ax = af$, one could write the solution as a continuous integral or as an interval type one:

$$x(t) = \int^{t} [-ax(\tau) + af(\tau)]d\tau \quad \textbf{(2.5a)}$$

$$x(kT) = \int^{kT-T} [-ax + af]d\tau + \int_{kT-T}^{kT} [-ax + af]d\tau = x(kT-T) + A_{rect} \quad \textbf{(2.5b)}$$

Here, A_{rect} is the area of $(-ax + af)$ over the interval $(kT - T) < \tau < kT$. One now has a choice looking back over the rectangular area or looking forward. The rectangular width is, of course, T. For the forward-looking case, a first approximation for x_1 is[2]

$$\begin{aligned} x_1(kT) &= x_1(kT-T) + T[ax_1(kT-T) + af(kT-T)] \\ &= (1-aT)x_1(kT-T) + aTf(kT-T) \end{aligned} \quad \textbf{(2.5c)}$$

Or, in general, for Euler's forward rectangle method, the integral may be approximated in its simplest form (using the notation $t_{k+1} - t_k$ for the width, instead of T, which is $kT–T$) as

$$\int_{t_k}^{t_{k+1}} x(\tau)\,d\tau \approx (t_{k+1} - t_k)x(t_k) \quad \textbf{(2.6)}$$

Trapezoidal Rule

This trapezoidal rule is based upon a straight-line approximation between the values of a function, $f(t)$, at t_0 and t_1. To find the area under the function, say a curve, is to evaluate the integral of the function between points a and b. The interval between these points is subdivided into subintervals; the area of each subinterval is approximated by a trapezoid between the end points. It will be necessary only to sum these individual trapezoids to get the whole area; by making the intervals all the same size, the solution will be simpler. For each interval of delta t (i.e., $t_{k+1} - t_k$), the area is then given by

$$\int_{t_k}^{t_{k+1}} x(\tau)\,d\tau \approx (1/2)(t_{k+1} - t_k)[x(t_{k+1}) + x(t_k)] \quad \textbf{(2.7)}$$

This equation gives good results if the delta t's are small, but it is for only one interval and is called the "local error." This error may be shown to be – (1/12) (delta t)$^3 f''(t = \xi_1)$, where ξ_1 is between t_0 and t_1. For a larger "global error" it may be shown that

---

[2] This method is as presented in Franklin & Powell, *Digital Control of Dynamic Systems*, Addison-Wesley, 1980, page 55.

$$\text{Global error} = -(1/12)(\text{delta } t)^3\,[f''(\xi_1)+f''(\xi_2)+\cdots+f''(\xi_n)] \tag{2.8}$$

Following through on Equation 2.8 allows one to predict the error for the trapezoidal integration. This technique is beyond the scope of this review or probably the examination; however, for those interested, please refer to pages 249–250 of the previously mentioned reference to Gerald & Wheatley.

NUMERICAL SOLUTIONS OF DIFFERENTIAL EQUATIONS

This solution will be based upon first-order ordinary differential equations. However, the method may be extended to higher-ordered equations by converting them to a matrix of first-ordered ones.

Integration routines produce values of system variables at specific points in time and update this information at each interval of delta time as T (delta $t = T = t_{k+1} - t_k$). Instead of a continuous function of time, $x(t)$, the variable x will be represented with discrete values such that $x(t)$ is represented by $x_0, x_1, x_2, \ldots, x_n$. Consider a simple differential equation as before as based upon Euler's method,

$$dx/dt + ax = f(t)$$

Now assume the delta time periods, T, are fixed (not all routines use fixed step sizes); then one writes the continuous equations as a difference equation where $dx/dt \approx (x_{k+1} - x_k)/T = -ax_k + f_k$ or, solving for the updated value, x_{k+1},

$$x_{k+1} = x_k - Tax_k + Tf_k \tag{2.9a}$$

For fixed increments by knowing the first value of $x_{k=0}$ (or the initial condition), one may calculate the solution for as many "next values" of x_{k+1} as desired for some value of T. The difference equation may be programmed in almost any high-level language on a digital computer; however, T must be small as compared to the shortest time constant of the equation (here, $1/a$).

The following equation—with the "f" term meaning "a function of" rather than as a "forcing function" term as used in Equation (2.5a)—is a more general form of Equation (2.9a). This equation is obtained by letting the notation x_{k+1} become $y\,[k+1\,\Delta t]$ and is written (perhaps somewhat more confusingly) as

$$y[(k+1)\Delta t] = y(k\Delta t) + \Delta t f[y(k\Delta t), k\Delta t)] \tag{2.9b}$$

Reduction of Differential Equation Order

To reduce the order of a linear time-dependent differential equation, the following technique is used. For example, assume a second-order equation: $x'' + ax' + bx = f(t)$. If we define $x = x_1$ and $x' = x_1' = x_2$, then

$$x_2' + ax_2 + bx_1 = f(t)$$
$$x_1' = x_2 \text{ (by definition)}$$
$$x_2' = -b_{x1} - ax_2 + f(t)$$

This technique can be extended to higher-order systems and, of course, be put into a matrix form (called the state variable form). And it can easily be set up as a matrix of first-order difference equations for solving digitally.

DIFFERENTIAL CALCULUS

Definition of a Function

Notation. A variable y is said to be a function of another variable x if, when x is given, y is determined. The symbols $f(x)$, $F(x)$, etc., represent various functions of x. The symbol $f(a)$ represents the value of $f(x)$ when $x = a$.

Definition of a Derivative

Let $y = f(x)$. If Δx is any increment (increase or decrease) given to x, and Δy is the corresponding increment in y, then the derivative of y with respect to x is the limit of the ratio of Δy to Δx as Δx approaches zero; that is,

$$\frac{dy}{dx} = \lim_{\Delta x \to 0} \frac{\Delta y}{\Delta x} = \lim_{\Delta x \to 0} \frac{f(x+\Delta x) - f(x)}{\Delta x} = f'(x)$$

Some Relations among Derivatives

If $x = f(y)$, then $\dfrac{dy}{dx} = 1 \div \dfrac{dx}{dy}$

If $x = f(t)$ and $y = F(t)$, then $\dfrac{dy}{dx} = \dfrac{dy}{dt} \div \dfrac{dx}{dt}$

If $y = f(u)$ and $u = F(x)$, then $\dfrac{dy}{dx} = \dfrac{dy}{du} \times \dfrac{du}{dx}$

If $x = f(t)$, then $f''(t) = \dfrac{d^2x}{dt^2} = \dfrac{d}{dt}\left(\dfrac{dx}{dt}\right)$

Table of Derivatives

Functions of x are represented by u and v, and constants are represented by a, n, and e.

$$\frac{d}{dx}(x) = 1 \qquad \frac{d}{dx}(a) = 0$$

$$\frac{d}{dx}(u \pm v \pm \ldots) = \frac{du}{dx} \pm \frac{dv}{dx} \pm \ldots \qquad \frac{d}{dx}(au) = a\frac{du}{dx}$$

$$\frac{d}{dx}(uv) = u\frac{dv}{dx} + v\frac{du}{dx} \qquad \frac{d}{dx}\left(\frac{u}{v}\right) = \frac{v\dfrac{du}{dx} - u\dfrac{dv}{dx}}{v^2}$$

$$\frac{d}{dx}(u^n) = nu^{n-1}\frac{du}{dx} \qquad \frac{d}{dx}\log_a u = \frac{\log_a e}{u}\frac{du}{dx}$$

$$\frac{d}{dx}a^u = a^u \ln a \frac{du}{dx}$$

$$\frac{d}{dx}e^u = e^u\frac{du}{dx} \qquad \frac{d}{dx}u^v = vu^{v-1}\frac{du}{dx} + u^v \ln u\frac{dv}{dx}$$

$$\frac{d}{dx}\sin u = \cos u\frac{du}{dx} \qquad \frac{d}{dx}\cot u = -\csc^2 u\frac{du}{dx}$$

$$\frac{d}{dx}\cos u = -\sin u\frac{du}{dx} \qquad \frac{d}{dx}\sec u = \sec u \tan u\frac{du}{dx}$$

$$\frac{d}{dx}\tan u = \sec^2 u\frac{du}{dx} \qquad \frac{d}{dx}\csc u = -\csc u \cot u\frac{du}{dx}$$

$$\frac{d}{dx}\sin^{-1} u = \frac{1}{\sqrt{1-u^2}}\frac{du}{dx} \quad \text{where} \quad -\pi/2 \le \sin^{-1} u \ge \pi/2$$

$$\frac{d}{dx}\cos^{-1} u = -\frac{1}{\sqrt{1-u^2}}\frac{du}{dx} \quad \text{where} \quad 0 \le \cos^{-1} u \ge \pi$$

$$\frac{d}{dx}\tan^{-1} u = \frac{1}{1+u^2}\frac{du}{dx}$$

$$\frac{d}{dx}\cot^{-1} u = -\frac{1}{1+u^2}\frac{du}{dx}$$

$$\frac{d}{dx}\sec^{-1} u = \frac{1}{u\sqrt{u^2-1}}\frac{du}{dx} \quad \text{where} \quad 0 \le \sec^{-1} u \le \pi/2 \text{ and } -\pi \le \sec^{-1} u \le -\pi/2$$

$$\frac{d}{dx}\csc^{-1} u = -\frac{1}{u\sqrt{u^2-1}}\frac{du}{dx} \quad \text{where} \quad -\pi < \csc^{-1} u \le -\pi/2 \text{ and } 0 < \csc^{-1} u \le \pi/2$$

Example 2.32

Find the derivatives of the following functions with respect to x.

(i) $x^3 + 4e^{-2x}$, (ii) $x^2 \sin x$, (iii) $2\sin^2 x$, (iv) $(\frac{e^{-x}}{x})$

Solution

(i) $\frac{d}{dx}(x^3 + 4e^{-2x}) = \frac{d}{dx}(x^3) + \frac{d}{dx}(4e^{-2x}) = 3x^2 + 4(-2)e^{-2x} = 3x^2 - 8e^{-2x}$

(ii) $\frac{d}{dx}(x^2 \sin x) = x^2\frac{d}{dx}(\sin x) + (\sin x)\frac{d}{dx}(x^2) = x^2 \cos x + 2x \sin x$

(iii) $\frac{d}{dx}(2\sin^2 x) = 2(2)(\sin^{2-1} x)\frac{d}{dx}(\sin x) = (4\sin x)\cos x$

(iv) $\frac{d}{dx}(\frac{e^{-x}}{x}) = \frac{x\frac{d}{dx}(e^{-x}) - e^{-x}\frac{d}{dx}(x)}{x^2} = \frac{x(-e^{-x}) - (e^{-x})1}{x^2} = \frac{-e^{-x}(x+1)}{x^2}$

Slope of a Curve: Tangent and Normal

The slope of the curve (slope of the tangent line to the curve) whose equation is $y = f(x)$ is

$$\text{Slope} = m = \tan\phi = \frac{dy}{dx} = f'(x)$$

$$\text{Slope at } x_1 \text{ is } m_1 = f'(x_1)$$

The equation of a tangent line at $P_1(x_1, y_1)$ is $y - y_1 = m_1(x - x_1)$. The equation of a normal at $P_1(x_1, y_1)$ is

$$y - y_1 = -\frac{1}{m_1}(x - x_1)$$

The angle β of the intersection of two curves whose slopes at a common point are m_1 and m_2 is

$$\beta = \tan^{-1}\frac{m_2 - m_1}{1 + m_1 m_2}$$

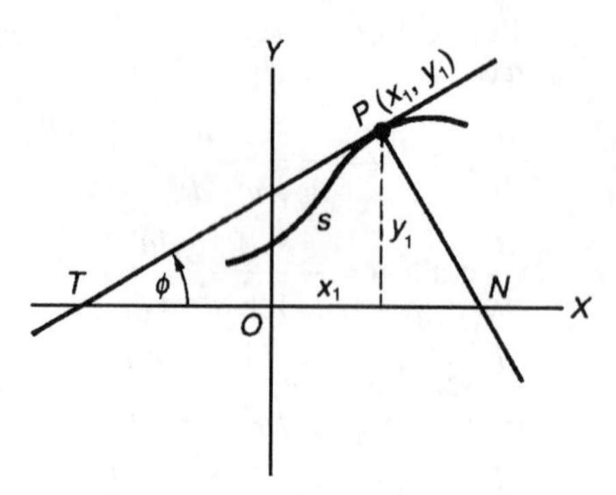

Figure 2.33

Example 2.34

Find the slope of the curve $y = (4x^2 - 8)$ at $x = 1$.

Solution

Slope $\frac{dy}{dx} = \frac{d}{dx}(4x^2 - 8) = \frac{d}{dx}(4x^2) - \frac{d}{dx}(8) = 4\frac{d}{dx}(x^2) - 0 = 8x$;

at $x = 1$, slope = 8

Example 2.35

Find the equation of the tangent to the curve $y = (x^2 - x - 4)$ at $(1, -4)$.

Solution

Slope of the curve $\frac{dy}{dx} = \frac{d}{dx}(x^2 - x - 4) = \frac{d}{dx}(x^2) - \frac{d}{dx}(x) - \frac{d}{dx}(4) = 2x - 1$;

at $(1, -4)$, slope $= 2(1) - 1 = 1$

Equation of the tangent is $(y + 4) = 1(x - 1)$ or $y = x - 5$.

Maximum and Minimum Values of a Function

The maximum or minimum value of a function $f(x)$ in an interval from $x = a$ to $x = b$ is the value of the function that is larger or smaller, respectively, than the values of the function in its immediate vicinity. Thus, the values of the function at M_1 and M_2 in Figure 2.34 are maxima, and its values at m_1 and m_2 are minima.

Test for a maximum at $x = x_1$: $f'(x_1) = 0$ or ∞, and $f''(x_1) < 0$

Test for minimum at $x = x_1$: $f'(x_1) = 0$ or ∞, and $f''(x_1) > 0$

If $f''(x_1) = 0$ or ∞, then for a maximum, $f'''(x_1) = 0$ or ∞ and $f^{IV}(x_1) < 0$; for a minimum, $f'''(x_1) = 0$ or ∞ and $f^{IV}(x_1) > 0$, and similarly if $f^{IV}(x_1) = 0$ or ∞, and so on, where f^{IV} represents the fourth derivative.

In a practical problem that suggests that the function $f(x)$ has a maximum or has a minimum in an interval from $x = a$ to $x = b$, simply equate $f'(x)$ to 0 and solve for the required value of x. To find the largest or smallest values of a function $f(x)$ in an interval from $x = a$ to $x = b$, find also the values $f(a)$ and $f(b)$. L and S may be the largest and smallest values, although they are not maximum or minimum values (see Figure 2.34).

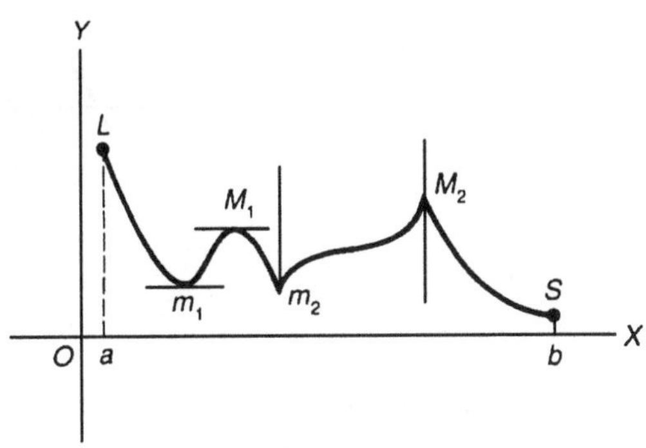

Figure 2.34

Points of Inflection of a Curve

Wherever $f''(x) < 0$, the curve is concave down.

Wherever $f''(x) > 0$, the curve is concave up.

The curve is said to have a point of inflection at $x = x_1$ if $f''(x_1) = 0$ or ∞, and the curve is concave up on one side of $x = x_1$ and concave down on the other (see points I_1 and I_2 in Figure 2.35).

Example 2.36

For the function $y = f(x) = (x^3 - 3x)$, find the maximum, minimum, and the point of inflection.

Solution

The derivative of $f(x)$, $f'(x) = 3x^2 - 3$.

Equating $f'(x)$ to 0, $3x^2 - 3 = 0$, which has roots at $x = 1$ and -1.

The derivative of $f'(x)$, $f''(x) = 6x$. At $x = 1$, $f''(1) = 6$ and at $x = -1$, $f''(-1) = -6$.

Minimum occurs at $x = 1$, since $f''(1) > 0$ and $f''(1) = 0$. Value of minimum $= f(1) = -2$.

Maximum occurs at $x = -1$, since $f'(-1) < 0$ and $f'(-1) = 0$. Value of maximum $= f(-1) = 2$.

$f''(x) = 0$ at $x = 0$; $f''(x)$ changes sign as x passes through 0; then $x = 0$ is the inflection point (it is between 1 and –1).

Taylor and Maclaurin Series

In general, any $f(x)$ may be expanded into a **Taylor series**:

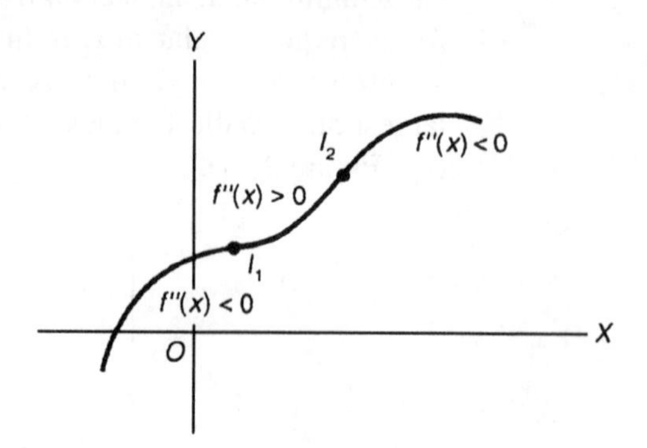

Figure 2.35

$$f(x) = f(a) + f'(a)\frac{x-a}{1} + f''(a)\frac{(x-a)^2}{2!} + f'''(a)\frac{(x-a)^3}{3!} + \ldots$$

$$+ f^{(n-1)}(a)\frac{(x-a)^{n-1}}{(n-1)!} + R_n$$

where a is any quantity whatever, so chosen that none of the expressions $f(a)$, $f'(a)$, $f''(a)$,... become infinite. If the series is to be used for the purpose of computing the approximate value of $f(x)$ for a given value of x, a should be chosen such that $(x - a)$ is numerically very small, and thus only a few terms of the series need be used. If $a = 0$, this series is called a Maclaurin series.

Example 2.37

Find the power series expansion of sin x about the point 0.

Solution

In the Taylor's series expansion, $f(x) = \sin(x)$ and $a = 0$. Derivatives of $f(x)$ are: $f'(x) = \cos(x)$, $f''(x) = -\sin(x)$, $f'''(x) = -\cos(x)$, and so on.

Substituting $f(x)$ and its derivatives at a in the expansion,

$$\sin(x) = \sin(0) + x\cos(0) + \frac{x^2(-\sin(0))}{2!} + \frac{x^3(-\cos(0))}{3!} + \ldots = x - \frac{x^3}{3!} + \frac{x^5}{5!} - \ldots$$

Evaluation of Indeterminate Forms

Let $f(x)$ and $F(x)$ be two functions of x, and let a be a value of x.

1. If $\frac{f(a)}{F(a)}=\frac{0}{0}$ or $\frac{\infty}{\infty}$, use $\frac{f'(a)}{F'(a)}$ for the value of this fraction.

   If $\frac{f'(a)}{F'(a)}=\frac{0}{0}$ or $\frac{\infty}{\infty}$, use $\frac{f''(a)}{F''(a)}$ for the value of this fraction, and so on.

2. If $f(a) \bullet F(a) = 0 \bullet \infty$ or if $f(a) - F(a) = \infty - \infty$, evaluate the expression by changing the product or difference to the form $\frac{0}{0}$ or $\frac{\infty}{\infty}$ and use the previous rule.

3. If $f(a)^{F(a)} = 0^0$ or ∞^0 or 1^∞, then form $e^{F(a) \bullet \ln f(a)}$, and the exponent, being of the form $0 \bullet \infty$, may be evaluated by rule 2.

Example **2.38**

Find the following limits: (i) $\lim\limits_{x \to 0} \frac{\sin x}{x}$ (ii) $\lim\limits_{x \to 0} \frac{1 - \cos x}{x^2}$

Solution

(i) Since $\frac{\sin(0)}{0} = \frac{0}{0}$, we apply the limiting theorem:

$$\lim_{x \to 0} \frac{\sin x}{x} = \lim_{x \to 0} \frac{d(\sin x)}{d(x)} = \lim_{x \to 0} \frac{\cos x}{1} = 1$$

(ii) Since $\frac{1 - \cos(0)}{0^2} = \frac{0}{0}$, we apply the limiting theorem:

$$\lim_{x \to 0} \frac{1 - \cos x}{x^2} = \lim_{x \to 0} \frac{\sin x}{2x} = \lim_{x \to 0} \frac{\cos x}{2} = \frac{1}{2}$$

Example **2.39**

The cubic $y = x^3 + x^2 - 3$ has one point of inflection. Where does it occur?

Solution

The answer requires knowing where y' changes sign. Now $y' = 3x^2 + 2x$ and $y'' = 6x + 2$, which is 0 where $x = -\frac{1}{3}$. Thus the only inflection point is at $x = -\frac{1}{3}$.

Example **2.40**

The function of Example 2.39 has one local maximum and one local minimum. Where are they?

Solution

Setting $y' = 0$ $(3x^2 + 2x = 0)$ yields $x = 0$ or $x = -\frac{2}{3}$. Since the second derivative is 2 at $x = 0$, this is the local minimum. At $x = -\frac{2}{3}$, $y'' = -2$, so $x = -\frac{2}{3}$ is the local maximum.

Differential of a Function

If $y = f(x)$ and Δx is an increment in x, then the differential of x equals the increment of x, or $dx = \Delta x$; and the differential of y is the derivative of y multiplied by the differential of x; thus

$$dy = \frac{dy}{dx}dx = \frac{df(x)}{dx}dx = f'(x)\,dx \quad \text{and} \quad \frac{dy}{dx} = dy \div dx$$

If $x = f_1(t)$ and $y = f_2(t)$, then $dx = f_1'(t)\,dt$, and $dy = f_2'(t)\,dt$.

Every derivative formula has a corresponding differential formula; thus, from the Table of Derivatives subsection, we have, for example,

$$d(uv) = u\,dv + v\,du; \quad d(\sin u) = \cos\,u\,du; \quad d(\tan^{-1} u) = \frac{du}{1+u^2}$$

Functions of Several Variables, Partial Derivatives, and Differentials

Let z be a function of two variables, $z = f(x, y)$; then its partial derivatives are

$$\frac{\partial z}{\partial x} = \frac{dz}{dx} \text{ when } y \text{ is kept constant} \qquad \frac{\partial z}{\partial y} = \frac{dz}{dy} \text{ when } x \text{ is kept constant}$$

Example 2.41

Find the partial derivatives of $f(x, y) = 4x^2y - 2y$ (i) with respect to x, and (ii) with respect to y.

Solution

(i) $\dfrac{\partial f}{\partial x} = \dfrac{\partial}{\partial x}(4x^2y - 2y) = 4y\dfrac{\partial(x^2)}{\partial x} - 0 = (4y)(2x) = 8xy$

(ii) $\dfrac{\partial f}{\partial y} = \dfrac{\partial}{\partial y}(4x^2y - 2y) = 4x^2\dfrac{\partial(y)}{\partial y} - \dfrac{\partial(2y)}{\partial y} = 4x^2 - 2$

Example 2.42

Two automobiles are approaching the origin. The first one is traveling from the left on the x-axis at 30 mph. The second is traveling from the top on the y-axis at 45 mph. How fast is the distance between them changing when the first is at (–5, 0) and the second is at (0, 10)? (Both coordinates are in miles.)

Solution

If $x(t)$ is taken as the position of the first auto at time t and $y(t)$ as the position of the second auto at time t, then the distance between them at time t is $s(t) = \sqrt{[x(t)]^2 + [y(t)]^2}$. Using the chain rule,

$$s'(t) = \frac{ds}{dt} = \frac{1}{d_2 s(t)} \frac{d}{dt}\{[x(t)]^2 + [y(t)]^2\} = \frac{1}{2s(t)}[2x(t)x'(t) + 2y(t)y'(t)]$$

Now $x'(t) = 30$ and $y'(t) = -45$ for all t; and when $t = t_0$, $x(t_0) = -5$ and $y(t_0) = 10$.

Therefore,

$$s'(t_0) = \frac{-2(5)(30) - 2(10)(45)}{2\sqrt{(-5)^2 + (10)^2}} = \frac{-1200}{2\sqrt{125}} = \frac{-120}{\sqrt{5}} = 24\sqrt{5} \approx 54$$

Thus, the two automobiles are “closing” at about 54 mph.

Example 2.43

How close do the two automobiles in Example 2.42 get?

Solution

One wants to minimize $s(t)$ in Example 2.42, so set $s'(t) = 0$. Thus,

$$\frac{xx' + yy'}{s} = 0 \quad \text{or} \quad xx' + yy' = 0$$

Since $x' = 30$ and $y' = -45$, $30x = 45y$. However, since $x'(t) = 30$, $x(t) = 30t + x_0$, and similarly $y(t) = -45t + y_0$. If one takes $t_0 = 0$ when the problem starts, $x_0 = -5$ and $y_0 = 10$, so $30(30t - 5) = 45(-45t + 10)$ gives time of minimum distance. Solving for t, factor out 75 from both sides to get $2(6t - 1) = 3(-9t + 2)$, or $39t = 8$.

Thus the minimum distance occurs at 8/39 of an hour after the initial conditions of Example 2.42. At this time $x = -5 + 240/39$ and $y = 10 - 360/39$, so $x = 45/39$ and $y = 30/39$. The minimum distance is

$$s\left(\frac{8}{39}\right) = \frac{\sqrt{(45)^2 + (30)^2}}{39} = \frac{15}{39}\sqrt{9+4} = \frac{15\sqrt{13}}{39} \approx 1.4 \text{ miles}$$

Example 2.44

In Example 2.42, which reaches the origin car first, Car 1 or 2?

Solution

This is obvious if, in Example 2.43, one notices that x is positive and y is (still) positive. Alternatively, notice that Car 1 takes 5/30 of an hour to reach the origin and Car 2 takes 10/45 of an hour. The time 5/30 < 10/45, so Car 1 gets there first.

INTEGRAL CALCULUS

Definition of an Integral

The function $F(x)$ is said to be the integral of $f(x)$ if the derivative of $F(x)$ is $f(x)$, or if the differential of $F(x)$ is $f(x)\,dx$. In symbols,

$$F(x) = \int f(x)\,dx \quad \text{if} \quad \frac{dF(x)}{dx} = f(x), \qquad \text{or} \quad dF(x) = f(x)dx$$

In general, $\int f(x)\,dx = F(x) + C$, where C is an arbitrary constant.

Fundamental Theorems on Integrals

$$\int df(x) = f(x) + C$$

$$\int df(x)\,dx = f(x)\,dx$$

$$\int [f_1(x) \pm f_2(x) \pm \cdots]\,dx = \int f_1(x)\,dx \pm \int f_2(x)\,dx \pm \cdots$$

$$\int af(x)\,dx = a\int f(x)\,dx, \text{ where } a \text{ is any constant}$$

$$\int u^n\,du = \frac{u^{n+1}}{n+1} + C \quad (n \neq -1), \text{ where } u \text{ is any function of } x$$

$$\int \frac{du}{u} = \ln u + C, \text{ where } u \text{ is any function of } x$$

$$\int u\,dv = uv - \int v\,du, \text{ where } u \text{ and } v \text{ are any functions of } x$$

$$\int [u(x) \pm v(x)]\,dx = \int u(x)\,dx \pm \int v(x)\,dx$$

$$\int \frac{dx}{ax+b} = \frac{1}{a}\ln|ax+b|$$

$$\int \frac{dx}{\sqrt{x}} = 2\sqrt{x}$$

$$\int a^x dx = \frac{a^x}{\ln a}$$

$$\int \sin x\, dx = -\cos x$$

$$\int \sin^2 x\, dx = \frac{x}{2} - \frac{\sin 2x}{4}$$

$$\int x \sin x\, dx = \sin x - x\cos x$$

$$\int \cos x\, dx = \sin x$$

$$\int \cos^2 x\, dx = \frac{x}{2} + \frac{\sin 2x}{4}$$

$$\int x\cos x\, dx = \cos x + \sin x$$

$$\int \sin x \cos x\, dx = (\sin^2 x)/2$$

$$\int \tan x\, dx = -\ln|\cos x| = \ln|\sec x|$$

$$\int \tan^2 x\, dx = \tan x - x$$

$$\int \cot x\, dx = -\ln|\csc x| = \ln|\sin x|$$

$$\int \cot^2 x\, dx = -\cot x - x$$

$$\int e^{ax} dx = (1/a)e^{ax}$$

$$\int \ln x\, dx = x[\ln(x) - 1] \qquad (x > 0)$$

Example 2.45

Evaluate the following integrals:

(i) $\int (x^3 + 4)dx$ (ii) $\int \sin 3x dx$ (iii) $\int \sqrt{(1-x)}dx$

Solution

(i)

$$\int (x^3 + 4)dx = \left(\frac{x^4}{4} + 4x\right) + \text{constant}$$

(ii)

$$\int \sin 3x dx;\ \text{let } t = 3x;\ \text{then } dt = 3dx \text{ and}$$

$$\int \frac{1}{3}\sin t dt = -\frac{\cos t}{3} = \frac{-\cos 3x}{3} + \text{constant}$$

(iii)

$$\int \sqrt{(1-x)}dx;\ \text{let } t = 1-x;\ \text{then } dt = -dx;$$

$$\int -\sqrt{t}dt = -\frac{2}{3}t^{\frac{3}{2}} = -\frac{2}{3}(1-x)^{\frac{3}{2}} + \text{constant}$$

Moment of Inertia

Moment of inertia J of a mass m:

About $OX: I_x = \int y^2 dm = \int r^2 \sin^2\theta \, dm$

About $OY: I_y = \int x^2 dm = \int r^2 \cos^2\theta \, dm$

About $O: J_0 = \int (x^2 + y^2) dm = \int r^2 \, dm$

Center of Gravity

Coordinates $(\bar{x}, \bar{y})$ of the center of gravity of a mass m:

$$\bar{x} = \frac{\int x \, dm}{\int dm}, \qquad \bar{y} = \frac{\int y \, dm}{\int dm}$$

The center of gravity of the differential element of area may be taken at its midpoint. In the above equations, x and y are the coordinates of the center of gravity of the element.

Work

The work W done in moving a particle from $s = a$ to $s = b$ against a force whose component in the direction of motion is F_s is

$$dW = F_s \, ds, \qquad \mathrm{W} = \int_a^b F_s \, ds$$

where F_s must be expressed as a function of s.

Example 2.46

Consider the function $y = x^2 + 1$ between $x = 0$ and $x = 2$. What is the area between the curve and the x-axis?

Solution

$$A = \int_0^2 (x^2 + 1) \, dx = \left(\frac{x^3}{3} + x \right) \Bigg|_0^2 = \frac{8}{3} + 2 = \frac{14}{3}$$

Example 2.47

Consider the area bounded by $x = y^2$, the x-axis, and the line $x = 4$. Find (1) the area; (2) the first moment of area with respect to the x-axis; (3) the first moment of area with respect to the y-axis; (4) the centroid; (5) the second moment of area with respect to the x-axis; (6) the second moment of area with respect to the y-axis; (7) the moment of inertia about the line $x = -2$; and (8) the moment of inertia about the line $y = 4$.

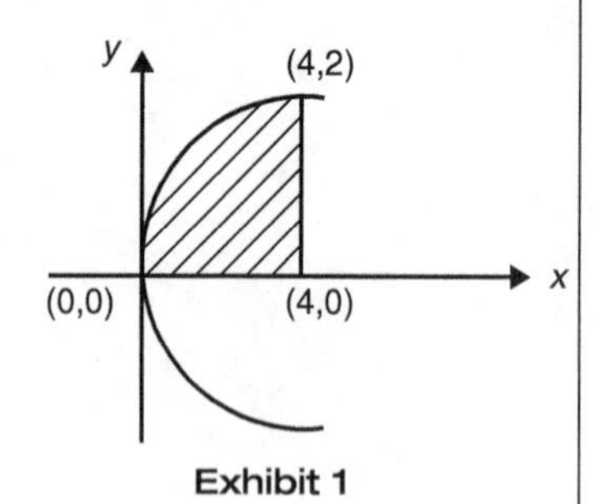

Exhibit 1

Solution

The area is shown shaded in Exhibit 1.

For the vertical strip shown in Exhibit 2, $dA = y\,dx = \sqrt{x}dx$.

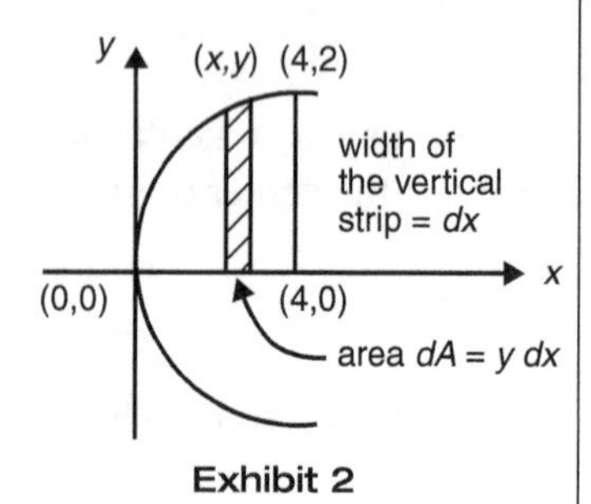

Exhibit 2

$$\text{(1) area } A = \int dA = \int y dx = \int_0^4 \sqrt{x} dx = \frac{16}{3} \text{cm}^2$$

$$\text{(3) first moment with respect to } y\text{-axis } M_y = \int x dA = \int_0^4 x\sqrt{x} dx = 12.8 \text{ cm}^3$$

(6) second moment with respect to y-axis

$$I_y = \int x^2 dA = \int_0^4 x^2 \sqrt{x} dx = 36.57 \text{ cm}^4$$

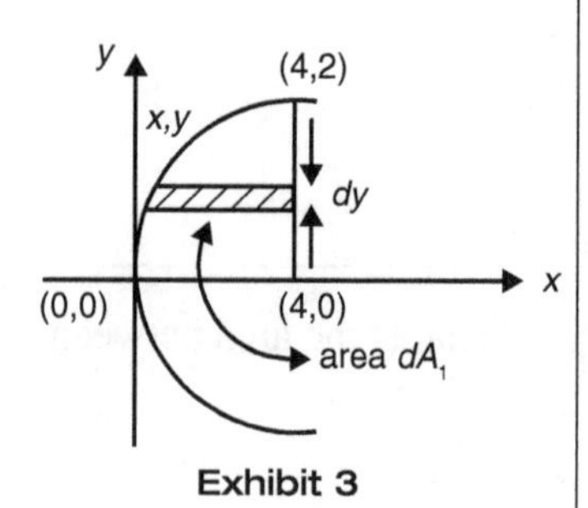

Exhibit 3

For the horizontal strip shown in Exhibit 3, $dA_1 = (4 - x)\,dy = (4 - y^2)\,dy$

$$\text{Note: area } A = \int_0^2 (4 - y^2) dy$$

$$\text{(2) first moment with respect to } x\text{-axis } M_x = \int y dA_1 = \int_0^2 y(4 - y^2) dy = 4 \text{ cm}^3$$

$$\text{(4) } x \text{ coordinate of centroid } x_c = \frac{M_y}{A} = \frac{12.8}{5.33} = 2.4 \text{ cm}$$

$$y \text{ coordinate of the centroid } y_c = \frac{M_x}{A} = \frac{4}{5.33} = 0.75 \text{ cm}$$

(5) second moment with respect to the x-axis

$$I_x = \int y^2 dA_1 = \int_0^2 y^2 (4 - y^2) dy = 4.27 \text{ cm}^4$$

Now, using parallel-axis theorem,

(7) moment of inertia about the line $x = -2$ is $I_y + Ad^2 = 36.57 + (5.333)(2 + 2.4)^2 = 139.8\ \text{cm}^4$

(8) moment of inertia about the line $y = 4$ is $I_x + Ad^2 = 4.27 + (5.333)(4 - 0.75)^2 = 60.6\ \text{cm}^4$

DIFFERENTIAL EQUATIONS

Definitions

A **differential equation** is an equation involving differentials or derivatives.

The **order** of a differential equation is the order of the derivative of highest order that it contains.

The **degree** of a differential equation is the power to which the derivative of highest order in the equation is raised, that derivative entering the equation free from radicals.

The **solution** of a differential equation is the relation involving only the variables (but not their derivatives) and arbitrary constants, consistent with the given differential equation.

The most **general solution** of a differential equation of the *n*th order contains *n* arbitrary constants. If particular values are assigned to these arbitrary constants, the solution is called a particular solution.

Notation

Symbol or Abbreviation	Definition
M, N	Functions of x and y
X	Function of x alone or a constant
Y	Function of y alone or a constant
C, c	Arbitrary constants of integration
$a, b, k, 1, m, n$	Given constants

Equations of First Order and First Degree: *M dx* 1 *N dy* 5 0

Variables Separable: $X_1 Y_1 dx + X_2 Y_2 dy = 0$

Solution

$$\int \frac{X_1}{X_2} dx + \int \frac{Y_2}{Y_1} dy = 0$$

Linear Equation: $dy + (X_1 Y - X_2)\, dx = 0$

Solution

$$y = e^{-\int X_1 dx} \left(\int X_2 e^{\int X_1 dx} dx + C \right)$$

Second-Order Differential Equations

A second-order differential expression, $L(x, y, y', y'')$, is linear if

$$L(x, ay_1 + by_2, ay_1' + by_2', ay_1'' + by_2'') = aL(x, y_1, y_1', y_1'') + bL(x, y_2, y_2', y_2'')$$

or, if it has the form

$$L(x, y_1, y_1', y_1'') = f(x)y + g(x)y' + h(x)y''$$

A second-order linear differential equation is

$$L(x, y, y', y'') = F(x)$$

If $F(x) \equiv 0$, it is homogeneous; if $F(x)$ is nonzero, it is inhomogeneous.

Constant Coefficients

If $L = ay'' + by' + cy$ where a, b, and c are constants with $a \neq 0$, the first step is to solve the associated homogeneous equation $ay'' + by' + cy = 0$. By replacing y by $1, y'$ by r, and y'' by r^2, one obtains the characteristic equation $ar^2 + bc + c = 0$ with roots r_1 and r_2 obtained from factoring or from the quadratic formula. There are three cases to consider:

Case 1: $r_1 \neq r_2$, both real; $y = c_1e^{r_1x} + c_2e^{r_2x}$, where c_1 and c_2 are arbitrary constants.

Case 2: $r_1 = r_2$; $y = c_1e^{r_1x} + c_2xe^{r_2x}$.

Case 3: $r_1 = \alpha + j\beta$, $r_2 = \alpha - j\beta$, where α and β are real and $j^2 = -1$;
$y = d_1e^{r_1x} + d_2e^{r_2x} = e^{\alpha x}(c_1 \sin \beta x + c_2 \cos \beta x)$. In particular, if $\alpha = 0$,
$y = c_1 \sin \beta x + c_2 \cos \beta x$.

After finding the two solutions to the associated homogeneous equation (y_1 is the result of setting $c_1 = 1$ and $c_2 = 0$, whereas y_2 has $c_1 = 0$ and $c_2 = 1$), one proceeds in either of the following two ways.

Variation of Parameters

If $L(y) = F(x)$ in which the coefficient of y'' is 1, and $W(x) = y_1(x)y_2'(x) - y_1'(x)y_2(x)$ in which y_1 and y_2 are those solutions found above, and if

$$u_1' = \frac{-F(x)y_2(x)}{W(x)} \quad \text{and} \quad u_2' = \frac{F(x)y_1(x)}{W(x)}$$

one solution to the inhomogeneous equation is $y_p = u_1(x)y_1(x) + u_2(x)y_2(x)$.

Undetermined Coefficients

In this technique, one guesses y_p by using the following patterns.

One guesses that the solution may be of the same form as the $F(x)$ function but with coefficients to be determined. This method requires modification if $F(x)$ and $c_1y_1 + c_2y_2$ from the associated homogeneous equation interfere, but it is frequently easier than the integration required in the Variation of Parameters technique to construct u_1 and u_2 from their derivatives.

To guess, classify $F(x)$. If it is a polynomial of degree k, the guess will be a polynomial of degree k. However, if $r = 0$ occurs in the homogeneous equation, increase the degree of the polynomial by one. If $F(x)$ is a polynomial times e^{Ax}, so will be the guess. Once again, the degree may have to be increased by one. If $F(x)$ contains sines and cosines, so should the guess.

After making the guess, differentiate it twice and put it into the equation. The coefficients may be determined at this time.

When mixed forms of functions are present in $F(x)$—for example, $x^2 + 2 + 3 \sin 2x$—treat the terms $x^2 + 2$ and $3 \sin 2x$ independently. The principle of *superposition* then permits you to add the results.

Now, after finding the solution $c_1y_1 + c_2y_2$ to the associated homogeneous equation, and the particular solution y_p to the inhomogeneous equation, form the general solution $y = c_1y_1 + c_2y_2 + y_p$. If initial values are required, such as $y(1) = 2$ and $y'(1) = 3$, the final step is to determine the values of c_1 and c_2 that fit the initial conditions.

Example 2.48

Specify the order of each of the following differential equations and also identify each as linear/nonlinear and homogeneous/nonhomogeneous.

Note: $y' = \frac{dy}{dt}$; $y'' = \frac{d^2y}{dt^2}$

(i) $y'' + 4y' + 4y = 0$ (ii) $3y' + 2y = 0$ (iii) $y'' + 2y' + 2y = e^{-t}$

(iv) $2y' + y + 2 = 0$ (v) $y' + y^2 + 4 = 0$

Solution

(i) second order, linear, homogeneous

(ii) first order, linear, homogeneous

(iii) second order, linear, nonhomogeneous

(iv) first order, linear, nonhomogeneous

(v) first order, nonlinear, nonhomogeneous

Example 2.49

Find the homogeneous solutions for each of the following differential equations and specify whether it belongs to either overdamped, underdamped, or critically damped case.

(i) $y'' + 3y' + 2y = 0$ (ii) $y''+ 2y' + y = 4$ (iii) $y'' + 2y' + 2y = 0$

Solution

(i) Characteristic equation is $s^2 + 3s + 2 = 0$, and the characteristic roots are -1 and -2. Homogeneous solution is $y_h(t) = C_1e^{-1t} + C_2e^{-2t}$. The roots are real and distinct, so the function is overdamped.

(ii) Characteristic equation is $s^2 + 2s + 1 = 0$, and the characteristic roots are -1 and -1. Homogeneous solution is $y_h(t) = (C_1 + C_2t)e^{-1t}$. The roots are real and equal, so the function is critically damped.

(iii) Characteristic root is $s^2 + 2s + 2 = 0$ and the characteristic roots are $-1 + i1$ and $-1 - i1$. Homogeneous solution is $y_h(t) = e^{-1t}$ $(C_1 \cos t + C_2 \sin t)$. As the roots are complex, the function is underdamped.

Example 2.50

Solve the differential equation $y' + 2y = 0$ with initial condition $y(0) = 3$.

Solution

This is a first-order, linear, homogeneous differential equation with constant coefficient.

Characteristic equation is $s + 2 = 0$, and the root is -2; solution is $y(t) = Ce^{-2t}$.

To find C, $y(0) = C = 3$; then, $y(t) = 3e^{-2t}$.

Example 2.51

Solve the differential equation $y'' + 2y' + y = 4$ with initial conditions $y(0) = 0$, $y'(0) = 1$.

Solution

Characteristic equation is $s^2 + 2s + 1 = 0$.

Characteristic roots are: -1 and -1 (roots are real and equal; the solution is critically damped).

Natural or homogeneous solution $y_h(t) = (C_1 + C_2t)e^{-t}$

Particular solution $y_p(t) = B$ due to forcing function $f(t) = 4$, a constant

Substituting $y_p(t)$ in the differential equation, $y_p'' + 2y_p' + y_p = 4$; $0 + 0 + B = 4$; or $B = 4$

Complete solution $y(t) = y_h(t) + y_p(t) = C_1e^{-t} + C_2te^{-t} + 4$

To find C_1 and C_2, use the initial conditions in $y(t)$ and $y'(t)$; $y'(t) = -C_1e^{-t} - C_2te^{-t} + C_2e^{-t}$

$y(0) = C_1 + B = 0$ and $y'(0) = -C_1 + C_2 = 1$; solving $C_1 = -4$; $C_2 = -3$

Complete solution is $y(t) = -4e^{-t} - 3te^{-t} + 4$.

Find the solution of $y'' + 2y' = x^2 + 2 + 3 \sin 2x$ subject to $y(0) = 1, y'(0) = 0$.

Solution

Begin with $y'' + 2y' = 0$. The characteristic equation is $r^2 + 2r = 0$, which has roots 0 and –2. Thus, the two solutions to the associated homogeneous equation are $y_1 = 1$ and $y_2 = e^{-2x}$. To use the method of Undetermined Coefficients, guess $(ax^2 + bx + c) \bullet x$ for the $x^2 + 2$ term (the x is needed because $y_1 = 1$). Differentiate twice and insert in the equation: $(6ax + 2b) + 2(3ax^2 + 2bx + c)$ should be the same as $x^2 + 2$. Thus, $6a = 1, 4b + 6a = 0$, and $2b + 2c = 2$. Consequently, $a = \frac{1}{6}$, $b = -\frac{1}{4}$, and $c = \frac{5}{4}$. Next, guess $c \sin 2x + d \cos 2x$ for the other term. Then $y'' + 2y' = -4c \sin 2x - 4d \cos 2x + 2(2c \cos 2x - 2d \sin 2x)$ should match $3 \sin 2x$, so $-4c - 4d = 3$ and $-4d + 4c = 0$. Thus $c = d = -\frac{3}{8}$.

Putting all this together, one has the general solution

$$y = A + Be^{-2x} + \frac{1}{6}x^3 - \frac{1}{4}x^2 + \frac{5}{4}x - \frac{3}{8}\sin 2x - \frac{3}{8}\cos 2x$$

Now to fit the initial conditions,

$$y(0) = 1 = A + B - \frac{3}{8} \quad \text{and} \quad y'(0) = 0 = -2B + \frac{5}{4} - \frac{3}{4}$$

Hence, $B = \frac{1}{4}$ and $A = \frac{9}{8}$, so

$$y = \frac{9}{8} + \frac{1}{4}e^{-2x} + \frac{1}{6}x^3 - \frac{1}{4}x^2 + \frac{5}{4}x - \frac{3}{8}\sin 2x - \frac{3}{8}\cos 2x$$

Euler Equations

An equation of the form $x^2 y'' + axy' + by = F(x)$, with a and b constants, may be solved as readily as the constant coefficient case. These are called **Euler equations.** Upon substituting $y = x^m$, one obtains the *indicial equation* $m(m - 1) + am + b = 0$. This quadratic equation has two roots, m_1 and m_2.

If $m_1 \neq m_2$, both real, then $y = c_1 |x|^{m_1} + c_2 |x|^{m_2}$.
If $m_1 = m_2$, then $y = |x|^{m_1} (c_1 + c_2 \ln |x|)$.

If $m_1 = p + jq$ and $m_2 = p - jq$, then $y = |x|^p [c_1 \cos (q \ln |x|) + c_2 \sin (q \ln |x|)]$.

Once y is determined, y_p for the inhomogeneous equation may be found by Variation of Parameters. The method of Undetermined Coefficients is not recommended for Euler equations.

Higher-order linear equations with constant coefficients or of Euler form may be solved analogously.

Laplace Transform

The Laplace transform is an operation that converts functions of x on the half-line $[0, \infty]$ into functions of p on some half-line (a, ∞). The damping power of e^{-xp} is the basis for this useful technique. If $f(x)$ is a piecewise continuous function on $[0, \infty]$ that does not grow too fast, $L(f)$ is the function of p defined by

$$L[f(p)] = \int_0^{\infty} e^{-xp} f(x)\, dx$$

for the values of p for which the integral converges. For example, if $f(x) \equiv 1$,

$$L(f) = L(1) = \int_0^\infty e^{-xp}\,dx = \frac{1}{p} \quad (\text{for } p > 0)$$

As a further example,

$$L(f) = L(e^{ax}) = \int_0^\infty e^{-xp} e^{ax}\,dx = \int_0^\infty e^{-x(p-a)}\,dx = \frac{1}{p-a} \quad (\text{for } p > a)$$

The basic connection between the Laplace transform and differential equations is the following result achieved by integration by parts:

$$L\left[y'(p)\right] = \int_0^\infty e^{-xp} y'(x)\,dx = y(x)e^{-xp}\Big|_0^\infty + p\int_0^\infty e^{-xp} y(x)\,dx = -y(0) + pL\left[y(p)\right]$$

Consequently, the solution to $ay'' + by' + cy = F(x)$ may be obtained by transforming $L[ay'' + by' + cy] = L(F)$, so $a[p^2L(y) - py(0) - y'(0)] + b[pL(y) - y(0)] + cL(y) = L(F)$. Solving for $L(y)$,

$$L(y) = \frac{L(F) + apy(0) + ay'(0) + by(0)}{ap^2 + bp + c}$$

If one were able to "invert" this result,

$$y = L^{-1}\left[\frac{L(F) + apy(0) + ay'(0) + by(0)}{ap^2 + bp + c}\right]$$

the solution would appear, complete with initial values. The Laplace transform is invertible, and the process of finding $L^{-1}[f(p)]$ as a function of x is much like the process of integration.

Table 3.1 presents a tabulation of selected transforms, where the transform of $f(x)$ is called $F(p)$. Line 3 in Table 3.1 reveals that the operation of multiplying by x corresponds to the negative of the operation of differentiating with respect to p. Lines 2 and 8 have been previously discussed. The δ in Line 1 is a *pseudo-function* with great utility defined by $\delta(x) = 0$ for all x except $x = 0$, and $\int_0^\infty \delta(x)\,dx = 1$, so $\delta(0) = +\infty$. The u in Line 7 is called the *Heaviside function*, and it is *zero* until $x - c > 0$. Thus $u(x-c)f(x-c)$ is $f(x)$ shifted right to the point $x = c$. For example, if $f(x) = x$ and $c = 1$, $u(x-1)f(x-1)$ has the graph shown in Figure 2.36, whereas f has the graph shown in Figure 2.37. The operation in the left column of Line 9 is a new way to multiply functions, called **convolution**.

Table 3.1 Selected transforms

$f(x)$	$F(p)$
1. δ	1
2. 1	p^{-1}
3. $xf(x)$	$\frac{-dF}{dp}$

4. $e^{ax} f(x)$	$F(p-a)$
5. $\sin ax$	$\frac{a}{p^2+a^2}$
6. $\cos ax$	$\frac{p}{p^2+a^2}$
7. $u(x-c) f(x-c)$	$e^{-cp}F(p)$
8. $f'(x)$	$pF(p)-f(0)$
	(continued)
9. $\int_0^x f(u)g(x-u)du$	$F(p)\bullet G(p)$
10. $f(x)$, if f is periodic, of period L	$\frac{\int_0^L e^{-px}f(x)dx}{1-e^{-pL}}$

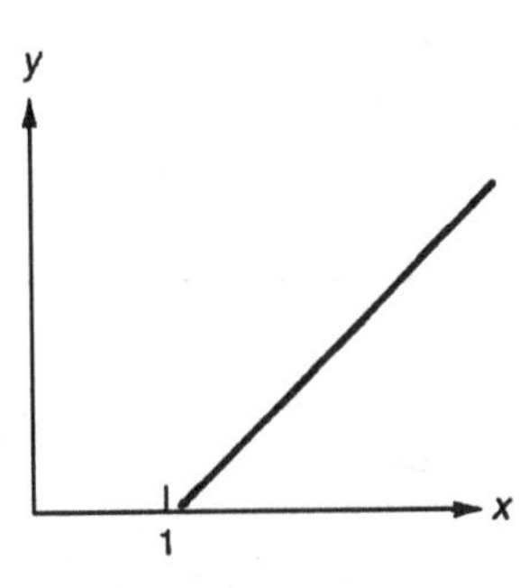

Figure 2.36

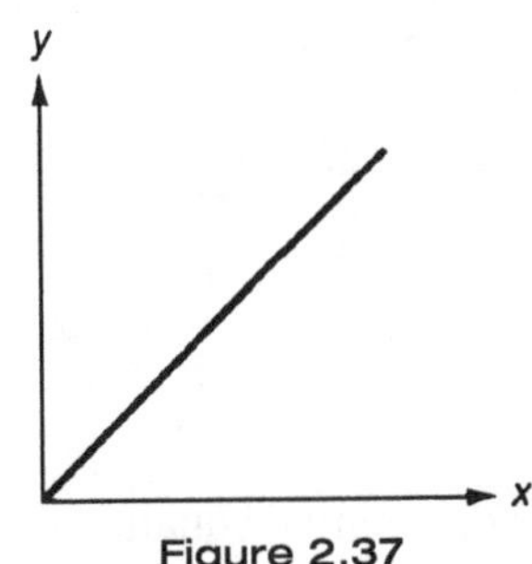

Figure 2.37

For an example of inversion of a transform, consider

$$F(p)=\frac{2p+3}{(p^2+1)(p-2)}$$

Begin by using partial fractions to write

$$F(p)=\frac{A}{p-2}+\frac{Bp+C}{p^2+1}$$

from which

$$F(p)=\frac{7/5}{p-2}+\frac{(-7/5)p-4/5}{p^2+1}$$

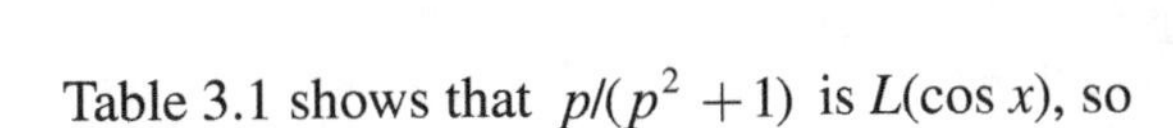

Table 3.1 shows that $p/(p^2+1)$ is $L(\cos x)$, so

$$L^{-1}\left[\frac{(-7/5)p}{p^2+1}\right]$$

is $-(7/5)\cos x$, and similarly

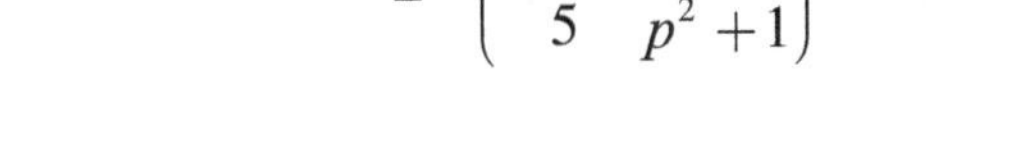

$$L^{-1}\left(-\frac{4}{5}\bullet\frac{1}{p^2+1}\right)$$

is $-(4/5)\sin x$. Since $L(1) = 1/p$, $L(e^{2x})$ is $1/(p-2)$ by Line 4, and combining these three terms yields

$$f(x)=\frac{7}{5}e^{2x}-\frac{7}{5}\cos x-\frac{4}{5}\sin x$$

The *Heaviside* operation in Line 7 leads to an easy solution of differential equations whose right-hand side is not continuous. For example, consider the response of $y''+y$ to a driving function $f(x)$ that is 1 for x between 0 and 2 and then becomes 0. Suppose $y(0) = 1$ and $y'(0) = 3$. By applying Line 8 in Table 3.1 twice,

$$L(y)=\frac{L(f)+p+3}{p^2+1}$$

and since $f(x) = 1 - u(x - 2)$,

$$L(f)=\frac{1}{p}-\frac{e^{-2p}}{p}$$

so

$$L(y)=\frac{p+3+\left(1-e^{-2p}\right)/p}{p^2+1}$$

The e^{-2p} portion represents delay, so write

$$L(y)=\frac{p^2+3p+1}{p(p^2+1)}-\frac{e^{-2p}}{p(p^2+1)}=\frac{1}{p}+\frac{3}{p^2+1}-e^{-2p}\left(\frac{1}{p}-\frac{p}{p^2+1}\right)$$

from which $y=1+3\sin x-u(x-2)+u(x-2)\cos(x-2)$.

Example 2.53

Find the Laplace transform of each of the following functions:

(i) $2\,u(t)+e^{-3t}$ (ii) $2te^{-t}$ (iii) $e^{-t}\sin 4t$ (iv) $\sin 4t$ (v) $\cos 2t$

Solution

(i) $\frac{2}{s}+\frac{1}{s+3}=\frac{2s+7}{s(s+3)}$ (ii) $\frac{2}{(s+1)^2}$ (iii) $\frac{4}{(s+1)^2+4^2}$

(iv) $\frac{4}{s^2+4^2}$ (v) $\frac{s}{s^2+2^2}$

Find an expression for $Y(s) = £\, y(t)$ for each of the following differential equations.

(i) $\frac{dy}{dt}+6\,y(t)=4\,u(t);\ y(0)=0$

(ii) $\frac{dy}{dt}+2\,y(t)=0;\ y(0)=-2$

(iii) $\frac{d^2y}{dt^2}+2\frac{dy}{dt}+3\,y(t)=\sin 2t$ with zero initial conditions

(iv) $\frac{d^2y}{dt^2}+2\frac{dy}{dt}+y(t)=e^{-2t}$ with initial conditions $y(0)=-1$ and $\frac{dy(0)}{dt}=1$

Solution

(i) $\left[sY(s)-0\right]+6Y(s)=\dfrac{4}{s};\ Y(s)=\dfrac{4}{s(s+6)}$

(ii) $\left[sY(s)-(-2)\right]+2Y(s)=0;\ Y(s)=-\dfrac{2}{s+2}$

(iii) $\left[s^2Y(s)-0-0\right]+2\left[sY(s)-0\right]+3Y(s)=\dfrac{2}{s^2+4};$

$$Y(s)=\frac{2}{(s^2+4)(s^2+2s+3)}$$

(iv) $\left[s^2Y(s)-(-s)-1\right]+2\left[sY(s)-(-1)\right]+Y(s)=\dfrac{1}{s+2}$

simplifying $Y(s)=\dfrac{(-s^2-3s-1)}{(s+2)(s^2+2s+1)}$

Example 2.55

Find the initial and final values of the functions whose Laplace transforms are given:

(i) $F(s)=\dfrac{2(s+1)}{s(s+4)(s+6)}$ (ii) $F(s)=\dfrac{4s}{s^2+2s+2}$

Solution

(i) Initial value is: $\lim\limits_{t\to 0} f(t)=\lim\limits_{s\to\infty} sF(s)=\lim\limits_{s\to\infty} s\dfrac{2(s+1)}{s(s+4)(s+6)}=0$

Final value is: $\lim\limits_{t\to\infty} f(t)=\lim\limits_{s\to 0} sF(s)=\lim\limits_{s\to 0} s\dfrac{2(s+1)}{s(s+4)(s+6)}=\dfrac{1}{12}$

(ii) Initial value is: $\lim\limits_{t\to 0} f(t)=\lim\limits_{s\to\infty} sF(s)=\lim\limits_{s\to\infty} s\dfrac{4s}{s^2+2s+2}=4$

Final value is: $\lim\limits_{t\to\infty} f(t)=\lim\limits_{s\to 0} sF(s)=\lim\limits_{s\to 0} s\dfrac{4s}{s^2+2s+2}=0$

FOURIER SERIES AND FOURIER TRANSFORM

Fourier Series

A periodic function $F(t)$ with period T can be expanded into Fourier series as

$$F(t)=a_0+\sum_{n=1}^{\infty}(a_n\cos n\omega_0 t+b_n\sin n\omega_0 t) \tag{2.10}$$

where $\omega_0=\dfrac{2\pi}{T}$ and the Fourier coefficients are defined as

$$a_0 = \left(\frac{1}{T}\right)\int_0^T F(t)dt \quad \textbf{(2.10a)}$$

$$a_n = \left(\frac{2}{T}\right)\int_0^T F(t)\cos(n\omega_0 t)dt \quad \textbf{(2.10b)}$$

$$b_n = \left(\frac{2}{T}\right)\int_0^T F(t)\sin(nw_0 t)dt \quad \textbf{(2.10c)}$$

For a truncated series, the root mean square (RMS) value F_N is defined as

$$F_N^{\ 2} = a_0^{\ 2} + \left(\frac{1}{2}\right)\sum_{n=1}^{N}(a_n^{\ 2} + b_n^{\ 2}) \quad \textbf{(2.11)}$$

Example 2.56

Find the Fourier coefficients of the periodic waveform shown in Exhibit 4 and the RMS value of the truncated Fourier series including five harmonics.

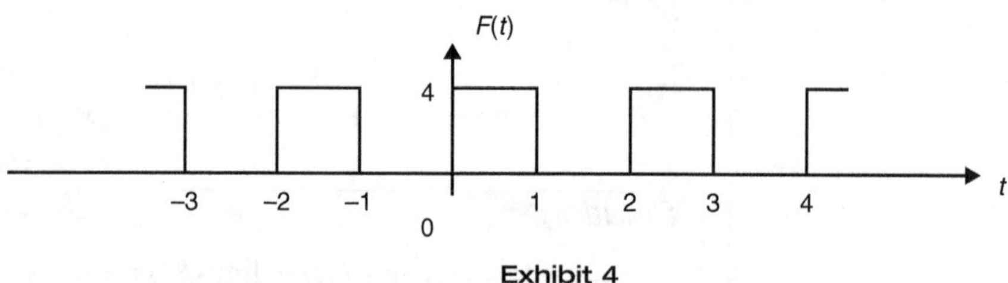

Exhibit 4

Solution

Period $T = 2\ s$; $w_0 = \dfrac{2\pi}{T} = \pi\ rad/s$

From Equation (2.10a) $a_0 = \dfrac{1}{2}\left[\int_0^1 4dt + \int_1^2 0dt\right] = 2$

From Equation (2.10b) $a_n = \dfrac{2}{2}\left[\int_0^1 4\cos n\omega_0 t dt + \int_1^2 0dt\right] = 0, \text{when } n \neq 0$

From Equation (2.10c) $b_n = \dfrac{2}{2}\int_0^1 4\sin n\omega_0 t dt + \int_1^2 0dt = \dfrac{4}{n\pi}(1 - \cos n\pi)$

$b_n = 0$ for all even n; $b_n = \dfrac{8}{n\pi}$ for all odd n

Fourier series is $F(t) = 2 + \dfrac{8}{\pi}\sin\ \omega_0 t + \dfrac{8}{3\pi}\sin\ 3\omega_0 t + \dfrac{8}{5\pi}\sin\ 5\omega_0 t + \ldots$

From Equation (2.11), for $N = 5$, $F_N^{\ 2} = 2^2 + \dfrac{1}{2}\left[(\dfrac{8}{\pi})^2 + (\dfrac{8}{3\pi})^2 + (\dfrac{8}{5\pi})^2\right] = 7.732;$

$$F_N = 2.7807$$

Fourier Transform

The Fourier transform of a function $x(t)$ and its inverse relation are:

$$X(f) = \int_{-\infty}^{+\infty} x(t)\exp(-j2\pi ft)\,dt \qquad \textbf{(2.12a)}$$

$$X(t) = \int_{-\infty}^{+\infty} x(f)\exp(j2\pi ft)\,dt \qquad \textbf{(2.12b)}$$

Table 3.2 lists the Fourier transforms for a few commonly used functions in communication systems. The NCEES *Fundamentals of Engineering Supplied-Reference Handbook* includes a more complete table of Fourier transform pairs.

Table 3.2 Fourier Transform Pairs

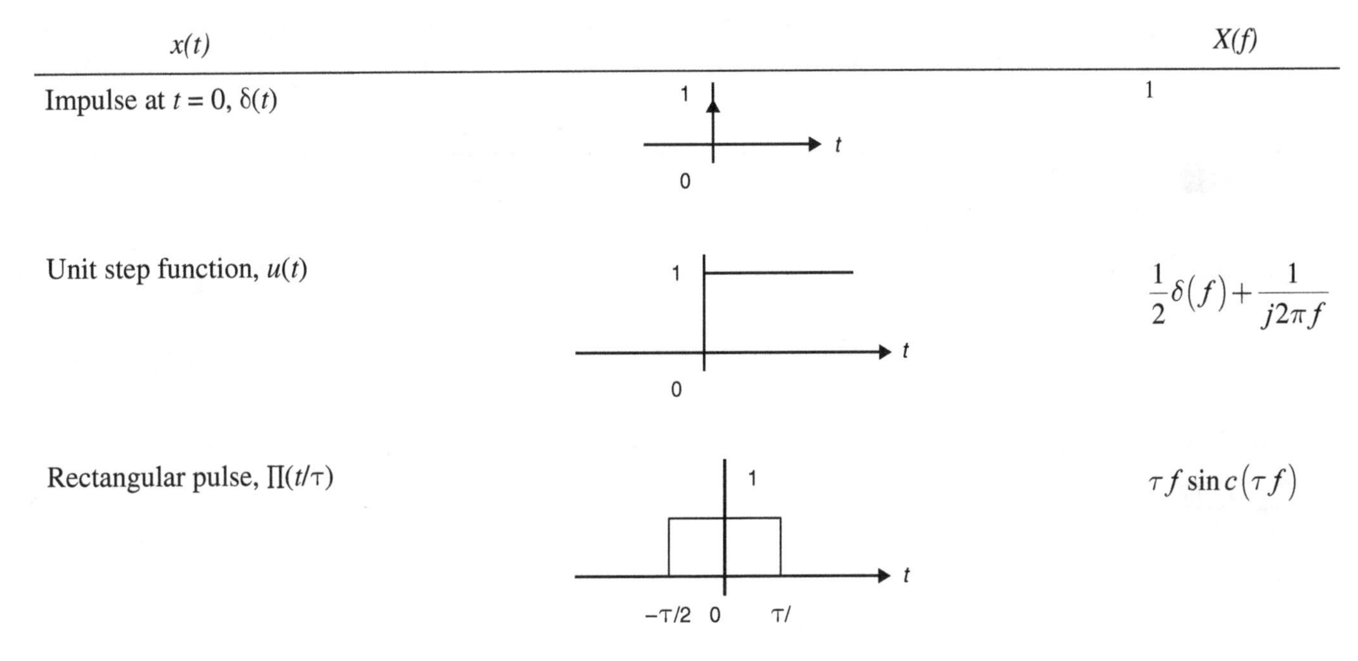

$x(t)$	$X(f)$
Impulse at $t = 0$, $\delta(t)$	1
Unit step function, $u(t)$	$\dfrac{1}{2}\delta(f) + \dfrac{1}{j2\pi f}$
Rectangular pulse, $\Pi(t/\tau)$	$\tau f \sin c(\tau f)$

Table 3.2 Fourier Transform Pairs

<table>
<tr><th>$x(t)$</th><th></th><th>$X(f)$</th></tr>
<tr><td>Sinc function, sinc(at)</td><td>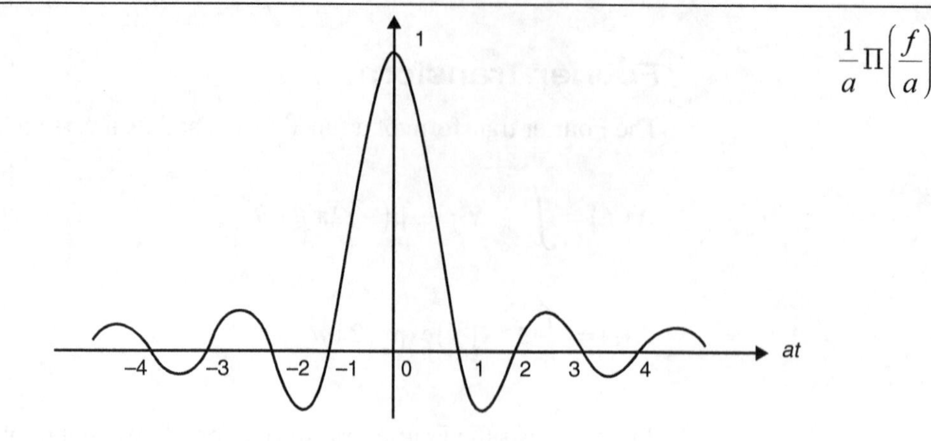
</td><td>$\frac{1}{a}\Pi\left(\frac{f}{a}\right)$</td></tr>
<tr><td>Triangular pulse, $\Lambda(t/\tau)$</td><td>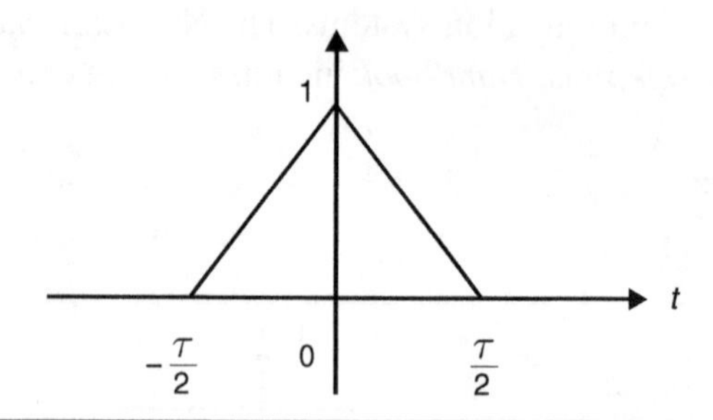
</td><td>$\tau f \sin c^2\left(\tau f\right)$</td></tr>
</table>

Example 2.57

Find the Fourier transform of (i) 4 $\delta(t)$, (ii) 2 $u(t)$, (iii) $\Pi(t/2)$, and (iv) 6 $\cos(100\pi t)$.

Solution

(i) 4 (ii) $\delta(f) + 1/j\pi f$ (iii) $2f\,\text{sinc}(2f)$ (iv) $3[\delta(f-50) + \delta(f+50)]$

DIFFERENCE EQUATIONS AND Z-TRANSFORMS

Difference equations are used to model discrete systems; they are analogous to differential equations that describe continuous systems. The equations $y(k) - y(k-1) = 10$ and $y(k+1) - y(k) = 5$ are some examples of first order linear difference equations; $y(k) = y(k-1) + y(k-2)$ is an example of second order difference equation.

Example 2.58

Find the values of $y(1)$, $y(2)$, and $y(3)$ of the equation $y(k) - 1.01y(k-1) = -50$ with the initial condition $y(0) = 1000$.

Solution

$y(1) - 1.01\,y(0) = -50;\ y(1) = \$960$

$y(2) - 1.01\,y(1) = -50;\ y(2) = \$919.60$

$y(3) - 1.01\,y(2) = -50;\ y(3) = \$878.80$ and so on

Z-Transforms

Z-transform of a discrete sequence is defined as

$$F(z) = \sum_{k=0}^{\infty} f[k]z^{-k} \tag{2.13}$$

For example, if the discrete sequence is $f[k] = 0, 1, 4$, then its z-transform is $(F_z) = 0 + z^{-1} + 4z^{-2}$.

Example 2.59

Find the z-transform of the function $f(k) = 3\,u(k) + 2^k$ for $k \ge 0$.

Solution

Using the Z-transform table from the *Fundamentals of Engineering Supplied-Reference Handbook*, $F(z)$ =

$$\frac{3}{1-z^{-1}} + \frac{1}{1-2z^{-1}}$$

Simplifying, $F(z) = \dfrac{3(1-2z^{-1}) + (1-z^{-1})}{(1-z^{-1})(1-2z^{-1})} = \dfrac{4-7z^{-1}}{(1-z^{-1})(1-2_z^{-1})}$

Multiplying by z^2, $F(z) = \dfrac{4z^2 - 7z}{z^2 - 3z + 2}$

PROBLEMS

1.1 The simplest value of $\frac{[(n+1)!]^2}{n!(n-1)!}$ is:

a. n^2
b. $n(n+1)$
c. $n+1$
d. $n(n+1)^2$

1.2 If $x^{3/4} = 8$, x equals:

a. 6 c. –9
b. 9 d. 16

1.3 If $\log_a 10 = 0.250$, $\log_{10} a$ equals:

a. 4 c. 2
b. 0.50 d. 0.25

1.4 If $\log_5 x = -1.8$, $\log_x 5$ is:

a. 0.35 c. –0.56
b. 0.79 d. undefined

1.5 If $\log x + \log (x - 10) - \log 2 = 1$, x is:

a. –1.708 c. 7.824
b. 5.213 d. 11.708

1.6 A right circular cone, cut parallel with the axis of symmetry, reveals a(n):

a. circle
b. hyperbola
c. eclipse
d. parabola

1.7 The expression $\frac{6!}{3!0!}$ is equal to:

a. ∞
b. 120
c. 2!
d. 0

1.8 To find the angles of a triangle, given only the lengths of the sides, one would use:

a. the law of cosines
b. the law of tangents
c. the law of sines
d. the inverse-square law

1.9 If $\sin \alpha = \dfrac{a}{\sqrt{a^2 + b^2}}$, which of the following equations is true?

a. $\tan^{-1} \dfrac{b}{a} = \dfrac{\pi}{2} - \alpha$

b. $\tan^{-1} \dfrac{b}{a} = -\alpha$

c. $\cos^{-1} \dfrac{b}{\sqrt{a^2 + b^2}} = \dfrac{\pi}{2} - \alpha$

d. $\cos^{-1} \dfrac{a}{\sqrt{a^2 + b^2}} = \alpha$

1.10 The sine of 840° equals:
a. –cos 30°
b. –cos 60°
c. sin 30°
d. sin 60°

1.11 One root of $x^3 - 8x - 3 = 0$ is:
a. 2
b. 3
c. 4
d. 5

1.12 Roots of the equation $3x^3 - 3x^2 - 18x = 0$ are:
a. –2, 3
b. 0, – 2, 3
c. $2 + 1i, 2 - 1i$
d. 0, 2, –3

1.13 The equation whose roots are $-1 + i1$ and $-1 - i1$ is given as:
a. $x^2 + 2x + 2 = 0$
b. $x^2 - 2x - 2 = 0$
c. $x^2 + 2 = 0$
d. $x^2 - 2 = 0$

1.14 Natural logarithms have a base of:
a. 3.1416
b. 2.171828
c. 10
d. 2.71828

1.15 $(5.743)^{1/30}$ equals:
a. 1.03
b. 1.04
c. 1.05
d. 1.06

1.16 The value of $\tan(A + B)$, where $\tan A = 1/3$ and $\tan B = 1/4$ (A and B are acute angles) is:
a. 7/12
b. 1/11
c. 7/11
d. 7/13

1.17 To cut a right circular cone in such a way as to reveal a parabola, it must be cut:
a. perpendicular to the axis of symmetry
b. at any acute angle to the axis of symmetry
c. at any obtuse angle to the axis of symmetry
d. none of these

1.18 The equation of the line perpendicular to $3y + 2x = 5$ and passing through $(-2, 5)$ is:
a. $2x = 3y$
b. $2y = 3x$
c. $2y = 3x + 16$
d. $3x = 2y + 8$

1.19 Equation of a line that has a slope of -2 and passes through $(2, 0)$ is:
a. $y = -2x$
b. $y = 2x + 4$
c. $y = -2x + 4$
d. $y = 2x - 4$

1.20 Equation of a line that intercepts the x-axis at $x = 4$ and the y-axis at $y = -6$ is:
a. $3x - 2y = 12$
b. $2x - 3y = 12$
c. $x + y = 6$
d. $x - y = 4$

1.21 The x-axis intercept and the y-axis intercept of the line $x + 3y + 9 = 0$ are:
a. 0 and 3
b. -9 and -3
c. -3 and -9
d. 9 and 0

1.22 The distance between the points $(1, 0, -2)$ and $(0, 2, 3)$ is:
a. 3.45
b. 5.39
c. 6.71
d. 7.48

1.23 The equation of a parabola with center at $(0, 0)$ and directrix at $x = -2$ is:
a. $y^2 = 4x$
b. $y = 2x^2$
c. $y^2 = 8x$
d. $y^2 = 2x$

1.24 The equation of the directrix of the parabola $y^2 = -4x$ is:
a. $y = 2$
b. $x = 1$
c. $x + y = 0$
d. $x = -2$

1.25 The equation of an ellipse with foci at $(\pm 2, 0)$ and directrix at $x = 6$ is:

a. $\dfrac{x^2}{12} + \dfrac{y^2}{8} = 1$

b. $\dfrac{x^2}{8} + \dfrac{y^2}{8} = 1$

c. $\dfrac{x^2}{12} + \dfrac{y^2}{12} = 1$

d. $\dfrac{x^2}{4} + \dfrac{y^2}{4} = 1$

1.26 The foci of the ellipse $\left(\dfrac{x}{3}\right)^2 + \left(\dfrac{y}{2}\right)^2 = 1$ are at:

a. $(\pm\sqrt{5}, 0)$ c. $(0, \pm\sqrt{2})$
b. $(\pm\sqrt{2}, 0)$ d. $(0, \pm\sqrt{5})$

1.27 The equation of a hyperbola with center at $(0, 0)$, foci at $(\pm 4, 0)$, and eccentricity of 3 is:

a. $\dfrac{x^2}{16} - \dfrac{y^2}{16} = 1$

b. $\dfrac{x^2}{16} - \dfrac{y^2}{128} = 1$

c. $\dfrac{x^2}{64} - \dfrac{y^2}{48} = 1$

d. $\dfrac{x^2}{128} - \dfrac{y^2}{128} = 1$

1.28 The equation of a circle with center at $(1, 2)$ and passing through the point $(4, 6)$ is:

a. $x^2 + y^2 = 25$

b. $(x + 1)^2 + (y - 2)^2 = 25$

c. $x^2 + (y - 2)^2 = 25$

d. $(x - 1)^2 + (y - 2)^2 = 25$

1.29 The length of the tangent from $(4, 8)$ to the circle $x^2 + (y - 1)^2 = 3^2$ is:

a. 3.81 c. 5.66
b. 4.14 d. 7.48

1.30 The conic section described by the equation $x^2 - 10xy + y^2 + x + y + 1 = 0$ is:

a. circle c. hyperbola

b. parabola d. ellipse

1.31 A triangle has sides of length 2, 3, and 4. The angle subtended by the sides of length 2 and 4 is:

a. 21.2° c. 46.6°
b. 35.0° d. 61.2°

1.32 Length a of one side of the triangle below is:

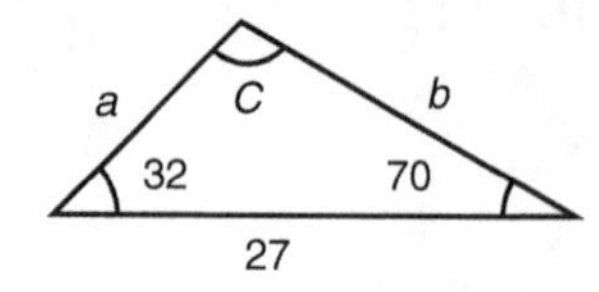

a. 25.9 c. 12.7
b. 19.1 d. 4.8

1.33 The relation $\sec\theta - (\sec\theta)(\sin^2\theta)$ can be simplified as:

a. $\sin\theta$ c. $\cot\theta$
b. $\tan\theta$ d. $\cos\theta$

1.34 If $\sin\theta = m$, $\cot\theta$ is:

a. $\dfrac{\sqrt{1-m^2}}{m}$

b. $\dfrac{m}{\sqrt{1-m^2}}$

c. $\sqrt{1-m^2}$

d. m

1.35 If vectors $\mathbf{A} = 3i - 6j + 2k$ and $\mathbf{B} = 10i + 4j - 6k$, their cross product $\mathbf{A} \times \mathbf{B}$ is:

a. $-12\mathbf{i} + 38\mathbf{j}$ c. $-12\mathbf{i} + 18\mathbf{j} + 24\mathbf{k}$
b. $12\mathbf{i} + 24\mathbf{j} + 36\mathbf{k}$ d. $28\mathbf{i} + 38\mathbf{j} + 72\mathbf{k}$

1.36 The sum of all integers from 10 to 50 (both inclusive) is:

a. 990 c. 1230
b. 1110 d. 1420

1.37 The 50th term of the series 10, 16, 22, 28, 34, 40. . . is:

a. 272 c. 428
b. 304 d. 584

1.38 Sum of the infinite series 4, 2, 1, 0.5, 0.25. . . is:

a. 8 c. 10,400
b. 128 d. ∞

1.39 If $\mathbf{A} = \begin{bmatrix} 1 & 2 & 3 \\ 1 & 2 & 9 \end{bmatrix}$ and $\mathbf{B} = \begin{bmatrix} 5 & 1 \\ 6 & 0 \\ 4 & 7 \end{bmatrix}$, the (2,1) entry of AB is:

a. 29 c. 33
b. 53 d. 64

1.40 The inverse of the matrix $\begin{bmatrix} 1 & 1 \\ 3 & 2 \end{bmatrix}$ is:

a. $\begin{bmatrix} 2 & -1 \\ -3 & 1 \end{bmatrix}$ b. $\begin{bmatrix} 2 & 3 \\ 1 & 1 \end{bmatrix}$ c. $\begin{bmatrix} 1 & 3 \\ 1 & 2 \end{bmatrix}$ d. $\begin{bmatrix} -2 & 1 \\ 3 & -1 \end{bmatrix}$

1.41 The determinant of the matrix $\begin{bmatrix} 1 & 2 & -1 \\ 3 & 0 & 2 \\ 2 & -2 & -1 \end{bmatrix}$ is:

a. 4 c. 24
b. 16 d. –16

1.42 In the system of equations

$$\begin{aligned} 3x_1 + 2x_2 - x_3 &= 5 \\ x_2 - x_3 &= 2 \\ x_1 + 2x_2 - 3x_3 &= -1 \end{aligned}$$

the value of x_2 = is:
a. 2 c. 4
b. –1 d. 6

1.43 What is the determinant of M?

$$\boldsymbol{M} = \begin{bmatrix} 0 & 1 & 1 & 1 \\ 1 & 1 & 1 & 1 \\ 1 & 1 & 3 & 1 \\ 2 & 1 & 3 & 4 \end{bmatrix}$$

a. –6 c. 0
b. 6 d. 7

1.44 $\int_{\pi/2}^{\pi} \sin 2x \, dx =$

a. 2
b. 1
c. 0
d. –1

1.45 $\int_0^2 x^2\sqrt{1+x^3}\,dx =$

a. 52/9
b. 0
c. 52/3
d. 26/3

1.46 $\int_1^e x(\ln x)dx =$

a. $\frac{1}{2}e^2+1$
b. $\frac{1}{2}e^2-e+\frac{1}{2}$
c. $\frac{1}{4}e^2+\frac{1}{4}$
d. $\frac{1}{4}e^2-\frac{1}{2}e+\frac{1}{4}$

1.47 If the first derivative of the equation of a curve is constant, the curve is a:
a. circle
b. hyperbola
c. parabola
d. straight line

1.48 Which of the following is a characteristic of all trigonometric functions?
a. The values of all functions repeat themselves every 45 degrees.
b. All functions have units of length or angular measure.
c. The graphs of all functions are continuous.
d. All functions have dimensionless units.

1.49 For a given curve $y = f(x)$ that is continuous between $x = a$ and $x = b$, the average value of the curve between the ordinates at $x = a$ and $x = b$ is represented by:

a. $\dfrac{\int_a^b x^2\,dy}{b-a}$
b. $\dfrac{\int_a^b y^2 dx}{b-a}$
c. $\dfrac{\int_a^b x\,dy}{a-b}$
d. $\dfrac{\int_a^b y\,dx}{b-a}$

1.50 If $y = \cos x$, $\dfrac{dy}{dx}$ is:

a. $\sin x$
b. $-\tan x \cos x$
c. $\dfrac{1}{\sec x}$
d. $\sec x \sin x$

1.51 The derivative of $\cos^3 5x$ is:

a. $3 \sin^2 5x$
b. $15 \sin^2 5x$
c. $\cos^2 5x \sin x$
d. $-15 \cos^2 5x \sin 5x$

1.52 The slope of the curve $y = 2x^3 - 3x$ at $x = 1$ is:

a. -1 c. 3

b. 0 d. 5

1.53 A stone is dropped from the top of a building at $t = 0$. The position of the stone is given by the equation $s(t) = 16\,t^2$ m. Acceleration of the stone (in m/s^2) 2 seconds after it is dropped is:

a. 64 c. 24

b. 32 d. 16

1.54 Maximum value of the function $f(x) = x^3 - 5x - 4$ occurs at:

a. 0 c. -0.30

b. 1.29 d. -1.29

1.55 The partial derivative with respect to x of the function $xy^2 - 5y + 6$ is:

a. xy c. y^2

b. $2y$ d. $-5y$

1.56 The power series expansion of $\cos(x)$ about the point $x = 0$ is:

a. $1-\frac{x^2}{2!}+\frac{x^4}{4!}-\frac{x^6}{6!}+\ldots$ c. $1-\frac{x}{2!}+\frac{x}{4!}-\frac{x}{6!}+\ldots$

b. $1+\frac{x^2}{2!}+\frac{x^4}{4!}+\frac{x^6}{6!}+\ldots$ d. $1+\frac{x}{2!}+\frac{x}{4!}+\frac{x}{6!}+\ldots$

1.57 The value of $\lim\limits_{x\to 2}\frac{x^2-4}{x-2}$ is:

a. 0 c. 1

b. ∞ d. 4

1.58 If $A = \int_0^{\frac{\pi}{4}} \sin^2\theta\, d\theta$, the value of A is:

a. 0.29 c. 1.75

b. 0.58 d. 3.14

1.59 If $x^3 + 3x^2y + y^3 = 4$ defines y implicitly, $dy/dx =$

a. $-\frac{x^2+2xy}{x^2+y^2}$ c. $-\frac{x^2+y^2}{x^2+2xy}$

b. $3x^2+3y^2$ d. $-\frac{x^2+2xy}{x^2+y^2}$

1.60 Estimate $\sqrt{34}$ using differentials. The answer is closest to:

a. $6+\frac{1}{6}$ c. 6

b. $6-\frac{1}{6}$ d. $6-\frac{1}{3}$

1.61 The only relative maximum of $f(x)=x^4-\frac{4}{3}x^3-12x^2+1$ is:

a. −1 c. 0

b. 1 d. −1

1.62 The area between $y = x^2$ and $y = 2x + 3$ is:

a. 9 c. $6\frac{1}{3}$

b. 20 d. $10\frac{2}{3}$

1.63 The area enclosed by the curve $r = 2(\sin\theta + \cos\theta)$ is:

a. π c. 2π

b. $\frac{\pi}{2}$ d. $\pi\sqrt{2}$

1.64 The curve in Exhibit 2.64 has the equation $y = f(x)$. At point A, what are the values of $\frac{dy}{dx}$ and $\frac{d^2y}{dx^2}$?

y
A
x

Exhibit 2.64

a. $\frac{dy}{dx}<0,\ \frac{d^2y}{dx^2}<0$ c. $\frac{dy}{dx}=0,\ \frac{d^2y}{dx^2}=0$

b. $\frac{dy}{dx}<0,\ \frac{d^2y}{dx^2}>0$ d. $\frac{dy}{dx}>0,\ \frac{d^2y}{dx^2}<0$

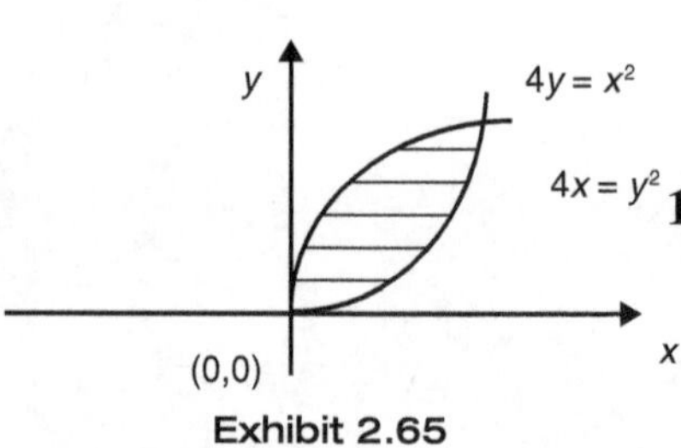

Exhibit 2.65

1.65 The area of the shaded region in Exhibit 2.65 is:

a. 1.37 c. 5.33

b. 3.82 d. 6.80

Refer to Exhibit 2.66 for problems 2.66 through 2.71.

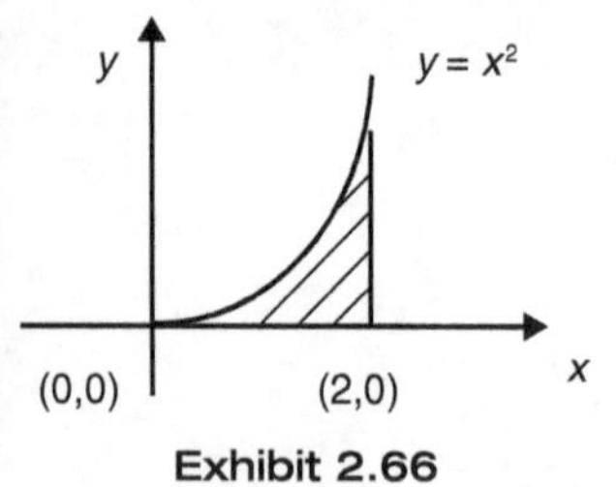

Exhibit 2.66

1.66 Area of the shaded region is:

a. 1.10 c. 4.02
b. 2.67 d. 6.80

1.67 The first moment of the shaded area with respect to the x-axis is:

a. 1.20 c. 3.20
b. 2.30 d. 4.60

1.68 The first moment of the shaded area with respect to the y-axis is:

a. 0.85 c. 3.07
b. 1.90 d. 4.00

1.69 The centroid of the shaded area is:

a. (1.5, 1.2) c. (0, 1.2)
b. (1.5, 0) d. (1.0, 1.0)

1.70 The second moment of the shaded area with respect to the x-axis is:

a. 6.10 c. 12.11
b. 9.05 d. 18.32

1.71 The second moment of the shaded area with respect to the y-axis is:

a. 1.98 c. 4.63
b. 3.12 d. 6.40

1.72 If the characteristic roots of a differential equation are $-4 - i4$ and $-4 + i4$, the homogeneous solution is:

a. $C_1 \cos 4x + C_2 \sin 4x$ c. $C_1 e^{-i4x} + C_2 e^{i4x}$
b. $e^{-4x}(C_1 \cos 4x + C_2 \sin 4x)$ d. $C_1 \cos (4x + \theta)$

1.73 Characteristic roots of a differential equation are -2 and -2. If the forcing function is e^{-2x}, the particular solution, $y_p(x)$ is:

a. Ax^2 c. Axe^{-2x}
b. Ae^{-2x} d. Ax^2e^{-2x}

1.74 The solution of the differential equation $y'' + 5y' + 6y = 2e^{-2x}$ with zero initial conditions is:

a. $y = -2e^{-2x} + 2e^{-3x} + 2xe^{-2x}$ c. $y = 2e^{-3x} + 2xe^{-2x}$
b. $y = x - 2e^{-2x} + 2e^{-3x}$ d. $y = -2e^{-2x} + 2xe^{-2x}$

1.75 $\lim\limits_{x \to 1} \dfrac{x^2 - 1}{x - 1} =$

a. 2 c. 0
b. ∞ d. 1

1.76 The solution to $xy' + 2y = e^{3x}$ is:

a. $y = e^{3x} - \dfrac{e^{3x}}{x} + \dfrac{c}{x}$

b. $y = \dfrac{xe^{3x} - 3e^{3x} + 3c}{3x^2}$

c. $y = xe^{3x} - 3e^{3x} + c$

d. $y + x = e^{3x} + c$

1.77 Solve $xy'' - 2(x + 1)y' + (x + 2)y = 0$.

a. $y = Ae^x + Bx^3e^x$
c. $y = A\sin(x + 1) + B\cos(x + 2)$
b. $y = Ae^x + Be^{2x}$
d. $y = Ae^x + Be^{-x}$

1.78 Solve $y'' + 4y = 8\sin x$.

a. $y = Ae^{2x} + Be^{-2x}$
b. $y = A\sin 2x + B\cos 2x$
c. $y = A\sin 2x + B\cos 2x + \sin x$
d. $y = A\sin 2x + B\cos 2x + \frac{8}{3}\sin x$

1.79 The family of trajectories orthogonal to the family $x^2 + y^2 = 2cy$ is:

a. $x - y = c$
c. $x^2 + y^2 = c$
b. $x^2 - y^2 = cx$
d. $x^2 + y^2 = 2cx$

1.80 The Laplace transform of the function $e^{-t}\cos(t)$ is:

a. $\dfrac{s}{s^2 + 1^2}$
c. $\dfrac{(s+1)}{s^2 + 1^2}$
b. $\dfrac{(s+1)}{(s+1)^2 + 1^2}$
d. $\dfrac{(s-1)}{(s-1)^2 + 1^2}$

1.81 For the differential equation $\dfrac{d^2y}{dt^2} + 2\dfrac{dy}{dt} + y(t) = \cos 2t$ with zero initial conditions, the Laplace transform $Y(s)$ of $y(t)$ is:

a. $\dfrac{s}{(s^2 + 2s + 1)}$
c. $\dfrac{s}{(s^2 + 4)(s^2 + 2s + 1)}$
b. $\dfrac{2}{(s^2 + 2)(s^2 + 2s + 1)}$
d. $\dfrac{s}{(s^2 + 4)(s^2 + 1)}$

1.82 The initial value of the function whose Laplace transform $F(s) = \dfrac{s+5}{s(s+1)(s+10)}$ is:

a. 2.0
c. 0.5
b. 1.0
d. 0

1.83 For the difference equation $y(k) = 2y(k-1) + 3y(k-2)$ with initial conditions $y(-1) = 1$; $y(-2) = 1$, the value of $y(1)$ is:

a. 1 c. 13
b. 5 d. 41

1.84 For the difference equation $y(k + 1) + y(k) = u(k)$ with $y(0) = 0$, find $Y(z)$, the z-transform of the function $y(k)$ is:

a. $\dfrac{1}{(z-1)(z+2)}$ c. $\dfrac{z}{(z+1)(z-2)}$

b. $\dfrac{z}{(z-1)(z+2)}$ d. $\dfrac{2}{(z+1)(z+2)}$

1.85 The initial value of the function whose Z-transform $F(z) = \dfrac{z+2}{z-4}$ is:

a. 0 c. 0.5
b. –0.5 d. 1.0

1.86 Find the Fourier coefficient b_3 of the waveform shown in Exhibit 2.86.

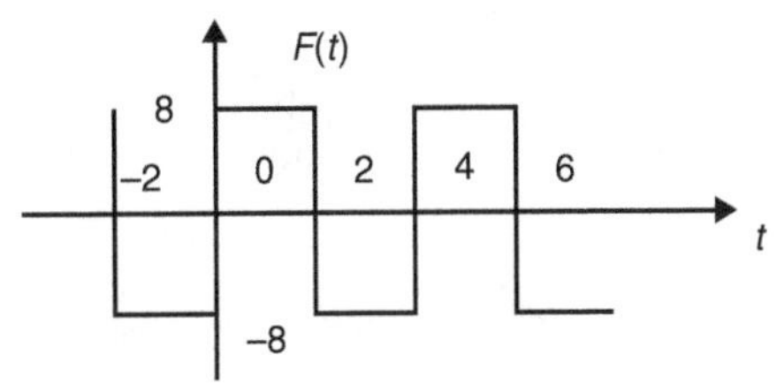

Exhibit 2.86

a. 1.17 c. 3.40
b. 2.23 d. 4.78

1.87 Find the Fourier transform, $F(f)$, of a triangular pulse of width 2.

a. $\text{sinc}(f)$ c. $\text{sinc}(2f)$
b. $\text{sinc}^2(f)$ d. $f\,\text{sinc}^2(f)$

1.88 Find the Fourier transform of $f(t) = \delta(t - 1) + \delta(t + 1)$.

a. $\exp(-j2\pi f)$ c. $2\cos(2\pi f)$
b. $\exp(j2\pi f)$ d. $j2\sin(2\pi f)$

1.89 Find $y(1)$ if $y(k) = 2y(k - 1) + 3y(k - 2)$ with $y(-1) = 1$; $y(-2) = 1$.

a. 1 c. 13
b. 5 d. 41

1.90 Find the final value of the function $f(k)$ whose Z-transform is $F(z) = \dfrac{2(z+1)}{(z-1)}$.

a. 1 c. 4
b. 2 d. ∞

SOLUTIONS

1.1 d. The value $(n + 1)!$ may be written as $(n +1)(n)\ [(n -1)!]$. It may be written also as $n!(n + 1)$. Hence the given expression may be written as follows:

$$\frac{\{(n+1)(n)\ [(n-1)!]\}\ \{n\ !(n+1)\}}{n\ !(n-1)!} = (n+1)^2 n$$

1.2 d. Raise both sides of the equation to the 4/3 power:

$$[x^{3/4}]^{4/3} = 8^{4/3}$$

$$x = \sqrt[3]{8^4} = \sqrt[3]{(2^3)^4} = 2^{\frac{3\bullet 4}{3}} = 2^4 = 16$$

1.3 a. $\log_a 10 = 0.250$ can be written as $10 = a^{0.250}$. Taking $\log_{10}$,

$$\log_{10} 10 = \log_{10} a^{0.250}$$
$$1 = 0.250 \log_{10} a$$

and

$$\log_{10} a = \frac{1}{0.250} = 4$$

1.4 c. Since $(\log_5 x)(\log_x 5)=1$, $\log_x 5= \dfrac{1}{-1.8} = -0.556$

1.5 d. Since $\log(a) + \log(b) - \log(c) = \log \dfrac{ab}{c}$, the given equation simplifies to $\log \dfrac{x(x-10)}{2} = 1$. Equivalently, $\dfrac{x(x-10)}{2} = 10$ or $x^2 - 10x - 20 = 0$. The roots are 11.708 and –1.708. Since log is not defined for negative values, $x = 11.708$.

1.6 b.

1.7 b. $\dfrac{6!}{3!0!} = \dfrac{6(5)(4)(3)!}{3!(1)} = 120$

1.8 a. The law of cosines is $a^2 = b^2 + c^2 - 2bc \cos A$ for any plane triangle with angles A, B, C and sides a, b, c, respectively.

This law can be applied to solve for the angles, given three sides in a plane triangle (Exhibit 2.8).

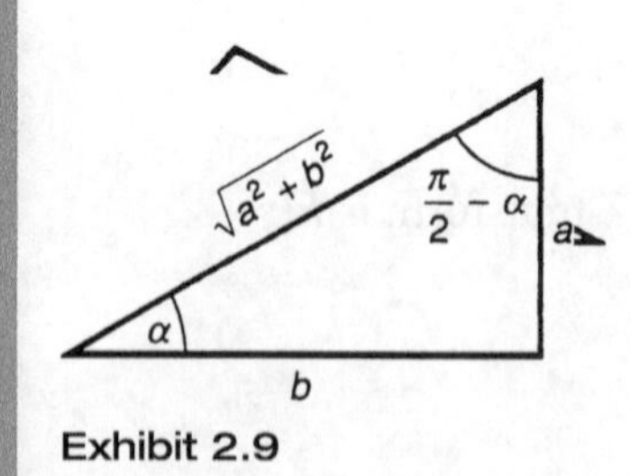

Exhibit 2.9

1.9 a. The triangle appears in Exhibit 2.9.

$$\tan\left(\frac{\pi}{2} - \alpha\right) = \frac{b}{a}$$

$$\tan^{-1}\frac{b}{a} = \frac{\pi}{2} - \alpha$$

1.10 d.

$$840° = 2(360) + 120 = 2(2\pi) \text{ rad} + 120°$$
$$\sin[2(2\pi) \text{ rad} + 120°] = \sin 120° = \sin 60°$$

1.11 b. The solution is obtained by seeing which of the five answers satisfies the equation.

x	$x^3 - 8x - 3$
2	–11
3	0
4	29
5	82
6	165

1.12 b. Any scientific calculator can be used to find the roots as 0, 3, and –2.

1.13 a. If x_1 and x_2 are the roots, the equation is $(x - x_1)(x - x_2) = 0$. Since the roots are $-1 + i$ and $-1 - i$, the equation is $(x - (-1 + i))(x - (-1 - i)) = 0$ or $(x + 1 - i)(x + 1 + i) = 0$ or $x^2 + 2x + 2 = 0$.

1.14 d. Common logarithms have base 10. Natural, or napierian, logarithms have base $e = 2.71828$.

1.15 d.

$$\log\ (5.743)^{1/30} = \frac{1}{30}\log 5.743 = \frac{1}{30}(0.7592) = 0.0253$$

The antilogarithm of 0.0253 is 1.06.

1.16 c.

$$\sin\ (A + B) = (\sin A \cos B) + (\cos A \sin B)$$

$$\cos\ (A + B) = (\cos A \cos B) - (\sin A \sin B)$$

$$\tan\ (A + B) = \frac{\sin(A + B)}{\cos(A + B)} = \frac{(\sin A \cos B) + (\cos A \sin B)}{(\cos A \cos B) - (\sin A \sin B)}$$

Dividing by $\cos A \cos B$,

$$\tan(A + B) = \frac{\dfrac{(\sin A \cos B)}{(\cos A \cos B)} + \dfrac{(\cos A \sin B)}{(\cos A \cos B)}}{\dfrac{(\cos A \cos B)}{(\cos A \cos B)} - \dfrac{(\sin A \sin B)}{(\cos A \cos B)}} = \frac{\tan A + \tan B}{1 - \tan A \tan B}$$

$$= \frac{\dfrac{1}{3} + \dfrac{1}{4}}{1 - \dfrac{1}{3} \times \dfrac{1}{4}} = \frac{\dfrac{4}{12} + \dfrac{3}{12}}{1 - \dfrac{1}{12}} = \frac{\dfrac{7}{12}}{\dfrac{11}{12}} = \frac{7}{11}$$

The problem could also be solved by determining angle A (whose tangent is 1/3) and angle B (whose tangent is 1/4). Then we could find the tangent of $(A + B)$.

$$\tan^{-1}\frac{1}{3} = 18.435° \quad \tan^{-1}\frac{1}{4} = 14.036°$$

$$\tan(18.435 + 14.036)° = \tan(32.471°) = 0.6364 = \frac{7}{11}$$

1.17 d. To reveal a parabola, a right circular cone must be cut parallel to an element of the cone and intersecting the axis of symmetry.

1.18 c. Rewriting the given line, $y = \frac{5}{3} - \frac{2}{3}x$. This line has slope $-\frac{2}{3}$, so a perpendicular line must have slope $\frac{3}{2}$. Using the point-slope form, $\frac{y-5}{x+3} = \frac{3}{2}$. Simplifying, $y = \frac{3}{2}x + 8$.

1.19 d. The equation of a straight line is $y = mx + b$ where the slope is m, (x, y) is any point on the line, and b is the y-axis intercept. Here, $m = -2$ and (2, 0) is a point. Substituting these values, $0 = -2(2) + b$ or $b = 4$. Then, the equation is $y = -2x + 4$ or $y + 2x = 4$.

1.20 a. (4,0) and (0, –6) are two points on the straight line. Then, the slope $m = \frac{y_2 - y_1}{x_2 - x_1} = \frac{0-(-6)}{4-0} = \frac{3}{2}$. Substituting one of the points, say, (0, –6) in the general equation $y = mx + b$; $-6 = 0 + b$. Then, $b = -6$ and the equation is $y = \left(\frac{3}{2}\right)x - 6$ or $3x - 2y = 12$.

1.21 b. For y-axis intercept, $x = 0$. Substituting this in the equation, $0 + 3y + 9 = 0$ or $y = -3$ is the y-axis intercept. For x-axis intercept, $y = 0$. Substituting this in the equation, $x + 0 + 9 = 0$ or $x = -9$ is the x-axis intercept.

1.22 b. Distance d between any two points (x_1, y_1, z_1) and (x_2, y_2, z_2) can be determined as $d^2 = (x_1 - x_2)^2 + (y_1 - y_2)^2 + (z_1 - z_2)^2$. Here, $d^2 = (1 - 0)^2 + (0 - 2)^2 + (-2 - 3)^2 = 30$; $d = 5.385$.

1.23 c. Directrix, $x = -p/2 = -2$ or $p = 4$. Equation of a parabola with center at origin is $y^2 = 2px$; as $p = 4$, $y^2 = 8x$.

1.24 b. Equation of a parabola with center at origin is $y^2 = 2px$. Since $y^2 = -4x$, $2p = -4$ or $p = -2$. Equation of a directrix is $x = (-p/2)$; since $p = 2$, the equation is $x = 1$.

1.25 d.

$$\int_{\pi/2}^{\pi} \sin 2x \, dx = -\frac{1}{2}\cos 2x = -\frac{1}{2}\cos 2\pi + \frac{1}{2}\cos \pi = -\frac{1}{2} - \frac{1}{2} = -1$$

1.26 a. Let $u = 1 + x^3$, so $du = 3x^2 dx$. The integral becomes $\frac{1}{3}\int_1^9 \sqrt{u}\,du$ $= \frac{2}{9}u^{3/2} = \frac{2}{9}(27-1)$.

1.27 c. Prepare to solve using $\int u\,dv = uv - \int v\,du$.

Let $u = \ln x$ and $dv = x\,dx$.

Therefore, $du = \frac{1}{x}dx$ and $v = \frac{1}{2}x^2$

$$\int_1^e x(\ln x)dx = \ln x\left(\frac{1}{2}x^2\right)\Big|_1^e - \int_1^e \frac{1}{2}x^2\left(\frac{1}{x}\right)dx$$

$$= \ln x\left(\frac{1}{2}x^2\right)\Big|_1^e - \frac{1}{4}x^2\Big|_1^e = \ln e\left(\frac{1}{2}e^2\right) - \ln 1\left(\frac{1}{2}1^2\right) - \left[\frac{1}{4}e^2 - \frac{1}{4}1^2\right]$$

$$= \frac{1}{4}e^2 + \frac{1}{4}$$

1.28 d. If $\frac{dy}{dx} = m$, $y = \int m\,dx = m\int dx = mx + b$, so $y = mx + b$ is a straight line.

1.29 d. All trigonometric functions are ratios of lengths, with the result that they are dimensionless.

1.30 d.

$$\text{Area} = \int_a^b y\,dx$$

$$\text{Average value} = \frac{\text{Area}}{\text{Base width}} = \frac{\int_a^b y\,dx}{b-a}$$

1.31 b. Since $\frac{dy}{dx} = -\sin x$ and $\tan x = \frac{\sin x}{\cos x}$, then $\sin x = \tan x \cos x$. Thus, the derivative is

$$\frac{dy}{dx} = -\tan x \cos x$$

1.32 d. Apply the chain rule. The "outside" function is u^3, so $y' = 3\cos^2 5x$ $(\cos 5x)' = 3\cos^2 5x\,(-\sin 5x)\,(5)$.

1.33 c. Derivative of y, $y' = 6x^2 - 3$. Since the slope is the derivative, at $x = 1$ the slope is $y'\,|_{x=1} = 3$.

1.34 b. Position $s(t) = 16t^2$; then, velocity $v(t) = s'(t) = 32t$ m/s and acceleration $a(t) = s''(t) = 32$ m/s^2; at any time the acceleration is 32 m/s^2.

1.35 d. Derivative of y is $y' = 3x^2 - 5$. Equating y' to 0, $3x^2 - 5 = 0$ has roots at $x = -1.29, +1.29$.

Second derivative of y is $y'' = 6x$.

$y''|_{x=-1.29} < 0$; then, the maximum of y occurs at $x = -1.29$.

(Note: value of $y_{max} = (-1.29)^3 - 5(-1.29) - 4 = 0.30$

$y''|_{x=1.29} > 0$; then, the minimum of y occurs at $x = 1.29$; $y_{min} = (1.29)^3 - 5(1.29) - 4 = -8.30$.

At the inflection point $y'' = 0$: $y'' = 6x = 0$; then, $x = 0$. Also, y'' changes sign at $x = 0$.)

1.36 c. y^2

1.37 a. $f(x) = \cos x$; $f'(x) = -\sin x$; $f''(x) = -\cos x$; $f'''(x) = \sin x$; $f^{iv}(x) = \cos x$
Substituting these values in the Taylor's series expansion,

$$f(x) = \cos(0) + \frac{-\sin 0}{1!}x + \frac{-\cos 0}{2!}x^2 + \frac{\sin 0}{3!}x^3 + \frac{\cos 0}{4!}x^4$$

$$= 1 - \frac{x^2}{2!} + \frac{x^4}{4!} - \frac{x^6}{6!} + \ldots$$

1.38 d. Using L'Hopital's rule, $\lim_{x \to 2} \frac{x^2 - 4}{x - 2} = \lim_{x \to 2} \frac{2x}{1} = 4$.

1.39 a. Using the integral table, the integral =

$$2\left[\frac{\theta}{2} - \frac{\sin 2\theta}{4}\right]_0^{\pi/4} = \left[\left(\frac{\pi}{4} - \frac{1}{2}\right) - 0\right] = 0.285$$

1.40 a. Taking the derivative with respect to x,

$$3x^2 + 6xy + 3x^2y' + 3y^2y' = 0, \text{ so } y' = -\frac{3x^2 + 6xy}{3x^2 + 3y^2}$$

1.41 b. Since $\sqrt{36} = 6$, take $x_0 = 36$ and $f(x) = \sqrt{x}$. In general,

$$\Delta y = f(x) - f(x_0) = f'(x_0)(x - x_0)$$

$$\sqrt{34} - \sqrt{36} = \frac{1}{2}\frac{1}{\sqrt{36}}(-2) = -\frac{1}{6}$$

1.42 b. Here,

$$f'(x) = 4x^3 - 4x^2 - 24x = x(4x^2 - 4x - 24) = 4x(x - 3)(x + 2) = 0,$$

so possible extrema are at 0, 3, and –2. Since $f''(0) = -24$, it is the maximum (3 and –2 are minima). Since $f(0) = 1$, the relative maximum is 1.

1.43 d. The line and the parabola intersect when $x^2 = 2x + 3$, or $x^2 - 2x + 1 = 4$, or $(x-1)^2 = 2^2$. The line is above the parabola, so

$$A = \int_{-1}^{3} \left(2x + 3 - x^2\right) dx = \left(x^2 + 3x - \frac{1}{3}x^3\right)\Bigg|_{-1}^{3} = 9 - \left(-\frac{5}{3}\right)$$

1.44 c. Multiply by r to obtain $x^2 + y^2 = 2y + 2x$, or $x^2 - 2x + y^2 - 2y = 0$, or $(x-1)^2 + (y-1)^2 = 2$ a circle centered at (1, 1) of radius $\sqrt{2}$. The area is $\pi(\sqrt{2})^2 = 2\pi$.

1.45 d. The first derivative $\frac{dy}{dx}$ is the slope of the curve. At point A the slope is positive. The second derivative $\frac{d^2y}{dx^2}$ gives the direction of bending. A negative value indicates the curve is concave downward.

1.46 c. Point of intersection of the curves is determined as $y^2 = 4x = 4\sqrt{4y}$, $y = 4$; $x = 4$.

Area of the strip in Exhibit 2.46a, $dA = (y_1 - y_2)dx = \left(\sqrt{4x} - \frac{x^2}{4}\right)dx$

$$\text{Area} = A = \int_{x=0}^{4} \left(\sqrt{4x} - \frac{x^2}{4}\right)dx = \frac{16}{3}$$

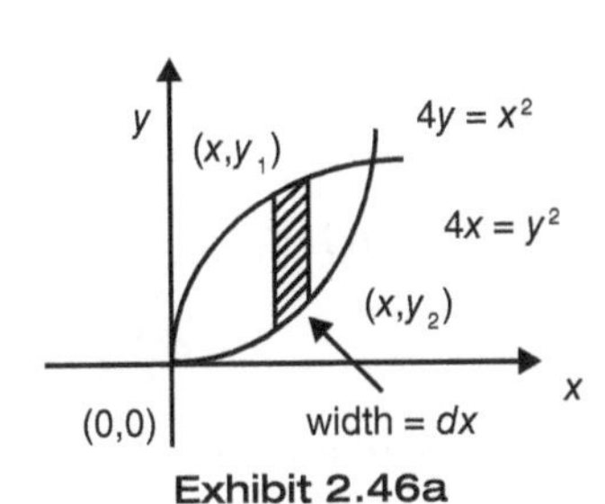

Exhibit 2.46a

1.47 b. In Exhibit 2.47a, area $dA_2 = ydx$.

$$\text{Shaded area} = \int dA_2 = \int_{x=0}^{2} y.dx = \int_{x=0}^{2} x^2 dx = \frac{8}{3}$$

y
y = x²
(x,y)
x
(0,0)
(2,0)
width = dx
Area, $dA_2 = y\,dx$

Exhibit 2.47a

1.48 c. In Exhibit 2.48, area $dA_1 = (2 - x)dy$.

First moment with respect to x-axis of this area $dM_x = y\,dA_1$

First moment with respect to x-axis of shaded area is

$$M_x = \int dM_x = \int_{y=0}^{4} y(2-x)dy = \int_{0}^{4} y(2-\sqrt{y})dy = 3.2$$

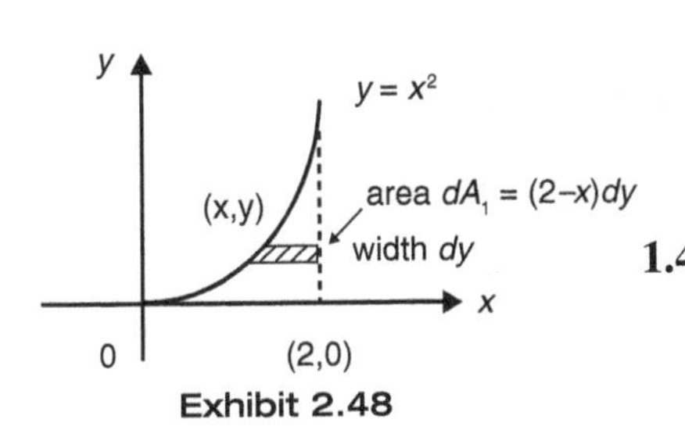

Exhibit 2.48

1.49 d. From Exhibit 2.47a, first moment with respect to y-axis of the shaded area dA_2 is $dM_y = x\,dA_2 = xy\,dx$; first moment with respect to y-axis of shaded area is

$$M_y = \int dM_y = \int_{x=0}^{2} x.y.dx = \int_{0}^{2} x.x^2 dx = 4$$

1.50 **a.** x-coordinate $x_c = \frac{M_y}{area} = \frac{4}{2.67} = 1.5$; y-coordinate y_c

$$= \frac{M_x}{area} = \frac{3.2}{2.67} = 1.2$$

1.51 **a.** From Exhibit 2.48, moment of inertia with respect to x-axis of area dA_1 is $dMI_x = y^2 dA_1$. Moment of inertia with respect to x-axis of shaded area:

$$MI_x = \int dMI_x = \int y^2 dA_1 = \int_{y=0}^{4} y^2(2-x)dy = \int_{0}^{4} y^2(2-\sqrt{y})dy$$

Integrating, $MI_x = 6.095$.

1.52 **d.** From Exhibit 2.47a, moment of inertia with respect to y-axis of area dA_2 is $dMI_y = x^2 dA_2$. Moment of inertia with respect to y-axis of shaded area:

$$MI_y = \int dMI_y = \int x^2 dA_2 = \int_{x=0}^{2} x^2.ydx = \int_{x=0}^{2} x^4 dx = 6.4$$

1.53 **b.** Homogeneous solution, $y_h(x) = e^{-4x}(C_1 \cos 4x + C_2 \sin 4x)$ or $e^{-4x} C_3 \cos(4x + \theta)$

1.54 **d.** Since the characteristic roots are –2 and –2, the particular solution is $y_p(x) = A\, x^2 e^{-2x}$.

1.55 **a.** The characteristic equation is $r^2 + 5r + 6 = 0$ and the characteristic roots are –2, –3. Then, the homogeneous solution is $y_h(x) = c_1 e^{-2x} + c_2 e^{-3x}$.

The particular solution due to e^{-2x} is $y_p(x) = Bxe^{-2x}$.

[As –2 is a characteristic root, e^{-2x} cannot be a particular solution.]

$y_p' = -2Be^{-2x}.x + Be^{-2x}$ and $y_p'' = 4Be^{-2x}.x - 2Be^{-2x} - 2Be^{-2x}$

$y_p'' + 5y_p' + 6y = 2e^{-2x}$ yields $B = 2$

$y(x) = 2e^{-2x}x + c_1 e^{-2x} + c_2 e^{-3x}$ $\quad y(0) = c_1 + c_2 = 0 \quad c_1 = -2$

$y'(x) = 2e^{-2x} - 4xe^{-2x} - 2c_1 e^{-2x} - 3c_2 2e^{-3x}$ $\quad y'(0) = 2 - 2c_1 - 3c_2 = 0 \quad c_2 = 2$

The solution is $y = -2e^{-2x} + 2e^{-3x} + 2xe^{-2x}$.

1.56 **a.**

$$\lim_{x\to 1} \frac{x^2 - 1}{x - 1} = \lim_{x\to 1} \frac{(x-1)(x+1)}{x-1} = \lim_{x\to 1}(x+1) = 2$$

1.57 **b.** This is linear equation, $y' + \frac{2}{x}y = \frac{1}{x}e^{3x}$. The integrating factor is $e^{\int \frac{2}{x}\,dx} = x^2$ so the equation becomes $d(x^2y) = xe^{3x}\,dx$. Integrating,

$$x^2 y = \frac{1}{3}xe^{3x} - \frac{1}{9}e^{3x} + c, \text{ or } y = \frac{1}{3x}e^{3x} - \frac{1}{9x^2}e^{3x} + \frac{c}{x^2}.$$

1.58 **a.** By inspection, $y_1 = e^x$ is one solution. Use reduction of order to obtain

$$\left(\frac{y_2}{y_1}\right)' = \frac{e^{\int \frac{-2(x+1)}{x}dx}}{y_1^2} = \frac{e^{2x+2\ln x}}{e^{2x}} = x^2$$

Hence $\frac{y_2}{y_1} = \frac{x^3}{3}$, so $y_2 = \frac{x^3}{3}y_1$. Suppressing the $\frac{1}{3}$, $y_2 = x^3 e^x$.

1.59 **d.** The associated homogeneous equation, $y'' + 4y = 0$, has the solution $y_h = A\sin 2x + B\cos 2x$. Using the method of undetermined coefficients,

$$y_p = a\sin x + b\cos x$$
$$y_p'' = -a\sin x - b\cos x$$
$$y_p'' + 4y_p = (-a + 4a)\sin x + (-b + 4b)\cos x = 8\sin x$$
$$3a\sin x = 8\sin x$$
$$3b\cos x = 0\cos x$$

Thus, $b = 0$ and $a = \frac{8}{3}$.

1.60 **d.** Begin by eliminating c by solving for it from the derivative of the given equation: $2x + 2yy' = 2cy'$, $c = \frac{x + yy'}{y'}$,

$$x^2 + y^2 = 2\frac{x + yy'}{y'}y', \; x^2y' + y^2y' = 2xy + 2y^2y', \text{ and}$$

$$y' = \frac{2xy}{x^2 - y^2}.$$

Now, the orthogonal family will have $y'_{\text{new}} = -\frac{1}{y'_{\text{old}}}$, so $y'_{\text{old}} = \frac{y^2 - x^2}{2xy}$.

Letting $u = \frac{y}{x}$, $xu' + u = \frac{u^2 - 1}{2u}$, $xu' = \frac{u^2 - 1 - 2u^2}{2u} = -\frac{1 + u^2}{2u}$,

and $\frac{2u\,du}{1 + u^2} = -\frac{dx}{x}$. Integrating, $\ln(1 + u^2) = -\ln|x| + c$, and

$$\ln\left\{\left[1 + \left(\frac{y}{x}\right)^2\right]|x|\right\} = c$$

$$\left|x + \frac{y^2}{x}\right| = e^c = c_1 > 0, \quad \text{or} \quad x + \frac{y^2}{x} = c_2\,(= \pm c_1)$$

Thus $x^2 + y^2 = c_2x$, or $x^2 + y^2 = 2cx$.

1.61 b. Using Laplace transform table, the transform is $\dfrac{(s+1)}{(s+1)^2+1^2}$.

1.62 c. Taking the Laplace transform of the differential equation and simplifying,

$$s^2Y(s)+2sY(s)+Y(s)=\frac{s}{s^2+2^2} \Rightarrow Y(s)=\frac{s}{(s^2+4)(s^2+2s+1)}$$

1.63 d. Initial value, $\lim_{t\to 0} f(t)=\lim_{s\to\infty} s\dfrac{(s+5)}{s(s+1)(s+10)}=0$

1.64 c. $y(k) = 2y(k-1) + 3y(k-2)$ with $y(-1) = 1$ and $y(-2) = 1$
For $k = 0$, $y(0) = 2y(-1) + 3y(-2) = 2(1) + 3(1) = 5$
For $k = 1$, $y(1) = 2y(0) + 3y(-1) = 2(5) + 3(1) = 13$

1.65 a. $y(k+1) + y(k) = u(k)$; taking Z-transform, $[z\,y(z) - 0] + y(z)$

$=\dfrac{1}{1-z^{-1}}$. Simplifying, $y(z)=\dfrac{z}{(z-1)(z+2)}$.

1.66 d. Using the initial value theorem, the initial value,

$$\lim_{k\to 0} f(k)=\lim_{z\to\infty}\frac{z+2}{z-4}=1$$

1.67 c. Period T = 4 s; $\omega_0 = \dfrac{2\pi}{T}=\dfrac{\pi}{2}$ rad/s;

$$b_3=\frac{2}{4}\left[\int_0^2 8\sin\left(\frac{3\pi t}{2}\right)dt+\int_2^4 -8\sin\left(\frac{3\pi t}{2}\right)dt\right]=\frac{16}{3\pi}(1-\cos 3\pi)=3.40$$

1.68 d. Since the pulse width is 2, $\tau = 1$ and from the Table, $X(f) = f\,\text{sinc}^2(f)$.

1.69 c. Using time shift theorem for Fourier transforms,
$X(f) = 1\exp(-j2\pi f) + 1\exp(+j2\pi f) = 2\cos(2\pi f)$

1.70 c. Substituting $k = 0$, $y(0) = 2y(-1) + 3y(-2) = 2 + 3 = 5$
Substituting $k = 1$, $y(1) = 2y(0) + 3y(-1) = 10 + 3 = 13$

1.71 c. Final value $=\lim_{k\to\infty} f(k)=\lim_{z\to 1}(1-z^{-1})F(z)=\lim_{z\to 1}(1-z^{-1})\dfrac{2(z+1)}{(z-1)}=4$

1.72 a. Focus $ae = 2$ and directrix $\dfrac{a}{e} = 6$. Solving, $a^2 = 12$ and $e = 1/\sqrt{3}$.

But $e = \sqrt{1 - \dfrac{b^2}{a^2}}$; solving, $b^2 = 8$. The equation of an ellipse is

$$\frac{x^2}{a^2} + \frac{y^2}{b^2} = 1 \text{ or } \frac{x^2}{12} + \frac{y^2}{8} = 1.$$

1.73 **a.** The equation of an ellipse is $\dfrac{x^2}{a^2} + \dfrac{y^2}{b^2} = 1$. Here, $a = 3$ and $b = 2$.

The eccentricity $e = \sqrt{1 - \dfrac{b^2}{a^2}} = \sqrt{\dfrac{5}{9}}$, and the foci $= (\pm ae, 0) = (\pm \sqrt{5}, 0)$.

1.74 **b.** Focus $ae = 4$ and eccentricity $e = 3$. Solving, $a = 4/3$.

Also, $b = a\sqrt{e^2 - 1} = \dfrac{4\sqrt{8}}{3}$. Equation for a hyperbola is

$$\frac{x^2}{a^2} - \frac{y^2}{b^2} = 1 \Rightarrow \frac{x^2}{16} - \frac{y^2}{128} = 1.$$

1.75 **d.** Radius of a circle with center at (h, k) and passing through a point (x, y) is $r^2 = (x - h)^2 + (y - k)^2$. Here, $r^2 = (4 - 1)^2 + (6 - 2)^2 = 25$. Equation of a circle with center at (h, k) is $(x - h)^2 + (y - k)^2 = r^2$; here, $(x - 1)^2 + (y - 2)^2 = 25$.

1.76 **d.** Length of the tangent from any point (x', y') outside a circle with center at (h, k) and radius r is given as $t^2 = (x' - h)^2 + (y' - k)^2 - r^2$. Here, $(x', y') = (4, 8)$, center $(h, k) = (0, 1)$, and $r^2 = 3^2$. Then, $t^2 = (4 - 0)^2 + (8 - 1)^2 - 3^2 = 56$ or $t = 7.483$.

1.77 **c.** General equation of a conic section is $Ax^2 + 2Bxy + Cy^2 + 2Dx + 2Ey + F = 0$, where both A and C are not zeros. Here, $A = 1$; $B = -5$; $C = 1$; then, $(B^2 - AC) > 0 \Rightarrow$ hyperbola.

1.78 **c.** Since three sides of a triangle are given, use the law of cosines; $3^2 = 2^2 + 4^2 - 2(2)(4)\cos(\theta)$ where θ is the angle opposite to side 3. Solving, $\theta = 46.6°$.

1.79 **a.** Angle $C = 180 - (70 + 32) = 78°$. Using the law of sines,

$\dfrac{a}{\sin 70} = \dfrac{b}{\sin 32} = \dfrac{27}{\sin 78}$. Solving, $a = 25.94$.

1.80 **d.** $(\sec \theta)(1 - \sin^2 \theta) = \sec \theta \cos^2 \theta = \dfrac{1}{\cos\theta} \cos^2 \theta = \cos \theta$

1.81 **a.** $\sin\theta = \dfrac{m}{1} = \dfrac{\text{opposite side}}{\text{hypotenuse}}$; then, adjacent side $= \sqrt{1 - m^2}$

and $\cos\theta = \dfrac{\text{adjacent side}}{\text{hypotenuse}} = \dfrac{\sqrt{1 - m^2}}{1}$; then, $\cot\theta = \dfrac{\cos\theta}{\sin\theta} = \dfrac{\sqrt{1 - m^2}}{m}$

1.82 d. Cross product, $\mathbf{A} \times \mathbf{B} = \begin{vmatrix} i & j & k \\ 3 & -6 & 2 \\ 10 & 4 & -6 \end{vmatrix}$

Expanding, $i[(-6)(-6) - (2)(4)] - j[(3)(-6) - (2)(10)] + k[(3)(4) - (-6)(10)] = 28\mathbf{i} + 38\mathbf{j} + 72\mathbf{k}$

1.83 c. This is an arithmetic series; first term $a = 10$, last term $l = 50$, and common difference $d = 1$. The number of terms, n, can be calculated as, $l = a + (n - 1)d$; $50 = 10 + (n - 1)1$; solving, $n = 41$.

$$\text{sum } S = \frac{n(a+1)}{2} = \frac{41}{2}(10 + 50) = 1230$$

1.84 b. This is an arithmetic series; first term $a = 10$ and the common difference $d = 6$. Taking the number of terms n as 50, the nth term (last term) can be calculated as, $l = a + (n - 1)d$. In this case, $l = 10 + (50 - 1)6 = 304$.

1.85 a. This is a geometric series; first term $a = 4$ and common ratio $r = 0.5$. Since $r < 1$, the series is convergent and the sum as the number of terms n tend to infinity is $S = \frac{a}{1-r} = \frac{4}{1-0.5} = 8$.

1.86 b. To compute the (2,1) entry, take [1 2 9] • [5 6 4] = 5 + 12 + 36 = 53.

1.87 d. To invert a 2×2 matrix,

$$\begin{bmatrix} a & b \\ c & d \end{bmatrix}^{-1} = \frac{1}{ad - bc}\begin{bmatrix} d & -b \\ -c & a \end{bmatrix} = \frac{1}{2-3}\begin{bmatrix} 2 & -1 \\ -3 & 1 \end{bmatrix}$$

1.88 c. This 3×3 determinant can be computed quickly be expanding it in minors, especially around the second column:

$$\begin{vmatrix} 1 & 2 & -1 \\ 3 & 0 & 2 \\ 2 & -2 & 1 \end{vmatrix} = (-1)^{1+2}(2)\begin{vmatrix} 3 & 2 \\ 2 & -1 \end{vmatrix} + (-1)^{2+2}(0)\begin{vmatrix} 1 & -1 \\ 2 & -1 \end{vmatrix} + (-1)^{3+2}(-2)\begin{vmatrix} 1 & -1 \\ 3 & 2 \end{vmatrix}$$

$$= -2(-3-4) + 0 + 2(2+3) = 24$$

1.89 d. By Cramer's Rule,

$$x_2 = \frac{\text{Det}\begin{bmatrix} 3 & 5 & -1 \\ 0 & 2 & -1 \\ 1 & -1 & -3 \end{bmatrix}}{\text{Det}\begin{bmatrix} 3 & 2 & -1 \\ 0 & 1 & -1 \\ 1 & 2 & -3 \end{bmatrix}} = \frac{3\begin{vmatrix} 2 & -1 \\ -1 & -3 \end{vmatrix} + 1\begin{vmatrix} 5 & -1 \\ 2 & -1 \end{vmatrix}}{3\begin{vmatrix} 1 & -1 \\ 2 & -3 \end{vmatrix} + 1\begin{vmatrix} 2 & -1 \\ 1 & -1 \end{vmatrix}} = \frac{3(-7) + (-3) = -24}{3(-1) + (-1) = -4} = 6$$

1.90 **a.** To evaluate a 4×4 matrix, one must do some row or column operations and expand by minors:

$$\begin{bmatrix} 0 & 1 & 1 & 1 \\ 1 & 1 & 1 & 1 \\ 1 & 1 & 3 & 1 \\ 2 & 1 & 3 & 4 \end{bmatrix} \sim \begin{bmatrix} 0 & 1 & 1 & 1 \\ 1 & 1 & 1 & 1 \\ 0 & 0 & 2 & 0 \\ 0 & -1 & 1 & 2 \end{bmatrix}$$

Taking minors of column 1,

$$\text{Det}(M) = (1)(-1)^{2+1}\,\text{Det}\begin{bmatrix} 1 & 1 & 1 \\ 0 & 2 & 0 \\ -1 & 1 & 2 \end{bmatrix} = -(4+2) = -6$$

CHAPTER 3

Probability and Statistics

OUTLINE

SETS AND SET OPERATIONS

A set is any well-defined list, collection, or class of objects. The objects in a set can be anything: numbers, letters, cards, people, and so on. They are called the **elements**, or **members**, of a set.

The name of a set is usually denoted by a capital letter, such as A, B, Y, Z. The elements of a set are usually denoted by small letters, such as a, b, y, z.

To specify that an element a is a member of a set B, we say "a is in B," which is written

$$a \in B$$

A set is called the **null**, or **empty**, set, denoted by $\varnothing$, if it has no elements. We say the set A is a subset of the set B, written $A \subset B$, if all the members of A are also in B. The universal set, denoted by U, is the set that contains all the members of the subsets. The **complement** of a set A, denoted as A', consists of all the elements in U that are not in A.

Sets are defined in either of two ways: (1) by listing the members in a tabular form (for example, if A consists of the numbers 1, 3, 5, and 7, we write $A = \{1, 3, 5, 7\}$), or (2) by stating the properties that the members must satisfy in a set-builder form. For example, a set B containing all the odd numbers is written $B = \{x|x \text{ is odd}\}$, where x is an element of the set and the vertical line "|" is read "such that." The full notation is read "B is the set of numbers x such that x is odd."

Example 3.1

Define the set R of all outcomes of the roll of a six-sided die.

Solution

$R = \{1, 2, 3, 4, 5, 6\}$

Example 3.2

Let U be the set of integers. Define the complement of $B = \{x|x \text{ is odd}\}$.

Solution

$B' = \{x|x \text{ is even}\}$

Set Operations

The **union** of two sets A and B is defined as the set of all elements in either A or B and is traditionally written as $A \cup B$ or, alternatively, as A or B.

The **intersection** of two sets A and B is defined as the set of all elements in both A and B and is traditionally written $A \cap B$.

Set operations are well-behaved mathematically and follow these laws:

identity
$$A \cup 0 = A \qquad A \cup U = U$$
$$A \cap 0 = 0 \qquad A \cap A = A$$

complement
$$A \cup A' = U \quad (A')' = A \quad A \cap A' = 0 \quad U' = 0$$

commutative
$$A \cup B = B \cup A \qquad A \cap B = B \cap A$$

associative
$$(A \cup B) \cup C = A \cup (B \cup C)$$
$$(A \cap B) \cap C = A \cap (B \cap C)$$

distributive
$$A \cup (B \cap C) = (A \cup B) \cap (A \cup C)$$
$$A \cap (B \cup C) = (A \cap B) \cup (A \cap C)$$

de Morgan's Law
$$(A \cup B)' = A' \cap B'$$
$$(A \cap B)' = A' \cup B'$$

For example, define the following sets:

$U = \{x|\, x \text{ is a person}\}$

$A = \{a|\, a \text{ is American}\}$

$F = \{f|\, f \text{ is French}\}$

$B = \{b\,|\, b \text{ is a person with dual French and American citizenship}\}$

Then the union of the sets A and F is the set of all people who are either American or French and is written $A \cup F$. The set of people who are not American is A'. The set of people who are not French is F'.

The set of people who are not French and not American is $A' \cap F'$. This is the same as $(A \cup F)'$, according to de Morgan's Law. In this example, de Morgan's Law means that all the people who are neither American nor French is the same as the set of all the people who are not French and not American.

Venn Diagrams

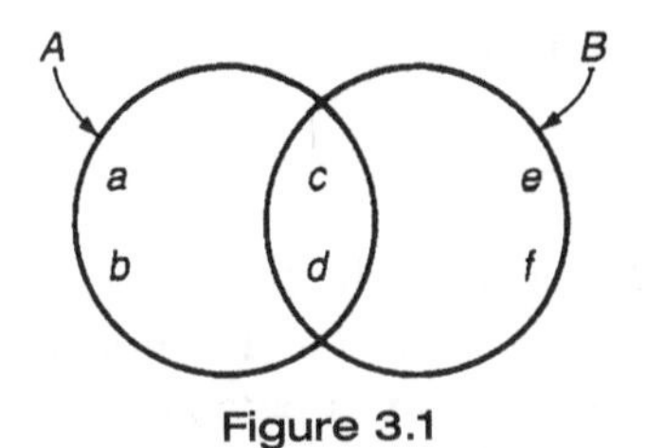

Figure 3.1

A simple and intuitive way to represent these set relationships is known as the **Venn diagram**. Here we represent a set by a closed area and an element in the set by a point in the area. For example, let $A = \{a, b, c, d\}$ and $B = \{c, d, e, f\}$; the sets are represented in a **Venn diagram** as shown in Figure 3.1.

Relational properties such as the intersection of A and $B = \{c, d\}$ are intuitively understood as the overlap of two regions.

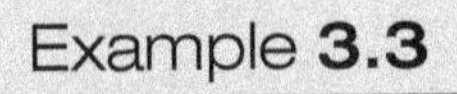

Draw a Venn diagram of the relationship between the sets U, A, F, and B as defined in the previous section.

Solution

1. $A \subset U$. Americans are a subset of people (Exhibit 1).
2. $A \cup F$ is the set of Americans and French (Exhibit 2).
3. $B = A \cap F$ is the intersection of two sets; it represents Americans who are also French (Exhibit 3).

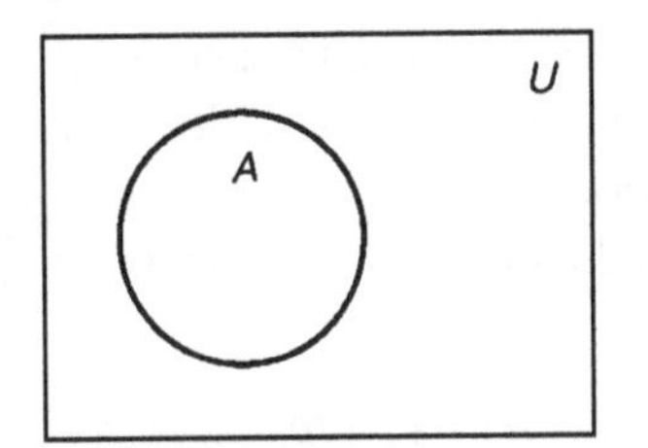

Exhibit 1

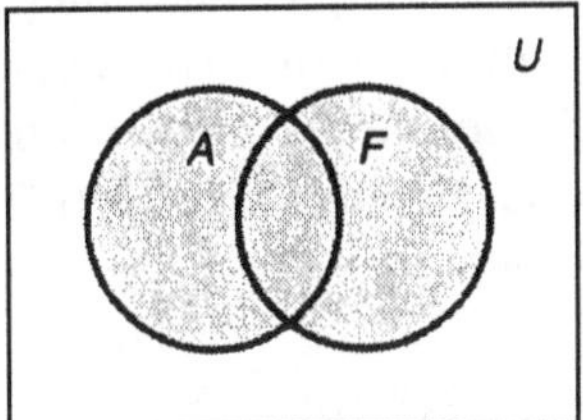
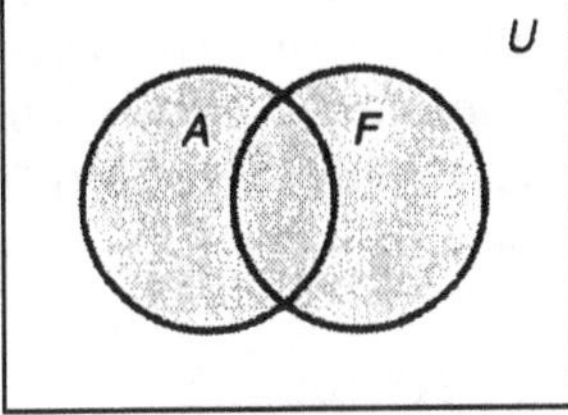

Exhibit 2

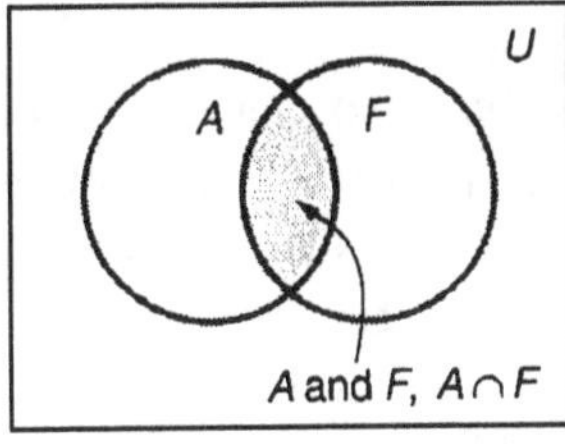

Exhibit 3

Product Sets

An ordered pair of elements a and b is denoted by (a, b). Two ordered pairs (a, b) and (c, d) are equal only if $a = c$ and $b = d$. For example, the ordered pairs (3, 4) and (4, 3) are different.

The product x of two sets A and B is denoted by $A \times B$ and consists of all ordered pairs of elements in A and B.

The two subsets of a product set can be treated like the intersection of two Cartesian coordinate axes. The subsets each act like a dimension, and the product set can be represented as a matrix where the elements of one subset are written along the row of the matrix and those of the other along the column.

For example, the set of results of a single coin toss is either heads (H) or tails (T) and is defined $R = \{H, T\}$. The result of two coin tosses is the product $R \times R$ and is defined as

$$R \times R = \{(H, H), (H, T), (T, H), (T, T)\}$$

COUNTING SETS

For discrete probability calculations it is important to count the number of elements in sets of possible outcomes. The primary method is simply to write down all the elements in a set and count them. For example, count the number of elements in the set S of all possible outcomes of a six-sided die throw. The set definition is $S = \{1, 2, 3, 4, 5, 6\}$. Simple counting gives six elements.

Most useful sets are large, and simple counting is too time-consuming. There are several methods for simplifying this task. One can use the product set concept introduced in the last section. If A is a set with n elements and B is a set with m elements, then the product set $A \times B$ has the arithmetic $n \times m$ number of elements. To simplify counting then, first count the sets making up the product set (usually containing a much smaller number of elements) and simply multiply these counts.

Example 3.4

Count the number of possible outcomes for tossing five coins.

Solution

The number of outcomes for a single toss defined by the set $R = \{H, T\}$ is 2. The result of five coin tosses is the product set $R \times R \times R \times R \times R$. The total number of possible outcomes is then the arithmetic product of the number of outcomes in each of the individual five tosses. This is $2 \times 2 \times 2 \times 2 \times 2 = 2^5 = 32$.

Permutations

If A is a set with n elements, a **permutation** of A is an ordered arrangement of A. Given the set $A = \{a, b, c\}$, the order a, b, c of the elements is one permutation. Any other order—for example, b, c, a—is another permutation.

The set B of all permutations of the set A is defined as the set of all arrangements of the three elements. These are

$$B = \{\{a, b, c\}, \{a, c, b\}, \{c, b, a\}, \{b, a, c\}, \{b, c, a\}, \{c, a, b\}\}$$

There are six permutations. This number also can be derived as follows. The number of ways an element can be chosen for the first space is three. Then there are two elements left. One of these can go in the second space. Then there is one element left. This must go in the third space. This gives the formula $3 \times 2 \times 1 = 6$. In general, the number of ways n distinct elements can be arranged is given by

$$n! = n \times (n-1) \times (n-2) \times \cdots 1$$

and is called the **factorial** of the number n. The factorial of 0 is 1 ($0! = 1$).

For example, count the number of ways a standard playing deck can be arranged. Since there are 52 distinct cards in a deck, there are 52! different arrangements, or permutations.

Now suppose we have the set of letters L in the word *obtuse* so that $L = \{o, b, t, u, s, e\}$. How many two-letter symbols could be made from this set? Notice that the letters are all distinct. We again count the number of ways the letters can be selected. For the first choice it is six; for the second choice it is five. The two selections are now complete. There are therefore $6 \times 5 = 30$ possibilities.

The general formula for the number of permutations, taking r items from a set of n, is given by

$$P(n, r) = n!/(n-r)!$$

Using this equation, one can express the previous example as $P(6, 2) = 6!/(6-2)! = 30$.

Example **3.5**

A jeweler has nine different beads and a bracelet design that requires four beads. To find out which looks the best, he decides to try all the permutations. How many different bracelets will he have to try?

Solution

There are $n = 9$ beads. He selects $r = 4$ at a time. The order is important, because each arrangement of r beads on the bracelet makes a different bracelet. So the number of different bracelets is

$$P(9,4) = 9!/(9-4)! = 9\times 8\times 7\times 6 = 3024$$

If the bracelet is a closed circle, there is no discernible difference when it is rotated. Then one observes four identical states for each unique bracelet. This is called ring permutation and is given by the formula

$$P_{\text{ring}}(n,r) = P(n,r)/r$$

There are only 3024/4 = 756 distinct ring bracelets the jeweler can make.

Example **3.6**

(i) In how many ways can four people be asked to form a line of three people?
(ii) In how many ways can the letters of the word BEAUTY be arranged?
(iii) In how many ways can the letters of the word GOOD be arranged?

Solution

(i) $P(4,3) = \dfrac{4!}{(4-3)!} = 24$

(ii) $P(6,6) = \dfrac{6!}{(6-6)!} = 720$

(iii) $P(4;1,1,2) = \dfrac{4!}{1!1!2!} = 12$

Combinations

When the order of the set of r things that are selected from the set of n things does not matter, we talk about combinations.

Again consider the standard playing deck of 52 cards. How many hands of 5 cards can we get from a deck of 52 cards? Count the number of ways the hands can be drawn. The first draw can be any of the 52 cards. The second draw can only be one of the remaining 51 cards. The third draw can only be one of the remaining 50, the next is one of 49, the last one of 48. So the result is

$$52 \times 51 \times 50 \times 49 \times 48 = 52!/(52-5)!$$

This is the formula for permutations discussed in the last section. But the order in which we receive the cards is not important, so many of the hands are the same. In fact there are 5! similar arrangements of cards that make the same hand. The number of distinct hands is

$$(52 \times 51 \times 50 \times 49 \times 48)/(5 \times 4 \times 3 \times 2 \times 1) = 52!/[5! \times (52-5)!]$$

The general form r items taken from a set of n items when order is not important is written as the binomial coefficient $C(n, r)$, also written $\binom{n}{r}$, and is given by the formula

$$C(n,r)=\frac{n!}{r!(n-r)!}$$

Example 3.7

There are six skiers staying in a cabin with four bunks. How many combinations of people will be able to sleep in beds?

Solution

$C(6, 4) = 6!/[4! \times (6-4)!] = (6 \times 5 \times 4 \times 3 \times 2 \times 1)/[(4 \times 3 \times 2 \times 1) \times (2 \times 1)] = 15$

PROBABILITY

Definitions

An **experiment**, or **trial**, is an action that can lead to a measurement.

Sampling is the act of taking a measurement. The **sample space** S is the set of all possible outcomes of an experiment (trial). An event e is one of the possible outcomes of the trial.

If an experiment can occur in n mutually exclusive and equally likely ways, and if m of these ways correspond to an event e, then the probability of the event is given by

$$P\{e\} = m/n$$

Example 3.8

A die is a cube of six faces designated as 1 through 6. The set of outcomes R of one die roll is defined as $R = \{1, 2, 3, 4, 5, 6\}$. If two dice are rolled, define trial, sample space, n, m, and the probability of rolling a seven when adding both dice together.

Solution

The trial is the rolling of two dice. The sample space is all possible outcomes of a two-dice roll, and the event is the outcome that the sum is 7.

The number of all possible outcomes, n, is the number of elements in the product set of the outcome of two dice when each is rolled independently. The product set is $R \times R$ and contains 36 elements.

The number of all possible ways, m, that the (7) event can occur is (1,6), (2,5), (3,4), (4,3), (5,2), and (6,1) for a total of six ways. The probability of rolling a 7 is $P\{7\} = \frac{6}{36} = \frac{1}{6}$.

Example 3.9

What is the probability of (i) a tail showing up when a fair coin is tossed, (ii) number 3 showing up when a fair die is tossed, and (iii) a red king is drawn from a deck of 52 cards.

Solution

(i) 1/2 (ii) 1/6 (iii) 2/52

General Character of Probability

The probability $P\{E\}$ of an event E is a real number in the range 0 through 1. Two theorems identify the range between which all probabilities are defined:

1. If ø is the null set, $P\{ø\} = 0$.
2. If S is the sample space, $P\{S\} = 1$.

The first states that the probability of an impossible event is zero, and the second states that, if an event is certain to occur, the probability is 1.

Complementary Probabilities

If E and E' are complementary events, $P\{E\} = 1 - P\{E'\}$. Complementary events are defined with respect to the sample space. The probability that an event E will happen is complementary to the probability that any of the other possible outcomes will happen.

Example 3.10

If the probability of throwing a 3 on a die is 1/6, what is the probability of not throwing a 3?

Solution

E is the probability of not throwing a 3, so $P\{E\} = 1 - P\{E'\} = 1 - \frac{1}{6} = \frac{5}{6}$.

Sometimes the complementary property of probabilities can be used to simplify calculations. This will happen when seeking the probability of an event that represents a larger fraction of the sample space than its complement.

Example 3.11

What is the probability $P\{E\}$ of getting at least one head in four coin tosses?

Solution

The complementary event $P\{E'\}$ to getting at least one head is getting no heads (or all tails) in four tosses. So the probability of getting at least one head is

$$P\{E\} = 1 - (0.5)^4 = 1 - 0.0625 = 0.9375$$

Joint Probability

The probability that a combination of events will occur is covered by joint probability rules. If E and F are two events, the joint probability is given by the rule

$$P\{E \cup F\} = P\{E\} + P\{F\} - P\{E \cap F\} \qquad \text{(Rule 1)}$$

A special case of the joint probability rule can be derived by considering two events, E and F, to be mutually exclusive. In this case the last term in Rule 1 is zero since $P\{E \cap F\} = P\{0\} = 0$. Thus, if E and F are mutually exclusive events,

$$P\{E \cup F\} = P\{E\} + P\{F\} \qquad \text{(Rule 2)}$$

Example 3.12

What is the probability of throwing a 7 or a 10 with two dice?

Solution

We will call the event of throwing a 7 A, and of throwing a 10 B. We know from previous examples that $P\{A\} = \frac{1}{6}$, and we can count outcomes to get $P\{B\} = \frac{1}{12}$. Applying the formula,

$$P\{A \cup B\} = P\{A\} + P\{B\} = \frac{1}{6} + \frac{1}{12} = \frac{1}{4}$$

If two events E and F are independent—that is, if they come from different sample spaces—then the probability that both will happen is given by the rule

$$P\{E \cap F\} = P\{E\} \times P\{F\} \qquad \text{(Rule 3)}$$

Example 3.13

What is the probability of throwing two heads in two coin tosses?

Solution

Call the throwing of one head E, the other F. The probability of throwing a single head is $P\{E\} = \frac{1}{2}$, and $P\{F\} = \frac{1}{2}$. The probability of throwing both heads is

$$P\{E \cap F\} = P\{E\} \times P\{F\} = \frac{1}{2} \times \frac{1}{2} = \frac{1}{4}$$

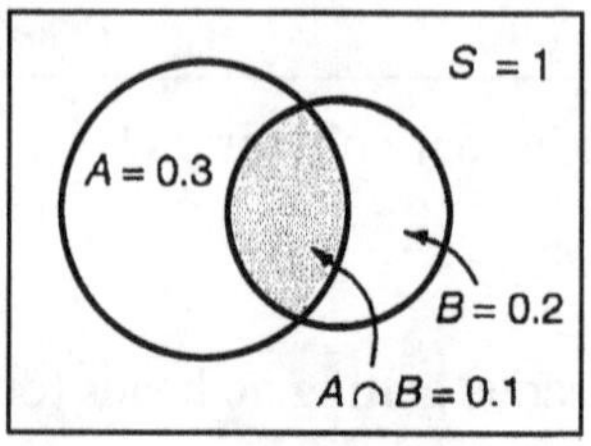

To visualize joint probabilities, we can use a Venn diagram showing two intersecting events, A and B, as shown in Figure 3.2. Let the normalized areas of each event represent the probability that the event will occur. For example, think of a random dart thrown at the Venn diagram: What are the chances of hitting one of the areas? Assume the areas correspond to probabilities and are given by $P\{S\} = 1$, $P\{A\} = 0.3$, $P\{B\} = 0.2$, and $P\{A \cap B\}$ is 0.6. The probability of hitting either area A or area B is calculated as the sum of the areas A and B minus the overlap area so it is not counted twice:

Figure 3.2
Venn diagram of joint probabilities

$$P\{A \cup B\} = 0.3 + 0.2 - .06 = .44$$

The result is also equal to the normalized area covered by A and B. The probability of hitting both A and B on one throw is simply the overlap area $P\{A \cap B\} = 0.1$.

If we throw two darts, the area S is used twice and represents two independent sample spaces. Hence Rule 3 applies.

Example 3.14

A die is tossed. Event A = {an odd number shows up}; event B = {a number > 4 shows up}.

(i) Find the probabilities, P(A) and P(B). (ii) What is the probability that either A or B or both occur?

Solution

(i) P(A) = P(1 or 3 or 5 showing up) = 3 (1/6) = 0.5
P(B) = P(5 or 6 showing up) = 2 (1/6) = 0.333

(ii) Event (A, B) = {5} and P(A, B) = 1/6;
then P(A + B) = P(A) + P(B) – P(AB) = 1/2 + 2/6 – 1/6 = 4/6
Check: Event (A + B) = {1,3,5,6}; then P(A + B) = 4/6

Conditional Probability

The conditional probability of an event E given an event F is denoted by $P\{E \mid F\}$ and is defined as

$$P\{E \mid F\} = P\{E \cap F\}/P\{F\} \qquad \text{for } P\{F\} \text{ not zero}$$

Example 3.15

Two six-sided dice, one red and one green, are tossed. What is the probability that the green die shows a 1, given that the sum of numbers on both dice is less than 4?

Solution

Let E be the event "green die shows 1" and let F be the event "sum of numbers shows less than four." Then

$$E = \{(1,1), (1,2), (1,3), (1,4), (1,5), (1,6)\}$$
$$F = \{(1,1), (1,2), (2,1)\}$$
$$E \cap F = \{(1,1), (1,2)\}$$
$$P\{E \mid F\} = P\{E \cap F\}/P\{F\} = (2/36)/(3/36) = 2/3$$

The generalized form of conditional probability is known as Bayes' theorem and is stated as follows: If $E_1, E_2, \ldots, E_n$ are n mutually exclusive events whose union is the sample space S, and E is any arbitrary event such that $P\{E\}$ is not zero, then

$$P\{E_k \mid E\} = \frac{P\{E_k\} \times P\{E \mid E_k\}}{\sum_{j=1}^{n} [P\{E_j\} \times P\{E \mid E_j\}]}$$

RANDOM VARIABLES

The method of random variables is a powerful concept. It casts the set-theory-based probability calculations of previous sections into a functional form and allows the application of standard mathematical tools to probability theory. It is often easy to solve fairly complex probability problems using random variables, although an approach different from the usual one is required.

A random variable, usually denoted by X, is a mapping of the sample space to some set of real numbers. The mapping transforms points of a sample space into points, or more accurately intervals, on the x-axis. The mapping, or random, variable X is called a discrete random variable if it assumes only a denumerable number of values on the x-axis. A random variable is called a continuous random variable if it assumes a continuum of values on the x-axis. The mapping is usually quite easy and intuitive for numerical events but provides no major advantage for nonnumerical discrete sample spaces, where counting remains the major tool.

Example 3.16

Cast the sample space of the outcomes of a roll of a die into random variable form.

Solution

The sample space is the set R defined by $R = \{1, 2, 3, 4, 5, 6\}$. These can easily by written along the x-axis as

$$R = \{1, 2, 3, 4, 5, 6\} \rightarrow 1 \mid 2 \mid 3 \mid 4 \mid 5 \mid 6 \mid x\text{-axis}$$

Probability Density Functions

A probability density function $f(x)$ is a mathematical rule that assigns a probability to the occurrence of the random variable x. Since the random variable is a mapping from trial outcomes, or events, to the numerical intervals on the x-axis, the probability that an event will occur is the area under the probability density function curve over the x interval defining the event.

For a continuous random variable the probability that an event E, mapped into an interval between x_1 and x_2, will occur is defined as

$$\int_{x_1}^{x_2} f(x)\,dx = P\{E\} \qquad \text{for } E \text{ mapped into } (x_1, x_2)$$

For a discrete case the formula is

$$\sum_{i=1}^{n} f(x_i) = P\{E\} \qquad \text{for } E \text{ containing } x_1, x_2, \ldots, x_n$$

It is assumed here that a step interval is associated with each value of x_i; therefore, the equivalent dx in the integral is 1 and is not required in the sum.

Example 3.17

The probability density function of a single six-sided die throw is shown graphically in Exhibit 4. The probability of throwing a 3 is given by the area under the curve over the interval assigned to the numeral 3, which is the step interval from 2.5 to 3.5.

Hence $P\{3\} = f(x) \times 1 = \frac{1}{6} \times 1 = \frac{1}{6}$.

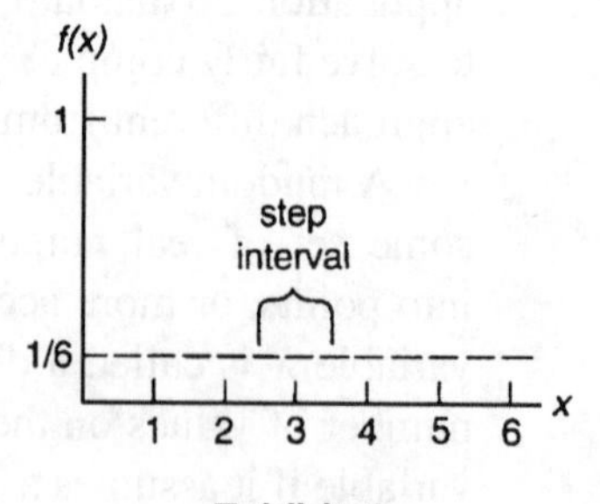

Exhibit 4

Example **3.18**

A probability density function is defined as $f(x) = Ax^2$ for $-1 < x < 2$ and zero elsewhere. Find the value of A so that it is a valid density function.

Solution

For a valid density function, $\int_{-\infty}^{\infty} f(x)\,dx = 1$

Then, $\int_{-1}^{2} Ax^2\,dx = A\frac{9}{3} = 3A = 1$; then $A = \frac{1}{3}$

Properties of Probability Density Functions

The expected value $E\{X\}$ of a probability density function is also called the mean, and for a discrete case it is given by

$$E\{X\} = \sum x_i \times f(x_i) = u$$

The expected value of a continuous random variable is

$$E\{X\} = \int_{-\inf}^{+\inf} x \times f(x) \times dx = u$$

The expected value for a discrete random variable of a function $g(X)$ is given by

$$E\{g(X)\} = \sum g(x_i) \times f(x_i)$$

The expected value of a continuous random variable is

$$E\{g(X)\} = \int_{-\inf}^{+\inf} g(x) \times f(x) \times dx$$

Of special interest are the functions of the form

$$g(x) = (x - u)^r$$

These are the powers of the random variables around the mean. The expected values of these power functions are called the rth moments about the mean of the distribution, where r is the power. The second moment about the mean is also known as the variance and is calculated as follows:

$$V\{X\} = E\{(x-u)^2\} = E\{(x^2 - 2xu + u^2)\} = E\{x^2\} - E\{2xu\} + E\{u^2\}$$

Since u is a constant, the second term is $2u^2$ and the third term evaluates to u^2; therefore, the second moment about the mean becomes

$$V\{X\} = E\{x^2\} - u^2 = \sigma^2$$

The square root of the variance is signified by the Greek letter sigma and is called the **standard deviation**.

Example 3.19

Calculate the mean and standard deviation of a single die throw.

Solution

This is a discrete function and can be calculated numerically by the discrete formulas given above. The mean, where $f(x_i) = \frac{1}{6}$ (all outcomes are equally likely), is given by

$$u = E\{X\} = \sum_{i=1}^{i=6} x_i \times f(x_i) = (1+2+3+4+5+6)/6 = \frac{21}{6} = 3.5$$

The standard deviation is given by

$$\sigma = \sqrt{V\{x\}} = \sqrt{E\{x^2\} - u^2} = \sqrt{[(1^2 + 2^2 + 3^2 + 4^2 + 5^2 + 6^2)/6] - 3.5^2} = 1.7$$

STATISTICAL TREATMENT OF DATA

Whether from the outcome of an experiment or trial, or simply the output of a number generator, we are constantly presented with numerical data. A statistical treatment of such data involves ordering, presentation, and analysis. The tools available for such treatment are generally applicable to a set of numbers and can be applied without much knowledge about the source of the data, although such knowledge is often necessary to make sensible use of the statistical results.

In its raw form, numerical data is simply a list of n numbers denoted by x_i, where $i = 1, 2, 3, \ldots, n$. There is no specific significance associated with the order implicit in the i numbers. They are names for the individuals in the list, although they are often associated with the order in which the raw data was recorded. For example, consider a box of 50 resistors. They are to be used in a sensitive circuit, and their resistances must be measured. The results of the 50 measurements are presented in the following table.

Table of Raw Measurements (Ω)

101	105	110	115	82
86	91	96	117	112
109	103	89	97	98
101	104	99	95	97
85	90	94	112	107
103	94	98	106	98
114	112	108	101	99
93	96	99	104	90
109	106	101	93	92
104	99	109	100	107

Each number is named by the variable x_i, and there are $n = 50$ of them. The numbers range from 82 to 117.

Frequency Distribution

A systematic tool used in ordering data is the frequency distribution. The method requires counting the number of occurrences of raw numbers whose values fall within step intervals. The step intervals (or bins) are usually chosen to (1) be of constant size, (2) cover the range of numbers in the raw data, (3) be small enough in quantity to limit the amount of writing yet not have many empty steps, and (4) be sufficient in quantity so that significant information is not lost.

For example, the aforementioned raw data of measured resistances may be ordered in a frequency distribution table such as Table 3.1. Here the step interval is the event E of a random variable that can be mapped onto the x-axis. The set of eight events is the sample space. If we take a number randomly from the raw measurement set, the probability that it will be in bin 5 is

$$f(E_5) = P\{E_5\} = 10/50 = 0.2$$

Table 3.1 Frequency and cumulative frequency table

Event, E_i	Range, Ω	Frequency	Cumulative Frequency	Probability Density Function, $f(E_i)$
1	80–84	1	1	0.02
2	85–89	3	4	0.06
3	90–94	8	12	0.16
4	95–99	12	24	0.24
5	100–104	10	34	0.20
6	105–109	9	43	0.18
7	110–114	5	48	0.10
8	115–119	2	50	0.04

The last column in Table 3.1 is the probability density function of the distribution. The probability table can be plotted along the x-axis in several ways, as shown in Figures 3.3 through 3.5.

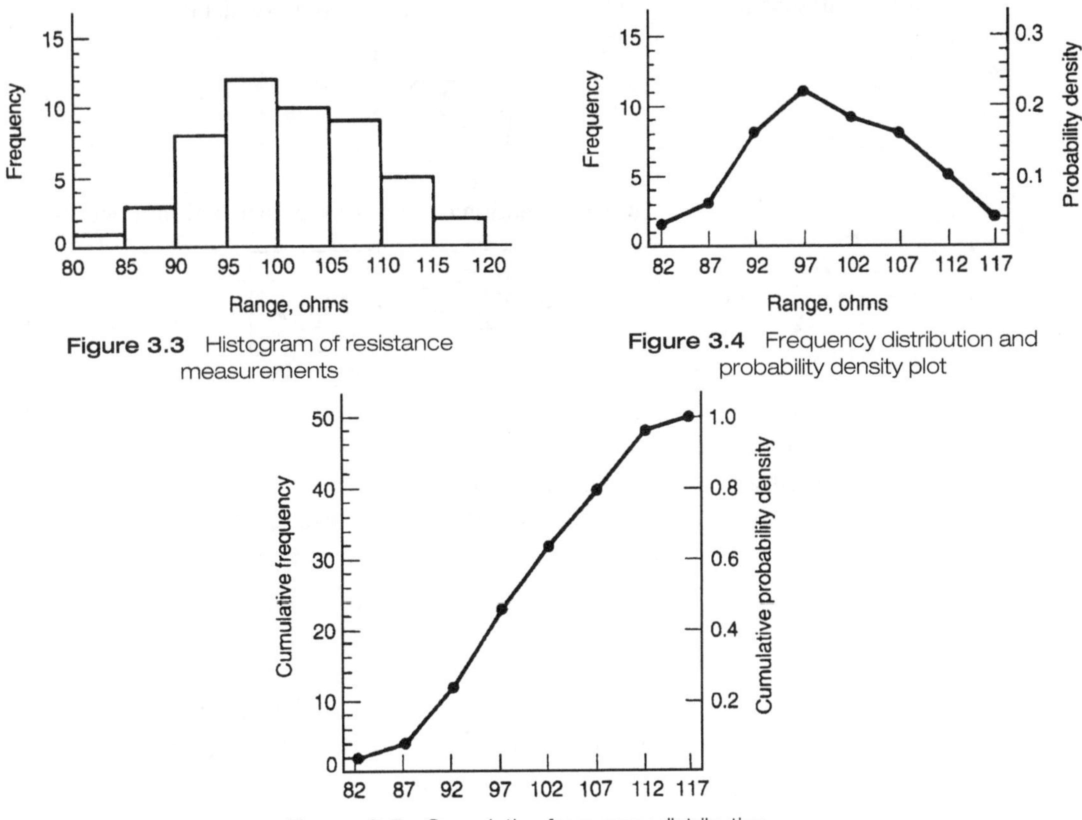

Figure 3.3 Histogram of resistance measurements

Figure 3.4 Frequency distribution and probability density plot

Figure 3.5 Cumulative frequency distribution and cumulative probability density

Standard Statistical Measures

There are several statistical quantities that can be calculated from a set of raw data and its distribution function. Some of the more important ones are listed here, together with the method of their calculation.

Mode The observed value that occurs most frequently; here the mode is bin 4 with a range of 95–99 Ω.

Median The point in the distribution that divides the number of observations such that half of the observations are above and half are below. The median is often the mean of the two middle values; here the median is 4.5 bins, 100 Ω.

Mean The arithmetic mean, or average, is calculated from raw data as

$$\mu = \frac{1}{n}\sum_{i=1}^{n} x_i = 100.6$$

It is calculated from the distribution function as

$$\mu = \sum_{i=1}^{m} b_i \times f(E_i) = 100.4$$

where b_i is the ith event value (for i = 1, b_i = 82) and m is the number of bins; $f(E_i)$ is the probability density function. (The two averages are not quite the same because of the information lost in assigning the step intervals.)

Standard deviation (a) Computational form for the raw data:

$$\sigma = \sqrt{\frac{1}{n}\left[\left(\sum_{i=1}^{n} x_i^2\right) - n \times \mu^2\right]} = 8.08$$

(b) Computational form for the distribution function:

$$\sigma = \sqrt{\left\{\left[\sum_{i=1}^{m} b_i^2 \times f(E_i)\right] - \mu^2\right\}} = 8.02$$

Sample standard deviation If the data set is a sample of a larger population, then the sample standard deviation is the best estimate of the standard deviation of the larger population.

The computational form for the raw data set is

$$\sigma = \sqrt{\frac{1}{n-1}\left[\left(\sum_{i=1}^{n} x_i^2\right) - n \times \mu^2\right]} = 8.166$$

Sample standard deviations and the use of $(n - 1)$ in the denominator are discussed in the section on sampling.

Skewness This is a measure of the frequency distribution asymmetry and is approximately

skewness ≅ 3(mean – median)/(standard deviation)

Example **3.20**

Two professors give the following scores to their students. What is the mode and arithmetic mean?

Frequency	1	3	6	11	13	10	2
Score	35	45	55	65	75	85	95

Solution

mode = 75; N = 1 + 3 + 6 + 11 + 13 + 10 + 2 = 46

weighted arithmetic mean = $\overline{X_w} = [35(1) + 45(3) + \ldots 95(2)] / 46 = 70$

STANDARD DISTRIBUTION FUNCTIONS

In the previous section, we calculated several general properties of probability distribution functions.

To know the appropriate probability density function for an actual situation, two general methods are available:

1. The probability density function is actually calculated, as was done in the last section, by analyzing the physical mechanism by which experimental events and outcomes are generated and counting the number of ways an individual event occurs.
2. Recognition of an overall similarity between the present experiment and another for which the probability density function is already known permits the known behavior of the function to be applied to the new experiment. This work-saving method is by far the more popular one. Of course, to apply this method, it is necessary to have a repertoire of known probability functions and to understand the problem characteristics to which they apply.

This section lists several popular probability density functions and their characteristics.

Binomial Distribution

The binomial distribution applies when there is a set of discrete binary alternative outcomes. Deriving this distribution function helps one understand the class of problems to which it applies. For example, given a set of n events, each with a probability p of occurring, what is the probability that r of the events will occur and $(n - r)$ not occur?

The probability of one event occurring is p.

The probability of r events occurring is p^r.

The probability of $(n - r)$ events not occurring is $(1 - p)^{n-r}$.

The probability of exactly r events occurring and $(n - r)$ not occurring in a trial is given by the joint probability Rule 3:

$$P[r \cap (n-r)] = p^r \times (1-p)^{n-r}$$

However, there are many ways of choosing r occurrences out of n events. In fact, the number of different ways of choosing r items from a set of n items when order is not important is given by the binomial coefficient $C(n, r)$. The total probability of r

occurrences from n trials, given an individual probability of occurrence as p, is thus given by

$$C(n,r)\times p^r \times (1-p)^{n-r} = f(r)$$

This is the **binomial probability density function.**

The mean of this density function is the first moment of the density function, or expected value, and is calculated as

$$E\{x\} = \sum_{r=0}^{n} r \times f(r) = \sum_{r=0}^{n} r \times \frac{n!}{(r)!(n-r)!} \times p^r \times (1-p)^{n-r}$$

This can be rewritten as

$$\sum_{r=1}^{n} \frac{n!}{(r-1)!(n-r)!} \times p^r \times (1-p)^{n-r}$$

We can now factor out the quantity $n \times p$ and let $r - 1 = y$. This can be rewritten as

$$n \times p \times \sum_{y=0}^{n-1} \frac{(n-1)!}{(y)!(n-1-y)!} \times p^y \times (1-p)^{n-1-y} = n \times p \times [p + (1-p)]^{n-1}$$

Since the sum is merely the expansion of a binomial raised to a power, and the number 1 raised to any power is 1, the mean is

$$\mu = n \times p$$

A similar calculation shows the variance is

$$\text{var} = n \times p \times (1-p)$$

The standard deviation is

$$\sigma = \sqrt{\text{var}} = \sqrt{n \times p \times (1-p)}$$

Example 3.21

A truck carrying dairy products and eggs damages its suspension and 5% of the eggs break.

(i) What is the probability that a carton of 12 eggs will have exactly one broken egg?

(ii) What is the probability that one or more eggs in a carton will be broken?

Solution

(i) Since an egg is either broken or not broken, the binomial distribution applies. The probability p that an egg is broken is 0.05 and that one is not broken is $(1-p) = 0.95$. From the equation for the binomial distribution, with $n = 12$ and $r = 1$,

$$p\{1\} = f(1) = C(12,1) \times 0.05^1 \times 0.95^{11} = 12 \times 0.05 \times 0.57 = 0.34$$

(ii) The probability that one or more eggs will be broken can be calculated as the sum of each individual probability:

$$p\{x > 0\} = p\{1\} + p\{2\} + \cdots + p\{12\}$$

However, this requires 12 calculations. The problem can also be solved using the complementary rule:

$$p\{x>0\}=1-p\{0\}=C(12,0)\times 0.05^0\times 0.95^{12}=0.95^{12}=0.54$$

Example **3.22**

A biased coin is tossed. Find the probability that a head appears once in three trials.

P(Head) = p = 0.6

Solution

Here q = P(Head not occurring) or P(Tail) = 1 – 0.6 = 0.4.

Then P(1 Head) = C(3,1) 0.6^1 0.4^2 = 0.2880.

Normal Distribution Function

The normal distribution, or Gaussian distribution, is widely used to represent the distribution of outcomes of experiments and measurements. It is popular because it can be derived from a few empirical assumptions about the errors presumed to cause the distribution of results about the mean. One assumption is that the error is the result of a combination of N elementary errors, each of magnitude e and equally likely to be positive or negative. The derivation then assumes $N\rightarrow\infty$ and $e\rightarrow 0$ in such a way as to leave the standard deviation constant. This error model is universal, since most experiments are analyzed to eliminate systematic errors. What remains is attributable to errors that are too small to explain systematically, so the normal probability distribution is evoked.

The form of the probability density and distribution functions for the **normal distribution** with a mean μ and variance σ^2 is given by

$$f(x)=\frac{e^{-(x-\mu)^2/2\sigma^2}}{\sigma\sqrt{2\pi}}\qquad -\infty<x<\infty$$

$$F(x)=\int_{-\infty}^{x}\frac{e^{-(x-\mu)^2/2\sigma^2}}{\sigma\sqrt{2\pi}}\,dt$$

The normal distribution is the typical bell-shaped curve shown in Figure 3.6. Here we see that the curve is symmetric about the mean μ. Its width and height are determined by the standard deviation σ. As σ increases, the curve becomes wider and lower.

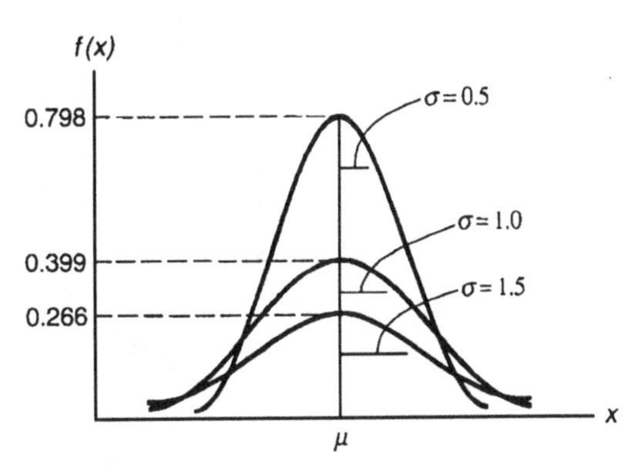

Figure 3.6 Normal distribution curve

Since this function is difficult to integrate, reference tables are used to calculate probabilities in a standard format; then the standard probabilities are converted to the actual variable required by the problem. The relation between the standard variable, z, and a typical problem variable, x, is

$$z = (x - \mu)/\sigma$$

Since μ and σ are constants, the standard probability at a value z is the same as the problem probability for the value at x.

The standard probability density function is

$$f(z) = \frac{1}{\sqrt{2\pi}} \times e^{-z^2/2}$$

Table 3.2 Standard probability table

z	$F(z)$	$f(z)$	z	$F(z)$	$f(z)$
0.0	0.5000	0.3989	2.0	0.9773	0.0540
0.1	0.5398	0.3970	2.1	0.9821	0.0440
0.2	0.5793	0.3910	2.2	0.9861	0.0355
0.3	0.6179	0.3814	2.3	0.9893	0.0283
0.4	0.6554	0.3683	2.4	0.9918	0.0224
0.5	0.6915	0.3521	2.5	0.9938	0.0175
0.6	0.7257	0.3332	2.6	0.9953	0.0136
0.7	0.7580	0.3123	2.7	0.9965	0.0104
0.8	0.7881	0.2897	2.8	0.9974	0.0079
0.9	0.8159	0.2661	2.9	0.9981	0.0060
1.0	0.8413	0.2420	3.0	0.9987	0.0044
1.1	0.8643	0.2179	3.1	0.9990	0.0033
1.2	0.8849	0.1942	3.2	0.9993	0.0024
1.3	0.9032	0.1714	3.3	0.9995	0.0017
1.4	0.9192	0.1497	3.4	0.9997	0.0012
1.5	0.9332	0.1295	3.5	0.9998	0.0009
1.6	0.9452	0.1109	3.6	0.9998	0.0006
1.7	0.9554	0.0940	3.7	0.9999	0.0004
1.8	0.9641	0.0790	3.8	0.9999	0.0003
1.9	0.9713	0.0656	3.9	1.0000	0.0002
			4.0	1.0000	0.0001

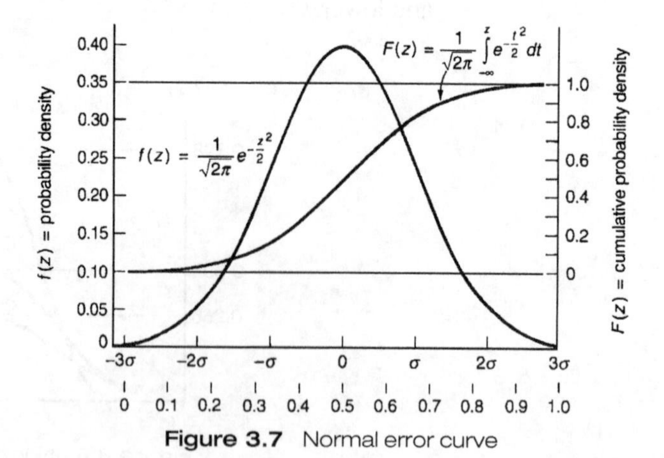

Figure 3.7 Normal error curve

The standard cumulative distribution function is

$$F(z) = \int_{-\infty}^{z} \frac{1}{\sqrt{2\pi}} \times e^{-t^2/2} \times dt$$

The standard probability function is shown graphically in Figure 3.7, and Table 3.2 shows the corresponding numerical values. The standard probability curve is symmetric about the origin and is given in terms of unit *sigma*. To use the table, remember that the function $F(z)$ is the area under the probability curve from minus infinity to the value z. The area under the curve up to $x = 0$ is therefore 0.5. Also, from symmetry,

$$F(-z) = 1 - F(z)$$

Example 3.23

Find the probability that the standard variable z lies within (i) 1σ, (ii) 2σ, and (iii) 3σ of the mean.

Solution

(i) The probability is $P_1 = F(1.0) - F(-1.0)$. From the symmetry of F, $F(-1.0) = 1 - F(1.0)$, so

$$P_1 = 2F(1.0) - 1 = 2(0.8413) - 1$$
$$= 0.6826$$

(ii) In this case, the probability is

$$P_2 = F(2.0) - F(-2.0)$$
$$= F(2.0) - [1 - F(2.0)]$$
$$= 2F(2.0) - 1 = 2(0.9773) - 1$$
$$= 0.9546$$

(iii) In the same way,

$$P_3 = 2F(3.0) - 1$$
$$= 2(0.9987) - 1$$
$$= 0.9974$$

Example 3.24

A Gaussian random variable has a mean of 1830 and standard deviation of 460. Find the probability that the variable will be more than 2750.

Solution

$P(X > 2750) = 1 - P(X \leq 2750) = 1 - F[(2750 - 1830)/460] = 1 - F(2.0) = 1 - 0.9772 = 0.0228$

t-Distribution

The t-distribution is often used to test an assumption about a population mean when the parent population is known to be normally distributed but its standard deviation is unknown. In this case, the inferences made about the parent mean will depend upon the size of the samples being taken.

It is customary to describe the t-distribution in terms of the standard variable t and the number of degrees of freedom ν. The number of degrees of freedom is a measure of the number of independent observations in a sample that can be used to estimate the standard deviation of the parent population; the number of degrees of freedom ν is one less than the sample size ($\nu = n - 1$).

The density function of the t-distribution is given by

$$f(t) = \frac{\Gamma\left(\frac{\nu+1}{2}\right)}{\sqrt{\nu\pi}\,\Gamma\left(\frac{\nu}{2}\right)\left(1+t^2/\nu\right)^{(\nu+1)/2}}$$

and is provided in Table 3.3. The mean is $m = 0$, and the standard deviation is

$$\sigma = \sqrt{\frac{\nu}{\nu - 2}}$$

Table 3.3 t-Distribution; values of $t_{\alpha,\nu}$

Degrees of Freedom, ν	Area of the Tail				
	$\alpha = 0.10$	$\alpha = 0.05$	$\alpha = 0.025$	$\alpha = 0.01$	$\alpha = 0.005$
1	3.078	6.314	12.706	31.821	63.657
2	1.886	2.920	4.303	6.965	9.925
3	1.638	2.353	3.182	4.541	5.841
4	1.533	2.132	2.776	3.747	4.604
5	1.476	2.015	2.571	3.365	4.032
6	1.440	1.943	2.447	3.143	3.707
7	1.415	1.895	2.365	2.998	3.499
8	1.397	1.860	2.306	2.896	3.355
9	1.383	1.833	2.262	2.821	3.250
10	1.372	1.812	2.228	2.764	3.169
11	1.363	1.796	2.201	2.718	3.106
12	1.356	1.782	2.179	2.681	3.055
13	1.350	1.771	2.160	2.650	3.012
14	1.345	1.761	2.145	2.624	2.977
15	1.341	1.753	2.131	2.602	2.947
16	1.337	1.746	2.120	2.583	2.921
17	1.333	1.740	2.110	2.567	2.898
18	1.330	1.734	2.101	2.552	2.878
19	1.328	1.729	2.093	2.539	2.861
20	1.325	1.725	2.086	2.528	2.845
21	1.323	1.721	2.080	2.518	2.831
22	1.321	1.717	2.074	2.508	2.819
23	1.319	1.714	2.069	2.500	2.807
24	1.318	1.711	2.064	2.492	2.797
25	1.316	1.708	2.060	2.485	2.787
26	1.315	1.706	2.056	2.479	2.779
27	1.314	1.703	2.052	2.473	2.771
28	1.313	1.701	2.048	2.467	2.763
29	1.311	1.699	2.045	2.462	2.756
inf.	1.282	1.645	1.960	2.326	2.576

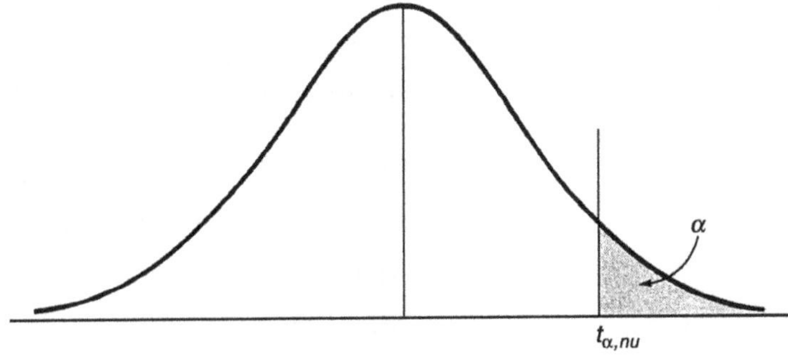

Figure 3.8

Probability questions involving the t-distribution can be answered by using the distribution function $t_{\alpha,\nu}$ shown in Figure 3.8. Table 3.3 gives the value of t as a function of the degrees of freedom v down the column and the area (α) of the tail across the top. The t-distribution is symmetric. As an example, the probability of t falling within ±3.0 when a sample size of 8 ($\nu = 7$) is selected is one minus twice the tail ($\alpha = 0.01$):

$$P\{-3.0 < t < 3.0\} = 1 - (2 \times 0.01) = 0.98$$

The t-distribution is a family of distributions that approaches the Gaussian distribution for large n.

X^2 DISTRIBUTION

In probability theory and statistics, the chi-square distribution (also chi-squared or x^2-distribution) is one of the most widely used theoretical probability distributions in inferential statistics (e.g., in statistical significance tests). It is useful because, under reasonable assumptions, easily calculated quantities can be proven to have distributions that approximate to the chi-square distribution if the null hypothesis is true.

The best known situations in which the chi-square distribution is used are the common chi-square tests for goodness of fit of an observed distribution to a theoretical one, and of the independence of two criteria of classification of qualitative data. Many other statistical tests also lead to a use of this distribution.

If $Z_1, Z_2, \ldots, Z_n$ are independent unit normal random variables, then

$$\chi^2 = Z_1^2 + Z_2^2 + \ldots + Z_n^2$$

is said to have a chi-square distribution with n degrees of freedom.

PROBLEMS

3.1 Define the set of all outcomes for the roll of two dice.

3.2 Draw a Venn diagram showing the following:

The universal set of all people in the United States as U

All the males as M

All the females as F

All the students of both sexes as S

All the students with grades above "B" as A

3.3 What is the probability of drawing a pair of aces in two cards when an ace has been drawn on the first card?

a. 1/13 c. 3/51
b. 1/26 d. 4/51

3.4 An auto manufacturer has three plants (A, B, C). Four out of 500 cars from Plant A must be recalled, 10 out of 800 from Plant B, and 10 out of 1000 from Plant C. Now a customer purchases a car from a dealer who gets 30% of his stock from Plant A, 40% from Plant B, and 30% from Plant C, and the car is recalled. What is the probability it was manufactured in Plant A?

a. 0.0008 c. 0.0125
b. 0.01 d. 0.2308

3.5 There are ten defectives per 1000 times of a product. What is the probability that there is one and only one defective in a random lot of 100?

a. 99×0.01^{99} c. 0.5
b. 0.01 d. 0.99^{99}

3.6 The probability that both stages of a two-stage missile will function correctly is 0.95. The probability that the first stage will function correctly is 0.98. What is the probability that the second stage will function correctly given that the first one does?

a. 0.99 c. 0.97
b. 0.98 d. 0.95

3.7 A standard deck of 52 playing cards is thoroughly shuffled. The probability that the first four cards dealt from the deck will be the four aces is closest to:

a. 2.0×10^{-1} c. 4.0×10^{-4}
b. 8.0×10^{-2} d. 4.0×10^{-6}

3.8 In statistics, the standard deviation measures:

a. a standard distance c. central tendency
b. a normal distance d. dispersion

3.9 There are three bins containing integrated circuits (ICs). One bin has two premium ICs, one has two regular ICs, and one has one premium IC and one regular IC. An IC is picked at random. It is found to be a premium IC. What is the probability that the remaining IC in that bin is also a premium IC?

a. $\frac{1}{5}$ c. $\frac{1}{3}$

b. $\frac{1}{4}$ d. $\frac{2}{3}$

3.10 How many teams of four can be formed from 35 people?

a. About 25,000 c. About 50,000

b. About 2,000,000 d. About 200,000

3.11 A bin contains 50 bolts, 10 of which are defective. If a worker grabs 5 bolts from the bin in one grab, what is the probability that no more than 2 of the 5 are bad?

a. About 0.5 c. About 0.90

b. About 0.75 d. About 0.95

3.12 How many three-letter codes may be formed from the English alphabet if no repetitions are allowed?

a. 26^3 c. $26 \times 25 \times 24$

b. 26/3 d. $26^3/3$

3.13 A widget has three parts, A, B, and C, with probabilities of 0.1, 0.2, and 0.25, respectively, of being defective. What is the probability that exactly one of these parts is defective?

a. 0.375 c. 0.95

b. 0.55 d. 0.005

3.14 If three students work on a certain math problem, student A has a probability of success of 0.5; student B, 0.4; and student C, 0.3. If they work independently, what is the probability that no one works the problem successfully?

a. 0.12 c. 0.32

b. 0.25 d. 0.21

3.15 A sample of 50 light bulbs is drawn from a large collection in which each bulb is good with a probability of 0.9. What is the approximate probability of having less than 3 bad bulbs in the 50?

a. 0.1 c. 0.3

b. 0.2 d. 0.4

3.16 The number of different 3-digit numbers that can be formed from the digits 1, 2, 3, 7, 8, 9 without reusing the digits is:

a. 10 c. 30

b. 20 d. 40

3.17 The number of different ways that a party of seven councilmen can be seated in a row is:

a. 1 c. 2080
b. 560 d. 5040

3.18 A student must answer six out of eight questions on an exam. The number of different ways in which he can do the exam is:

a. 8 c. 28
b. 18 d. 48

3.19 Repeat Problem 3.18 if the first two questions are mandatory.

a. 4 c. 12
b. 8 d. 15

3.20 A group of five women wishes to form a subcommittee consisting of two of them. The number of possible ways to do so is:

a. 5 c. 15
b. 10 d. 20

3.21 An integer has to be chosen from numbers between 1 and 100 (both inclusive). The probability of choosing a number divisible by 9 (with a remainder of 0) is:

a. 0 c. 0.11
b. 0.01 d. 0.91

3.22 Four fair coins are tossed. The probability of either one head or two heads showing up is:

a. 1/8 c. 4/8
b. 2/8 d. 5/8

3.23 Two identical bags contain ten apples and five oranges each. The probability of selecting an apple from the first bag and an orange from the second bag is:

a. 1/9 c. 3/9
b. 2/9 d. 4/9

3.24 A bag contains 5 red, 10 orange, 15 green, 20 violet, and 25 black cards. The probability that you will get a black card or a red card if you remove a card from the bag is:

a. 5/85 c. 25/85
b. 15/85 d. 35/85

3.25 Two bags each contain two orange balls, five white balls, and three red balls. The probability of selecting an orange ball from the first bag or a white ball from the other bag is:

a. 0 c. 0.6
b. 0.2 d. 1.0

3.26 A bag contains 100 balls numbered 1 to 100. One ball is drawn from the bag. What is the probability that the number on the ball will be even or greater than 72?

a. 0.64 c. 0.28
b. 0.50 d. 0.14

3.27 A circuit has two switches connected in series. For a signal to pass through, both switches must be closed. The probability that the first switch is closed is 0.95, and the probability that a signal passes through is 0.90. The probability that the second switch is closed is:

a. 0.8545 c. 0.9474
b. 0.9000 d. 0.9871

3.28 Four probability density distributions are shown in Exhibit 3.28. The only valid distribution is:

a. $f_1(x)$ c. $f_3(x)$
b. $f_2(x)$ d. $f_4(x)$

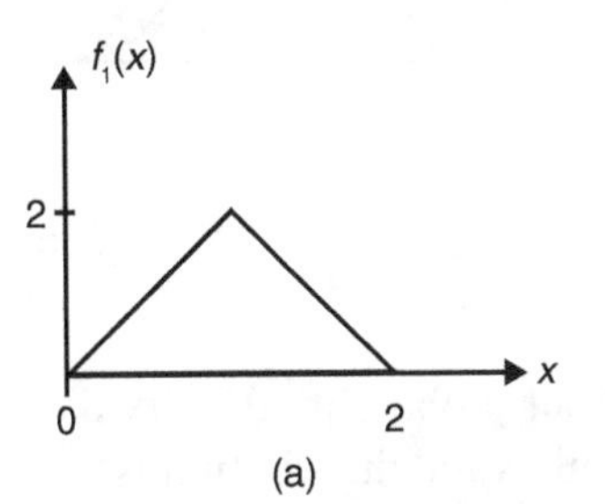

(a)

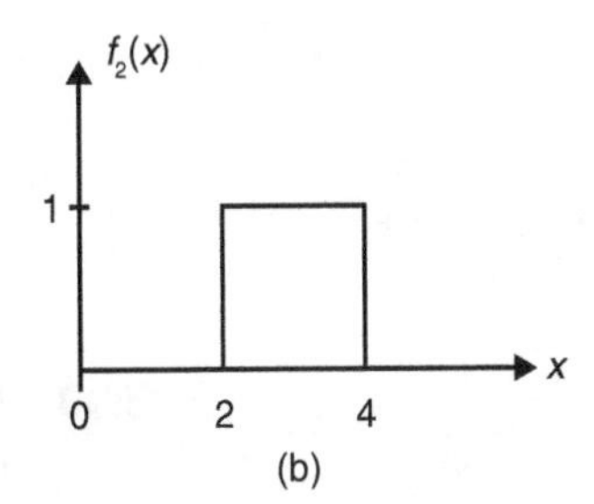

(b)

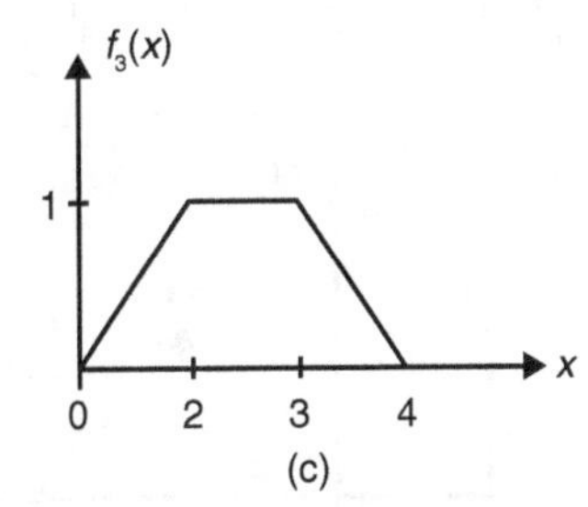

(c)

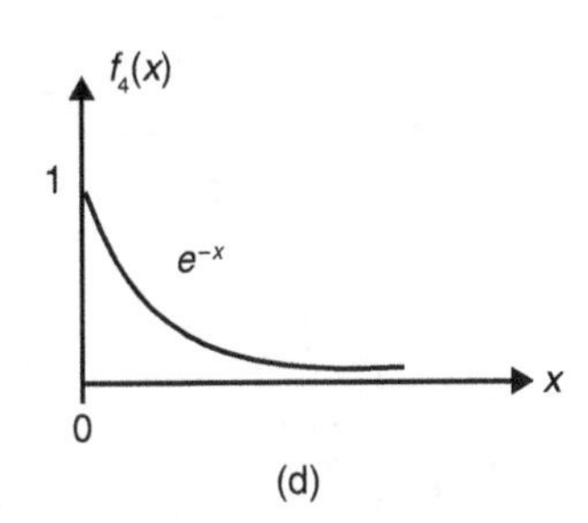

(d)

Exhibit 3.28

3.29 Four probability density distributions are shown in Exhibit 3.29. The only valid distribution is:

a. $f_1(x)$ c. $f_3(x)$
b. $f_2(x)$ d. $f_4(x)$

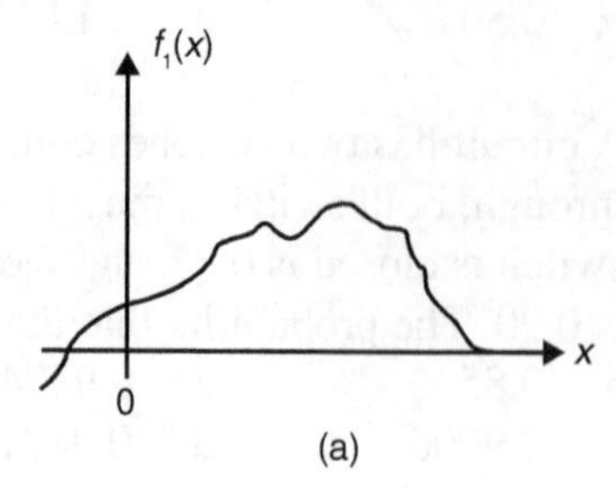

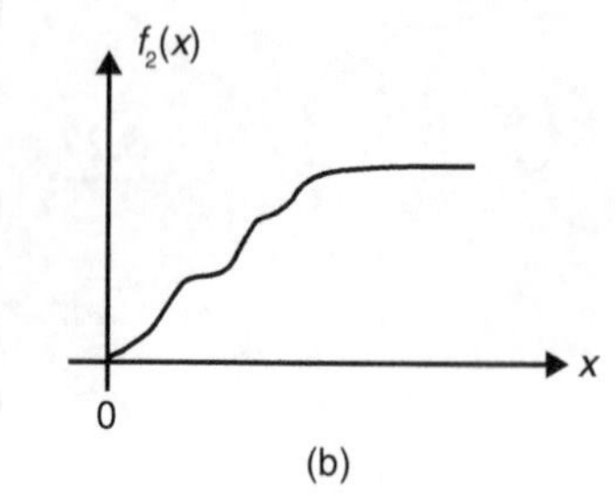

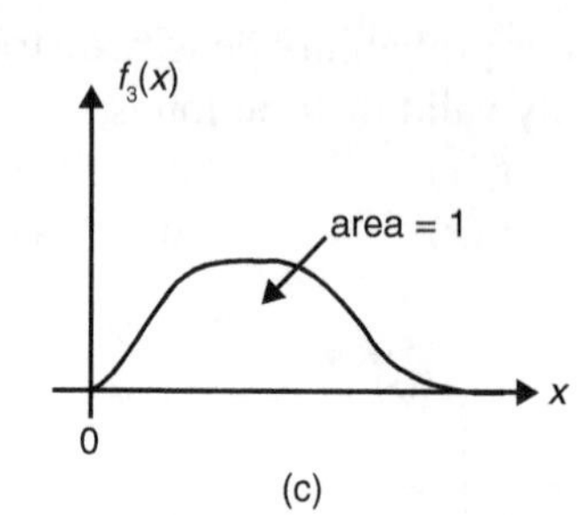

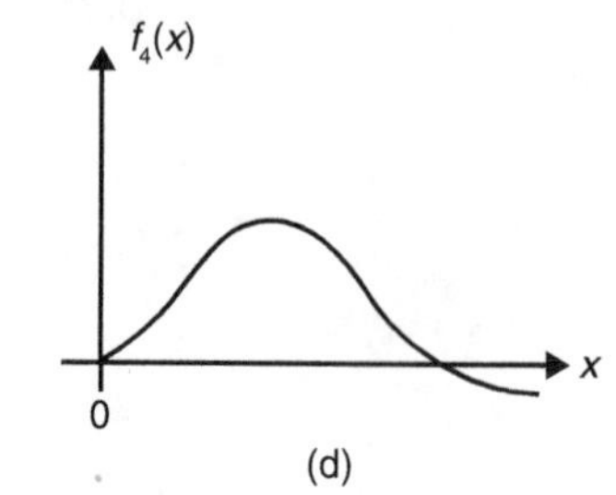

Exhibit 3.29

3.30 Four probability density distributions are shown in Exhibit 3.30. The only valid distribution is:

a. $f_1(x)$ c. $f_3(x)$
b. $f_2(x)$ d. $f_4(x)$

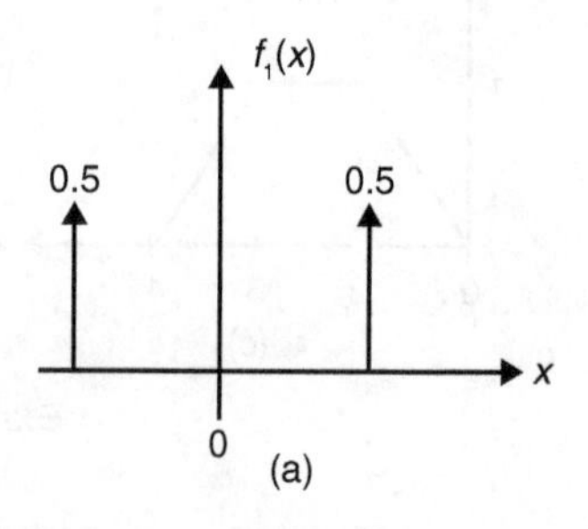

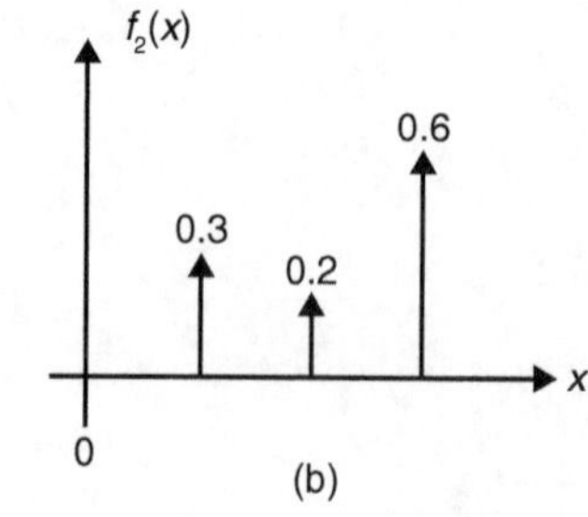

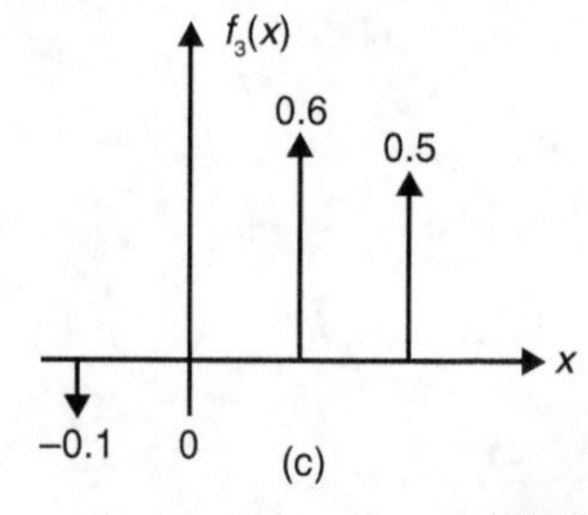

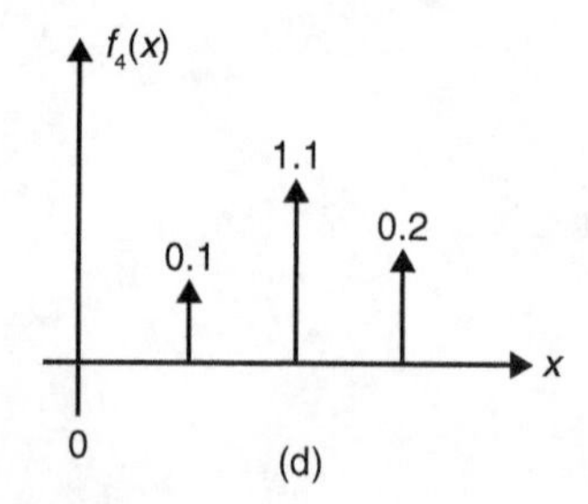

Exhibit 3.30

3.31 Four cumulative probability distribution functions are shown in Exhibit 3.31. The only valid distribution is:

a. $F_1(x)$ c. $F_3(x)$
b. $F_2(x)$ d. $F_4(x)$

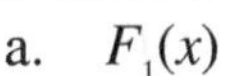

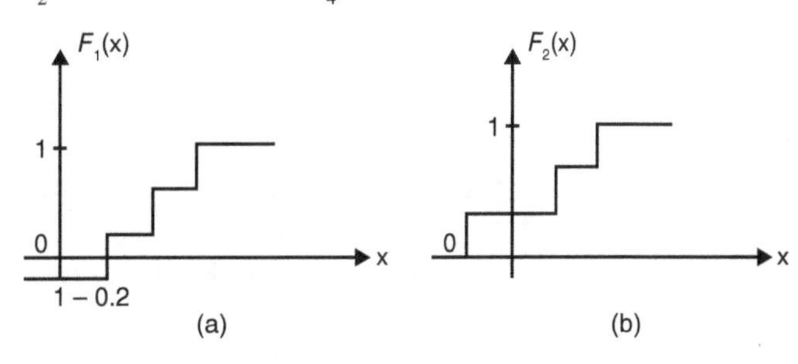

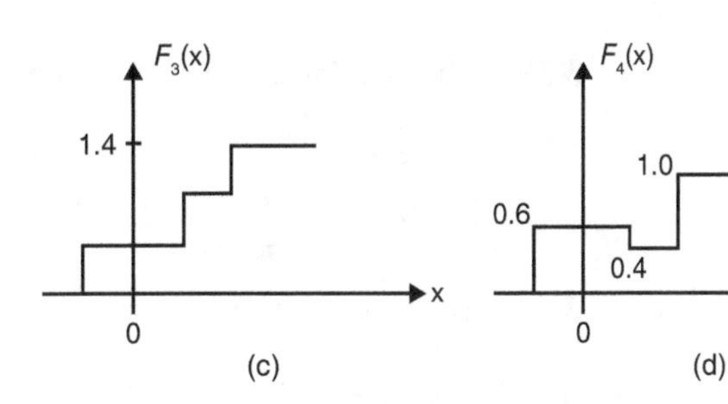

Exhibit 3.31

3.32 A coin is weighted so that heads is twice as likely to appear as tails. The probability that two heads occur in four tosses is:

a. 0.15 c. 0.45
b. 0.35 d. 0.75

3.33 It is given that 20% of all employees leave their jobs after one year. A company hired seven new employees. The probability that nobody will leave the company after one year is:

a. 0.1335 c. 0.3815
b. 0.2315 d. 0.6510

3.34 If four fair coins are tossed simultaneously, the probability that at least one head appears is:

a. 0.1335 c. 0.7815
b. 0.5635 d. 0.9375

3.35 For unit normal distribution, the probability that $(x > 3)$ is:

a. 0.0013 c. 0.1807
b. 0.0178 d. 0.5402

3.36 Scores in a particular game have a normal distribution with a mean of 30 and a standard deviation of 5. Contestants must score more than 26 to qualify for the finals. The probability of being disqualified in the qualifying round is:

a. 0.121 c. 0.304
b. 0.212 d. 0.540

3.37 The radial distance to the impact points for shells fired by a cannon is approximated by a normal Gaussian random variable with a mean of 2000 m and standard deviation of 40 m. When a target is located at 1980 m distance, the probability that shells will fall within ± 68 m of the target is:

a. 0.2341 c. 0.5847
b. 0.3248 d. 0.8710

3.38 The chance of a car being stolen from a residential area is 1 in 120. In one area there are five cars parked in front of the houses. The probability that none will be stolen is:

a. 0.0131 c. 0.5847
b. 0.3248 d. 0.9590

3.39 The standard deviation of the sequence 3, 4, 4, 5, 8, 8, 8, 10, 11, 15, 18, 20 is:

a. 5.36 c. 15.62
b. 9.35 d. 28.75

3.40 Weighted arithmetic mean of the following 50 data points is:

Frequency	3	8	18	12	9
Score	1.5	2.5	3.5	4.5	5.5

a. 1.56 c. 5.62
b. 3.82 d. 8.75

SOLUTIONS

3.1 We will write this in ordered pairs:

$$\begin{aligned} R = \{&(1,1),\\ &(1,2), (2,1),\\ &(1,3), (3,1), (2,2),\\ &(1,4), (4,1), (2,3), (3,2)\\ &(1,5), (5,1), (2,3), (3,2), (3,3),\\ &(1,6), (6,1), (5,2), (2,5), (3,4), (4,3),\\ &(2,6), (6,2), (5,3), (3,5), (4,4),\\ &(3,6), (6,3), (5,4), (4,5),\\ &(4,6), (6,4), (5,5),\\ &(5,6), (6,5),\\ &(6,6)\} \end{aligned}$$

3.2 See Exhibit 3.2.

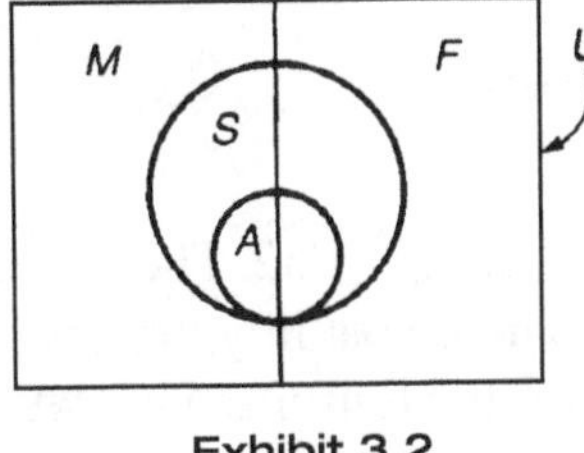

Exhibit 3.2

3.3 **c.** This is a conditional probability problem. Let B be "draw an ace," and let A be "draw a second ace": $P\{B\} = 4/52$ (1/13) and $P\{A\} = 3/51$. Then $P\{A|B\} = P\{A\} \times P\{B\}/P\{B\} = 3/51$.

3.4 **d.** This is a Bayes' theorem problem application because partitions are involved. The event E is a recall, with E_1 = Plant A, E_2 = Plant B, and E_3 = Plant C. The conditional probabilities of a recall from Plants E_1, E_2, and E_3 are

$$P(E \mid E_1) = 4/500 = 0.008$$
$$P(E \mid E_2) = 10/800 = 0.0125$$
$$P(E \mid E_3) = 10/1000 = 0.01$$

The probabilities that the dealer had a car from E_1, E_2, or E_3 are $P(E_1) = 0.3$, $P(E_2) = 0.4$, and $P(E_3) = 0.3$. Now applying Bayes' formula gives the probability that the recall was built in Plant A (E_1) as

$$P\{E_1|\text{recall}\} = \frac{P\{E_1\} \times P\{E|E_1\}}{P\{E_1\} \times P\{E \mid E_1\} + P\{E_2\} \times P\{E|E_2\} + P\{E_3\} \times P\{E \mid E_3\}}$$
$$= \frac{0.3 \times 0.008}{0.3 \times 0.008 + 0.4 \times 0.0125 + 0.3 \times 0.01} = 0.2308$$

3.5 **d.** The problem involves binomial probability. The probability that one item, selected at random, is defective is

$$p_{\text{defective}} = \frac{10}{1000} = 0.01$$

and the probability that one item is good (not defective) is

$$p_{\text{good}} = 1 - p_{\text{defective}} = 0.99$$

The probability that exactly one defective item will be found in a random sample of 100 items is given by the binomial $b(1, 100, 0.01)$, in which

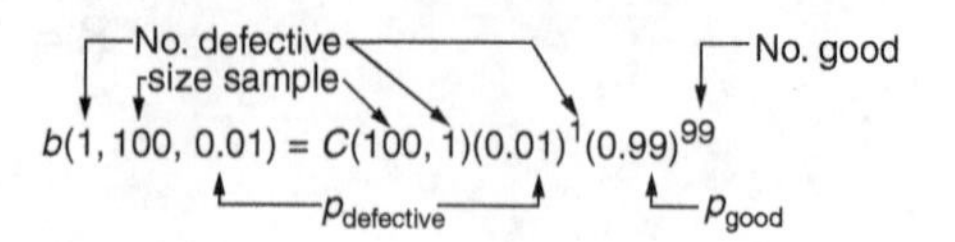

$C(n,r)=\binom{n}{r}=\dfrac{n!}{(n-r)!r!}$ is the number of combinations of n objects taken r at a time without concern for the order of arrangement.

$C(100, 1) = \dfrac{100!}{99!1!} = 100$, so $b(1, 100, 0.01) = 100(0.01)(0.99)^{99}$ $= 0.99^{99} = 0.3697$.

3.6 c. Here, $P(S_1) = 0.98$ and $P(S_2 \cap S_1) = 0.95$ are given. Hence the conditional probability $P(S_2 | S_1)$ is

$$P(S_2|S_1)=\frac{P(S_2 \cap S_1)}{P(S_1)}=\frac{0.95}{0.98}=0.97$$

3.7 d. The probability of drawing an ace on the first card is 4/52. The probability that the second card is an ace is 3/51. The probability that the third card is an ace is 2/50, and probability for the fourth ace is 1/49. The probability that the first four cards will all be aces is

$$P=\frac{4}{52}\bullet\frac{3}{51}\bullet\frac{2}{50}\bullet\frac{1}{49}=0.00\ 0037=3.7\times10^{-6}$$

3.8 d.

3.9 d. Since the first IC that is picked is a premium IC, it was drawn from either bin 1 or bin 3. From the distribution of premium ICs, the probability that the premium IC came from bin 1 is $\frac{2}{3}$, and from bin 3 is $\frac{1}{3}$.

In bin 1, the probability that the remaining IC is a premium IC is 1; in bin 3, the probability is 0. Thus, the probability that the remaining IC is a premium IC is

$$\frac{2}{3}(1)+\frac{1}{3}(0)=\frac{2}{3}$$

An alternative solution using Bayes' theorem for conditional probability is

$$P(\text{bin 1} \mid \text{drew premium}) = \frac{P(\text{bin 1 and premium})}{P(\text{premium})}$$

$$= \frac{P(\text{premium} \mid \text{bin 1}) \bullet P(\text{bin 1})}{\sum_{i=1}^{3} P(\text{premium} \mid \text{bin 1}) P(\text{bin 1})}$$

$$= \frac{1\left(\frac{1}{3}\right)}{1\left(\frac{1}{3}\right) + 0\left(\frac{1}{3}\right) + \frac{1}{2}\left(\frac{1}{3}\right)} = \frac{2}{3}$$

3.10 c. The answer is the binomial coefficient

$$\binom{35}{4} = \frac{35 \bullet 34 \bullet 33 \bullet 32}{4 \bullet 3 \bullet 2 \bullet 1} = 35 \bullet 34 \bullet 11 \bullet 4 = 52,360$$

3.11 d. The total number of choices of 5 is $\binom{50}{5}$. Of these, $\binom{40}{5}$ have no bad bolts, $\binom{40}{4} \times \binom{10}{1}$ have one bad bolt, and $\binom{40}{3}\binom{10}{2}$ have two bad bolts. Thus,

$$\frac{\binom{40}{5} + \binom{40}{4}\binom{10}{1} + \binom{40}{3}\binom{10}{2}}{\binom{50}{5}}$$

$$= \frac{\frac{40 \bullet 39 \bullet 38 \bullet 37 \bullet 36}{5 \bullet 4 \bullet 3 \bullet 2} + \frac{40 \bullet 39 \bullet 38 \bullet 37}{4 \bullet 3 \bullet 2} \bullet 10 + \frac{40 \bullet 39 \bullet 38}{3 \bullet 2} \bullet \frac{10 \bullet 9}{2}}{\frac{50 \bullet 49 \bullet 48 \bullet 47 \bullet 46}{5 \bullet 4 \bullet 3 \bullet 2}}$$

$$= \frac{658,008 + 913,900 + 444,600}{2,118,760} = 0.9517$$

3.12 c. There are 26 choices for the first letter; 25 remain for the second, and 24 for the third.

3.13 a. The probability that only A is defective is

$$0.1 \times (1 - 0.2) \times (1 - 0.25) = 0.06$$

The probability that only B is defective is

$$(1 - 0.1) \times (0.2) \times (1 - 0.25) = 0.135$$

The probability that only C is defective is

$$(1 - 0.1) \times (1 - 0.2) \times (0.25) = 0.18$$

Now add to find the final probability, which is

$$0.06 + 0.135 + 0.18 = 0.375$$

3.14 d. Simply multiply the complementary probabilities $(1 - 0.5) \times (1 - 0.4) \times (1 - 0.3) = 0.21$.

3.15 a. Apply the binomial distribution. The probability of 0 bad is $(0.9)^{50}$; of 1 bad, $\binom{50}{1}(0.1)(0.9)^{49}$; and of 2 bad, $\binom{50}{1}(0.1)^2(0.9)^{48}$. Adding these, $(0.9)^{48}[(0.9)^2 + 5.0(0.9) + 1225(0.1)^2] = 0.112$.

3.16 b. This is the permutation of arranging 3 objects out of 6:

$$P(6,3) = \frac{6!}{(6-3)!} = 20$$

3.17 d. This is the permutation of arranging 7 persons out of 7:

$$P(7,7) = \frac{7!}{(7-7)!} = 5040$$

3.18 c. This is the selection (or combination) of 6 out of 8:

$$C(8,6) = \frac{8!}{(8-6)6!} = 28$$

3.19 d. Since two questions are mandatory, only four questions have to be selected out of six.

Then, $C(6,4) = \frac{6!}{(6-4)!4!} = 15$.

3.20 b. This is the selection (or combination) of 2 out of 5:

$$C(5,2) = \frac{5!}{3!2!} = 10$$

3.21 c. Since there are 11 integers that are exactly divisible by 9, probability = 11/100 = 0.11.

3.22 d. $P(1 \text{ head}) = \frac{C(4,1)}{2^4} = \frac{4}{16}$; $P(2 \text{ heads}) = \frac{C(4,2)}{2^4} = \frac{6}{16}$

Since P(1 head AND 2 heads) is 0, P(1 head or 2 heads) = (4/16) + (6/16) = 5/8.

3.23 b. Let event A = (Apple from a bag) and event B = (Orange from the other bag).

Then, P(A) = 10/15 and P(B) = 5/15.

Since the events are independent, P(A AND B) = P(A) P(B) = (10/15) (5/15) = 2/9.

3.24 d. P(Black OR Red) = P(Black) + P(Red) – P(Black AND Red) = (5/85) + (30/85) – 0 = 35/85.

3.25 **c.** Let event A = (Orange from a bag) and event B = (White from the other bag).

P(A) = 2/10 = 0.2 and P(B) = 5/10 = 0.5

P(A OR B) = P(A) + P(B) – P(A AND B) = 0.2 + 0.5 – (0.2)(0.5) = 0.6

Note: P(A AND B) = P(A) P(B) as events A and B are independent.

3.26 **a.** Let event A = (number is even) and event B = (number > 72).

P(A) = 50/100 = 0.5 and P(B) = 28/100 = 0.28

Event (A and B) = (number is odd and > 72); P(A AND B) = 14/100 = 0.14

P(A OR B) = P(A) + P(B) – P(A AND B) = 0.50 + 0.28 – 0.14 = 0.64

3.27 **c.** P(both closed) = P(1 is closed) P(2 is closed)

0.90 = (0.95) P(2 is closed); then, P(2 is closed) = 0.90/0.95 = 0.9474

3.28 **d.** For a valid probability density function, $\int_{-\infty}^{\infty} f(x)dx = 1,$ or the total area under the curve should be 1. For $f_1(x)$, the area is (1/2)(2)(2) = 2; for $f_2(x)$, it is 1(4 – 2) = 2; for $f_3(x)$, it is (1/2)(4 + 1) = 2.5. For $f_4(x)$, $\int_0^{\infty} e^{-x}\,dx = 1.$ So, $f_4(x)$ is the only valid distribution.

3.29 **c.** For a valid probability distribution function, both $f(\infty)$ and $f(-\infty)$ should be zero and $f(x) \geq 0$ for all x. Also, $\int_{-\infty}^{\infty} f(x)dx = 1.$ Then, $f_1(x)$ and $f_4(x)$ are not valid since $f(x) < 0$ for certain values of x. $f_2(x)$ is not valid because $f(\infty)$ is not 0. $f_3(x)$ is the valid function as it satisfies all the conditions.

3.30 **a.** The distribution is discrete, but the rules are similar to those of Problem 3.29. $f_1(x)$ is the only valid distribution. $f_2(x)$ is not valid since the sum of the densities (equivalent to integrating) is more than 1. $f_3(x)$ has a negative value. $f_4(x)$ has a value more than 1.

3.31 **b.** For a valid cumulative probability distribution function $F(x)$, the following rules apply: $F(-\infty) = 0$, $F(\infty) = 1$, $0 \leq F(x) \leq 1$, and $F(x_1) \leq F(x_2)$ if $x_1 < x_2$. For $F_1(x)$, $F(x)$ has a negative value; for $F_3(x)$, $F(\infty)$ is more than 1; for $F_4(x)$, the rule $F(x_1) \leq F(x_2)$ if $x_1 < x_2$ fails. Only $F_2(x)$ obeys all the rules.

3.32 **a.** Let p = P(head on the first toss); then, P(tail on the first toss) = $1 - p$

But, $p = 2(1 - p)$; solving, $p = 0.667$.

P(two heads in four tosses) = C(4,2)$(0.667)^2(1 - 0.667)^2$ = 0.1481

3.33 **a.** P(leaving the job) = 0.25; P(none will leave the job) = C(7,0)$(0.25)^0$ $(1 - 0.25)^7$ = 0.1335

3.34 **d.** P(head) = p = 0.5; P(at least one head) = 1 – P(no head) = 1 – C(4,0) $(0.5)^0(1 - 0.5)^4$ = 0.9375

3.35 **a.** Using the normal distribution table, P(X > 3) = 1 – F(3) = 0.0013.

3.36 **b.** $P\{X \le 26\} = F(26) = F\left(\frac{26-30}{5}\right) = F(-0.8) = 1 - F(0.8) = 1 - 0.7881$ $= 0.2119$

3.37 **d.** $P\{1980 - 68 < x \le 1980 + 68\} = F(2048) - F(1912)$

$$= F\left(\frac{2048-2000}{40}\right) - F\left(\frac{1912-2000}{40}\right) = F(1.20) - F(-2.2)$$

$$= 0.8849 - \{1 - 0.9861) = 0.8710$$

3.38 **d.** $C(5,0)(1/120)^0(119/120)^5 = 0.9590$

3.39 **a.** $\text{mean} = \overline{X} = \frac{\sum x}{n} = \frac{114}{12} = 9.5$

$$\text{variance, } \sigma^2 = (1/12)\left[(3-9.5)^2 + (4-9.5)^2 + \ldots\right] = 28.75$$

standard deviation $\sigma = 5.36$

3.40 **b.** $\frac{3(1.5) + 8(2.5) + \ldots + 9(5.5)}{3 + 8 + 18 + \ldots + 9} = 3.82$

CHAPTER 4

Computational Tools

OUTLINE

INTRODUCTION

Current information for the FE exam indicates that approximately 7% of the *morning session* of the exam will contain questions related to computers, regardless of exam discipline. This means that there will be approximately eight or nine questions related to computers in the morning session for all FE examinees. There may be additional questions related to *computer systems* in the afternoon for examinees taking the *electrical* FE exam. All examinees should have basic familiarity with the following computer topics:

- Terminology (e.g., memory types, CPU, baud rates, Internet)
- Spreadsheets (e.g., addresses, interpretation, what if, copying formulas)
- Structured programming (e.g., assignment statements, loops and branches, function calls)

These three broad areas are specifically identified as topic areas addressed on the FE exam by NCEES. These and similar topics will be covered in this review chapter. More advanced computer topics, such as computer architecture, interfacing, microprocessors, and software design, which could be covered in the *afternoon section* of the *electrical* exam, should be reviewed using materials appropriate to that exam.

TERMINOLOGY

Types of Computers and Networks

There are many types of computers. The predecessors of all modern day computers were **mainframe** computers. Mainframe computers are large computers that typically fill an entire room, require air conditioning, and support many hundreds of users. Large and especially fast computers are known as **supercomputers**. Supercomputers are used to model very complex phenomena such as the operation of the human brain, air currents resulting from a fire, or nuclear explosions. Supercomputers are sometimes referred to as **Cray** computers because Cray Research, Inc. (founded by Seymour Cray) is a noted manufacturer of supercomputers. Mainframe and supercomputers tend to be expensive to buy and maintain.

Slightly smaller computers, which might still fill a room but support only 10 to 100 users, are known as **minicomputers**. Minicomputers are not in much use today, having been displaced by networks of smaller computers.

Smaller and moderately less powerful computers used for engineering and scientific calculations are often referred to as **workstations**. Workstations typically are designed for one user but can support multiple users. Computers that are less powerful than workstations and that are used in the home or office are typically known as **PCs** (personal computers). PCs derive their name from the IBM-PC, but other similarly used computers, such as Apple's Mac family of computers, also are typically referred to as PCs. Both workstations and PCs are **microcomputers**. Over time, the prefix "micro" has tended to be dropped from the term, and most people simply refer to microcomputers as computers. Similarly, in recent years, the distinction between workstations and PCs has blurred, and low-end workstations may also be referred to simply as PCs.

The computer chips that go into consumer devices, such as microwave ovens, washing machines, and cellular telephones, are known as **microprocessors**. If the chip has advanced systems on it such as analog-to-digital converters, timer systems, and significant memory, then it is known as a **microcontroller**.

To obtain large computing power at small cost, computers are sometimes networked together such that individual computers can process different parts of one problem. This is sometimes referred to as **cluster computing** or **grid computing**. Cluster computing typically takes place on computers located in close proximity to one another. A common cluster computing arrangement is a **Beowulf Cluster**. Grid computing typically uses computers more widely dispersed over the **Internet**, a worldwide network of computers for the purpose of sharing information. An arrangement in which computing services, such as network support, hardware and software upgrades, and data storage, take place over the Internet is known as **cloud computing** or **computing in the cloud**.

A lot of information available on the Internet can be viewed using an **Internet browser**, a computer program written to easily access information provided on the Internet using **http (hypertext transfer protocol)**, a standard method for publishing information on the Internet. A key feature of http is the use of a **URL**, or **universal resource locator**. A URL provides a standard format for providing a computer address of where information resides on the Internet. **Search engines**, such as Google, Yahoo, and Bing, can provide URLs for sought-after information.

Computer Communication

Computer communication has always been important. Typically a monitor and keyboard were used to access mainframe computers. The monitor and keyboard were known as a **terminal**. Today, a PC is often used as a terminal for large computers. If the terminal has very low processing power, it is known as a **thin client**. If the terminal has significant processing power, it is known as a **fat client** or **thick client**.

Computers communicate by sending **bits** of data. A bit is the simplest piece of information that can be stored in a computer. It is represented by a one or a zero. Representing information in only one of two possible states is known as **binary** representation.

Early computers communicated to terminals and other equipment at slow rates typically measured in hundreds or thousands of **baud**. The baud rate is the greatest number of state transitions per second in a communication system. In older technology, this is equivalent to the **bit rate**, the maximum number of bits that can be transmitted per second. However, modern communication systems may send more than one bit of information per state transition, so baud and bit rate should not be used interchangeably. The speed of **RS-232** communication, a common serial communication standard, has typically been expressed in baud. Its maximum data rate is about 20,000 baud.

Serial communication is a method of communication in which only one bit of information is sent at a time. Sometimes the bit that is sent is not received correctly. To prevent this type of error, serial communication systems often send a **parity bit**. A parity bit is an extra bit sent with a group of other bits (usually numbering seven or eight) such that the number of 1s or 0s sent is always even or odd. If the receiver detects that it has received the wrong number of even or odd bits, it can request retransmission of the data. Serial communication tends to be much slower than **parallel** communication in which many bits are sent simultaneously over separate physical lines.

RS-232 serial ports have been common on many PCs for many years; however, communication on PCs now typically takes place over **Ethernet, firewire**, or **USB**. Ethernet is a communication standard primarily used for **local area networks** (LANs), physically connected computers sharing a small geographic region. Ethernet now supports data speeds up to 100 billion bits per second. Firewire and USB standards support speeds up to about 5 billion bits per second and are primarily used to connect computers to other noncomputer devices, such as digital video cameras, external hard drives, and other personal electronic devices. **Wide area networks** (WANs) are used to link LANs and other computer networks together over far-reaching geographical areas.

Example **4.1**

Which of the following communication protocols would be most suitable for a new data communication center's local area network (LAN)?

(i) RS-232 (ii) Ethernet (iii) firewire (iv) USB

Solution

The correct choice is (ii) Ethernet. All other options are considerably slower than Ethernet and more likely to be used to connect computers to other noncomputer devices rather than to other computers. As more information is being sent over networks, a new data communication center needs the fastest communication standard.

Computer Memory

Computer memory and communication speeds are typically expressed in bits (b) or **bytes** (B) and bits per second (bps) or bytes per second (B/s). A byte consists of eight bits.

As computer memory has increased in size, prefixes have been used to indicate larger units of computer memory. This has led to some confusion in specifying computer memory. Because computer memory is utilized in sizes consisting of powers of two, a **kilobyte (KB)** is not 1000 bytes, but is actually 1024 (2^{10}) bytes. The confusion exists because the prefix kilo means 1000 in SI units. The confusion is even more pronounced because in computer information storage (such as in hard drives, tape drives, and removable disks) and in other computer-related usage (such as in clock speeds, operations per second, or data transfer rates), the standard SI meaning of the prefix is *typically* used. Consequently, a **megabyte (MB)** is 1,048,576 (2^{20}) bytes if referring to computer memory and is 1,000,000 bytes if referring to the storage capacity of a removable disk. A **gigabyte (GB)** indicates 2^{30} bytes of computer memory or 10^9 bytes of disk storage capacity. Likewise, a **terabyte (TB)** consists of 2^{40} bytes if referring to computer memory or 10^{12} bytes if referring to the total amount of data to be transferred to a computer over the Internet.

If the base 2 usage of the prefix is meant, especially when using it in reference to something other than computer memory, then the prefixes should be altered to kibi (2^{10}), mebi (2^{20}), gibi (2^{30}), and tebi (2^{40}) to reduce confusion. The **kibibyte, mebibyte**, **gibibyte**, and **tebibyte** are then abbreviated **KiB**, **MiB**, **GiB**, and **TiB**, respectively. Thus, a 200 GiB hard drive has $200 \times 2^{30} = 214{,}748{,}364{,}800$ bytes of storage capacity, whereas a 200 GB hard drive has $200 \times 10^9 = 200{,}000{,}000{,}000$ bytes of storage capacity.

Example 4.2

A supercomputer has 200 TB of computer memory. How many bits does this represent?

Solution

Because the reference is to computer memory, the TB abbreviation indicates the nonstandard use of the prefix *tera* to mean 2^{40}. Additionally, the number of bits must be found from the number of bytes (8 bits = 1 byte).

$$200 \text{ TB} \times \frac{2^{40} \text{ B}}{1 \text{ TB}} \times \frac{8 \text{ bits}}{1 \text{ B}} = 1{,}759{,}218{,}604{,}441{,}600 \text{ bits}$$

Computer memory is typically referred to as belonging to **RAM** or **ROM**. For historical reasons, the name RAM was created to indicate *random access memory*, and ROM indicated *read-only memory*. However, the term *RAM* now indicates memory that is **volatile**, and *ROM* indicates memory that is **nonvolatile**. Volatile memory is memory in which the data is lost if power is removed, whereas nonvolatile memory retains its data even if power is removed. RAM is used to temporarily store data and program code while a computer is in operation. ROM is used to permanently store programs and other information that doesn't change (or at least doesn't change very often).

Although there are other types of RAM, the two most common types are **static RAM (SRAM)** and **dynamic RAM (DRAM)**. SRAM is faster than DRAM but is more expensive, takes up more space, and consumes more energy. It is typically used for a computer's **cache**. A cache is a separate memory structure that is used to rapidly access frequently used data or program code. Because of the disadvantages of SRAM, cache memory is typically much smaller than main memory. DRAM is typically used for main memory. DRAM is so named because the information is stored as a charge on a capacitor that gradually loses its charge. To keep from losing the data, the memory system must provide a periodic refresh charge. RAM chips are built on silicon wafers using complementary metal oxide semiconductor **(CMOS)** circuits. CMOS circuits are an especially energy-efficient means of storing data in RAM, especially if the transitions between the binary states occur infrequently.

Read-only memory (ROM) was initially just as the name states—memory that could only be read. The data or program code was placed into the memory chip during manufacture. This memory evolved into programmable ROM **(PROM)** memory, which was memory that could be programmed after manufacture but only programmed once. After a PROM was programmed, it could only be read and not written to again. This was followed by **EPROM** memory. EPROM stands for erasable programmable ROM. This memory type could be written to many times but only after erasing it via exposure to ultraviolet light through a quartz crystal window manufactured into the chip housing. **EEPROM** followed EPROM. EEPROM stands for electrically erasable PROM. Instead of needing ultraviolet light to erase the memory, this memory could be erased and reprogrammed by applying appropriate voltages to the chip.

Flash memory is a type of EEPROM and is commonly accessed via a USB port. This arrangement is referred to as a **pen drive**, **thumb drive**, **flash drive**, or **jump drive**. Sometimes hard drives and **CD** (compact disc) or **DVD** (digital video disc or digital versatile disc) drives are considered to be ROM, especially if the CD or DVD disc can only be written to once.

Data transmission is slower in ROM as compared to RAM. Data transmission to or from a hard drive, CD drive, or DVD drive is extremely slow as compared to accessing other memory types. Although considerably slower, these drives have substantially more memory capacity than internal RAM or ROM. Storage capacity for hard drives is now commonly measured in hundreds of GB or tens of TB.

Example 4.3

Which of the following memory types would most likely be used for a computer's cache memory?

(i) SRAM (ii) DRAM (iii) EPROM (iv) FLASH

Solution

The correct answer is (i) SRAM because it is the fastest type of RAM. ROM types of memory would never be used for a cache.

Example 4.4

To what do the first two letters in the name of the memory type known as EEPROM refer?

(i) Enhanced Emitter (ii) Energy Efficient (iii) Exception Event

(iv) Electrically Erasable

Solution

The correct answer is (iv) Electrically Erasable. EEPROM is a type of read only memory (ROM) that can be erased and reprogrammed by applying appropriate voltages to the memory chip.

Program Execution

Execution of computer programs to process data takes place in the **central processing unit (CPU)**. This is a specialized circuit that once consisted of vacuum tubes or mechanical relays but is now built on silicon as an **integrated circuit (IC)**. An IC is a complex electrical circuit consisting of many transistors to implement one or more functions on a single silicon wafer. CPUs have many substructures, such as **arithmetic and logic units (ALUs)**, registers (also commonly known as **accumulators)**, **buses**, and **control units**.

- **ALUs** are responsible for implementing data manipulation via arithmetic instructions such as add, subtract, and multiply. They also implement logic instructions such as logical AND, OR, and XOR (exclusive or).
- **Registers** or **accumulators** are special memory locations within the CPU used for holding data and intermediate results that are currently being accessed and manipulated by the ALU. They are also used for keeping track of which instruction in memory is to be executed next and for keeping track of the CPU's current operational state.

- **Buses** are the electrically conductive paths on which code, data, control information, and memory locations (addresses) are sent and received.
- The **control unit** is the circuit that makes all the subsystems work together appropriately. The control unit ensures data is written to or read from the appropriate registers at the appropriate times. The control unit may also assist with getting the next instruction from memory and other similar tasks.

The four main functions of the CPU are to *fetch*, *decode*, *execute*, and *write back* program code and data. The speed at which a CPU can accomplish these functions is often measured in **MIPS, million instructions per second**, also sometimes referred to as **MOPS, million operations per second.**

A single integrated circuit having more than one CPU is referred to as a **multicore processor**. Multicore processors can increase processing speed by sharing operations between two or more CPUs. Splitting processing tasks so that they can be shared between processors is known as **parallel programming**. An arrangement in which a CPU accesses both program code and data from a single memory structure is known as the **von Neumann architecture**. The **Harvard architecture** is an arrangement in which the CPU accesses program code from a memory structure that is separate and distinct from another memory structure used for holding data.

Number Systems and Data Types

CPUs are designed to work with code and data as binary numbers (0s and 1s). A computer program represented in binary is often referred to as **machine code**. As the digits of a binary number are examined from right to left, each digit represents a higher consecutive integer power of two. For example, the binary number 1101 represents the decimal number 13 because $1 \times 2^3 + 1 \times 2^2 + 0 \times 2^1 + 1 \times 2^0 = 8 + 4 + 0 + 1 = 13$.

Example 4.5

The binary number 110 corresponds to what decimal (base 10) number?

Solution

$$1 \times 2^2 + 1 \times 2^1 + 0 \times 2^0 = 4 + 2 = 6$$

The conversion of a decimal number to a binary number can be achieved by the method of remainders as follows. A decimal integer is divided by 2, giving an integer quotient and a remainder. This process is repeated until the quotient becomes 0. The remainders (in the reverse order) form the binary number. The following example illustrates this process.

Example 4.6

Convert decimal number 43 to a binary number.

Solution

	Quotient		Remainder
43 ÷ 2 =	21	+	1
21 ÷ 2 =	10	+	1
10 ÷ 2 =	5	+	0
5 ÷ 2 =	2	+	1
2 ÷ 2 =	1	+	0
1 ÷ 2 =	0	+	1

Answer: $(43)_{10} = (101011)_2$

Although the conversion from decimal to binary is straightforward, binary numbers are typically difficult for humans to read and understand, especially as the number of digits increase. A small improvement in readability was made by introducing **hexadecimal** (or **hex**) numbers. Hex numbers are formed by taking four binary digits at a time and using them to represent a base 16 number. This is accomplished by using the letters A, B, C, D, E, and F to represent decimal numbers 10, 11, 12, 13, 14, and 15, respectively. Thus, the binary number 1101, hexadecimal number D, and decimal number 13 are all equivalent. The equivalent decimal number can be found for a hexadecimal number in a manner similar to that for a binary number.

Example 4.7

Find the equivalent decimal value for the hexadecimal number 25E6.

Solution

$2 \times 16^3 + 5 \times 16^2 + 14 \times 16^1 + 6 \times 16^0 = 8192 + 1280 + 224 + 6 = 9702$

Fractional binary values can be represented by including digits to the right of a radix point (commonly understood as the decimal point for decimal numbers). For instance, the binary number 1101.101 is equivalent to the decimal number 13.625 because $1101.101 = 1 \times 2^3 + 1 \times 2^2 + 0 \times 2^1 + 1 \times 2^0 + 1 \times 2^{-1} + 0 \times 2^{-2} + 1 \times 2^{-3}$, which is equivalent to $8 + 4 + 0 + 1 + 0.5 + 0 + 0.125 = 13.625$.

A similar procedure can be followed for hex numbers. To find equivalent binary or hex numbers from a decimal number, repeatedly divide the decimal number and all remainders by integer powers of 2 or 16, respectively. For instance, the decimal number 2672 is A70 in hex because 16^2 divides into 2672 ten times and 16^1 divides into the remainder seven times and 16^0 divides into the final remainder zero times.

Alternatively, conversion from decimal to hexadecimal may be carried out in a manner similar to that for decimal to binary conversion, with the divisor 2 (or multiplier in the case of fractions) replaced by 16.

Example **4.8**

Convert the base 10 integer 458.75 to base 16 equivalent value.

	Quotient		Remainder	Hexadecimal
458 ÷ 16	28	+	10	A
28 ÷ 16	1	+	12	C
1 ÷ 16	0	+	1	1
	Integer		**Remainder**	**Hexadecimal**
.75 × 16	12		0	C
458.75 = 1CA.C				

If fractional values of a number are never needed, all of the available bits can be used to represent an **integer** data type. Because there must always be a finite number of bits used to represent a number, fractional numbers will often suffer from **round off**, or **rounding error**. For instance, if a binary number only had four digits available to represent a fractional value, any fraction that couldn't be exactly represented as an integer multiple of $1/2^4$ would be represented as the nearest integer multiple of $1/2^4$. This creates a rounding error of as much as $1/2^5$ = 0.03125. A limited number of digits for the fractional portion creates a limit to the precision of a binary number.

Similarly, a limited number of digits for the integer portion creates a limit to the range of a binary number. To help alleviate these limitations, the **floating point** data type was created. Floating point data permits the radix point to move (float) among the binary digits depending on whether greater range of the number is needed or whether increased precision is needed. It can be thought of as a type of scientific notation representation of numbers for computers. Common floating point number standards consist of **single precision**, **double precision**, and **quadruple precision**. These standards are able to accurately represent about 7, 16, and 34 significant figures, respectively. If a computation is attempted that cannot be represented by the computer, a **NaN** (Not a Number) data type might result. Examples of these types of computation might be trying to divide by zero or taking the square root of a negative number.

Character data types are often used to represent letters of the alphabet or the character representation of the numerals zero through nine or punctuation. Some other special symbols and control characters may also be represented by the character data type. Some of these include the symbol for a new line, carriage return, and bell sound. Common codes for representing character data include **ASCII (American Standard Code for Information Interchange), EBCDIC (Extended Binary Coded Decimal Information Code), and Unicode**. ASCII and EBCDIC are used for the English language alphabet and symbols. Unicode is an evolving standard used to represent alphabets and writing symbols of many different languages of the world. When character data is put together, as in forming a name, it is often referred to as a **string**.

Example 4.9

Which of the following would most likely be an example of a character string?

(i) 241.38
(ii) 2.4138e2
(iii) 241
(iv) Two hundred forty-one and thirty-eight hundredths

Solution

The correct answer is (iv) Two hundred forty-one and thirty-eight hundredths. Although all of the answers could be strings if the character representation of each of the digits were being displayed, only (iv) is the *most likely* example of a character string. This is because answers (i) and (ii) could also be examples of floating point data types, and answer (iii) could be an example of an integer data type. Only answer (iv) uses a significant number of alphabetic letters and dashes to guarantee it to be a character string.

Placing a number of data values in a continuous section of memory with one data value sequentially following the next is often referred to as a **data array**. This arrangement often makes processing of the data faster because the data can be easily loaded into the computer's cache memory for ready access. Data arrays may be one-dimensional arrays (as in a single column or row) or two-dimensional arrays in which there are multiple columns and rows of data (as in a matrix). Higher dimensional arrays are possible but not as common. A data array often uses a **pointer**, a variable that "points" to the location of the array, to provide access to the data within the array.

Computer Languages

To process data, computer programs are necessary. Originally, programs were written by physical switches or circuits that were opened or closed to represent the 0s or 1s of machine code. However, this was very difficult for humans to do without making mistakes, so **assembly languages** (or **assembler languages**) were developed. Assembly languages use short **mnemonics** to correspond to a specific instruction available to the CPU. For example, the instruction LDAA might stand for Load Accumulator A. Once the program is written in assembly language, a separate program known as an **assembler** converts the assembly code into machine code. The resulting file created is known as an **executable** because it can be directly loaded and executed on the computer.

Even though assembly languages were a vast improvement over writing programs in machine code, they were still difficult to use, especially as data processing tasks became more complicated. For this reason, **high-level languages** were created to make programming easier. Examples of high-level languages include Fortran, Basic, C, Java, Matlab, and Python. Similar to assemblers, **compilers** convert high-level languages into machine code so that executables are available to work directly on the computer. As long as a compiler is available for any given computer, high-level languages can usually be compiled to work on the computer with little or no modification to the original programming code. **Compiler errors** occur when the program is compiled if there are errors in the code that the compiler cannot process. A syntax error is a common error made by programmers. A **syntax error** occurs when the rules for how the language is to be used are not followed. An example of a syntax error is trying to use a variable before the variable is **declared** in a statement that names the variable and describes its data type.

Interpreted languages are also considered high-level languages. Interpreted languages do not need to be compiled and an executable is not generated. Interpreted languages are "interpreted" into machine code at runtime. This has some advantages in that variables and arrays can be created and altered "on the fly," but also has disadvantages in that the code executes more slowly, and an **interpreter** must exist on the machine in order for the interpreted language to run. Fortran and C are usually compiled into executables; whereas Matlab and Python are usually interpreted.

High-level languages use **functions** (which may also be known as **subroutines**, **procedures**, or **methods**) to break code into smaller reusable sections. Functions often need **parameters** (the datum or data) passed to them to execute properly. Parameters may be passed using **call-by-value** or **call-by-reference**. Call-by-value sends the data values directly to the function, whereas call-by-reference sends the function the memory location of where the data is stored. Functions may or may not return values to the program that originally called it.

Structured programs use three main structures to process data. These are **sequence**, **looping** (or **repetition**), and **decision** (or **choice**). Sequence is the process by which code executes only one instruction at a time, and each instruction executed follows in sequential order as determined by the programmer. Looping involves execution of an instruction or a sequential list of instructions repeatedly until some condition is satisfied. The determination for looping can be made prior to the loop beginning (pre-test) or after the loop has started execution (post-test). In structured programming, loops should only have one entry point and exit point. A decision is used to determine whether a given condition is true or false or what condition exists so that a program **branch** (a change in the flow of program execution) can take place.

As programs have become larger, **object oriented programming (OOP)** has become a major programming paradigm. Object oriented programming organizes programs around the data as objects rather than organizing around the tasks that must be performed on the data. Many believe OOP makes maintenance of programs easier, especially as the programming code becomes very large.

Operating Systems

A PC needs programs in order to interface with its own input and output devices, control those devices, and manage resources. These programs are known as **operating systems**. Some common operating systems include Unix, Linux, Mac OS, and Microsoft Windows. Operating system functions are generally accessed through an **API** (an **application program interface**). An API ensures that the operating system is in control of resources and that computer hardware is not *directly* controlled by the user or other programs. Operating systems permit other work productivity programs (**application programs**) to also run on the computer. Common application programs include programs for word processing, e-mail, spreadsheets, database management, presentation graphics, Internet browsers, and accounting programs. In addition to these, engineers often use application programs for **CAD (computer aided design)** and **CAM (computer aided manufacturing)**. Many application programs work with the user using a **GUI (graphical user interface)**. A GUI makes interacting with an application program (or operating system) easier by visually displaying options available and making those options easily selectable by a mouse or other input device.

Computer Security

Many operating systems and networking application programs implement security features to protect computer data and prevent unauthorized use of computers. A **password** is a sequence of letters, numbers, and other characters that must be provided before the operating system or networking program will permit access to computer data and programs. Typically, passwords should be changed often and consist of an unobvious sequence of character data involving both upper and lowercase letters, numbers, and other symbols. In some cases, passwords may be necessary to access computer data.

Encryption is also used to protect passwords and sensitive data. Encryption scrambles and encodes the data or password such that it is not readily seen or understandable without a key. A common method of providing encryption over the Internet is by using **SSL, secure socket layer**. The URL for Web sites using SSL begin with https rather than http.

Damage to data can occur due to a **computer virus**, **malware**, **spyware**, or other deliberate destruction by an unauthorized user. A computer virus is software written to cause damage to another computer's files or operation and to replicate itself so that it can be spread to other computers. **Antivirus software** is software that attempts to recognize computer viruses and inactivate them before they can infect a computer or replicate. Malware is any software written by unscrupulous people to cause damage to a computer's data, whether or not it is self-replicating. Spyware is software written to record or search for sensitive information on a computer to send to the spyware's programmer in order to steal information such as credit card numbers or computer passwords. People who write such programs are known as **crackers**, and sometimes **hackers**, although the term *hacker* is sometimes used to denote skilled programmers who don't necessarily have any malicious intent for the code they write.

Often a **firewall** is used to prevent unauthorized access to a computer over the Internet. A firewall carefully checks transmitted information over a network to make sure only authorized information is being conveyed. Some computers require **biometric data** to gain access to the computer. Biometric data is information about an individual that is uneasy to copy and is often used to provide a higher level of security for gaining computer access. Examples of biometric data would include a fingerprint or iris scan.

Some unauthorized people will try to gain access to computers by **spoofing**. This is an attempt to gain access by convincing the computer's security features that the unauthorized person is someone else or another computer that does have authorized access. Another commonly used method to try to gain unauthorized access to a computer is by **phishing**. Phishing is an attempt, usually via e-mail, to convince a user that the person sending the e-mail is someone who can be trusted and that the user should send confidential information such as passwords or banking information.

STRUCTURED PROGRAMMING

To solve a problem by using a computer program, an algorithm is usually developed. An **algorithm** is an ordered sequence of steps to take to arrive at a solution.

In the early days of computers and computer programs, there were no clear and concise rules to follow in implementing an algorithm in a computer program. Two of the earliest computer languages, assembly and BASIC, permitted the use of the *jump* and *goto* instructions, respectively. Although a computer could implement these instructions without any problems, human programmers overused them to their own detriment. Because no structured rules existed, programmers were left to implement the algorithm in whatever way they could to arrive at a solution. The resulting code solutions often were difficult to read and understand by people other than the original programmer. Such code also was prone to error because the programmer had not considered all the possible circumstances in which program execution could be altered by a goto statement. If the code ever needed to be upgraded, it was often difficult to follow the logical sequence and know what parts of the code should be changed and what should remain the same. Thus, program maintenance was difficult to achieve. Large blocks of code were often written to implement the algorithm all at once. This too made it difficult for others to keep track of all the variables that were being used, what the variables' current values were, and when program execution left the current sequence or jumped back in. As programming tasks became more complicated and larger code needed to be written, a real need for code that was easy to read and understand became apparent.

Structured programming is a concept generally credited to Edsger Dijkstra. He advocated much less dependence on the *goto* statement, even that it should be eliminated. Dijkstra articulated simple rules to make programming easier to read, understand, and maintain. To eliminate the rat's nest of code generated by overuse of the *goto* statement, structured programming specifies that there should be only one entry point and exit point within a loop, function call, or sequence of instructions. The entry or exit point should be determined by a decision statement that clearly indicates the condition or conditions under which the program execution could change (that is, under what conditions the program could *branch* to execute other instructions). Similarly, structured programming specifies that code should be modular. This means that large problems should be broken into smaller problems that can be solved first; the smaller solutions are then pieced together to solve the large problem. This top-down design approach means that function calls should be used to solve smaller portions of the problem. This keeps the code modular so that it is easy to see what small problem is being solved, the logical steps taken to solve it, and how it relates to solving the big problem without the confusion of solving too much at one time. Once the function is developed and well tested, it can be reused in other code, making new code development times shorter and facilitating code upgrades and maintenance. Indenting distinct chunks of code improves its readability and is another important contribution of structured programming. For example, the programmer might indent a section of instructions all pertaining to the same loop function.

In structured programming, only three main control structures are needed:

1. Statements for sequential execution of instructions, such as assignment statements
2. Decision statements, such as an *if-then-else*, for causing program branching
3. Repetition statements, such as the *while-do* or *for* looping commands

To also help develop algorithms and make programs easier to read and understand, programmers use flowcharts and pseudocode to illustrate the algorithm.

Example 4.10

Which of the following is considered to be good structured programming technique?

(i) Frequent use of the *goto* instruction

(ii) Frequent use of the *jump* instruction

(iii) An absence of function calls

(iv) Code that is broken into small sections

Solution

The correct answer is (iv) Code that is broken into small sections. This is the idea of modularity that makes the code easily readable and understandable as subsections that solve a small part of the larger problem rather than trying to solve the entire problem in one large block of code.

Flowcharts

An algorithmic flowchart is a pictorial representation of the step-by-step solution of a problem using standard symbols. Some of the commonly used shapes are shown in Figure 4.1. Consider the simple problem in Example 4.11.

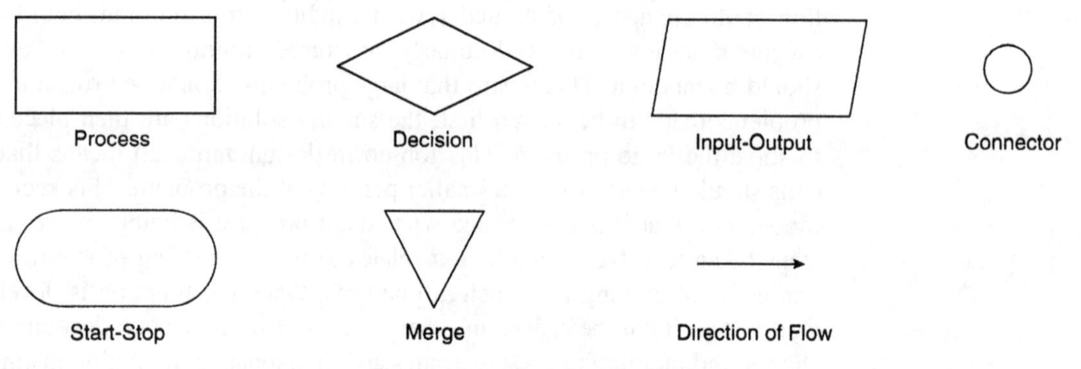

Figure 4.1 Flowchart Symbols

Example 4.11

A present sum of money (P) at an annual interest rate (I), if kept in a bank for N years, would amount to a future sum (F) at the end of that time according to the equation $F = P(1 + I)^N$. Prepare a flowchart for $P = \$100$, $I = 0.07$, and $N = 5$ years. Then compute and output the values of F for all values of N from 1 to 5.

Solution

Exhibit 1 shows a flowchart for this situation.

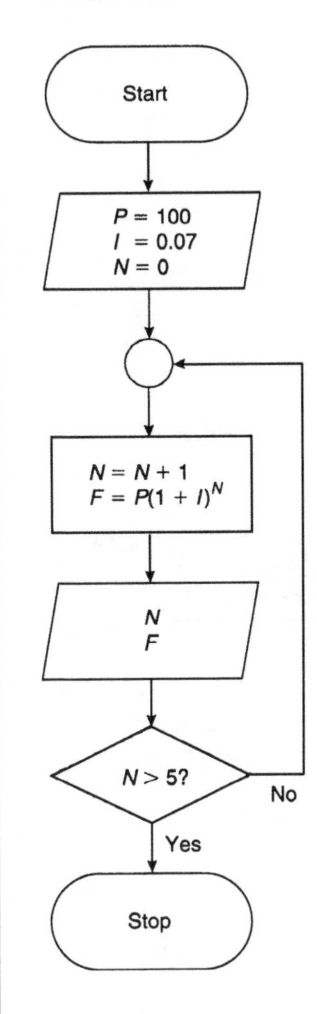

Exhibit 1
Algorithmic Flowchart

Example 4.12

Consider the flowchart in Exhibit 2.
The computation does which of the following?

(i) Inputs hours worked and hourly pay and outputs the weekly paycheck for 40 hours or less.

(ii) Inputs hours worked and hourly pay and outputs the weekly paycheck for hours worked including over 40 hours at premium pay.

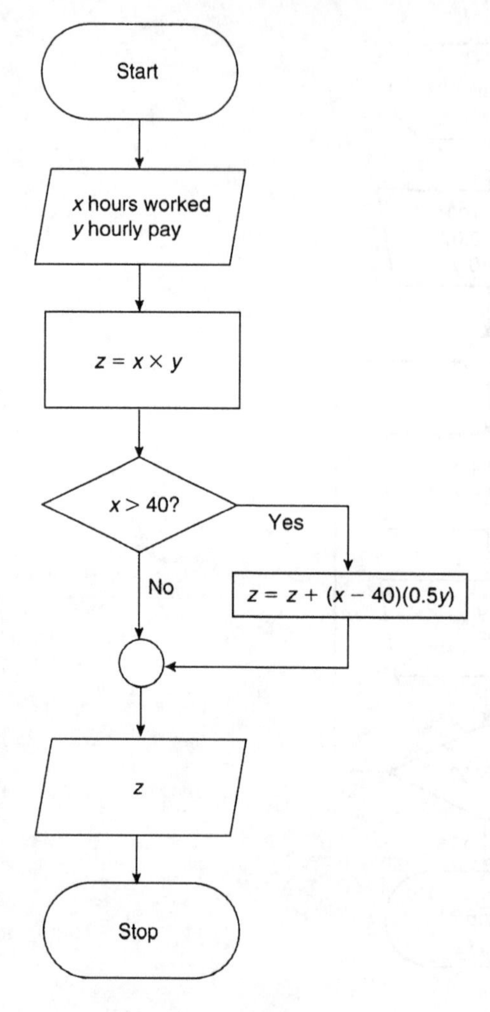

Exhibit 2

Solution

The answer is (ii).

Pseudocode

Pseudocode is an English-like language representation of computer programming. It is carefully organized to be more precise than a simple statement, but may lack the detailed precision of a flowchart or of the computer program itself.

Example 4.13

Prepare pseudocode for the computer problem described in Example 4.11.

Solution

```
INPUT P, I, and N = 0
DOWHILE N < 5
COMPUTE N = N + 1
   F = P (1 + I)^N
OUTPUT N, F
IF N > 5 THEN ENDDO
```

SPREADSHEETS

For today's engineers the ability to create and use spreadsheets is essential. Although several spreadsheet application programs exist, the most popular is Microsoft's Excel. The following explanations and examples are applicable to any spreadsheet package.

As noted in Figure 4.2, three types of information may be entered into a spreadsheet: text, values, and formulas. Text includes labels, headings, and explanatory text (columns A and B of Figure 4.2). Values are numbers, times, or dates (columns C, D, and E of Figure 4.2). Formulas combine operators and values in an algebraic expression (column F of Figure 4.2).

Figure 4.2 Example Spreadsheet

	A	B	C	D	E	F
1	**Last Name**	**First Name**	**Date**	**Age**	**Answers Correct**	**Percentage Correct**
2	Dempsey	Lou	4/17/2010	22	138	76.7
3	Johnson	Nathan	10/30/2010	26	105	58.3
4	Smith	Julie	10/30/2010	21	162	90.0

A **cell** is the intercept of a column and a row. Its location is based upon its column-row location; for example, B3 is the intercept of column B and row 3. Columns are labeled with letters across the top of the spreadsheet, and rows are labeled with numbers on the side. To change a cell entry, the cell must be highlighted using either an address or a pointer.

A group of cells may be called out by using a **range**. Cells A1, A2, A3, A4 could be called out using the range reference A1:A4 (or A1..A4). Similarly, the range A2, B2, C2, D2 could use the range reference A2:D2 (or A2..D2).

In order to call out a block of cells, a range callout might be A2:C4 (or A2..C4) and would reference the following cells:

A2 B2 C2
A3 B3 C3
A4 B4 C4

Formulas may include cell references, operators (such as + ,–,*,/), and functions (such as SUM, AVERAGE). The formula SUM(A2:A6) or SUM(A2..A6) would be evaluated as equal to A2 + A3 + A4 + A5 + A6. Built-in formulas such as SUM or AVERAGE must be preceded by the @ symbol or an equals sign (=) in order to be recognized as a formula as opposed to text.

Relational References

Most spreadsheet references are relative to the cell's position. For example, if the content of cell A5 contains B4, then the value of A5 is the value of the cell up one and over one. The relational reference is most frequently used in tabulations, as in the following example for an inventory where cost times quantity equals value and the sum of the values yields the total inventory cost.

Inventory		Valuation	
	A	B	C
1 Item	Cost	Quantity	Value
2 box	5.2	2	10.4
3 tie	3.4	3	10.2
4 shoe	2.4	2	4.8
5 hat	1.0	1	1.0
6 Sum			26.4

In C2, the formula is A2*B2, in C3 A3*B3, and so forth. For the summation the function SUM is used; for example, SUM(C2:C5) in C6.

Instead of typing in each cell's formula, the formula can be copied from the first cell to all of the subsequent cells by first highlighting cell C2 then dragging the mouse to include cell C5. The first active cell, C2, would be displayed in the edit window. Typing the formula for C2 as = A2*B2 and holding the control key down and pressing the enter key will copy the relational formula to each of the highlighted cells. Since the call is relational, the formula in cell C3 is evaluated as A3*B3. In C4 the cell is evaluated A4*B4. Similarly, edit operations to include copy or fill operations simplify the duplication of relational formulas from previously filled-in cells.

Arithmetic Order of Operation

Operations in equations use the following sequence for precedence: exponentiation, multiplication, or division, followed by addition or subtraction. Parentheses in formulas supercede normal operator order.

Absolute References

Sometimes one must use a reference to a cell that should not be changed, such as a data variable. An absolute reference can be specified by inserting a dollar sign ($) before the column-row reference. If B2 is the data entry cell, then by using B2 as its reference in another cell, the call will always be evaluated to cell B2, regardless of which cell a formula might be copied to. Mixed reference can be made by using the dollar sign for only one of the elements of the reference. For example, the reference B$2 is a mixed reference in that the row does not change but the column remains a relational reference. Thus, a formula copied using this mixed reference would be relative to column B but unchanging in its reference to row 2.

The power of spreadsheets makes repetitive calculations very easy. Changing a value in one cell automatically updates every other cell that references it. Furthermore, program-like commands can be used in spreadsheets by executing a macro. Macros permit the execution of a large number of spreadsheet commands that have been previously recorded or programmed by the user.

All spreadsheet programs allow for changes in the appearance of the spreadsheet. Headings, borders, or type fonts are usual customizing tools. Spreadsheets also permit presentation of data as line graphs, bar graphs, pie charts, and other figures. This makes spreadsheets especially valuable for report generation and presentation.

Example 4.14

Which of the following tasks is possible within a spreadsheet?

(i) Creating a graph
(ii) Creating reports
(iii) Analyzing data
(iv) All of these are possible.

Solution

The correct answer is (iv). All of these are possible within a spreadsheet. All of these tasks are easily done in a spreadsheet. There are very few things that can't be done, especially if the user knows how to use macros.

Example 4.15

A spreadsheet has the values 5, 10, and 15 in cells A1, A2, and A3, respectively; and the formula A1 + A2 in cell B2. What value will display in cell B3 if the formula from cell B2 is copied to cell B3?

(i) 15
(ii) 20
(iii) 25
(iv) 30

Solution

The correct answer is (ii) 20. The $ symbol represents absolute addressing within spreadsheets, and its absence indicates relative addressing. Thus, when the formula is copied from cell B2 to B3, the values added are the value 5 in cell A1 and the value 15 in cell A3. Thus, the formula in cell B3 will be A1 + A3, and the value displayed will be 20.

Example 4.16

Which of the following would likely be most difficult to incorporate into a spreadsheet cell?

(i) A jpeg image
(ii) Text representing a name
(iii) Value representing currency
(iv) Cell shading and/or borders

Solution

The correct answer is (i) a jpeg image. Although it is quite easy to insert an image into a spreadsheet, inserting one into an individual cell is not possible. There are ways to make it seem as if an image is inserted into a cell, but this is not easily done. The other answer choices are all easily inserted or performed within a spreadsheet cell.

Example 4.17

A spreadsheet has the values 2, 4, and 6 in cells A1, A2, and A3, respectively. It also has values 3, 5, and 7 in cells B1, B2, and B3, respectively. What is displayed in cell D8 if it contains the formula @SUM(A1:B3)?

(i) 12
(ii) 15
(iii) 27
(iv) An error would result.

Solution

The correct answer is (iii) 27. Referencing cells in this manner provides the sum of all the values in the block. Where the formula is located (in this case, cell D8) is irrelevant unless the formula is copied to another cell. The format of the formula is correct in that it must start with the @ or = symbol, and the argument for the formula must either be specified with a colon (A1:B3) or two periods (A1..B3).

GLOSSARY OF COMPUTER TERMS

Accumulators	Registers that hold data, addresses, or instructions for further manipulations in the ALU
Address bus	Two-way parallel path that connects processors and memory containing addresses
AI	Artificial intelligence
Algorithm	A sequence of steps applied to a given data set that solves the intended problem
Alphanumeric data	Data containing the characters a, b, c, ..., z, 0, 1, 2, ..., 9
ALU	Arithmetic and logic unit
ASCII	American Standard Code for Information Interchange, 7 bit/character (Pronounced AS-key)
Asynchronous	Form of communications in which message data transfer is not synchronous, with the basic transfer rate requiring start/stop protocol
Baud rate	Number of state transitions that can be realized per second. Often equivalent to bits per second
BIOS	Basic input/output system
Bit	0 or 1
Buffer	Temporary storage device
Byte	8 bits
Cache memory	Fast look-ahead memory connecting processors with memory, offering faster access to often-used data
Channel	Logic path for signals or data
CISC	Complex instruction-set computing
Clock rate	Cycles per second
Control bus	Separate physical path for control and status information
Control unit	Ensures all subsystems within CPU work together appropriately
CPU	Central Processing Unit, the primary processor
Data buffer	Temporary storage of data
Data bus	Separate physical path dedicated for data
Digital	Discrete level or valued quantification, as opposed to analog or continuous valued
Duplex communication	Communications mode where data is transmitted in both directions at the same time
Dynamic memory	Storage that must be continually hardware-refreshed to retain valid information
EBCDIC	Extended Binary Coded Decimal Interchange Code—8 bits/character (pronounced EB-see-dick)
EPROM	Erasable programmable read-only memory
Expert systems	Programs with AI, which imitate the knowledge of a human expert
Floppy disk	Removable disk media in various sizes, $5^1/_4$", $3^1/_2$"
Flowchart	Graphical depiction of logic using shapes and lines
Gbyte (GB)	Gigabytes: 1,073,741,824 or 2^{30} bytes; a measurement of computer memory

Half-duplex communication	Two-way communications path in which only one direction operates at a time (transmit or receive)
Handshaking	Communications protocol to start/stop data transfer
Hard drive	Disk that has nonremovable media
Hardware	Physical elements of a system
Hexadecimal	Numbering system (base 16) that uses 0–9, A, B, ..., F
Hierarchical database	Database organization containing hierarchy of indexes/keys to records
I/O	Input/output devices such as terminal, keyboard, mouse, printer
IR	Instruction register
Kbytes (KB)	Kilobytes: 1024 or 2^{10} bytes; a measurement of computer memory
LAN	Local area network
LIFO	Last in–first out
LSI	Large scale integration
Main memory	That memory seen by the CPU
Mbytes (MB)	Megabytes: 1,048, 576 or 2^{20} bytes; a measurement of computer memory
Memory	Generic term for random access storage
Microprocessor	Computer architecture with Central Processing Unit in one LSI chip
MODEM	Modulator-demodulator
MOS	Metal oxide semiconductor
Multiplexer	Device that switches several input sources, one at a time, to an output
Nibble	Four bits
Nonvolatile memory	As opposed to volatile memory, does not need power to retain its present state
OCR	Optical character recognition
OS	Operating system
OS memory	Memory dedicated to the OS, not usable for other functions
Parallel interface	A character (8-bit) or word (16-bit) interface with as many wires as bits in interface plus data clock wire.
Parity	Method for detecting errors in data: one extra bit carried with data, to make the sum of one bit in a data stream even or odd
PC	Program counter or personal computer
Peripheral devices	Input/output devices not contained in main processing hardware
Program	A sequence of computer instructions
PROM	Programmable read-only memory
Protocols	Established set of handshaking rules enabling communications
Pseudocode	An English-like way of representing programming control structures
RAM	Random access memory
Real time/Batch	Method of program execution: real-time implies immediate execution; batch mode is postponed until run on a group of related activities
RISC	Reduced instruction set computing
ROM	Read-only memory
Sequential storage	Memory (usually tape) accessed only in sequential order (n, $n + 1$, ...)
Serial interface	Single data stream that encodes data by individual bit per clock period
Simplex communication	One-way communication
Static memory	Memory that does not require intermediate refresh cycles to retain state
Structured programming	Use of programming constructs, such as Do-While or If-Then-Else, to produce code that is logical and easy to follow
Synchronous	Communications mode in which data and clock are at same rate
Transmission speed	Rate at which data is moved, in baud (bits per second, bps)
Virtual memory	Addressable memory outside physical address bus limits through use of memory mapped pages
Volatile memory	Memory whose contents are lost when power is removed
Words	8, 16, or 32 bits
WYSIWYG	What you see is what you get
16-bit	Basic organization of data with 2 bytes per word
32-bit	Basic organization of data with 4 bytes per word
64-bit	Basic organization of data with 8 bytes per word

PROBLEMS

4.1 In spreadsheets, what is the easier way to write B1 + B2 + B3 + B4 + B5?

a. Sum (B1:B5) c. @B1..B5SUM
b. (B1..B5) Sum d. @SUMB2..B5

4.2 The address of the cell located at row 23 and column C is:

a. 23C c. C.23
b. C23 d. 23.C

4.3 Which of the following is *FALSE*?

a. Flowcharts use symbols to represent input/output, decision branches, process statements, and other operations.
b. Pseudocode is an English-like description of a program.
c. Pseudocode uses symbols to represent steps in a program.
d. Structured programming breaks a program into logical steps or calls to subprograms.

4.4 In pseudocode using DOWHILE, which of the following is *TRUE*?

a. DOWHILE is normally used for decision branching.
b. The DOWHILE test condition must be false to continue the loop.
c. The DOWHILE test condition tests at the beginning of the loop.
d. The DOWHILE test condition tests at the end of the loop.

4.5 A spreadsheet contains the following formulas in the cells:

	A	B	C
1		A1 +1	B1 +1
2	A1 ^2	B1^2	C1^2
3	Sum (A1:A2)	Sum (B1:B2)	Sum (C1:C2)

If 2 is placed in cell A1, what is the value in cell C3?

a. 12 c. 8
b. 20 d. 28

4.6 A spreadsheet contains the following:

	A	B	C	D
1		3	4	5
2	2	A$2		
3	4			
4	6			

If you copy the formula from B2 into D4, what is the equivalent formula in D4?

a. A$2 c. C4
b. C4 d. C$2

4.7 The hexadecimal number 2DB.A is most nearly equivalent to which decimal number?

a. 731.625
b. 731.10
c. 453.625
d. 341.10

4.8 The decimal number 1938.25 is most nearly equivalent to which hexadecimal number?

a. $(792.25)_{16}$
b. $(792.4)_{16}$
c. $(279.4)_{16}$
d. $(279.04)_{16}$

4.9 The type of office computer that would most likely be found at an engineer's desk is a:

a. mainframe
b. minicomputer
c. workstation
d. microcontroller

4.10 Which of the following is *NOT* considered an Internet search engine?

a. Bing
b. Google
c. Unix
d. Yahoo

4.11 Which term describes a worldwide network of computers for sharing information?

a. Beowulf cluster
b. Cloud computing
c. Internet
d. LAN

4.12 The baud rate of an old serial communication system is the same as the bit rate. If the baud rate is 9600, transmitting a 500 KB file would take:

a. 1 minute
b. 7 minutes
c. 2 minutes
d. 18 minutes

4.13 If the data rate for USB 3.0 is 4.8 billion bps, the file size that can be sent in 30 seconds is:

a. 1800 KB
b. 18 MB
c. 1.8 GB
d. 18 GB

4.14 Which of the following types of memory is considered to be nonvolatile?

a. Cache
b. DRAM
c. Flash
d. SRAM

4.15 Which of the following drives typically uses EEPROM memory?

a. CD
b. DVD
c. Hard drive
d. Thumb drive

4.16 Which of the following is *NOT* likely to be found as a subunit of a CPU?

a. ALU
b. Control unit
c. Registers
d. Transformer

4.17 Which of the following data types would give the *BEST* precision?

a. Floating point
b. Integer
c. Long integer
d. Unsigned integer

4.18 Which of the following is *NOT* a standard for encoding character data types?
a. ALPHA
b. ASCII
c. EBCDIC
d. Unicode

4.19 Which of the following programming language types would be the hardest for a human to code, read, and understand?
a. Assembly
b. High-level
c. Interpreted
d. Machine

4.20 Each of the following is another term for "function" *EXCEPT*:
a. declarative
b. method
c. procedure
d. subroutine

4.21 An application program is *MOST* likely to use which of the following to interact with the computer's operating system?
a. API
b. CAM
c. GUI
d. URL

4.22 Which of the following is *NOT* used as a safeguard for preventing unauthorized computer access?
a. Antivirus software
b. Biometrics
c. Firewall
d. Spoofing

4.23 A spreadsheet has the values 15, 18, and 32 in cells A1, A2, and A3, respectively. If cell B1 has the formula A1 + A2 in it, and cell B2 has the formula B1 + A3 in it, what will be displayed in cell B2?
a. 32
b. 55
c. 65
d. 98

4.24 If the formulas in cells B1 and B2 from Problem 4.23 are now copied to cells C1 and C2, respectively, what value will be displayed in cell C2?
a. 55
b. 65
c. 98
d. 130

4.25 If the conditions described in problems 4.23 and 4.24 exist, and if the value in cell A1 is changed from 15 to 10, which cells would show a change in values displayed?
a. All cell values would change.
b. Only cell A1would change.
c. Only cells A2 and A3 do not change.
d. Only row 1 would change.

4.26 The operation used within a spreadsheet to execute many previously recorded commands in a manner similar to running a program is a:
a. functional
b. macro
c. system tool
d. vector tool

4.27 In structured programming, how many possible exit points should there be for a loop structure?
a. One
b. Two
c. None
d. As many as desired

4.28 A change in the order of program execution is known as:
a. branching
b. inclusion
c. slippage
d. yielding

4.29 Giving a variable an initial value or modifying its value is known as:
a. an assignment
b. a disclosure
c. a fixture
d. a method

4.30 In a flowchart, an oval-shaped symbol represents:
a. a decision
b. initialization
c. input or output
d. program start or stop

4.31 Pseudocode is useful for developing an algorithm because:
a. it can be directly compiled into machine language
b. it has a very easy-to-read, English-like structure
c. it is highly structured, making mistakes unlikely
d. it requires the programmer to think about all parts of a problem simultaneously

SOLUTIONS

4.1 **a.** Sum (B1:B5) or @Sum (B1..B5)

4.2 **b.**

4.3 **c.** Pseudocode does not use symbols but uses English-like statements such as IF-THEN and DOWHILE.

4.4 **c.** IF-THEN is normally used for branching. The DOWHILE test condition must be true to continue branching, and the test is done at the beginning of the loop. The DOUNTIL test is done at the end of the loop.

4.5 **b.** Plugging 2 into cell A1 of the spreadsheet produces the following spreadsheet display:

	A	B	C
1	2	3	4
2	4	9	16
3	6	12	20

The value of C3 is 20.

4.6 **d.** The formula contains mixed references. The $ implies absolute row reference, whereas the column is relative. The result of any copy would eliminate any answer except for the absolute row 2 entry. The relative column reference A gets replaced by C. The cell contains C$2.

4.7 **a.**

$$(2DB)_{16} = 2 \times 16^2 + 13 \times 16^1 + 11 \times 16^0 = 512 + 208 + 11 = 731$$
$$(.A)_{16} = 10 \times 16^{-1} = .625$$
$$(2DB.A)_{16} = 731.625$$

4.8 **b.**

		Quotient		Remainder	Hexadecimal Digit
1938 ÷ 16	=	121	+	2	2
121 ÷ 16	=	7	+	9	9
7 ÷ 16	=	0	+	7	7

$$(1938)_{10} = (792)_{16}$$

		Integer		Fraction	Hexadecimal Digit
0.25 × 16	=	4	+	0.00	4

$$(.25)_{10} = (.4)_{16}$$
$$(1938.25)_{10} = (792.4)_{16}$$

4.9 **c.** The key terms are *office* and *desk* and *most likely*. Only a workstation or PC (microcomputer) would be most likely to be at an engineer's desk. A microcontroller could conceivably be at an engineer's desk, but it is only likely if the engineer works with embedded systems.

4.10 **c.** Unix is an operating system, not an Internet search engine.

4.11 **c.** The key term is *worldwide*. Only the Internet fits this description.

4.12 **b.** In this case, the problem explicitly states that the baud rate and the bit rate are the same. This means that the communication system can transmit 9600 *bits* per second. To find the total number of seconds needed to send the file, multiply the file size by 8 bits per byte and divide this result by the baud rate. Finally, convert the number of seconds to minutes by dividing by 60 seconds per minute.

$$\text{Total time} = \frac{\left(500\times10^{3}\ \text{B}\right)\left(\dfrac{8\ \text{bits}}{1\ \text{B}}\right)}{\dfrac{9600\ \text{bits}}{\text{s}}}\times\frac{1\ \text{minute}}{60\ \text{s}} = 6.94\ \text{minutes}$$

Note that had the file been 500 KiB, the transmission would still have been about 7 minutes.

4.13 **d.** To find the file size (in bits), multiply the data transmission rate by the total time. Then, convert the answer to bytes by multiplying 1 B/8 bits. Finally, use the standard SI definitions for KB, MB, or GB to determine the file size as provided in the answer choices.

$$\text{File size} = \frac{4.8\times10^{9}\ \text{b}}{\text{s}}\times 30\ \text{s}\times\frac{1\ \text{B}}{8\ \text{b}}\times\frac{1\ \text{GB}}{1\times10^{9}\ \text{B}} = 18\ \text{GB}$$

4.14 **c.** All of the memory types are RAM except for Flash memory. Flash memory is a type of EEPROM, which is nonvolatile memory.

4.15 **d.** A thumb drive uses Flash memory, which is a type of EEPROM.

4.16 **d.** Transformers are used to increase or decrease alternating currents and voltages. They are not found in CPUs.

4.17 **a.** Answers b, c, and d are all forms of integer data types; they cannot represent fractional values. The floating point data type permits the radix point to "float," or move, so that all or most of the significant figures can be used to represent a fractional value. For instance, the integer data types have poor precision in representing π because they can only represent π as the integer 3. The single precision floating point data type can represent π as 3.141593.

4.18 **a.** ASCII, EBCDIC, and Unicode are all standards for encoding character data.

4.19 **d.** Machine language consists of 1s and 0s. It would be the hardest for a human to read and understand.

4.20 **a.** Methods, procedures, and subroutines are all synonymous terms for function.

4.21 **a.** An application program is most likely to use an API, an application program interface, to interact with the operating system.

4.22 **d.** Spoofing refers to a method of attempting to gain unauthorized computer access. It is not a preventative safeguard.

4.23 **c.** With values 15, 18, and 32 in cells A1, A2, and A3, respectively, and with the formulas A1 + A2 in cell B1 and B1 + A3 in cell B2, cell B1 will have 15 + 18 = 33 in it, and cell B2 will have 33 + 32 = 65. In this case, the absolute addressing is irrelevant to the final calculation. In other words, the answer would have been the same if cell B2 had contained the formula B1 + A3.

4.24 **d.** In this problem, the relative or absolute addressing scheme is important to the copying procedure. After copying, cell C1 will have the relative addressing formula B1 + B2. Cell C2 will have a mixture of relative addressing and absolute addressing. Cell C2 will contain C1 + A3. With these formulas, C1 will have the value 33 + 65 = 98, and C2 will have 98 + 32 = 130.

4.25 **c.** Given that the conditions in problems 4.23 and 4.24 exist, changing the value of A1 will change every cell that directly or indirectly references A1. Changing the value of A1 from 15 to 10 would give the following results: A1 = 10, A2 = 18, A3 = 32, B1 = 10 + 18 = 28, B2 = 28 + 32 = 60, C1 = 28 + 60 = 88, and C2 = 88 + 32 = 120. Thus, all cells change except for cells A2 and A3.

4.26 **b.** A macro is a sequence of spreadsheet commands that can be recorded and executed, as is done in traditional computer programming.

4.27 **a.** In structured programming, there should be only one exit point for each loop structure. Having more than one exit point makes debugging and maintaining code very difficult.

4.28 **a.** The term branch or branching indicates a change in the order of program execution.

4.29 **a.** An assignment statement assigns or changes a value within a variable.

4.30 **d.** An oval shape denotes program initiation (start) or termination (stop). Decisions are represented by diamonds. Inputs and outputs are represented by rhomboids. Processing steps (which might include an initialization procedure) are presented by rectangles.

4.31 **b.** Pseudocode is useful for developing an algorithm because of its natural, easy-to-read, English-like structure. It permits the programmer to capture the basic ideas and sequences of the algorithm on paper without having to worry about programming syntax and other rules.

CHAPTER 5

Ethics and Professional Practice

OUTLINE

The topic of ethics and business practices represents about 7% of the morning FE/EIT exam. This is about the same fraction as represented by morning-section questions on subjects such as probability and statistics, material properties, strength of materials, and fluid mechanics, so it is clearly an important subject. Questions in the section on ethics and business practices of the exam may cover the following topics:

- Code of ethics (professional and technical societies)
- Agreements and contracts
- Ethical versus legal behavior
- Professional liability
- Public protection issues (for example, licensing boards)

MORALS, PERSONAL ETHICS, AND PROFESSIONAL ETHICS

To put professional ethics for engineers in perspective, it is helpful to distinguish it from morals and personal ethics. *Morals* are beliefs about right and wrong behaviors that are widely held by significant portions of a given culture. Obviously, morals will vary from culture to culture, and though some are common across different cultures, there seems to be no universal moral code.

Personal ethics are the beliefs that individuals hold that often are more restrictive than and sometimes contradictory to the morals of the culture. An example of personal ethics that might be more restrictive than morals might be the belief of an individual that alcohol should not be consumed in a culture that accepts the use of alcohol.

Professional ethics, on the other hand, is the formally adopted code of behavior by a group of professionals held out to society as that profession's pledge about how the profession will interact with society. Such rules or codes represent the agreed-on basis for a successful relationship between the profession and the society it serves.

The engineering profession has adopted several such codes of ethics, and different practitioners may adhere to or be bound by codes that vary by professional discipline but are similar in their basics. Codes adopted by the state boards of registration are typically codified into law and are legally binding for licensed engineers in the respective state. Codes adopted by professional societies are not legally binding but are voluntarily adhered to by members of those societies. The successful understanding of professional ethics for engineers requires an understanding of various codes; for purposes of examining registration applicants, the NCEES has adopted a "model code" that includes many canons common to most codes adopted nationwide.

CODES OF ETHICS

Codes of ethics are published by professional and technical societies and by licensing boards. Why are codes published and why are they important? These fundamental questions are at the heart of the definition of a "profession." Some important aspects of the definition of a profession might include skills and knowledge vital to society; extensive and intellectual education and training important for proper practice in the profession; an importance of autonomous action by practitioners; a recognition by society of these aspects, leading to a governmentally endorsed monopoly on the practice of the profession; and a reliance on published standards of ethical conduct, usually in the form of a code of ethics (Harris, et al., 2005). Such codes are published and followed to maintain a high standard of confidence in the profession by the public served by the practicing professionals, because without high standards of confidence, the ability of a profession to serve the public need may be seriously impaired.

The FE exam questions on ethics and business practices are based on the NCEES code of ethics, a concise body of model rules designed to guide state boards and practitioners as a model of good practice in the regulation of engineering. These rules do not bind any engineer, but the codes of ethics published by individual state boards and of professional societies will be very similar to these in principle.

NCEES Model Rules of Professional Conduct

A. Licensee's Obligation to Society

1. Licensees, in the performance of their services for clients, employers, and customers, shall be cognizant that their first and foremost responsibility is to the public welfare.
2. Licensees shall approve and seal only those design documents and surveys that conform to accepted engineering and surveying standards and safeguard the life, health, property, and welfare of the public.
3. Licensees shall notify their employer or client and such other authority as may be appropriate when their professional judgment is overruled under circumstances where the life, health, property, or welfare of the public is endangered.
4. Licensees shall be objective and truthful in professional reports, statements, or testimony. They shall include all relevant and pertinent information in such reports, statements, or testimony.
5. Licensees shall express a professional opinion publicly only when it is founded upon an adequate knowledge of the facts and a competent evaluation of the subject matter.
6. Licensees shall issue no statements, criticisms, or arguments on technical matters which are inspired or paid for by interested parties, unless they explicitly identify the interested parties on whose behalf they are speaking and reveal any interest they have in the matters.
7. Licensees shall not permit the use of their name or firm name by, nor associate in the business ventures with, any person or firm which is engaging in fraudulent or dishonest business or professional practices.
8. Licensees having knowledge of possible violations of any of these Rules of Professional Conduct shall provide the board with the information and assistance necessary to make the final determination of such violation. (Section 150, Disciplinary Action, NCEES Model Law)

B. Licensee's Obligation to Employer and Clients

1. Licensees shall undertake assignments only when qualified by education or experience in the specific technical fields of engineering or surveying involved.
2. Licensees shall not affix their signatures or seals to any plans or documents dealing with subject matter in which they lack competence, nor to any such plan or document not prepared under their direct control and personal supervision.
3. Licensees may accept assignments for coordination of an entire project, provided that each design segment is signed and sealed by the licensee responsible for preparation of that design segment.
4. Licensees shall not reveal facts, data, or information obtained in a professional capacity without the prior consent of the client or employer except as authorized or required by law. Licensees shall not solicit or accept gratuities, directly or indirectly, from contractors, their agents, or other parties in connection with work for employers or clients.

5. Licensees shall make full prior disclosures to their employers or clients of potential conflicts of interest or other circumstances which could influence or appear to influence their judgment or the quality of their service.
6. Licensees shall not accept compensation, financial or otherwise, from more than one party for services pertaining to the same project, unless the circumstances are fully disclosed and agreed to by all interested parties.
7. Licensees shall not solicit or accept a professional contract from a governmental body on which a principal or officer of their organization serves as a member. Conversely, licensees serving as members, advisors, or employees of a government body or department, who are the principals or employees of a private concern, shall not participate in decisions with respect to professional services offered or provided by said concern to the governmental body which they serve. (Section 150, Disciplinary Action, NCEES Model Law)

C. Licensee's Obligation to Other Licensees

1. Licensees shall not falsify or permit misrepresentation of their, or their associates', academic or professional qualifications. They shall not misrepresent or exaggerate their degree of responsibility in prior assignments nor the complexity of said assignments. Presentations incident to the solicitation of employment or business shall not misrepresent pertinent facts concerning employers, employees, associates, joint ventures, or past accomplishments.
2. Licensees shall not offer, give, solicit, or receive, either directly or indirectly, any commission, or gift, or other valuable consideration in order to secure work, and shall not make any political contribution with the intent to influence the award of a contract by public authority.
3. Licensees shall not attempt to injure, maliciously or falsely, directly or indirectly, the professional reputation, prospects, practice, or employment of other licensees, nor indiscriminately criticize other licensees' work. (Section 150, Disciplinary Action, NCEES Model Law)

Many ethical questions arise in the formulation of business practices. Professionals should appreciate that expressions like "all is fair in business" and "let the buyer beware" can conflict with fundamental ideas about how a professional engineer should practice. The reputation of the profession, not only the individual professional, is critically important to the ability of all engineers to discharge their duty to protect the public health, safety, and welfare.

The *NCEES Model Rules* addressing a licensee's obligation to other licensees prohibit misrepresentation or exaggeration of academic or professional qualifications, experience, level of responsibility, prior projects, or any other pertinent facts that might be used by a potential client or employer to choose an engineer.

The *Model Rules* also prohibit gifts, commissions, or other valuable consideration to secure work. Political contributions intended to influence public authorities responsible for awarding contracts are also prohibited.

Often, these rules are misunderstood in the arena of foreign practice. Increasingly, engineering is practiced globally, and engineers must deal with foreign clients and foreign governmental officials, many times on foreign soil where laws and especially cultural practices vary greatly. In the United States, the federal Foreign Corrupt Practices Act (FCPA) is a relatively recent recognition and regulation of this problem. Among other purposes, it provides clearer legal boundaries for U.S. engineers involved with international projects.

According to the FCPA, it is not illegal for a U.S. engineer to make petty extortion payments ("grease payments," "expediting payments," and "facilitating payments" are common expressions) to governmental officials when progress of otherwise legitimate projects is delayed by demands for such payments consistent with prevailing practice in that country. It is illegal, however, for U.S. engineers to give valuable gifts or payments to develop contracts for *new business*. In some cultures, reciprocal, expensive gift giving is an important part of business relationships, and the reciprocal nature of this practice can make it acceptable under the FCPA. Most commonly, when the engineer's responsibilities include interactions with foreign clients or partners, the engineer's corporate employer will publish detailed and conservative guidelines intended to guide the engineer in these ethical questions.

The practicing engineer should always be watchful of established and, especially, new business practices to be sure the practices are consistent with the codes of ethics he or she is following.

Example 5.1

The *NCEES Model Rules of Professional Conduct* allow an engineer to do which one of the following?

(i) Accept money from contractors in connection with work for an employer or client

(ii) Compete with other engineers in seeking to provide professional services

(iii) Accept a professional contract from a governmental body even though a principal or officer of the engineer's firm serves as a member of the governmental body

(iv) Sign or seal all design segments of the project as the coordinator of an entire project

Solution

Although the other items are not allowed by the *Model Rules*, nowhere does it say that an engineer cannot compete with other engineers in seeking to provide professional services. But, of course, he or she should conduct business in an ethical manner. The correct answer is (ii).

AGREEMENTS AND CONTRACTS

One aspect of business practice is understanding the concepts and terminology of agreements and contracts.

Elements of a Contract

Contracts may be formed by two or more parties; that is, there must be a party to make an offer and a party to accept.

To be enforceable in a court of law, a contract must contain the following five essential elements:

1. There must be a mutual agreement.
2. The subject matter must be lawful.

3. There must be a valid consideration.
4. The parties must be legally competent.
5. To be a formal contract, the contract must comply with the provisions of the law with regard to form.

A *formal contract* depends on a particular form or mode of expression for legal efficacy. All other contracts are called *informal contracts* since they do not depend on mere formality for their legal existence.

Contract and Related Legal Terminology

Case law—the body of law created by courts interpreting statute law. Judges use precedents, the outcome of similar cases, to construct logically their decision in a given issue.

Changed or concealed conditions—in construction contracting, it is important to specify how changed or concealed conditions will be handled, usually by changes in the contract terms. For example, if an excavation project is slowed by a difficult soil pocket between soil corings, the excavation contractor may be able to support a claim for increased costs due to these unforeseen conditions. When the concealed conditions are such that they should have been foreseen, such claims are more difficult to support.

Common law—the body of rules of action and principles that derive their authority solely from usage and customs.

Damages for delays—in many contracts, completion time is an important concern, and contractual clauses addressing penalties for delays (or rewards for early completion) are often incorporated.

Equal or approved equivalent—terms used in specifications for materials to permit use of alternative but equal material when an original material is not available or an equivalent material can be obtained at lower cost. The engineer is responsible for approving the alternative material.

Equity—system of doctrines supplementing common and statute law, such as the Maxims of Equity.

Errors and omissions—term used to describe the kind of mistakes that can be made by engineers and architects leading to damage to the client. Often, this risk is protected by liability insurance policies.

Force account—a method of work by which the owner elects to do work with his or her own forces instead of employing a construction contractor. Under this method, the owner maintains direct supervision over the work, furnishes all materials and equipment, and employs workers on his or her own payroll.

Hold harmless—clauses are often included requiring one party to agree not to make a claim against the other and sometimes to cooperate in the defense of the other party if a claim is made by a third party.

Incorporate by reference—the act of making a document legally binding by referencing it within a contract, although it is not attached to or reproduced in the contract. This is done to eliminate unnecessary repetition.

Indemnify—to protect another person against loss or damage, as with an insurance policy.

Liquidated damages—a specific sum of money expressly stipulated as the amount of damages to be recovered by either party for a breach of the agreement by the other.

Mechanics' liens—legal mechanism by which unpaid contractors, suppliers, mechanics, or laborers are allowed to claim or repossess construction materials that have been delivered to the worksite in lieu of payment.

Plans—the drawings that show the physical characteristics of the work to be done. The plans and specifications form the guide and standards of performance that will be required.

Punitive damages—a sum of money used to punish the defendant in certain situations involving willful, wanton, malicious, or negligent torts.

Specifications—written instructions that accompany and supplement the plans. The specifications cover the quality of the materials, workmanship, and other technical requirements. The plans and specifications form the guide and standards of performance that will be required.

Statute law—acts or rules established by legislative action.

Statute of limitations—a time limit on claims resulting from design or construction errors, usually beginning with the date the work was performed, but in some cases beginning on the date the deficiency could first have been discovered.

Surety bond—bonds issued by a third party to guarantee the faithful performance of the contractor. Surety bonds are normally used in connection with competitive-bid contracts, namely, bid bonds, performance bonds, and payment bonds.

Workers' compensation—insurance protecting laborers and subcontractors in case of an on-the-job injury; it is often required of contractors.

ETHICAL VERSUS LEGAL BEHAVIOR

Engineers have a clear obligation to adhere to all laws and regulations in their work—what they do must be done legally. But the obligation goes beyond this. Unlike the world of business where cutthroat but legal practices are commonly condoned and frequently rewarded, engineers assume important obligations to the public and to the profession that restrict how they must practice and that often are much more stringent than law or regulation.

When you realize that restricting the practice of engineering to certain licensed professionals by the state is essentially a state-provided monopoly, you may begin to see why there is a difference. Competitive businesses compete in many ways to gain the kind of advantage in their field that engineers and other licensed professionals are given by the state.

Aggressive advertising is one example of a business practice that engineers avoid, even though it is not illegal or prohibited. Before 1978, it was common for professional societies to prohibit or narrowly restrict advertising by their practitioners; however, in 1978 the U.S. Supreme Court ruled such broad restrictions unconstitutional, allowing only reasonable restraints on advertising by professional societies. Since that time, engineering societies have adopted guidelines on advertising. Other professions have been less successful in regulating advertising. For example, the profusion of television advertising by lawyers, and the language of those advertisements, contrasts with the practice of engineering professionals where advertising is more commonly seen in technical journals or trade literature. Many believe the legal profession has suffered a loss of respect as a result, while the profession of engineering still is held in high regard by the public. It is in the interest of the engineering profession to avoid this kind of advertising, even though it is legal, because it can damage the reputation of the profession.

Another example of the importance of self-regulation is the engineer's responsibility to the environment. Although many laws and regulations restrict engineering practices that might damage the environment, there are still many legal ways to accomplish engineering projects that can have adverse environmental effects. Increasingly, codes of ethics are adding requirements for the engineer to consider the environment or the "sustainability" of proposed engineering projects. The engineer's ethical responsibility to work toward sustainable development may go beyond any legal requirements intended to prevent environmental damage.

Conflicts of Interest

A conflict of interest is any situation where the decision of an engineer can have some significant effect on his or her financial situation. It would be a clear conflict of interest for a designing engineer to specify exclusively some component that is only available from a supplier in which that engineer has a significant financial interest, when other components from other suppliers would serve equally well. Engineers must avoid even the *appearance* of a conflict of interest. This is critically important for the reputation of the profession, which the engineer is charged with protecting, in order for engineers to effectively serve the public interest.

An apparent conflict of interest is any situation that might appear to an outside observer to be an actual conflict of interest. For example, if the engineer in the case mentioned above had subsequently divested himself of all interest in the supplier, there is no longer an actual conflict of interest. However, to an outside observer with imperfect information, there might be the appearance of a conflict, resulting in the perception of unethical behavior in the public eye.

The usual remedy for conflicts of interest and apparent conflicts of interest is disclosure and, often, recusal. The engineer's interest must be disclosed in advance, generally to a supervisor, and recusal must at least be discussed. In many cases, recusal may not be necessary, but disclosure is vitally important. In every case, the public perception of the conflict must be considered, with the goal of protecting the reputation of the individuals and the profession.

PROFESSIONAL LIABILITY

Good engineering practice includes numerous checks and conservative principles of design to protect against blunders, but occasional errors and omissions can result in damage or injury. The engineer is responsible for such damage or injury, and it is good practice to carry errors and omissions insurance to provide appropriate compensation to any injured party, whether a client or a member of the public. Such insurance can be a significant cost in some fields of engineering, but it represents a cost of doing business that should be reflected in the fees charged. The most important factor in preventing errors and blunders is to provide adequate time for careful review of all steps in the project by knowledgeable senior licensed engineers. Frantic schedules and unrealistic deadlines can significantly increase the risk.

PUBLIC PROTECTION ISSUES

State boards in all 50 states and the District of Columbia are charged by their states with the responsibility for the licensing of engineers and the regulation of the practice of engineering to protect the health, safety, and welfare of the public. Licensed engineers in each state are legally bound by laws and regulations published by the respective state board. The boards are generally made up of engineers appointed by the state governor; sometimes nonengineering members also are appointed to make sure the public is adequately represented.

State boards commonly issue cease and desist letters to nonengineers who have firms or businesses with names that imply engineering services are being offered to the public or who may actually be offering engineering services without the required state license. These boards also regulate the practice of engineering by their registrants, often sanctioning registrants for inappropriate business practices or engineering design decisions. Many boards require continuing education by registrants for maintenance of proficiency. A weakness of many boards is in the area of discipline for incompetent practices, but this weakness is often offset by tort law whereby incompetent practitioners who cause damage or injury are commonly subject to significant legal damages.

CONTRACTS AND CONTRACT LAW

Bidding

The bid process begins with a Notice to Bidders, indicating the scope and location of the project, client, availability of plans and specifications to be used, date and location of submission, and bond requirements. Bidders may be required to attend a prebid meeting/site visit to ask questions, clarify scope, and see the actual project site.

Interested contractors will estimate material and labor costs, time and equipment requirements, and overhead and profit desired. Upon completion of the bid, it must be submitted at the proper location no later than the time required. Late submissions will be rejected.

Bonding

On all public construction projects, and most private projects, three types of bonds are obtained by contractors prior to submitting a bid. These are obtained from a surety company, and consist of the following:

- *Bid bond.* Typically 5% to 20% of the estimated project cost. This guarantees that the bidder will enter into a contract with the client if the bidder is the low bidder. If the low bidder does not sign a contract, the amount of the bid bond is used by the client to either offset the cost of the next lowest bid or to offset the cost of rebidding the project.
- *Performance bond.* Typically 100% of the project cost. This guarantees that the contractor will perform the specified work in accordance with the contract. If the contractor defaults on the contract, the bond is the upper amount that the surety company will incur to arrange completion of the project.
- *Payment (labor and material payment) bond.* Typically 100% of the project cost. This guarantees that the contractor will pay for all materials and labor used on the project, protecting the client from liens against the project by third parties.

REFERENCES

Harris, Charles E., Jr., Michael S. Pritchard, and Michael J. Rabins. *Engineering Ethics: Concepts and Cases*. Thompson Wadsworth, 2005.

National Council of Examiners for Engineering and Surveying. *Model Rules*, September 2006.

PROBLEMS

5.1 Jim is a PE working for an HVAC designer who often must specify compressors and other equipment for his many clients. He reports to Joan, the VP of engineering. Jim specifies compressors from several different manufacturers and suppliers based on the technical specifications and on his experience with those products in past projects. Joan's long-time friend Charlie, who has been working in technical sales of construction materials, takes a new job with one of the compressor suppliers that Jim deals with from time to time. Charlie calls on Joan, inviting her and any of her HVAC designers to lunch to discuss a new line of high-efficiency compressors; Joan invites Jim to come along. Jim should:

a. decline to attend the lunch, citing concerns about conflict of interest
b. agree to attend the lunch but insist on paying for his own meal
c. agree to attend the lunch and learn about the new line of compressors
d. report Joan to the state board and never specify compressors from that supplier again

5.2 Harry C. is an experienced geotechnical engineer who has many years' experience as a PE designing geotechnical projects and who is very familiar with the rules regarding the requirement for trench shoring and trench boxes to protect construction workers during excavations. During a vacation visit to a neighboring state, he observes a city sewer construction project with several workers in an unprotected deep trench, which, to Harry's experienced eye, is probably not safe without a trench box or shoring. Harry should:

a. remember that he is not licensed in the neighboring state and has no authority to interfere
b. approach the contractor's construction foreman and insist that work be halted until the safety of the trench is investigated
c. advise all the workers in the trench that they are in danger and encourage them to go on strike for safer conditions
d. contact the city engineer to report his concerns

5.3 Engineering student Travis is eagerly anticipating his graduation in three months and has interviewed with several firms for entry-level employment as an electrical engineer. He has received two offers to work for firms A and B in a nearby city, and after comparing the jobs, salaries, and benefits and discussing the choice with his faculty advisor, he telephones firm A whose offer is more appealing and advises them he will accept their job offer. Two weeks later he is contacted by firm C in a different city with a job offer that includes a salary more than 15% higher than the offer he has accepted plus a generous relocation allowance. Travis should:

a. decline the offer from firm C, explaining that he has already accepted a position
b. contact firm A and ask if he can reconsider his decision
c. contact firm A to give them a fair chance to match the offer from firm C
d. advise firm C that he can accept their offer if they will contact firm A to inform them of this change

5.4 EIT Jerry works for a small civil engineering firm that provides general civil engineering design services for several municipalities in the region. He has become concerned that his PE supervisor Eddie is not giving careful reviews to Jerry's work before sealing the drawings and approving them for construction. Jerry asks Eddie to review with him the design assumptions from Jerry's latest design, a steel fire exit staircase to be added to an elementary school building, because he has concerns about the appropriate design loadings. But before the design assumptions are reviewed, Jerry notices the drawings have been approved and released to the fabricator. Jerry should:

a. quit his job and find another employer
b. take a review course in live loadings for steel structures
c. in the future mark each drawing he prepares "Not Approved for Construction"
d. None of the above

5.5 Dr. Willis Hemmings, PE, is an engineering professor whose research in fire protection engineering is nationally recognized. He is retained as an expert witness for the defendant, a structural engineering design firm, in a lawsuit filed by a firefighter who was injured while fighting a fire in a steel structure that collapsed during the fire. The plaintiff's lawyer alleges that the original design of certain components of the fire protection system protecting the steel structural members was inadequate. Hemmings reviews the original design documents, which call for a protective coating that is slightly thinner than is required by the local building code. Hemmings testifies that even though the specified coating is thinner than required, he believes that the design was sound because the product used is applied by a new process that is probably more efficient and the thinner coating probably gave the same level of protection. He bases his testimony on his national reputation as an expert in this field. Such expert testimony is:

a. a commonly accepted method of certifying good engineering design in tort law
b. legal only when given by a licensed professional engineer like Hemmings
c. unethical because it contradicts accepted practice without supporting tests or other data
d. effective only because Hemmings is involved in cutting-edge research

5.6 Jackie is a young PE who works for a garden tool manufacturer that has produced about 100,000 shovels, rakes, and other garden implements annually for more than 20 years. The company recently won a contract to manufacture and supply 5000 folding entrenching tools of an existing design to a Central American military client. The vice president of marketing has been working to develop contracts with other military clients and asks Jackie to prepare a statement of qualifications (SOQ). Jackie is asked to describe the design group (consisting of two engineers, one EIT, one student intern, three CAD technicians, and one IT technician) as a "team of eight tool design engineers," and to describe the company as "experienced in the design, testing, and manufacturing of military equipment, with a recent production history of over two million entrenching tools and related hardware." Jackie should:

a. check the production records to be sure the figures cited are accurate
b. ask the vice president to sign off on the draft of the SOQ
c. object to describing the qualifications and experiences of her group in an exaggerated way
d. be sure to mention that she is a PE and list the states in which she is licensed

5.7 The Ford Motor Company paid millions of dollars to individuals injured and killed in crashes of the Ford Pinto, which had a fuel tank and filler system that sometimes ruptured in rear-end collisions, spilling gasoline and causing fires. While many considered the filler system design deficient because of this tendency, one important factor played a role in the lawsuits. An internal Ford memo was discovered that included the cost-benefit calculations Ford managers used in making the decision not to improve the tank/filler system design. This memo was significant because:

a. it is unethical to use the cost-benefit method for safety-related decisions
b. it is illegal to estimate the value of human life in cost-benefit calculations
c. state law requires estimates of the value of human life be at least $500,000 in such calculations
d. None of the above

5.8 Charles is tasked to write specifications for electric motors and pumps for a new sanitary sewage treatment plant his employer is designing for a municipal client. Charles is concerned that he doesn't have a very good knowledge about current pump design standards but is willing to learn. Charles's fiancée is an accountant employed by a pump distributor and offers to provide Charles with a binder of specifications for all the pumps her firm distributes. Charles should:

a. decline to accept the binder, citing concerns about conflict of interest
b. accept the binder but turn it over to his employer's technical librarian without reviewing it
c. accept the binder and study the materials to gain a better understanding of pump design and specifications
d. ask his fiancée if she knows an applications engineer at her firm who would draft specifications for him

5.9 Professor Martinez is a PE who teaches chemical engineering classes at a small engineering school. His student, Erica, recently graduated and took a job with WECHO, a small firm that provides chemicals and support to oil well drilling operations. WECHO has never employed an engineer and has hired Erica, partly on Prof. Martinez's strong recommendation, in hopes that she will one day become their chief engineer. After she has worked at WECHO for about two years, her supervisor Harry calls Prof. Martinez to explain that WECHO has been required to complete an environmental assessment before deploying a new surfactant, and the assessment must be sealed by a PE. Harry explains that Erica has done all the research to collect data and answer questions on the assessment, and everyone at WECHO agrees that she has done a superb job in completing the assessment, but it still requires the seal of a PE before submission. Harry asks Prof. Martinez if he can review Erica's work and seal the report, reminding him that he had given a glowing recommendation of Erica at the time WECHO hired her. Prof. Martinez should:

a. negotiate a consulting contract to allow him sufficient time and funding to review the report before sealing it
b. require Erica to first sign the report as an EIT and graduate engineer before reviewing it
c. require WECHO to purchase a bond against environmental damage before sealing the report
d. decline to review or seal the report, citing responsible charge issues

5.10 Jack Krompten, PE, is an experienced civil engineer working for a land development firm that has completed several successful residential subdivision developments in WoodAcres, a suburban bedroom community of a large, sprawling, and rapidly growing city. The WoodAcres city engineer, who also served half-time as the mayor, has retired, and the city council realizes that with rapid growth ahead it will be important to hire a new city engineer. They approach Krompton with an offer of half-time city engineer, suggesting that he can keep his current job while discharging the responsibilities of the city engineer—primarily reviewing plans for future residential subdivision developments in WoodAcres. Krompton should:

a. recognize that by holding two jobs he is being paid by two parties for the same work
b. insist that he can only accept the offer if his present employer agrees to reduce his responsibilities to half-time
c. recognize that a 60-hour workweek schedule will take time away from his family
d. recognize that this arrangement will probably create a conflict of interest and refuse the offer

5.11 Willis is an aerospace engineering lab test engineer who works for a space systems contractor certifying components for spacecraft service. He is in charge of a team of technicians testing a new circuit breaker design made of lighter weight materials intended for service in unpressurized compartments in rockets and spacecraft. The new design has passed all tests except for some minor overheating during certain rare electrical load conditions. The lead technician notices that this overheating does not occur when a fan is used to cool the test apparatus and proposes to run the test with the fan to complete the certification process. He points out that the load conditions will only occur during thruster operation in space, which is a much colder environment. Willis should:

a. agree to the lead technician's suggestion, since he has many years of experience in testing and certification
b. agree to run the test as suggested but include a footnote explaining the use of the fan
c. insist on running the test as specified without the use of the fan
d. report the technician to the state board for falsifying test reports

5.12 Shamar is a registered PE mechanical engineer assigned as a project manager on a new transmission line project. He is tasked to build a project team to include several engineers and EITs that will be responsible for design and construction of 7.6 miles of high tension transmission lines consisting of steel towers and aluminum conductors in an existing right of way. He realizes that foundation design and soil mechanics will be an important technical area to his project, and he has never studied these subjects. He wonders if he is qualified to supervise such a project. He should:

a. meet with his supervisor to decline the assignment
b. decline the assignment and contact the state board to report that he is being asked to take responsibility for tasks he is not knowledgeable about
c. accept the assignment and check out an introductory soil mechanics textbook from the firm's technical library
d. accept the assignment and be sure his team includes licensed engineers with expertise in these areas

5.13 Julio is a design engineer working for a sheet metal fabricating firm. He is tasked with the design of a portable steel tank for compressed air to be mass produced and sold to consumers for pressurizing automobile tires. He designs a cylindrical tank to be manufactured by rolling sheet metal into a cylinder, closing with a longitudinal weld along the top, and welding on two elliptical heads. His design drawings are approved by his supervisor, Sonja, a licensed engineer, and by the vice president of manufacturing, but when the client reviews the designs, he asks the VP to change the design so that the longitudinal weld along the top is moved to the bottom where it will not be visible to improve the esthetics and marketability of the product. The VP agrees with this change. Julio learns of this change and objects, citing concerns about corrosion at the weld if it is on the bottom. Sonja forwards Julio's objection with a recommendation against the change to the VP, with a copy to the client, but the VP insists, saying esthetics is very important in this product. Julio should:

a. accept the fact that esthetics governs this aspect of the design
b. write a letter to the client stating his objections
c. put a clear disclaimer on the drawing indicating his objections
d. contact the state board to report that his recommendation has been overruled by the VP

5.14 William is a PE who designs industrial incineration systems. He is working on a system to incinerate toxic wastes, and his employer has developed advanced technology using higher temperatures and chemical-specific catalyst systems that minimize the risk to the environment, workers, and the public. A public hearing is scheduled to address questions of safety and environmental risk posed by the project, and William is briefed by the corporate VP for public affairs about how to handle questions from the public. He is told to buy a new suit, project an air of technical competence, point out that his firm is the industry leader with many successful projects around the world, and describe the proposed system as one with "zero risk" to the public. William should:

a. follow his instructions to the letter
b. insist that his old suit is adequate, because he refuses to appear more successful than he really is, but follow the other instructions
c. follow all instructions, except use the term "minimal risk" rather than "zero risk"
d. resign from his position and look for a different employer who won't ask him to face the public

5.15 Darlene is a metallurgical engineering EIT who works for a firm that manufactures automotive body panels. She has been tasked with improvements to the design of inner fender and trunk floor panels to reduce corrosion damage. After several weeks of study and comparison of alternatives, she submits a new trunk floor panel design utilizing a weldable stainless steel that will significantly reduce corrosion compared to the galvanized carbon steel alternatives she has been considering. The new panels will cost more, however, and after much study and debate, the VP of manufacturing rejects her design and approves an alternative made of a cheaper material. Darlene should:

a. accept the decision and work to finalize a workable design
b. resign from her position, since her employer has lost confidence in her
c. contact the state board to advise that her design decision has been overturned by a nonengineering manager
d. None of the above

5.16 Matt is a young PE who has just started his own consulting practice after six years of work with a small consulting firm providing structural engineering design services to architects. He has worked on steel and timber framed churches, prestressed and reinforced concrete parking structures, and many tilt-up strip center buildings. His expertise has been in the area of design of tilt-up concrete construction, where he has developed some innovative details regarding reinforcement at lifting points. His building designs, when constructed by experienced contractors, have reduced construction times and costs. Because of his expertise, he is approached by lawyer Marlene, who tells Matt that she represents a construction worker who is suing a project owner, contractor, and designer over a construction accident in which a tilt-up wall was dropped during construction, seriously injuring several workers. Marlene asks Matt to serve as an expert witness to assess the design and construction practices in the project and testify as to the causes of the accident. Marlene has taken the case on a contingency fee basis, in which she will earn 40% of any settlement, and she asks Matt if he would rather be paid by the hour for his study and testimony or instead accept 5% of any settlement, which she believes could be as high as $25 million. Matt should:

a. compare the 5% contingency fee with an expected fee based on his hourly rates, realizing that there is some chance he will earn nothing
b. be sure to have Marlene put the contingency fee arrangement in the form of a legal contract
c. decline the contingency fee arrangement and bill on an hourly basis
d. accept the contingency fee arrangement but donate the difference over his hourly rate to charity

5.17 Victor is a consulting engineer who is also in charge of a crew providing land surveying and subdivision design services to developers. He has been contracted to provide a survey of a 14-acre tract where a local developer is contemplating a subdivision, and he realizes that his crew had surveyed this same tract last month for another developer who has abandoned the project. He reprints the survey drawings, changing the title block for the new client. With respect to billing for the drawings, Victor should:

a. bill the new client the same as he billed the original client to be fair to both
b. bill the new client for half of the amount billed to the original client
c. bill the new client only for any work he did to change the drawings and reprint them
d. provide the new client with the drawings without any charge

5.18 Frank is a PE who works for ELEC, an electrical engineering design and construction firm. Frank's job is estimating construction costs, bidding construction projects, and supervision of design of electrical systems for buildings. Frank's bright EIT of five years, Linda, has just received her PE license and has resigned her position to open her own consulting business in a nearby community. Until she is replaced, Frank will have to also do all detail design of electrical systems for their projects. Frank receives a request for proposal (RFP) from a general contractor regarding design of electrical systems for a local independent school district. He realizes that his firm may be in competition with Linda for the engineering design, and knowing that Linda's salary was about half of his, he expects she may have a competitive advantage. Frank should:

a. ask Linda not to bid on this project
b. remind his contact with the general contractor that Linda has just left his firm, that she is inexperienced, and hint that she was sometimes slow to complete her design assignments
c. emphasize his 18 years of experience and subsequent design efficiency in his proposal
d. promote a CAD technician to a designer position so he can show a lower billing rate for engineering design hours

5.19 It is important to avoid the appearance of a conflict of interest because:

a. the engineer's judgment might be adversely affected
b. the engineer's client might suffer financial damages
c. the appearance of a conflict of interest is a misdemeanor
d. the appearance of a conflict of interest damages the reputation of the profession

5.20 The code of ethics published by the American Society of Civil Engineers is:

a. legally binding on all licensed engineers practicing civil engineering
b. adhered to voluntarily by members of the ASCE as a condition of membership
c. legally binding on all engineers with a degree in civil engineering
d. published only as a training guideline for young civil engineers

5.21 Which statement *MOST* accurately describes an engineer's responsibility to the environment?

a. The engineer has a legal obligation to make sure all development is sustainable.
b. The engineer has no obligation to the environment beyond protecting public health and safety.
c. The engineer has a moral obligation to consider the impact of his or her work on the environment.
d. The engineer's environmental responsibility is primarily governed by specific state laws.

SOLUTIONS

5.1 **c.** Jim can accept this invitation. We can assume there is no corporate policy prohibiting or restricting lunch invitations since VP Joan has accepted the invitation; therefore, there is no reason for Jim to decline the invitation. The opportunity to learn more about the new product is useful to him, his employer, and his clients; the cost of the lunch presumably would not be considered a "valuable" gift; and the lunch would not create either a conflict of interest or the appearance of a conflict to a reasonable person. If instead of lunch the offer involved a 10-day elk hunting trip or a vacation in the south of France, the solution would be very different because of the obvious "value" of the gift.

5.2 **d.** Doing nothing (a) is not an option if Harry really believes the trench represents a serious hazard to the workers. His code of ethics requires him to remember that his first and foremost responsibility is the public welfare, which includes the safety of the construction workers. Answers (b) and (c) are not the best way to proceed; his concerns should be reported to an engineer with some authority over the project. Since this is a city-contracted sewer improvement project, the city engineer will have project responsibility and will be the appropriate individual for Harry to take his concerns to.

5.3 **a.** Travis should decline the offer from firm C, explaining that he has already accepted an offer to work for firm A. While the *NCEES Model Rules* don't specifically address issues of personal integrity, it is clear that the engineer's obligation to employer and clients will not be satisfied by any decision that ignores Travis's verbal agreement to employment with firm A. Furthermore, such actions will tarnish his integrity in the eyes of firm A and by implication will harm the reputation and credibility of other students and the profession.

5.4 **d.** None of the first three solutions will address the concerns Jerry has raised about the safety of the particular project in question, so (d) is the correct solution. He should instead meet with Eddie to discuss the details of his design and make a determination if the design is completed safely. If it isn't, he will need to take further action to stop fabrication and construction while the design is reviewed and possibly modified to address any deficiencies. After this is done, he might want to consider all three of the other choices for his future. If he is thwarted in these responsibilities by Eddie, he should contact the state board with his concerns.

5.5 **c.** Hemmings cannot offer expert opinion that is contrary to accepted engineering practice without supporting that opinion with computer modeling, lab test results, or study of the literature. He can offer expert opinion that the design is not in line with accepted engineering practice without any supporting calculations, but he can't maintain that a substandard practice is acceptable without rational supporting evidence. Hemmings's credentials and experience may qualify him as an expert, but they do not relieve him of the requirement to base his professional opinion on facts.

5.6 **c.** Jackie should object to the request to exaggerate the size, qualifications, and experience level of her design group. The *Model Rules* require engineers to "be objective and truthful" in all professional matters, and the suggested exaggerations are clearly in conflict with this requirement.

5.7 **d.** It is not unethical or illegal to use the cost-benefit method, nor are certain values for human life prescribed by law; the answer is none of the above. The assumed values and calculations in the memo may have appeared callous or inflammatory to the juries in the resulting lawsuits, but they were not unethical or illegal. They may have been imprudent—an important lesson is that the public (jury) apparently objected to a design decision that increased the risk of a post-crash fire for such a small net benefit. Even though risk of death had been considered by the designers, the mode of death (burning to death in otherwise survivable crashes) was a factor in the strong reactions by the juries.

5.8 **c.** Studying products from a distributor is perfectly acceptable and can be a good way to gain a better understanding of pump equipment on the market. Accepting the binder does not represent a conflict of interest as implied by choices (a) and (b). Option (d) would be clearly setting himself up for an apparent conflict of interest.

5.9 **d.** Since Prof. Martinez has not been in responsible charge of the development of the assessment, he can't seal it, regardless of the level of review or the capabilities of the EIT who has done the work. Engineering students should be cautious in accepting a position as the sole engineering employee in a small firm; they are first encouraged to gain experience as an EIT under the guidance of an experienced PE and qualify for licensure as a PE before taking a position where licensure might be needed.

5.10 **d.** Krompton should recognize that the proposed arrangement will create a conflict of interest by placing him in charge of reviewing and approving plans from developers and potential developers that are in direct competition with his own employer. Any decision by him that might tend to make development less profitable for other developers could be advantageous to his employer by making their services more cost effective. Even if he were able to make all decisions rationally and without bias, the appearance of a conflict of interest would be very real and would cause a loss of credibility in the city engineer's actions and damage the reputation of the engineering profession.

5.11 **c.** Willis should insist on running the test as specified. The technician's suggestion to use a fan to "fudge" the test is technically indefensible, as well as unethical, in any case—the fan simulates convection cooling, which does not occur in the vacuum of space. Even if the technician had suggested a way to simulate an increased radiative heat flux, changes to a specified test procedure are not made casually. Much more study, documentation, and higher level approvals are involved.

5.12 d. Shamar should accept the assignment and select team members so that all areas of needed expertise are represented by licensed individuals who can seal appropriate portions of the plans. Option (a) is not recommended for an engineer's career advancement—he is expected to accept assignments of increasing responsibility; (b) is detrimental to his career—he will cause unnecessary concern with the board and with his supervisors; and (c) studying an introductory soil mechanics textbook may give him a better understanding of the problem but probably won't qualify him to seal foundation plans for the project described.

5.13 d. Julio is objecting because he knows water accumulates in compressed air storage tanks and causes corrosion, particularly at the bottom where the water will collect. Because of the metallurgy at the weld, corrosion is more aggressive at the weld site, and Julio considers this fundamentally a bad design if the weld is at the bottom. Since his recommendation (based on technical reasons and an increased risk to the public) is overruled under circumstances where the safety of the public is endangered, he is obligated (by the NCEES rules) to "notify his employer or client and other such authority." Having already notified his employer (through Sonja and the VP of manufacturing) and the client, all of whom except Sonja are part of the problem, his next logical step is to contact the state board with his concerns. This is rarely necessary; in most cases, the client and VP will be very interested in Julio's objection as it is based on public safety and will serve to reduce the company's own liability. However, if the employer and VP of manufacturing persist as described here, Julio must take additional action. It is not unreasonable to expect that Julio may face some sort of sanction from a management team that has put him in this position. He may even need a lawyer as this unpleasant situation deteriorates. Whistle-blowing should be considered the solution of last resort, as in this case.

5.14 c. William should avoid the use of the term "zero risk"; he knows that no project has zero risk. He should instead look for ways to quantify the risk that will be informative and meaningful to the public and try to convey the attitude of concern for minimizing the risk consistent with the potential public benefits of the project (increased employment and tax base). He should not do (a), and (b) does not address the problem of misinforming the public about the risk. There is no need to resign; to do so will not help his career or the project.

5.15 a. Darlene should accept the decision and work to make the chosen design successful and profitable for her employer. She does not need to resign—the business decision probably does not reflect a lack of confidence. It was made based on costs and profitability, not public safety, so she should not contact authorities to complain that a manager has overruled her. This kind of decision should be considered a "management decision" because it affects business and profits. Decisions that adversely affect public health, safety, or welfare should be considered "engineering decisions," and when these are overturned by nontechnical managers for other reasons, the engineer may be justified or even obligated to report this to authorities.

5.16 **c.** Matt should decline the contingency fee arrangement. While lawyers commonly work on contingency fee arrangements, engineers can't do this. Any contingency fee arrangement would put the engineer in a conflict of interest situation, where his engineering judgment can influence his income. In such a situation, his engineering judgment may not be sound or will at least appear to be conflicted to an outside observer.

5.17 **c.** Victor should be cautious in making sure that no additional surveying or resurveying is needed because of any changes to the tract. If additional fieldwork is not needed, he should bill the new client only for the work required to edit and reprint the drawings. He should not bill the new client the same as the original client, because that would be billing two clients for the same work, which is specifically prohibited by the *Model Rules*. Billing for half of the original amount is the same, just for an arbitrary amount. When two clients require the same survey simultaneously, it may make sense for Victor to facilitate a partnership in the project, but this is not always feasible when one client has already been billed for work done.

5.18 **c.** Option (c) is the only ethical and practical solution listed. Option (a) asking a competitor not to bid is not practical. Option (b) starting rumors about Linda's capabilities is clearly unethical. And option (d) is troublesome—experienced CAD technicians can do some aspects of design if closely supervised by a PE, but it isn't really necessary to make such a promotion just to show a lower billing rate. The billing rate for engineering design could be maintained at the same (competitive) level as when Linda was employed, even if the actual design work is done by Frank at twice the salary until a replacement for Linda can be hired. This in itself is not unethical, but if Frank does not budget sufficient time to do the design work in addition to his other work, it becomes a question of ethics. An engineer must allow sufficient time to do a professional job. The most practical and desirable solution is not listed—Frank should expedite hiring a qualified replacement for Linda, and he might have to consider declining the opportunity to bid on some projects until she is replaced.

5.19 **d.** Even the appearance of a conflict of interest can damage the reputation of the individuals involved and the profession as a whole, and such situations should be avoided. If there is no actual conflict of interest, the engineer's judgment will not be affected and the client will not suffer damages; nor is it criminal.

5.20 **b.** Codes of ethics published by professional societies are voluntarily adhered to by membership. They don't carry the weight of law but are much more than training guidelines—members who do not adhere to the society's code can be sanctioned by the society or forfeit their membership.

5.21 **c.** Many professional societies require the engineer to "consider" the impact on the environment. Some say he or she should consider whether the development is "sustainable." Most legal restrictions only require the engineer to prevent certain kinds of environmental damage and do not require sustainable development. There are, of course, specific state laws that must be followed, but many federal laws and regulations also apply. The engineer's responsibility is broader than laws and regulations in any event, making choice (c) the best answer.

CHAPTER 6

Engineering Economics

OUTLINE

This is a review of the field known as **engineering economics**, **engineering economy**, or **engineering economic analysis**. Since engineering economics is straightforward and logical, even people who have not had a formal course should be able to gain sufficient knowledge from this chapter to successfully solve most engineering economics problems.

There are 30 example problems throughout this review. These examples are an integral part of the review and should be examined as you come to them.

The field of engineering economics uses mathematical and economic techniques to systematically analyze situations that pose alternative courses of action. The initial step in engineering economics problems is to resolve a situation, or each possible alternative in a given situation, into its favorable and unfavorable consequences or factors. These are then measured in some common unit, usually money. Factors that cannot readily be equated to money are called **intangible** or **irreducible** factors. Such factors are considered in conjunction with the monetary analysis when making the final decision on proposed courses of action.

CASH FLOW

A cash flow table shows the *money consequences* of a situation and its timing. For example, a simple problem might be to list the year-by-year consequences of purchasing and owning a used car:

Year	Cash Flow	
Beginning of first year 0	–$4500	Car purchased now for $4500 cash. (The minus sign indicates a disbursement.)
End of year 1	−350	Maintenance costs are $350 per year.
End of year 2	−350	
End of year 3	−350	
End of year 4	−350 +2000	This car is sold at the end of the fourth year for $2000. (The plus sign represents the receipt of money.)

This same cash flow may be represented graphically, as shown in Figure 6.1. The upward arrow represents a receipt of money, and the downward arrows represent disbursements. The horizontal axis represents the passage of time.

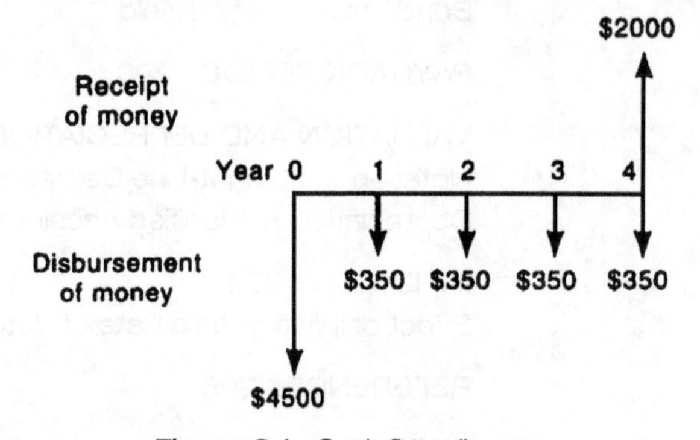

Figure 6.1 Cash flow diagram

Example **6.1**

In January 2003, a firm purchased a used copier for $500. Repairs cost nothing in 2003 or 2004. Repairs were $85 in 2005, $130 in 2006, and $140 in 2007. The machine was sold in 2007 for $300. Complete the cash flow table.

Solution

Unless otherwise stated, the customary assumption is a beginning-of-year purchase, followed by end-of-year receipts or disbursements, and an end-of-year resale or salvage value. Thus the copier repairs and the copier sale are assumed to occur at the end of the year. Letting a minus sign represent a disbursement of money and a plus sign a receipt of money, we are able to set up the cash flow table:

Year	Cash Flow
Beginning of 2003	–$500
End of 2003	0
End of 2004	0
End of 2005	–85
End of 2006	–130
End of 2007	+160

Notice that at the end of 2007, the cash flow table shows +160, which is the net sum of –140 and +300. If we define year 0 as the beginning of 2003, the cash flow table becomes:

Year	Cash Flow
0	–$500
1	0
2	0
3	–85
4	–130
5	+160

From this cash flow table, the definitions of year 0 and year 1 become clear. Year 0 is defined as the *beginning* of year 1. Year 1 is the *end* of year 1, and so forth.

TIME VALUE OF MONEY

When the money consequences of an alternative occur in a short period of time—say, less than one year—we might simply add up the various sums of money and obtain the net result. But we cannot treat money this way over longer periods of time. This is because money today does not have the same value as money at some future time.

Consider this question: Which would you prefer, $100 today or the assurance of receiving $100 a year from now? Clearly, you would prefer the $100 today. If you had the money today, rather than a year from now, you could use it for the year. And if you had no use for it, you could lend it to someone who would pay interest for the privilege of using your money for the year.

EQUIVALENCE

In the preceding section we saw that money at different points in time (for example, $100 today or $100 one year hence) may be equal in the sense that it is $100, but $100 a year hence is *not* an acceptable substitute for $100 today. When we have acceptable substitutes, we say they are *equivalent* to each other. Thus at 8% interest, $108 a year hence is equivalent to $100 today.

Example 6.2

At a 10% per year (compound) interest rate, $500 now is *equivalent* to how much three years hence?

Solution

A value of $500 now will increase by 10% in each of the three years.

$$\text{Now} = \$500.00$$
$$\text{End of 1st year} = 500 + 10\%(500) = \$550.00$$
$$\text{End of 2nd year} = 550 + 10\%(550) = \$605.00$$
$$\text{End of 3rd year} = 605 + 10\%(605) = \$665.50$$

Thus $500 now is *equivalent* to $665.50 at the end of three years. Note that interest is charged each year on the original $500 plus the unpaid interest.

Equivalence is an essential factor in engineering economics. Suppose we wish to select the better of two alternatives. First, we must compute their cash flows. For example:

	Alternative	
Year	***A***	***B***
0	–$2000	–$2800
1	+800	+1100
2	+800	+1100
3	+800	+1100

The larger investment in alternative *B* results in larger subsequent benefits, but we have no direct way of knowing whether it is better than alternative *A*. So we do not know which to select. To make a decision, we must resolve the alternatives into *equivalent* sums so that they may be compared accurately.

COMPOUND INTEREST

To facilitate equivalence computations, a series of compound interest factors will be derived here, and their use will be illustrated in examples.

Symbols and Functional Notation

i = effective interest rate per interest period. In equations, the interest rate is stated as a decimal (i.e., 8% interest is 0.08).

n = number of interest periods. Usually, the interest period is one year, in which case n would be number of years.

P = a present sum of money.

F = a future sum of money. The future sum F is an amount n interest periods from the present that is equivalent to P at interest rate i.

A = an end-of-period cash receipt or disbursement (annuity) in a uniform series continuing for n periods. The entire series is equivalent to P or F at interest rate i.

G = uniform period-by-period increase in cash flows; the uniform gradient.

r = nominal annual interest rate.

From Table 6.1 we can see that the functional notation scheme is based on writing (to find/given, i, n). Thus, if we wished to find the future sum F, given a uniform series of receipts A, the proper compound interest factor to use would be $(F/A, i, n)$.

Table 6.1 Periodic compounding: Functional notation and formulas

Factor	Given	To Find	Functional Notation	Formula
Single payment				
Compound amount factor	P	F	$(F/P, i\%, n)$	$F = P(1 + i)^n$
Present worth factor	F	P	$(P/F, i\%, n)$	$P = F(1 + i)^{-n}$
Uniform payment series				
Sinking fund factor	F	A	$(A/F, i\%, n)$	$A = F\left[\frac{i}{(1+i)^n - 1}\right]$
Capital recovery factor	P	A	$(A/P, i\%, n)$	$A = P\left[\frac{i(1+i)^n}{(1+i)^n - 1}\right]$
Compound amount factor	A	F	$(F/A, i\%, n)$	$F = A\left[\frac{(1+i)^n - 1}{i}\right]$
Present worth factor	A	P	$(P/A, i\%, n)$	$P = A\left[\frac{(1+i)^n - 1}{i(1+i)^n}\right]$
Uniform gradient				
Gradient present worth	G	P	$(P/G, i\%, n)$	$P = G\left[\frac{(1+i)^n - 1}{i^2(1+i)^n} - \frac{n}{i(1+i)^n}\right]$
Gradient future worth	G	F	$(F/G, i\%, n)$	$F = G\left[\frac{(1+i)^n - 1}{i^2} - \frac{n}{1}\right]$
Gradient uniform series	G	A	$(A/G, i\%, n)$	$A = G\left[\frac{1}{i} - \frac{n}{(1+i)^n - 1}\right]$

Single-Payment Formulas

Suppose a present sum of money P is invested for one year at interest rate i. At the end of the year, the initial investment P is received together with interest equal to Pi, or a total amount $P + Pi$. Factoring P, the sum at the end of one year is $P(1 + i)$. If the investment is allowed to remain for subsequent years, the progression is as follows:

Amount at Beginning of the Period	+	Interest for the Period	=	Amount at End of the Period
1st year, P	+	Pi	=	$P(1 + i)$
2nd year, $P(1 + i)$	+	$Pi(1 + i)$	=	$P(1 + i)^2$
3rd year, $P(1 + i)^2$	+	$Pi(1 + i)^2$	=	$P(1 + i)^3$
nth year, $P(1 + i)^{n-1}$	+	$Pi(1 + i)^{n-1}$	=	$P(1 + i)^n$

The present sum P increases in n periods to $P(1 + i)^n$. This gives a relation between a present sum P and its equivalent future sum F:

$$\text{Future sum} = (\text{present sum})(1 + i)^n$$
$$F = P(1 + i)^n$$

This is the **single-payment compound amount formula**. In functional notation it is written

$$F = P(F/P, i, n)$$

The relationship may be rewritten as

$$\text{Present sum} = (\text{Future sum})\ (1 + i)^{-n}$$
$$P = F(1 + i)^{-n}$$

This is the **single-payment present worth formula**. It is written

$$P = F(P/F, i, n)$$

Example 6.3

At a 10% per year interest rate, \$500 now is *equivalent* to how much three years hence?

Solution

This problem was solved in Example 6.2. Now it can be solved using a single-payment formula. $P = \$500$, $n = 3$ years, $i = 10\%$, and F = unknown:

$$F = P(1 + i)^n = 500(1 + 0.10)^3 = \$665.50$$

This problem also may be solved using a compound interest table:

$$F = P(F/P, i, n) = 500(F/P, 10\%, 3)$$

From the 10% compound interest table, read $(F/P, 10\%, 3) = 1.331$.

$$F = 500(F/P, 10\%, 3) = 500(1.331) = \$665.50$$

Example 6.4

To raise money for a new business, a small startup company asks you to lend it some money. The entrepreneur offers to pay you \$3000 at the end of four years. How much should you give the company now if you want to realize 12% interest per year?

Solution

P = unknown, F = \$3000, n = 4 years, and i = 12%:

$$P = F(1 + i)^{-n} = 3000(1 + 0.12)^{-4} = \$1906.55$$

Alternative computation using a compound interest table:

$$P = F(P/F, i, n) = 3000(P/F, 12\%, 4) = 3000(0.6355) = \$1906.50$$

Note that the solution based on the compound interest table is slightly different from the exact solution using a hand-held calculator. In engineering economics, the compound interest tables are considered to be sufficiently accurate.

Uniform Payment Series Formulas

Consider the situation shown in Figure 6.2. Using the single-payment compound amount factor, we can write an equation for F in terms of A:

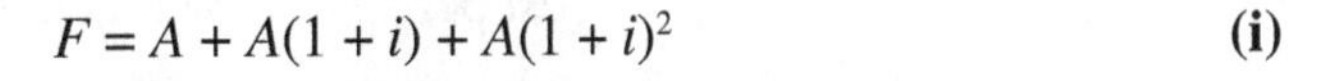

$$F = A + A(1 + i) + A(1 + i)^2 \qquad \text{(i)}$$

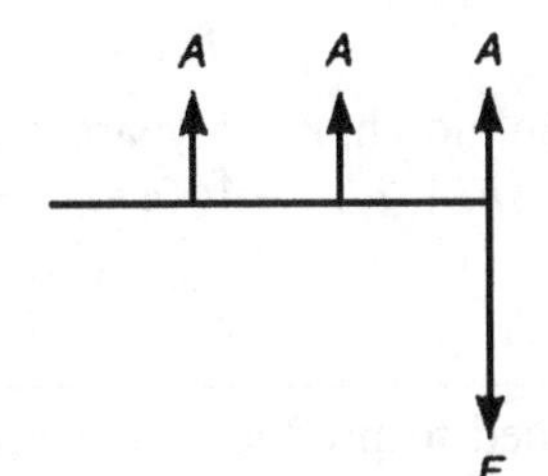

A = End-of-period cash receipt or disbursement in a uniform series continuing for n periods

F = A future sum of money

Figure 6.2 Cash flow diagram—uniform payment series

In this situation, with n = 3, Equation (i) may be written in a more general form:

$$F = A + A(1 + i) + A(1 + i)^{n-1} \qquad \text{(ii)}$$

Multiply Eq. (ii) by $(1 + i)$ $\quad (1 + i)F = A(1 + i) + A(1 + i)^{n-1} + A(1 + i)^n \qquad$ **(iii)**

Subtract Eq. (ii) yields: $\quad iF = -A + A(1 + i)^n$

This produces the **uniform series compound amount formula**:

$$F = A\left[\frac{(1+i)^n - 1}{i}\right].$$

Solving this equation for A produces the **uniform series sinking fund formula**:

$$R = \frac{8.31}{\text{m.w.}} = \frac{8.31}{18} =$$

Since $F = P(1 + i)^n$, we can substitute this expression for F in the equation and obtain the **uniform series capital recovery formula**:

$$A = P\left[\frac{i(1+i)^n}{(1+i)^n - 1}\right].$$

Solving the equation for P produces the **uniform series present worth formula**:

$$P = A\left[\frac{(1+i)^n - 1}{i(1+i)^n}\right].$$

In functional notation, the uniform series factors are:

Compound amount $(F/A, i, n)$

Sinking fund $(A/F, i, n)$

Capital recovery $(A/P, i, n)$

Present worth $(P/A, i, n)$

Example 6.5

If \$100 is deposited at the end of each year in a savings account that pays 6% interest per year, how much will be in the account at the end of five years?

Solution

$A = \$100$, $F =$ unknown, $n = 5$ years, and $i = 6\%$:

$$F = A(F/A, i, n) = 100(F/A, 6\%, 5) = 100(5.637) = \$563.70$$

Example 6.6

A fund established to produce a desired amount at the end of a given period, by means of a series of payments throughout the period, is called a **sinking fund**. A sinking fund is to be established to accumulate money to replace a \$10,000 machine. If the machine is to be replaced at the end of 12 years, how much should be deposited in the sinking fund each year? Assume the fund earns 10% annual interest.

Solution

$$\text{Annual sinking fund deposit } A = 10{,}000(A/F, 10\%, 12) = 10{,}000(0.0468) = \$468$$

Example 6.7

An individual is considering the purchase of a used automobile. The total price is \$6200. With \$1240 as a down payment, and the balance paid in 48 equal monthly payments with interest at 1% per month, compute the monthly payment. The payments are due at the end of each month.

Solution

The amount to be repaid by the 48 monthly payments is the cost of the automobile *minus* the \$1240 down payment.

$P = \$4960$, $A =$ unknown, $n = 48$ monthly payments, and $i = 1\%$ per month:

$$A = P(A/P, 1\%, 48) = 4960(0.0263) = \$130.45$$

Example 6.8

A couple sell their home. In addition to cash, they take a mortgage on the house. The mortgage will be paid off by monthly payments of \$450 for 50 months. The couple decides to sell the mortgage to a local bank. The bank will buy the mortgage, but it requires a 1% per month interest rate on its investment. How much will the bank pay for the mortgage?

Solution

$A = \$450$, $n = 50$ months, $i = 1\%$ per month, and $P =$ unknown:

$$P = A(P/A, i, n) = 450(P/A, 1\%, 50) = 450(39.196) = \$17{,}638.20$$

Uniform Gradient

At times, one will encounter a situation where the cash flow series is not a constant amount A; instead, it is an increasing series. The cash flow shown in Figure 6.3 may be resolved into two components (Figure 6.4). We can compute the value of P^* as equal to P' plus P, and we already have the equation for P':$P' = A(P/A, i, n)$.

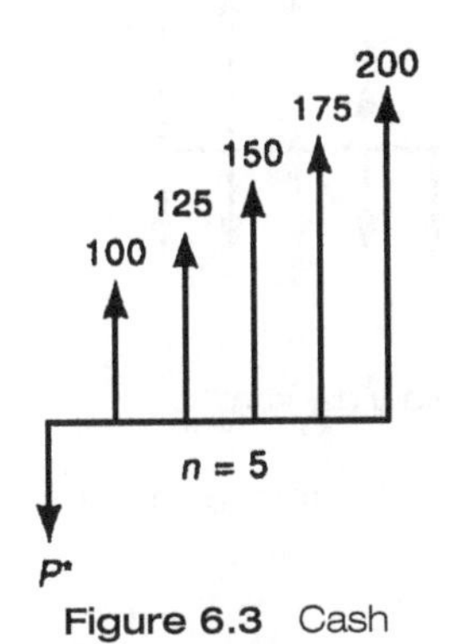

Figure 6.3 Cash flow diagram—uniform gradient

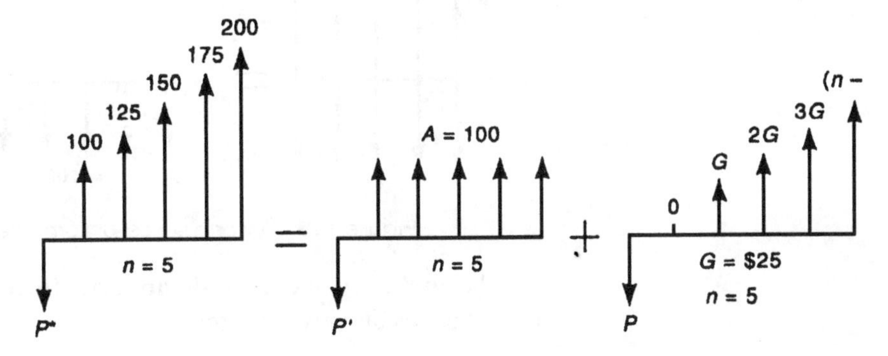

Figure 6.4 Uniform gradient diagram resolved

The value for P in the right-hand diagram is

$$P = G\left[\frac{(1+i)^n - 1}{i^2(1+i)^n} - \frac{n}{i(1+i)^n}\right]$$

This is the **uniform gradient present worth formula**. In functional notation, the relationship is $P = G(P/G, i, n)$.

Example 6.9

The maintenance on a machine is expected to be \$155 at the end of the first year, and it is expected to increase \$35 each year for the following seven years (Exhibit 1). What sum of money should be set aside now to pay the maintenance for the eight-year period? Assume 6% interest.

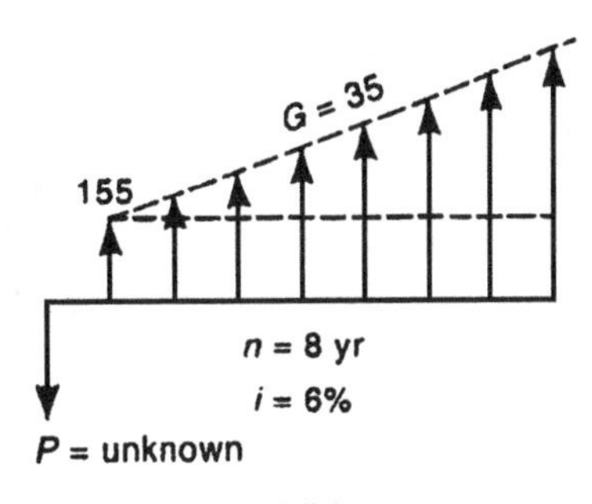

Exhibit 1

Solution

$$P = 155(P/A, 6\%, 8) + 35(P/G, 6\%, 8)$$
$$= 155(6.210) + 35(19.841) = \$1656.99$$

In the gradient series, if—instead of the present sum, P—an equivalent uniform series A is desired, the problem might appear as shown in Figure 6.5. The relationship between A' and G in the right-hand diagram is

$$A' = G\left[\frac{1}{i} - \frac{n}{(1+i)^n - 1}\right]$$

In functional notation, the uniform gradient (to) uniform series factor is: $A' = G(A/G, i, n)$.

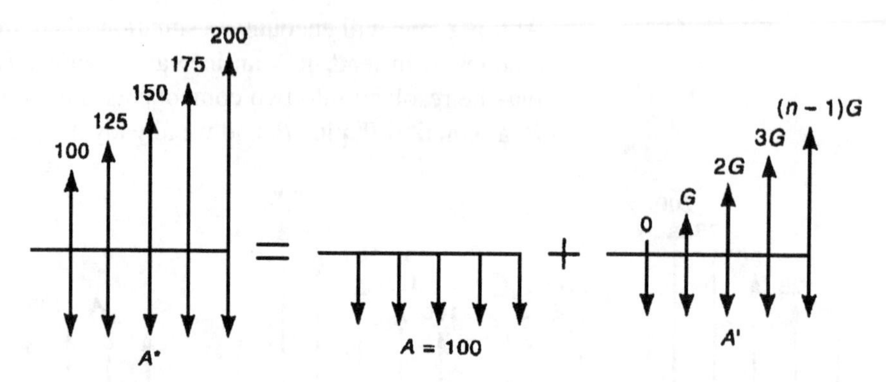

Figure 6.5 Uniform series, uniform gradient cash flow diagram

The **uniform gradient uniform series factor** may be read from the compound interest tables directly, or computed as

$$(A/G, i, n) = \frac{1 - n(A/F, t, n)}{i}$$

Note carefully the diagrams for the uniform gradient factors. The first term in the uniform gradient is zero and the last term is $(n - 1)G$. But we use n in the equations and function notation. The derivations (not shown here) were done on this basis, and the uniform gradient compound interest tables are computed this way.

Example **6.10**

For the situation in Example 6.9, we wish now to know the uniform annual maintenance cost. Compute an equivalent A for the maintenance costs.

Solution

Refer to Exhibit 2. The equivalent uniform annual maintenance cost is

$$A = 155 + 35(A/G, 6\%, 8) = 155 + 35(3.195) = \$266.83$$

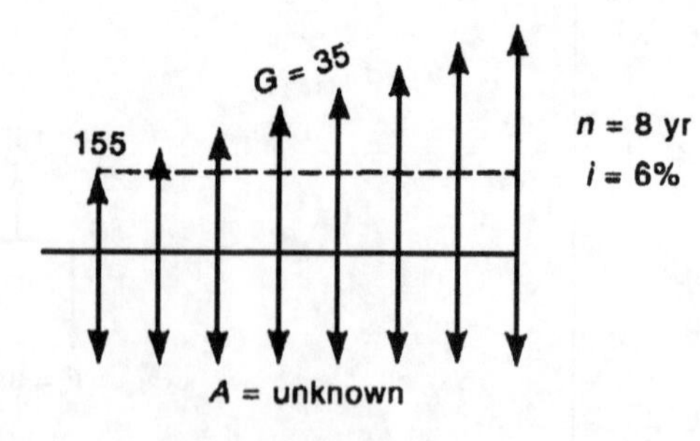

Exhibit 2

Standard compound interest tables give values for eight interest factors: two single payments, four uniform payment series, and two uniform gradients. The tables do *not* give the uniform gradient future worth factor, $(F/G, i, n)$. If it is needed, it may be computed from two tabulated factors:

$$(F/G, i, n) = (P/G, i, n)(F/P, i, n)$$

For example, if $i = 10\%$ and $n = 12$ years, then $(F/G, 10\%, 12) = (P/G, 10\%, 12)\,(F/P, 10\%, 12) = (29.901)(3.138) = 93.83$.

A second method of computing the uniform gradient future worth factor is

$$(F/G,i,n) = \frac{(F/G,i,n)-n}{i}$$

Using this equation for $i = 10\%$ and $n = 12$ years, $(F/G, 10\%, 12) = [(F/A, 10\%, 12) - 12]/0.10 = (21.384 - 12)/0.10 = 93.84$.

Table 6.2 Continuous compounding: Functional notation and formulas

Factor	Given	To Find	Functional Notation	Formula
Single payment				
Compound amount factor	P	F	$(F/P, r\%, n)$	$F = P\,[e^{rn}]$
Present worth factor	F	P	$(P/F, r\%, n)$	$P = F\,[e^{-rn}]$
Uniform payment series				
Sinking fund factor	F	A	$(A/F, r\%, n)$	$A = F\left[\frac{e^{r}-1}{e^{rn}-1}\right]$
Capital recovery factor	P	A	$(A/P, r\%, n)$	$A = P\left[\frac{e^{r}-1}{1-e^{-rn}}\right]$
Compound amount factor	A	F	$(F/A, r\%, n)$	$F = A\left[\frac{e^{rn}-1}{e^{r}-1}\right]$
Present worth factor	A	P	$(P/A, r\%, n)$	$P = A\left[\frac{1-e^{-rn}}{e^{r}-1}\right]$

r = nominal annual interest rate, n = number of years.

Example 6.11

Five hundred dollars is deposited each year into a savings bank account that pays 5% nominal interest, compounded continuously. How much will be in the account at the end of five years?

Solution

$A = \$500$, $r = 0.05$, $n = 5$ years.

$$F = A(F/A, r\%, n) = A\left[\frac{e^{rn}-1}{e^{r}-1}\right] = 500\left[\frac{e^{0.05(5)}-1}{e^{0.05}-1}\right] = \$2769.84$$

NOMINAL AND EFFECTIVE INTEREST

Nominal interest is the annual interest rate without considering the effect of any compounding. **Effective interest** is the annual interest rate taking into account the effect of any compounding during the year.

Non-annual Compounding

Frequently an interest rate is described as an annual rate, even though the interest period may be something other than one year. A bank may pay 1% interest on the amount in a savings account every three months. The *nominal* interest rate in this situation is $4 \times 1\% = 4\%$. But if you deposited \$1000 in such an account, would you have 104%(1000) = \$1040 in the account at the end of one year? The answer is no, you would have more. The amount in the account would increase as follows.

Amount in Account

Beginning of year:	1000.00
End of three months:	1000.00 + 1%(1000.00) = 1010.00
End of six months:	1010.00 + 1%(1010.00) = 1020.10
End of nine months:	1020.10 + 1%(1020.10) = 1030.30
End of one year:	1030.30 + 1%(1030.30) = 1040.60

At the end of one year, the interest of \$40.60, divided by the original \$1000, gives a rate of 4.06 percent. This is the *effective* interest rate.

$$\text{Effective interest rate per year:} \quad i_{eff} = (1 + r/m)^m - 1$$

where r = nominal annual interest rate
m = number of compound periods per year
r/m = effective interest rate per period

Example **6.12**

A bank charges 1.5% interest per month on the unpaid balance for purchases made on its credit card. What nominal interest rate is it charging? What is the effective interest rate?

Solution

The nominal interest rate is simply the annual interest ignoring compounding, or 12(1.5%) = 18%.

$$\text{Effective interest rate} = (1 + 0.015)^{12} - 1 = 0.1956 = 19.56\%$$

SOLVING ENGINEERING ECONOMICS PROBLEMS

The techniques presented so far illustrate how to convert single amounts of money, and uniform or gradient series of money, into some equivalent sum at another point in time. These compound interest computations are an essential part of engineering economics problems.

The typical situation is that we have a number of alternatives; the question is, which alternative should we select? The customary method of solution is to express each alternative in some common form and then choose the best, taking both the monetary and intangible factors into account. In most computations an interest rate must be used. It is often called the **minimum attractive rate of return (MARR)**, to indicate that this is the smallest interest rate, or rate of return, at which one is willing to invest money.

Criteria

Engineering economics problems inevitably fall into one of three categories:

1. *Fixed input.* The amount of money or other input resources is fixed. *Example*: A project engineer has a budget of $450,000 to overhaul a plant.
2. *Fixed output.* There is a fixed task or other output to be accomplished. *Example*: A mechanical contractor has been awarded a fixed-price contract to air-condition a building.
3. *Neither input nor output fixed.* This is the general situation, where neither the amount of money (or other inputs) nor the amount of benefits (or other outputs) is fixed. *Example*: A consulting engineering firm has more work available than it can handle. It is considering paying the staff to work evenings to increase the amount of design work it can perform.

There are five major methods of comparing alternatives: present worth, future worth, annual cost, rate of return, and benefit-cost analysis. These are presented in the sections that follow.

Present Worth

Present worth analysis converts all of the money consequences of an alternative into an equivalent present sum. The criteria are:

Category	Present Worth Criterion
Fixed input	Maximize the present worth of benefits or other outputs
Fixed output	Minimize the present worth of costs or other inputs
Neither input nor output fixed	Maximize present worth of benefits minus present worth of costs, or maximize net present worth

Appropriate Problems

Present worth analysis is most frequently used to determine the present value of future money receipts and disbursements. We might want to know, for example, the present worth of an income-producing property, such as an oil well. This should provide an estimate of the price at which the property could be bought or sold.

An important restriction in the use of present worth calculation is that there must be a common analysis period for comparing alternatives. It would be incorrect, for example, to compare the present worth (PW) of cost of pump A, expected to last 6 years, with the PW of cost of pump B, expected to last 12 years

(Figure 6.6). In situations like this, the solution is either to use some other analysis technique (generally, the annual cost method is suitable in these situations) or to restructure the problem so that there is a common analysis period.

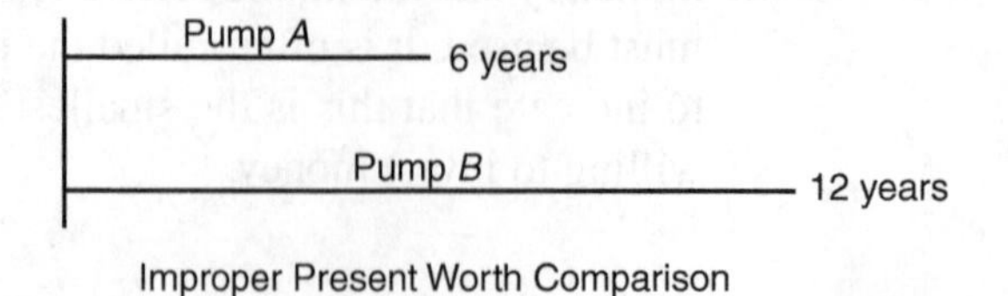

Figure 6.6 Improper present worth comparison

In this example, a customary assumption would be that a pump is needed for 12 years and that pump A will be replaced by an identical pump A at the end of 6 years. This gives a 12-year common analysis period (Figure 6.7). This approach is easy to use when the different lives of the alternatives have a practical least-common-multiple life. When this is not true (for example, the life of J equals 7 years and the life of K equals 11 years), some assumptions must be made to select a suitable common analysis period, or the present worth method should not be used.

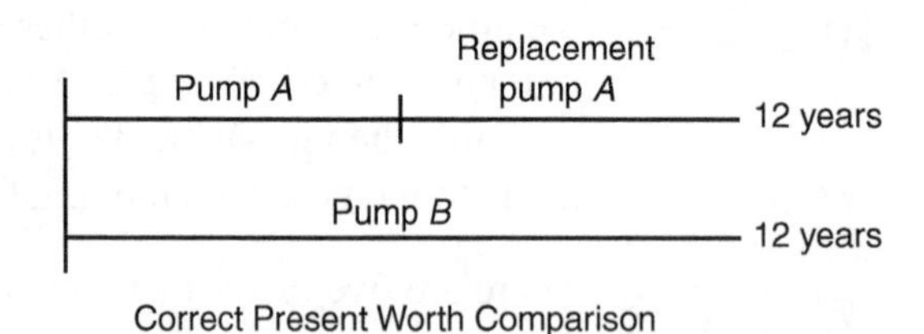

Figure 6.7 Proper present worth comparison

Example 6.13

Machine X has an initial cost of \$10,000, an annual maintenance cost of \$500 per year, and no salvage value at the end of its 4-year useful life. Machine Y costs \$20,000, and the first year there is no maintenance cost. Maintenance is \$100 the second year, and it increases \$100 per year thereafter. The machine has an anticipated \$5000 salvage value at the end of its 12-year useful life. If the minimum attractive rate of return (MARR) is 8%, which machine should be selected?

Solution

The analysis period is not stated in the problem. Therefore, we select the least common multiple of the lives, or 12 years, as the analysis period.

Present worth of cost of 12 years of machine X:

$$\begin{aligned} PW &= 10{,}000 + 10{,}000(P/F, 8\%, 4) + 10{,}000(P/F, 8\%, 8) + 500(P/A, 8\%, 12) \\ &= 10{,}000 + 10{,}000(0.7350) + 10{,}000(0.5403) + 500(7.536) = \$26{,}521 \end{aligned}$$

Present worth of cost of 12 years of machine Y:

$$\begin{aligned} PW &= 20{,}000 + 100(P/G, 8\%, 12) - 5000(P/F, 8\%, 12) \\ &= 20{,}000 + 100(34.634) - 5000(0.3971) = \$21{,}478 \end{aligned}$$

Choose machine Y, with its smaller PW of cost.

Example 6.14

Two alternatives have the following cash flows:

	Alternative	
Year	*A*	*B*
0	–\$2000	–\$2800
1	+800	+1100
2	+800	+1100
3	+800	+1100

At a 4% interest rate, which alternative should be selected?

Solution

The net present worth of each alternative is computed:

$$\text{Net present worth (NPW)} = \text{PW of benefit} - \text{PW of cost}$$
$$\text{NPW}_A = 800(P/A, 4\%, 3) - 2000 = 800(2.775) - 2000 = \$220.00$$
$$\text{NPW}_B = 1100(P/A, 4\%, 3) - 2800 = 1100(2.775) - 2800 = \$252.50$$

To maximize NPW, choose alternative *B*.

Infinite Life and Capitalized Cost

In the special situation where the analysis period is infinite ($n = \infty$), an analysis of the present worth of cost is called **capitalized cost**. There are a few public projects where the analysis period is infinity. Other examples are permanent endowments and cemetery perpetual care.

When n equals infinity, a present sum P will accrue interest of Pi for every future interest period. For the principal sum P to continue undiminished (an essential requirement for n equal to infinity), the end-of-period sum A that can be disbursed is Pi (Figure 6.8).

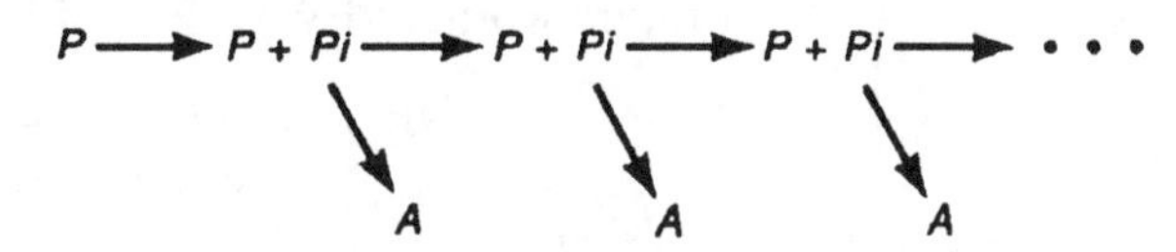

Figure 6.8 Infinite life, capitalized cost diagram

When $n = \infty$, the fundamental relationship is

$$A = Pi$$

Some form of this equation is used whenever there is a problem involving an infinite analysis period.

Example 6.15

In his will, a man wishes to establish a perpetual trust to provide for the maintenance of a small local park. If the annual maintenance is \$7500 per year and the trust account can earn 5% interest, how much money must be set aside in the trust?

Solution

When $n = \infty$, $A = Pi$ or $P = A/i$. The capitalized cost is $P = A/i = \$7500/0.05 = \$150{,}000$.

Future Worth or Value

In present worth analysis, the comparison is made in terms of the equivalent present costs and benefits. But the analysis need not be made in terms of the present—it can be made in terms of a past, present, or future time. Although the numerical calculations may look different, the decision is unaffected by the selected point in time. Often we do want to know what the future situation will be if we take some particular course of action now. An analysis based on some future point in time is called **future worth analysis.**

Category	Future Worth Criterion
Fixed input	Maximize the future worth of benefits or other outputs
Fixed output	Minimize the future worth of costs or other inputs
Neither input nor output fixed	Maximize future worth of benefits minus future worth of costs, or maximize net future worth

Example 6.16

Two alternatives have the following cash flows:

	Alternative	
Year	***A***	***B***
0	–$2000	–$2800
1	+800	+1100
2	+800	+1100
3	+800	+1100

At a 4% interest rate, which alternative should be selected?

Solution

In Example 6.14, this problem was solved by present worth analysis at year 0. Here it will be solved by future worth analysis at the end of year 3.

Net future worth (NFW) = FW of benefits – FW of cost

$$\begin{aligned}\text{NFW}_A &= 800(F/A, 4\%, 3) - 2000(F/P, 4\%, 3)\\ &= 800(3.122) - 2000(1.125) = +\$247.60\end{aligned}$$

$$\begin{aligned}\text{NFW}_B &= 1100(F/A, 4\%, 3) - 2800(F/P, 4\%, 3)\\ &= 1100(3.122) - 2800(1.125) = +\$284.20\end{aligned}$$

To maximize NFW, choose alternative *B*.

Annual Cost

The annual cost method is more accurately described as the method of equivalent uniform annual cost (EUAC). Where the computation is of benefits, it is called the method of equivalent uniform annual benefits (EUAB).

Criteria

For each of the three possible categories of problems, there is an annual cost criterion for economic efficiency.

Category	Annual Cost Criterion
Fixed input	Maximize the equivalent uniform annual benefits (EUAB)
Fixed output	Minimize the equivalent uniform annual cost (EUAC)
Neither input nor output fixed	Maximize EUAB – EUAC

Application of Annual Cost Analysis

In the section on present worth, we pointed out that the present worth method requires a common analysis period for all alternatives. This restriction does not apply in all annual cost calculations, but it is important to understand the circumstances that justify comparing alternatives with different service lives.

Frequently, an analysis is done to provide for a more-or-less continuing requirement. For example, one might need to pump water from a well on a continuing basis. Regardless of whether each of two pumps has a useful service life of 6 years or 12 years, we would select the alternative whose annual cost is a minimum. And this still would be the case if the pumps' useful lives were the more troublesome 7 and 11 years. Thus, if we can assume a continuing need for an item, an annual cost comparison among alternatives of differing service lives is valid. The underlying assumption in these situations is that the shorter-lived alternative can be replaced with an identical item with identical costs, when it has reached the end of its useful life. This means that the EUAC of the initial alternative is equal to the EUAC for the continuing series of replacements.

On the other hand, if there is a specific requirement to pump water for ten years, then each pump must be evaluated to see what costs will be incurred during the analysis period and what salvage value, if any, may be recovered at the end of the analysis period. The annual cost comparison needs to consider the actual circumstances of the situation.

Examination problems are often readily solved using the annual cost method. And the underlying *continuing requirement* is usually present, so an annual cost comparison of unequal-lived alternatives is an appropriate method of analysis.

Example 6.17

Consider the following alternatives:

	A	*B*
First cost	$5000	$10,000
Annual maintenance	500	200
End-of-useful-life salvage value	600	1000
Useful life	5 years	15 years

Based on an 8% interest rate, which alternative should be selected?

Solution

Assuming both alternatives perform the same task and there is a continuing requirement, the goal is to minimize EUAC.

Alternative *A*:

$$\begin{aligned} \text{EUAC} &= 5000(A/P, 8\%, 5) + 500 - 600(A/F, 8\%, 5) \\ &= 5000(0.2505) + 500 - 600(0.1705) = \$1650 \end{aligned}$$

Alternative *B*:

$$\begin{aligned} \text{EUAC} &= 10{,}000(A/P, 8\%, 15) + 200 - 1000(A/F, 8\%, 15) \\ &= 10{,}000(0.1168) + 200 - 1000(0.0368) = \$1331 \end{aligned}$$

To minimize EUAC, select alternative *B*.

Rate of Return Analysis

A typical situation is a cash flow representing the costs and benefits. The rate of return may be defined as the interest rate where PW of cost = PW of benefits, EUAC = EUAB, or PW of cost – PW of benefits = 0.

Example 6.18

Compute the rate of return for the investment represented by the following cash flow table.

Year:	0	1	2	3	4	5
Cash flow:	–\$595	+250	+200	+150	+100	+50

Solution

This declining uniform gradient series may be separated into two cash flows (Exhibit 3) for which compound interest factors are available.

Note that the gradient series factors are based on an *increasing* gradient. Here the declining cash flow is solved by subtracting an increasing uniform gradient, as indicated in the figure.

PW of cost – PW of benefits = 0

$$595 - [250(P/A, i, 5)] - 50(P/G, i, 5) = 0$$

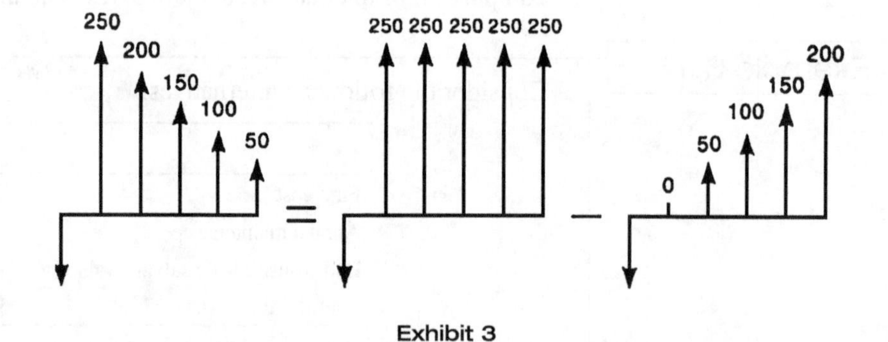

Exhibit 3

Try $i = 10\%$:

$$595 - [250(3.791) - 50(6.862)] = -9.65$$

Try $i = 12\%$:

$$595 - [250(3.605) - 50(6.397)] = +13.60$$

The rate of return is between 10 and 12%. It may be computed more accurately by linear interpolation:

$$\text{Rate of return} = 10\% + (2\%)\left(\frac{9.65 - 0}{13.60 + 9.65}\right) = 10.83\%$$

Two Alternatives

Compute the incremental rate of return on the cash flow representing the difference between the two alternatives. Since we want to look at increments of *investment*, the cash flow for the difference between the alternatives is computed by taking the higher initial-cost alternative minus the lower initial-cost alternative. If the incremental rate of return is greater than or equal to the predetermined minimum attractive rate of return (MARR), choose the higher-cost alternative; otherwise, choose the lower-cost alternative.

Example **6.19**

Two alternatives have the following cash flows:

	Alternative	
Year	*A*	*B*
0	–$2000	–$2800
1	+800	+1100
2	+800	+1100
3	+800	+1100

If 4% is considered the minimum attractive rate of return (MARR), which alternative should be selected?

Solution

These two alternatives were previously examined in Examples 15.14 and 15.16 by present worth and future worth analysis. This time, the alternatives will be resolved using a rate-of-return analysis.

Note that the problem statement specifies a 4% MARR, whereas Examples 15.14 and 15.16 referred to a 4% interest rate. These are really two different ways of saying the same thing: the minimum acceptable time value of money is 4%.

First, tabulate the cash flow that represents the increment of investment between the alternatives. This is done by taking the higher initial-cost alternative minus the lower initial-cost alternative:

	Alternative		Difference between alternatives
Year	*A*	*B*	*B – A*
0	–$2000	–$2800	–$800
1	+800	+1100	+300
2	+800	+1100	+300
3	+800	+1100	+300

Then compute the rate of return on the increment of investment represented by the difference between the alternatives:

$$\text{PW of cost} = \text{PW of benefits}$$
$$800 = 300(P/A, i, 3)$$
$$(P/A, i, 3) = 800/300 = 2.67$$
$$i = 6.1\%$$

Since the incremental rate of return exceeds the 4% MARR, the increment of investment is desirable. Choose the higher-cost alternative *B*.

Before leaving this example, one should note something that relates to the rates of return on alternative *A* and on alternative *B*. These rates of return, if calculated, are:

	Rate of Return
Alternative *A*	9.7%
Alternative *B*	8.7%

The correct answer to this problem has been shown to be alternative *B*, even though alternative *A* has a higher rate of return. The higher-cost alternative may be thought of as the lower-cost alternative plus the increment of investment between them. Viewed this way, the higher-cost alternative *B* is equal to the desirable lower-cost alternative *A* plus the difference between the alternatives.

The important conclusion is that computing the rate of return for each alternative does *not* provide the basis for choosing between alternatives. Instead, incremental analysis is required.

Example 6.20

Consider the following:

	Alternative	
Year	***A***	***B***
0	−$200.0	−$131.0
1	+77.6	+48.1
2	+77.6	+48.1
3	+77.6	+48.1

If the MARR is 10%, which alternative should be selected?

Solution

To examine the increment of investment between the alternatives, we will examine the higher initial-cost alternative minus the lower initial-cost alternative, or *A* – *B*.

	Alternative		Increment
Year	***A***	***B***	***A* – *B***
0	−$200.0	−$131.0	−$69.0
1	+77.6	+48.1	+29.5
2	+77.6	+48.1	+29.5
3	+77.6	+48.1	+29.5

Solve for the incremental rate of return:

$$\text{PW of cost} = \text{PW of benefits}$$
$$69.0 = 29.5(P/A, i, 3)$$
$$(P/A, i, 3) = 69.0/29.5 = 2.339$$

From compound interest tables, the incremental rate of return is between 12 and 18%. This is a desirable increment of investment; hence we select the higher-initial-cost alternative *A*.

Three or More Alternatives

When there are three or more mutually exclusive alternatives, proceed with the same logic presented for two alternatives. The components of incremental analysis are listed below.

Step 1. Compute the rate of return for each alternative. Reject any alternative where the rate of return is less than the desired MARR. (This step is not essential, but helps to immediately identify unacceptable alternatives.)

Step 2. Rank the remaining alternatives in order of increasing initial cost.

Step 3. Examine the increment of investment between the two lowest-cost alternatives as described for the two-alternative problem. Select the better of the two alternatives and reject the other one.

Step 4. Take the preferred alternative from step 3. Consider the next higher initial-cost alternative, and proceed with another two-alternative comparison.

Step 5. Continue until all alternatives have been examined and the best of the multiple alternatives has been identified.

Example 6.21

Consider the following:

	Alternative	
Year	***A***	***B***
0	–\$200.0	–\$131.0
1	+77.6	+48.1
2	+77.6	+48.1
3	+77.6	+48.1

If the MARR is 10%, which alternative, if any, should be selected?

Solution

One should carefully note that this is a *three-alternative* problem, where the alternatives are *A*, *B*, and *Do nothing*. In this solution we will skip step 1. Reorganize the problem by placing the alternatives in order of increasing initial cost:

	Alternative		
Year	***Do Nothing***	***B***	***A***
0	0	– \$131.0	–\$200.0
1	0	+ 48.1	+77.6
2	0	+ 48.1	+77.6
3	0	+ 48.1	+77.6

Examine the *B – Do nothing* increment of investment:

Year	*B – Do Nothing*
0	–\$131.0 – 0 = – \$131.0
1	+ 48.1 – 0 = + 48.1
2	+ 48.1 – 0 = + 48.1
3	+ 48.1 – 0 = + 48.1

Solve for the incremental rate of return:

$$\text{PW of cost} = \text{PW of benefits}$$
$$131.0 = 48.1(P/A, i, 3)$$
$$(P/A, i, 3) = 131.0/48.1 = 2.723$$

From compound interest tables, the incremental rate of return is about 5%. Since the incremental rate of return is less than 10%, the *B – Do nothing* increment is not desirable. Reject alternative *B*.

Year	*A – Do Nothing*
0	–\$200.0 – 0 = –\$200.0
1	+77.6 – 0 = +77.6
2	+77.6 – 0 = +77.6
3	+77.6 – 0 = +77.6

Next, consider the increment of investment between the two remaining alternatives. Solve for the incremental rate of return:

$$\text{PW of cost} = \text{PW of benefits}$$
$$200.0 = 77.6(P/A, i, 3)$$
$$(P/A, i, 3) = 200.0/77.6 = 2.577$$

The incremental rate of return is 8%, less than the desired 10%. Reject the increment and select the remaining alternative: *Do nothing*.

If you have not already done so, you should go back to Example 6.20 and see how the slightly changed wording of the problem has radically altered it. Example 6.20 required a choice between two undesirable alternatives. This example adds the *Do nothing* alternative, which is superior to *A* and *B*.

BENEFIT-COST ANALYSIS

Generally, in public works and governmental economic analyses, the dominant method of analysis is the **benefit-cost ratio (B/C)**. It is simply the ratio of benefits divided by costs, taking into account the time value of money.

$$B/C = \frac{\text{PW of benefits}}{\text{PW of cost}} = \frac{\text{Equivalent uniform annual benefits}}{\text{Equivalent uniform annual cost}}$$

For a given interest rate, a B/C ratio ≥1 reflects an acceptable project. The B/C analysis method is parallel to rate-of-return analysis. The same kind of incremental analysis is required.

Example **6.22**

Solve Example 6.20 by benefit-cost analysis.

Solution

	Alternative		Increment
Year	***A***	***B***	***A* – *B***
0	–$200.0	–$131.0	–$69.0
1	+77.6	+48.1	+29.5
2	+77.6	+48.1	+29.5
3	+77.6	+48.1	+29.5

The benefit-cost ratio for the $A - B$ increment is

$$\text{B/C} = \frac{\text{PW of benefits}}{\text{PW of cost}} = \frac{29.5(P/A, 10\%, 3)}{69.0} = \frac{73.37}{69.0} = 1.06$$

Since the B/C ratio exceeds 1, the increment of investment is desirable. Select the higher-cost alternative A.

Breakeven Analysis

In business, *breakeven* is defined as the point where income just covers costs. In engineering economics, the breakeven point is defined as the point where two alternatives are equivalent.

Example **6.23**

A city is considering a new $50,000 snowplow. The new machine will operate at a savings of $600 per day compared with the present equipment. Assume that the MARR is 12%, and the machine's life is ten years with zero resale value at that time. How many days per year must the machine be used to justify the investment?

Solution

This breakeven problem may be readily solved by annual cost computations. We will set the equivalent uniform annual cost (EUAC) of the snowplow equal to its annual benefit and solve for the required annual utilization. Let X = breakeven point = days of operation per year.

$$\text{EUAC} = \text{EUAB}$$

$$50{,}000(A/P, 12\%, 10) = 600X$$

$$X = 50{,}000(0.1770)/600 = 14.8 \text{ days/year}$$

BONDS

A **bond** is a form of debt represented by a certificate. The bond will specify the amount of the debt, the interest rate of the bond, how often the interest is paid, and when the debt will be repaid.

Bond Value

Bond value is the present worth of all the future interest payments plus the future repayment of the debt, computed at some selected interest rate.

Example **6.24**

A \$5000 bond is being offered for sale. It has a stated interest rate of 7%, paid annually. At the end of eight years the \$5000 debt will be repaid along with the last interest payment. It you want an 8% rate of return on this investment (bond yield), how much would you be willing to pay for the bond (bond value)?

Solution

The bond pays 7% × \$5000 = \$350 at the end of every year and will repay the \$5000 debt at the end of eight years.

$$\begin{aligned}\text{Bond value} &= \text{PW of all future benefits}\\ &= 350(P/A,8\%,8) + 5000(P/F,8\%,8)\\ &= 350(5.747) + 5000(0.5403) = \$4712.95\end{aligned}$$

Bond Yield

Bond yield is the interest rate at which the benefits of owning the bond are equivalent to the cost of the bond.

Example **6.25**

If the bond in Example 6.24 can actually be purchased for \$4200, what is the bond yield?

Solution

Set the cost of the bond equal to the PW of the bond benefits and solve for the unknown interest rate. The resulting i^* is the bond yield.

$$\$4200 = 350(P/A,i,8) + 5000(P/F,i,8)$$

The equation must be solved by trial and error. Try $i = 10\%$.

$$\$4200 \stackrel{?}{=} 350(5.335) + 5000(0.467) = 4202.25$$

We see that i^* is very close to 10%. No further computations are required. The bond yield is very close to 10%.

PAYBACK PERIOD

Payback period is the period of time required for the profit or other benefits of an investment to equal the cost of the investment.

Example **6.26**

A project has the following costs and benefits.

Year	Costs	Benefits
0	$1400	
1	500	$200
2	300	100
3–10		400/year

What is the payback period?

Solution

The total cost is $2200. At the end of year 6 the total benefits will be 200 + 100 + 400 + 400 + 400 + 400 = $1900. And at the end of year 7 benefits will be 1900 + 400 = $2300. The payback period is where benefits equal cost. Since a cash flow table is normally based on the end-of-year convention (See Example 6.1), the payback period is at the end of year 7 and the answer is 7 years.

If, on the other hand, the problem had been stated in words something like ". . .and the benefits from the third year on are $400/year," you would assume the benefits occur uniformly throughout the year. In this situation, the correct answer would be 6.75 years.

VALUATION AND DEPRECIATION

Depreciation of capital equipment is an important component of many after-tax economic analyses. For this reason, one must understand the fundamentals of depreciation accounting.

Notation

BV = book value
C = cost of the property (basis)
D_j = depreciation in year j
S_n = salvage value in year n

Depreciation is the systematic allocation of the cost of a capital asset over its useful life. **Book value** is the original cost of an asset minus the accumulated depreciation of the asset.

$$\text{Book value (BV)} = C - \Sigma(D_j)$$

In computing a schedule of depreciation charges four items are considered:

1. Cost of the property, C (called the *basis* in tax law)
2. Type of property. Property is classified either as **tangible** (such as machinery) or **intangible** (such as a franchise or a copyright) and as either **real property** (real estate) or **personal property** (everything not real property)
3. Depreciable life in years, n
4. Salvage value of the property at the end of its depreciable (usable) life, S_n

Straight-Line Depreciation

Depreciation charge in any year is given by

$$D_j = \frac{C - S_n}{n}$$

An alternate computation of the depreciation charge in year j is

$$D_j = \frac{C - \text{Depreciation taken to beginning of year } j - S_n}{\text{Remaining useful life at beginning of year } j}$$

Double-Declining-Balance Depreciation

DDB depreciation in any year, $D_j = \frac{2}{n}$ (C – depreciation in years prior to j).

For 150% declining-balance depreciation, replace the 2 in the equation with 1.5.

Modified Accelerated Cost Recovery System Depreciation

The modified accelerated cost recovery system (MACRS) depreciation method generally applies to property placed in service after 1986. To compute the MACRS depreciation for an item one must know the following:

1. Cost (basis) of the item
2. Property class. All tangible property is classified in one of six classes (3, 5, 7, 10, 15, and 20 years), based on the life over which it is depreciated (see Table 6.3). Residential real estate and nonresidential real estate are in two separate real property classes of 27.5 years and 39 years, respectively.
3. Depreciation computation
    - The 3-, 5-, 7-, and 10-year property classes use double-declining-balance depreciation with conversion to straight-line depreciation in the year that increases the deduction.
    - The 15- and 20-year property classes use 150% declining-balance depreciation with conversion to straight-line depreciation in the year that increases the deduction.
    - In MACRS the salvage value is assumed to be zero.

Half-Year Convention

Except for real property, a half-year convention is used. Under this convention all property is considered to be placed in service in the middle of the tax year, and a half year of depreciation is allowed in the first year. For each of the remaining years, one is allowed a full year of depreciation. If the property is disposed of prior to the end of the recovery period (property class life), a half year of depreciation is allowed in that year. If the property is held for the entire recovery period, a half year of depreciation is allowed for the year following the end of the recovery period (see Table 6.3).

Table 6.3 Modified ACRS (MACRS) depreciation for personal property—half-year convention

Recovery Year	3-year Recovery	Applicable Percentage for the Class of Property 5-year Recovery	7-year Recovery	10-year Recovery
1	33.33	20.00	14.29	10.00
2	44.45	32.00	24.49	18.00
3	14.81†	19.20	17.49	14.40
4	7.41	11.52†	12.49	11.52
5		11.52	8.93†	9.22
6		5.76	8.92	7.37
7			8.93	6.55†
8			4.46	6.55
9				6.56
10				6.55
11				3.28

† Use straight-line depreciation for the year marked and all subsequent years.

Example 6.27

A $5000 computer has an anticipated $500 salvage value at the end of its five-year depreciable life. Compute the depreciation schedule by MACRS depreciation. Do the MACRS computation by hand, and then compare the results with the values from Table 6.3.

Solution

The depreciation method is double declining balance with conversion to straight line for the computer's five-year property class and the half-year convention is used. Salvage value S_n is assumed to be zero for MACRS. Using the equation for DDB depreciation in any year:

Year

$$1\left(\frac{1}{2}\text{year}\right)\quad D_1 = \frac{1}{2}\times\frac{2}{5}(5000-0) = \$1000$$

$$2\quad D_2 = \frac{2}{5}(5000-1000) = 1600$$

$$3\quad D_3 = \frac{2}{5}(5000-2600) = 960$$

$$4\quad D_4 = \frac{2}{5}(5000-3560) = 576$$

$$5\quad D_5 = \frac{2}{5}(5000-4136) = 346$$

$$6\left(\frac{1}{2}\text{year}\right)\quad D_6 = \frac{1}{2}\times\frac{2}{5}(5000-4482) = 104$$

$$\overline{\$4586}$$

The computation must now be modified to convert to straight-line depreciation at the point where the straight-line depreciation will be larger. Using the alternate straight-line computation,

$$D_5 = \frac{5000 - 4136 - 0}{1.5 \text{ years remaining}} = \$576$$

This is more than the $346 computed using DDB; hence, switch to the straight-line method for year 5 and beyond.

$$D_6\left(\frac{1}{2}\text{year}\right) = \frac{1}{2}(576) = \$288$$

Answers:

Year	Depreciation (MACRS)
1	$1000
2	1600
3	960
4	576
5	576
6	288
	$5000

The computed MACRS depreciation is identical with that obtained from Table 6.3.

INFLATION

Inflation is characterized by rising prices for goods and services, while deflation produces a decrease in prices. An inflationary trend makes future dollars have less purchasing power than present dollars. This benefits long-term borrowers of money because they may repay a loan of present dollars in the future with dollars of reduced buying power. The help to borrowers is at the expense of lenders. Deflation has the opposite effect. Money borrowed at a point in time followed by a deflationary period subjects the borrower to loan repayment with dollars of greater purchasing power than those he borrowed. This is to the lenders' advantage at the expense of borrowers.

Price changes occur in a variety of ways. One method of stating a price change is a uniform rate of price change per year.

f = general inflation rate per interest period
i = effective interest rate per interest period

The following example problem will illustrate the computations.

Example 6.28

A mortgage will be repaid in three equal payments of $5000 at the end of years 1, 2, and 3. If the annual inflation rate, f, is 8% during this period, and the investor wishes a 12% annual interest rate (i), what is the maximum amount he would be willing to pay for the mortgage?

Solution

The computation is a two-step process. First, the three future payments must be converted into dollars with the same purchasing power as today's (year 0) dollars.

Year	Actual Cash Flow		Multiplied by		Cash Flow Adjusted to Today's (Year 0) Dollars
0	—	×	—	=	—
1	+5000	×	$(1 + 0.08)^{-1}$	=	+ 4630
2	+5000	×	$(1 + 0.08)^{-2}$	=	+ 4286
3	+5000	×	$(1 + 0.08)^{-3}$	=	+ 3969

The general form of the adjusting multiplier is

$$(1 + f)^{-n} \text{ or } (P/F, f, n)$$

Now that the problem has been converted to dollars of the same purchasing power (today's dollars in this example), we can proceed to compute the present worth of the future payments at the desired 12% interest rate.

Year	Actual Cash Flow		Multiplied by		Present Worth
0	—	×	—	=	—
1	+4630	×	$(1 + 0.12)^{-1}$	=	+4134
2	+4286	×	$(1 + 0.12)^{-2}$	=	+3417
3	+3969	×	$(1 + 0.12)^{-3}$	=	+2825
					\$10,376

The investor would pay \$10,376.

Alternate Solution

Instead of doing the inflation and interest rate computations separately, one can compute a combined equivalent interest rate per interest period, d.

$$d = (1 + f)(1 + i) - 1 = i + f + (i \times f)$$

For this cash flow, $d = 0.12 + 0.08 + 0.12(0.08) = 0.2096$. Since we do not have 20.96% interest tables, the problem must be calculated using present-worth equations.

$$\begin{aligned} \text{PW} &= 5000(1 + 0.2096)^{-1} + 5000(1 + 0.2096)^{-2} + 5000(1 + 0.2096)^{-3} \\ &= 4134 + 3417 + 2825 = \$10{,}376 \end{aligned}$$

Example 6.29

One economist has predicted that there will be a 7% per year inflation of prices during the next ten years. If this proves to be correct, an item that presently sells for \$10 would sell for what price ten years hence?

Solution

$f = 7\%$, $P = \$10$

$F = ?$, $n = 10$ years

Here, the computation is to find the future worth F, rather than the present worth, P.

$$F = P(1 + f)10 = 10(1 + 0.07)10 = \$19.67$$

Effect of Inflation on a Rate of Return

The effect of inflation on the computed rate of return for an investment depends on how future benefits respond to the inflation. If benefits produce constant dollars, which are not increased by inflation, the effect of inflation is to reduce the before-tax rate of return on the investment. If, on the other hand, the dollar benefits increase to keep up with the inflation, the before-tax rate of return will not be adversely affected by the inflation.

This is not true when an after-tax analysis is made. Even if the future benefits increase to match the inflation rate, the allowable depreciation schedule does not increase. The result will be increased taxable income and income tax payments. This reduces the available after-tax benefits and, therefore, the after-tax rate of return.

Example 6.30

A man bought a 5% tax-free municipal bond. It cost \$1000 and will pay \$50 interest each year for 20 years. The bond will mature at the end of 20 years and return the original \$1000. If there is 2% annual inflation during this period, what rate of return will the investor receive after considering the effect of inflation?

Solution

$$d = 0.05,\ i = \text{unknown},\ f = 0.02$$

Combined effective interest rate/interest period,

$$d = i + f + (i \times f)$$
$$0.05 = i + 0.02 + 0.02i$$
$$1.02i = 0.03,\ i = 0.294 = 2.94\%$$

REFERENCE

Newnan, D. G., et al. *Engineering Economic Analysis*, 6th ed. San Jose, CA: Engineering Press, 2000.

PROBLEMS

6.1 A retirement fund earns 8% interest, compounded quarterly. If $400 is deposited every three months for 25 years, the amount in the fund at the end of 25 years is nearest to:

a. $50,000
b. $75,000
c. $100,000
d. $125,000

6.2 The repair costs for some handheld equipment are estimated to be $120 the first year, increasing by $30 per year in subsequent years. The amount a person needs to deposit into a bank account paying 4% interest to provide for the repair costs for the next five years is nearest to:

a. $500
b. $600
c. $700
d. $800

6.3 One thousand dollars is borrowed for one year at an interest rate of 1% per month. If this same sum of money were borrowed for the same period at an interest rate of 12% per year, the saving in interest charges would be closest to:

a. $0
b. $3
c. $5
d. $7

6.4 How much should a person invest in a fund that will pay 9%, compounded continuously, if he wishes to have $10,000 in the fund at the end of ten years?

a. $4000
b. $5000
c. $6000
d. $7000

6.5 A store charges 1.5% interest per month on credit purchases. This is equivalent to a nominal annual interest rate of:

a. 1.5%
b. 15.0%
c. 18.0%
d. 19.6%

6.6 A small company borrowed $10,000 to expand its business. The entire principal of $10,000 will be repaid in two years, but quarterly interest of $330 must be paid every three months. The nominal annual interest rate the company is paying is closest to:

a. 3.3%
b. 5.0%
c. 6.6%
d. 13.2%

6.7 A store's policy is to charge 3% interest every two months on the unpaid balance in charge accounts. The effective interest rate is closest to:

a. 6%
b. 12%
c. 15%
d. 19%

6.8 The effective interest rate on a loan is 19.56%. If there are 12 compounding periods per year, the nominal interest rate is closest to:

a. 1.5%
b. 4.5%
c. 9.0%
d. 18.0%

6.9 A deposit of $300 was made one year ago into an account paying monthly interest. If the account now has $320.52, the effective annual interest rate is closest to:

a. 7% c. 12%
b. 10% d. 15%

6.10 If the effective interest rate per year is 12%, based on monthly compounding, the nominal interest rate per year is closest to:

a. 8.5% c. 10.0%
b. 9.3% d. 11.4%

6.11 If 10% nominal annual interest is compounded daily, the effective annual interest rate is nearest to:

a. 10.00% c. 10.50%
b. 10.38% d. 10.75%

6.12 An individual wishes to deposit a certain quantity of money now so that he will have $500 at the end of five years. With interest at 4% per year, compounded semiannually, the amount of the deposit is nearest to:

a. $340 c. $410
b. $400 d. $416

6.13 A steam boiler is purchased on the basis of guaranteed performance. A test indicates that the operating cost will be $300 more per year than the manufacturer guaranteed. If the expected life of the boiler is 20 years, and the time value of money is 8%, the amount the purchaser should deduct from the purchase price to compensate for the extra operating cost is nearest to:

a. $2950 c. $4100
b. $3320 d. $5520

6.14 A consulting engineer bought a fax machine with one year's free maintenance. In the second year the maintenance cost is estimated at $20. In subsequent years the maintenance cost will increase $20 per year (that is, third year maintenance will be $40, fourth year maintenance will be $60, and so forth). The amount that must be set aside now at 6% interest to pay the maintenance costs on the fax machine for the first six years of ownership is nearest to:

a. $101 c. $229
b. $164 d. $284

6.15 An investor is considering buying a 20-year corporate bond. The bond has a face value of $1000 and pays 6% interest per year in two semiannual payments. Thus the purchaser of the bond will receive $30 every six months, and in addition he will receive $1000 at the end of 20 years, along with the last $30 interest payment. If the investor believes he should receive 8% annual interest, compounded semiannually, the amount he is willing to pay for the bond (bond value) is closest to:

a. $500 c. $700
b. $600 d. $800

6.16 Annual maintenance costs for a particular section of highway pavement are \$2000. The placement of a new surface would reduce the annual maintenance cost to \$500 per year for the first five years and to \$1000 per year for the next five years. The annual maintenance after ten years would again be \$2000. If maintenance costs are the only saving, the maximum investment that can be justified for the new surface, with interest at 4%, is closest to:

a. \$5500 c. \$10,000
b. \$7170 d. \$10,340

6.17 A project has an initial cost of \$10,000, uniform annual benefits of \$2400, and a salvage value of \$3000 at the end of its ten-year useful life. At 12% interest the net present worth (NPW) of the project is closest to:

a. \$2500 c. \$4500
b. \$3500 d. \$5500

6.18 A person borrows \$5000 at an interest rate of 18%, compounded monthly. Monthly payments of \$167.10 are agreed upon. The length of the loan is closest to:

a. 12 months c. 24 months
b. 20 months d. 40 months

6.19 A machine costing \$2000 to buy and \$300 per year to operate will save labor expenses of \$650 per year for eight years. The machine will be purchased if its salvage value at the end of eight years is sufficiently large to make the investment economically attractive. If an interest rate of 10% is used, the minimum salvage value must be closest to:

a. \$100 c. \$300
b. \$200 d. \$400

6.20 The amount of money deposited 50 years ago at 8% interest that would now provide a perpetual payment of \$10,000 per year is nearest to:

a. \$3000 c. \$50,000
b. \$8000 d. \$70,000

6.21 An industrial firm must pay a local jurisdiction the cost to expand its sewage treatment plant. In addition, the firm must pay \$12,000 annually toward the plant operating costs. The industrial firm will pay sufficient money into a fund that earns 5% per year to pay its share of the plant operating costs forever. The amount to be paid to the fund is nearest to:

a. \$15,000 c. \$60,000
b. \$30,000 d. \$240,000

6.22 At an interest rate of 2% per month, money will double in value in how many months?

a. 20 months c. 24 months
b. 22 months d. 35 months

6.23 A woman deposited $10,000 into an account at her credit union. The money was left on deposit for 80 months. During the first 50 months the woman earned 12% interest, compounded monthly. The credit union then changed its interest policy so that the woman earned 8% interest compounded quarterly during the next 30 months. The amount of money in the account at the end of 80 months is nearest to:

a. $10,000 c. $15,000
b. $12,500 d. $20,000

6.24 An engineer deposited $200 quarterly in her savings account for three years at 6% interest, compounded quarterly. Then for five years she made no deposits or withdrawals. The amount in the account after eight years is closest to:

a. $1200 c. $2400
b. $1800 d. $3600

6.25 A sum of money, Q, will be received six years from now. At 6% annual interest the present worth of Q is $60. At this same interest rate the value of Q ten years from now is closest to:

a. $60 c. $90
b. $77 d. $107

6.26 If $200 is deposited in a savings account at the beginning of each year for 15 years and the account earns interest at 6%, compounded annually, the value of the account at the end of 15 years will be most nearly:

a. $4500 c. $4900
b. $4700 d. $5100

6.27 The maintenance expense on a piece of machinery is estimated as follows:

Year	1	2	3	4
Maintenance	$150	$300	$450	$600

If interest is 8%, the equivalent uniform annual maintenance cost is closest to:

a. $250 c. $350
b. $300 d. $400

6.28 A payment of $12,000 six years from now is equivalent, at 10% interest, to an annual payment for eight years starting at the end of this year. The annual payment is closest to:

a. $1000 c. $1400
b. $1200 d. $1600

6.29 A manufacturer purchased $15,000 worth of equipment with a useful life of six years and a $2000 salvage value at the end of the six years. Assuming a 12% interest rate, the equivalent uniform annual cost (EUAC) is nearest to:

a. $1500 c. $3500
b. $2500 d. $4500

6.30 Consider a machine as follows:

Initial cost: $80,000
End-of-useful-life salvage value: $20,000
Annual operating cost: $18,000
Useful life: 20 years

Based on 10% interest, the equivalent uniform annual cost for the machine is closest to:

a. $21,000 c. $25,000
b. $23,000 d. $27,000

6.31 Consider a machine as follows:

Initial cost: $80,000
Annual operating cost: $18,000
Useful life: 20 years

What must be the salvage value of the machine at the end of 20 years for the machine to have an equivalent uniform annual cost of $27,000? Assume a 10% interest rate. The salvage value S_{20} is closest to:

a. $10,000 c. $30,000
b. $20,000 d. $40,000

6.32 Twenty-five thousand dollars is deposited in a savings account that pays 5% interest, compounded semiannually. Equal annual withdrawals are to be made from the account beginning one year from now and continuing forever. The maximum amount of the equal annual withdrawals is closest to:

a. $625 c. $1250
b. $1000 d. $1265

6.33 An investor is considering the investment of $10,000 in a piece of land. The property taxes are $100 per year. The lowest selling price the investor must receive if she wishes to earn a 10% interest rate after keeping the land for ten years is:

a. $20,000 c. $23,000
b. $21,000 d. $27,000

6.34 The rate of return for a $10,000 investment that will yield $1000 per year for 20 years is closest to:

a. 1% c. 8%
b. 4% d. 12%

6.35 An engineer invested $10,000 in a company. In return he received $600 per year for six years and his $10,000 investment back at the end of the six years. His rate of return on the investment was closest to:

a. 6% c. 12%
b. 10% d. 15%

6.36 An engineer made ten annual end-of-year purchases of $1000 of common stock. At the end of the tenth year, just after the last purchase, the engineer sold all the stock for $12,000. The rate of return received on the investment is closest to:

a. 2% c. 8%
b. 4% d. 10%

6.37 A company is considering buying a new piece of machinery.

Initial cost: $80,000
End-of-useful-life salvage value: $20,000
Annual operating cost: $18,000
Useful life: 20 years

The machine will produce an annual saving in material of $25,700. What is the before-tax rate of return if the machine is installed? The rate of return is closest to:

a. 6% c. 10%
b. 8% d. 15%

6.38 Consider the following situation: Invest $100 now and receive two payments of $102.15—one at the end of year 3 and one at the end of year 6. The rate of return is nearest to:

a. 6% c. 10%
b. 8% d. 18%

6.39 Two mutually exclusive alternatives are being considered:

Year	A	B
0	–$2500	–$6000
1	+746	+1664
2	+746	+1664
3	+746	+1664
4	+746	+1664
5	+746	+1664

The rate of return on the difference between the alternatives is closest to:

a. 6% c. 10%
b. 8% d. 12%

6.40 A project will cost $50,000. The benefits at the end of the first year are estimated to be $10,000, increasing $1000 per year in subsequent years. Assuming a 12% interest rate, no salvage value, and an eight-year analysis period, the benefit-cost ratio is closest to:

a. 0.78 c. 1.28
b. 1.00 d. 1.45

6.41 Two alternatives are being considered:

	A	B
Initial cost	\$500	\$800
Uniform annual benefit	\$140	\$200
Useful life, years	8	8

The benefit-cost ratio of the difference between the alternatives, based on a 12% interest rate, is closest to:

a. 0.60
b. 0.80
c. 1.00
d. 1.20

6.42 An engineer will invest in a mining project if the benefit-cost ratio is greater than 1.00, based on an 18% interest rate. The project cost is \$57,000. The net annual return is estimated at \$14,000 for each of the next eight years. At the end of eight years the mining project will be worthless. The benefit-cost ratio is closest to:

a. 0.60
b. 0.80
c. 1.00
d. 1.20

6.43 A city has retained your firm to do a benefit-cost analysis of the following project:

Project cost: \$60,000,000
Gross income: \$20,000,000 per year
Operating costs: \$5,500,000 per year
Salvage value after ten years: None

The project life is ten years. Use 8% interest in the analysis. The computed benefit-cost ratio is closest to:

a. 0.80
b. 1.00
c. 1.20
d. 1.60

6.44 A piece of property is purchased for \$10,000 and yields a \$1000 yearly profit. If the property is sold after five years, the minimum price to break even, with interest at 6%, is closest to:

a. \$5000
b. \$6500
c. \$7700
d. \$8300

6.45 Given two machines:

	A	B
Initial cost	\$55,000	\$75,000
Total annual costs	\$16,200	\$12,450

With interest at 10% per year, at what service life do these two machines have the same equivalent uniform annual cost? The service life is closest to:

a. four years
b. five years
c. six years
d. eight years

6.46 A machine part that is operating in a corrosive atmosphere is made of low-carbon steel. It costs $350 installed, and lasts six years. If the part is treated for corrosion resistance it will cost $700 installed. How long must the treated part last to be as economical as the untreated part, if money is worth 6%?

a. 8 years
b. 11 years
c. 15 years
d. 17 years

6.47 A firm has determined that the two best paints for its machinery are Tuff-Coat at $45 per gallon and Quick at $22 per gallon. The Quick paint is expected to prevent rust for five years. Both paints take $40 of labor per gallon to apply, and both cover the same area. If a 12% interest rate is used, how long must the Tuff-Coat paint prevent rust to justify its use?

a. Five years
b. Six years
c. Seven years
d. Eight years

6.48 Two alternatives are being considered:

	A	B
Cost	$1000	$2000
Useful life in years	10	10
End-of-useful-life salvage value	$100	$400

The net annual benefit of alternative A is $150. If interest is 8%, what must be the net annual benefit of alternative B for the two alternatives to be equally desirable?

a. $150
b. $200
c. $225
d. $275

6.49 A $5000 municipal bond is offered for sale. It will provide 8% annual interest by paying $200 to the bondholder every six months. At the end of ten years, the $5000 will be paid to the bondholder along with the final $200 interest payment. If you consider 12% nominal annual interest, compounded semiannually, an appropriate bond yield, the amount you would be willing to pay for the bond is closest to:

a. $2750
b. $3850
c. $5000
d. $7400

6.50 A municipal bond is being offered for sale for $10,000. It is a zero-coupon bond, that is, the bond pays no interest during its 15-year life. At the end of 15 years the owner of the bond will receive a single payment of $26,639. The bond yield is closest to:

a. 4%
b. 5%
c. 6%
d. 7%

6.51 A firm is considering purchasing $8000 of small hand tools for use on a production line. It is estimated that the tools will reduce the amount of required overtime work by $2000 the first year, with this amount increasing by $1000 per year thereafter. The payback period for the hand tools is closest to:

a. 2.00 years
b. 2.50 years
c. 2.75 years
d. 3.00 years

6.52 Special tools for the manufacture of finished plastic products cost $15,000 and have an estimated $1000 salvage value at the end of an estimated three-year useful life and recovery period. The third-year straight-line depreciation is closest to:

a. $3000
b. $3500
c. $4000
d. $4500

6.53 Refer to the facts of Problem 6.52. The first-year MACRS depreciation is closest to:

a. $3000
b. $3500
c. $4000
d. $5000

6.54 An engineer is considering the purchase of an annuity that will pay $1000 per year for ten years. The engineer feels he should obtain a 5% rate of return on the annuity after considering the effect of an estimated 6% inflation per year. The amount he would be willing to pay to purchase the annuity is closest to:

a. $1500
b. $3000
c. $4500
d. $6000

6.55 An automobile costs $20,000 today. You can earn 12% tax-free on an *auto purchase account*. If you expect the cost of the auto to increase by 10% per year, the amount you would need to deposit in the account to provide for the purchase of the auto five years from now is closest to:

a. $12,000
b. $14,000
c. $16,000
d. $18,000

6.56 An engineer purchases a building lot for $40,000 cash and plans to sell it after five years. If he wants an 18% before-tax rate of return, after taking the 6% annual inflation rate into account, the selling price must be nearest to:

a. $55,000
b. $65,000
c. $75,000
d. $125,000

6.57 A piece of equipment with a list price of $450 can actually be purchased for either $400 cash or $50 immediately plus four additional annual payments of $115.25. All values are in dollars of current purchasing power. If the typical customer considered a 5% interest rate appropriate, the inflation rate at which the two purchase alternatives are equivalent is nearest to:

a. 5%
b. 6%
c. 8%
d. 10%

SOLUTIONS

6.1 d.

$$F = A(F/A,i,n) = 400(F/A,2\%,100)$$
$$= 400(312.23) = \$124{,}890$$

6.2 d.

$$P = A(P/A,i,n) + G(P/G,i,n)$$
$$= 120(P/A,4\%,5) + 30(P/G,4\%,5)$$
$$= 120(4.452) + 30(8.555) = \$791$$

6.3 d.

At $i = 1\%$/month: $F = 1000(1 + 0.01)^{12} = \$1126.83$
At $i = 12\%$/year: $F = 1000(1 + 0.12)^{1} = 1120.00$
Saving in interesting charges = 1126.83 – 1120.00 = \$6.83

6.4 a.

$$P = Fe^{-rn} = 10{,}000e^{-0.09(10)} = 4066$$

6.5 c. The nominal interest rate is the annual interest rate ignoring the effect of any compounding. Nominal interest rate = 1.5% × 12 = 18%.

6.6 d. The interest paid per year = 330 × 4 = 1320. The nominal annual interest rate = 1320/10,000 = 0.132 = 13.2%.

6.7 d.

$$i_e = (1 + r/m)^m - 1 = (1 + 0.03)6 - 1 = 0.194 = 19.4\%$$

6.8 d.

$$i_e = (1 + r/m)^m - 1$$
$$r/m = (1 + i^e)^{1/m} - 1 = (1 + 0.1956)1/12 - 1 = 0.015$$
$$r = 0.015(m) = 0.015 \times 12 = 0.18 = 18\%$$

6.9 a.

$$i_e = 20.52/300 = 0.0684 = 6.84\%$$

6.10 d.

$$i^e = (1 + r/m)^m - 1$$
$$0.12 = (1 + r/12)12 - 1$$
$$(1.12)1/12 = (1 + r/12)$$
$$1.00949 = (1 + r/12)$$
$$r = 0.00949 \times 12 = 0.1138 = 11.38\%$$

6.11 c.

$$i^e = (1 + r/m)^m - 1 = (1 + 0.10/365)^{365} - 1 = 0.1052 = 10.52\%$$

6.12 c.

$$P = F(P/F,i,n) = 500(P/F,2\%,10) = 500(0.8203) = \$410$$

6.13 a.

$$P = 300(P/A,8\%,20) = 300(9.818) = \$2945$$

6.14 c. Using single payment present worth factors:

$$P = 20(P/F,6\%,2) + 40(P/F,6\%,3) + 60(P/F,6\%,4) + 80(P/F,6\%,5) + 100(P/F,6\%,6) = \$229$$

Alternate solution using the gradient present worth factor:

$$P = 20(P/G,6\%,6) = 20(11.459) = \$229$$

6.15 d.

$$\begin{aligned} \text{PW} &= 30\,(P/A,4\%,40) + 1000(P/F,4\%,40) \\ &= 30(19.793) + 1000(0.2083) = \$802 \end{aligned}$$

6.16 d. Benefits are \$1500 per year for the first five years and \$1000 per year for the subsequent five years.

As Exhibit 6.16 indicates, the benefits may be considered as \$1000 per year for ten years, plus an additional \$500 benefit in each of the first five years.

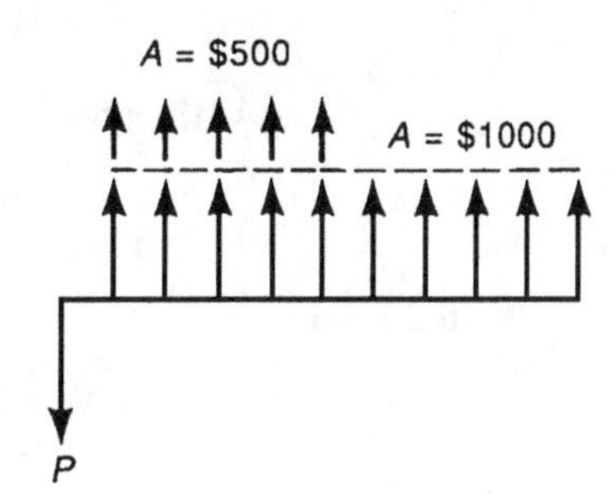

Exhibit 6.16

$$\begin{aligned} \text{maximum investment} &= \text{present worth of benefits} \\ &= 1000(P/A,4\%,10) + 500(P/A,4\%,5) \\ &= 1000(8.111) + 500(4.452) = \$10{,}337 \end{aligned}$$

6.17 c.

$$\begin{aligned} \text{NPW} &= \text{PW of benefits} - \text{PW of cost} \\ &= 2400(P/A,12\%,10) + 3000(P/F,12\%,10) - 10{,}000 = \$4526 \\ &= 2400(5.65) + 3000(.32) - 10{,}000 = \$4526 \end{aligned}$$

6.18 d.

$$\begin{aligned} \text{PW of benefits} &= \text{PW of cost} \\ 5000 &= 167.10(P/A,1.5\%,n) \\ (P/A,1.5\%,n) &= 5000/167.10 = 29.92 \end{aligned}$$

From the $1\frac{1}{2}\%$ interest table, $n = 40$.

6.19 c.

$$\begin{aligned} \text{NPW} &= \text{PW of benefits} - \text{PW of cost} = 0 \\ &= (650 - 300)(P/A,10\%,8) + S_8\,(\text{P/F},10\%,8) - 2000 = 0 \\ &= 350(5.335) + S_8(0.4665) - 2000 = 0 \\ S_8 &= 132.75/0.4665 = \$285 \end{aligned}$$

6.20 a. The amount of money needed now to begin the perpetual payments is $P' = A/i = 10{,}000/0.08 = 125{,}000$. From this we can compute the amount of money, P, that would need to have been deposited 50 years ago:

$$P = 125{,}000(P/F,8\%,50) = 125{,}000(0.0213) = \$2663$$

6.21 d.

$$P = A/i = 12{,}000/0.5 = \$240{,}000$$

6.22 d.

$$\begin{aligned} 2 &= 1(F/P,i,n) \\ (F/P,2\%,n) &= 2 \end{aligned}$$

From the 2% interest table, n = about 35 months.

6.23 d. At the end of 50 months

$$F = 10{,}000(F/P,1\%,50) = 10{,}000(1.645) = \$16{,}450$$

At the end of 80 months

$$F = 16{,}450(F/P,2\%,10) = 16{,}450(1.219) = \$20{,}053$$

6.24 d.

$$\begin{aligned} \text{FW} &= 200(\text{F/A},1.5\%,12)(F/P,\ 1.5\%,20) \\ &= 200(13.041)(1.347) = \$3513 \end{aligned}$$

6.25 d. The present amount $P = 60$ is equivalent to Q six years hence at 6% interest. The future sum F may be calculated by either of two methods:

$$F = Q(F/P,6\%,4) \text{ and } Q = 60(F/P,6\%,6)$$

or

$$F = P(F/P,6\%,10)$$

Since P is known, the second equation may be solved directly.

$$F = P(F/P,6\%,10) = 60(1.791) = \$107$$

6.26 c.

$$F' = A(F/A,i,n) = 200(F/A,6\%,15) = 200(23.276) = \$4655.20$$
$$F = F'(F/P,i,n) = 4655.20(F/P,6\%,1) = 4655.20(1.06) = \$4935$$

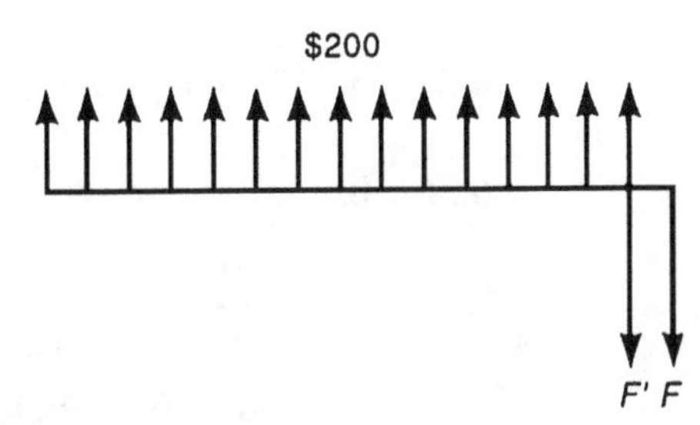

Exhibit 6.26

6.27 c.

$$\text{EUAC} = 150 + 150(A/G,8\%,4) = 150 + 150(1.404) = \$361$$

6.28 b.

$$\text{Annual payment} = 12{,}000(P/F,10\%,6)(A/P,10\%,8)$$
$$= 12{,}000(0.5645)(0.1874) = \$1269$$

6.29 c.

$$\text{EUAC} = 15{,}000(A/P,12\%,6) - 2000(A/F,12\%,6)$$
$$= 15{,}000(0.2432) - 2000(0.1232) = \$3402$$

6.30 d.

$$\text{EUAC} = 80{,}000(A/P,10\%,20) - 20{,}000(A/F,10\%,20)$$
$$+ \text{annual operating cost}$$
$$= 80{,}000(0.1175) - 20{,}000(0.0175) + 18{,}000$$
$$= 9400 - 350 + 18{,}000 = \$27{,}050$$

6.31 b.

$$\text{EUAC} = \text{EUAB}$$
$$27{,}000 = 80{,}000(A/P,10\%,20) + 18{,}000 - S_{20}(A/F,10\%,20)$$
$$= 80{,}000(0.1175) + 18{,}000 - S_{20}(0.0175)$$
$$S_{20} = (27{,}400 - 27{,}000)/0.0175 = \$22{,}857$$

6.32 d. The general equation for an infinite life, $P = A/i$, must be used to solve the problem.

$$ie = (1 + 0.025)2 - 1 = 0.050625$$

The maximum annual withdrawal will be $A = Pi = 25{,}000(0.050625) = \$1266$.

6.33 d.

$$\text{Minimum sale price} = 10{,}000(F/P,10\%,10) + 100(F/A,10\%,10)$$
$$= 10{,}000(2.594) + 100(15.937) = \$27{,}530$$

6.34 **c.**

$$NPW = 1000(P/A,i,20) - 10{,}000 = 0$$
$$(P/A,i,20) = 10{,}000/1000 = 10$$

From interest tables: $6\% < i < 8\%$.

6.35 **a.** The rate of return was $600/10{,}000 = 0.06 = 6\%$.

6.36 **b.**

$$F = A(F/A,i,n)$$
$$12{,}000 = 1000(F/A,i,10)$$
$$(F/A,i,10) = 12{,}000/1000 = 12$$

In the 4% interest table: $(F/A,4\%,10) = 12.006$, so $i = 4\%$.

6.37 **b.**

$$\text{PW of cost} = \text{PW of benefits}$$
$$80{,}000 = (25{,}700 - 18{,}000)(P/A,i,20) + 20{,}000(P/F,i,20)$$

Try $i = 8\%$.

$$80{,}000 \stackrel{?}{=} 7700(9.818) + 20{,}000(0.2145) = 79{,}889$$

Therefore, the rate of return is very close to 8%.

6.38 **d.**

$$\text{PW of cost} = \text{PW of benefits}$$
$$100 = 102.15(P/F,i,3) + 102.15(P/F,i,6)$$

Solve by trial and error. Try $i = 12\%$.

$$100 \stackrel{?}{=} 102.15(0.7118) + 102.15(0.5066) = 124.46$$

The PW of benefits exceeds the PW of cost. This indicates that the interest rate i is too low. Try $i = 18\%$.

$$100 \stackrel{?}{=} 102.15(0.6086) + 102.15(0.3704) = 100.00$$

Therefore, the rate of return is 18%.

6.39 **c.** The difference between the alternatives:

$$\text{Incremental cost} = 6000 - 2500 = \$3500$$

$$\text{Incremental annual benefit} = 1664 - 746 = \$918$$
$$\text{PW of cost} = \text{PW of benefits}$$
$$3500 = 918(P/A,i,5)$$
$$(P/A,i,5) = 3500/918 = 3.81$$

From the interest tables, i is very close to 10%.

6.40 **c.**

$$B/C = \frac{\text{PW of benefits}}{\text{PW of cost}} = \frac{10{,}000\ (P/A,\ 12\%,\ 8) + 1000\ (P/G,\ 12\%,\ 8)}{50{,}000}$$

$$= \frac{10{,}000\ (4.968) + 1000\ (14.471)}{50{,}000} = 1.28$$

6.41 c.

$$\text{B/C} = \frac{\text{PW of benefits}}{\text{PW of cost}} = \frac{60(P/A,12\%,8)}{300} = \frac{60(4.968)}{300} = 0.99$$

Alternate solution:

$$\text{B/C} = \frac{\text{EUAB}}{\text{EUAC}} = \frac{60}{300(A/P,12\%,8)} = \frac{60}{300(0.2013)} = 0.99$$

6.42 a.

$$\text{B/C} = \frac{\text{PW of benefits}}{\text{PW of cost}} = \frac{14{,}000(P/A,18\%,8)}{57{,}000} = \frac{14{,}000(4.078)}{57{,}000} = 1.00$$

6.43 d.

$$\text{B/C} = \frac{\text{EUAB}}{\text{EUAC}} = \frac{20{,}000{,}000 - 5{,}500{,}000}{60{,}000{,}000(A/P,8\%,10)} = 1.62$$

6.44 c.

$$\begin{aligned} F &= 10{,}000(F/P,6\%,5) - 1000(F/A,6\%,5) \\ &= 10{,}000(1.338) - 1000(5.637) = \$774 \end{aligned}$$

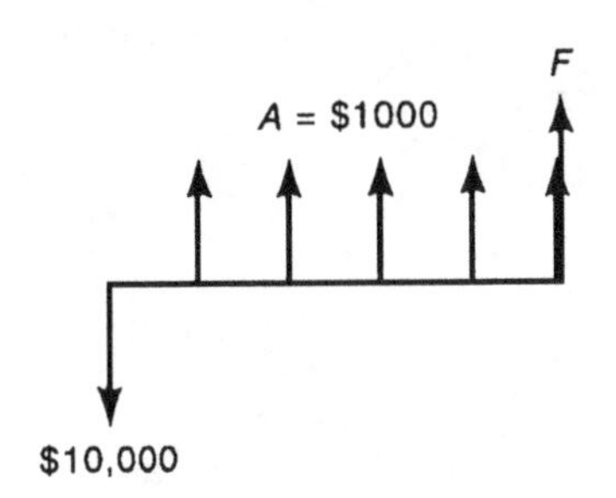

Exhibit 6.44

6.45 d.

$$\begin{aligned} \text{PW of cost}_A &= \text{PW of cost}_B \\ 55{,}000 + 16{,}200(P/A,10\%,n) &= 75{,}000 + 12{,}450(P/A,10\%,n) \\ (P/A,10\%,n) &= (75{,}000 - 55{,}000)/(16{,}200 - 12{,}450) \\ &= 5.33 \end{aligned}$$

From the 10% interest tables, $n = 8$ years.

6.46 c.

$$\begin{aligned} \text{EUAC}_{\text{untreated}} &= \text{EUAC}_{\text{treated}} \\ 350(A/P,6\%,6) &= 700(A/P,6\%,n) \\ 350(0.2034) &= 700(A/P,6\%,n) \\ (A/P,6\%,n) &= 71.19/700 = 0.1017 \end{aligned}$$

From the 6% interest table, $n = 15+$ years.

6.47 d.

$$\text{EUAC}_{\text{T-C}} = \text{EUAC}_{\text{Quick}}$$
$$(45 + 40)(A/P,12\%,n) = (22 + 40)(A/P,12\%,5)$$
$$(A/P,12\%,n) = 17.20/85 = 0.202$$

From the 12% interest table, $n = 8$.

6.48 d. At breakeven,

$$\text{NPWA} = \text{NPWB}$$
$$150(P/A,8\%,10) + 100(P/F,8\%,10) - 1000 = \text{NAB}(P/A,8\%,10) + 400(P/F,8\%,10) - 2000$$
$$52.82 = 6.71(\text{NAB}) - 1814.72$$

Net annual benefit (NAB) = (1814.72 + 52.82)/6.71 = \$278

6.49 b. The number of six-month compounding periods in this problem is 20. So $n = 20$ and 12%/2 = 6% is the interest rate for the six-month interest period.

$$\begin{aligned}\text{Bond value} &= \text{PW of all future benefits}\\ &= 200(P/A,6\%,20) + 5000(P/F,6\%,20)\\ &= 200(11.470) + 5000(0.3118) = \$3853\end{aligned}$$

6.50 d. We know $P = 10{,}000$, $F = 26{,}639$, $n = 15$, and i = bond yield. Using the equation for the single payment compound amount:

$$F = P(1 + i)^n$$
$$26{,}639 = 10{,}000(1 + i)^{15}$$
$$2.66391/15 = (1 + i)$$
$$1.0675 = 1 + i$$
$$i = 0.0675 = 6.75\%$$

6.51 c. The annual benefits are \$2000, \$3000, \$4000, \$5000, and so on. The payback period is the time when \$8000 of benefits are received. This will occur in 2.75 years.

6.52 d.

$$D_3 = (C - S)/n = (15{,}000 - 1000)/3 = \$4666$$

6.53 d. From the modified ACRS table (Table 6.3) read for the first recovery year and three-year recovery the MACRS depreciation is 33.33% × 15,000 = \$5000.

6.54 d.

$$d = i + f + (i \times f) = 0.05 + 0.06 + 0.05(0.06) = 0.113 = 11.3\%$$

$$P = A(P/A,11.3\%,10) = 1000\left[\frac{(1+0.113)^{10} - 1}{0.113(1+0.113)^{10}}\right]$$

$$= 1000\left[\frac{1.9171}{0.3296}\right] = \$5816$$

6.55 d.

Cost of auto five years hence $(F) = P(1 + \text{inflation rate})^n$

$= 20{,}000(1 + 0.10)5 = 32{,}210$

Amount to deposit now to have $32,210 available five years hence:

$P = F(P/F,i,n) = 32{,}210\ (P/F,12\%,5) = 32{,}210(0.5674) = \$18{,}276$

6.56 d.

$$\text{Selling price } (F) = 40{,}000(F/P,18\%,5)(F/P,6\%,5)$$
$$= 40{,}000(2.288)(1.338) = \$122{,}500$$

6.57 b.

$$\text{PW of cash purchase} = \text{PW of installment purchase}$$
$$400 = 50 + 115.25(P/A,d,4)$$
$$(P/A,d,4) = 350/11.25 = 3.037$$

From the interest tables, $d = 12\%$.

$$d = i + f + i\,(f)$$
$$0.12 = 0.05 + f + 0.05f$$
$$f = 0.07/1.05 = 0.0667 = 6.67\%$$

CHAPTER 7

Statics

OUTLINE

Statics is concerned with the forces of interaction between bodies or within bodies of mechanical systems that have no significant accelerations. Typical engineering problems require the analyst to predict forces induced at certain points by known forces applied at other points.

INTRODUCTORY CONCEPTS IN MECHANICS

Newton's Laws of Motion

Every element of a mechanical system must satisfy **Newton's second law of motion**, which states that the resultant force f acting on the element is related to the acceleration a of the element by

$$f = ma$$

where m represents the mass of the element. This entire chapter deals with the special case in which $a = 0$. **Newton's third law** requires that the force exerted on a body A by a body B is of equal magnitude and opposite direction to the force exerted on body B by body A. A careful, unambiguous account of this law is essential for successful analysis of forces; rules for ensuring that such an analysis is done properly will be reviewed in the Equilibrium section.

Newton's Law of Gravitation

Every pair of material elements is attracted toward one another by a pair of *gravitational* forces, the magnitude of which is given by

$$f_g = \frac{\gamma m_1 m_2}{r^2} \tag{7.1}$$

where γ is the **universal gravitational constant** (about 6.7×10^{-11} N•m^2/kg^2), m_1 and m_2 are the masses of the elements, and r is the distance between them. Because very large masses are necessary to make these forces significant, f_g can often be neglected. A notable exception is the force exerted by the earth (which has a mass of about 6×10^{24} kg) on objects near its surface. In this case, the gravitational force has a magnitude given by

$$f_g = mg \tag{7.2}$$

where m is the mass of the attracted object and g (which is related to the earth's mass and radius) has a value that varies between 9.78 and 9.83 N/kg with geographic location. In technically correct terminology, the word *weight* refers to the force of gravity by the earth. However, *weight* has other, closely related meanings that can be a source of serious confusion to the analyst of mechanical systems.

Dimensions and Units of Measurement

Every quantity in mechanics can be expressed in terms of three fundamental quantities. In the SI system of units, these are *mass*, *length*, and *time*, and units are the kilogram (kg), the meter (m), and the second (s), respectively. In the SI system, Newton's second law provides a definition of a fourth unit in terms of the three fundamental ones. The **newton** (N) is defined as the force required to accelerate a 1-kilogram body at the rate of one meter per second per second (m/s^2). This can be expressed symbolically as N = kg • m/s^2. See Table 7.1.

Considerable confusion results from the introduction and widespread use (especially outside the United States) of the **kilogram-force (kgf)**, defined to be 9.80665 N.

Table 7.1 Units common to mechanics

Quantity	SI Unit	Coherent U.S. Engineering Unit	Common Noncoherent Unit
Mass	kilogram (kg)	slug = lbf • s²/ft ≈ 14.594 kg	pound-mass (lbm) $\approx \frac{1}{32.174}$ slug ≈ 0.4536 kg
Length	meter (m)	foot (ft) = 0.3048 m	inch (in.) = 25.4 mm
Time	second (s)	second (s)	minute (min) = 60 s
Force	newton (N) N = kg • m/s²	pound-force (lbf) ≈ 4.448 N	kilogram-force (kgf) or kilopound (kp) = 9.80665 N

Errors stemming from force values given in this noncoherent unit can be avoided by converting to newtons before doing further calculation. For example, an estimate of the acceleration imparted by a force of 217 kgf to a 100-kg body would proceed as follows:

$$f = (217 \text{ kgf})\,(9.80665 \text{ N/kgf}) = 2128 \text{ N}$$

$$a = \frac{f}{m} = \frac{2128 \text{ N}}{100 \text{ kg}} = 21.28 \left(\text{N/kg} = \frac{\text{kg} \bullet \text{m/s}^2}{\text{kg}} = \frac{\text{m}}{\text{s}^2} \right)$$

Errors in unit conversion, and many errors of analysis, will be revealed by the practice of appending unit symbols to *every* number and algebraically reducing combinations of symbols resulting from multiplication and division.

VECTOR GEOMETRY AND ALGEBRA

Handling many of the problems that arise in mechanics can be greatly simplified by means of the operations of vector analysis. Their use will be successful if attention is given to the geometric meaning of each operation.

A **vector** represents a physical quantity that can be characterized by a magnitude and a direction in space, which will be taken to be three-dimensional here. We use an arrow to depict each vector, with the length of the arrow proportional to the magnitude represented, and the orientation representing the direction of the vector. Boldface letters are used in text and equations to represent vectors. The **magnitude** of vector **a** is written as $|\mathbf{a}|$ and sometimes as a.

Addition and Subtraction

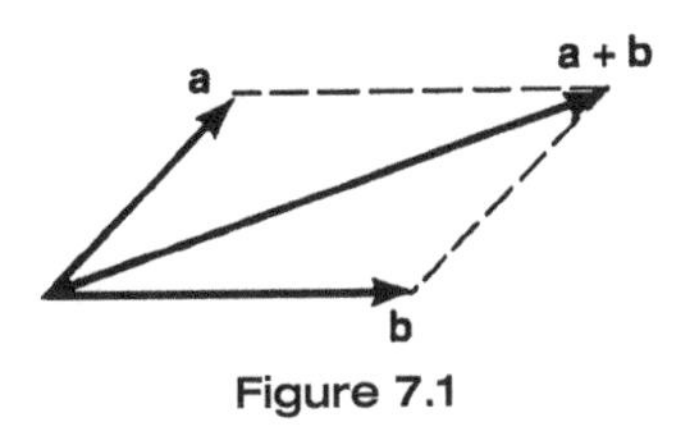

Figure 7.1

The **sum of two vectors** is a vector determined according to the so-called parallelogram law, as illustrated in Figure 7.1. The sum is the vector represented by the diagonal of the parallelogram formed by the two vectors placed with their tails coincident. The commutative law,

$$\mathbf{a} + \mathbf{b} = \mathbf{b} + \mathbf{a}$$

and the associative law,

$$\mathbf{a} + (\mathbf{b} + \mathbf{c}) = (\mathbf{a} + \mathbf{b}) + \mathbf{c}$$

both follow from this definition.

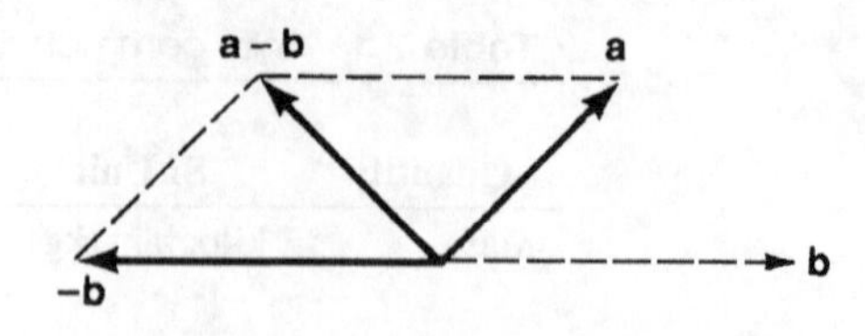

Figure 7.2

The negative of a vector **b** is defined to be of the same magnitude as, but of the opposite direction to, **b**, and it is written as –**b**. Subtraction of two vectors is then defined by

$$\mathbf{a} - \mathbf{b} = \mathbf{a} + (-\mathbf{b})$$

as indicated in Figure 7.2.

Multiplication by a Scalar

The product of a scalar, p, and a vector, **a**, is the vector written as $p\mathbf{a}$, and it is defined to have a magnitude of $p|\mathbf{a}|$, and direction the same as, or the opposite of, **a**, depending on whether p is positive or negative. The following laws can be readily verified from these definitions:

$$p(q\mathbf{a}) = (pq)\mathbf{a}$$
$$(p+q)\mathbf{a} = p\mathbf{a} + q\mathbf{a}$$
$$p(\mathbf{a}+\mathbf{b}) = p\mathbf{a} + p\mathbf{b}$$

Addition of two or more vectors is sometimes called the **composition** of the vectors. A reversal of this process is called the **resolution** of a vector, that is, determining a set of vectors (usually in prescribed directions) the sum of which will be the given vector.

Example 7.1

Resolve the vector **f** of magnitude 75 kN into two vectors (components) in the directions of lines L_1 and L_2 shown in Exhibit 1.

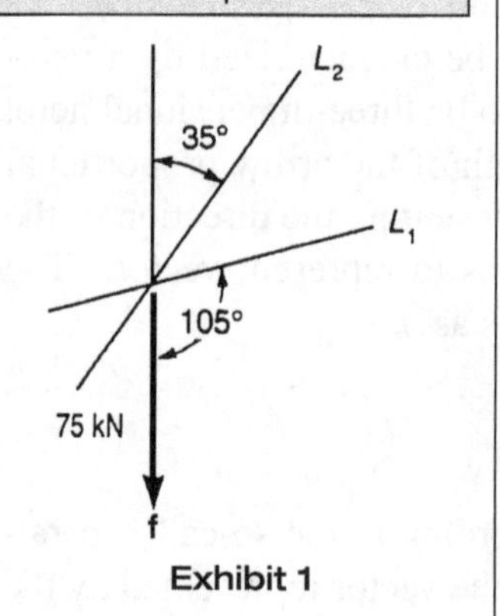

Exhibit 1

Solution

The given vector is to be the diagonal of the parallelogram having the desired components as sides. The parallelogram can be completed by drawing lines through the head of the vector, parallel to the given lines. The magnitudes of the two components can then be determined by applying the trigonometric law of sines:

$$|\mathbf{f}_1| = \frac{\sin 35°}{\sin 40°}(75\text{ kN}) = 66.9\text{ kN}$$

$$|\mathbf{f}_2| = \frac{\sin 105°}{\sin 40°}(75\text{ kN}) = 112.7\text{ kN}$$

Dot Product

The **dot product** of two vectors **a** and **b** is a scalar (sometimes called the scalar product or inner product) that is equal to the product of the magnitudes of the vectors and the cosine of the angle θ between the vectors. It is written as

$$\mathbf{a} \cdot \mathbf{b} = |\mathbf{a}||\mathbf{b}| \cos\theta \tag{7.3}$$

A special case is the dot product of a vector with itself,

$$\mathbf{a} \bullet \mathbf{a} = |\mathbf{a}||\mathbf{a}| \cos\theta$$

which provides a way of expressing the magnitude of a vector:

$$|\mathbf{a}| = \sqrt{\mathbf{a} \bullet \mathbf{a}} \tag{7.4}$$

The commutative law,

$$\mathbf{a} \bullet \mathbf{b} = \mathbf{b} \bullet \mathbf{a}$$

and the distributive law,

$$\mathbf{a} \cdot (\mathbf{b} + \mathbf{c}) = \mathbf{a} \cdot \mathbf{b} + \mathbf{a} \cdot \mathbf{c}$$

can both be verified from the above definitions.

Unit Vectors and Projections

An extremely useful tool is the **unit vector**. It has a magnitude of 1 and is designated here by the symbol **e**. Unit vectors can be introduced (defined) by giving the direction in terms of the geometry of the application, or defined in terms of a specified vector by multiplying the vector by the reciprocal of its magnitude. For example, the unit vector in the direction of **a** is given by

$$\mathbf{e}_a = \frac{1}{|\mathbf{a}|}\mathbf{a}$$

Rearranged, this relationship expresses the vector **a** in terms of its magnitude and a unit vector that gives its direction:

$$\mathbf{a} = |\mathbf{a}|\mathbf{e}_a \tag{7.5}$$

The **projection** of vector **a** onto a line L is the vector from the projection of the tail of **a** onto L to the projection of the head of **a** onto L, as indicated in Figure 7.3.

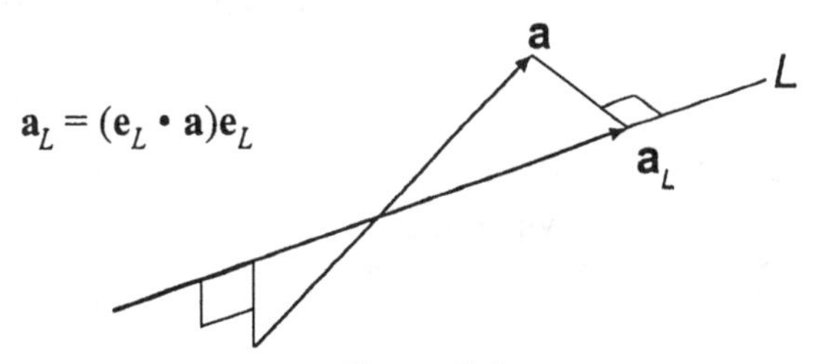

Figure 7.3

The magnitude of this projection will be the product of $|\mathbf{a}|$ and the cosine of the angle between $\mathbf{a}$ and L. With unit vector $\mathbf{e}_L$, defined to be parallel to L, the projection of $\mathbf{a}$ onto L can be expressed as

$$\mathbf{a}_L = (\mathbf{e}_L \bullet \mathbf{a})\mathbf{e}_L$$

That is, vector projection onto a line in a selected direction can be evaluated by dot-multiplying the vector by a unit vector in that direction. As indicated in the next section, this approach provides the most direct means of obtaining equivalent scalar relationships from a vector relationship.

Vector and Scalar Equations

Many physical laws, such as Newton's laws of motion, are best expressed by vector equations. In general, a vector equation can provide up to three independent scalar equations. In Example 7.1, the directions of the lines L_1 and L_2 might be determined by the orientations of two members of a structural truss, and the direction of $\mathbf{f}$ by a gravitational force. Then the relationship $\mathbf{f}_1 + \mathbf{f}_2 = \mathbf{f}$ might be a requirement of equilibrium and geometry, from which the magnitudes of $\mathbf{f}_1$ and $\mathbf{f}_2$ are to be determined in terms of the magnitude of $\mathbf{f}$. With the unit vectors $\mathbf{e}_1$ and $\mathbf{e}_2$ introduced as shown, the equation

$$f_1\mathbf{e}_1 + f_2\mathbf{e}_2 = \mathbf{f}$$

brings the unknowns f_1 and f_2 into evidence and makes available two scalar equations for their determination. A corresponding scalar equation can be obtained by dot-multiplying each member of the vector equation by a selected vector; for example, with the unit vectors already introduced, we can write the two equations

$$f_1\mathbf{e}_1 \bullet \mathbf{e}_1 + f_2\mathbf{e}_1 \bullet \mathbf{e}_2 = \mathbf{e}_1 \bullet \mathbf{f}$$
$$f_1\mathbf{e}_2 \bullet \mathbf{e}_1 + f_2\mathbf{e}_2 \bullet \mathbf{e}_2 = \mathbf{e}_2 \bullet \mathbf{f}$$

reduce them to

$$f_1 + (\cos 40°)f_2 = (75 \text{ kN}) \cos 105°$$
$$(\cos 40°)f_1 + f_2 = (75 \text{ kN}) \cos 145°$$

and solve these for

$$f_1 = 66.9 \text{ kN} \qquad f_2 = -112.7 \text{ kN}$$

The vector equation could also be dot-multiplied by the unit vectors $\mathbf{e}_a$ and $\mathbf{e}_b$ defined to be perpendicular to f_1 and f_2, as shown in Figure 7.4. This gives

$$f_1\mathbf{e}_a \bullet \mathbf{e}_1 + f_2\mathbf{e}_a \bullet \mathbf{e}_2 = \mathbf{e}_a \bullet \mathbf{f}$$
$$f_1\mathbf{e}_b \bullet \mathbf{e}_1 + f_2\mathbf{e}_b \bullet \mathbf{e}_2 = \mathbf{e}_b \bullet \mathbf{f}$$

which reduce to

$$(\cos 130°)f_2 = (75 \text{ kN}) \cos 15°$$
$$(\cos 50°)f_1 = (75 \text{ kN}) \cos 55°$$

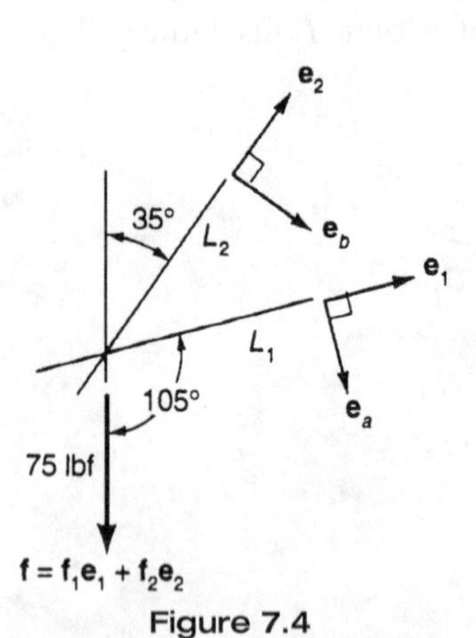

Figure 7.4

The unit vectors $\mathbf{e}_a$ and $\mathbf{e}_b$ are better choices for the purpose of evaluating f_1 and f_2 because each is perpendicular to one of the vectors with unknown magnitudes, so that we can eliminate an unknown from each equation.

The Cross Product

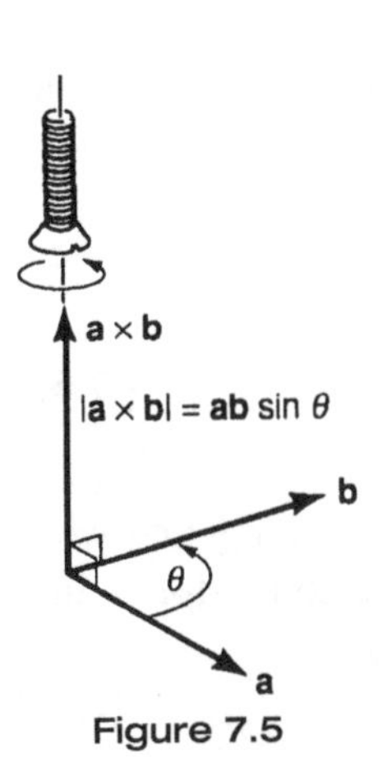

Figure 7.5

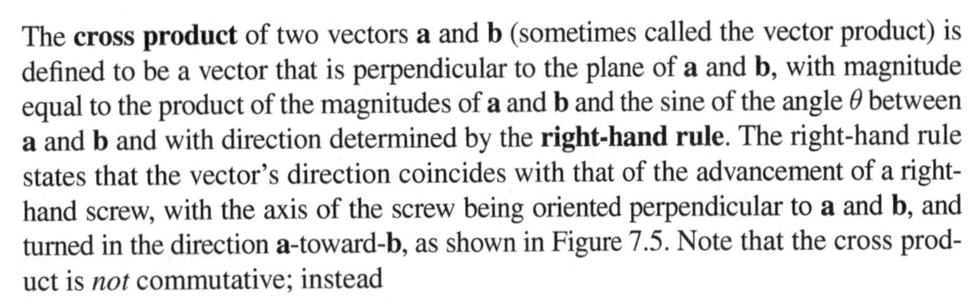

The **cross product** of two vectors **a** and **b** (sometimes called the vector product) is defined to be a vector that is perpendicular to the plane of **a** and **b**, with magnitude equal to the product of the magnitudes of **a** and **b** and the sine of the angle θ between **a** and **b** and with direction determined by the **right-hand rule**. The right-hand rule states that the vector's direction coincides with that of the advancement of a right-hand screw, with the axis of the screw being oriented perpendicular to **a** and **b**, and turned in the direction **a**-toward-**b**, as shown in Figure 7.5. Note that the cross product is *not* commutative; instead

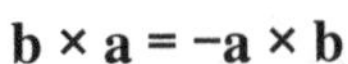

$$\mathbf{b} \times \mathbf{a} = -\mathbf{a} \times \mathbf{b}$$

However, the definitions of cross-multiplication and addition can be used to show that the distributive law

$$\mathbf{a} \times (\mathbf{b} + \mathbf{c}) = \mathbf{a} \times \mathbf{b} + \mathbf{a} \times \mathbf{c}$$

is valid, and that the cross product is associative with respect to multiplication by scalars:

$$(p\mathbf{a}) \times (q\mathbf{b}) = (pq)(\mathbf{a} \times \mathbf{b})$$

Example 7.2

It is desired to resolve a given vector **a** into two components, one parallel to a second given vector **b** and one perpendicular to **b**. The results are to be expressed in terms of **a** and **b**, using vector operations defined in the preceding pages.

Solution

The component parallel to **b** will be the projection of **a** onto the line parallel to **b**, which is expressible in terms of a unit vector in the direction of **b**:

$$\mathbf{a}_{\parallel} = (\mathbf{e}_b \bullet \mathbf{a})\mathbf{e}_b = \frac{(\mathbf{b} \bullet \mathbf{a})\mathbf{b}}{\mathbf{b} \bullet \mathbf{b}}$$

The component perpendicular to **b** can be determined from the fact that **a** is to be the sum of this and the component just calculated:

$$\mathbf{a}_{\perp} = \mathbf{a} - \mathbf{a}_{\parallel}$$

Alternatively, the perpendicular component can be calculated by

$$\mathbf{a}_{\perp} = \frac{(\mathbf{b} \times \mathbf{a}) \times \mathbf{b}}{\mathbf{b} \bullet \mathbf{b}}$$

Verifying this last relationship from the definition of the cross product provides useful practice in relating the operation to geometry. The vector **a** can now be expressed in terms of the components parallel and perpendicular to **b**:

$$\mathbf{a} = \frac{(\mathbf{b} \bullet \mathbf{a})\mathbf{b}}{\mathbf{b} \bullet \mathbf{b}} + \frac{(\mathbf{b} \times \mathbf{a}) \times \mathbf{b}}{\mathbf{b} \bullet \mathbf{b}}$$

Rectangular Cartesian Components

A special way of resolving vectors consists of forming three mutually perpendicular components. The directions are chosen (usually with consideration of the geometry of the problem at hand) and three mutually perpendicular unit vectors $\mathbf{e}_x$, $\mathbf{e}_y$, and $\mathbf{e}_z$ are defined to be parallel to these directions. The rectangular Cartesian components of a vector **a** are then the projections of **a** onto lines in the selected directions, expressed as

$$\mathbf{a} = a_x\mathbf{e}_x + a_y\mathbf{e}_y + a_z\mathbf{e}_z \qquad (7.6)$$

in which $a_x = \mathbf{e}_x \bullet \mathbf{a}$, $a_y = \mathbf{e}_y \bullet \mathbf{a}$, and $a_z = \mathbf{e}_z \bullet \mathbf{a}$. These relationships, together with the associative and distributive laws mentioned previously, lead to the following formulas:

$$\mathbf{a} \bullet \mathbf{b} = a_x b_x + a_y b_y + a_z b_z \qquad (7.7)$$

$$|\mathbf{a}| = \sqrt{a_x^2 + a_y^2 + a_z^2} \qquad (7.8)$$

$$\begin{aligned} \mathbf{a} \times \mathbf{b} = &(a_y b_z - a_z b_y)\mathbf{e}_x \\ &+ (a_z b_x - a_x b_z)\mathbf{e}_y \\ &+ (a_x b_y - a_y b_x)\mathbf{e}_z \end{aligned} \qquad (7.9)$$

Each of these equations depends on the fact that the unit vectors are mutually perpendicular; in addition, the expression for the cross product is valid only if the unit vectors form a *right-handed* set; that is,

$$\mathbf{e}_x = \mathbf{e}_y \times \mathbf{e}_z, \quad \mathbf{e}_y = \mathbf{e}_z \times \mathbf{e}_x, \quad \text{and} \quad \mathbf{e}_z = \mathbf{e}_x \times \mathbf{e}_y$$

Unit vectors defining the three dimensions in Cartesian coordinates are also often represented as **i**, **j**, and **k**.

Example 7.3

Determine the lengths of the guylines $O'P$ and $O'Q$, shown in Exhibit 2, and the angle between them.

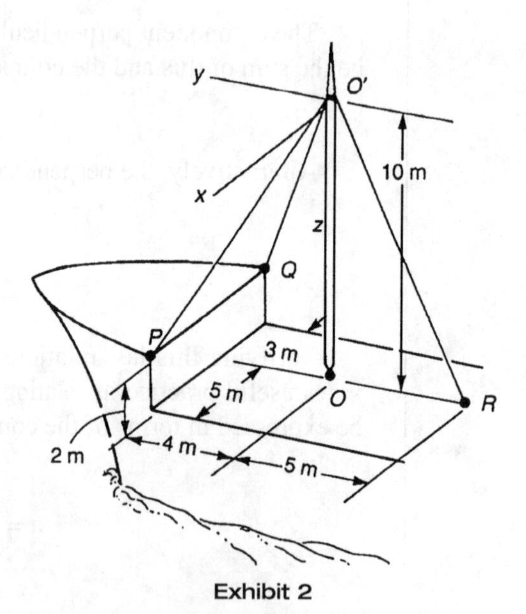

Exhibit 2

Solution

Let position vectors **a** and **b** extend from point O' to points P and Q, respectively, and resolve these into components along the x, y, and z axes shown:

$$\mathbf{a} = (5\text{ m})\mathbf{e}_x + (4\text{ m})\mathbf{e}_y + (8\text{ m})\mathbf{e}_z$$
$$\mathbf{b} = (-3\text{ m})\mathbf{e}_x + (4\text{ m})\mathbf{e}_y + (8\text{ m})\mathbf{e}_z$$

The required lengths can then be determined from Equation (7.8):

$$|\mathbf{a}| = \sqrt{(5\text{ m})^2 + (4\text{ m})^2 + (8\text{ m})^2} = 10.25\text{ m}$$

$$|\mathbf{b}| = \sqrt{(-3\text{ m})^2 + (4\text{ m})^2 + (8\text{ m})^2} = 9.43\text{ m}$$

The required angle is a factor in the definition of $\mathbf{a} \bullet \mathbf{b}$, and in fact it will be the only unknown in this equation after $\mathbf{a} \bullet \mathbf{b}$ is evaluated. From Equation (7.7), we have

$$\mathbf{a} \bullet \mathbf{b} = (5\text{ m})(-3\text{ m}) + (4\text{ m})(4\text{ m}) + (8\text{ m})(8\text{ m}) = 65\text{ m}^2$$

Then Equation (7.3) gives

$$\cos\theta = \frac{\mathbf{a} \bullet \mathbf{b}}{|\mathbf{a}||\mathbf{b}|} = \frac{65\text{ m}^2}{(10.25\text{ m})(9.43\text{ m})} = 0.672$$

from which

$$\theta = 47.7°$$

FORCE SYSTEMS

A body may have several forces acting on it simultaneously. To account for these forces in an organized way, some general properties of a set of forces, or a force system, will prove useful.

Types of Forces

Normally, forces are distributed over some region of the body they act upon; however, some simplification can often be gained without significant loss of accuracy by considering that a force is concentrated at a single point called the **point of application** of the force. However, in some circumstances, it may be necessary to account for the way the forces are distributed over a region of the body. Thus, we make a distinction between **concentrated forces** and **distributed forces**.

A second distinction that is often important is between **surface forces**, or actions that take place where surfaces contact, and **body forces**, which are distributed throughout a body, as in the case of gravity.

Point of Application and Line of Action

In addition to the vector value of a force (that is, magnitude and direction), the point at which a force acts on the body is important to the way the body responds. For this reason, analysis usually requires not only a **force** vector to specify the magnitude and direction of the force, but also a **position** vector to specify the location of the *point of application* of the force. The **line of action** of a force is the line parallel to the force vector and through the point of application. This line is important to the understanding of *moments* of forces.

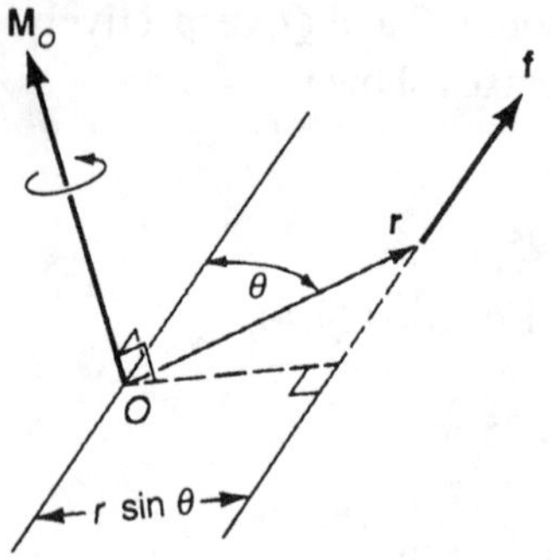

Figure 7.6

Moments of Forces

The **moment** about a point O of a force $\mathbf{f}$ is the vector defined as

$$\mathbf{M}_O = \mathbf{r} \times \mathbf{f} \tag{7.10}$$

where $\mathbf{r}$ is a position vector from O to any point on the line of action of $\mathbf{f}$. Reference to Figure 7.6 and the definition of the cross product reveal that the magnitude of the moment is

$$|\mathbf{M}_O| = df \tag{7.11}$$

where f is the magnitude of $\mathbf{f}$ and $d = r \sin\theta$ is the perpendicular distance from O to the line of action of $\mathbf{f}$.

Example 7.4

Suppose the guyline $O'P$ in Exhibit 3 has a tension of 800 N. What is the moment about O of the force from this cable acting on the mast?

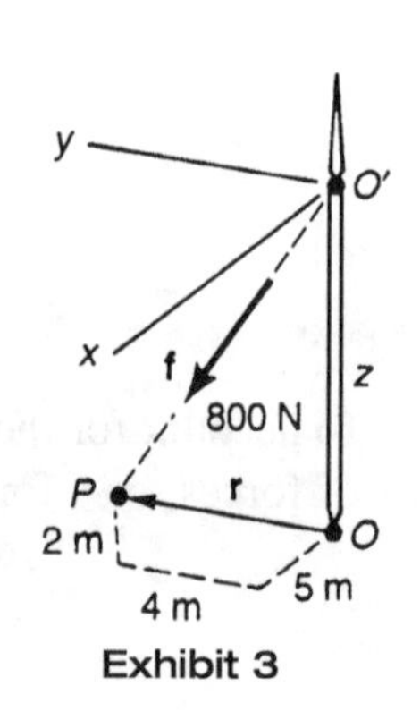

Exhibit 3

Solution

Because point P is on the line of action of this force, a position vector, $\mathbf{r}$, from O to P can be used for evaluation of the moment about O. Referring to the axes shown in Exhibit 3, the x-y-z resolution of this vector may be written as

$$\mathbf{r} = (5\text{ m})\mathbf{e}_x + (4\text{ m})\mathbf{e}_y + (-2\text{ m})\mathbf{e}_z$$

The resolution of the force can be determined by multiplying its magnitude by the unit vector in the direction of $O'P$, components of which can be obtained by dividing the vector $\mathbf{a}$ from the preceding example by its magnitude:

$$\mathbf{f} = 800\text{N}\left(\frac{5}{10.25}\mathbf{e}_x + \frac{4}{10.25}\mathbf{e}_y + \frac{8}{10.25}\mathbf{e}_z\right)$$

$$= (390.4\mathbf{e}_x + 312.3\mathbf{e}_y + 624.6\mathbf{e}_z)\text{N}$$

The moment can now be evaluated with reference to Equation (7.9):

$$\begin{aligned}\mathbf{M}_O &= [(4\text{ m})(624.6\text{ N}) - (-2\text{ m})(312.3\text{ N})]\mathbf{e}_x \\ &\quad + [(-2\text{ m})(390.4\text{ N}) - (5\text{ m})(624.6\text{ N})]\mathbf{e}_y \\ &\quad + [(5\text{ m})(312.3\text{ N}) - (4\text{ m})(390.4\text{ N})]\mathbf{e}_z \\ &= (3123\mathbf{e}_x - 3904\mathbf{e}_y)\text{ N}\bullet\text{m}\end{aligned}$$

Observe that point O' is also on the line of action of $\mathbf{f}$, so that a position vector $\mathbf{r} = (-10\text{ m})\mathbf{e}_z$ could have been used instead of the one above, and with less arithmetic.

The moment about an axis Oi is defined as the projection onto the axis of the moment about some point on the axis. To express this, we define the positive sense along the axis with the unit vector $\mathbf{e}_i$ and write

$$\mathbf{M}_{Oi} = (\mathbf{e}_i \bullet \mathbf{M}_O)\,\mathbf{e}_i \tag{7.12}$$

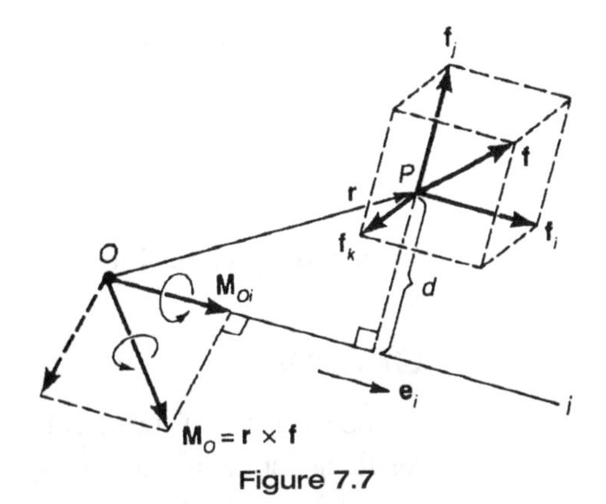

Figure 7.7

Although the moment about the axis can be computed according to this definition, an alternative form is often easier to use and provides a different interpretation. Substitution of Equation (7.10) into Equation (7.12) and use of the vector identity $\mathbf{a} \bullet (\mathbf{b} \times \mathbf{c}) = (\mathbf{a} \times \mathbf{b}) \bullet \mathbf{c}$ leads to $\mathbf{M}_{Oi} = [(\mathbf{e}_i \times \mathbf{r}) \bullet \mathbf{f}\,]\,\mathbf{e}_i$.

Now if $\mathbf{f}$ is resolved into a component $\mathbf{f}_i$ parallel to the axis Oi, a component $\mathbf{f}_j$ perpendicular to O_i and in the plane of $\mathbf{r}$ and O_i, and a component $\mathbf{f}_k$ perpendicular to this plane, several important facts become apparent. Referring to Figure 7.7, note that because both $\mathbf{f}_i$ and $\mathbf{f}_j$ are perpendicular to $\mathbf{e}_i \times \mathbf{r}$, neither of these components contributes to $\mathbf{M}_{Oi}$. Also, because $\mathbf{e}_i \times \mathbf{r}$ has the magnitude $d = |\mathbf{r}| \sin \angle \frac{\mathbf{r}}{\mathbf{i}}$, we can express the magnitude of the moment component as

$$|\mathbf{M}_{Oi}| = df_k$$

where d is the perpendicular distance from point P to the axis Oi.

The *sense* of $\mathbf{M}_{Oi}$ (that is, whether it is directed in the positive or negative i-direction) is readily determined from the direction of $\mathbf{f}_k$ and the right-hand rule. Alternatively, the sense may be determined by the sign of the factor $\mathbf{e}_i \bullet \mathbf{M}_O$ in Equation (7.12). Note that the same value of d would be obtained regardless of where the point O is on the i-axis, and recall that the position vector $\mathbf{r}$ in the definition $\mathbf{M}_O = \mathbf{r} \times \mathbf{f}$ can be from O to any point on the line of action of $\mathbf{f}$. This means that Equation (7.12) will yield the value $\mathbf{M}_{Oi}$ with $\mathbf{r}$ as a position vector from *any* point on the axis O_i to *any* point on the line of action of $\mathbf{f}$.

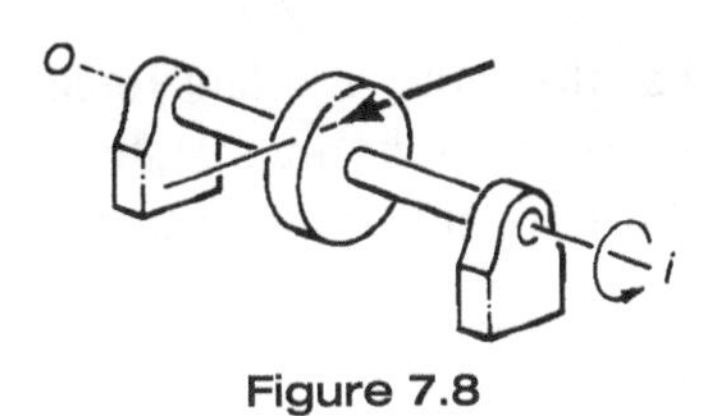

Figure 7.8

The moment about an axis is a measure of the tendency of the force(s) to cause rotation about the axis. For example, if a rotor is mounted in bearings and subjected to a set of forces, the moment of these forces about the axis of the bearings is found to be directly related to the rate of change of rotational speed. Neither forces parallel to the axis nor forces with lines of action passing through the axis will affect the rotation.

Resultant Forces and Moments

If there are several forces $\mathbf{f}_1, \mathbf{f}_2, \ldots, \mathbf{f}_n$, each with its own line of action, the **resultant force** is defined as

$$\mathbf{f} = \sum_{i=1}^{n} \mathbf{f}_i$$

and the **resultant moment** about a point O is defined as

$$\mathbf{M}_O = \sum_{i=1}^{n} \mathbf{r}_i \times \mathbf{f}_i$$

where $\mathbf{r}_i$ is a position vector from O to any point on the line of action of $\mathbf{f}_i$.

Couples

A special set of forces, called a **couple**, has zero resultant force but a nonzero resultant moment. An example is a pair of forces of equal magnitude, opposite directions, and separate lines of action.

Moments about Different Points

The resultant moment of a set of forces about two different points, O and O', are related as follows. With $\mathbf{r}_{oo'}$ designating the position vector from O to O', the position vectors from these two points to a point on the line of action of the force $\mathbf{f}_i$ are shown in Figure 7.9 and are related by

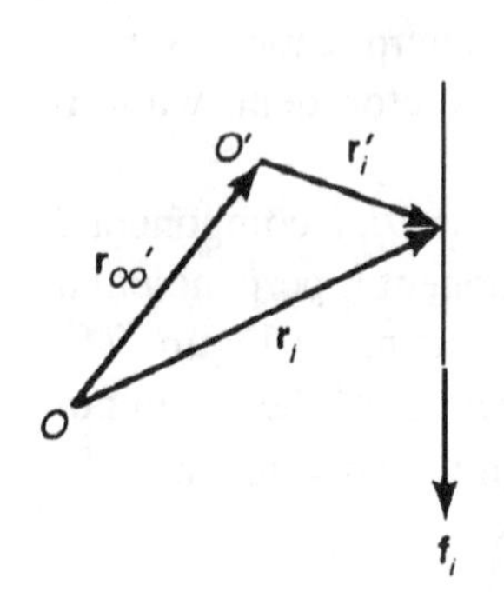

Figure 7.9

$$\mathbf{r}_i = \mathbf{r}_{i'} + \mathbf{r}_{oo'}$$

The moment about O can then be expressed as

$$\begin{aligned}\mathbf{M}_O &= \mathbf{r}_1 \times \mathbf{f}_1 + \mathbf{r}_2 \times \mathbf{f}_2 + \ldots + \mathbf{r}_n \times \mathbf{f}_n \\ &= (\mathbf{r}'_1 + \mathbf{r}_{OO'}) \times \mathbf{f}_1 + (\mathbf{r}'_2 + \mathbf{r}_{OO'}) \times \mathbf{f}_2 + \ldots + (\mathbf{r}_n + \mathbf{r}_{OO'}) \times \mathbf{f}_n \\ &= \mathbf{r}'_1 \times \mathbf{f}_1 + \mathbf{r}'_2 \times \mathbf{f}_2 + \ldots + \mathbf{r}'_n \times \mathbf{f}_n + \mathbf{r}_{OO'} \times (\mathbf{f}_1 + \mathbf{f}_1 + \ldots + \mathbf{f}_n) \\ &= \mathbf{M}_{O'} + \mathbf{r}_{OO'} \times \mathbf{f} \end{aligned} \qquad (7.13)$$

That is, the moment about O is equal to that about O' plus the moment that a force equal to the resultant of the given forces would have about O if the line of action of this force passed through O'. For the special case in which $\mathbf{f} = 0$, Equation (7.13) shows that the moment of a couple about every point is the same.

Equivalent Force Systems

Two sets of forces are said to be **equivalent** if each has the same resultant force and the same resultant moment about some point. If these conditions are met, then Equation (7.13) can be used to show that the sets will have the same resultant moment about *any* point.

Example 7.5

A 7-kN force acts on the end of the beam, its line of action passing along the center of the web of the channel section. If a bracket were attached to the end of the beam, allowing this force to be applied 80 mm to the left of the center of the web (see Exhibit 4), what horizontal forces applied along the flanges would need to be added so that, together with the displaced 7-kN force, they would form a set equivalent to the 7-kN force acting along the web?

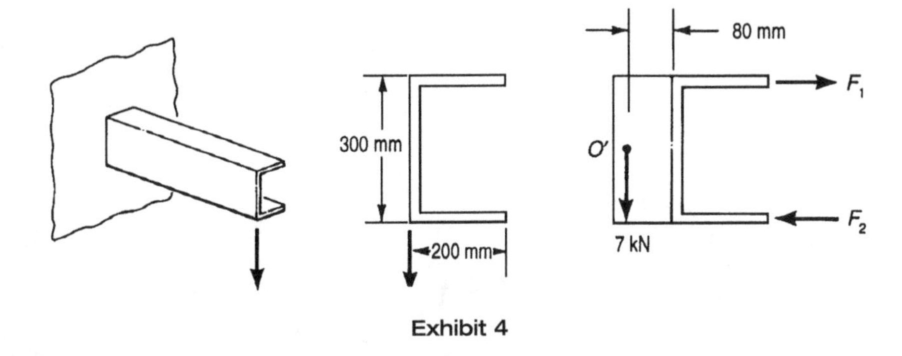

Exhibit 4

Solution

Since the resultants of the two sets must be the same (7-kN downward), the resultant of F_1 and F_2 must be zero, which in turn means that F_1 and F_2 must have equal magnitudes and opposite directions. To determine the magnitude, the resultant moment of each set can be equated; a convenient point about which to evaluate these moments is located where the lines of action of F_2 and the force through O' intersect. The moment about this point of the original force is (0.08 m)(7 kN) = 0.56 kN• m acting clockwise. The resultant moment of the equivalent set about this point is (0.3 m)F_1, also in the clockwise direction. Equivalence requires that (0.3 m)F_1 = 0.56 kN• m from which F_1 = 1.867 kN. Thus, the equivalent set consists of the vertical, 7-kN force through O', together with a couple that has a clockwise-acting moment of 0.56 kN• m.

EQUILIBRIUM

Newton's Laws require for every body or system of bodies that

$$\sum_i \mathbf{f}_i = \sum_j m_j \mathbf{a}_j$$

and

$$\sum_i \mathbf{r}_i \times \mathbf{f}_i = \sum_j \mathbf{r}_j \times m_j \mathbf{a}_j$$

in which $\mathbf{f}_i$ is one of the forces acting on the system (from an *external* source); m_j and $\mathbf{a}_j$ are the mass and acceleration, respectively, of the *j*th material element; $\mathbf{r}_i$ and $\mathbf{r}_j$ are position vectors from *any* selected point to, respectively, a point on the line of action of $\mathbf{f}_i$ and the *j*th material element; and the sums are to include *all* the external forces and material elements. A system will be in static equilibrium whenever the accelerations are all zero. In these cases the laws require that the resultant of all forces from external sources be zero, and that the resultant moment of these forces about any point be zero.

Free-Body Diagrams

In spite of the simplicity of equilibrium relationship, they can easily be misapplied. Experience has repeatedly shown that nearly all such errors stem from lack of attention to the appropriate **free-body diagram**. The free-body diagram must show clearly what body, bodies, or parts thereof are being considered as the system,

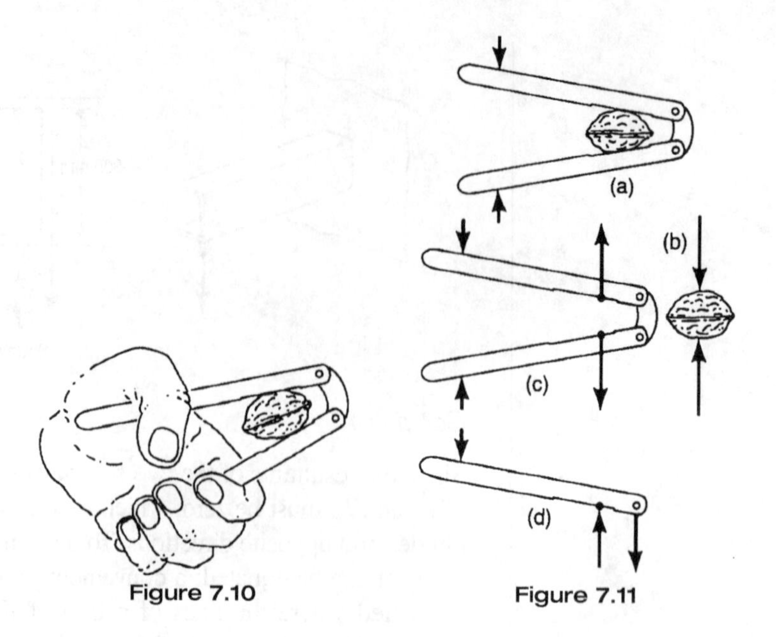

Figure 7.10

Figure 7.11

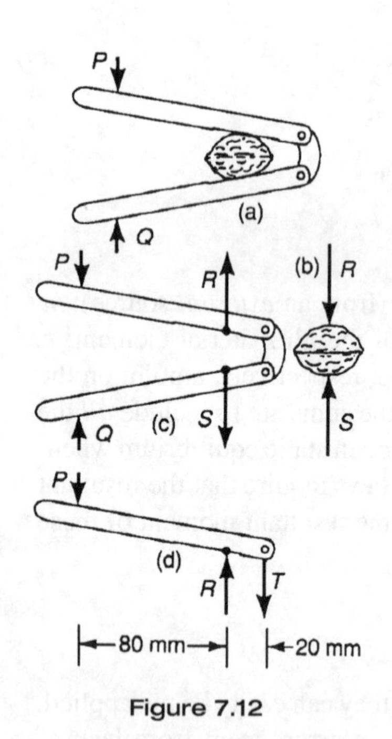

Figure 7.12

and all of the forces acting *on* the system from sources *outside* the system. For illustration, consider the device shown in Figure 7.10. Several different free-bodies are possibly useful and will be constructed.

First, consider the system consisting of the nutcracker together with the walnut. This system is shown isolated from all other objects, with arrows depicting the forces that come from objects *external* to the system. Assuming forces of gravity to be negligible, the only external body that exerts forces on this system is the hand. These forces are shown in Figure 7.11(a). The interaction between the nut and the cracker is *not* shown on this free body, since it is internal to this system.

To expose the force tending to break the nut, we might consider free-bodies of the nut and of the nutcracker, shown in Figures 7.11(b) and (c). In Figure 7.11(c) the hand and the walnut are external to the nutcracker. Therefore, arrows depicting the forces from the nut as well as from the hand are included in this free body.

Another possibly useful free body is that of the upper handle. Objects external to this are the hand, the walnut, and the connecting pin. This free body appears in Figure 7.11(d).

Here is a summary of the procedure: First, a sketch must be made clearly showing the system to be considered for equilibrium. The system boundary is normally chosen so that it passes through a point where a force interaction of particular interest occurs. Next, *all* forces acting on the system, from bodies *external* to the system, must be properly represented. Force interactions between bodies *within* the system are *not* considered.

To show clearly the physical significance of quantities in equilibrium equations, symbols must accompany the arrows representing the forces that appear in these equations. In Figure 7.12, the letters P, Q, R, S, and T have been chosen to indicate the magnitudes of several forces. Each letter represents a **scalar** multiplier of a unit vector in the direction of the arrow; since this scalar can take on a positive or negative value, a force in the same or opposite direction indicated by the arrow can be represented. If an analysis leads to the values, say, $P = 80$ N and $R = 400$ N, the forces acting on the upper handle would be 80 N *downward* on the right-hand end and 400 N *upward* where the walnut makes contact at R. If different circumstances led to, say, $P = -15$ N and $R = -75$, these values would imply that the forces acting on the upper handle are 15 N *upward* on the right-hand end and 75 N *downward* where the walnut makes contact. (A little adhesive between the walnut and the nutcracker would make this possible.)

Observe that the forces on the walnut have the same labels as their counterparts on the nutcracker, and the arrows have opposite directions. We have in this way implied satisfaction of Newton's third law without further fuss. Other relatively simple aspects of force analysis can be treated as the free-bodies are constructed; for example, unless the walnut is to accelerate, it is evident from a glance at its free body that $R = S$. Writing this equation could be circumvented by simply labeling both arrows with the same letter.

Equations of Equilibrium

With free-body diagrams properly drawn and labeled, equations of equilibrium can be written for any of the chosen bodies. For example, the vertical force equilibrium of the free body of Figure 7.12(a) requires that $Q - P = 0$. Each of the other two equations of force equilibrium, involving components in the horizontal direction and components perpendicular to the plane of the sketch, is the trivial equation $0 = 0$. Moment equilibrium shows that the lines of action of the two forces must coincide, a fact that has already been incorporated into the diagram. Similar analyses of the free-bodies of Figure 7.12(b) and (c) lead to

$$S - R = 0$$

and

$$Q - P + R - S = 0$$

Now suppose that the reason for this analysis is to obtain an estimate of how hard one must squeeze in order to crack a nut, given the cracking requires 245 N applied to the nut. Then $R = 245$ N, and the three equations above contain the three unknowns, P, Q, and S. Unfortunately, attempts to solve these for P or Q will fail, because the three equations are not independent, because the last equation can be deduced from the first two by addition. Equilibrium of still another body must be considered in order to obtain an independent equation. The free body of the handle in Figure 7.12(d) can provide two more equations: the vertical component for force equilibrium,

$$T + P = 245 \text{ N} \qquad \textbf{(a)}$$

and an equation of moment equilibrium. Summing moments about the point of contact with the walnut leads to

$$-(20 \text{ mm})T + (80 \text{ mm})P = 0 \qquad \textbf{(b)}$$

To solve for P, we can multiply Equation (a) by 20 mm and add Equation (b) to the result, yielding

$$(100 \text{ mm})P = (20 \text{ mm})\,(245 \text{ N}) \qquad \textbf{(c)}$$

This gives $P = 49$ N.

A more direct analysis stems from considering moments about a different point on the handle, resulting in Equation (c) as the first equation written.

The following examples provide further illustration of the use of the basic laws of static equilibrium.

Example 7.6

Neglecting gravity forces in Exhibit 5(a), except those on the 300-kg load, determine the forces in the cable and in the boom.

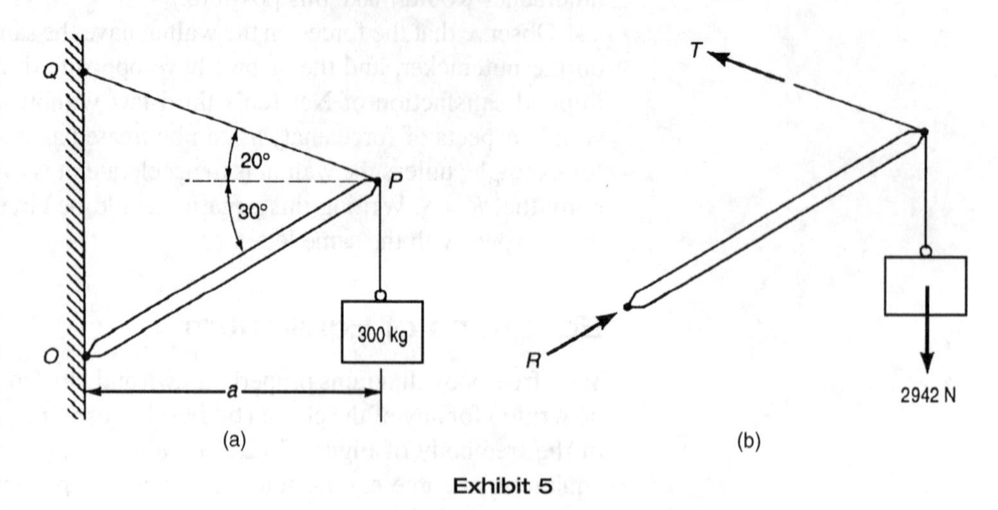

Exhibit 5

Solution

First, a free-body diagram is drawn (Exhibit 5(b)), with the system boundary passing through the support point O and through the cable segment PQ. Vanishing of moment about point P requires that the reaction at O must be directed along the boom. The force **T** can be separated into its x and y components. These two force components generate additive moments, $\mathbf{r} \times \mathbf{f}$, both of which are positive due to the right-hand rule. No moment is added for the force **R** because it is applied at the pivot point (the position vector at which **R** is applied is of length 0). The equation of moments about O can be expressed as

$$(T \sin 20° - 2942 \text{ N})a + (T \cos 20°)(a \tan 30°) = 0$$

from which

$$T = \frac{2942 \text{ kN}}{\sin 20° + \cos 20° \tan 30°} = 3.33 \text{ kN}$$

Equilibrium of horizontal forces,

$$R \cos 30° - T \cos 20° = 0$$

leads to the magnitude of the reaction at O:

$$R = \frac{(3.33 \text{ N}) \cos 20°}{\cos 30°} = 3.61 \text{ kN}$$

Example 7.7

Neglecting gravity forces except those on the 2-Mg load shown in Exhibit 6(a), determine the tension in cable AB, which is holding up the crane boom.

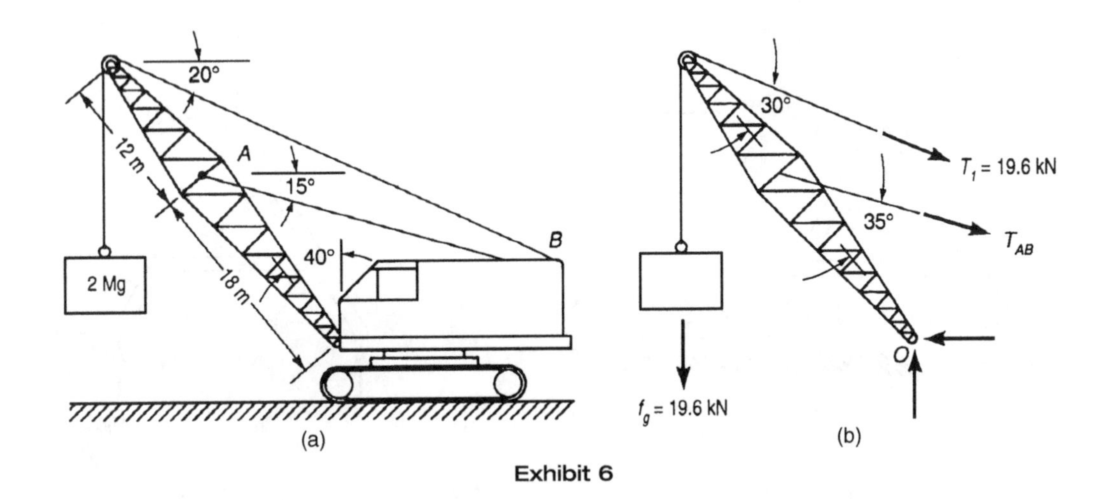

Exhibit 6

Solution

The boundary of the free body passes through the two upper cable segments and the support point, labeled O in Exhibit 6(b). By considering moment equilibrium of a system consisting of the pulley and a portion of the cable, including the section that is in contact with the pulley (not shown), we find that $T_1 = f_g = 19.6$ kN.

To avoid introducing the unknown reaction at O into the analysis, consider moments of forces about this point. With the radius of the pulley denoted as r, the equation of moments about O is

$$f_g[(12 \text{ m} + 18 \text{ m})\sin 40° + r] - T_1[(12 \text{ m} + 18 \text{ m})\sin 30° + r] - T_{AB} \sin 35°(18 \text{ m}) = 0$$

With the value of $T_1 = f_g$ substituted, this is readily solved for the tension in the supporting cable:

$$T_{AB} = \frac{(19.6 \text{ kN})(30 \text{ m})(\sin 40° - \sin 30°)}{(18 \text{ m})\sin 35°} = 8.13 \text{ kN}$$

Example 7.8

Gravity forces on the structural members are negligible compared with P and Q in Exhibit 7(a). Evaluate all the forces acting on each of the three members in the A-frame.

Solution

Free-bodies of the entire frame and of each individual member are shown in Exhibit 7(b)–(e). The roller support at D means that no horizontal force can be transmitted from the ground at that point. From the free body in view (b) we can consider moments about point E,

$$(3a)Q + (3a \tan 30°)P - (6a \tan 30°)R_D = 0$$

horizontal forces,

$$R_{Ex} - Q = 0$$

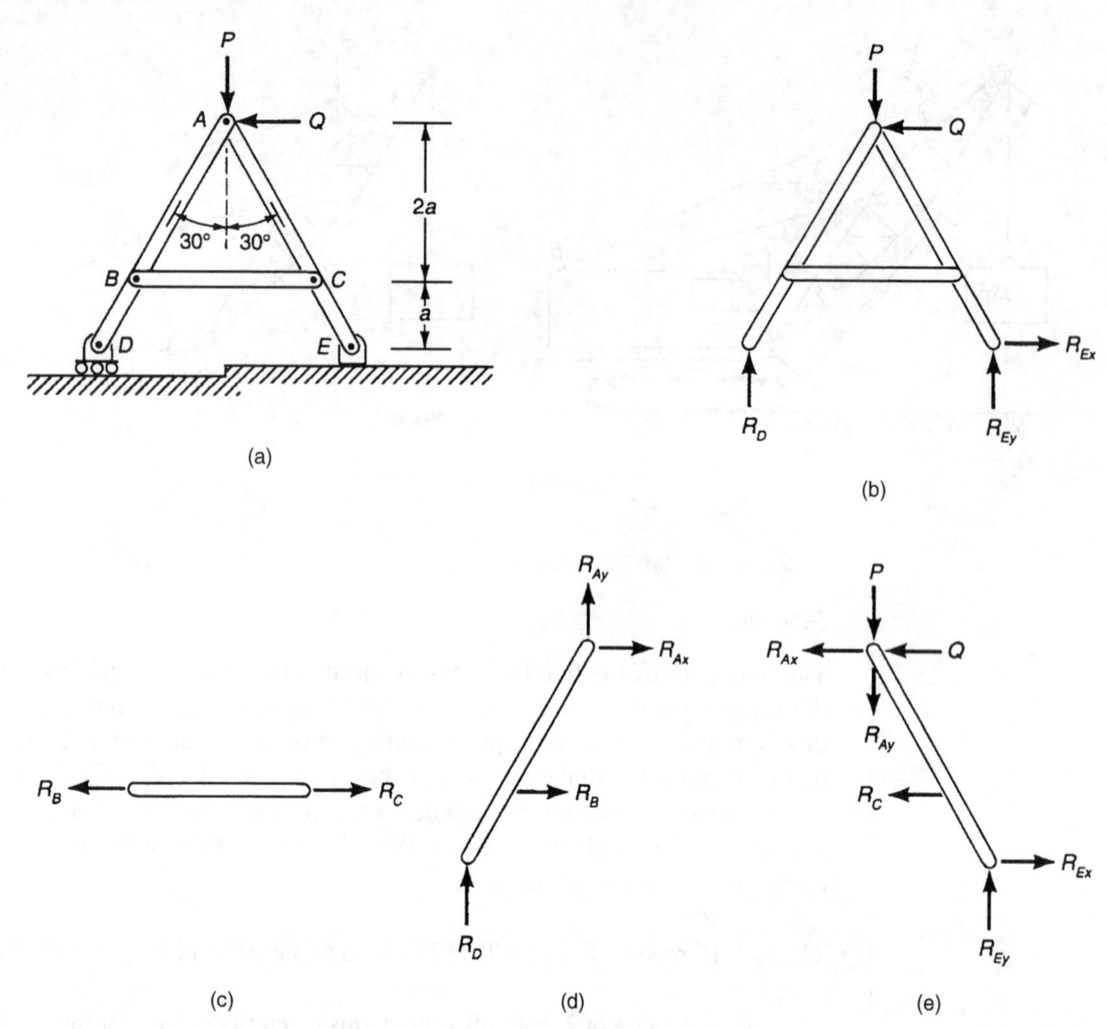

Exhibit 7

and moments about point D:

$$(6a \tan 30°)R_{Ey} + (3a)Q - (3a \tan 30°)P = 0$$

Note that $6a \tan 30°$ equals the distance from point O to point E.

These equations can then be solved for the support reactions:

$$R_D = \frac{1}{2}(P + \cot 30° Q)$$

$$R_{Ex} = Q$$

$$R_{Ey} = \frac{1}{2}(P - \cot 30° Q)$$

As a check, it might be a good idea to consider vertical forces. Moment equilibrium of the free body in view (c) implies that the lines of action of R_B and R_C are along the bar. Its horizontal equilibrium gives us

$$R_C - R_B = 0$$

Now, turning to the free body in view (d), we can write equations of moment equilibrium about points A as

$$(2a)R_B - (3a \tan 30°)R_D = 0$$

and horizontal and vertical force equilibrium as

$$R_{Ax} + R_B = 0$$
$$R_{Ay} + R_D = 0$$

With values of R_D, R_{Ax}, and R_{Ay} above, these equations give the following values of the remaining unknown reactions:

$$R_B = R_C = \frac{3}{4}(\tan 30° P + Q)$$

$$R_{Ax} = -\frac{3}{4}(\tan 30° P + Q)$$

$$R_{Ay} = -\frac{1}{2}(P + \cot 30°\ Q)$$

Example 7.9

The cables OA, OB, and OC support a suspended block shown in Exhibit 8. Determine the tension in each cable in terms of the gravitational force f_g.

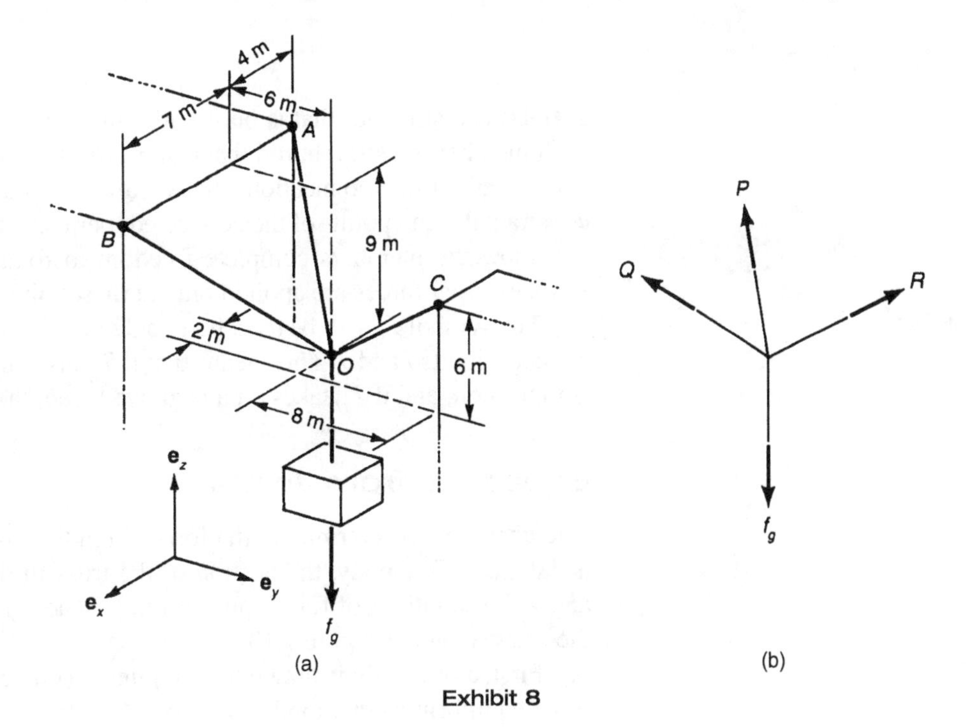

Exhibit 8

Solution

In Exhibit 8(b), the desired tensions P, Q, and R are shown on the free body of the portion of the structure in the neighborhood of point O. To resolve the forces in the directions of $\mathbf{e}_x$, $\mathbf{e}_y$, and $\mathbf{e}_z$, we need to find the direction cosines between OA, OB, and OC and these directions. This is done, as in Example 7.3, by dividing the projection of the cable onto the axis by the length of the cable; for instance, the direction cosine between OA and $\mathbf{e}_x$ is

$$\cos\angle_{OA}^{e_x} = \frac{-4m}{\sqrt{(4m)^2 + (6m)^2 + (9m)^2}} = -0.347$$

The results are summarized in the following equations:

$$\mathbf{e}_{OA} = -0.347\mathbf{e}_x - 0.520\mathbf{e}_y + 0.780\mathbf{e}_z$$
$$\mathbf{e}_{OB} = 0.543\mathbf{e}_x - 0.466\mathbf{e}_y + 0.699\mathbf{e}_z$$
$$\mathbf{e}_{OC} = 0.196\mathbf{e}_x + 0.784\mathbf{e}_y + 0.588\mathbf{e}_z$$

Force equilibrium requires that

$$P\mathbf{e}_{OA} + Q\mathbf{e}_{OB} + R\mathbf{e}_{OC} - f_g\mathbf{e}_z = 0$$

Dot-multiplying this equation by $\mathbf{e}_x$, $\mathbf{e}_y$, and $\mathbf{e}_z$ yields the following relations:

$$-0.347P + 0.543Q + 0.196R = 0$$
$$-0.520P - 0.466Q + 0.784R = 0$$
$$0.780P + 0.699Q + 0.588R = f_g$$

which can be solved for the desired forces:

$$P = 0.660f_g$$
$$Q = 0.217f_g$$
$$R = 0.567f_g$$

TRUSSES

A **truss** is a structure that is built with interconnected axial force members. Each such member is a straight rod that can transmit force along its axis. This limitation is the result of interconnections that are all of the ball-and-socket type; that is, they constrain the end points of the connected members against relative position change but allow the members complete freedom to rotate about the connection point. Also, external forces are applied only at these joints.

The symbol T_{IJ} will be used here to denote the tensile force in the member that connects joints I and J. This means that if T_{IJ} takes on a positive value, the member is in tension, and if T_{IJ} takes on a negative value, the member is in compression.

Equations from Joints

One approach to determining the forces in individual members within a truss is to isolate, as a free body, the portion of the truss in the neighborhood of each joint and write equations of force equilibrium for each. This procedure is illustrated for the truss shown in Figure 7.13.

First, a free body is drawn for the joint G (Figure 7.14), and the corresponding force equilibrium equation is written:

$$T_{GE}\mathbf{e}_{GE} + T_{GF}\mathbf{e}_{GF} - (50\text{ kN})\mathbf{e}_y = 0$$

To evaluate T_{GE}, the equation may be dot-multiplied by $\mathbf{e}_a$, which is defined to be perpendicular to the other unknown force, as shown in Figure 7.14:

$$\mathbf{e}_a \bullet \mathbf{e}_{GE}T_{GE} + \mathbf{e}_a \bullet \mathbf{e}_{GF}T_{GF} - \mathbf{e}_a \bullet \mathbf{e}_y(50\text{ kN}) = 0$$

$$\left[\left(\frac{1}{\sqrt{17}}\right)\left(-\frac{4}{\sqrt{5}}\right) + \left(\frac{4}{\sqrt{17}}\right)\left(\frac{3}{5}\right)\right]T_{GE} - \frac{4}{\sqrt{17}}(50\text{ kN}) = 0$$

$$T_{GE} = 125\text{ kN}$$

Figure 7.13

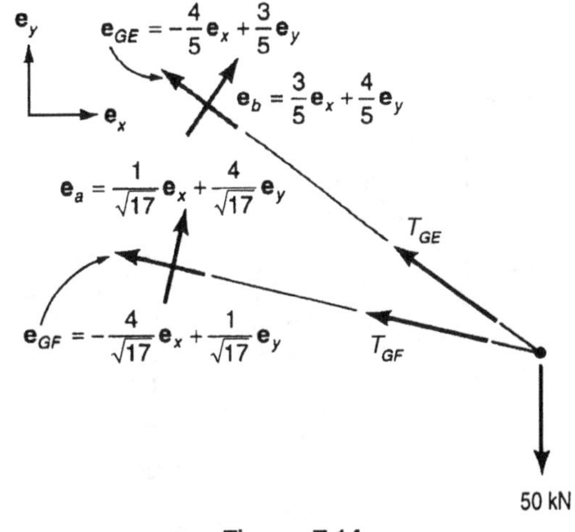

Figure 7.14

Dot-multiplication by $\mathbf{e}_b$ will similarly yield the value of T_{GF}:

$$\mathbf{e}_b \bullet \mathbf{e}_{GE} T_{GE} + \mathbf{e}_b \bullet \mathbf{e}_{GF} T_{GF} - \mathbf{e}_b \bullet \mathbf{e}_y (50 \text{ kN}) = 0$$

$$\left[\left(\frac{3}{5}\right)\left(-\frac{4}{\sqrt{17}}\right) + \left(\frac{4}{5}\right)\left(\frac{1}{\sqrt{17}}\right)\right] T_{GF} - \frac{4}{5}(50 \text{ kN}) = 0$$

$$T_{GF} = -103.1 \text{ kN}$$

Next, consider a free-body diagram of the neighborhood of joint F (Figure 7.15). Considering projections of forces perpendicular to the line DFG, it becomes evident without writing equations that $T_{FE} = 0$. In view of this, and considering forces parallel to DFG, it becomes evident that $T_{FD} = T_{FG} = -103.1$ kN.

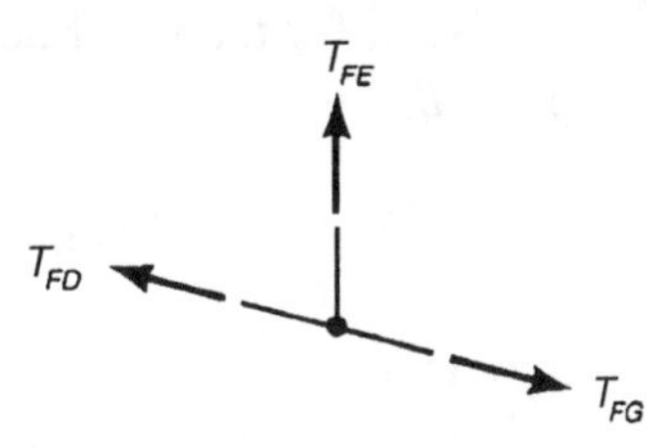

Figure 7.15

We can next proceed to joint E, where T_{ED} and T_{EC} are now the only unknowns. Once these forces are evaluated, the equilibrium equations for joint D contain only two unknowns. Proceeding in this manner, we can evaluate the remainder of the internal forces and the reactions at the supports.

By proper selection of the order in which joints of the truss are considered, it is usually possible to work through a planar truss in the manner indicated. If the values of all the forces are not required, however, use of the equations from joints may be much less efficient than the approach explained next.

Equations from Sections

The forces external to *any* portion of a system in equilibrium have zero resultants of force and moment. This is the basis for the procedure illustrated now. Figure 7.16 shows free-body diagrams of three different portions of the truss in

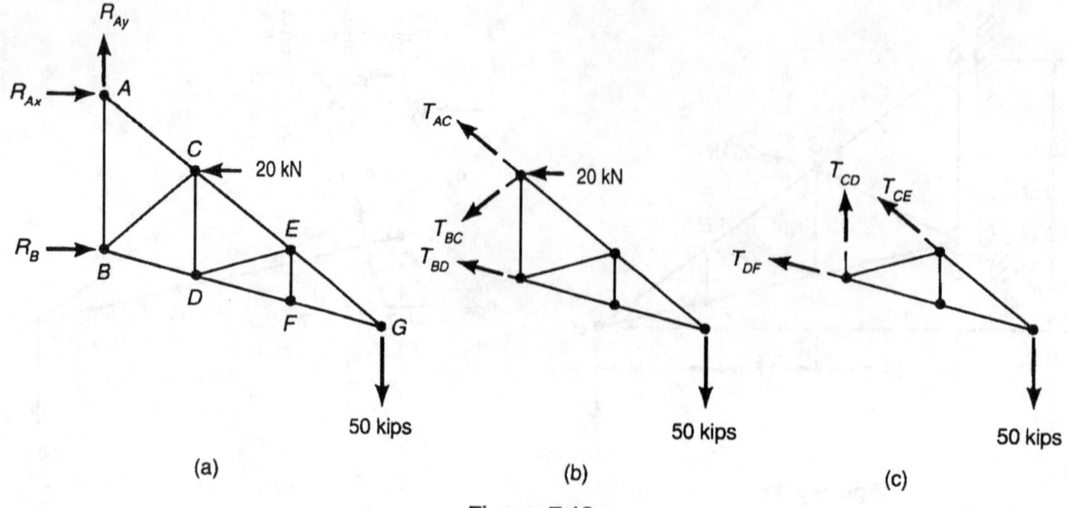

Figure 7.16

Figure 7.13. Referring to the free body of the entire structure, Figure 7.16(a), we can write the resultant moment about A as

$$(3.0\text{ m})R_B - (1.8\text{ m})(20\text{ kN}) - (6.0\text{ m})\,(50\text{ kN}) = 0$$

from which

$$R_B = 112\text{ kN}$$

Then summation of horizontal force components gives

$$R_{Ax} = -92\text{ kN}$$

and summation of vertical force components gives

$$R_{Ay} = 50\text{ kN}$$

Next, for the free body shown in Figure 7.16(b), we can sum moments about point B,

$$(3.0\text{ m})\frac{4}{5}T_{AC} + (1.2\text{ m})(20\text{ kN}) - (6.0\text{ m})(50\text{ kN}) = 0$$

to obtain

$$T_{AC} = 115\text{ kN}$$

Then we can sum moments about point G,

$$(3.6\text{ m})\left(\frac{1}{\sqrt{5}}T_{BC}\right) + (2.7\text{ m})\left(\frac{2}{\sqrt{5}}T_{BC} + 20\text{ kN}\right) = 0$$

to obtain

$$T_{BC} = 13.42\text{ kN}$$

and sum moments about point C,

$$-(1.8\text{ m})\frac{4}{\sqrt{17}}T_{BD} - (3.6\text{ m})(50\text{ kN}) = 0$$

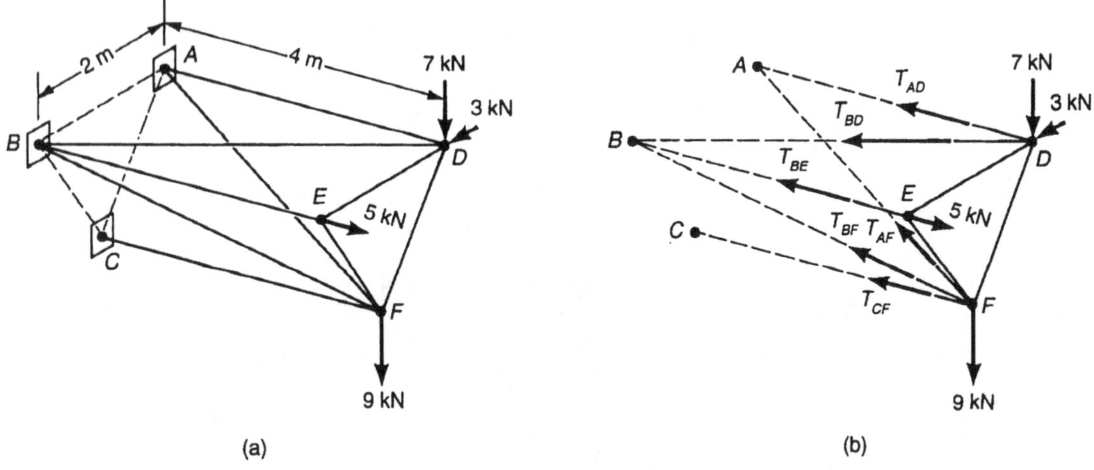

Figure 7.17

to obtain

$$T_{BD} = -103.1 \text{ kN}$$

Now, considering the free body in Figure 7.16(c), we can obtain T_{CD} by summing moments about point G:

$$T_{CD} = 0$$

By continuing in this fashion, we can evaluate the remaining forces in a fairly efficient manner.

The strategy in this method is to isolate a portion of the structure, with the boundary passing through the member in which a force is to be evaluated, and, with the free body completed, to find a direction for force reckoning or an axis for moment reckoning that will yield an equation with as few unknown forces as possible.

For further illustration, consider the truss shown in Figure 7.17(a). The vertical triangles ABC and DEF are both equilateral, and the rectangle $ABED$ is in a horizontal plane. The force in each of the nine members is to be evaluated.

Consider first the free body of the portion of the truss shown in Figure 7.17(b). To analyze this section, we select various axes, about which moments involve only one unknown force. The moment about each axis is evaluated most readily by resolving each force as indicated in Figure 7.17, keeping in mind that the moment of a force about an axis is zero if the line of action intersects the axis or is parallel to the axis:

$$\begin{aligned} M_{DF} &= \left(\sqrt{3} \text{ m}\right)(T_{BE} - 5 \text{ kN}) = 0 \\ T_{BE} &= 5 \text{ kN} \\ M_{CF} &= \left(\sqrt{3} \text{ m}\right)\left(\frac{2}{\sqrt{20}} T_{BD} + 3 \text{ kN}\right) - (1 \text{ m})(7 \text{ kN}) = 0 \\ T_{BD} &= \sqrt{5}\left(\frac{7}{\sqrt{3}} - 3\right) \text{kN} = 2.33 \text{ kN} \end{aligned}$$

$$M_{FE} = \left(\sqrt{3}\ \text{m}\right)\left(T_{AD} + \frac{4}{\sqrt{20}} T_{BD}\right) = 0$$

$$T_{AD} = -2\left(\frac{7}{\sqrt{3}} - 3\right)\text{kN} = 2.08\ \text{kN}$$

$$M_{BE} = \left(\sqrt{3}\ \text{m}\right)\left(\frac{2}{\sqrt{20}} T_{AF}\right) - (1\ \text{m})(9\ \text{kN}) - (2\ \text{m})(7\ \text{kN}) = 0$$

$$T_{AF} = 23\sqrt{\frac{5}{3}}\ \text{kN} = 29.7\ \text{kN}$$

$$M_{DA} = \left(\sqrt{3}\ \text{m}\right)\left(\frac{2}{\sqrt{20}} T_{BF}\right) - (1\ \text{m})(9\ \text{kN}) = 0$$

$$T_{BF} = 9\sqrt{\frac{5}{3}}\ \text{kN} = 11.62\ \text{kN}$$

$$M_{BA} = \left(\sqrt{3}\ \text{m}\right) T_{CF} + (4\ \text{m})(7\ \text{kN}) + (4\ \text{m})(9\ \text{kN}) = 0$$

$$T_{CF} = -\frac{64}{\sqrt{3}}\ \text{kN} = -37\ \text{kN}$$

To determine the forces in the remaining three members, it is a straightforward matter to isolate joints D and E and sum forces:

$$T_{DF} = -\frac{14}{\sqrt{3}}\ \text{kN} = -8.08\ \text{kN}$$

$$T_{DE} = T_{EF} = 0$$

As with planar trusses, three-dimensional trusses can be analyzed by writing equations of force equilibrium for each joint, or by considering a larger portion of the truss. The procedure for writing equilibrium equations for a joint is illustrated in the previous section and typically leads to a set of simultaneous equations for the unknown forces. Often, as in the preceding example, the task of solving simultaneous equations can be avoided by considering an entire section of the truss and finding axes about which only one unknown force has a nonzero moment.

COUPLE-SUPPORTING MEMBERS

The loads that a rigid bar can carry are not limited to axial forces. If lateral forces are applied, equilibrium requires that forces across a section have a nonzero moment about the center of the section, as demonstrated in Figure 7.18. This figure shows a sketch and free-body diagrams: one of the entire and others of two separated portions of the beam. Equilibrium of the portion on the left indicates at once that the forces across the plane of separation must form a lateral force and a couple. The detailed distribution of interaction forces may be fairly complicated and cannot be deduced from equilibrium alone; however, it is often possible to evaluate the resultant force and the moment of the equivalent couple from statics.

Twisting and Bending Moments

In general, the moment of the couple at a section can have any direction, depending on how the external loads are applied. For example, the bracket in Figure 7.19 has moment components at section A in each of three directions aligned with the axis

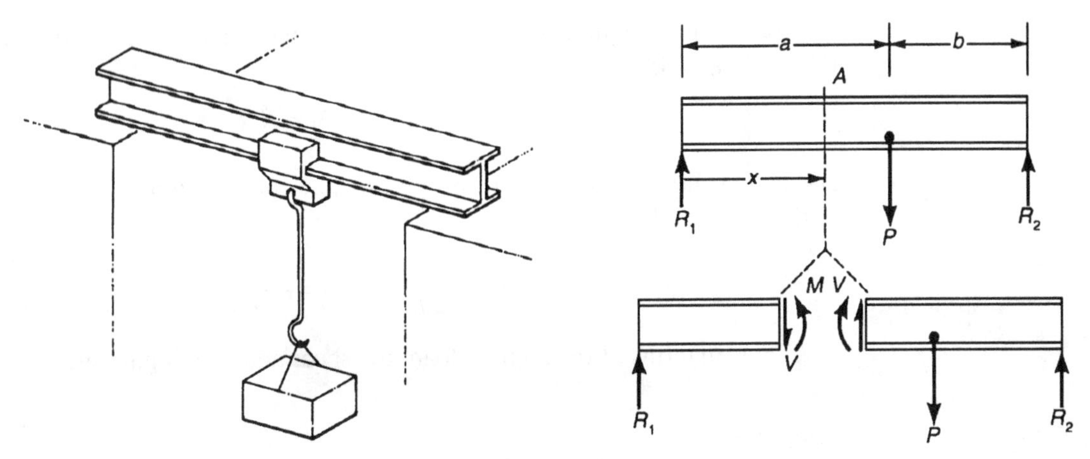

Figure 7.18

of the bracket. The moment vector is usually resolved into a component parallel to the axis of the bar as well as one or two components perpendicular to the axis. (This facilitates the analysis of strength and deformation of the bar.) The component of moment parallel to the axis of the bar is called the **twisting moment**, and the components perpendicular to the axis are called **bending moments**, after the types of deformation they produce, as shown in Figure 7.20.

The resultant force acting at a section of a bar is similarly resolved. The component parallel to the axis of the rod is called the **axial force**, and the components perpendicular to the axis are called **shearing forces**.

Evaluation of these force and moment components is accomplished by using the same basic ideas already examined: A free body of a portion of the member on either side of the section of interest is isolated and properly labeled, and equations of equilibrium are written and solved.

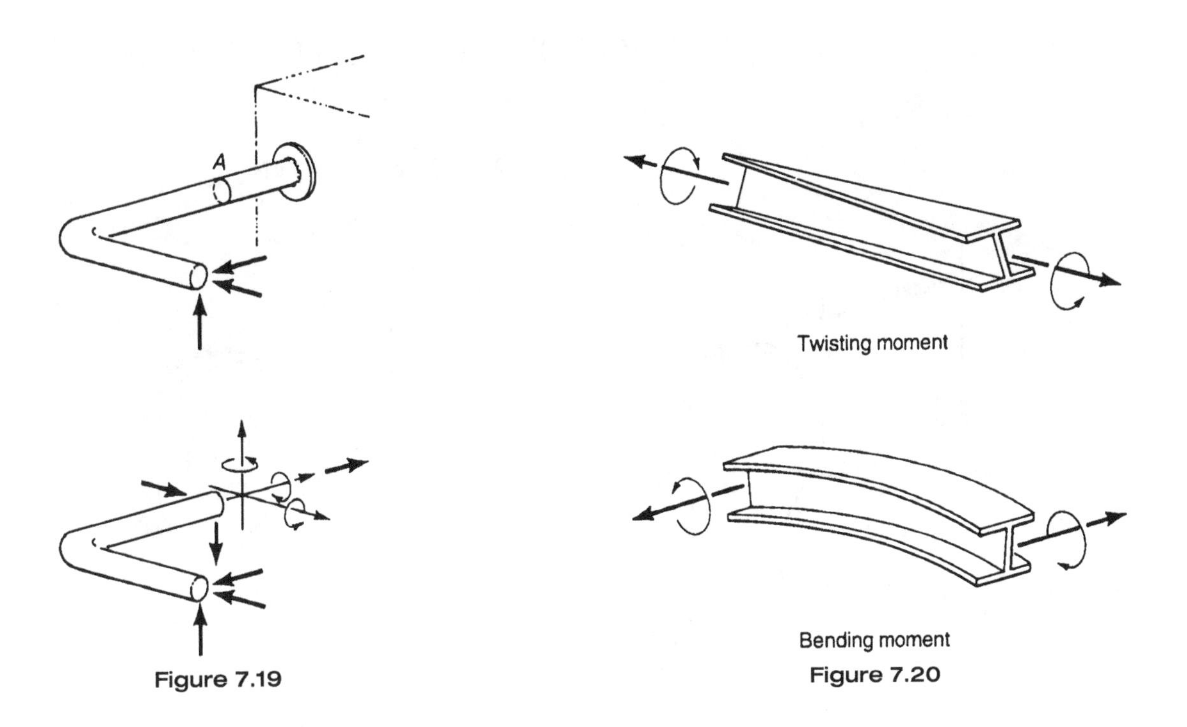

Figure 7.19

Figure 7.20

For example, equilibrium of the portion of the beam to the left of section A in Figure 7.18 yields

$$V = R_1 \quad \text{and} \quad M = R_1 x$$

and equilibrium of the entire beam gives values of the support reactions in terms of their applied load P as

$$R_1 = \frac{bP}{a+b} \qquad R_2 = \frac{aP}{a+b}$$

Thus, in terms of the applied load, the shear and bending moment are

$$V = \frac{bP}{a+b} \qquad x < a$$

$$M = \frac{bPx}{a+b} \qquad x < a$$

The qualification $x < a$ is necessary because the analysis was done for sections to the left of the applied load. A similar analysis for sections on the other side of the load results in

$$V = \frac{-aP}{a+b} \qquad x > a$$

$$M = \frac{a(a+b-x)P}{a+b} \qquad x > a$$

As another example, consider the force and moment components at the support O for the automobile torsion bar in Figure 7.21. Force equilibrium of the free body requires that

$$R + (6.8\text{ kN})\left(-\frac{8}{17}\mathbf{e}_x + \frac{15}{17}\mathbf{e}_z\right) + (2.5\text{ kN})\mathbf{e}_y = 0$$

or

$$R = (3.2\mathbf{e}_x - 2.5\mathbf{e}_y - 6.0\mathbf{e}_z)\text{ kN}$$

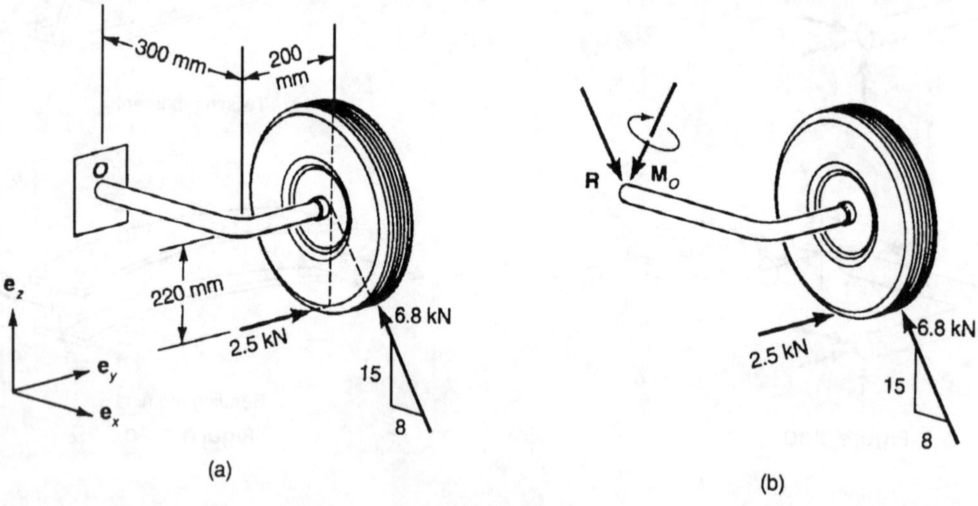

Figure 7.21

Therefore, there is a compressive axial force of 3.2 kN and a resultant shearing force of $\sqrt{(2.5)^2+(6.0)^2}$ kN $= 6.5$ kN at the section O. Next, moment equilibrium requires that

$$\begin{aligned}\mathbf{M}_O &+ [(0.3\text{ m})\mathbf{e}_x + (0.2\text{ m})\mathbf{e}_y] \times [(6.0\text{ kN})\mathbf{e}_z - (3.2\text{ kN})\mathbf{e}_x] \\ &+ [(0.3\text{ m})\mathbf{e}_x + (0.2\text{ m})\mathbf{e}_y - (0.22\text{ m})\mathbf{e}_z] \times (2.5\text{ kN})\mathbf{e}_y = 0\end{aligned}$$

or

$$\mathbf{M}_O = (-1.75\mathbf{e}_x + 1.80\mathbf{e}_y - 1.39\mathbf{e}_z)\text{ kN}\bullet\text{m}$$

This indicates the presence of a twisting moment

$$\mathbf{M}_t = 1.75\text{ kN}\bullet\text{m}$$

and a bending moment resultant

$$\begin{aligned}M_b &= \sqrt{(1.80)^2+(1.39)^2}\text{ kN}\bullet\text{m} \\ &= 2.27\text{ kN}\bullet\text{m}\end{aligned}$$

SYSTEMS WITH FRICTION

Friction forces act tangentially to the surfaces on which two objects make contact. The ratio of the tangential force to the normal force at the contact surfaces is called the **coefficient of friction**. In general, this ratio depends on several variables, such as the surface materials, the surface finishes, the presence of any surface films, the velocity of sliding, and the temperature. The ratio of tangential force necessary to initiate sliding from a state of rest to the normal force is called the **coefficient of static friction**, whereas the force ratio as sliding continues is called the **coefficient of sliding friction**. Typically, the coefficient of static friction is somewhat greater than the coefficient of sliding friction. Furthermore, a decrease in sliding friction force with an increase in the speed of sliding has been observed for some materials, although this variation is usually small enough that it can be neglected.

In spite of the complexity of the mechanism of friction (Figure 7.22), the approximation known as *Coulomb friction* has been found to lead to predictions

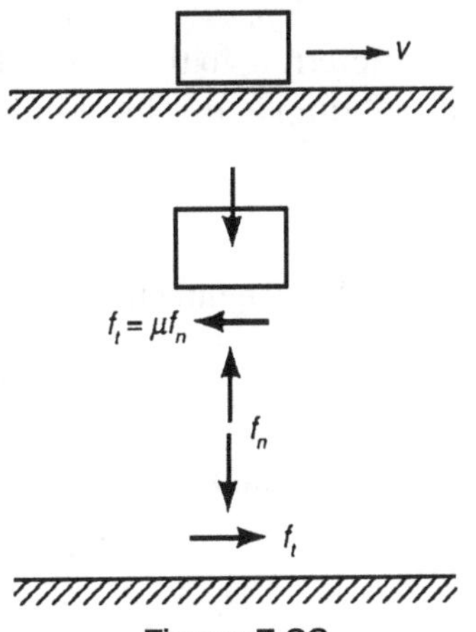

Figure 7.22

of acceptable accuracy for many dry surfaces. The so-called **Coulomb's Law of Friction** states that whenever sliding takes place, the tangential component of force between the surfaces is proportional to the normal component and acts in a direction to oppose the motion; that is,

$$\mathbf{f}_t = \mu_1 f_n \mathbf{e}_v \qquad v \neq 0 \tag{7.14a}$$

in which $\mathbf{e}_v$ is a unit vector in the direction on the relative velocity, v, of the objects on which $\mathbf{f}_t$ acts, and the coefficient of sliding friction μ_1 depends on the surfaces in contact but not on the magnitude of the normal force or on the velocity. When the surfaces are not sliding, the friction force can have any direction required for equilibrium, but its magnitude is limited by

$$f_t \leq \mu_0 f_n \qquad v = 0 \tag{7.14b}$$

in which μ_0 is the coefficient of static friction. When applied forces induce a friction force that reaches this limit, motion is incipient, meaning that any change tending to increase the friction force will cause acceleration.

Example 7.10

What is the magnitude P of the force required to move the 40-kg block up the 15° incline in Exhibit 9? Also, in the absence of P, will the block remain stationary on the incline? The coefficients of static and sliding friction are both 0.3.

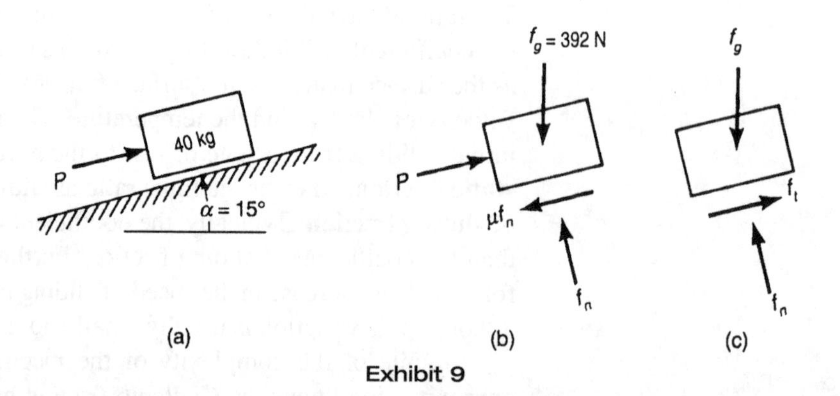

Exhibit 9

Solution

Referring to the free-body diagram (Exhibit 9(b)), we can write equations of equilibrium in the direction normal to the incline,

$$f_n - f_g \cos \alpha = 0$$

and in the direction along the incline,

$$P - f_g \sin \alpha - \mu f_n = 0$$

Elimination of f_n from these two equations results in

$$\begin{aligned} P &= f_g(\sin\alpha + \mu \cos\alpha) \\ &= (392 \text{ N})(\sin 15° + 0.30 \cos 15°) \\ &= 215 \text{ N} \end{aligned}$$

The free-body diagram in Exhibit 9(c) depicts the situation in the absence of the force P. Observe the reversal of the direction of the friction force. Now, if the block is to remain at rest,

$$f_n = f_g \cos\alpha$$
$$f_t = f_g \sin\alpha$$

But the friction force is limited by

$$f_t \le \mu f_n$$

Substitution of the equilibrium equations into this inequality gives

$$f_g \sin\alpha \le \mu f_g \cos\alpha$$

or

$$\tan\alpha \le \mu$$

Because $\tan\alpha = 0.27$, which is less than $\mu = 0.3$, the block will remain at rest.

Example 7.11

In the absence of P, the angle of the incline in the previous example is slowly increased until the block begins to slide downward. If the coefficient of static friction is $\mu_0 = 0.47$ and the coefficient of sliding friction is $\mu_1 = 0.44$, what will be the acceleration of the block after it breaks loose?

Solution

Prior to breakaway of the block,

$$f_t \le \mu_0 f_n$$

Or, with the equilibrium relationships from the free-body diagram (Exhibit 9(c)),

$$f_g \sin\alpha \le \mu_0 f_g \cos\alpha$$

The equality occurs when the critical angle, α_c, is reached:

$$\tan\alpha_c = \mu_0 = 0.47$$

from which

$$\alpha_c = 25.2°$$

After the block breaks loose,

$$f_t = \mu_1 f_n$$

Now, the component of acceleration perpendicular to the inclined plane remains zero, so that

$$f_n - f_g \cos\alpha_c = 0$$

But, in the direction parallel to the plane, the static equilibrium relationship must be replaced with Newton's second law,

$$f_g \sin\alpha_c - f_t = ma$$

in which a is the downward tangential acceleration. Combining these relationships leads to

$$a = \frac{f_g}{m}(\sin \alpha_c - \mu_1 \cos \alpha_c)$$

But, since $f_g = mg$,

$$a = g \sin \alpha_c \left(1 - \frac{\mu}{\tan \alpha_c}\right) = g \sin \alpha_c \left(1 - \frac{\mu_1}{\mu_2}\right) = (9.81 \text{ m/s}^2) \sin 25.2° \left(1 - \frac{0.44}{0.47}\right)$$

$$= 0.27 \text{ m/s}^2$$

DISTRIBUTED FORCES

When forces are distributed throughout some region (as in the case of gravity) or over a surface (as in the case of pressure on the wall of a water tank), the fundamental ideas illustrated previously apply. A feature not yet illustrated, however, is the computational detail of summation, which takes the form of integration.

Example 7.12

The beam in Exhibit 10 supports a load that varies in intensity along the length as indicated. The intensity (force per unit length of beam) has the values w_A and w_B at the two ends and varies linearly between these points.

In terms of w_A, w_B, and L, what is the resultant of the downward forces, and what are the magnitudes R_A and R_B of the reactions at the supports?

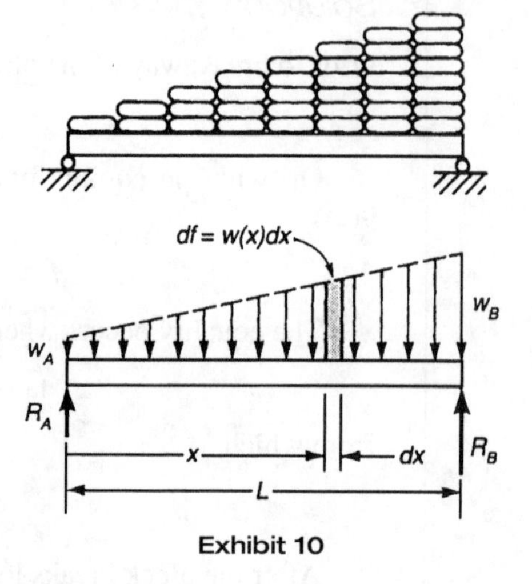

Exhibit 10

Solution

First, we write an expression for the load intensity as a function x, the distance along the span measured from the left-hand support:

$$w(x) = w_A + (w_B - w_A)\frac{x}{L}$$

Next, consider the force in the shaded portion of the load diagram, acting between the points given by x and $x + dx$. The force in this region will be the product of the intensity (force per unit length) and the length dx:

$$df = w(x)\, dx$$

Summing all such forces gives their resultant:

$$
\begin{aligned}
f &= \int w(x)\,dx \\
&= \int_0^L \left[w_A + (w_B - w_A)\frac{x}{L} \right] dx \\
&= \frac{1}{2}(w_A + w_B)L
\end{aligned}
$$

To evaluate R_B, we equate to zero the sum of moments about A of all the forces:

$$
\begin{aligned}
M_A &= R_B L - \int x\,[w(x)]dx \\
&= R_B L - \int_0^L x \left[w_A + (w_B - w_A)\frac{x}{L} \right] dx \\
&= R_B L - \frac{1}{6}(w_A + 2w_B)L^2 = 0
\end{aligned}
$$

This yields

$$
R_B = \frac{1}{6}(w_A + 2w_B)L
$$

To evaluate R_A, we can use the fact that the sum of all vertical forces must be zero:

$$
\begin{aligned}
R_A &= f - R_B \\
&= \frac{1}{2}(w_A + w_B)L - \frac{1}{6}(w_A + 2w_B)L \\
&= \frac{1}{6}(2w_A + w_B)
\end{aligned}
$$

Example 7.13

The uniform, slender, semicircular arch in Exhibit 11 is acted on by gravity and the reactions from the supports.

The free-body diagram shows the desired bending moment, M_b, as the reaction from the other half of the arch. Because of horizontal equilibrium, no axial force exists at this section. The vertical shearing force is also zero at this section because Newton's third law would require a shearing force in the opposite direction on the other half of the arch, and this pair of forces would be inconsistent with the symmetry of the system.

Because the arch is slender, the forces of gravity may be treated as distributed along a circular *line*. Let the cross-sectional area be denoted by A and the density (mass per unit volume) by ρ. Then the volume of the shaded element of the arch will be equal to $Aa\,d\theta$, and the magnitude of the force of gravity acting on it will be

$$
df_g = \rho(Aa\,d\theta)g
$$

The resultant of the gravitational forces then has the magnitude

$$
\begin{aligned}
f_g &= \int_0^{\pi/2} \rho Aga\,d\theta \\
&= \frac{1}{2}\pi a\rho Ag
\end{aligned}
$$

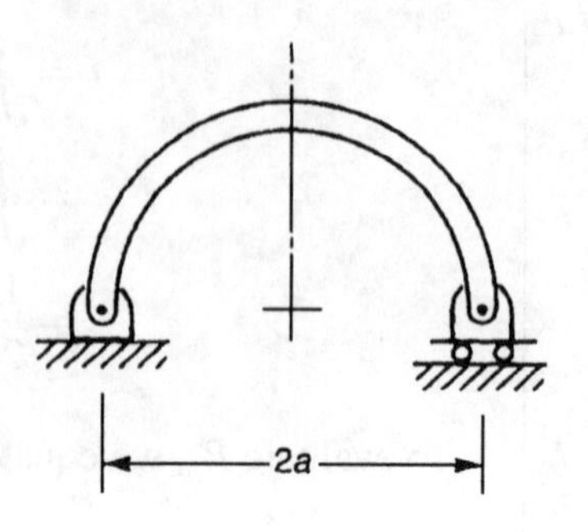

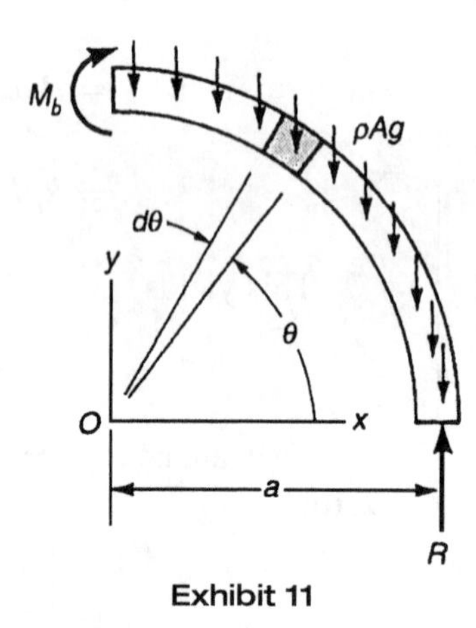

Exhibit 11

and vertical force equilibrium gives the support reaction as

$$R = \frac{1}{2}\pi a \rho A g$$

Now, the resultant moment of the forces of gravity about the point O will be in the clockwise direction and of magnitude

$$\begin{aligned} M_{Og} &= \int a\cos\theta \, df_g \\ &= \int_0^{\pi/2} a\cos\theta \, \rho A g a \, d\theta \\ &= a^2 \rho A g \end{aligned}$$

Finally, moment equilibrium about point O requires that

$$aR - M_{Og} - M_b = 0$$

which, together with the above results, gives

$$M_b = \left(\frac{\pi}{2} - 1\right) a^2 \rho A g$$

In an arch with a large radius, this bending moment could well cause failure of the structure.

Single Force Equivalents

The construction and evaluation of integrals such as those in Examples 7.12 and 7.13 can be circumvented if knowledge of an equivalent discrete force is available. This information, for a variety of special cases, is available in tabulated formulas, these formulas having been determined by integration. Proper use of such formulas requires an understanding of the following concepts.

Center of Mass and Center of Gravity

Consider the forces of gravity distributed throughout an arbitrary body, as shown in Figure 7.23. The force acting on the element with mass *dm* will be

$$d\mathbf{f}_g = dm\, g\mathbf{e}_g$$

and the resultant force will be given by

$$\begin{aligned}\mathbf{f}_g &= \int d\mathbf{f}_g \\ &= \int dm\, g\mathbf{e}_g \\ &= \left(\int dm\right) g\mathbf{e}_g = mg\mathbf{e}_g\end{aligned}$$

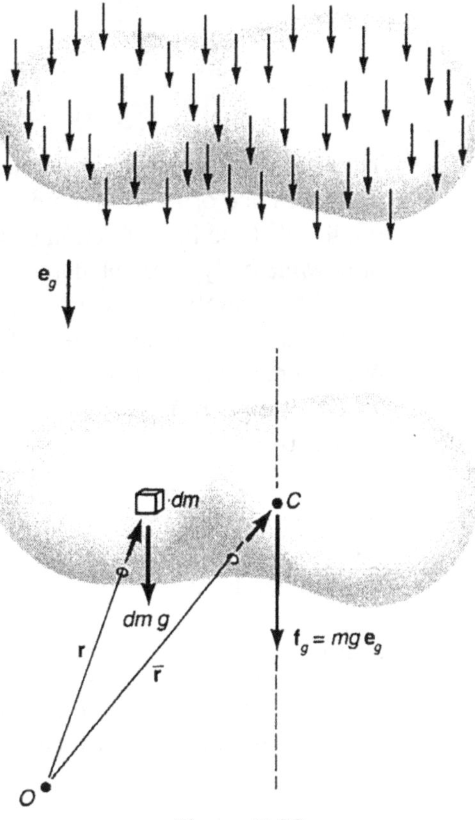

Figure 7.23

where m is the total mass of the body. The resultant moment about a point O is given by

$$M_{og} = \int \mathbf{r} \times d\mathbf{f}_g$$

$$= \int \mathbf{r} \times (g\mathbf{e}_g \, dm)$$

$$= \left(\int \mathbf{r} \, dm \right) \times g\mathbf{e}_g$$

where $\mathbf{r}$ is the position vector locating the mass element. Now, to compose the forces into a single, equivalent resultant, the line of action of the equivalent force $\mathbf{f}_g$ must pass through a point located by the position vector $\mathbf{r}_f$, satisfying the moment equivalence:

$$\mathbf{r}_f \times \mathbf{f}_g = \mathbf{M}_{Og}$$

Thus

$$\mathbf{r}_f \times (mg\mathbf{e}_g) = \left(\int \mathbf{r} \, dm \right) \times g\mathbf{e}_g$$

$$\mathbf{r}_f \times \mathbf{e}_g = \bar{\mathbf{r}} \times \mathbf{e}_g$$

where

$$\bar{\mathbf{r}} = \frac{1}{m} \int \mathbf{r} \, dm \qquad \textbf{(7.15)}$$

The vector $\bar{\mathbf{r}}$ locates an important point C, called the **center of mass** of the body. By the choice of $\mathbf{r}_f = \bar{\mathbf{r}}$, moment equivalence will be satisfied for any $\mathbf{e}_g$, which implies that the line of action of the equivalent force passes through C regardless of how the body is oriented.

The center of mass C is of fundamental importance in the study of dynamics. In statics its significance stems from the fact that the resultant moment about C, of the forces of uniform gravity, is zero. That is, the body could be statically balanced by supporting it at this point only. For this reason it is also called the center of gravity.

Example **7.14**

Find the location of the center for gravity for the portion of the arch isolated as a free body in Example 7.13.

Solution

Using the center of the circle as a reference point, we can locate the shaded mass element with the position vector

$$\mathbf{r} = a \cos\theta \, \mathbf{e}_x + a \sin\theta \, \mathbf{e}_y$$

Then, from the definition for the center of mass, Equation (7.15),

$$\bar{r} = \frac{1}{m}\int \left(a \cos\theta\, \mathbf{e}_x + a \sin\theta\, \mathbf{e}_y\right)\left(\rho A a\, d\theta\right)$$

$$= \frac{\rho A a^2}{m}\int_0^{\pi/2} \left(\cos\theta\, \mathbf{e}_x + \sin\theta\, \mathbf{e}_y\right) d\theta$$

$$= \frac{2a}{\pi}\left(\mathbf{e}_x + \mathbf{e}_y\right)$$

Centroids

The mass, dm, of the element used in the preceding integrals can be expressed in terms of the density, ρ, and the corresponding element of volume, dV, as $dm = \rho\, dV$. Then the position vector locating the center of mass can be written as

$$\bar{\mathbf{r}} = \frac{\int \mathbf{r}\rho\, dV}{\int \rho\, dV}$$

Now, if the density is uniform throughout the body, ρ can be brought outside the integrals with the result

$$\bar{\mathbf{r}} = \frac{1}{V}\int \mathbf{r}\, dV$$

The location of the point C here depends entirely on geometry, because all contributions having to do with material have been canceled. The point C, located according to this equation, is called the **centroid of the volume** V. Similarly, the **centroid of a surface area** A is defined as the point located by the position vector

$$\bar{\mathbf{r}} = \frac{1}{A}\int \mathbf{r}\, dA$$

and the **centroid of a line segment** of length L is defined as the point located by the position vector

$$\bar{\mathbf{r}} = \frac{1}{L}\int \mathbf{r}\, dL$$

The calculation in Example 7.14 was for a uniform mass per unit length of the arch. With this density canceled out, the center of mass of the arch segment is also the centroid of a quarter-segment of a circular line.

The integrals $\int \mathbf{r}\, dV, \int \mathbf{r}\, dA,$ and $\int \mathbf{r}\, dL$ are called the first moments of the volume, area, and line, respectively, about the reference point O. In carrying out the calculations, it is often convenient to work with one rectangular Cartesian component of the position vector at a time, that is, to evaluate separately the component equivalents obtained by dot-multiplying the vector definition by a unit vector in each direction:

$$\bar{x} = \mathbf{e}_x \bullet \bar{\mathbf{r}} = \frac{1}{V}\int x\,dv$$

$$\bar{y} = \mathbf{e}_y \bullet \bar{\mathbf{r}} = \frac{1}{V}\int y\,dv$$

$$\bar{z} = \mathbf{e}_z \bullet \bar{\mathbf{r}} = \frac{1}{V}\int z\,dv$$

Example 7.15

Determine the location of the centroid of the shaded triangle of Exhibit 12 in terms of the dimensions *a*, *b*, and *c*.

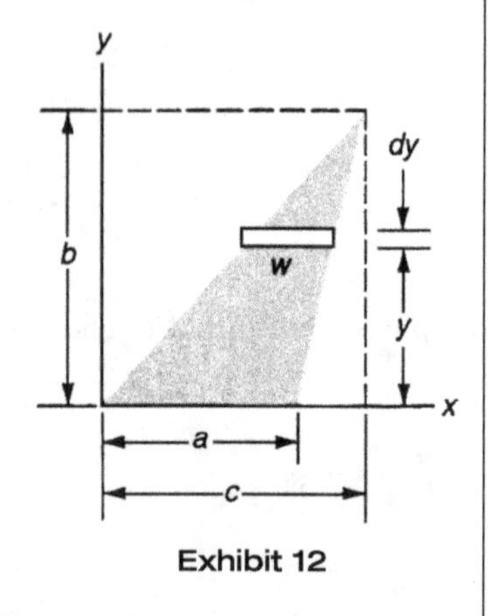

Exhibit 12

Solution

The area and first moment of area can be computed by evaluating the contribution to these quantities from the unshaded element in Exhibit 12 and summing these contributions by integration with respect to *y*. The width, *w*, of the element can be expressed in terms of *y* after observing that the entire triangle is similar to the triangle that lies above the unshaded element:

$$\frac{w}{b-y} = \frac{a}{b}$$

or

$$w = a\left(1 - \frac{y}{b}\right)$$

Thus the area of the element is

$$dA = w\,dy = a\left(1 - \frac{y}{b}\right)dy$$

and the area of the triangle is

$$A = \int_0^b a\left(1 - \frac{y}{b}\right)dy = \frac{1}{2}ab$$

This result is very well known. The center of the element is located on the straight line that connects the apex of the triangle with the midpoint of the base, that is, on the line that has the equation

$$x = \frac{a}{2} + \left(c - \frac{a}{2}\right)\frac{y}{b}$$

Therefore, the *x*-component of the first moment of area of the triangle is

$$\int x\,dA = \int_0^b \left[\frac{a}{2} + \left(c - \frac{a}{2}\right)\frac{y}{b}\right] a\left(1 - \frac{y}{b}\right)dy = \frac{1}{6}ab(a+c)$$

With this and the previously discussed value of area, the x-coordinate of the centroid is determined as

$$\bar{x} = \frac{1}{A}\int x\,dA = \frac{ab(a+c)/6}{ab/2} = \frac{1}{3}(a+c)$$

A similar calculation yields the y-component of the first moment of area as

$$\int y\,dA = \int_0^b ya\left(1-\frac{y}{b}\right)dy = \frac{ab^2}{6}$$

from which we find the y-coordinate of the centroid to be

$$\bar{y} = \frac{1}{A}\int y\,dA = \frac{ab^2/6}{ab/2} = \frac{1}{3}b$$

A composite volume, area, or line may be built up from several parts, for which the location of each centroid is known. In this case, a summation having the same form as the integral definition can readily be shown as valid. In the case of an area, A, that is a composite formed from n areas $A_1, A_2, \ldots, A_n$ and having centroids located by $\bar{\mathbf{r}}_1, \bar{\mathbf{r}}_2, \ldots, \bar{\mathbf{r}}_n$, the centroid of the composite is located at

$$\bar{\mathbf{r}} = \frac{1}{A}\sum_{i=1}^{n} A_i \bar{\mathbf{r}}_i$$

where

$$A = \sum_{i=1}^{n} A_i$$

Example 7.16

Locate the centroid of the plane area shown shaded in Exhibit 13.

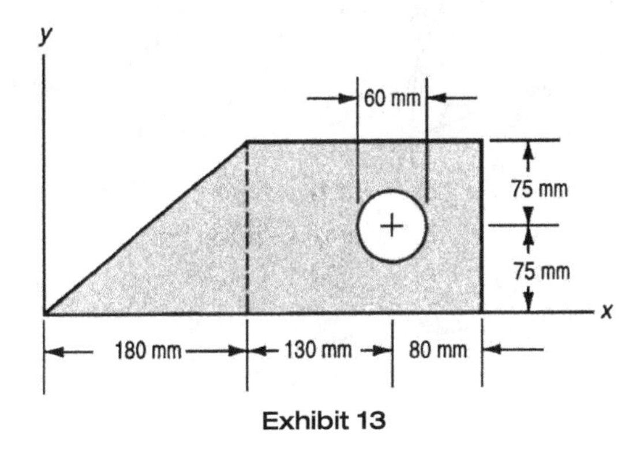

Exhibit 13

Solution

The areas and coordinates of centroids of individual parts are as follows:

	A, m²	$\bar{x}$**, mm**	$\bar{y}$**, mm**
Triangle	0.0135	120	50
Rectangle	0.0315	285	75
Circle	–0.00283	310	75

The coordinates locating the centroid are then

$$\bar{x} = \frac{(120)(0.0135)+(285)(0.0315)+(310)(-0.00283)}{0.0135+0.0315-0.00283} = 231 \text{ mm}$$

$$\bar{x} = \frac{(50)(0.0135)+(75)(0.0315)+(75)(-0.00283)}{0.0135+0.0315-0.00283} = 67 \text{ mm}$$

Second Moments of Area

The geometric properties defined in the previous two sections are associated with uniformly distributed forces. Forces that vary *linearly* over a plane area lead to additional geometric properties that are useful in evaluating moment resultants. One example is fluid pressure acting on submerged, plane surfaces; another is the bending stress induced in beams. The latter is covered in the chapter on mechanics of materials. Consider the flat surfaces shown in Figure 7.24 with a force intensity that varies linearly with the distance from the line *Op*. With the force per unit area denoted as σ_z and the distance from *Op* as *q*, the linear variation is expressed as

$$\sigma_z = kq$$

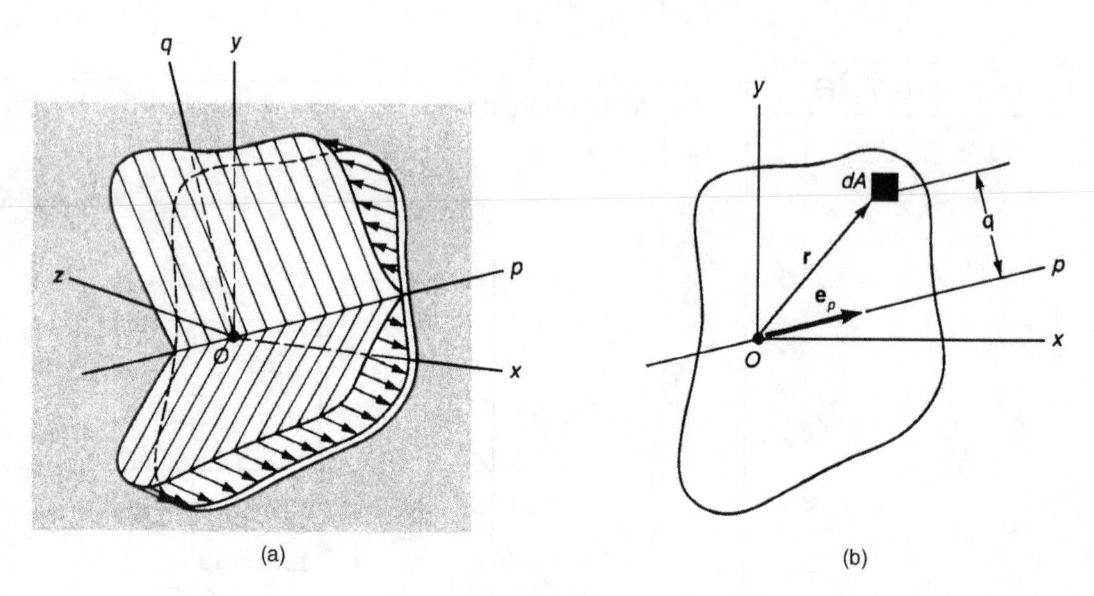

Figure 7.24

where k is a proportionality constant. (In the analysis of beam bending, this constant is the product of Young's modulus and the curvature of the deformed axis of the beam.) Observe from Figure 7.24(b) that the distance from axis Op to the area element dA is given by

$$q = r \sin \angle^{\mathbf{r}} \mathbf{e}_p = \left(\mathbf{e}_p \times \mathbf{r}\right) \bullet \mathbf{e}_z$$

so that the force acting on the area element can be expressed as

$$d\mathbf{f} = \sigma_z \, dA \, \mathbf{e}_z = k\left(\mathbf{e}_p \times \mathbf{r}\right) dA$$

The resultant moment about O is then

$$\mathbf{M}_O = \int_A \mathbf{r} \times d\mathbf{f}$$

$$= k\int_A \mathbf{r} \times \left(\mathbf{e}_p \times \mathbf{r}\right) dA$$

$$= k\int_A \left[\left(\mathbf{r} \bullet \mathbf{r}\right)\mathbf{e}_p - \left(\mathbf{e}_p \bullet \mathbf{r}\right)\mathbf{r}\right] dA$$

In terms of the rectangular Cartesian components in the directions of x and y,

$$\mathbf{r} = x\mathbf{e}_x + y\mathbf{e}_y$$
$$\mathbf{e}_p = p_x\mathbf{e}_x + p_y\mathbf{e}_y$$

The moment can be expressed as

$$\mathbf{M}_O = k\int_A [(x^2 + y^2)\mathbf{e}_p - (xp_x + yp_y)\mathbf{r}]dA$$

Components of the moment are

$$\mathbf{M}_{Ox} = \mathbf{e}_x \bullet \mathbf{M}_O = k\int [(x^2 + y^2)p_x - (xp_x + yp_y)x]dA$$

and

$$M_{Oy} = k\left[\left(-\int_A xy \, dA\right)p_x + \left(\int_A x^2 \, dA\right)p_y\right]$$

$$= k\left[\left(\int_A y^2 \, dA\right)p_x + \left(-\int_A xy \, dA\right)p_y\right]$$

The integrals that appear in these expressions are called **second moments of area** and are fundamental to the relation between the moment and the orientation of the zero-force line Op. The integrals

$$I_{xx} = \int_A y^2 dA \quad \text{and} \quad I_{yy} = \int_A x^2 dA \qquad \textbf{(7.16)}$$

are often called **moments of inertia** of the area about the x and y axes, respectively. (This terminology stems from analogous integrals related to mass distribution in the study of the kinetics of rigid bodies.) The integral

$$I_{xy} = -\int_A xy\, dA \tag{7.17}$$

is called the product of inertia of the area with respect to the x and y axes. (Many define it as the negative of this definition; then the corresponding terms in the next equation require opposite signs.) The quantities I_{xx}, I_{yy}, and I_{xy} are collectively called *second moments of area*.

In terms of the notation just introduced above, the relationship that gives the moment resultant in terms of the placement of the zero-force line Op is

$$\begin{Bmatrix} M_{Ox} \\ M_{Oy} \end{Bmatrix} = k \begin{bmatrix} I_{xx} & I_{xy} \\ I_{yx} & I_{yy} \end{bmatrix} \begin{Bmatrix} p_x \\ p_y \end{Bmatrix} \tag{7.18}$$

This is fundamental to the analyses of bending moment and deformation of elastic beams that are usually studied in mechanics of materials. Here, we examine only the evaluation of the moments and products of inertia.

The values of I_{xx}, I_{yy}, and I_{xy} depend on the placement of the origin O and the orientation of the x and y axes with respect to the area concerned. For special orientations of the x and y axes, called *principal directions*, the value of I_{xy} is zero. This situation will occur if the orientation is such that the area is symmetric with respect to either the x or y axis. (When no such symmetry exists, the determination of the principal directions requires computations that will not be addressed here.)

Parallel Axis Formulas

Often a value of one of the second moments of area is known for a given placement of the origin coordinates, but another value is needed for a different origin. Substituting the coordinate change indicated in Figure 7.25 into Equations (7.16) and (7.17) leads to the following *parallel axis formulas*, which make this evaluation much easier than carrying out an integration:

$$\begin{aligned} I_{xx}^O &= I_{xx}^C + A\bar{y}^2 \\ I_{yy}^O &= I_{yy}^C + A\bar{x}^2 \\ I_{xy}^O &= I_{xy}^C + A\bar{x}\bar{y}^2 \end{aligned} \tag{7.19}$$

Here, the superscript C indicates the value with the origin at the centroid, and the superscript O indicates an arbitrary origin. A is the area, and (x, y) are the coordinates of C with respect to O.

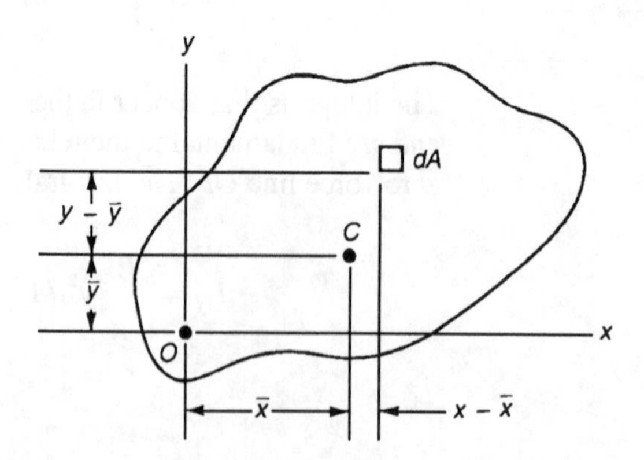

Figure 7.25

Example 7.17

In terms of the dimensions shown in Exhibit 14, evaluate the second moments of area of the triangle with respect to its centroid.

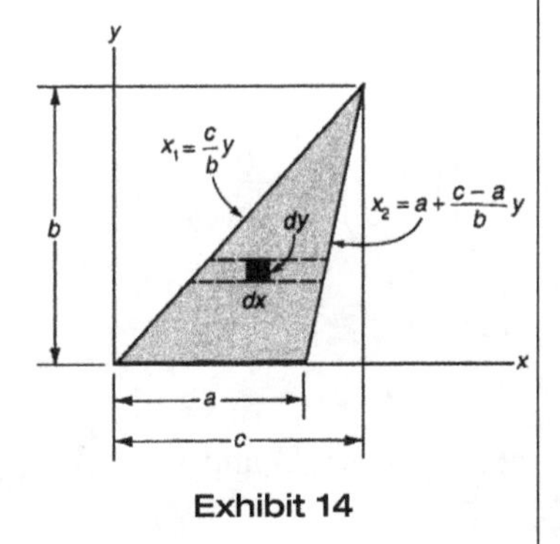

Exhibit 14

Solution

The area of the shaded element is *dx dy*. The line that bounds the triangle on the left is

$$x_1 = \frac{c}{b}y$$

and the line that bounds it on the right is

$$x_2 = a + \frac{c-a}{b}y$$

These two equations will form the limits of the first integration (with respect to *x*) in each case. The moments and product of inertia, with respect to the axes with origin at 0, have the values

$$I_{xx} = \int_0^b \int_{x_1}^{x_2} y^2\, dx\, dy$$

$$= \int_0^b [y^2 x]_{cy/b}^{a+(c-a)y/b}\, dy$$

$$= \int_0^b y^2\left(a - \frac{ay}{b}\right) dy = \frac{ab^3}{12}$$

$$I_{xy} = -\int_0^b \int_{x_1}^{x_2} xy\, dx\, dy$$

$$= -\int_0^b \left[\frac{x^2}{2}\right]_{cy/b}^{a+(c-a)y/b} y\, dy$$

$$= -\frac{a}{2}\int_0^b \left[a + 2(c-a)\frac{y}{b} - (2c-a)\frac{y^2}{b^2}\right] y\, dy$$

$$= -\frac{a}{2}\left[\frac{ab^2}{2} + \frac{2}{3}(c-a)b^2 - \frac{1}{4}(2c-a)b^2\right]$$

$$= -\frac{ab^2}{24}(a+2c)$$

$$I_{yy} = \int_0^b \int_{x_1}^{x_2} x^2 \, dx \, dy$$

$$= \frac{1}{3} \int_0^b [x^3]_{cy/b}^{a+(c-a)y/b} \, dy$$

$$= \frac{ab}{3} \int_0^b \left[a^2 \left(1 - \frac{y}{b}\right)^3 + 3ac\left(1 - \frac{y}{b}\right)^2 \frac{y}{b} + 3c^2\left(1 - \frac{y}{b}\right)\frac{y^2}{b^2} \right] \frac{dy}{b}$$

$$= \frac{ab}{12}(a^2 + ac + c^2)$$

As in Example 7.15, the coordinates of the centroid are $x = (a + c)/3$ and $y = b/3$. Use of the parallel axis formulas (7.19) then gives

$$I_{xx}^C = I_{xx}^O - A\bar{y}^2$$

$$= \frac{ab^3}{12} - \frac{ab}{2}\left(\frac{b}{3}\right)^2 = \frac{ab^3}{36}$$

$$I_{yy}^C = I_{yy}^O - A\bar{x}^2$$

$$= \frac{ab}{12}\left(a^2 + ac + c^2\right) - \frac{ab}{c}\left(\frac{a+c}{3}\right)^2 = \frac{ab}{36}\left(a^2 - ac + c^2\right)$$

$$I_{xy}^C = I_{xy}^O - A\overline{xy}$$

$$= \frac{ab^2}{24}(a + 2c) + \frac{ab}{c}\left(\frac{a+c}{3}\right)\left(\frac{b}{3}\right) = \frac{ab^2}{72}(a - 2c)$$

PROBLEMS

7.1 A 70-kg astronaut is "floating" inside a spaceship that is in a circular orbit at an altitude of 207 km above the earth, where the gravitational field intensity is 9.2 N/kg. What is the magnitude of the force of gravity on the astronaut?

a. Zero
b. 70 N
c. 70 kgf
d. 644 N

7.2 In the SI system of units, a pressure of 14.7 lbf/in.^{2} is:

a. 101 kPa
b. 101 Pa
c. 6.67 kg/in.^{2}
d. 4.77×10^4 kg/m^2

7.3 In the SI system of units, a fuel economy of 29 mi/gal is:

a. 12.33 km/L
b. 68.2 km/L
c. 46.7 km/L
d. 46.7 km/gal

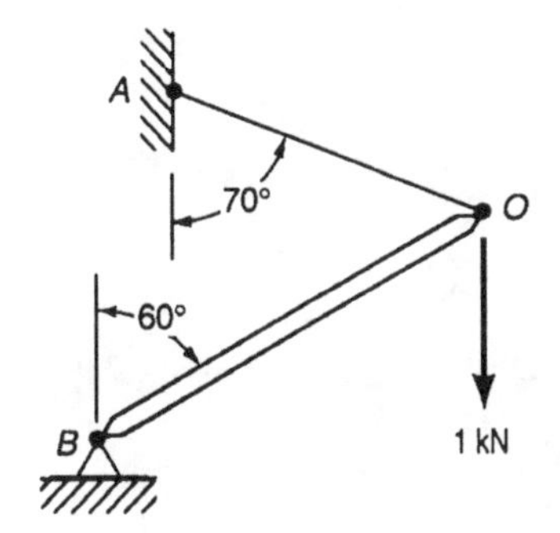

Exhibit 7.4

7.4 The components parallel to *OA* and *OB*, of the vertical 1-kN force in Exhibit 7.4 have magnitudes of:

a. 0.34 and 0.50 kN
b. 0.34 and 0.66 kN
c. 1.13 and 1.23 kN
d. 0.5 and 0.5 kN

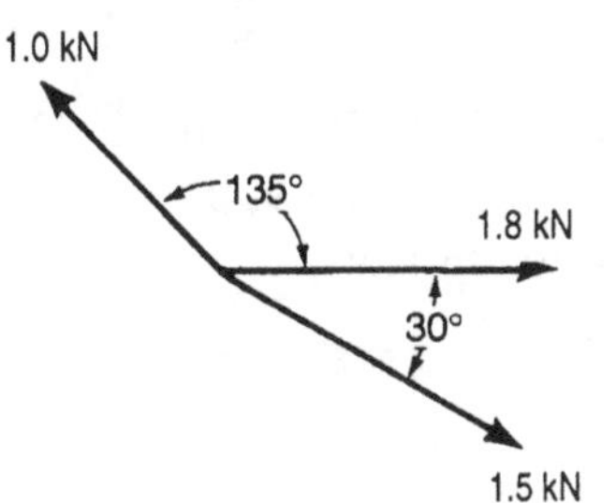

Exhibit 7.5

7.5 The resultant of the three forces in Exhibit 7.5 has a magnitude of:

a. 4.3 kN
b. 3.0 kN
c. 2.3 kN
d. 2.39 kN

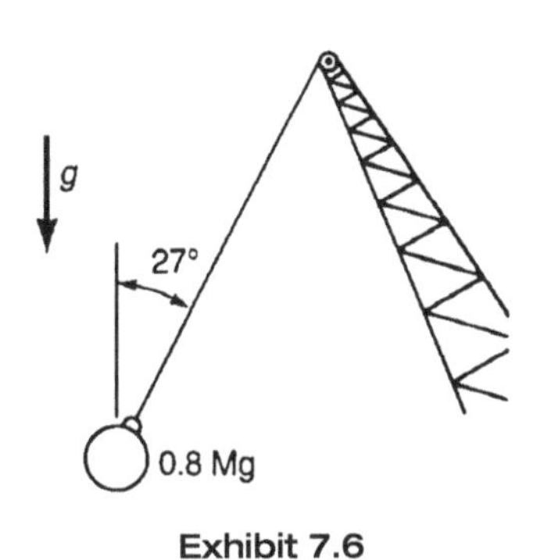

Exhibit 7.6

7.6 At a certain instant, the tension in the cable on which the destruction ball is suspended is 9.3 kN (see Exhibit 7.6). At this instant, the acceleration of the ball is:

a. 10.36 m/s^2
b. 5.30 m/s^2
c. 1.21 m/s^2
d. 9.81 m/s^2

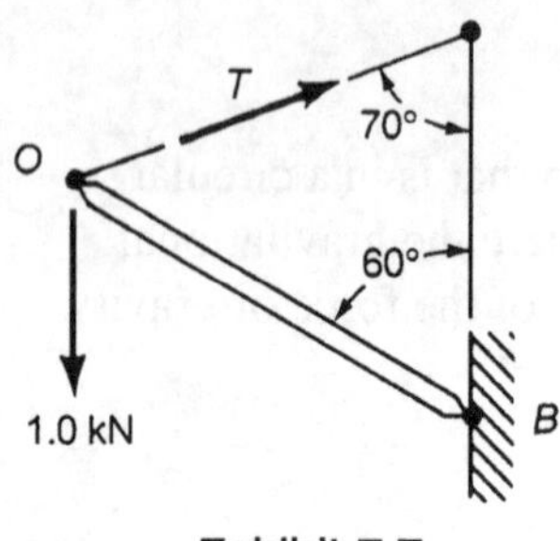

Exhibit 7.7

7.7 If the resultant of the 1-kN vertical force and the tensile force T from the cable is to be in the direction of the boom OB (see Exhibit 7.7), what must be the magnitude of T?

a. 2.9 kN c. 1.1 kN
b. 1.0 kN d. 0.94 kN

7.8 Evaluate the magnitude of the resultant force on the doorknob in Exhibit 7.8. The three components are mutually perpendicular.

a. 13 N c. 17 N
b. 19 N d. 5 N

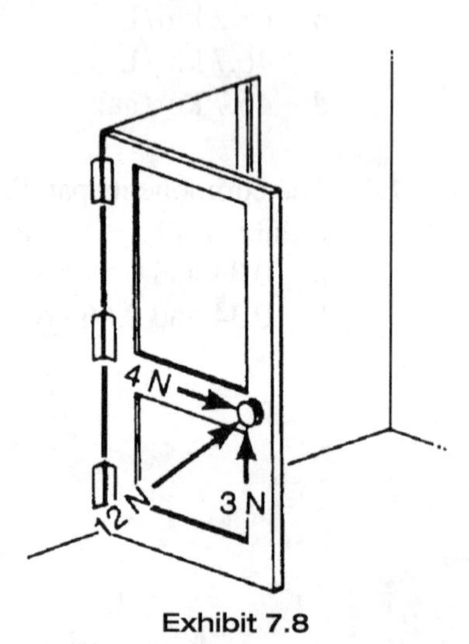

Exhibit 7.8

7.9 To raise the load, the hydraulic cylinder exerts a force of 50 kN in the direction of its axis, AB. For the position shown in Exhibit 7.9, the components of this force parallel and perpendicular to OB are:

a. 43.8 and 24.2 kN c. 23.1 and 26.9 kN
b. 41 and 28.7 kN d. 25 and 25 kN

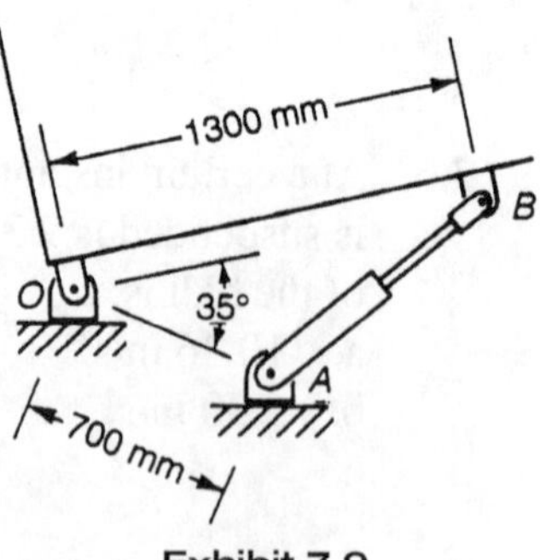

Exhibit 7.9

7.10 Refer to Exhibit 7.10. The moment about O of the 250-N force has a magnitude of:

a. 63.0 N•m　　c. 79.5 N•m
b. 77.0 N•m　　d. 23 N•m

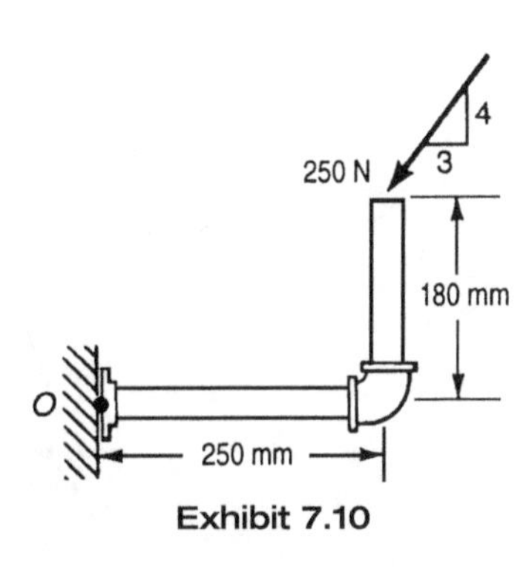

Exhibit 7.10

7.11 The connecting rod in Exhibit 7.11 exerts a force of 4.5 kN on the crank. The moment about O of this force has a magnitude of:

a. 408 N•m
b. 450 N•m
c. 318 N•m
d. 190 N•m

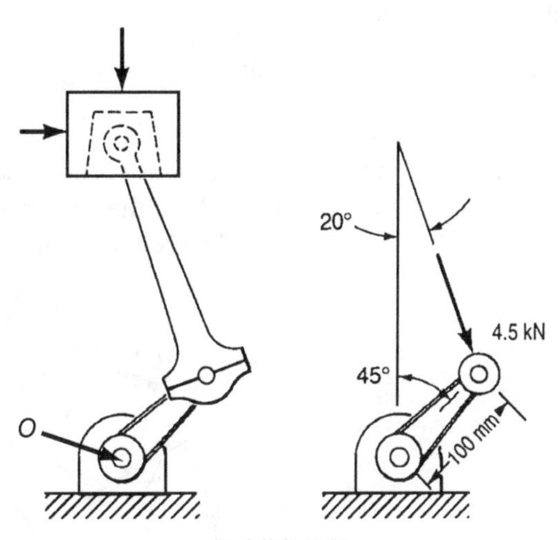

Exhibit 7.11

7.12 The tension in the line AB of Exhibit 7.12 is 3.50 kN. What must be the tension in the line BC if the moment about O of the force that the cable BC exerts on the spreader bar OB is equal and opposite to that of the force that the cable AB exerts on the spreader bar?

a. 3.50 kN　　c. 6.58 kN
b. 2.86 kN　　d. 4.29 kN

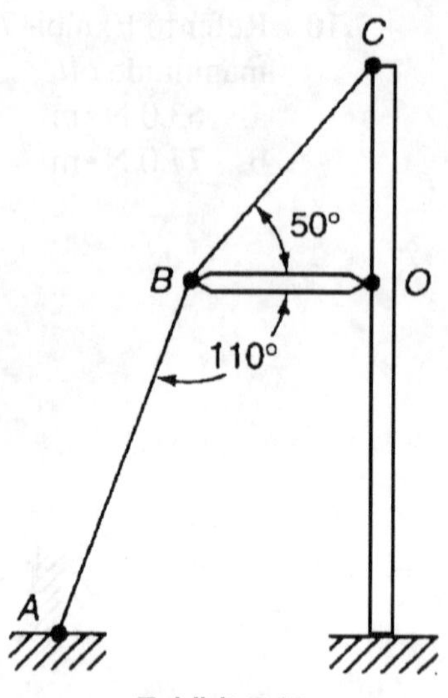

Exhibit 7.12

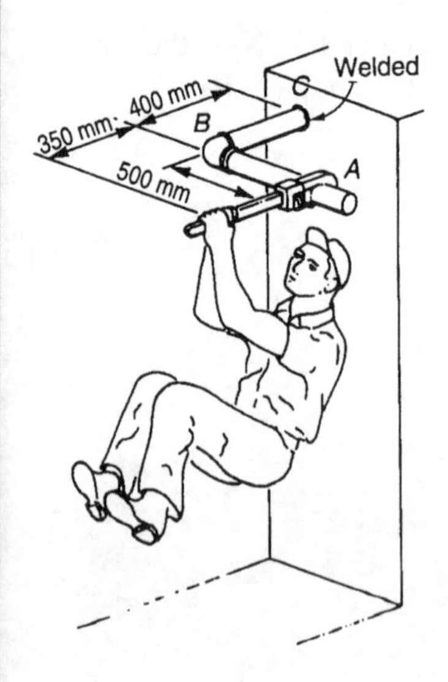

Exhibit 7.13

7.13 The plumber in Exhibit 7.13 exerts a vertical downward force of 1 kN on the wrench handle. The moment about *C* of this force has a magnitude of:

a. 500 N•m c. 900 N•m
b. 750 N•m d. 1250 N•m

7.14 The moment about the axis *CB* of the previous problem has a magnitude of:

a. 500 N•m c. 900 N•m
b. 750 N•m d. 1250 N•m

7.15 The brake is set on the wheel in Exhibit 7.15, and it will not slip until the moment about the center of the wheel of forces acting on the lug wrench reaches 150 N • m. Will the brake slip?

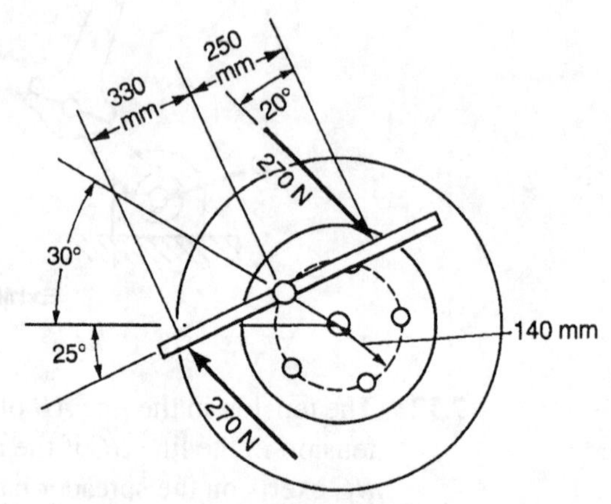

Exhibit 7.15

a. 147.2 N• m; No c. 1335.6 N• m; No
b. 156.6 N• m; Yes d. 313.2 N• m; Yes

7.16 The tension in the vertical line AC is 2 kN and that in the line BC is 6 kN (see Exhibit 7.16). The magnitude of the resultant force exerted by the two lines at C is:

a. 8.0 kN
b. 6.8 kN
c. 4.0 kN
d. 6.3 kN

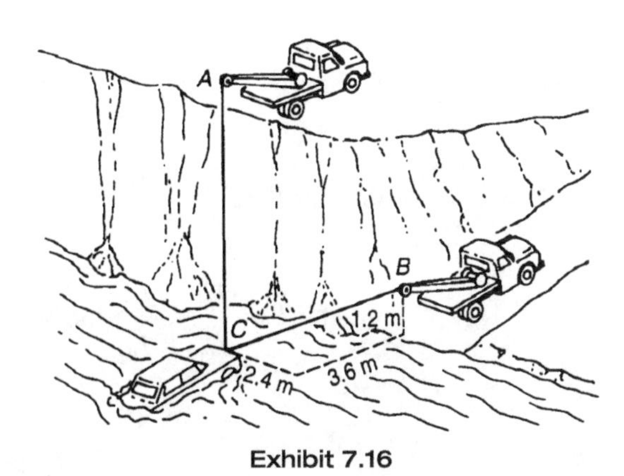

Exhibit 7.16

7.17 The moment of **f** about the axis AB in Exhibit 7.17 has the magnitude:

a. (144/65 m)f
b. (12/5 m)f
c. (29/13 m)f
d. (12 m)f

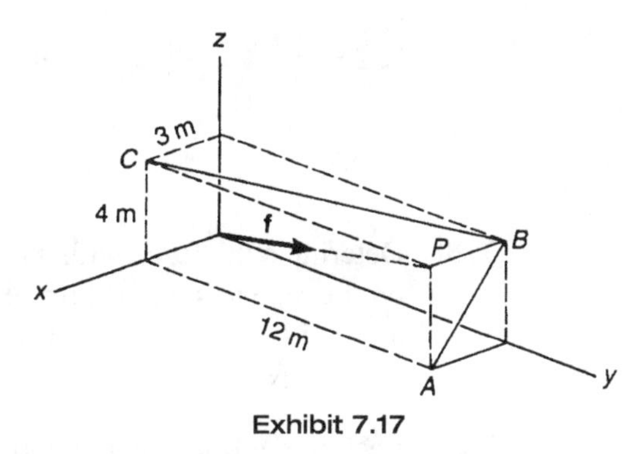

Exhibit 7.17

7.18 The surfaces are smooth where the drum makes contact (see Exhibit 7.18). The reaction at the contact point on the right is:

a. 1.77 kN
b. 3.06 kN
c. 2.29 kN
d. 2.70 kN

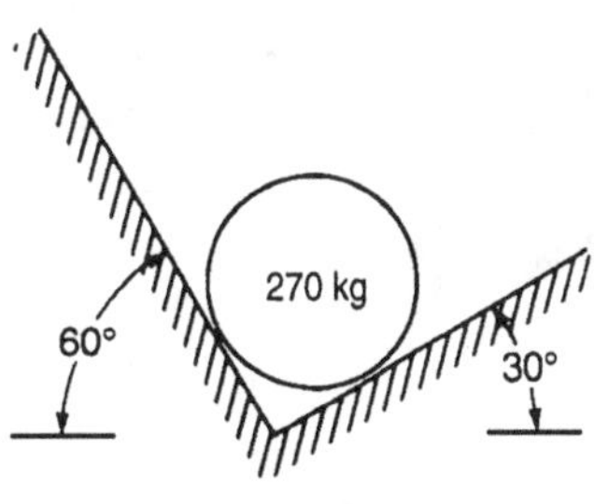

Exhibit 7.18

7.19 Two wheels, each of radius a but of different mass, are connected by the rod of length R in Exhibit 7.19. The assembly is free to roll in the circular trough. The angle θ for equilibrium is given by:

a. $\tan\theta = \dfrac{f_{g_1} - f_{g_2}}{f_{g_1} + f_{g_2}} \tan 30°$

b. $\sin\theta = \dfrac{f_{g_1} - f_{g_2}}{f_{g_1} + f_{g_2}} \sin 30°$

c. $\tan\theta = \dfrac{f_{g_1} - f_{g_2}}{f_{g_1} + f_{g_2}} \sin 30°$

d. $\sin\theta = \dfrac{f_{g_1} - f_{g_2}}{f_{g_1} + f_{g_2}} \tan 30°$

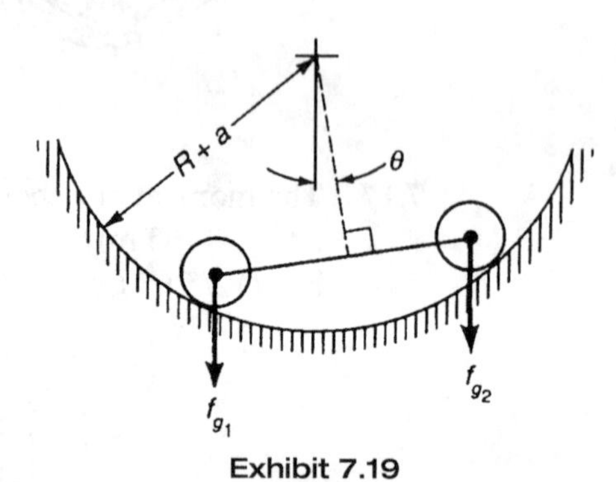

Exhibit 7.19

Exhibit 7.20

7.20 Determine the force with which the 80-kg man in Exhibit 7.20 must pull on the rope to support himself. The force is closest to:

a. 785 N
b. 628 N
c. 471 N
d. 157 N

7.21 Each of the tracks in the upper pulley unit is recessed to fit the chain, so as to prevent slipping (see Exhibit 7.21). The smaller track has a radius equal to 0.9 times that of the larger track. Evaluate the force P necessary to lift the block by means of the differential chain hoist. The force is nearest to:

a. 0.392 kN
b. 0.784 kN
c. 3.92 kN
d. 1.96 kN

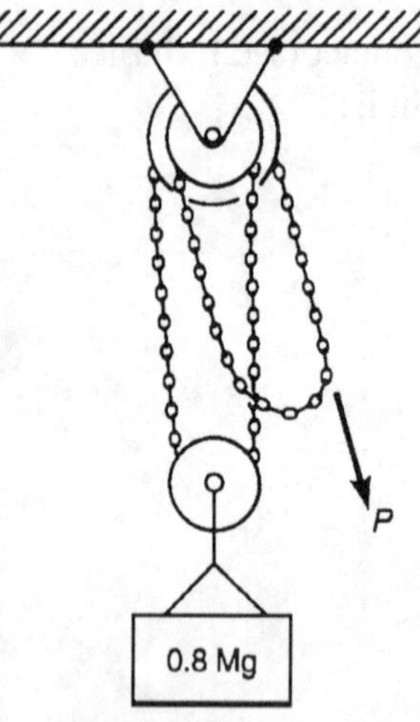

Exhibit 7.21

7.22 The weight of the linkage in Exhibit 7.22 is negligible compared with f_g. What is the value of P necessary to maintain equilibrium?

a. $0.333f_g$ c. $0.577f_g$
b. $0.500f_g$ d. $0.144f_g$

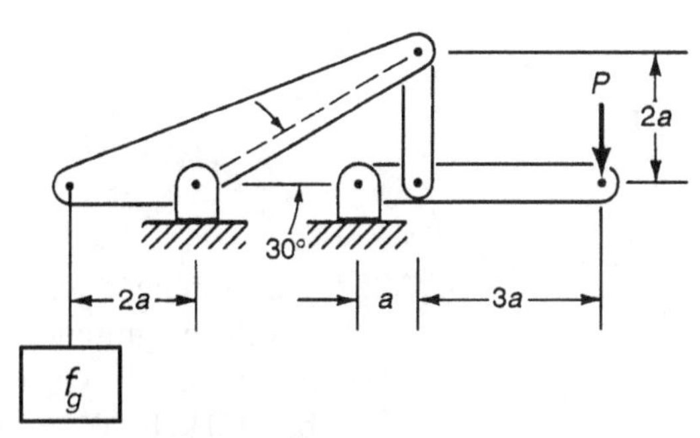

Exhibit 7.22

7.23 Neglecting the mass of the structure of Exhibit 7.23, the tension in the bar AB, induced by the 320-kg lifeboat, is nearest to:

a. 3.14 kN c. 1.57 kN
b. 0.32 kN d. 12.55 kN

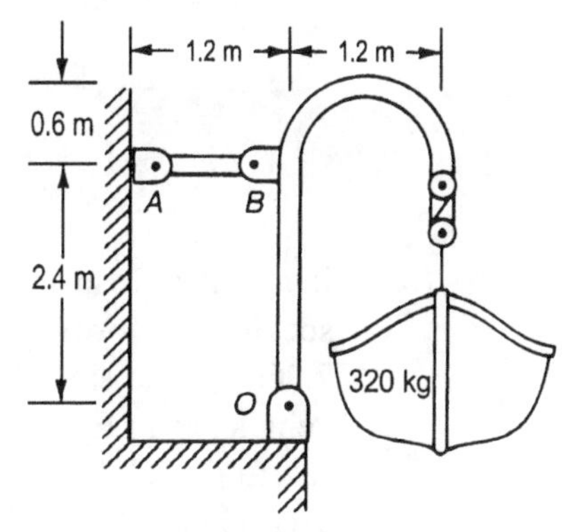

Exhibit 7.23

7.24 Until a clamp is tightened, the drill press table is free to slide along the column (see Exhibit 7.24). Estimate the coefficient of friction required so that the collar will be self-locking against the column under the action of the thrust from the drill. Neglect gravity.

a. 0.21 c. 0.63
b. 0.42 d. 0.84

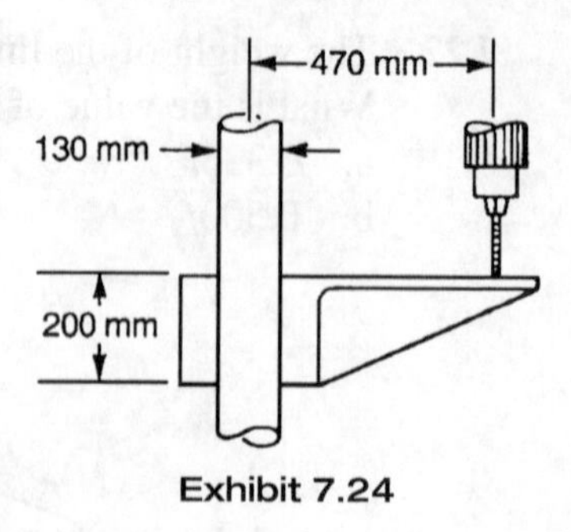

Exhibit 7.24

7.25 The combined reaction at the two rear wheels of the car in Exhibit 7.25 has the magnitude:

a. 5.0 kN c. 7.0 kN
b. 4.0 kN d. 6.3 kN

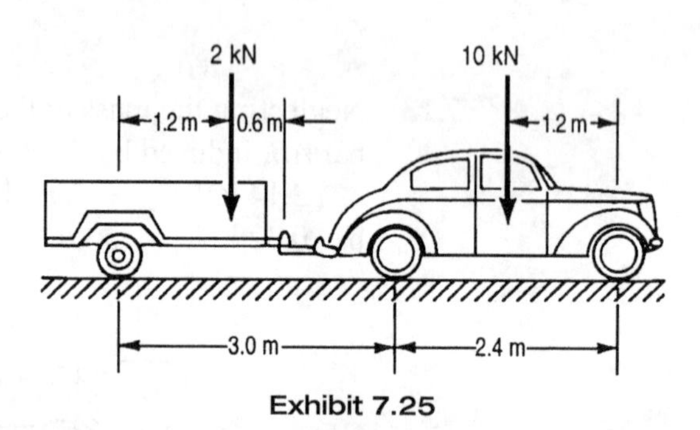

Exhibit 7.25

7.26 In a trailer "load-leveler" hitch, the angle bar slips into the cylindrical socket at *A*, forming a thrust bearing where the bar bottoms (see Exhibit 7.26(a)). The end, *B*, is then attached by a short chain to the towing vehicle, as in Exhibit 7.26(b). The pretension in the chain is 1.7 kN. The reaction at *C* (both trailer wheels) has magnitude:

a. 8.34 kN c. 6.20 kN
b. 8.05 kN d. 5.25 kN

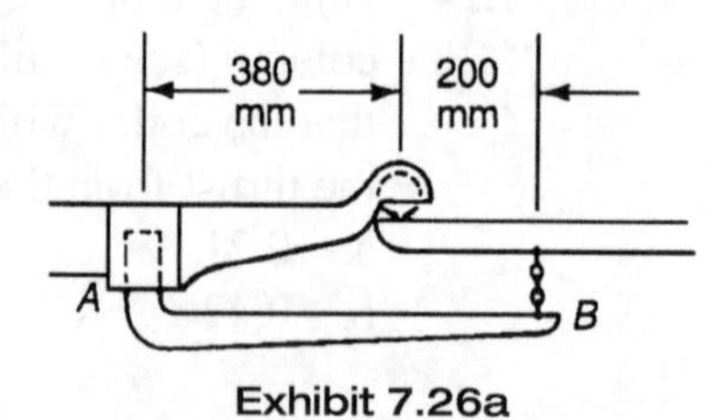

Exhibit 7.26a

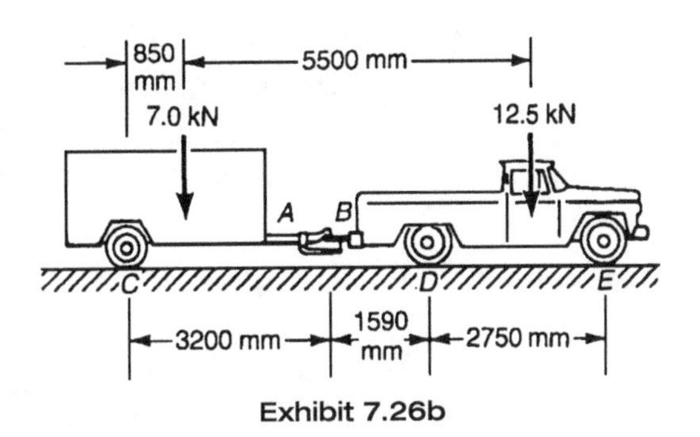

Exhibit 7.26b

7.27 The reaction at the near wheels, D, of the vehicle of Problem 7.26 has magnitude:

a. 8.34 kN
b. 8.05 kN
c. 6.20 kN
d. 5.25 kN

7.28 Evaluate the cutting force at C in terms of the force P on the handles of the compound snips (Exhibit 7.28).

a. $2P$
b. $4P$
c. $6P$
d. $8P$

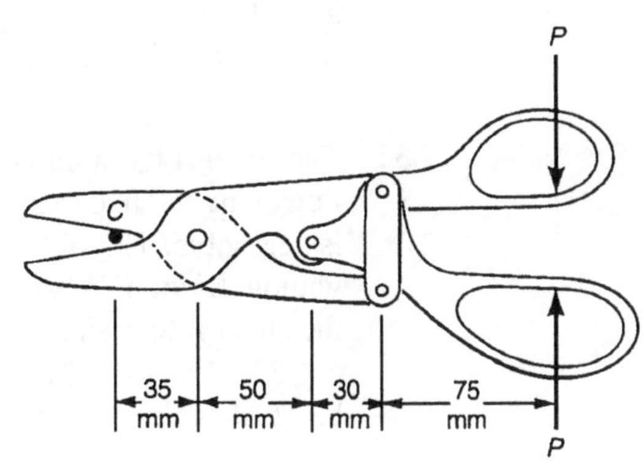

Exhibit 7.28

7.29 The magnitude of the force on the nut at C exerted by the jaws of the self-locking pliers (Exhibit 7.29) is:

a. 240 N
b. 480 N
c. 1600 N
d. 2020 N

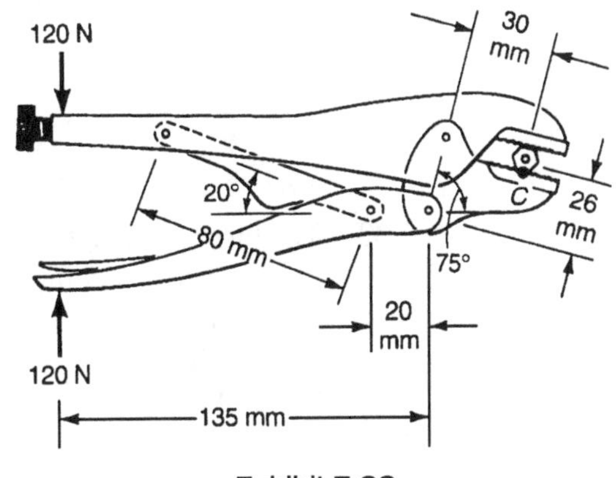

Exhibit 7.29

7.30 The moment of the couple transmitted through the shaft at section *A* (Exhibit 7.30) has magnitude:

a. 9.60 N•m c. zero
b. 15.0 N•m d. 30 N•m

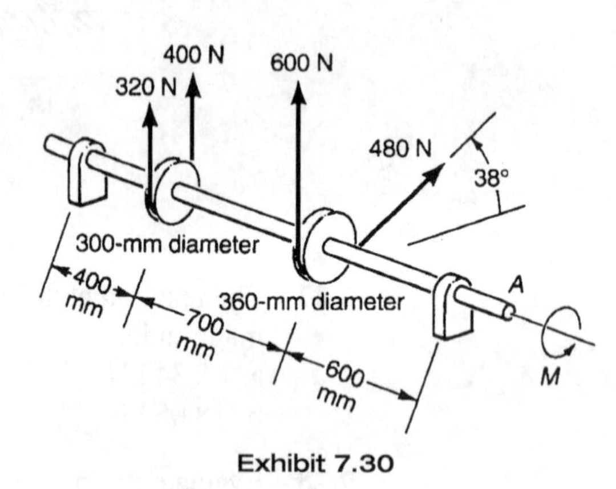

Exhibit 7.30

7.31 The reaction at the bearing near *A* in Problem 7.30 has magnitude:

a. 245 N c. 788 N
b. 745 N d. 1296 N

7.32 The slider *A* has a mass of 4 kg and is constrained to slide without friction along the fixed vertical rod (Exhibit 7.32). The mass of the wire *AB* is negligible. The slider *B* is constrained to slide along the horizontal rod without friction. What must be the magnitude *F* of the force applied to the slider *B* to maintain equilibrium?

a. 39.2 N c. 98 N
b. 19.6 N d. 49 N

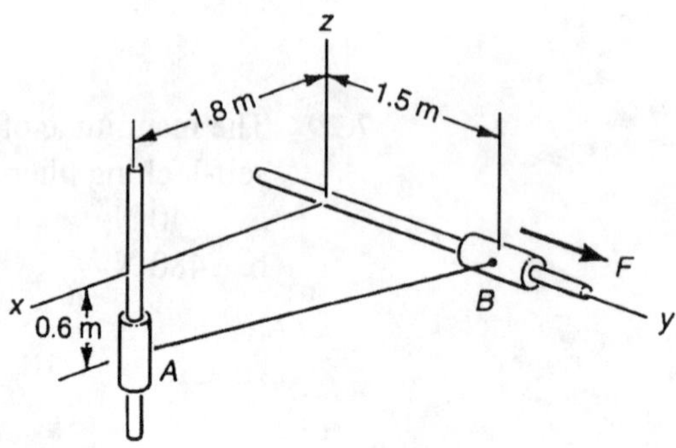

Exhibit 7.32

7.33 The turnbuckle in the guyline *AE* of Exhibit 7.33 is to be tightened such that the vertical component of the reaction at *O* is 800 N. What must be the tension in *AE*?

a. 424 N c. 636 N
b. 212 N d. 267 N

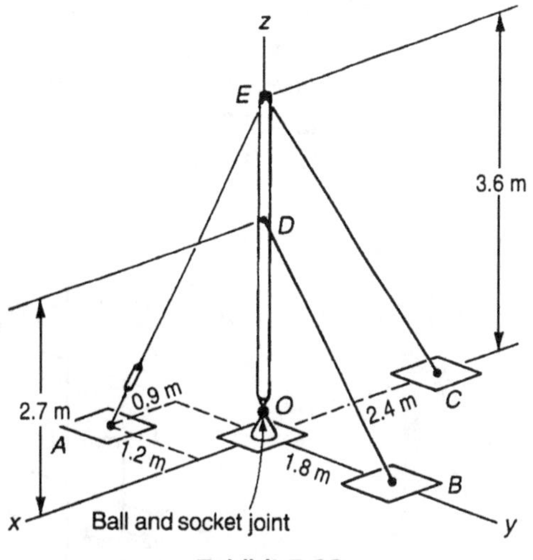

Exhibit 7.33

7.34 Determine the tensions in the cables AB and CD of Exhibit 7.34. They are closest to:

a. 2.27 kN, 1.45 kN c. 0.45 kN, 1.05 kN
b. 0.75 kN, 0.75 kN d. 1.87 kN, 1.20 kN

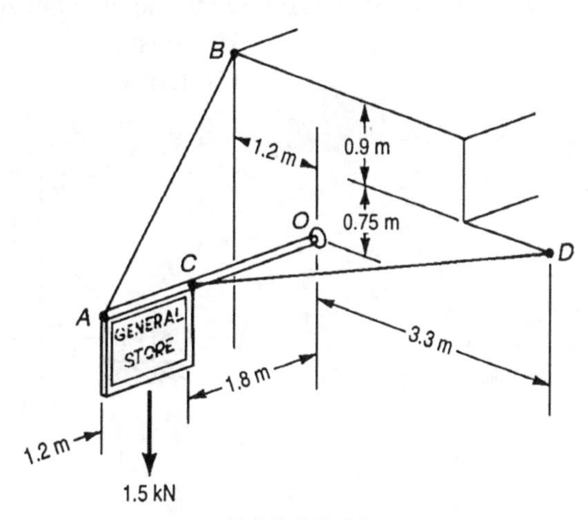

Exhibit 7.34

7.35 The 40-kg rectangular plate in Exhibit 7.35 is held by hinges along its edge *OA* and by the wire *BD*. The gravity force f_g acts through the geometric center of the plate. What is the tension of the wire?

a. 392 N
b. 196 N
c. 36.5 N
d. 358 N

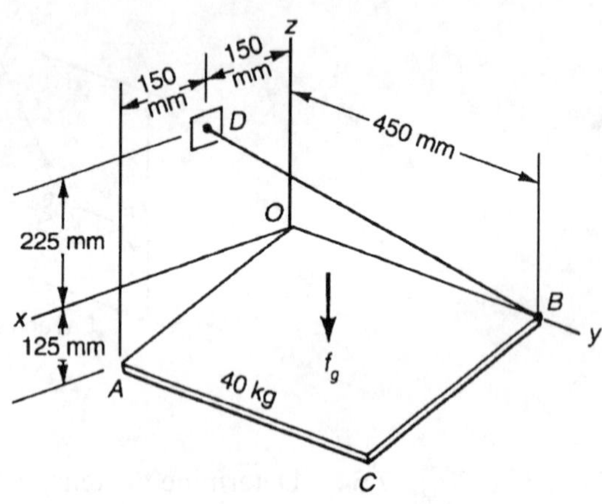

Exhibit 7.35

7.36 The force in the member *BC* of Exhibit 7.36 is:

a. 12 kN, tension
b. 17 kN, tension
c. 17 kN, compression
d. 12 kN, compression

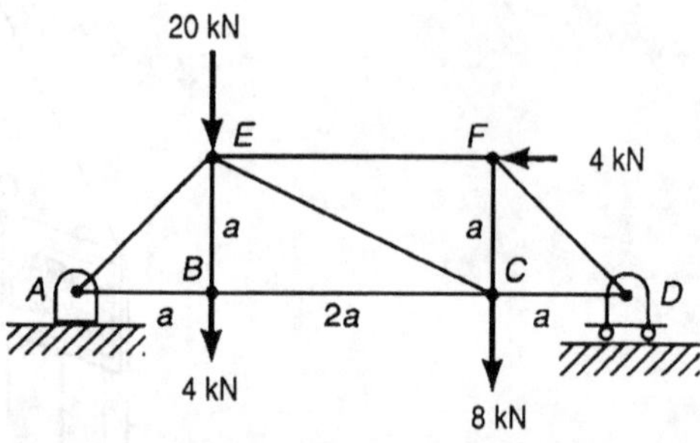

Exhibit 7.36

7.37 The force in the member *ML* of Exhibit 7.37 is:

a. 131 kN, tension
b. 131 kN, compression
c. 100 kN, tension
d. 106 kN, tension

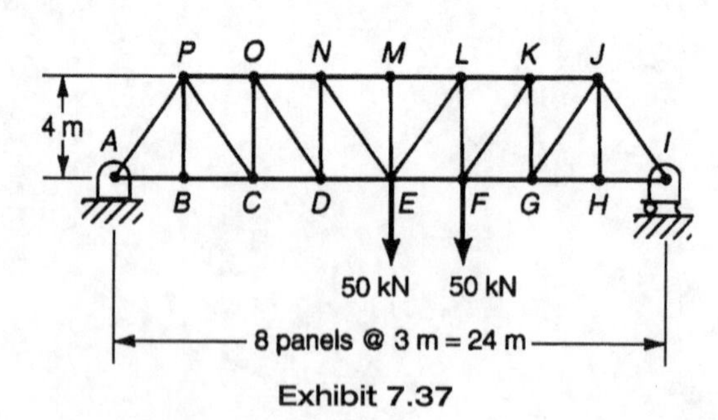

Exhibit 7.37

7.38 The force in the member *CF* of Exhibit 7.38 is:

a. 60.6 kN, tension
b. 60.6 kN, compression
c. 43.3 kN, tension
d. 43.3 kN, compression

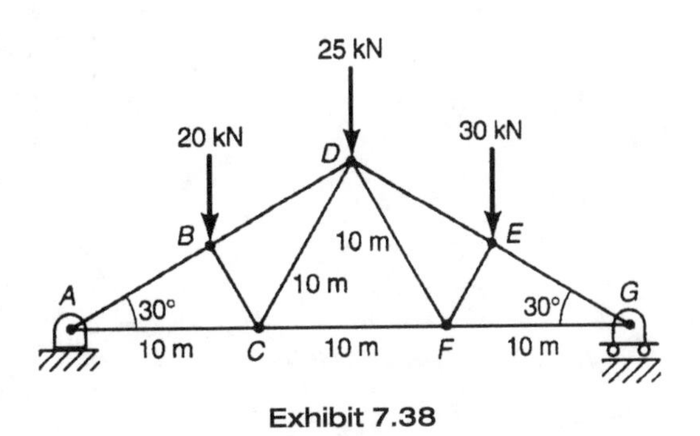

Exhibit 7.38

7.39 Members *AD*, *BE*, and *CF* are perpendicular to the plane *ABC* (see Exhibit 7.39). The force in the member *AD* is:

a. 6 kN, compression
b. 6 kN, tension
c. 8 kN, tension
d. 8 kN, compression

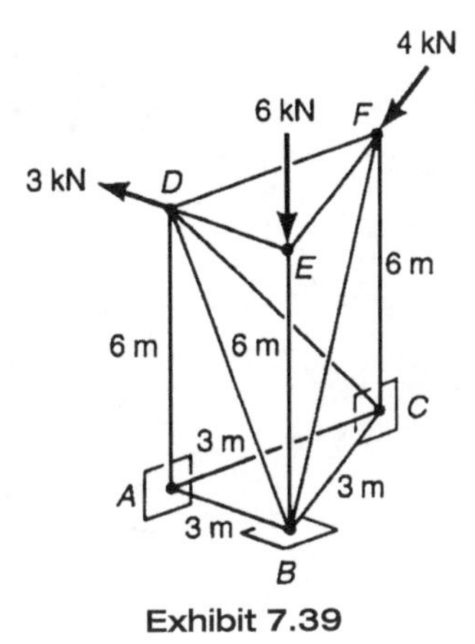

Exhibit 7.39

7.40 The force in member BE, shown in Exhibit 7.40, is:

a. $0.87P$, compression
b. P, compression
c. $0.87P$, tension
d. P, tension

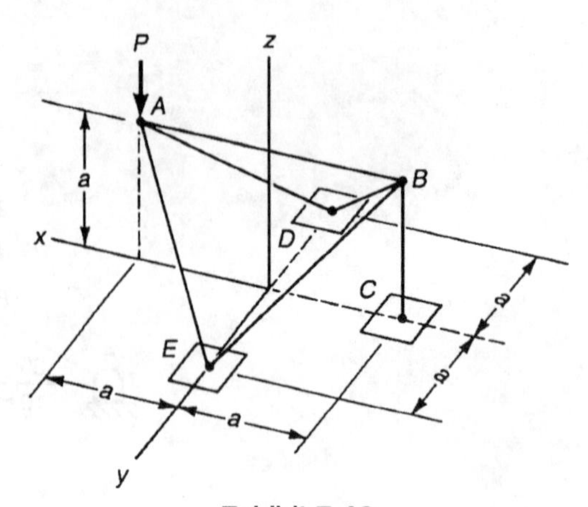

Exhibit 7.40

7.41 The contents of the crate in Exhibit 7.41 are such that the center of gravity is at the geometric center. In the absence of P, the crate will:

a. remain stationary
b. slide downward
c. tip over
d. slide upward

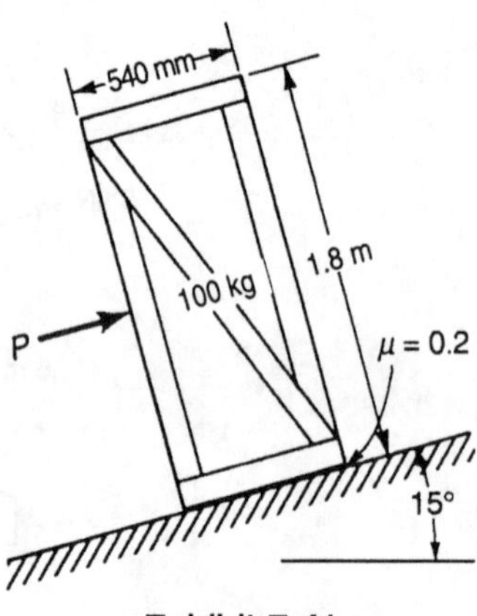

Exhibit 7.41

7.42 In Exhibit 7.41, the greatest distance from the incline to the line of action of P, such that the force can slide the crate up the incline without it, is:

a. 1.8 m
b. 0.9 m
c. 1.35 m
d. 1.09 m

7.43 The contents of the crate in Exhibit 7.43 are such that the center of gravity coincides with the geometric center. The greatest value of the angle that will allow the force to slide the crate without tipping it is given by:

a. $\tan^{-1}\left(\frac{h}{b}\right)$ c. $\tan^{-1}\left(\frac{1}{\mu}-\frac{2h}{b}\right)$

b. $\tan^{-1}\left(\frac{b}{h}\right)$ d. $\tan^{-1}\left(\frac{1}{\mu}+\frac{2h}{b}\right)$

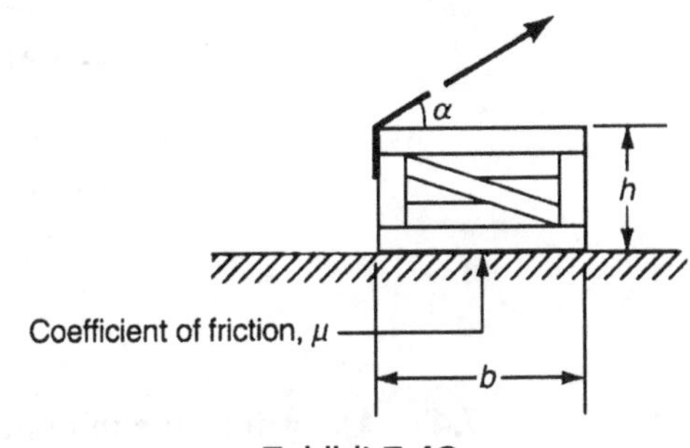

Exhibit 7.43

7.44 If the coefficient of friction between all surfaces in Exhibit 7.44 is 0.27, what will be the minimum value of P necessary to initiate motion?

a. 103 N c. 40.5 N
b. 81.0 N d. 51.5 N

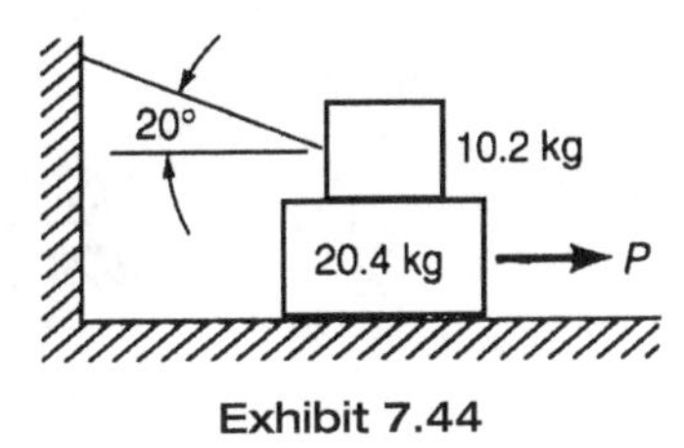

Exhibit 7.44

7.45 If the coefficient of friction between all surfaces in Exhibit 7.45 is 0.27, what will be the minimum value of P necessary to initiate motion?

a. 205 N c. 94.5 N
b. 103 N d. 9.63 N

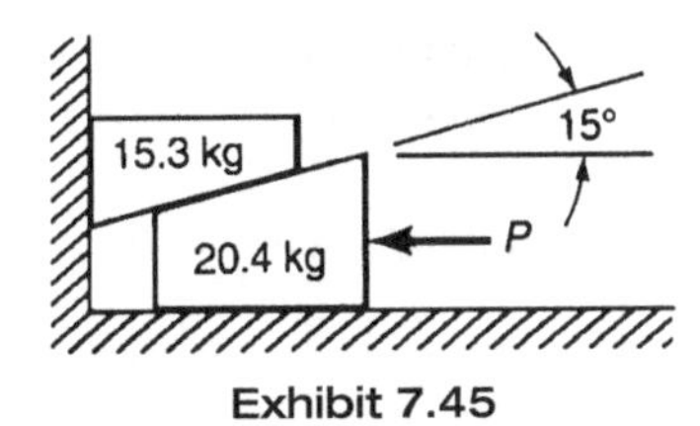

Exhibit 7.45

7.46 The forklift in Exhibit 7.46 is being used to roll the 2-Mg drum up the 40-degree incline while the height of the forks remains constant. The coefficient of friction is 0.45 between the vertical rails and the drum, and 0.30 between the incline and the drum. What horizontal thrust must the vehicle apply to the drum to move it?

a. 85 kN c. 78.2 kN
b. 62.7 kN d. 103 kN

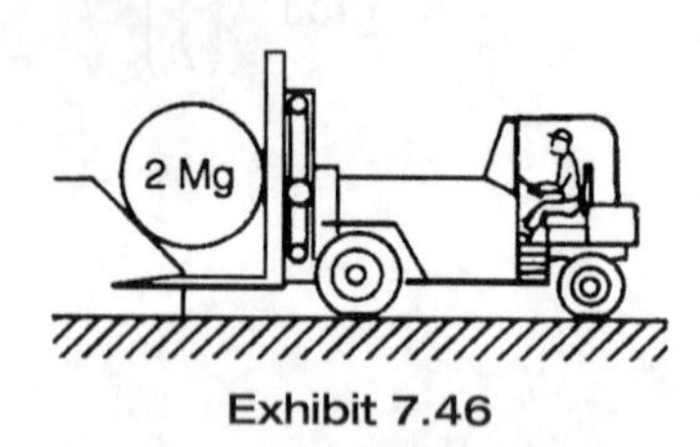

Exhibit 7.46

7.47 Determine the minimum coefficient of friction between the block and the weightless bar in Exhibit 7.47 necessary to prevent collapse.

a. 0.15 c. 0.42
b. 0.27 d. 0.72

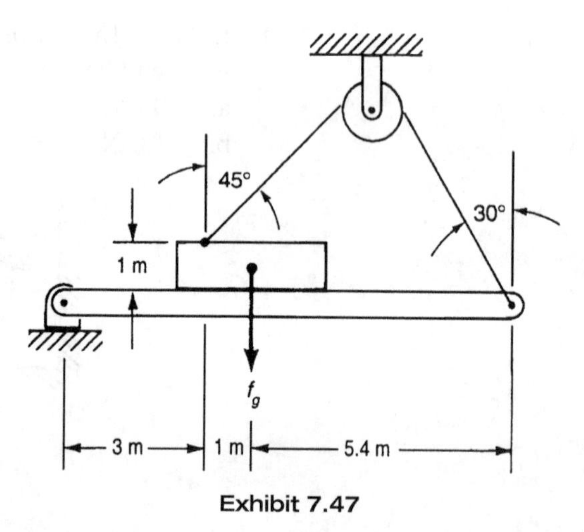

Exhibit 7.47

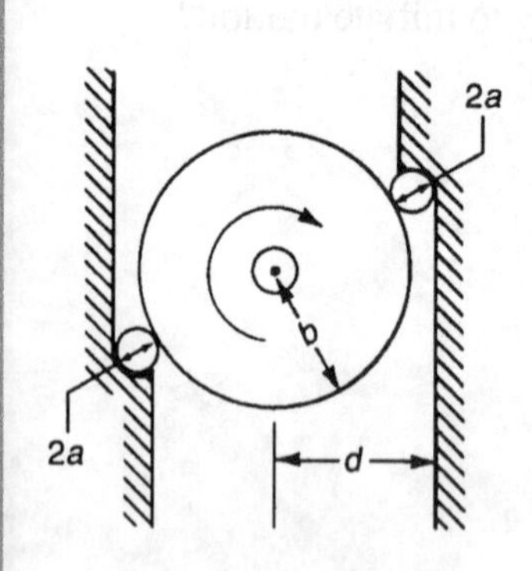

Exhibit 7.48

7.48 The small rollers in Exhibit 7.48 are intended to prevent clockwise rotation of the large drum. The coefficient of friction between the rollers and drum, and between the rollers and walls, is μ. Determine the minimum distance d such that the friction will effect a self-locking mechanism against clockwise rotation. Gravity is negligible.

a. $d > (a+b)\mu$ c. $d > \dfrac{2a+(1-\mu^2)b}{1+\mu^2}$

b. $d > \dfrac{a+b}{\mu}$ d. $d > \dfrac{(a+b)\mu}{1+\mu^2}$

SOLUTIONS

7.1 d. $f_g = (9.2\text{ N/kg})(70\text{ kg}) = 644\text{ N}$

7.2 a. $P = (14.7\text{ lbf/in.}^2)(39.37\text{ in./m})^2(4.448\text{ N/lbf}) = 101{,}000\text{ N/m}^2$
$= 101\text{ kPa}$

7.3 a. Fuel economy = (29 mi/gal)

$$\frac{(5280\text{ ft/mi})(0.3048\text{ m/ft})}{(1000\text{ m/km})(3.7854\text{ L/gal})} = 12.33\text{ km/L}$$

7.4 c.

Exhibit 7.4a

$$\frac{F_A}{\sin 60°} = \frac{F_B}{\sin 70°} = \frac{1\text{ kN}}{\sin 50°}$$

$$F_A = \frac{\sin 60°}{\sin 50°}(1\text{ kN}) = 1.13\text{ kN}$$

$$F_B = \frac{\sin 70°}{\sin 50°}(1\text{ kN}) = 1.23\text{ kN}$$

7.5 d.

$$F_x = 1.8\text{ kN} + (1.5\text{ kN})\cos 30° + (1.0\text{ kN})\cos 135° = 2.39\text{ kN}$$
$$F_y = (1.0\text{ kN})\sin 135° - (1.5\text{ kN})\sin 30° = -0.043\text{ kN}$$
$$F = \sqrt{F_x^2 + F_y^2} = 2.39\text{ kN}$$

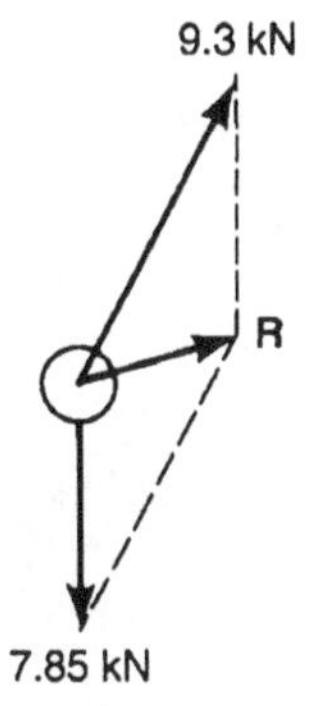

Exhibit 7.6a

7.6 b.

The force of gravity on the ball is $f_g = (800\text{ kg})(9.81\text{ N/kg}) = 7.85\text{ kN}$.
The net force, $\mathbf{R} = x\,\mathbf{e}_x + y\,\mathbf{e}_y = (9.3\text{ kN}\sin 27°)\,\mathbf{e}_x + (9.3\text{ kN}\cos 27° - 7.85\text{ kN})\,\mathbf{e}_y$

$$|\mathbf{R}| = R = \sqrt{x^2 + y^2}$$

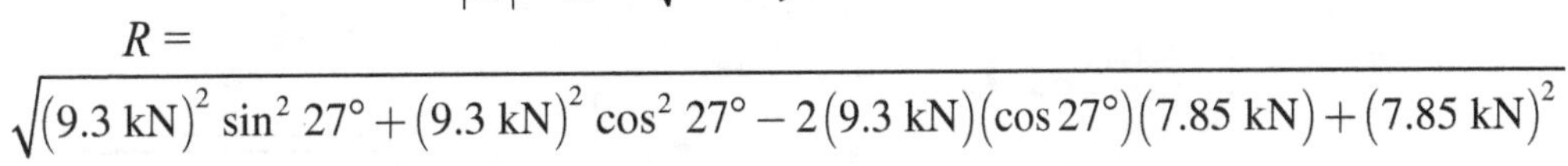

$$R = \sqrt{(9.3\text{ kN})^2\sin^2 27° + (9.3\text{ kN})^2\cos^2 27° - 2(9.3\text{ kN})(\cos 27°)(7.85\text{ kN}) + (7.85\text{ kN})^2}$$

$$R = 4.24\text{ kN}$$

$$a = \frac{4.24\text{ kN}}{0.8\text{ Mg}} = 5.30\text{ m/s}^2$$

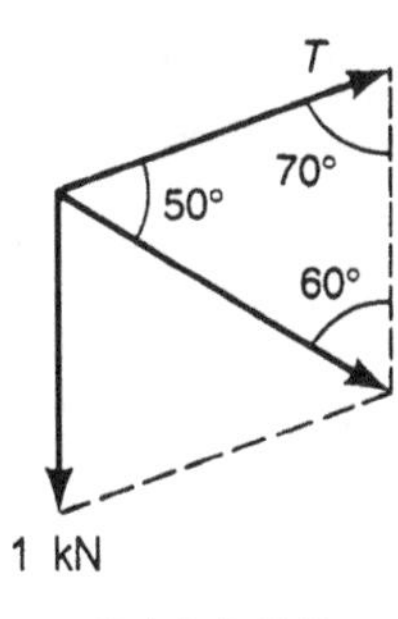

Exhibit 7.7a

7.7 c.

$$\frac{T}{\sin 60°} = \frac{1\text{ kN}}{\sin 50°}$$

$$T = 1.13\text{ kN}$$

7.8 a.

$$F = \sqrt{(4)^2 + (12)^2 + (3)^2} = 13\text{ N}$$

7.9 a.

First use geometry to solve the triangle.

$$\overline{AB} = \sqrt{(0.7)^2 + (1.3)^2 - 2(0.7)(1.3)\cos 35^\circ}\ \text{m} = 0.83\ \text{m}$$

$$\frac{\sin_B \angle_A^O}{0.7\ \text{m}} = \frac{\sin 35^\circ}{0.83\ \text{m}};\ {}_B\angle_A^O = 28.92^\circ$$

Then, compute the forces.

$$(50\ \text{kN})\cos\ 28.9^\circ = 43.8\ \text{kN}$$

$$(50\ \text{kN})\sin\ 28.9^\circ = 24.2\ \text{kN}$$

7.10 d.

Note that the direction of the 250-N force, **f**, is described by a 3-4-5 triangle. Separate **f** into its x and y components.

$$\mathbf{f} = (-250\ \text{N}\ (3/5))\ \mathbf{e}_x + (-250\ \text{N}\ (4/5))\ \mathbf{e}_y$$
$$M_0 = \mathbf{r} \times \mathbf{f}$$
$$\mathbf{r} = 0.25\ \text{m}\ \mathbf{e}_x + 0.18\ \text{m}\ \mathbf{e}_y$$
$$M_0 = (0.18\ \text{m})(150\ \text{N}) - (0.25\ \text{m})(200\ \text{N}) = -23.0\ \text{N}\bullet\text{m}$$

7.11 a.

$$M_0 = (0.1\ \text{m})\cos 25^\circ\ (4.5\ \text{kN}) = 408\ \text{N}\bullet\text{m}$$

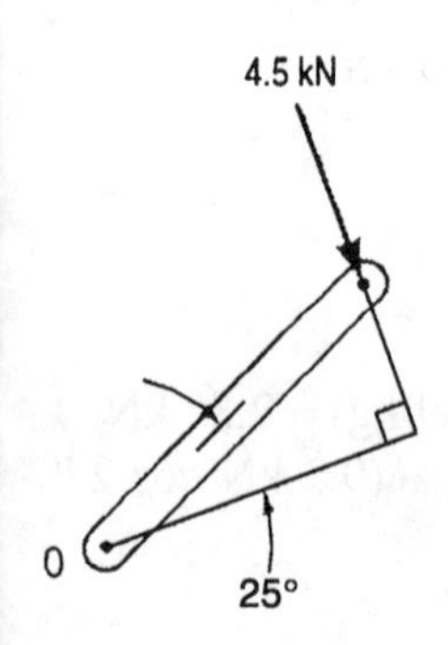

Exhibit 7.11a

7.12 d.

$$lT_{BC}\sin 50^\circ = l(3.5\ \text{kN})\sin 110^\circ$$
$$T_{BC} = 4.29\ \text{kN}$$

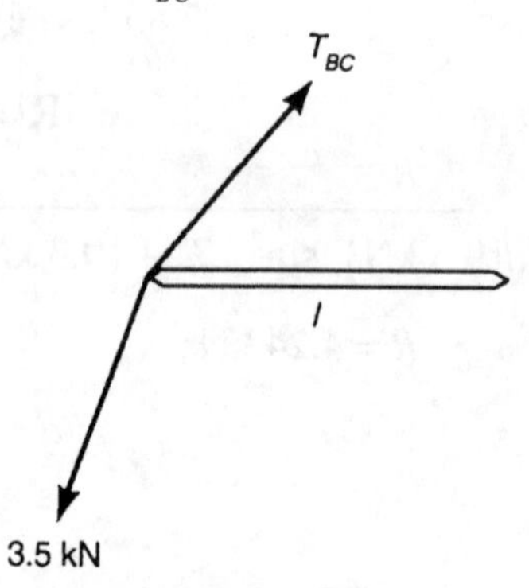

Exhibit 7.12a

7.13 c.

$$r = \sqrt{(0.75\ \text{m})^2 + (0.5\ \text{m})^2} = 0.9014\ \text{m}$$
$$M_C = (0.9014\ \text{m})(1\ \text{kN}) = 901.4\ \text{N}\bullet\text{m}$$

7.14 a.

$$M_{CB} = (0.5\ \text{m})(1\ \text{kN}) = 500\ \text{N}\bullet\text{m}$$

7.15 a.

$$M = (250 \text{ mm} + 330 \text{ mm}) \cos 20° \, (270 \text{ N}) = 147.2 \text{ N} \bullet \text{m}$$

7.16 b.

$$\overline{CB} = \sqrt{(2.4 \text{ m})^2 + (3.6 \text{ m})^2 + (1.2 \text{ m})^2} = 4.49 \text{ m}$$

$$R_x = \frac{2.4}{4.49}(6 \text{ kN}) = 3.207 \text{ kN}$$

$$R_y = \frac{3.6}{4.49}(6 \text{ kN}) = 4.811 \text{ kN}$$

$$R_z = \frac{1.2}{4.49}(6 \text{ kN}) + 2 \text{ kN} = 3.604 \text{ kN}$$

$$R = \sqrt{R_x^2 + R_y^2 + R_z^2} = 6.81 \text{ kN}$$

7.17 a.

$$M_{AB} = \mathbf{e}_{AB} \bullet (\mathbf{r}_{AP} \times \mathbf{f})$$

$$\mathbf{e}_{AB} = \frac{-3\mathbf{e}_x + 4\mathbf{e}_z}{\sqrt{(-3)^2 + 4^2}}$$

$$\mathbf{r}_{AP} = (4 \text{ m})\mathbf{e}_z$$

$$\mathbf{f} = \left(\frac{3\mathbf{e}_x + 12\mathbf{e}_y + 4\mathbf{e}_z}{\sqrt{3^2 + 12^2 + 4^2}}\right) f$$

$$M_{AB} = \left(-\frac{3}{5}\mathbf{e}_x + \frac{4}{5}\mathbf{e}_z\right) \bullet \left[(4 \text{ m})\mathbf{e}_z \times \left(\frac{3}{13}\mathbf{e}_x + \frac{12}{13}\mathbf{e}_y + \frac{4}{13}\mathbf{e}_z\right) f\right]$$

$$= \begin{vmatrix} -\frac{3}{5} & 0 & \frac{4}{5} \\ 0 & 0 & 4 \text{ m} \\ \frac{3}{13} & \frac{12}{13} & \frac{4}{13} \end{vmatrix} f = \left(\frac{144}{65} \text{ m}\right) f$$

7.18 c.

$$R_R - (2648 \text{ N}) \cos 30° = 0$$

$$R_R = 2.29 \text{ kN}$$

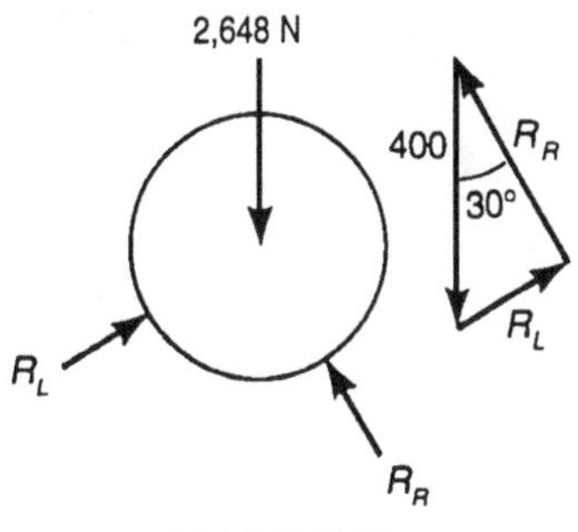

Exhibit 7.18a

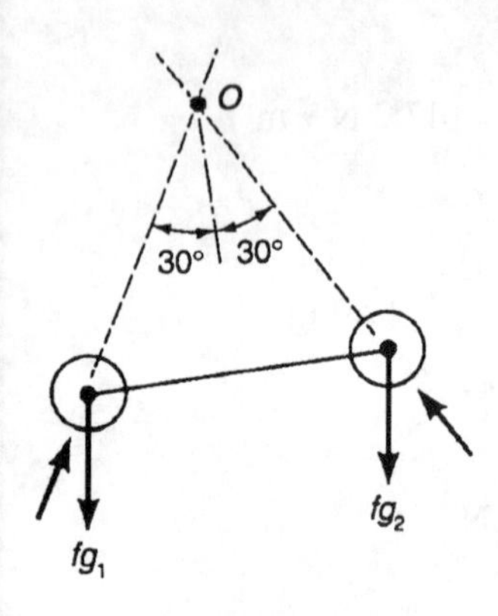

Exhibit 7.19a

7.19 a.

$$M_0 = R \sin(30° - \theta) f_{g_1} - R \sin(30° + \theta) f_{g_2} = 0$$

$$f_{g_1} (\sin 30° \cos\theta - \cos 30° \sin\theta) - f_{g_2} (\sin 30° \cos\theta + \cos\theta + \cos 30° \sin\theta) = 0$$

Dividing by cos 30° cos θ,

$$f_{g_1} (\tan 30° - \tan\theta) = f_{g_2} (\tan 30° + \tan\theta)$$

$$\tan\theta = \frac{f_{g_1} - f_{g_2}}{f_{g_1} + f_{g_2}} \tan 30°$$

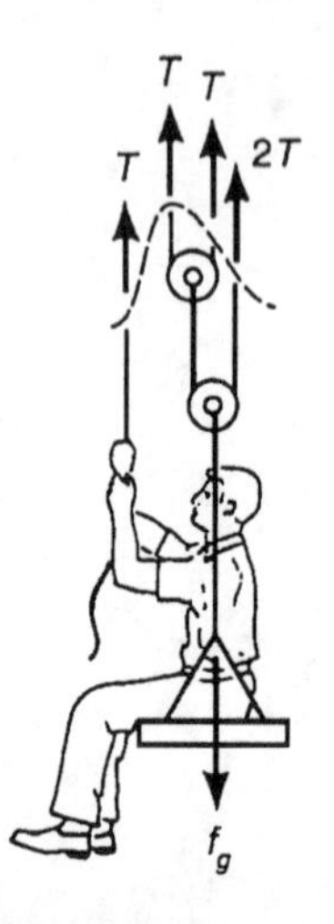

Exhibit 7.20a

7.20 d.

$$T + T + T + 2T - f_g = 0$$

$$T = \frac{1}{5} f_g$$

$$= \frac{1}{5} (80 \text{ kg})(9.8 \text{ N/kg}) = 157 \text{ N}$$

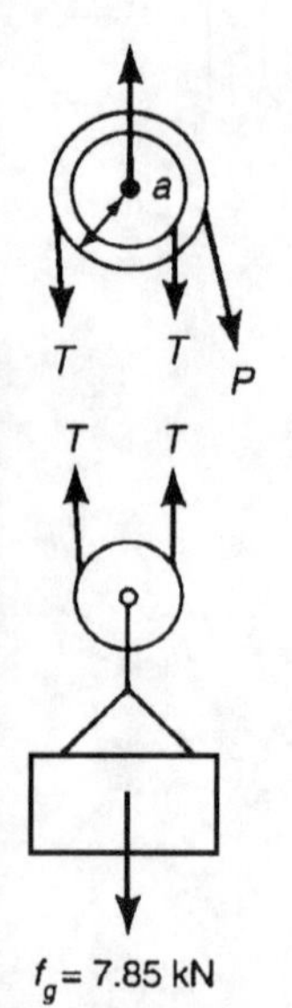

Exhibit 7.21a

7.21 a. The force of gravity on the block is

$$f_g = 0.8 \text{ Mg } (9.81 \text{ m/s}^2) = 7.85 \text{ kN}$$

$$aT - 0.9aT - aP = 0$$

$$P = 0.1T$$

$$2T = f_g$$

$$P = 0.1\left(\frac{1}{2} f_g\right) = 0.05 f_g = 0.392 \text{ kN}$$

7.22 d.

$$\sum M_A = aQ - 4aP = 0 \quad Q = 4P$$

$$\sum M_B = 2af_g - 2a \cot 30° \quad Q = 0$$

$$f_g = \sqrt{3}\, Q$$

$$P = \frac{Q}{4} = \frac{f_g}{4\sqrt{3}} = 0.144 f_g$$

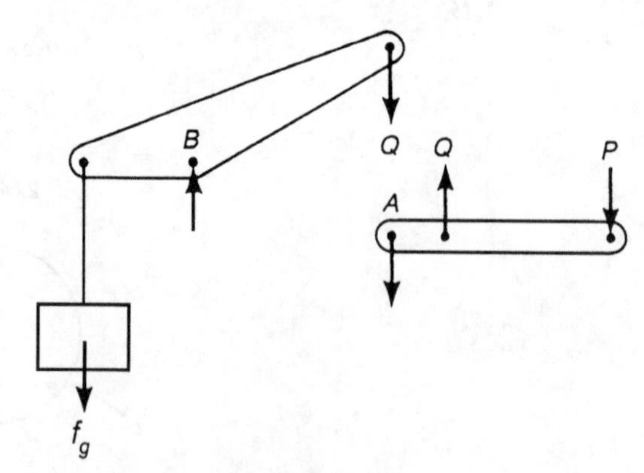

Exhibit 7.22a

7.23 **c.**

$$\sum M_O = (2.4 \text{ m})T_{AB} - (1.2 \text{ m})f_g = 0$$

$$T_{AB} = \frac{1.2}{2.4}(320 \text{ kg})(9.8 \text{ N/kg}) = 1.57 \text{ kN}$$

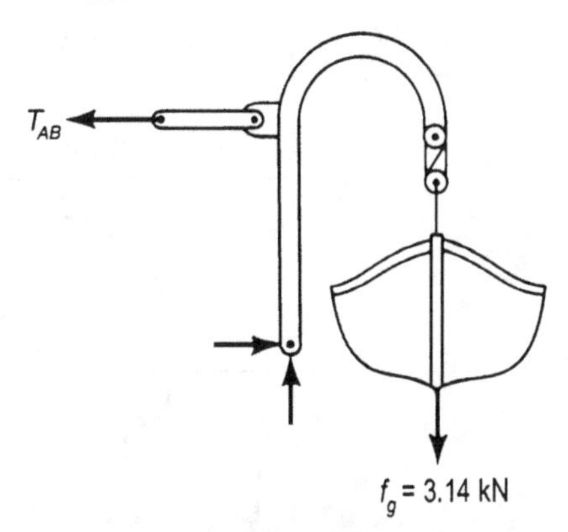

Exhibit 7.23a

7.24 **a.** Neglecting the force of gravity on the table, N(200 mm) + μN(130 mm) = P(535 mm) by summing moments about point A. Thus,

$$2\mu N \geq P = \left(\frac{200}{535} + \frac{130\mu}{535}\right)N$$

$$\mu \geq \frac{\frac{200}{535}}{2 - \frac{130}{535}} = \frac{10}{47} = 0.213$$

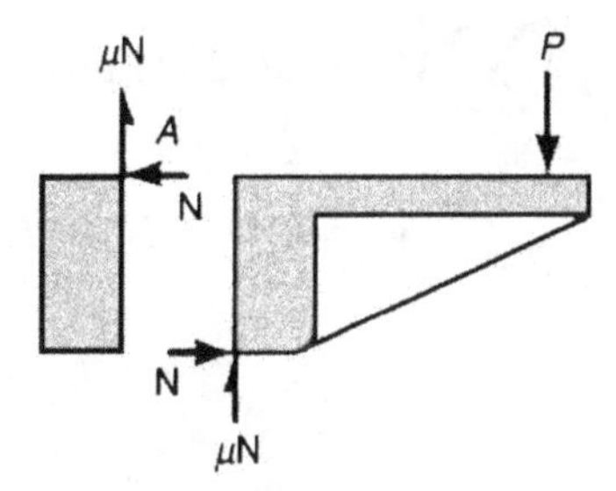

Exhibit 7.24a

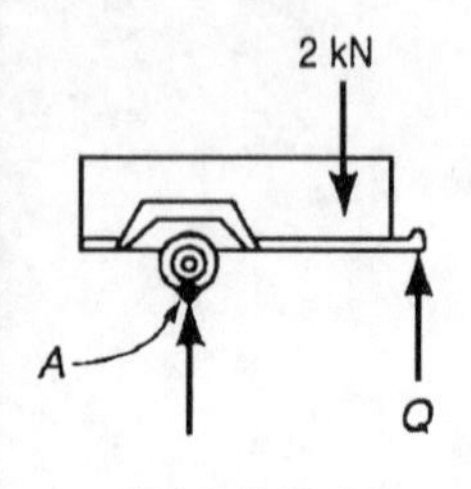

Exhibit 7.25a

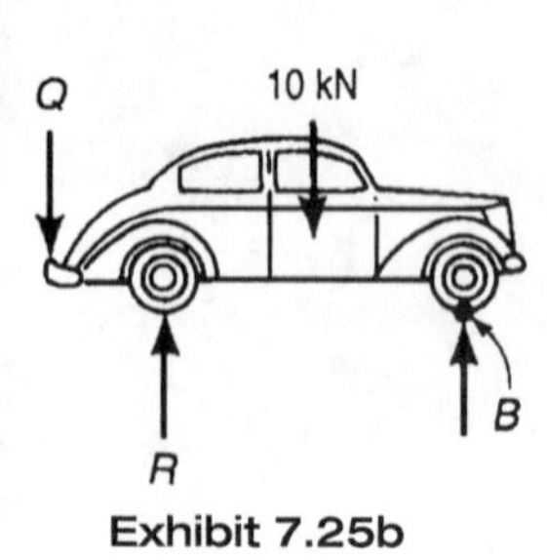

Exhibit 7.25b

7.25 c.

$$\sum M_A = (1.8\text{ m})Q - (1.2\text{ m})(2\text{ kN}) = 0$$

$$Q = \frac{1.2}{1.8}(2\text{ kN}) = 1.33\text{ kN}$$

$$\sum M_B = (3.6\text{ m})Q - (2.4\text{ m})R + (1.2\text{ m})(10\text{ kN}) = 0$$

$$R = \frac{(3.6\text{ m})(1.33\text{ kN}) + (1.2\text{ m})(10\text{ kN})}{2.4\text{ m}} = 7.0\text{ kN}$$

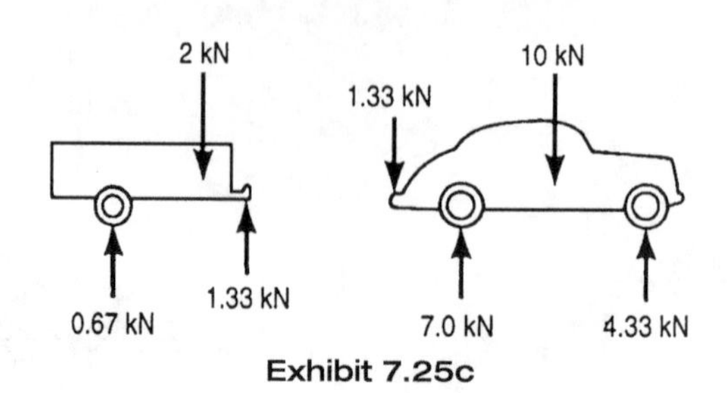

Exhibit 7.25c

7.26 d.

$$\sum M_A = (0.58\text{ m})(1.7\text{ kN}) - M = 0$$

$$M = 0.986\text{ kN}\bullet\text{m}$$

$$\sum F_U = 1.7\text{ kN} - P = 0$$

$$P = 1.7\text{ kN}$$

$$\sum M_C = R(3200\text{ mm}) + 986\text{ N}\bullet\text{m} + (1.7\text{ kN})(2820\text{ mm}) - (7.0\text{ kN})(850\text{ mm}) = 0$$

$$R = 0.053\text{ kN}$$

$$R_C = 7.0\text{ kN} - 1.7\text{ kN} - \text{R} = 5.25\text{ kN}$$

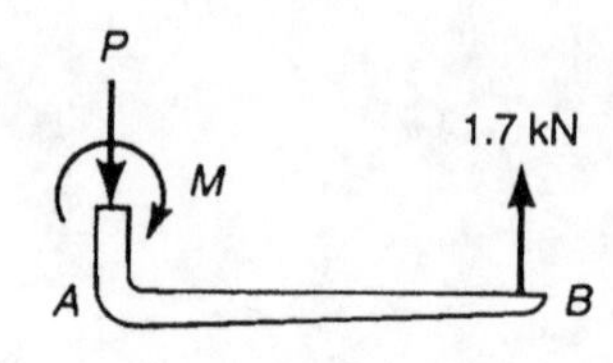

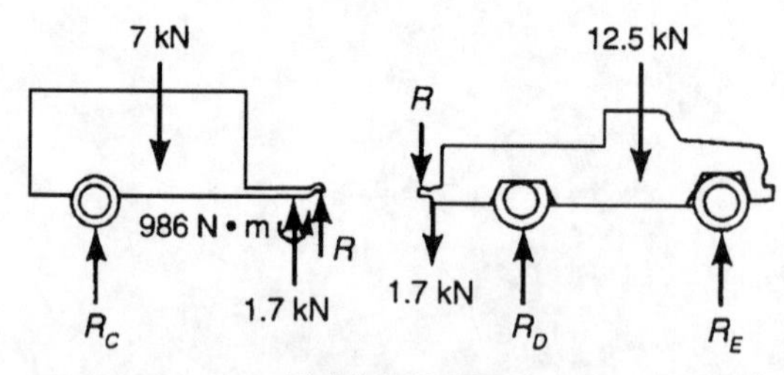

Exhibit 7.26c

7.27 c.

$$\sum M_E = (12.5\text{ kN})(1190\text{ mm}) + (1.7\text{ kN})(4140\text{ mm}) + R(4340\text{ mm}) - R_D(2750\text{ mm}) = 0$$

$$R_D = 8.05\text{ kN}$$

7.28 d.

$$\sum M_A = (105\text{ mm})P - (30\text{ mm})Q = 0$$

$$Q = 3.5\text{P}$$

$$\sum M_B = (80\text{ mm})Q - (35\text{ mm})R = 0$$

$$R = \frac{80}{35}(3.5P) = 8.0P$$

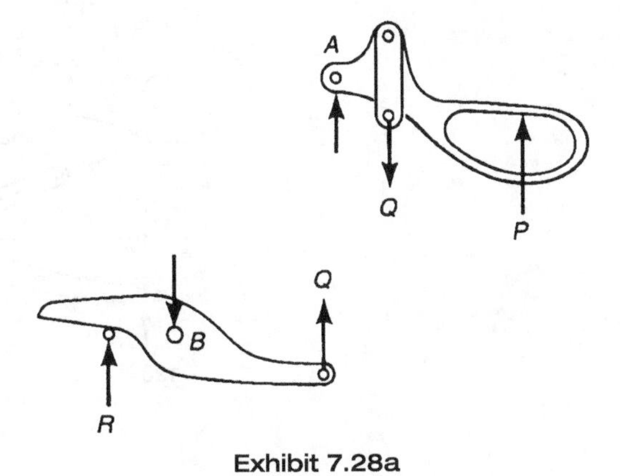

Exhibit 7.28a

7.29 d.

$$h = 115 \tan 20° = 135 \tan(20° - \phi)$$

$$\phi = 20° - \tan^{-1}\left(\frac{115}{135}\tan 20°\right) = 2.77°$$

$$\overline{OA} = \frac{115 \text{ mm}}{\cos 20°} = 122.4 \text{ mm}$$

$$\sum M_A = \overline{OA} \sin \phi \, R - (115 \text{ mm})(120 \text{ N}) = 0$$

$$R = \frac{115(120)}{122.4 \ \sin 2.77°} = 2330 \text{ N}$$

$$\sum M_B = (26 \text{ mm}) R \sin(85° + \phi) - (30 \text{ mm}) Q = 0$$

$$Q = \frac{26}{30}(2330 \text{ N}) \sin 87.77° = 2020 \text{ N}$$

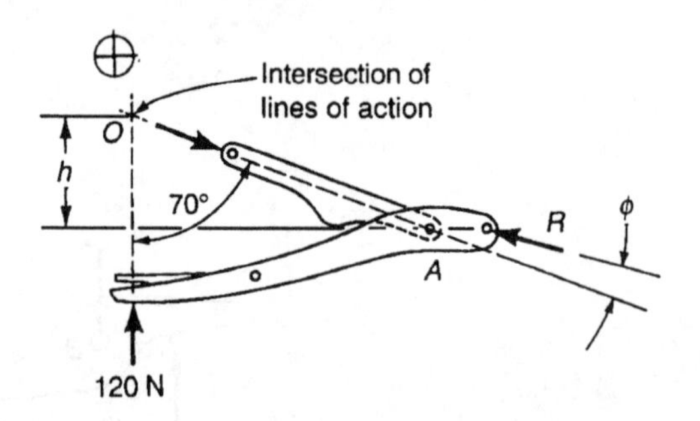

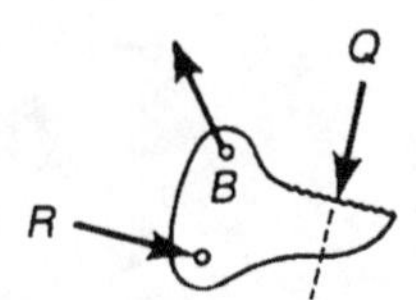

Exhibit 7.29a

7.30 a.

$$M = (600 \text{ N})(0.18 \text{ m}) - (480 \text{ N})(0.18 \text{ m}) + (320 \text{ N})(0.15 \text{ m}) - (400 \text{ N})(0.15 \text{ m}) = 9.60 \text{ N}\bullet\text{m}$$

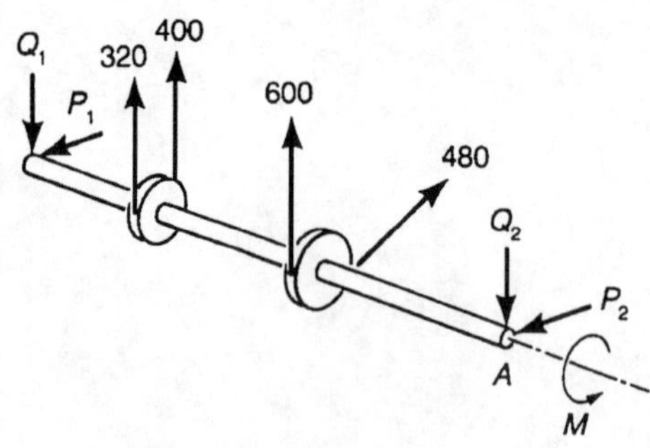

Exhibit 7.30a

7.31 c.

$$P_2(1.7\text{ m}) = (480\text{ N})(\cos 38°)(1.1\text{ m}),\ P_2 = 245\text{ N}$$

$$Q_2(1.7\text{ m}) = (720\text{ N})(0.4\text{ m}) + (600\text{ N} + 480\text{ N}\sin 38°)(1.1\text{ m}),\ Q_2 = 749\text{ N}$$

$$R_2 = \sqrt{P_2^2 + Q_2^2} = 788\text{ N}$$

7.32 c.

$$\mathbf{T} = T\mathbf{e}_{AB} = T\,\frac{-1.8\mathbf{e}_x + 1.5\mathbf{e}_y + 0.6\mathbf{e}_z}{\sqrt{5.85}}$$

Vertical forces at A: $\dfrac{0.6}{\sqrt{5.85}}T = mg$

y-forces at B: $F\,\dfrac{1.5}{\sqrt{5.85}}T$

Combine: $F = \dfrac{1.5}{0.6}mg = 98\text{ N}$

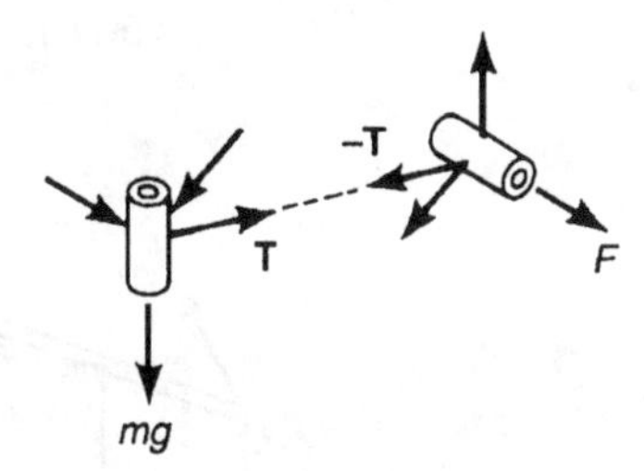

Exhibit 7.32a

7.33 a.

$$d = \frac{4}{5}(1.8\text{ m}) = 1.44\text{ m}$$

$$\mathbf{T}_{AE} = T_{AE}\left(\frac{3}{13}\mathbf{e}_x - \frac{4}{13}\mathbf{e}_y - \frac{12}{13}\mathbf{e}_z\right)$$

$$\sum M_{BC} = -(2.94\text{ m})\left(\frac{12}{13}T_{AE}\right) + (1.44\text{ m})R_z = 0$$

$$T_{AE} = \frac{26}{49}R_z = 424\text{ N}$$

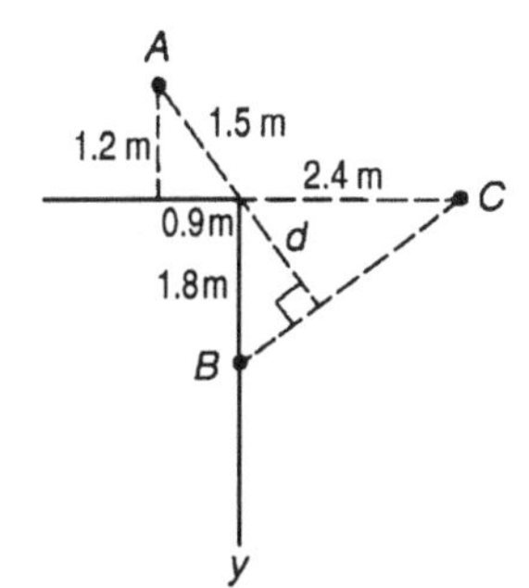

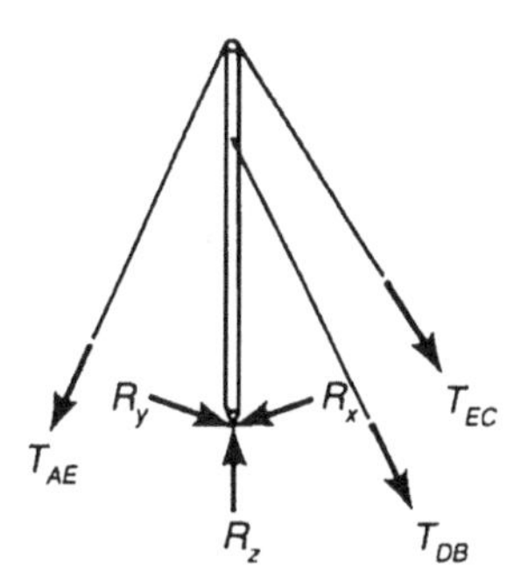

Exhibit 7.33a

7.34 a.

$$\mathbf{e}_{OD} = \frac{3.3\mathbf{e}_y + 0.75\mathbf{e}_z}{\sqrt{11.4525}} \qquad \mathbf{T}_{CD} = T_{CD}\frac{-1.8\mathbf{e}_x + 3.3\mathbf{e}_y + 0.75\mathbf{e}_z}{\sqrt{14.6925}}$$

$$\mathbf{e}_{OB} = \frac{-1.2\mathbf{e}_y + 1.65\mathbf{e}_z}{\sqrt{4.1625}} \qquad \mathbf{T}_{AB} = T_{AB}\frac{-3\mathbf{e}_x + 1.2\mathbf{e}_y + 1.65\mathbf{e}_z}{\sqrt{13.1625}}$$

$$M_{OD} = \mathbf{e}_{OD} \bullet \left[(3\text{ m})\mathbf{e}_x \times \mathbf{T}_{AB}\right] + \mathbf{e}_{OD} \bullet \left[(2.4\text{ m})\mathbf{e}_x \times (-1.5\text{ kN }\mathbf{e}_z)\right]$$

$$= (-1.5484\text{ m})T_{AB} + 3.5128\text{ kN}\bullet\text{m} = 0$$

$$T_{AB} = \frac{3.5128\text{ kN}\bullet\text{m}}{1.5484\text{ m}} = 2.27\text{ kN}$$

$$M_{OB} = \mathbf{e}_{OB} \bullet \left[(1.8\text{ m})\mathbf{e}_x \times \mathbf{T}_{CD}\right] + \mathbf{e}_{OB} \bullet \left[(2.4\text{ m})\mathbf{e}_x \times (-1.5\text{ kN }\mathbf{e}_z)\right]$$

$$= (-1.4604\text{ m})T_{CD} + 2.1174\text{ kN}\bullet\text{m} = 0$$

$$\mathbf{T}_{CD} = \frac{2.1174\text{ kN}\bullet\text{m}}{1.4604\text{ m}} = 1.45\text{ kN}$$

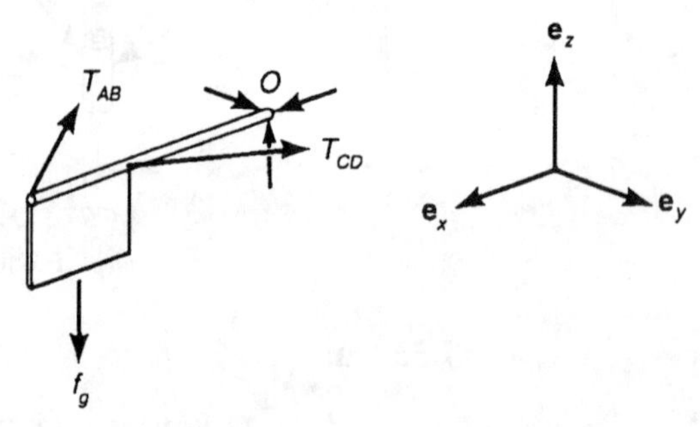

Exhibit 7.34a

7.35 d.

$$M_{OA} = \mathbf{e}_{OA} \bullet (0.45 \text{ m } \mathbf{e}_y \times \mathbf{T}) - (0.225 \text{ m})\frac{12}{13} f_g$$

$$= \frac{0.45 \text{ m}}{13}\left(\frac{46T}{7} - 6 f_g\right) = 0$$

$$T = \frac{21}{23} f_g = \frac{21}{23} (40 \text{ kg})(9.8 \text{ N/kg}) = 358 \text{ N}$$

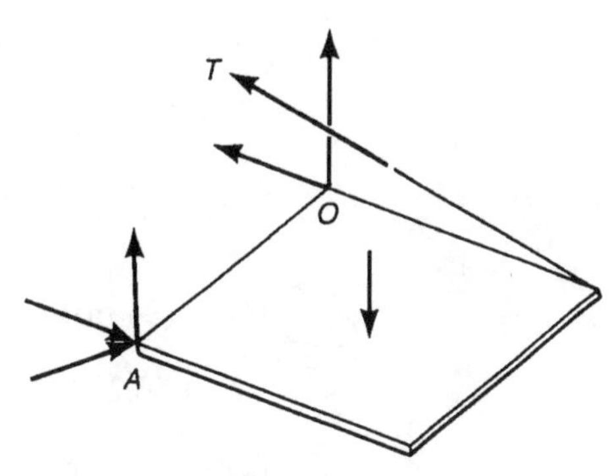

Exhibit 7.35a

7.36 b.

$$\sum M_A = a(4 \text{ kN}) + 4aR_D - a(24 \text{ kN}) - 3a(8 \text{ kN}) = 0$$

$$R_D = 11.0 \text{ kN}$$

$$\sum M_E = 3a(11 \text{ kN}) - 2a(8 \text{ kN}) - aT_{BC} = 0$$

$$T_{BC} = 17 \text{ kN (tension)}$$

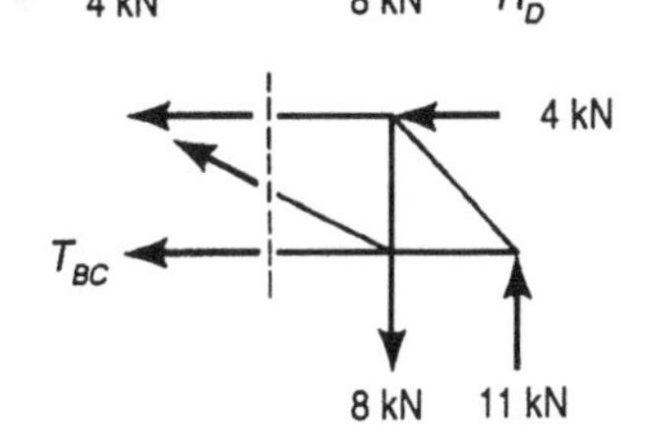

Exhibit 7.36a

7.37 b.

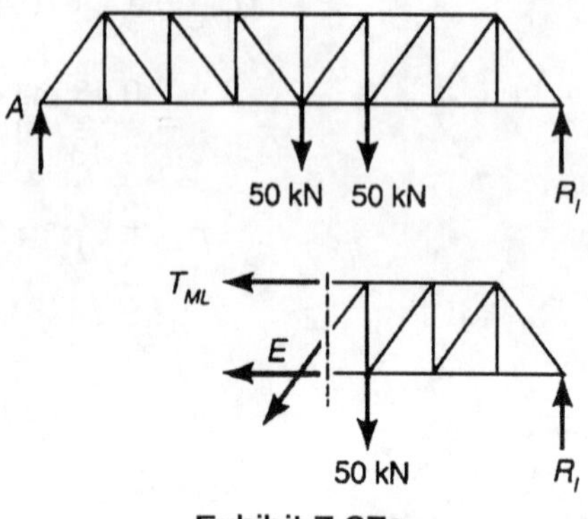

Exhibit 7.37a

$$\sum M_A = (24\text{ m})R_I - (12\text{ m})(50\text{ kN}) - (15\text{ m})(50\text{ kN}) = 0$$

$$R_I = 56.25\text{ kN}$$

$$\sum M_E = (4\text{ m})T_{ML} + (12\text{ m})(56.25\text{ kN}) - (3\text{ m})(50\text{ kN}) = 0$$

$$T_{ML} = -131.25\text{ kN}$$

7.38 c.

$$\sum M_A = (30\text{ m})R_G - (22.5\text{ m})(30\text{ kN}) - (15\text{ m})(25\text{ kN}) - (7.5\text{ m})(20\text{ kN}) = 0$$

$$R_G = 40\text{ kN}$$

$$\sum M_D = (15\text{ m})(40\text{ kN}) - (7.5\text{ m})(30\text{ kN}) - (5\sqrt{3}\text{ m})T_{CF} = 0$$

$$T_{CF} = 43.3\text{ kN}$$

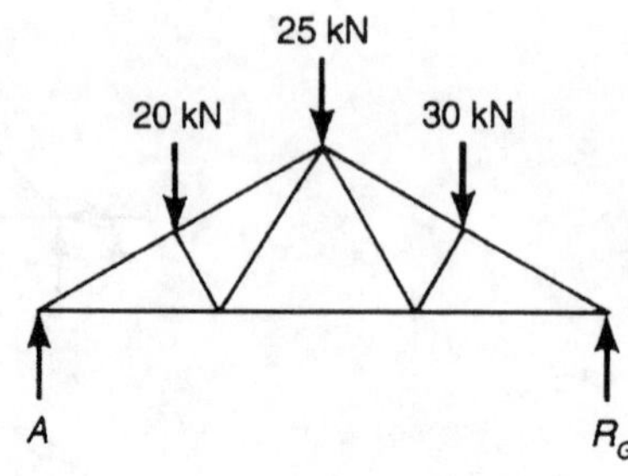

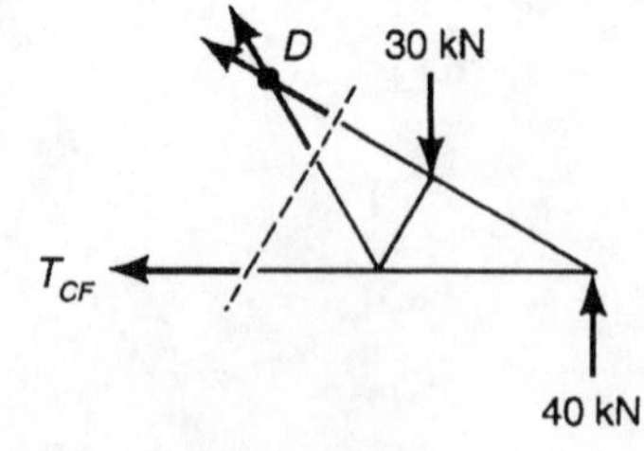

Exhibit 7.38a

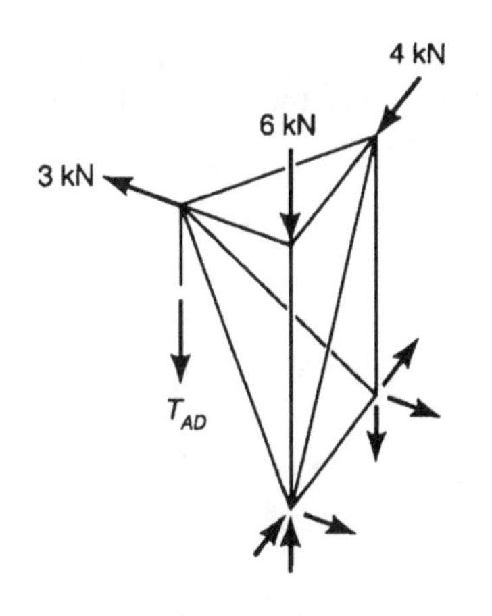

Exhibit 7.39a

7.39 a.

$$\sum M_{CB} = (6\text{ m})\frac{\sqrt{3}}{2}(3\text{ kN}) + \left(\frac{3\sqrt{3}}{2}\text{ m}\right)T_{AD} = 0$$

$$T_{AD} = -6\text{ kN (compression)}$$

7.40 a.

$$\sum M_{DE} = aP - aT_{AB} = 0; \qquad T_{AB} = P$$

$$\sum M_{DC} = \mathbf{e}_{CD} \bullet [a\mathbf{e}_z \times (P\mathbf{e}_x + \mathbf{T}_{BE})] = 0$$

$$\mathbf{T}_{BE} = T_{BE}\frac{\mathbf{e}_x + \mathbf{e}_y - \mathbf{e}_z}{\sqrt{3}}$$

$$(\mathbf{e}_{CD} \bullet \mathbf{e}_y)P = -\mathbf{e}_{CD} \bullet \left(\mathbf{e}_z \times T_{BE}\frac{\mathbf{e}_x + \mathbf{e}_y - \mathbf{e}_z}{\sqrt{3}}\right)$$

$$\frac{\mathbf{e}_x - \mathbf{e}_y}{\sqrt{2}} \bullet \mathbf{e}_y P = \frac{\mathbf{e}_x - \mathbf{e}_y}{\sqrt{2}} \bullet \frac{\mathbf{e}_x - \mathbf{e}_y}{\sqrt{3}} T_{BE}$$

$$T_{BE} = -\frac{\sqrt{3}}{2}P, \text{ or } -0.87P \text{ (compression)}$$

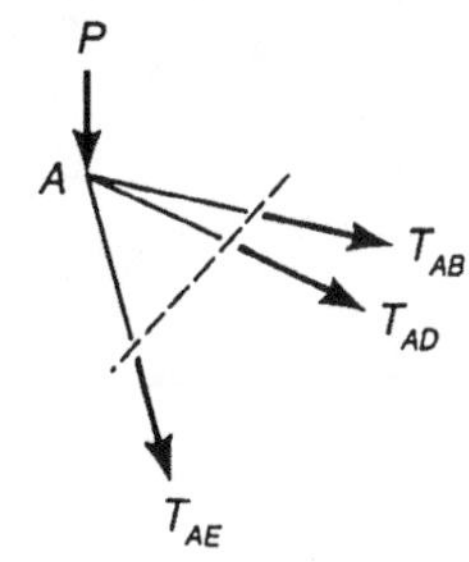

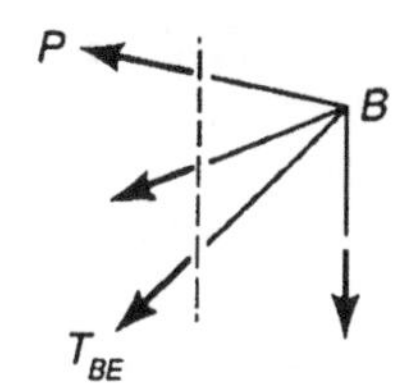

Exhibit 7.40a

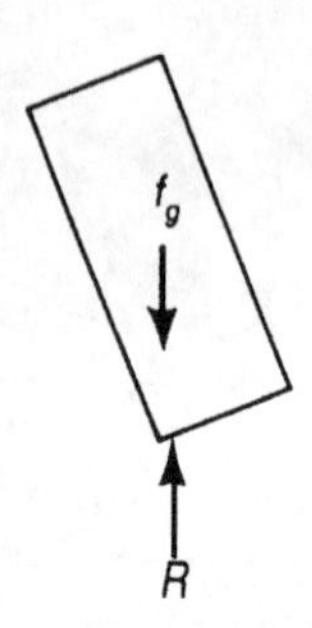

Exhibit 7.41a

7.41 b. The line of action of f_g (and hence of R) passes through the bottom at a distance of 28.8 mm from the lower corner. Hence the crate will not tip over: tan 15° = 0.27 > μ, so the crate will slide.

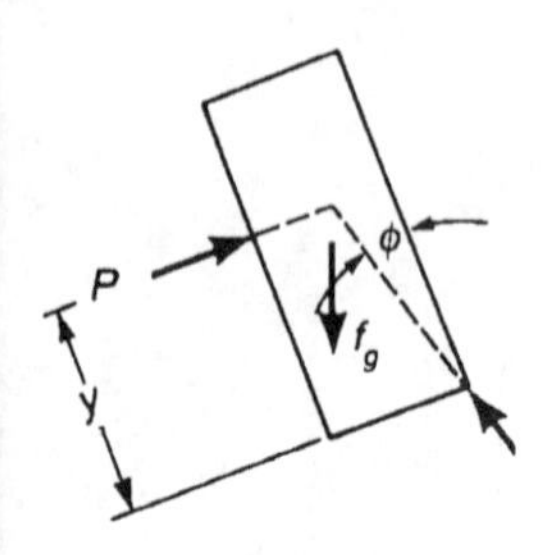

Exhibit 7.42

7.42 d. Vanishing resultant moment requires that the lines of action of the three forces intersect, so that

$$\tan\phi = \mu = \frac{0.27\text{ m} - (y - 0.9\text{ m})\tan 15°}{y}$$

$$y = \frac{0.27\text{ m} + (0.9\text{ m})\tan 15°}{\mu + \tan 15°} = 1.09\text{ m}$$

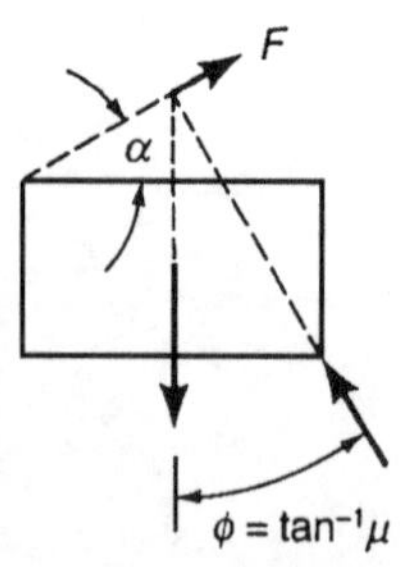

Exhibit 7.43a

7.43 c. Vanishing resultant moment requires that the lines of action of the three forces intersect, so that

$$\frac{b}{2}\tan\alpha = \frac{b}{2}\cot\phi - h = \frac{b}{2\mu} - h$$

$$\alpha = \tan^{-1}\left(\frac{1}{\mu} - \frac{2h}{b}\right)$$

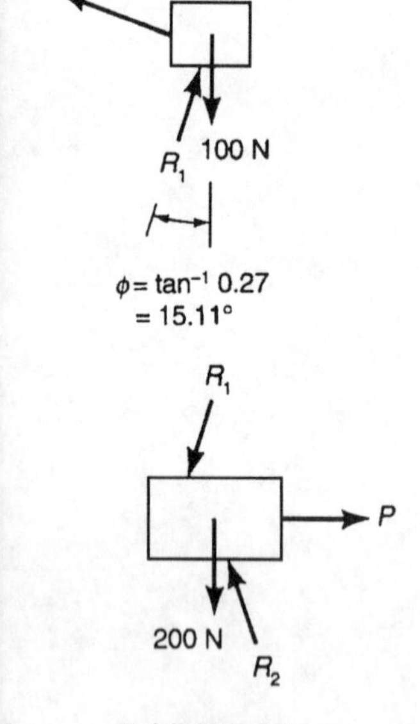

Exhibit 7.44a

7.44 a. Sum forces that are parallel to T on the upper block:

$$R_1\cos(20° - 15.11°) = (100\text{ N})\cos 20°;\ R_1 = 94.3\text{ N}$$

Sum forces that are perpendicular to R_2 on the lower block:

$$P\cos 15.11° - (200\text{ N})\sin 15.11° - (94.3\text{ N})\sin 30.22° = 0$$

$$P = 0.27(200\text{ N}) + \frac{\sin 45.22°}{\cos 15.11°}(206\text{ N}) = 103\text{ N}$$

7.45 a.

$$\phi = \tan^{-1}(0.27) = 15.11°$$

Sum forces that are perpendicular to R_0:

$$R_1 \cos(15° + 2\phi) = (150 \text{ N}) \cos \phi;\ R_1 = 206 \text{ N}$$

Sum forces that are perpendicular to R_2:

$$P \cos \phi - (200 \text{ N}) \sin \phi - R_1 \sin(15° + 2\phi) = 0$$

$$P = 0.27(200 \text{ N}) + \frac{\sin 45.22°}{\cos 15.11°}(206 \text{ N}) = 103 \text{ N s}$$

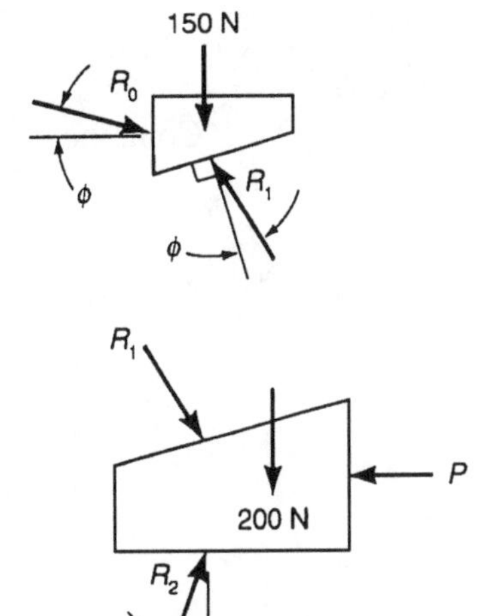

Exhibit 7.45a

7.46 b. Vanishing resultant moment requires that the lines of action of the three forces intersect, so that

$$a \tan \phi_1 = a \cos \alpha - \frac{a \sin \alpha}{\tan(\alpha + \phi_2)}$$

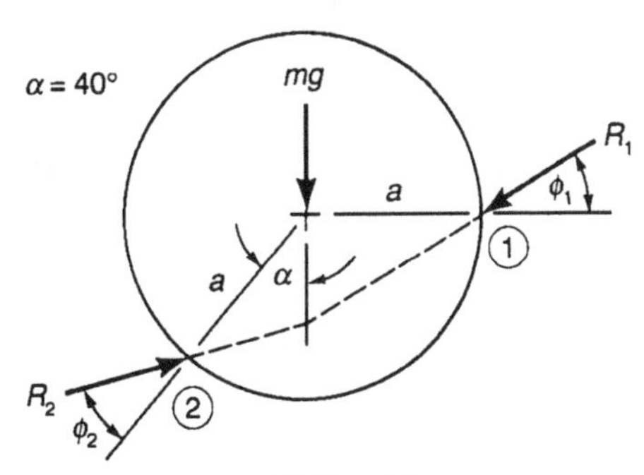

Exhibit 7.46a

If slip occurs at ②, $\phi_2 = \tan^{-1} 0.3 = 16.70°$.

$$\tan \phi_1 = \cos 40° - \frac{\sin 40°}{\tan 56.70°} = 0.344$$

Since this is less than $\mu_1 = 0.45$, slip *does* occur at ②.

$$\Sigma M_2 = a \cos\alpha\, R_1 \cos\phi_1 - (a + a \sin\alpha) R_1 \sin\phi_1 - a \sin\alpha\, \text{mg} = 0$$

$$R_1 \cos\phi_1 = \frac{\text{mg} \sin\alpha}{\cos\alpha - (1 + \sin\alpha)\tan\phi_1} = 3.20 \text{ mg} = 62.7 \text{ kN}$$

7.47 c. $\sum F_x = 0$ on the block leads to

$$T \sin 45° - \mu N = 0$$
$$-T\sqrt{2} + Ne = 0$$

Combine:

$$e = 2\mu$$

$$(T \cos 30°)(9.4) - N(4 + e) = 0$$

$$\frac{T}{N} = \frac{(4+e)}{(9.4 \cos 30°)} = \frac{\mu}{\sin 45°}$$

$$\mu = \frac{2}{\frac{9.4 \cos 30°}{2 \sin 45°} - 1} = 0.42$$

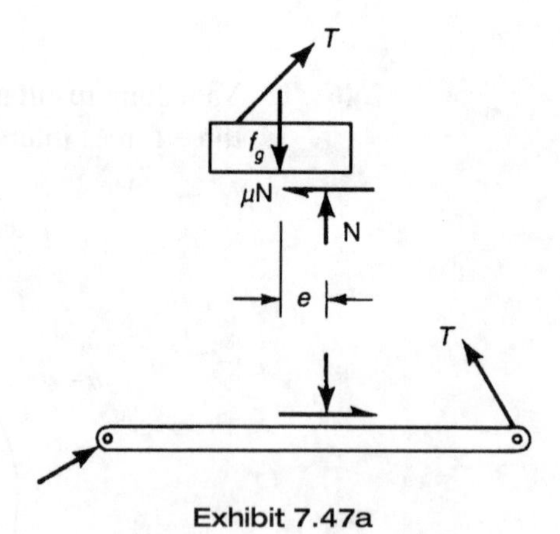

Exhibit 7.47a

7.48 c.

$$d = a + (b + a)\cos 2\phi$$

$$\cos 2\phi = \frac{1 - \tan^2 \phi}{1 + \tan^2 \phi}$$

To prevent slip, $\tan \phi < \mu$, so that

$$d > a + (b + a)\frac{1 - \mu^2}{1 + \mu^2} = \frac{2a + (1 - \mu^2)b}{1 + \mu^2}$$

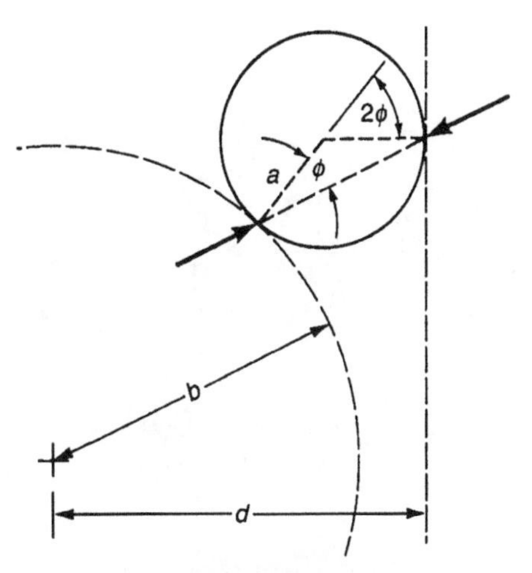

Exhibit 7.48a

CHAPTER 8

Dynamics

OUTLINE

MASS MOMENTS OF INERTIA

The mass moment of inertia of a body is a property that determines how much resistance to angular acceleration it has. This mass moment is called the second moment about an axis. The equation is:

$$I = \int r^2 \, dm$$

We can determine the mass moment of a cylinder as shown in Figure 8.1.

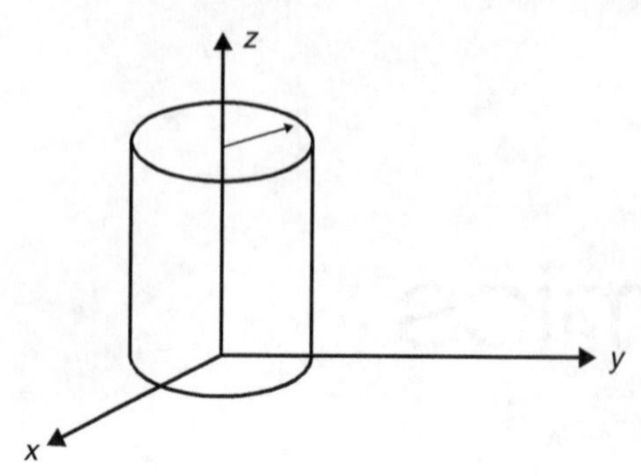

Figure 8.1

Consider an elemental moment of inertia of the cylinder:

$$d\,I_z = \int r^2 dm = \rho 2\pi h r^3 dr$$

And for uniform density of the cylinder, the equation becomes:

$$I_z = \rho 2\pi h \int_0^R r^3 dr = 2\rho\pi h \frac{R^4}{4} = \frac{\rho\pi R^4 h}{2}$$

Then, to find the mass of the circular cylinder:

$$m = \int_m dm = 2\pi\rho h \int^R rdr = \rho\pi h R^2$$

Therefore:

$$I_Z = \frac{1}{2} mR^2$$

Example 8.1

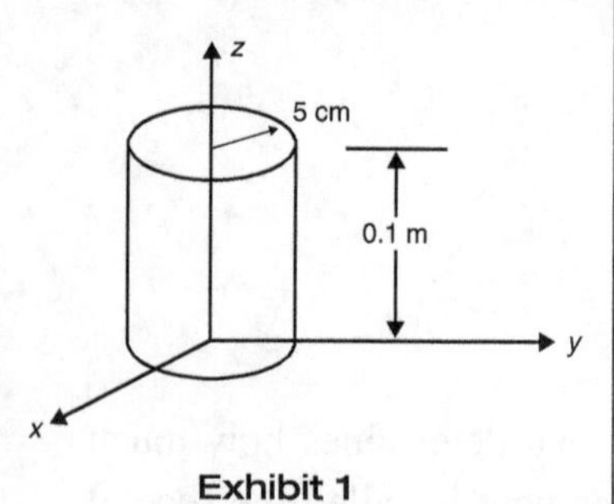

Exhibit 1

Consider a cylinder made with an alloy having a density of 400 kg/m³. The dimensions of the body are shown in Exhibit 1. Determine the mass moment of inertia of the cylinder.

Solution

Determine the mass of the cylinder.

$$\text{mass} = \rho\pi h R^2 = \left(400 \frac{\text{kg}}{\text{m}^3}\right)(\pi)(0.1\text{ m})(0.05\text{ m})^2 = 0.314\text{ kg}$$

Determine the mass moment of inertia about the Z axis.

$$I_z = \frac{1}{2}\text{ mass} \times R^2 = \frac{1}{2}(0.314\text{ kg})(0.05\text{ m})^2 = 3.9 \times 10^{-4}\text{ kg–m}^2$$

Example 8.2

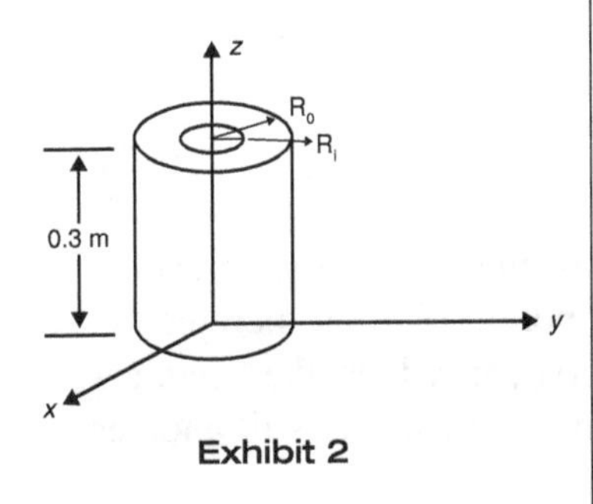

Exhibit 2

Consider the homogeneous hollow right circular cylinder, as shown in Exhibit 2, with the following dimensions:

$\rho = 7000\ \text{kg/m}^3$

Inner radius = 5 cm

Outer radius = 8 cm

$h = 0.3$ m

Determine the mass moment of inertia with respect to the geometric axis.

Solution

Determine the mass of the cylinder = mass of outer cylinder – mass of inner cylinder.

$$\text{mass} = \pi\rho h(R_0^2 - R_i^2) = \left(7000\frac{\text{kg}}{\text{m}^3}\right)(\pi)(0.3\ \text{m})(0.08^2 - 0.05^2)$$

$$\text{mass} = 25.73\ \text{kg}$$

Determine the mass moment of inertia with respect to the geometric axis.

$$I_z = \frac{1}{2}\text{mass}\left(R_0^2 - R_1^2\right) = \frac{25.73}{2}\text{kg}\left(0.08^2 - 0.05^2\right)\text{m}^2 = 0.05017\ \text{kg–m}^2$$

The analysis of a mechanical system having elements under acceleration must consider these accelerations along with the related forces. In such analysis, the force side of Newton's second law, $f = ma$, and the third law of action and reaction are dealt with in exactly the same manner as in statics. But it is the relationships among positions, velocities, and accelerations that complete the discipline of dynamic analysis. The following two sections review these relationships, and the remainder of the chapter deals with their incorporation into Newton's laws of motion.

KINEMATICS OF A PARTICLE

Consider a point P that moves along a smooth path as indicated in Figure 8.2. The position of the point may be specified by the vector $\mathbf{r}(t)$, defined to extend from an arbitrarily selected, fixed point O to the moving point P. The **velocity v** of the point is defined to be the derivative with respect to t of $\mathbf{r}(t)$, written as

$$\mathbf{v} = \frac{d\mathbf{r}}{dt} \tag{8.1}$$

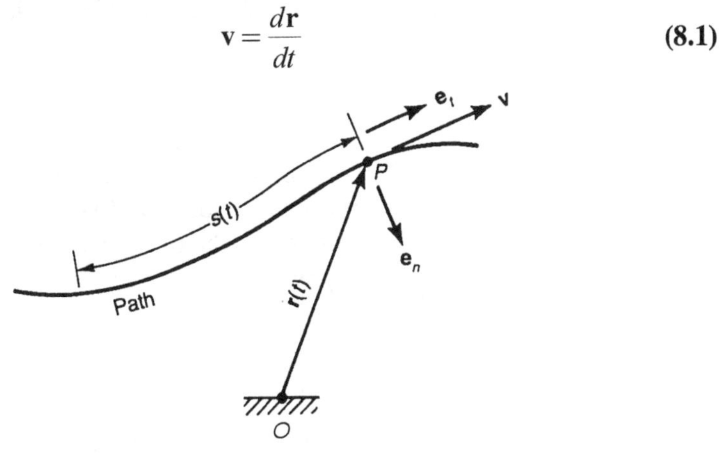

Figure 8.2

Although this definition is sometimes used for evaluation [that is, by differentiating a specific expression for $\mathbf{r}(t)$], it will often be more direct to use other relationships. It follows from the above definition that the velocity vector is tangent to the path of the particle; thus, upon introduction of a unit vector $\mathbf{e}_t$, defined to be tangent to the path, the velocity can also be expressed as

$$\mathbf{v} = v\mathbf{e}_t \tag{8.2}$$

The position of P can also be specified in terms of the distance $s(t)$ traveled along the path from an arbitrarily selected reference point. Then an incremental change in position may be approximated as $\Delta\mathbf{r} \approx \Delta s\mathbf{e}_t$, in which the accuracy increases as the increments Δt and $\Delta\mathbf{r}$ approach zero. This leads to still another way of expressing the velocity as

$$\mathbf{v} = \frac{ds}{dt}\mathbf{e}_t \tag{8.3}$$

The scalar

$$v = \frac{ds}{dt} = \dot{s} \tag{8.4}$$

can be either positive or negative, depending on whether the motion is in the same or the opposite direction as that selected in the definition of $\mathbf{e}_t$.

The **acceleration** of the point is defined as the derivative of the velocity with respect to time:

$$\mathbf{a} = \frac{d\mathbf{v}}{dt} \tag{8.5}$$

A useful relationship follows from application to Equation (8.2) of the rules for differentiating products and functions of functions:

$$\frac{d\mathbf{v}}{dt} = \dot{v}\mathbf{e}_t + \mathbf{v}\frac{ds}{dt}\frac{d\mathbf{e}_t}{ds}$$

As the direction of $\mathbf{e}_t$ varies, the square of its magnitude, $|\mathbf{e}_t|^2 = \mathbf{e}_t \bullet \mathbf{e}_t$, remains fixed and equal to 1, so that

$$\frac{d}{ds}|\mathbf{e}_t|^2 = \frac{d}{ds}(\mathbf{e}_t \bullet \mathbf{e}_t) = 2\mathbf{e}_t \bullet \frac{d\mathbf{e}_t}{ds} = 0$$

This shows that $d\mathbf{e}_t/ds$ is either zero or perpendicular to $\mathbf{e}_t$. With another unit vector $\mathbf{e}_n$ defined to be in the direction of $d\mathbf{e}_t/ds$, this vector may be expressed as

$$\frac{d\mathbf{e}_t}{ds} = \kappa\mathbf{e}_n$$

The scalar κ is called the local **curvature** of the path; its reciprocal, $\rho = 1/\kappa$, is called the local **radius of curvature** of the path. In the special case in which the path is straight, the curvature and hence $d\mathbf{e}_t/ds$ are zero. These lead to the following expression for the **acceleration** of the point:

$$\mathbf{a} = \dot{v}\mathbf{e}_t + \frac{v^2}{\rho}\mathbf{e}_n \tag{8.6}$$

The two terms express the **tangential** and **normal** (or **centripetal**) components of acceleration.

If a driver of a car with sufficient capability "steps on the gas," a positive value of $\dot{v}$ is induced, whereas if he "steps on the brake," a negative value is induced. If the path of the car is straight (zero curvature or "infinite" radius of curvature), the entire acceleration is $\dot{v}\mathbf{e}_t$. If the car is rounding a curve, there is an additional component of acceleration directed laterally, toward the center of curvature of the path. These components are indicated in Figure 8.3, a view of the plane of $\mathbf{e}_t$ and $d\mathbf{e}_t/ds$.

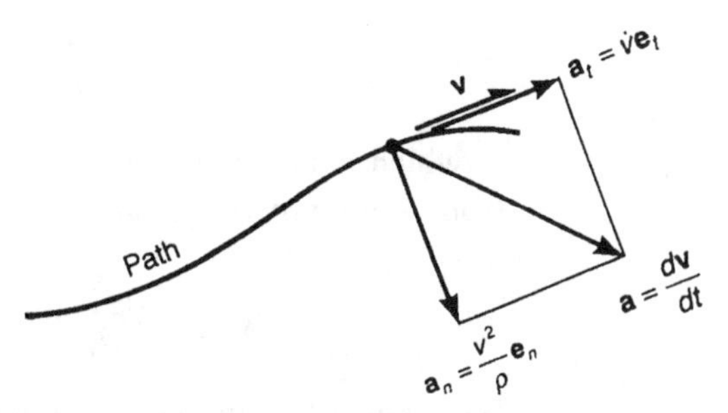

Figure 8.3

Example 8.3

At a certain instant, the velocity and acceleration of a point have the rectangular Cartesian components given by

$$\mathbf{v} = (3.5\mathbf{e}_x - 7.2\mathbf{e}_y + 9.6\mathbf{e}_z) \text{ m/s}$$
$$\mathbf{a} = (-20\mathbf{e}_x + 20\mathbf{e}_y + 10\mathbf{e}_z) \text{ m/s}^2$$

At this instant, what are the rate of change of speed dv/dt and the local radius of curvature of the path?

Solution

The rectangular Cartesian components of the unit tangent vector can be determined by dividing the velocity vector by its magnitude:

$$\mathbf{e}_t = \frac{\mathbf{v}}{|\mathbf{v}|} = \frac{3.5\mathbf{e}_x - 7.2\mathbf{e}_y + 9.6\mathbf{e}_z}{\sqrt{(3.5)^2 + (-7.2)^2 + (9.6)^2}} = 0.280\mathbf{e}_x - 0.576\mathbf{e}_y + 0.768\mathbf{e}_z$$

The rate of change of speed can then be determined as the projection of the acceleration vector onto the tangent to the path:

$$\dot{v} = \mathbf{e}_t \bullet \mathbf{a} = [(0.280)(-20) + (-0.576)(20) + (0.768)(10)] \text{ m/s}^2$$
$$= -9.44 \text{ m/s}^2$$

The negative sign indicates the projection is opposite to $\mathbf{e}_t$ (which was defined by the above equation to be in the same direction as the velocity). This means that the speed is *decreasing* at 9.44 m/s. One sees from Figure 8.3 that the normal component of acceleration has magnitude

$$a_n = \sqrt{|\mathbf{a}|^2 - \dot{v}^2} = \sqrt{(-20)^2 + (20)^2 + (10)^2 - (-9.44)^2 \text{m/s}^2} = 28.5 \text{ m/s}^2$$

which, from Equation (8.6), is related to the speed and radius of curvature by $a_n = v^2/\rho$. Rearrangement of this equation gives the radius of curvature as

$$\rho = \frac{v^2}{a_n} = \frac{[(3.5)^2 + (-7.2)^2 + (9.6)^2]\text{ m}^2/\text{s}^2}{28.5 \text{ m/s}^2} = 5.48 \text{ m}$$

Relating Distance, Velocity, and the Tangential Component of Acceleration

The basic relationships among tangential acceleration a_t, velocity $v\mathbf{e}_t$, and distance s are

$$\frac{dv}{dt} = a_t \quad \text{or} \quad v = v_0 + \int a_t \, dt \tag{8.7}$$

$$\frac{ds}{dt} = v \quad \text{or} \quad s = s_0 + \int v \, dt \tag{8.8}$$

in which v_0 and s_0 are constants of integration. An alternative relationship comes from writing $dv/dt = (ds/dt)(dv/ds) = v\,dv/ds$:

$$v\frac{dv}{ds} = a_t \quad \text{or} \quad v^2 = v_0^2 + 2\int a_t \, ds \tag{8.9}$$

Equations (8.7) and (8.8) are useful in dealing with the *time* histories of acceleration, velocity, and distance, whereas Equation (8.9) is helpful in dealing with the manner in which velocity and acceleration vary with distance.

Example 8.4

The variation of tangential acceleration with time is given in Exhibit 3. If a point with an initial velocity of 24 m/s is subjected to this acceleration, what will be its velocity at $t = 6$ s, 10 s, and 15 s, and what will be the values of s at $t = 4$ s, 7.6 s, and 15 s?

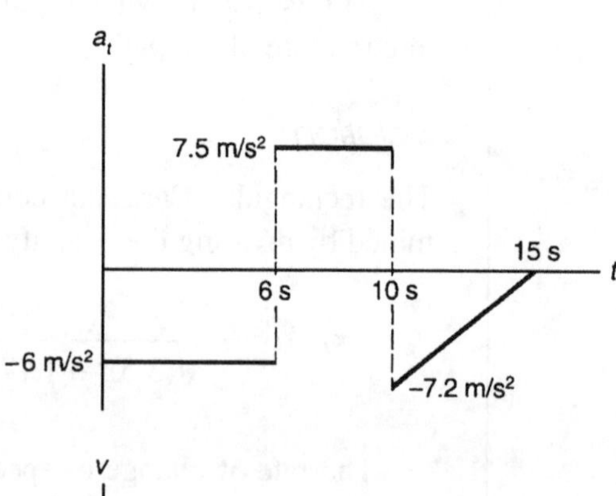

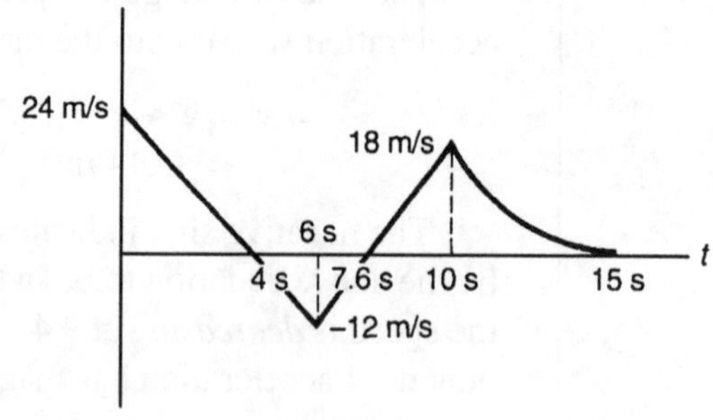

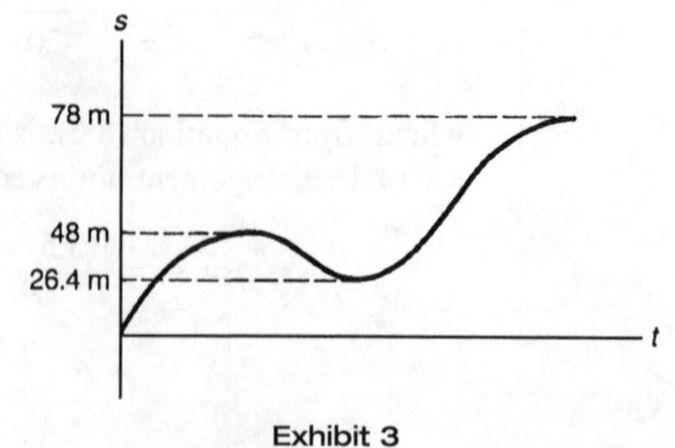

Exhibit 3

Solution

Equation (8.7) has the following graphical interpretations: At each point, the slope of the *v-t* curve is equal to the ordinate on the a_t-*t* curve. During any interval, the change in the value of *v* is equal to the area under the a_t-*t* curve for the same interval. With these rules and the given initial value of *v*, the variation of *v* with *t* can be plotted, and values of *v* can be calculated for each point.

The reader should use these rules to verify all details of the *v-t* curve shown. Equation (8.8) indicates that identical rules for slopes, ordinates, and areas relate the curve of distance *s* to that of velocity *v*, so the same procedure can be used to construct the *s-t* curve from the *v-t* curve. Again, the reader should verify all details of this curve.

Example 8.5

The tangential acceleration of the pendulum bob shown in Exhibit 4 varies with position according to $a = -g \sin(s/l)$, in which *g* is the local acceleration of gravity. If a speed v_0 is imparted at the vertical position (where $s = 0$), what will be the maximum value of *s* reached?

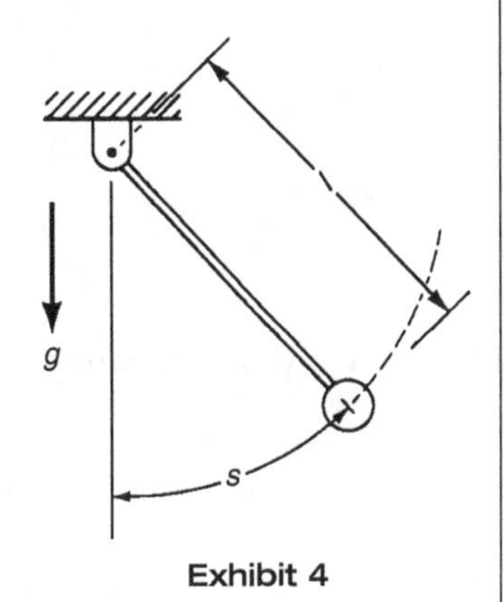

Exhibit 4

Solution

Because the relationship between tangential acceleration and *position* is given, Equation (8.9) will prove useful. The integrated form leads to

$$v^2 = v_0^2 + 2\int_0^s g \sin\left(\frac{s}{l}\right) ds$$
$$= v_0^2 + 2gl\left(1 - \cos\frac{s}{l}\right)$$

which gives the velocity *v* in the terms of any position *s*. Since $v = \dot{s}$, the maximum *s* will occur when $v = 0$, and the corresponding *s* is easily isolated from the above equation after setting $v = 0$:

$$s_{\max} = l \cos^{-1}\left(1 - \frac{v_0^2}{2gl}\right)$$

Observe that if $v_0^2 > 2gl$, no real value of $s_{\max}$ exists, because *v* never reaches zero in that case.

Constant Tangential Acceleration

When the tangential acceleration is constant, Equations (8.7) through (8.9) reduce to

$$v = v_0 + a_t t \qquad \textbf{(8.10)}$$

$$s = s_0 + v_0 t + \frac{1}{2} a_t t^2 \qquad \textbf{(8.11)}$$

$$v^2 = v_0^2 + 2a_t s \qquad \textbf{(8.12)}$$

Rectilinear Motion

In the special case in which the path is a straight line, the unit tangent vector $\mathbf{e}_t$ is constant, and the curvature $1/\rho$ is zero throughout. The acceleration is then given by $(dv/dt)\mathbf{e}_t$, and the subscript on the symbol a_t may be dropped without ambiguity.

Example 8.6

A particle is launched vertically upward with an initial speed of 10 m/s and subsequently moves with constant downward acceleration of magnitude 9.8 m/s². What is the maximum height reached by the particle? How long does it take to return to the original launch position? And how fast is it traveling at its return to the launch position?

Solution

In this case the path will be straight and the acceleration is constant. With $\mathbf{e}_t$ defined as upward, the constant scalars appearing in Equations (8.10) through (8.12) have the values $v_0 = 10$ m/s and $a_t = a = -9.8$ m/s², so that these equations become

$$v = 10\text{ m/s} - (9.8\text{ m/s}^2)t \qquad \textbf{(i)}$$

$$s = (10\text{ m/s})t - \frac{1}{2}(9.8\text{ m/s}^2)t^2 \qquad \textbf{(ii)}$$

$$v^2 = (10\text{ m/s})^2 - 2(9.8\text{ m/s}^2)s \qquad \textbf{(iii)}$$

The maximum height reached can be obtained by setting $v = 0$ in (iii), which gives

$$s_{max} = \frac{(10\text{ m/s})^2}{2(9.8\text{ m/s}^2)} = 5.1\text{ m}$$

The time required to reach this height can be obtained by setting $v = 0$ in (i), which gives

$$t_1 = \frac{10\text{ m/s}}{9.8\text{ m/s}^2} = 1.02\text{ s}$$

Finally, setting $s = 0$ in (iii) yields the two values of v that specify the velocity at the launch position:

$$v = \pm 10\text{ m/s}$$

The positive value gives the upward initial velocity, and the negative value gives the equal-magnitude, downward velocity of the particle when it returns to the launch position.

Rectangular Cartesian Coordinates

Multidimensional motion can be analyzed in terms of components associated with a set of fixed unit vectors $\mathbf{e}_x$, $\mathbf{e}_y$, and $\mathbf{e}_z$, which are defined to be mutually perpendicular. For some aspects of analysis, it is also important that they form a "right-handed" set, or $\mathbf{e}_z = \mathbf{e}_x \times \mathbf{e}_y$, $\mathbf{e}_x = \mathbf{e}_y \times \mathbf{e}_z$, and $\mathbf{e}_y = \mathbf{e}_z \times \mathbf{e}_x$. In terms of these unit vectors, the position, velocity, and acceleration can be expressed as

$$\mathbf{r} = x\mathbf{e}_x + y\mathbf{e}_y + z\mathbf{e}_z$$
$$\mathbf{V} = v_x\mathbf{e}_x + v_y\mathbf{e}_y + v_z\mathbf{e}_z$$
$$\mathbf{A} = a_x\mathbf{e}_x + a_y\mathbf{e}_y + a_z\mathbf{e}_z$$

with

$$v_x = \dot{x}$$
$$a_x = \dot{v}_x = \ddot{x}, \quad \text{etc.}$$

Example **8.7**

A wheel rolls without slipping along a straight surface with the orientation of the wheel given in terms of the angle $\theta(t)$. See Exhibit 5. Express the velocity and acceleration of the point P on the rim of the wheel in terms of this angle, its derivatives, and the radius b of the wheel.

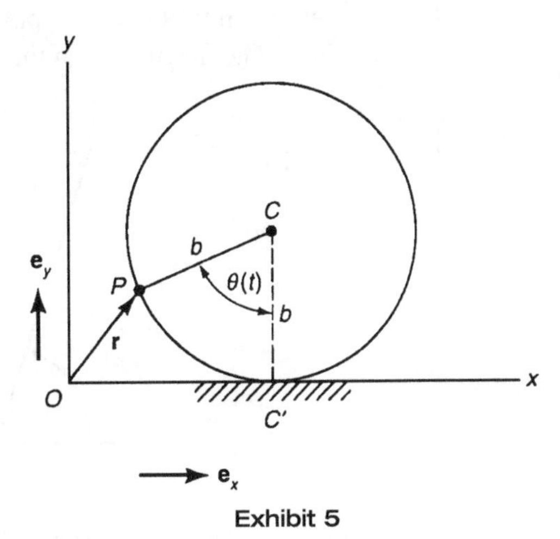

Exhibit 5

Solution

The origin for the x-y coordinates of P is the location of P when $\theta = 0$. Because the wheel rolls without slipping, the distance OC' is equal to the length of the circular arc PC'. The x-coordinate of P is then $OC' - b\sin\theta = b\theta - b\sin\theta$. The y-coordinate of P is that of C (i.e., b) minus $b\cos\theta$. In terms of these coordinates, the position vector from O to P may be expressed as

$$\mathbf{r} = b(\theta - \sin\theta)\mathbf{e}_x + b(1 - \cos\theta)\mathbf{e}_y$$

The velocity is then determined by differentiation of this expression:

$$\mathbf{v} = b\dot{\theta}[(1 - \cos\theta)\mathbf{e}_x + \sin\theta\mathbf{e}_y]$$

The acceleration is determined by another differentiation:

$$\mathbf{a} = b\ddot{\theta}[(1 - \cos\theta)\mathbf{e}_x + \sin\theta\mathbf{e}_y] + b\dot{\theta}^2(\sin\theta\mathbf{e}_x + \cos\theta\mathbf{e}_y)$$

These expressions may be simplified somewhat by rewriting them in terms of the unit vectors $\mathbf{e}_r$ and $\mathbf{e}_\theta$ as defined in Exhibit 6. These unit vectors are given in terms of the original horizontal and vertical unit vectors by

$$\mathbf{e}_r = -\sin\theta\mathbf{e}_x - \cos\theta\mathbf{e}_y$$
$$\mathbf{e}_\theta = -\cos\theta\mathbf{e}_x + \sin\theta\mathbf{e}_y$$

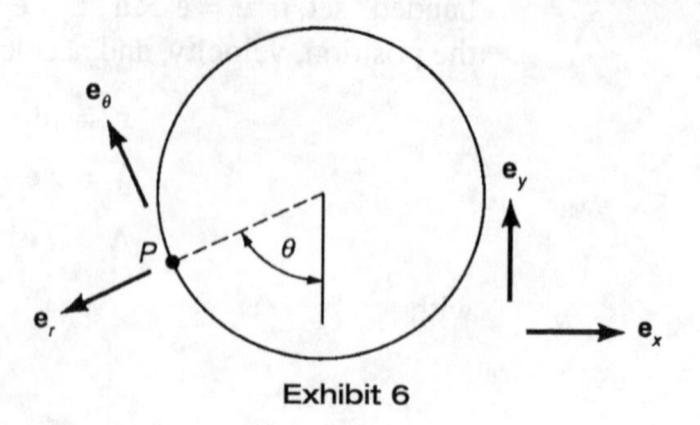

Exhibit 6

The above expressions for velocity and acceleration can now be written as

$$\mathbf{v} = b\dot{\theta}(\mathbf{e}_x + \mathbf{e}_\theta)$$
$$\mathbf{a} = b\ddot{\theta}(\mathbf{e}_x + \mathbf{e}_\theta)b\dot{\theta}^2\mathbf{e}_r$$

Further simplification is possible upon examination of the sum $\mathbf{e}_x + \mathbf{e}_\theta$, shown in Exhibit 7. The magnitude of this sum is $2\sin(\theta/2)$, and its direction is perpendicular

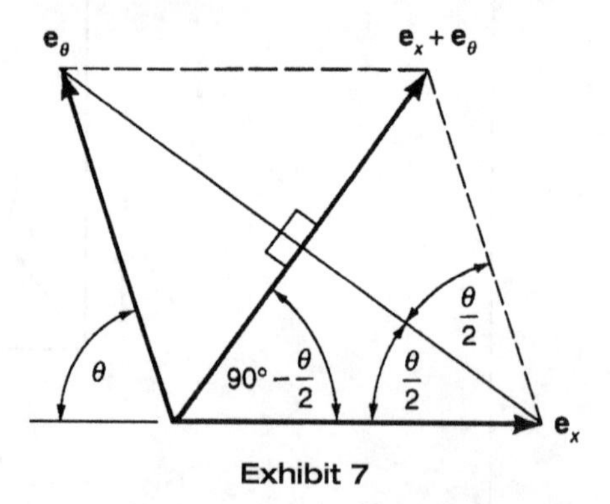

Exhibit 7

to the line connecting points P and C'. The velocity can thus be expressed as $\mathbf{v} = (2b\sin\theta/2\,\dot{\theta})\mathbf{e}_t$ in which the unit tangent vector is perpendicular to the line PC'. The acceleration can be simplified correspondingly to

$$\mathbf{a} = 2b\sin\frac{\theta}{2}\ddot{\theta}\mathbf{e}_t - b\dot{\theta}^2\mathbf{e}_r$$

Several steps were taken to reach the results in Example 8.7. The position vector was expressed in terms of the geometric constraints on the rolling of the wheel, differentiation led to expressions for the velocity and acceleration, and the introduction of auxiliary unit vectors and several trigonometric relationships simplified several expressions.

As mentioned earlier, direct use of the definitions expressed by Equations (8.1) and (8.5) may not be the easiest means of evaluating velocities and accelerations. Indeed, we will now review some kinematic relationships for rigid bodies that will make much shorter work of this example.

Circular Cylindrical Coordinates

Figure 8.4 shows a coordinate system that is useful for a number of problems in particle kinematics. The x and y coordinates of the rectangular Cartesian system are replaced with the distance r and the angle ϕ, while the definition of the z-coordinate remains unchanged. Two of the unit vectors associated with the rectangular Cartesian system are also replaced with $\mathbf{e}_r = \cos\phi\,\mathbf{e}_x + \sin\phi\,\mathbf{e}_y$ and $\mathbf{e}_\phi = -\sin\phi\,\mathbf{e}_x + \cos\phi\,\mathbf{e}_y$.

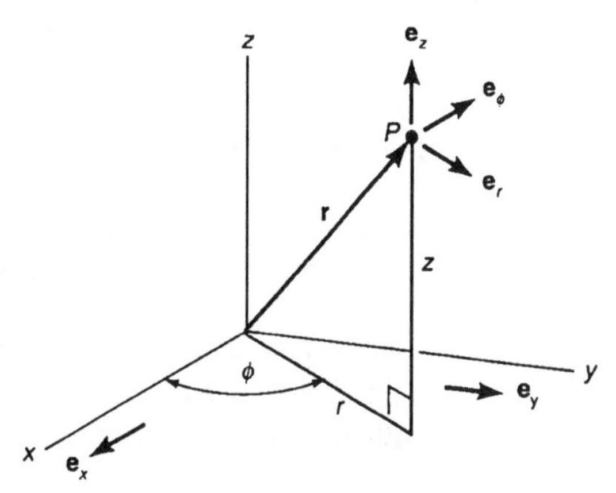

Figure 8.4

Since the angle ϕ varies, these two unit vectors also vary; their derivatives may be obtained by differentiating the above expressions:

$$\frac{d\mathbf{e}_r}{dt} = (-\mathbf{e}_x \sin\phi + \mathbf{e}_y \cos\phi)\frac{d\phi}{dt} = \dot{\phi}\mathbf{e}_\phi$$
$$\frac{d\mathbf{e}_\phi}{dt} = (-\mathbf{e}_x \cos\phi - \mathbf{e}_y \sin\phi)\frac{d\phi}{dt} = -\dot{\phi}\mathbf{e}_r$$

These are used along with the expression $\mathbf{r} = r\,\mathbf{e}_r + z\,\mathbf{e}_z$ for position to obtain expressions for velocity and acceleration:

$$\begin{aligned}\mathbf{v} = \dot{\mathbf{r}} &= \dot{r}\mathbf{e}_r + r\dot{\mathbf{e}}_r + \dot{z}\mathbf{e}_z \\ &= \dot{r}\mathbf{e}_r + r\dot{\phi}\mathbf{e}_\phi + \dot{z}\mathbf{e}_z\end{aligned} \tag{8.13}$$

$$\begin{aligned}\mathbf{a} = \dot{\mathbf{v}} &= \ddot{r}\mathbf{e}_r + \dot{r}\dot{\mathbf{e}}_r + (\dot{r}\dot{\phi} + r\ddot{\phi})\mathbf{e}_\phi + r\dot{\phi}\dot{\mathbf{e}}_\phi + \ddot{z}\mathbf{e}_z \\ &= (\ddot{r} - r\dot{\phi}^2)\mathbf{e}_r + (r\ddot{\phi} + 2\dot{r}\dot{\phi})\mathbf{e}_\phi + \ddot{z}\mathbf{e}_z\end{aligned} \tag{8.14}$$

Example 8.8

In Exhibit 8, the slider moves along the rod as it rotates about the fixed point *O*. At a particular instant, the slider is 200 mm from *O*, moving outward at 3 m/s relative to the rod; this relative speed is increasing at 130 m/s². At the same instant, the rod is rotating at a constant rate of 191 rpm. Evaluate the velocity and acceleration of the slider, and determine the rate of change of speed of the slider.

Exhibit 8

Solution

The angular speed of the rod is

$$\dot{\phi} = (191 \text{ rpm})\,\frac{2\pi \text{ rad/rev}}{60 \text{ s/min}} = 20.00 \text{ rad/s}$$

and its angular acceleration $\ddot{\phi}$ is zero. Other values to be substituted into Equations (8.13) and (8.14) are $r = 0.2$ m, $\dot{r} = 3$ m/s, and $\ddot{r} = 130$ m/s².

Substitution into Equations (8.13) and (8.14) leads directly to the following radial and transverse components of velocity and acceleration:

$$\mathbf{v} = (3\ \mathbf{e}_r + 4\ \mathbf{e}_\phi)\ \text{m/s}$$
$$\mathbf{a} = (50\ \mathbf{e}_r + 120\ \mathbf{e}_\phi)\ \text{m/s}^2$$

Now the radial and transverse components of the unit vector tangent to the path can be obtained by dividing the velocity vector by its magnitude:

$$\mathbf{e}_t = \frac{3\mathbf{e}_r + 4\mathbf{e}_\phi}{\sqrt{(3)^2 + (4)^2}} = 0.6\mathbf{e}_r + 0.8\mathbf{e}_\phi$$

The rate of change of speed is the projection of the acceleration vector onto the tangent to the path, which can be obtained by dot-multiplying the acceleration with the unit tangent vector:

$$\begin{aligned}\dot{v} = a_t &= \mathbf{e}_t \bullet \mathbf{a} \\ &= (0.6)(50\ \text{m/s}^2) + (0.8)(120\ \text{m/s}^2) \\ &= 126\ \text{m/s}^2\end{aligned}$$

Circular Path

When the path is circular, r is constant, and Equations (8.13) and (8.14) reduce to

$$\mathbf{v} = r\dot{\phi}\mathbf{e}_\phi$$
$$\mathbf{a} = -r\dot{\phi}^2\mathbf{e}_r + r\ddot{\phi}\mathbf{e}_\phi$$

Comparing these with Equations (8.2) and (8.6) (with $\rho = r$),

$$\mathbf{v} = v\mathbf{e}_t$$

$$\mathbf{a} = \dot{v}\mathbf{e}_t + \frac{v^2}{r}\mathbf{e}_n$$

we see that, for circular path motion, $\mathbf{e}_t = \mathbf{e}_\phi$, $\mathbf{e}_n = -\mathbf{e}_r$, and

$$v = r\dot{\phi} \tag{8.15}$$

$$a_n = r\dot{\phi}^2 \tag{8.16}$$

Example 8.9

A satellite is to be placed in a circular orbit over the equator at such an altitude that it makes one revolution around the earth per sidereal day (23.9345 hours). The gravitational acceleration is $(3.99 \times 10^{14}\ \text{m}^3/\text{s}^2)/r^2$, where r is the distance from the center of the earth. What is the altitude at which the satellite must be placed to achieve this period of orbit?

Solution

The angular speed of the line from the center of the earth to the satellite is

$$\dot{\phi} = \frac{2\pi \text{ rad}}{(23.9345 \text{ h})(3600 \text{ s/h})} = 7.292 \times 10^{-5} \text{ rad/s}$$

The acceleration has no tangential component, but the radial component in terms of the orbit radius and the angular speed will be $a_n = r(7.292 \times 10^{-5}\text{s}^{-1})^2$. This acceleration is imparted by the earth's gravitational attraction, so that $r(7.292 \times 10^{-5}\text{s}^{-1})^2 = (3.99 \times 10^{14} \text{ m}^3/\text{s}^2)/r^2$. This equation is readily solved for r, resulting in

$$r = \sqrt[3]{\frac{3.99 \times 10^{14} \text{m}^3/\text{s}^2}{(7.292 \times 10^{-5} \text{s}^{-1})^2}} = 42.2 \times 10^6 \text{ m}$$

The altitude will then be the difference between this value and the size of earth's radius, which is about 6.4×10^6 m: altitude = 35.8×10^6 m.

RIGID BODY KINEMATICS

The analysis of numerous mechanical systems rests on the assumption that the bodies making up the system are *rigid*. If the forces involved and the materials and geometry of the bodies are such that there is little deformation, the resulting predictions can be expected to be quite accurate.

The Constraint of Rigidity

If a body is **rigid**, the distance between each pair of points remains constant as the body moves. This constraint may be expressed in terms of a position vector $\mathbf{r}_{PQ}$ from a point P of the body to a point Q of the body, as indicated in Figure 8.5. If the magnitude of $\mathbf{r}_{PQ}$ is constant, then

$$\frac{d}{dt}|\mathbf{r}_{PQ}|^2 = \frac{d}{dt}(\mathbf{r}_{PQ} \bullet \mathbf{r}_{PQ}) = 2\mathbf{r}_{PQ} \bullet \frac{d\mathbf{r}_{PQ}}{dt} = 0 \qquad \textbf{(i)}$$

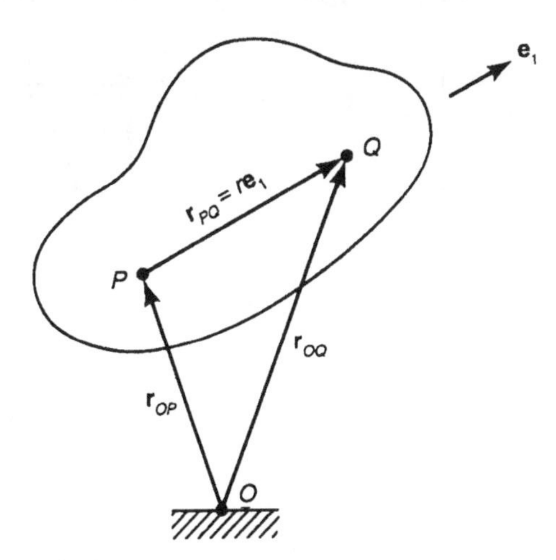

Figure 8.5

which indicates that $\dot{\mathbf{r}}_{PQ}$ is perpendicular to $\mathbf{r}_{PQ}$. Now, with a selected, fixed point designated as O, and vectors $\mathbf{r}_{OP}$ and $\mathbf{r}_{OQ}$ defined as indicated in Figure 8.5, differentiation of the vector relationship $\mathbf{r}_{PQ} = \mathbf{r}_{OQ} - \mathbf{r}_{OP}$ leads to the relationship

$$\frac{d\mathbf{r}_{PQ}}{dt} = \mathbf{v}_Q - \mathbf{v}_P \tag{ii}$$

in which $\mathbf{v}_P$ and $\mathbf{v}_Q$ designate the velocities of P and Q, respectively. Finally, if we define $\mathbf{e}_1$ to be the unit vector in the direction of $\mathbf{r}_{PQ}$, so that

$$\mathbf{r}_{PQ} = r\mathbf{e}_1 \tag{iii}$$

then substitution of (ii) and (iii) into (i) leads to

$$2r\mathbf{e}_1 \bullet (\mathbf{v}_Q - \mathbf{v}_P) = 0$$

or

$$\mathbf{e}_1 \bullet \mathbf{v}_Q = \mathbf{e}_1 \bullet \mathbf{v}_P \tag{8.17}$$

This shows that *the projections of the velocities of any two points of a rigid body onto the line connecting the two points must be equal.* This is intuitively plausible; otherwise the distance between the points would be changing. This frequently provides the most direct way of evaluating the velocities of various points within a mechanism.

Example 8.10

As the crank OQ in Exhibit 9 rotates clockwise at 200 rad/s, the piston P moves vertically. What will be the velocity of the piston at the instant when the angle θ is 50 degrees?

Solution

Since point Q must follow a circular path, its speed may be determined from Equation (8.15): $v_Q = (0.075\text{ m})(200\text{ s}^{-1}) = 15$ m/s, with the direction of $\mathbf{v}_Q$ as indicated in the figure. Because the cylinder wall constrains the piston, its velocity is vertical. The connecting rod PQ is rigid, so the velocities of the points P and Q must

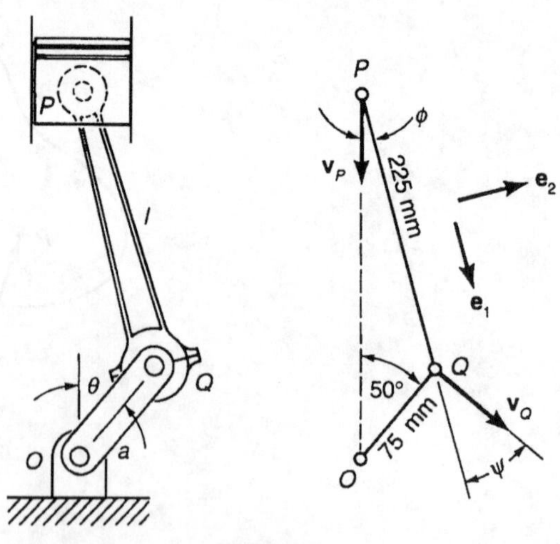

Exhibit 9

satisfy $v_P \cos\phi = v_Q \cos\psi$. The trigonometric rule of sines, applied to the triangle *OPQ*, gives

$$\sin\phi = \frac{a}{l}\sin\theta = \frac{75}{225}\sin 50^\circ$$

which yields $\phi = 14.8^\circ$. The other required angle is then $\psi = 90^\circ - \theta - \phi = 25.2^\circ$. Once these angles are determined, the constraint equation yields the speed of the piston:

$$v_P = \frac{\cos\psi}{\cos\phi} v_Q = 14.04 \text{ m/s}$$

The Angular Velocity Vector

If a rigid body is in *plane motion*, that is, if the velocities of all points of the body lie in a fixed plane, then its orientation may be specified by the angle θ between two fixed lines, one of which passes through the body, as indicated in Figure 8.6. The rate of change of this angle is central to the analysis of the velocities of various points of the body.

To determine this relationship, consider Figure 8.7, which shows a position vector from the point P to point Q, both fixed in the moving body. Two configurations are shown, one at time t and another after an arbitrary change during a time increment Δt. $\mathbf{e}_1$ is defined to be the unit vector in the direction of $\mathbf{r}_{PQ}(t)$, and $\mathbf{e}_2$ is

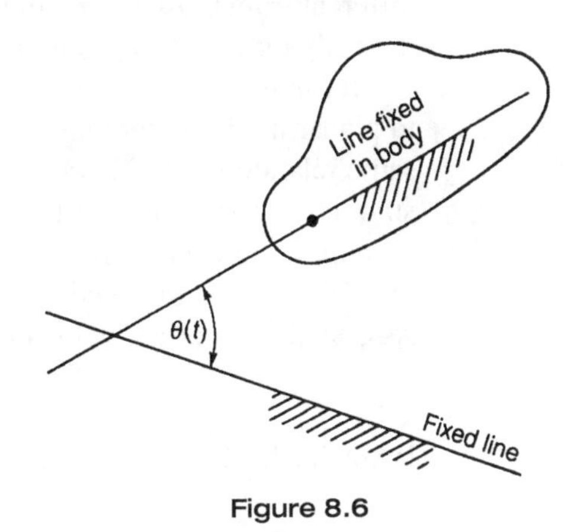

Figure 8.6

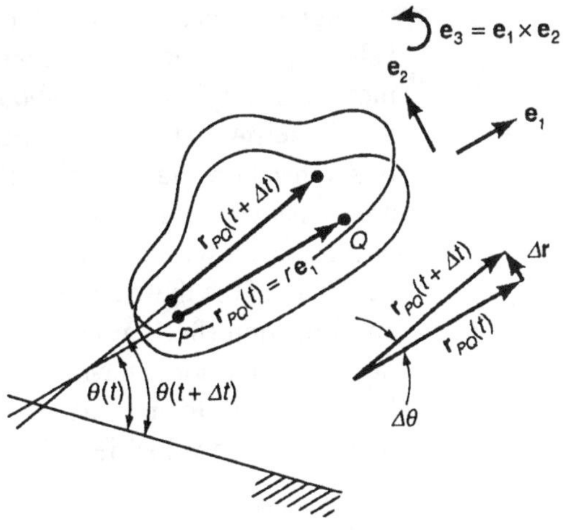

Figure 8.7

defined to be the unit vector of P and Q, 90° counterclockwise from $\mathbf{e}_1$. Both $\mathbf{e}_1$ and $\mathbf{e}_2$ are further assumed to lie in the plane of motion; this assumption is convenient but not limiting. The vector diagram in Figure 8.7 shows the change in $\mathbf{r}_{PQ}$ to be given by the approximation $\Delta\mathbf{r} \approx r\Delta\mathbf{e}_2$. Dividing both sides by the time increment Δt and letting this increment approach zero leads to the relation

$$\frac{d\mathbf{r}_{PQ}}{dt} = r\dot{\theta}\mathbf{e}_2$$

The scalar $\dot{\theta}$ will be denoted also by ω. Note from the definition of θ that a positive value of ω indicates a counterclockwise rotation, whereas a negative value of ω indicates a clockwise rotation. Another useful form of this relation may be written in terms of the **angular velocity vector,**

$$\boldsymbol{\omega} = \omega\,\mathbf{e}_3$$

where $\mathbf{e}_3$ is defined to be $\mathbf{e}_1 \times \mathbf{e}_2$, oriented perpendicular to the plane of Figure 8.7. With this definition,

$$\frac{d\mathbf{r}_{PQ}}{dt} = \boldsymbol{\omega} \times \mathbf{r}_{PQ}$$

Note that this relation is valid for *any* two points in the body. (A pair of points different from those shown in Figure 8.7 might give rise to a different angle, but as the body moves, *changes* in this angle would equal *changes* in θ.)

It can be shown that for the most general motion of a rigid body (not restricted to planar motion) there also exists a unique angular velocity vector for which the same relation holds. However, in nonplanar motion, the angular velocity $\boldsymbol{\omega}$ is not straightforwardly related to the rate of change of an angle, and its calculation requires a more extensive analysis than in the case of planar motion.

When $\dot{\mathbf{r}}_{PQ}$ is replaced with $\mathbf{v}_Q - \mathbf{v}_P$ according to Equation (8.ii) of the previous section, the important velocity relationship

$$\mathbf{v}_Q = \mathbf{v}_P + \boldsymbol{\omega} \times \mathbf{r}_{PQ} \qquad \textbf{(8.20)}$$

is obtained, which, for planar motion, becomes

$$\mathbf{v}_Q = \mathbf{v}_P + r\omega\mathbf{e}_2 \qquad \textbf{(8.21)}$$

In the special case in which $\omega = 0$, this indicates that all points have the same velocity, a motion called **translation**. In the special case in which $\mathbf{v}_P = \mathbf{0}$, the motion is simply rotation about a fixed axis through P. Thus, in the general case, the two terms on the right of Equation (8.20) can be seen to express a superposition of a translation and a rotation about P. But since P can be selected *arbitrarily*, there are as many combinations of a translation and a corresponding "center of rotation" as the analyst wishes to consider!

In all of the these cases, the angular velocity is a property of the *body's* motion, and Equation (8.21) relates the velocities of *any* two points of a body experiencing planar motion. Dot-multiplication of each member of Equation (8.21) with $\mathbf{e}_2$ leads to the following means of evaluating the angular velocity of a plane motion in terms of the velocities of two points:

$$\omega = \frac{\mathbf{e}_2 \bullet \mathbf{v}_Q - \mathbf{e}_2 \bullet \mathbf{v}_P}{r} \qquad \textbf{(8.22)}$$

That is, ω will be the difference between the magnitudes of the projections of the velocities of P and Q onto the perpendicular to the line connecting P and Q, divided by the distance between P and Q.

Example **8.11**

What will be the angular velocity of the connecting rod in Example 8.10, at the instant when the angle θ is 50 degrees?

Solution

Referring to Exhibit 9 for the definition of $\mathbf{e}_2$, we see that

$$\omega = \frac{v_Q \sin\psi + v_Q \sin\phi}{l}$$

$$= \frac{(15 \text{ m/s})\sin 25.2° + (14.04 \text{ m/s})\sin 14.8°}{0.225 \text{ m}} = 44.3 \text{ rad/s}$$

The positive value indicates that the rotation is counterclockwise at this instant.

Instantaneous Center of Zero Velocity

For planar motion with $\omega \neq 0$, there always exists a point C' of the body (or an imagined extension of the body) that has zero velocity. If point P of Equation (8.21) is selected to be this special point, the equation reduces to $\mathbf{v}_Q = \mathbf{v}_{C'} + \omega \times \mathbf{r}_{C'Q} = r\omega\mathbf{e}_2$ where r is now the distance from C' to Q and $\mathbf{e}_2$ is perpendicular to the line connecting C' and Q. This latter property can be used to locate C' if the directions of the velocities of two points of the body are known.

Example **8.12**

What is the location of the instantaneous center C' of the connecting rod in Examples 8.10 and 8.11? Use this to verify the previously determined values of the angular velocity of the connecting rod and the velocity of point P.

Solution

The velocity of any point of the connecting rod must be perpendicular to the line from C' to that point. Hence C' must lie at the point of intersection of the horizontal line through P and the line through Q perpendicular to $\mathbf{v}_Q$ (i.e., on the line through O and Q), as shown in Exhibit 10. The pertinent distances can be found as follows:

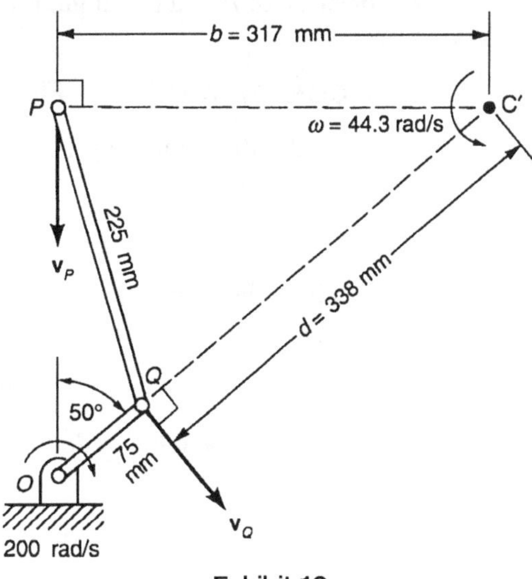

Exhibit 10

$$OP = (75 \text{ mm}) \cos 50° + (225 \text{ mm}) \cos 14.8° = 266 \text{ mm}$$
$$PC' = OP \tan 50° = 317 \text{ mm}$$
$$QC' = OP \sec 50° - 75 \text{ mm} = 338 \text{ mm}$$

The angular velocity of the connecting rod is then

$$\omega = \frac{v_Q}{QC'} = \frac{15 \text{ m/s}}{0.339 \text{ m}} = 44.3 \text{ rad/s}$$

and the velocity of P is then

$$v_P = PC'\,\omega = (0.317 \text{ m})(44.3 \text{ s}^{-1}) = 14.04 \text{ m/s}$$

in agreement with values the previously obtained.

Example 8.13

Using the properties of the instantaneous center, determine the velocity of the point P on the rim of the rolling wheel in Example 8.7.

Solution

Since the wheel rolls without slipping, the point of the wheel in contact with the flat surface has zero velocity and is therefore its instantaneous center. The angular speed of the wheel is $\dot{\theta}$, and the distance from C' to P is readily determined from Exhibit 11:

$$r = 2b \sin\frac{\theta}{2}$$

The velocity of point P then has the magnitude

$$v_P = r\omega = 2b \sin\frac{\theta}{2}\dot{\theta}$$

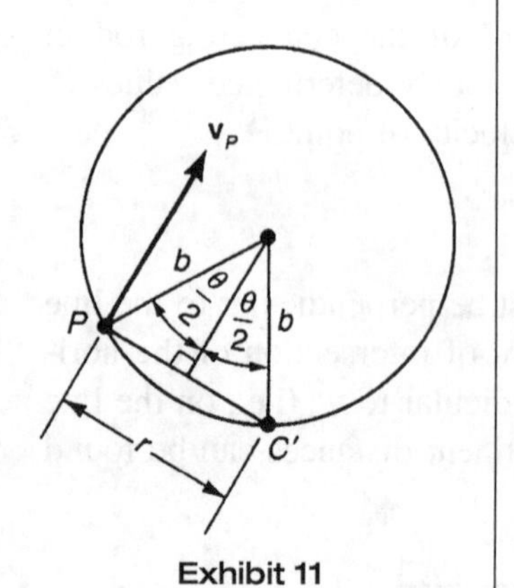

Exhibit 11

and the direction shown in Exhibit 11. This direction should be evident by inspection once it is realized that a positive $\dot{\theta}$ corresponds to clockwise rotation. The reader may find it instructive to recall the conventions for the choice of $\mathbf{e}_2$ and positive ω used in the derivation leading to Equation (8.21) and verify the agreement. Note the simplicity of this analysis as compared with the one expressing the position of P in a rectangular Cartesian coordinate system.

Accelerations in Rigid Bodies

Formally differentiating Equation (8.20) and substituting for $\dot{\mathbf{r}}_{PQ}$ using Equation (8.19) leads to

$$\mathbf{a}_Q = \mathbf{a}_P + (\alpha \times \mathbf{r}_{PQ}) + \omega \times (\omega \times \mathbf{r}_{PQ}) \quad \textbf{(8.23)}$$

in which the vector $\alpha = d\omega/dt$ is called the **angular acceleration** of the body. For planar motion, $\alpha = \alpha\mathbf{e}_3 = \dot{\omega}\mathbf{e}_3$ and $\omega \times (\omega \times \mathbf{r}) = -\omega^2\mathbf{r}$ so that

$$\mathbf{a}_Q = \mathbf{a}_P + r\alpha\mathbf{e}_2 - r\omega^2\mathbf{e}_1 \quad \textbf{(8.24)}$$

where $\mathbf{e}_1$ and $\mathbf{e}_2$ are defined as indicated in Figure 8.7.

Equivalent relationships, analogous to Equation (8.17) and Equation (8.22) for velocity, can be obtained by dot-multiplying this equation by $\mathbf{e}_1$ and by $\mathbf{e}_2$:

$$\mathbf{e}_1 \bullet \mathbf{a}_Q = \mathbf{e}_1 \bullet \mathbf{a}_P - r\omega^2 \tag{8.25}$$

$$\alpha = \frac{\mathbf{e}_2 \bullet \mathbf{a}_Q - \mathbf{e}_2 \bullet \mathbf{a}_P}{r} \tag{8.26}$$

Example 8.14

If the speed of the crank in Examples 8.8 through 8.10 is constant, what are the acceleration $\mathbf{a}_P$ of the piston and the angular acceleration α of the connecting rod at the instant when the angle θ is 50 degrees (Exhibit 12)?

Solution

When the crank speed is constant, the acceleration of Q is entirely centripetal, of magnitude

$$a_Q = r\omega^2$$

$$a_Q = (0.075\text{ m})(200\text{ s}^{-1})^2 = 3000\text{ m/s}^2$$

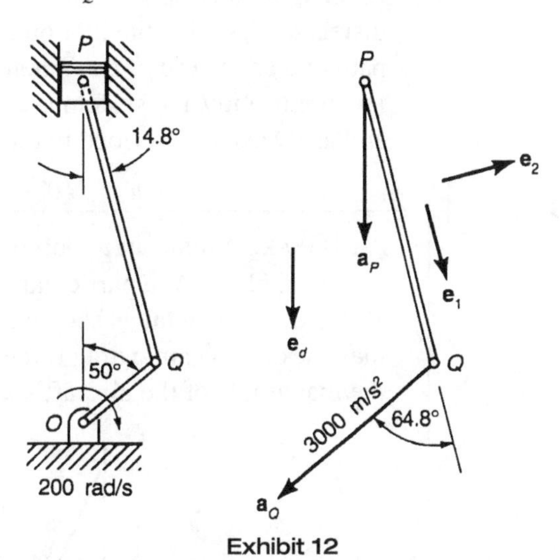

Exhibit 12

and directed toward the center of curvature O of the path of Q. The acceleration of P is vertically upward or downward. To determine the direction, we define a downward unit vector $\mathbf{e}_d$ and let $\mathbf{a}_p = a_p\mathbf{e}_d$ (see Exhibit 12). A positive value of a_P then indicates a downward acceleration and a negative value an upward acceleration. These expressions for a_Q and $\mathbf{a}_P$ are substituted into Equation (8.25), along with the previously determined angular velocity of the rod, giving

$$(3000\text{ m/s}^2)\cos 64.8° = a_P \cos 14.8° - (0.225\text{ m})(44.3\text{ s}^{-1})^2$$

which yields

$$a_P = 1779\text{ m/s}^2$$

The angular acceleration α of the rod can then be determined from Equation (8.26):

$$\alpha = \frac{(3000\text{ m/s}^2)\cos 154.8° - (1779\text{ m/s}^2)\cos 104.8°}{0.225\text{ m}} = -10{,}050\text{ rad/s}^2$$

The negative value indicates that the angular acceleration is clockwise; that is, the 44.3-rad/s counterclockwise angular velocity is rapidly decreasing at this instant.

NEWTON'S LAWS OF MOTION

Every element of a mechanical system must satisfy Newton's second law of motion; that is, the resultant force **f** acting on the element is related to the acceleration **a** of the element by

$$\mathbf{f} = m\mathbf{a} \tag{8.27}$$

in which m represents the mass of the element. Newton's third law requires that the force exerted on a body A by a body B is of equal magnitude and opposite direction to the force exerted on body B by body A. These laws and their logical consequences provide the basis for relating motions to the forces that cause them.

Applications to a Particle

A **particle** is an idealization of a material element in which its spatial extent is disregarded, so that the motion of all of its parts is completely characterized by the path of a geometric *point*. When the accelerations of various parts of a system differ significantly, the system is considered to be composed of a number of particles and analyzed as described in the next section.

Example 8.15

An 1800-kg aircraft in a loop maneuver follows a circular path of radius 3 km in a vertical plane. At a particular instant, its velocity is 210 m/s directed 25 degrees above the horizontal as shown in Exhibit 13. If the engine thrust is 16 kN greater than the aerodynamic drag force, what is the rate of change of the aircraft's speed, the magnitude of the aircraft's acceleration, and the aerodynamic lift force?

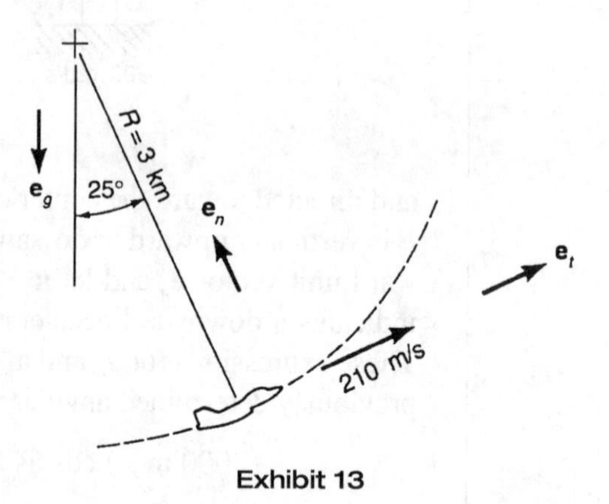

Exhibit 13

Solution

Since the dimensions of the aircraft are small compared with the radius of the path, all of its material elements can be considered to have essentially the same motion, so treating the aircraft as a particle as described above is reasonable.

The forces acting on the aircraft are shown on the free-body diagram, Exhibit 14. The thrust **T**, the drag **D**, and the lift **L** all result from aerodynamic

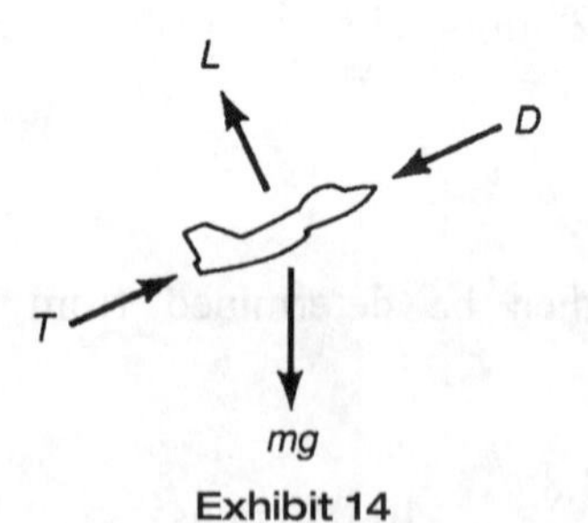

Exhibit 14

pressure from the surrounding air and engine gas. The lift is defined to be the component of the total force that is perpendicular to the flight path, and arises primarily from the wings. The force of gravity, mg, is the only other force arising from a source external to the free body. The left-hand side of Equation (8.27) is the resultant of these forces, whereas the right-hand side is obtained from Equation (8.6). Thus, Newton's second law is written in this case as

$$(T - D)\mathbf{e}_t + L\mathbf{e}_n + mg\mathbf{e}_g = m\left(\dot{v}\mathbf{e}_t + \frac{v^2}{R}\mathbf{e}_n\right)$$

Two independent equations arise from this two-dimensional vector equation. Dot multiplication with $\mathbf{e}_t$ yields $(T - D) + mg\mathbf{e}_t \bullet \mathbf{e}_g = m\dot{v}$ because $\mathbf{e}_t$ and $\mathbf{e}_n$ are always perpendicular vectors and thus their dot product is 0. This equation, rearranged, leads to the rate of change of speed:

$$\dot{v} = \frac{T - D}{m} - g\sin 25° = \frac{16{,}000 \text{ N}}{1800 \text{ kg}} - (9.81 \text{ m/s}^2)\sin 25° = 4.74 \text{ m/s}^2$$

Dot multiplication with $\mathbf{e}_n$ yields

$$L + mg\mathbf{e}_n \bullet \mathbf{e}_g = \frac{mv^2}{R}$$

which then allows us to determine the magnitude of lift force,

$$L = m\left(g\cos 25° + \frac{v^2}{R}\right) = (1800 \text{ kg})\left[(9.81 \text{ m/s}^2)\cos 25° + \frac{(210 \text{ m/s})^2}{3000 \text{ m}}\right] = 42.5 \text{ kN}$$

The magnitude of the acceleration is then determined by combining the tangential and normal components found above:

$$|\mathbf{a}| = \sqrt{(4.74 \text{ m/s}^2)^2 + \left[\frac{(210 \text{ m/s})^2}{3000 \text{ m}}\right]^2} = 15.45 \text{ m/s}^2$$

Example 8.16

Two blocks are interconnected by an inextensible, massless line through the pulley arrangement shown in Exhibit 15. The inertia and friction of the pulleys are negligible. The coefficient of friction between the block of mass m_1 and the horizontal surface is μ. What is the acceleration of the block of mass m_2 as it moves downward?

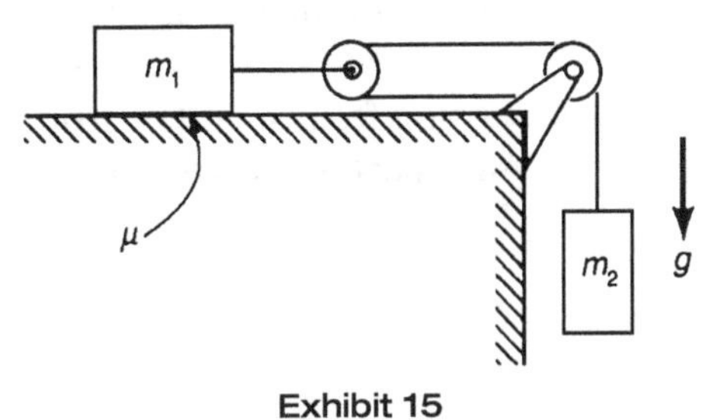

Exhibit 15

Solution

Since the motion of each block is a translation, the acceleration of each element in a block is the same; hence their spatial extensions may be ignored and each block may be treated as a particle.

Moment equilibrium of each pulley requires that the tension be the same in each part of the longer line around the pulleys. Denoting this tension by T, the two free-body diagrams in Exhibit 16 are used to write expressions for Newton's second law. Since the acceleration of the block on the left has no vertical component, $R_1 - m_1 g = 0$. Denoting its rightward acceleration by a_1, consideration of the horizontal forces leads to $2T - \mu R_1 = m_1 a_1$. Denoting the downward acceleration of the other block by a_2, application of Newton's second law to the other free body yields $m_2 g - T = m_2 a_2$. To relate the two accelerations, consider the pulley connected to the horizontal block. The instantaneous center of this pulley is at its rim and directly below its center. Thus, the speed of the pulley's upper rim is twice that of its center, so that the speed of the vertically moving block is twice that of the horizontally moving block. Since this is true at all times, we have $a_2 = 2a_1$.

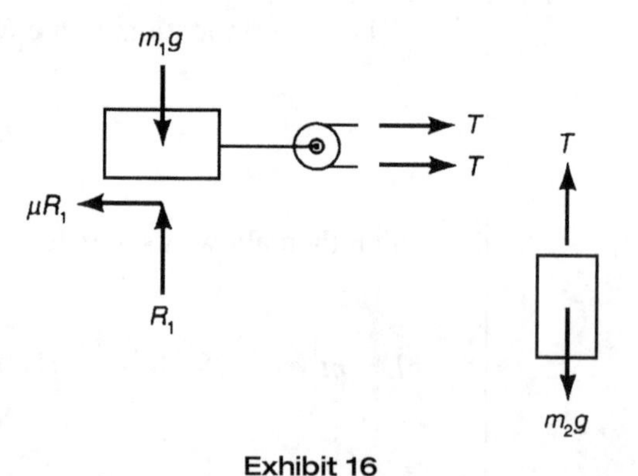

Exhibit 16

Eliminating R, T, and a_1 from these four equations now leads to

$$a_2 = \frac{4m_2 - 2\mu m_1}{4m_2 + m_1} g$$

Observe that if $m_2 > \mu m_1/2$, the acceleration is downward, whereas if $m_2 < \mu m_1/2$, the acceleration is upward, implying that the downward velocity will reach zero, after which time the friction force will no longer equal μR_1.

Systems of Particles

A mechanical **system** is any collection of material elements of fixed identity whose motion we may wish to consider. Such a system is treated as a collection of particles in which the individual particles must obey Newton's laws of motion.

It proves to be very useful to separate the forces acting on a system into those arising from sources outside the system and arising from the interaction between members of the system, as shown in Figure 8.8. That is, the resultant force on the ith particle is written as

$$\mathbf{f}_i = \mathbf{f}_{ie} + \sum_j \mathbf{f}_{i/j}$$

in which $\mathbf{f}_{ie}$ represents the resultant of all forces on the ith particle arising from sources external to the system, and $\mathbf{f}_{i/j}$ represents the force exerted on ith particle by the jth particle. With this notation, Newton's third law may be expressed as $\mathbf{f}_{j/i} = -\mathbf{f}_{i/j}$.

Now, each particle moves according to Newton's second law:

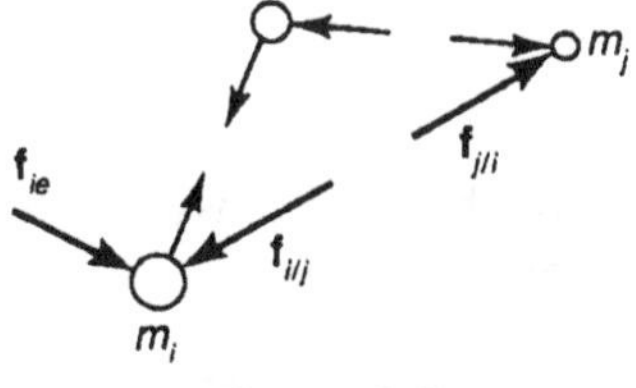

Figure 8.8

$$\mathbf{f}_{ie} + \sum_j \mathbf{f}_{i/j}$$

in which m_i denotes the mass of the ith particle and $\mathbf{a}_i$ its acceleration. There are as many such equations as there are particles in the system; if all such equations are added, the result is

$$\sum_i \mathbf{f}_{ie} + \sum_i \sum_j \mathbf{f}_{i/j} = \sum_i m_i \mathbf{a}_i$$

In view of Newton's third law, the internal forces can be grouped as pairs of oppositely directed forces of equal magnitude, and so their sum vanishes, leaving

$$\sum_i \mathbf{f}_{ie} = \sum_i m_i \mathbf{a}_i \quad \textbf{(8.28a)}$$

That is, the resultant of all *external* forces is equal to the sum of the products of the individual masses and their corresponding accelerations.

Linear Momentum and Center of Mass

The right-hand member of Equation (8.28a) can be expressed alternatively in terms of the **linear momentum** of the system, which is defined as

$$\mathbf{p} = \sum_i m_i \mathbf{v}_i \quad \textbf{(8.29)}$$

in which $\mathbf{v}_i$ denotes the velocity of the ith particle. Differentiation of this equation results in

$$\frac{d\mathbf{p}}{dt} = \sum_i m_i \mathbf{a}_i$$

which is the same expression appearing in Equation (8.28a). Hence an alternative to Equation (8.28a) is

$$\sum_i \mathbf{f}_{ie} = \frac{d\mathbf{p}}{dt} \quad \textbf{(8.28b)}$$

which states that the sum of the external forces is equal to the time rate of change of the linear momentum of the system.

The **center of mass** of the system is a point C located, relative to an arbitrarily selected reference point O, by the position vector $\mathbf{r}_C$, which satisfies the defining equation

$$m\mathbf{r}_C = \sum_i m_i \mathbf{r}_i \quad \textbf{(8.30)}$$

in which m denotes the total mass of the system and $\mathbf{r}_i$ is a position vector from O to the ith particle. Differentiation of this equation leads to

$$m\mathbf{v}_C = \sum_i m_i \mathbf{v}_i \tag{8.31}$$

which shows that the linear momentum is the product of the total mass and the velocity of the center of mass. Another differentiation yields

$$m\mathbf{a}_C = \sum_i m_i \mathbf{a}_i$$

which provides still another way of expressing Equation (8.28a):

$$\sum_i \mathbf{f}_{ie} = m\mathbf{a}_C \tag{8.28c}$$

This is sometimes called the **principle of motion of the mass center**. It indicates that the mass center responds to the resultant of external forces exactly as would a single particle having a mass equal to the total mass of the system.

Example **8.17**

A motor inside the case shown in Exhibit 17 drives the eccentric rotor at a constant angular speed ω. The distance from the rotor bearing to its center of mass is e, the mass of the rotor is m_r, and the mass of the nonrotating housing is $m - m_r$. (That is, the total mass of the rotor and housing together is m.) The housing is free to translate horizontally, constrained by the rollers, and under the influence of a spring of stiffness k and a dashpot that transmits a force to the housing of magnitude c times the speed of the housing in the direction opposite to that of the velocity of the housing. Write the differential equation that governs the extension $x(t)$ of the spring from its relaxed position.

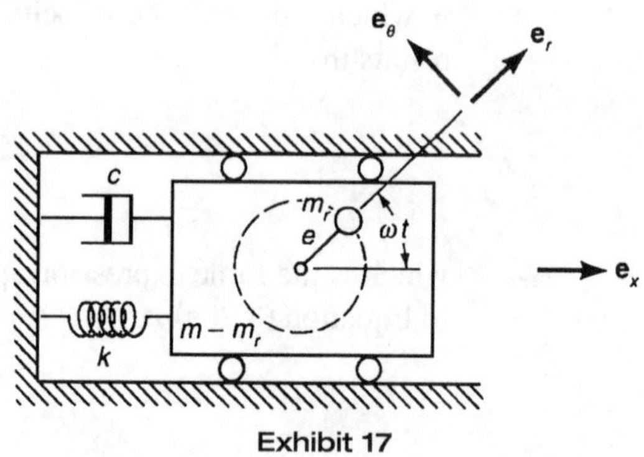

Exhibit 17

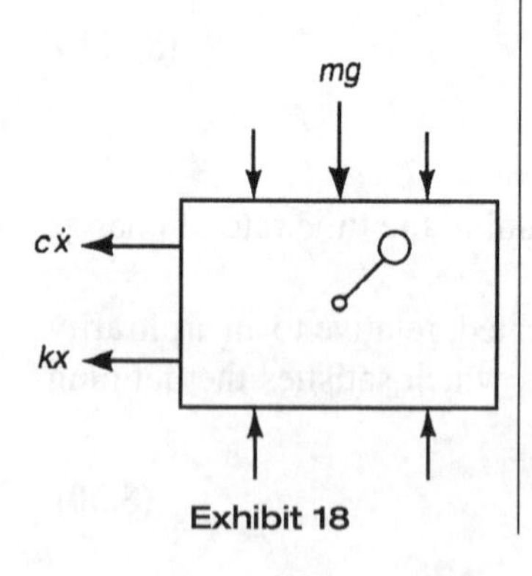

Exhibit 18

Solution

Consider the system consisting of the housing and rotor together. The free-body diagram (Exhibit 18) shows forces acting on this system from sources *external* to it. It does *not* include the torque necessary to maintain constant rotor speed nor the reaction at the bearing, these being internal, action-reaction pairs.

Note that when x is positive (the spring extended), the force exerted by the spring on the housing acts to the left, and when x is negative (the spring compressed), this force acts to the right. Both situations are depicted properly by the label kx on the arrow; that is, this indicates that the force equals $-kx\mathbf{e}_x$ in all cases. The same consideration applies to the arrow and label representing the force from the dashpot. The sum of all external forces then is

$$\sum_i \mathbf{f}_{ie} = -(kx + c\dot{x})\mathbf{e}_x + f_y\mathbf{e}_y$$

The acceleration of the housing is simply $\ddot{x}\mathbf{e}_x$, while that of the mass center of the rotor can be most readily determined by using Equation (8.24), letting P be the center of the bearing and noting that $\alpha = 0$. This gives the acceleration of the mass center of the rotor as

$$\mathbf{a} = \ddot{x}\mathbf{e}_x - e\omega^2\mathbf{e}_r$$

Substitution into Equation (8.28a) results in

$$-(kx + c\dot{x})\mathbf{e}_x + f_y\mathbf{e}_y = (m - m_r)\ddot{x}\mathbf{e}_x + m_r(\ddot{x}\mathbf{e}_x - e\omega^2\mathbf{e}_r)$$

The forces f_y are neither known nor of interest for our purpose; they may be eliminated from the equation by dot-multiplying each member with $\mathbf{e}_x$, which leads to

$$-(kx + c\dot{x}) = m\ddot{x} - m_r e\omega^2 \cos\omega t$$

A "standard" form of this equation is obtained by placing the dependent variable and its derivatives on one side and the known function of time on the other:

$$m\ddot{x} + c\dot{x} + kx = m_r e\omega^2 \cos\omega t$$

The same result can be obtained using either Equation (8.28b) or (8.28c).

Impulse and Momentum

The integral with respect to time of the resultant of external forces is called the **impulse** of this resultant force:

$$\mathbf{g} = \int_{t_1}^{t_2} \sum_i \mathbf{f}_{ie}\, dt$$

Integration of both members of Equation (8.28b) results in the following integrated form:

$$\mathbf{g} = \mathbf{p}(t_2) - \mathbf{p}(t_1) \qquad \textbf{(8.32)}$$

This states that impulse of the resultant external force is equal to the change in momentum of the system. Since it is a vector equation, we may obtain up to three independent relationships from it, one for each of three dimensions.

A special case occasionally arises, in which one or more components of the impulse are absent. The corresponding components of momentum then remain constant, and are said to be **conserved**.

Example 8.18

A rocket is simulated by a vehicle that is accelerated by the action of the passenger throwing rocks in the rearward direction as the vehicle moves along the roadway as shown in Exhibit 19. At a certain time, the mass of the vehicle, passenger, and supply of rocks is m, and all are moving at speed v. The passenger then launches a rock of mass m_1 with a rearward velocity of magnitude v_e *relative to the vehicle.* What is the increase Δv in the speed of the vehicle resulting from this action?

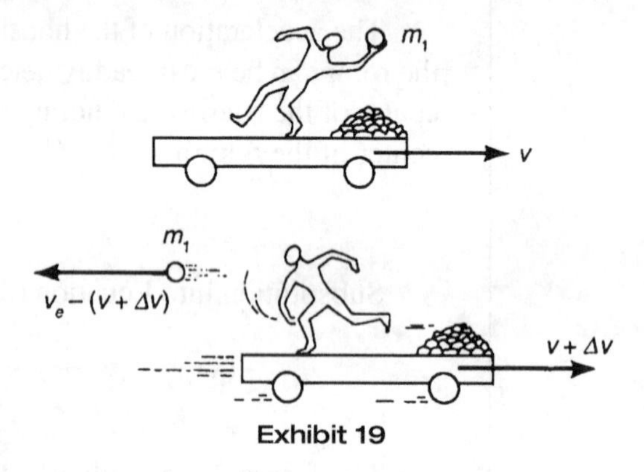

Exhibit 19

Solution

Assuming there is negligible friction at the wheels, the system consisting of passenger, vehicle, and rocks has no external forces acting on it in the direction of travel. The horizontal component of momentum is therefore conserved; that is, it is the same after the rock is launched as it was prior to the launching. After this is written in detail as

$$\mathbf{p}_{\text{final}} - \mathbf{p}_{\text{rock}} = \mathbf{p}_{\text{initial}}$$

$$(m - m_1)(v + \Delta v) - m_1[v_e - (v + \Delta v)] = mv$$

the equation can then be solved for the increase in vehicle speed:

$$\Delta v = \frac{m_1}{m} v_e$$

Moments of Force and Momentum

Equations (8.28) and (8.29) are valid regardless of the lines of action of the forces $\mathbf{f}_{ie}$. For example, the acceleration of the center of mass and the change in momentum of the system shown in Figure 8.9 will be the same for each of the different points of application of the force. However, there are characteristics of the motions induced by these forces that *do* depend on the lines of action, some of which are revealed by considering *moments* of forces.

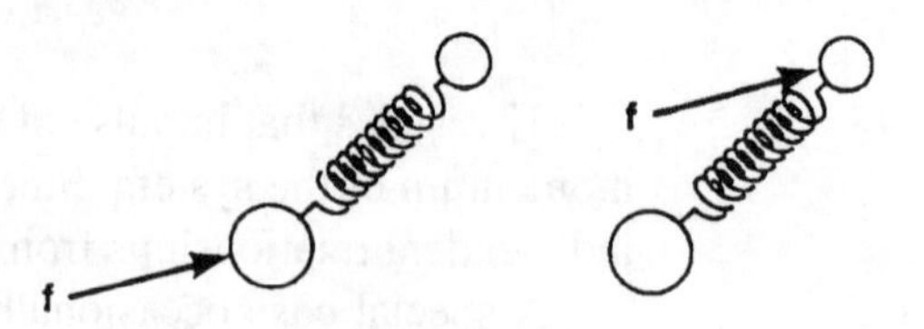

Figure 8.9

Consider again the external and internal forces acting on two typical particles of a system (Figure 8.10). Let the position vectors $\mathbf{r}_i$ and $\mathbf{r}_j$ locate the ith and jth particles, respectively, with respect to a selected point O. If the equation expressing Newton's second law for the ith particle is cross-multiplied by $\mathbf{r}_i$ and all such equations are added, the result is

$$\sum_i \mathbf{r}_i \times \mathbf{f}_{ie} + \sum_i \sum_j \mathbf{f}_{i/j} = \sum_i \mathbf{r}_i \times m_i \mathbf{a}_i$$

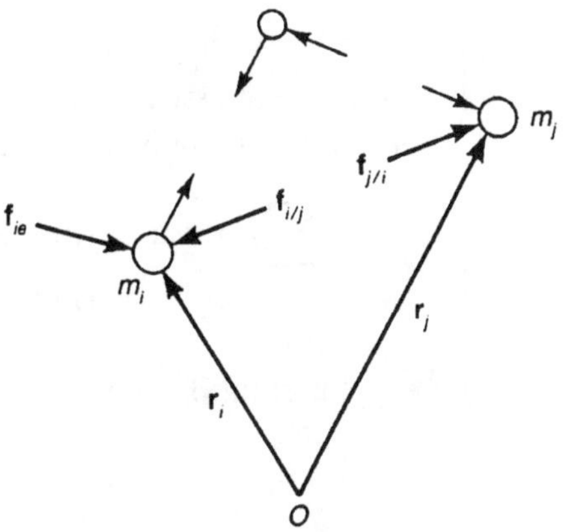

Figure 8.10

Now, if the forces $\mathbf{f}_{i/j}$ and $\mathbf{f}_{j/i} = -\mathbf{f}_{i/j}$ have a common line of action, then

$$\mathbf{r}_i \times \mathbf{f}_{i/j} + \mathbf{r}_j \times \mathbf{f}_{j/i} = 0$$

That is, the **moments** of the members of each action-reaction pair cancel one another, leaving the **moment equation** for the system:

$$\mathbf{M}_O = \sum_i \mathbf{r}_i \times m_i \mathbf{a}_i \qquad \textbf{(8.33a)}$$

in which the moment of the external forces is evaluated as in the previous chapter:

$$\mathbf{M}_O = \sum_i \mathbf{r}_i \times \mathbf{f}_{ie}$$

Example 8.19

The pendulum in Exhibit 20 consists of a stiff rod of negligible mass with two masses attached, and it swings in the vertical plane about the frictionless hinge at O under the influence of gravity. What will be the angular acceleration of the pendulum in terms of angular displacement θ and the other parameters indicated in the sketch?

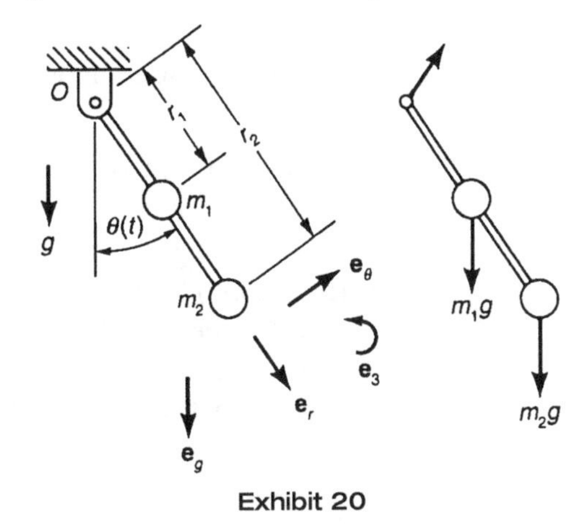

Exhibit 20

Solution

The free-body diagram shows the forces external to the system consisting of the rod together with the two particles. Since the reaction at the support is unknown and is of no interest for our purpose, a good strategy would be to consider moments about this point. Referring to the free-body diagram, we evaluate the resultant moment as usual:

$$\mathbf{M}_O = (r_i\mathbf{e}_r)\times(m_1 g\mathbf{e}_g) + (r_2\mathbf{e}_r)\times(m_2 g\mathbf{e}_g) = -(m_1 r_1 + m_2 r_2) g \sin\theta \mathbf{e}_3$$

Since each particle follows a circular path with center at O, their accelerations may be expressed as $\mathbf{a}_i = r_i\dot{\theta}^2\mathbf{e}_r + r_i\ddot{\theta}\mathbf{e}_\theta$. The right-hand member of the moment law, Equation (8.33a), is then evaluated in this case as

$$\begin{aligned}\sum \mathbf{r}_i \times m_i\mathbf{a}_i &= (r_1\mathbf{e}_r)\times m_1 r_1(-\dot{\theta}^2\mathbf{e}_r + \ddot{\theta}\mathbf{e}_\theta) + (r_2\mathbf{e}_r)\times m_2 r_2(-\dot{\theta}^2\mathbf{e}_r + \ddot{\theta}\mathbf{e}_\theta)\\ &= (m_1 r_1^2 + m_2 r_2^2)\ddot{\theta}\mathbf{e}_3\end{aligned}$$

Substitution into the moment law, Equation (8.33a), results in

$$-(m_1 r_1 + m_2 r_2) g \sin\theta\, \mathbf{e}_3 = (m_1 r_1^2 + m_2 r_2^2)\ddot{\theta}\, \mathbf{e}_3$$

or

$$\ddot{\theta} = -\frac{m_1 r_1 + m_2 r_2}{m_1 r_1^2 + m_2 r_2^2} g \sin\theta$$

The **moment of momentum** or **angular momentum about point** O is defined as

$$\mathbf{H}_O = \sum_i \mathbf{r}_i \times m_i\mathbf{v}_i \tag{8.34}$$

Now, if O is fixed in the inertial frame, then $\mathbf{v}_i = \mathbf{r}_i$, and it follows that

$$\frac{d\mathbf{H}_O}{dt} = \sum_i (\dot{\mathbf{r}}_i \times m_i\mathbf{v}_i + \mathbf{r}_i \times m_i\dot{\mathbf{v}}_i) = \sum_i \mathbf{r}_i \times m_i\mathbf{a}_i$$

This provides an alternative way of writing the moment law, as

$$\mathbf{M}_O = \frac{d\mathbf{H}_O}{dt} \tag{8.33b}$$

In the preceding example, the angular momentum about O is

$$\mathbf{H}_O = (r_1\mathbf{e}_r)\times(m_1 r_1\dot{\theta}\mathbf{e}_\theta) + (r_2\mathbf{e}_r)\times(m_2 r_2\dot{\theta}\mathbf{e}_\theta) = \left(m_1 r_1^2 + m_2 r_2^2\right)\dot{\theta}\mathbf{e}_3$$

Differentiating this expression and substituting for the right-hand member of Equation (8.33b) leads to the result achieved using Equation (8.33a).

As with forces and linear momentum, there are situations in which the moment of external forces vanishes. Then, Equation (8.33b) implies that the angular momentum about O remains constant, or is *conserved*.

Example 8.20

Suppose the pendulum of Example 8.19 is suspended at rest when it is struck by a small projectile, which becomes imbedded in the lower ball (Exhibit 21). What angular velocity ω is imparted to the pendulum?

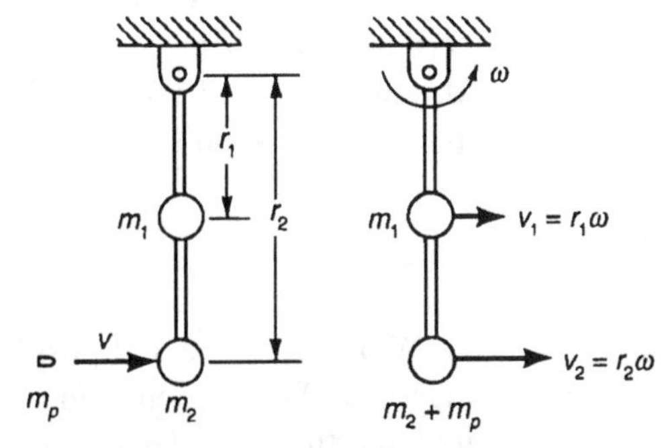

Exhibit 21

Solution

During the collision, which can induce a large reaction at the support as well as between the projectile and ball, the moment of forces external to the system—the pendulum and the projectile—will be zero. Hence, the angular momentum of this system prior to impact will equal that immediately after impact:

$$r_2 m_p v\mathbf{e}_3 = r_1 m_1 v_1 \mathbf{e}_3 + r_2 (m_2 + m_p) v_2 \mathbf{e}_3 = [m_1 r_1^2 + (m_2 + m_p) r_2^2]\omega \mathbf{e}_3$$

Hence,

$$\omega = \frac{r_2 m_p v}{m_1 r_1^2 + (m_2 + m_p) r_2^2}$$

WORK AND KINETIC ENERGY

The integration in Equation (8.9) has extensive implications that will be examined in this section.

A Single Particle

If the form of the tangential acceleration indicated in Equation (8.9) is merged with Equation (8.6), the result can be used to express Newton's second law as

$$\mathbf{f} = m\left(v\frac{dv}{ds}\mathbf{e}_t + \frac{v^2}{\rho}\mathbf{e}_n \right) \tag{i}$$

Now, if each member is dot-multiplied by an increment of change of position, $d\mathbf{r} = ds\mathbf{e}_t$, and the resulting scalars are integrated, they become

$$\int_{\mathbf{r}_1}^{\mathbf{r}_2} \mathbf{f} \bullet d\mathbf{r} = \int_{s_1}^{s_2} f_t \, ds = W_{1-2}$$

and

$$m\int_{v_1}^{v_2} v\,dv = \tfrac{1}{2}mv_2^{\,2} - \tfrac{1}{2}mv_1^{\,2}$$

The integral W_{1-2} is called the **work** done on the particle by the force **f** as the particle moves from position 1 to position 2. The scalar $T = ½\,mv^2$ is called the **kinetic energy** of the particle. Since (i) holds throughout any interval, a consequence of Newton's second law is the **work-kinetic energy relationship**:

$$W_{1-2} = T_2 - T_1 \qquad \textbf{(8.35)}$$

Work is considered positive when the kinetic energy of the system has been increased. When enough information is available to permit evaluation of the work integral, this provides a useful way of predicting the change in the speed of the particle.

Example 8.21

A 3.5-Mg airplane is to be launched from the deck of an aircraft carrier with the aid of a steam-powered catapult. The force that the catapult exerts on the aircraft varies with the distance s along the deck as shown in Exhibit 22. If other forces are negligible, what value of the constant f_0 is necessary for the catapult to accelerate the aircraft from rest to a speed of 160 km/h at the end of the 30-m travel?

$$f(s) = \frac{f_0}{1+\dfrac{s}{30\text{m}}}$$

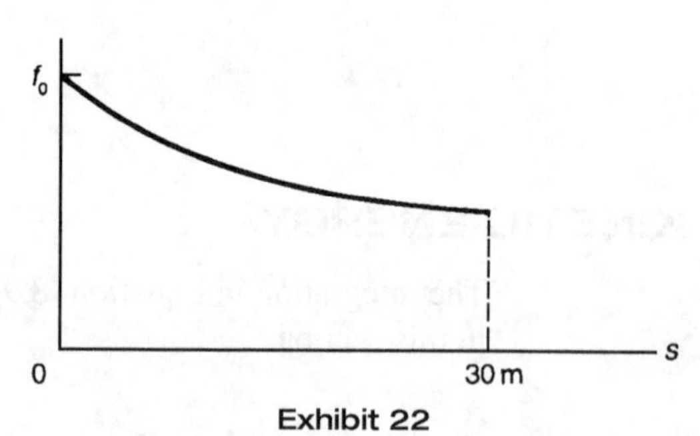

Exhibit 22

Solution

Letting d stand for the 30-m travel, the work done on the aircraft will be

$$W = \int_0^d \frac{f_0 ds}{1+\frac{s}{d}} = (f_0 d)\ln\left(1+\frac{s}{d}\right)\Bigg|_0^d = (f_0 d)\ln 2 = (20.8\,\text{m})f_0$$

This will equal the change in kinetic energy, which is initially zero.

$$\Delta T = (0.5)mv^2 - 0$$

$$\Delta T = \frac{1}{2}(3500\text{ kg})\left[\left(160\frac{\text{km}}{\text{h}}\right)\frac{1000\text{ m/km}}{3600\text{ s/h}}\right]^2 = 3.46\times10^6\text{ J}$$

The work-kinetic energy relationship

$$W = \Delta T$$

$$(20.8 \text{ m}) f_0 = 3.46 \times 10^6 \text{ N•m}$$

implies that the constant f_0 must have the value

$$f_0 = 166 \text{ kN}$$

Work of a Constant Force

A commonly encountered force of constant magnitude and direction is that of gravity near the earth's surface. When a constant force acts on a particle as it moves, the work done by the force can be evaluated as indicated in Figure 8.11. The increment of work as the particle undergoes an increment $d\mathbf{r}$ of displacement can be expressed as

$$dW = \mathbf{f}_0 \bullet d\mathbf{r} = |\mathbf{f}_0||d\mathbf{r}|\cos \measuredangle_{\mathbf{f}_0}^{d\mathbf{r}} = f_0 dq$$

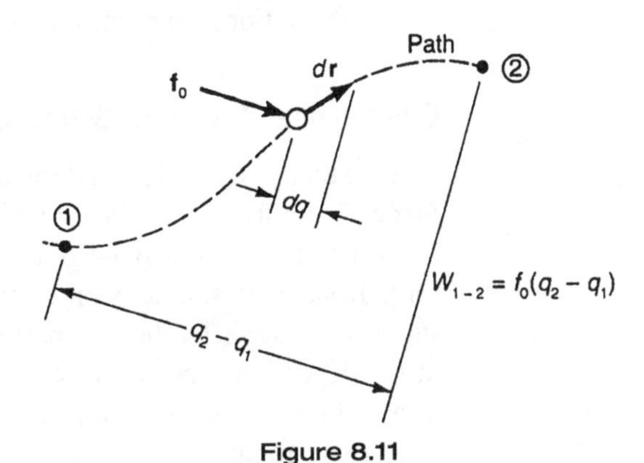

Figure 8.11

in which dq is the component of the displacement increment that is parallel to the force. Since the force is constant,

$$W_{1-2} = f_0 \int_1^2 dq = f_0(q_2 - q_1) \quad \textbf{(8.36)}$$

Thus, any movement of the particle that is perpendicular to the direction of the force has no effect on the work. In other words, the work is the same as would have been done if the particle had moved rectilinearly through a distance of $(q_2 - q_1)$ in a direction parallel to the force.

Example 8.22

How fast must the toy race car be traveling at the bottom of the hill to be able to coast to the top of the hill (see Exhibit 23)?

Solution

As the car moves up the hill, the work done by the force of gravity will be $W_g = -mgh$. The work done by friction forces may be negligible if the wheels are well made. If this is the case, the work done by all forces is approximately that due to

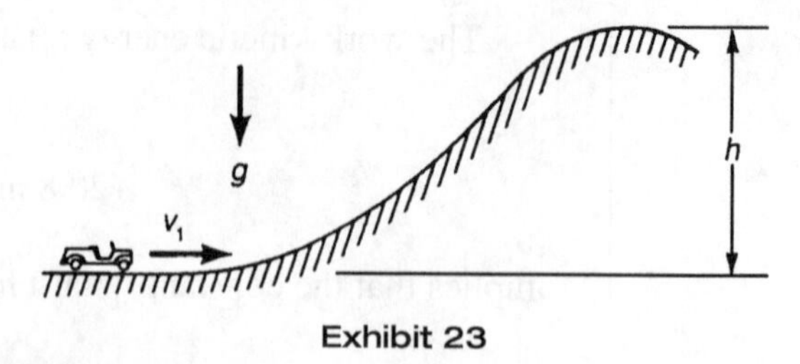

Exhibit 23

gravity. The speed of the car as it nears the hilltop can approach zero, so the work-kinetic energy equation may be written as

$$-mgh = T_2 - T_1 = 0 - \frac{1}{2}mv_1^2$$

which implies a minimum required speed of

$$v_1 = \sqrt{2gh}$$

With friction, the required speed will be somewhat greater.

Distance-Dependent Central Force

A force that remains directed toward or away from a fixed point is called a **central force**. Examples of forces for which the magnitude depends only on the distance from the particle to a fixed point are the force of gravitational attraction and the force from an elastic, tension-compression member with one end pinned to a fixed support. Figure 8.12 shows a particle *P* moving with such a central force acting on it; the dependence on distance is expressed by the function *f*(*r*), with the convention that a positive value of *f* indicates an attractive force and a negative value of *f* a repulsive force.

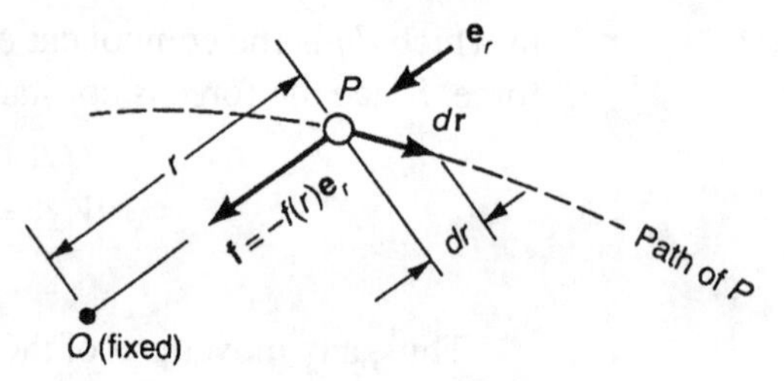

Figure 8.12

Now, the increment of work may be written as

$$dW = \mathbf{f} \bullet d\mathbf{r} = -f(r)\mathbf{e}_r \bullet d\mathbf{r}$$

Referring to the figure, we see that $\mathbf{e}_r \bullet d\mathbf{r} = |d\mathbf{r}| \cos \measuredangle_{er}^{dr}$ is equal to the change dr in radial distance r. Thus,

$$W_{1-2} = -\int_{r_1}^{r_2} f(r)\,dr \qquad \textbf{(8.37)}$$

Similar to the case of the constant force, the work done by a central force through an arbitrary motion is the same as would be done for a rectilinear motion, but in the radial direction.

Example 8.23

The elastic spring in Exhibit 24 has a linear force-displacement characteristic; that is, it exerts a force equal to the stiffness k times the amount it is stretched from its relaxed length l_0. As the particle moves from position 1 to position 2, what is the work done by the spring force on the particle?

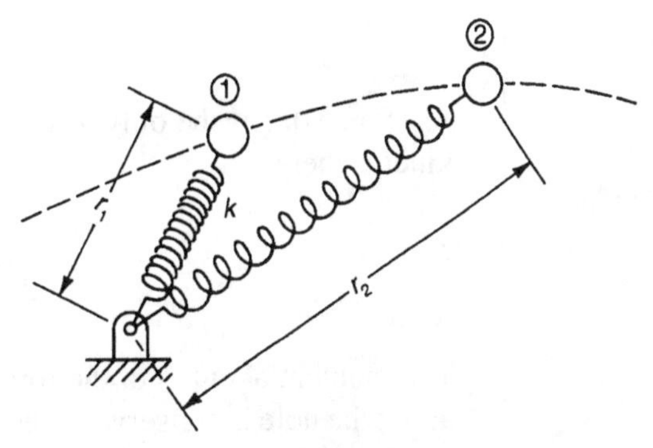

Exhibit 24

Solution

This is a case of a central distance-dependent force with $f(r) = -k(r - l_0)$. Equation (8.37) then becomes

$$W_{1-2} = -k\int_{r_1}^{r_2} (r - l_0)\,dr$$

A more convenient form results if we introduce the amount of spring extension as $\delta = (r - l_0)$. The integral then becomes

$$W_{1-2} = -k\int_{\delta_1}^{\delta_2} \delta\,d\delta = -\frac{k}{2}\left(\delta_2^2 - \delta_1^2\right)$$

Example 8.24

A 0.6-kg puck slides on a horizontal surface without friction under the influence of the tension in a light cord that passes through a small hole at O (see Exhibit 25). A spring under the surface imparts a tension in the cord that is proportional to the distance from the hole to the puck; its stiffness is $k = 30$ N/m. At a certain instant, the puck is 200 mm from the hole and moving at 2 m/s in the direction indicated in the top view. If the spring is in its relaxed position when the puck is at the hole, what is the maximum distance from the hole reached by the puck?

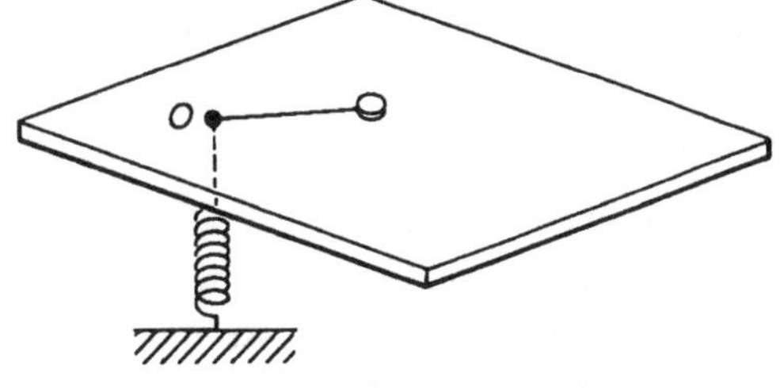

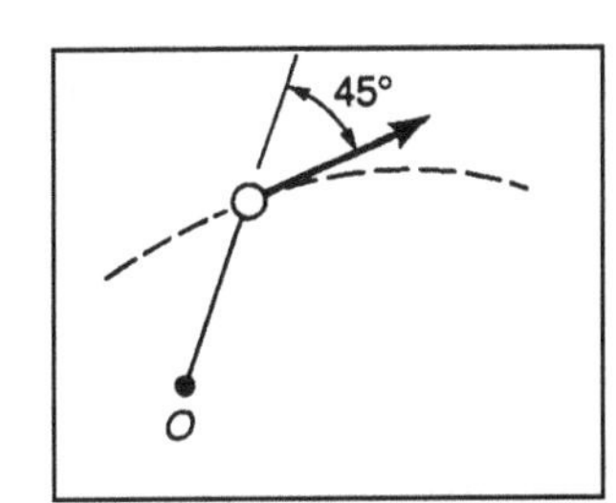

Exhibit 25

Solution

With the initial and maximum distances denoted by r_1 and r_2, respectively, the work done on the puck by the force from the cord from the initial position to that of maximum distance will be

$$W_{1-2} = -\frac{k}{2}\left(r_2^2 - r_1^2\right)$$

Since this is the only force that does work, this value must equal the change in kinetic energy:

$$-\frac{k}{2}\left(r_2^2 - r_1^2\right) = \frac{m}{2}\left(v_2^2 - v_1^2\right)$$

The moment about O of the force from the cord is zero, so the angular momentum about the hole is conserved. Because there is no radial component of velocity at the maximum distance, the angular momentum there is simply $r_2 m v_2$. Hence,

$$r_1\ m v_1\ \sin\ 45° = r_2 m v_2$$

These two equations contain the unknowns r_2 and v_2. Isolating v_2 from the latter, substituting this expression into the energy equation, and rearranging leads to the equation

$$\left(\frac{r_2}{r_1}\right)^4 - \left(1 + \frac{m v_1^2}{k r_1^2}\right)\left(\frac{r_2}{r_1}\right)^2 + \frac{m v_1^2}{k r_1^2}\sin^2 45° = 0$$

When the given values are substituted, the quadratic formula yields

$$\left(\frac{r_2}{r_1}\right) = 1.618$$

as the largest root so that the maximum distance reached is r_2 = 1.618 (200 mm) = 324 mm.

Example 8.25

A torpedo expulsion device operates by means of gas expanding within a tube that holds the torpedo. When test-fired with the tube firmly anchored, a 550-kg torpedo leaves the tube at 20 m/s. In operation, a 30-Mg submarine is traveling at 5 m/s when it expels a 550-kg torpedo in the forward direction. What are the speeds of the submarine and torpedo immediately after expulsion?

Solution

Considering the two-body system consisting of the submarine and torpedo, let us assume that the external forces remain in balance during expulsion. Then $W_e = 0$. Assuming also that the gas pressure depends only on the position of the torpedo relative to the submarine, the work W_{12} done by the internal forces will be the same during actual operation as during the test-firing. The work-kinetic energy relation ship for the test-firing yields $W_{12} = ½ (550\text{ kg})(20\text{ m/s})^2 = 110\text{ kJ}$, and for the operating condition the relationship is

$$110\text{ kJ}=\frac{1}{2}(30{,}000\text{ kg})\left[v_1^2-(5\text{ m/s})^2\right]+\tfrac{1}{2}(550\text{ kg})\left[v_2^2-(5\text{ m/s})^2\right]$$

Also, if the external forces are in balance, momentum will be conserved:

$$(30{,}550\text{ kg})(5\text{ m/s}) = (30{,}000\text{ kg})v_1 + (550\text{ kg})v_2$$

These two relationships give the desired speeds as

$$v_1 = 4.6\text{ m/s},\ v_2 = 24.8\text{ m/s}$$

The 19.8-m/s boost in speed given the torpedo is slightly less than when it is fired from the firmly anchored tube. However, the speed of the torpedo relative to the submarine is

$$v_2 - v_1 = 20.2\text{ m/s}$$

or slightly higher than in the fixed-tube test.

Two special cases of the work done by internal forces are of interest. The simpler is that in which the particles are constrained so that the distances between all pairs remain fixed, that is, the case of a rigid body. In this case, all the $dr_{i/j}$ are zero and the work-kinetic energy equation, Equation (8.38), reduces to $W_e = \Delta T$.

Another special case occurs when the force T_{ij} depends only on the distance $r_{i/j}$. That is, the force is not a function of relative velocity or previous history of deformation. Then the work integral $-\int T_{ij}dr_{i/j}$ is a function only of the distance between the particles and does not depend on the manner in which the particles move to reach a particular configuration. This would be the case, for example, with elastic spring interconnections or gravitational interactions. In this case, we can define the potential functions as

$$V_{ij}(r_{i/j})=\int_{(r_{i/j})_0}^{r_{i/j}} T_{ij}(\rho_{i/j})d\rho_{i/j}$$

and if their sum is denoted by

$$V=\sum_{i-j}V_{ij}$$

the work-energy integral becomes

$$W_e=\Delta T+\Delta V$$

That is, when the internal forces are all conservative, the work done by the external forces is equal to the change in total mechanical energy within the system.

KINETICS OF RIGID BODIES

If a system of particles is structurally constrained so that the distance between every pair of particles remains constant as the system moves, it forms a rigid body. Thus, the laws of kinetics in the previous section are applicable, along with the kinematics relationships reviewed earlier. Of the kinematics relationships, Equation (8.28c) will be useful in the form given, whereas the moment equation, Equation (8.33a), must be specialized to relate accelerations to angular velocity and angular acceleration.

In the general (three-dimensional) case, both the moment equation and kinematics relationships become considerably more complicated than they are for planar motion, and will be outside the scope of this review.

Moment Relationships for Planar Motion

Figure 8.13 shows the plane of the motion of a rigid body, with a point *P* (to be selected by the analyst for moment reference) along with an element of mass *dm*, which in the following is analogous to the mass m_i in the earlier analysis of a set of particles. The summation appearing in Equation (8.33a) will be written as an integral in this case, because the body is viewed as having continuously distributed mass. The element of mass is located relative to *P* by the position vector $\mathbf{r} = r\mathbf{e}_1 + z\mathbf{e}_3$. Its acceleration is related to that of point *P* through Equation (8.24), and the moment equation, Equation (8.33a), may be written as

$$\mathbf{M}_P = \int_m (r\mathbf{e}_1 + z\mathbf{e}_3)\times(\mathbf{a}_P + r\alpha\mathbf{e}_2 - r\omega^2\mathbf{e}_1)\,dm$$
$$= \left(\int_m \mathbf{r}\,dm\right)\times\mathbf{a}_P + \left(\int_m r^2\,dm\right)\alpha\mathbf{e}_3 - \alpha\int_m zr\mathbf{e}_1\,dm - \omega^2\int_m zr\mathbf{e}_2\,dm$$

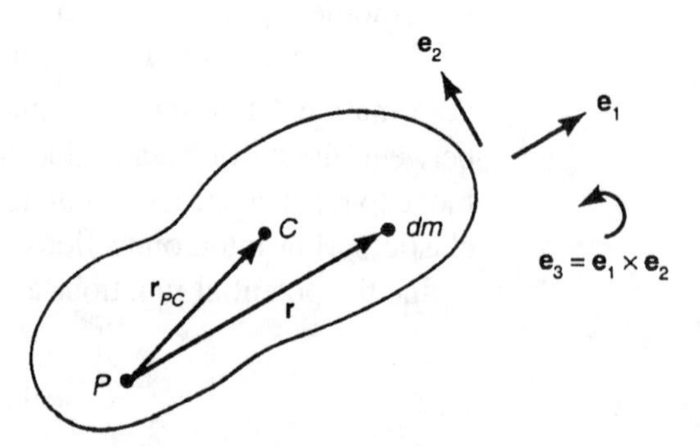

Figure 8.13

If the body's mass is distributed symmetrically with respect to the plane of motion through *P*, the last two integrals in the last line will vanish; without such symmetry, these terms indicate the possibility of components of moment in the plane of motion. Thus, even for *plane* motion, the distribution of mass may imply forces *perpendicular* to the plane of motion. These will not be pursued in detail here, but the reader must be aware of this possibility. The other two integrals will be of concern. The first is exactly the expression one would write to determine the location of the center of mass from point *P*:

$$\int_m \mathbf{r}\,dm = m\mathbf{r}_{PC}$$

The second integral is called the **moment of inertia** of the body about the axis through *P* and perpendicular to the plane of motion:

$$\int_m r^2\,dm = I_P$$

The moment about this axis is thus related to accelerations through

$$M_{P3} = \mathbf{e}_3 \bullet \mathbf{M}_P = I_P\alpha + \mathbf{e}_3 \bullet (m\mathbf{r}_{PC} \times \mathbf{a}_P) \quad \textbf{(8.40a)}$$

By using Equation (8.24) to relate the acceleration of P to that of the mass center C, it is possible to express this moment law in the alternative form

$$M_{P3} = I_C\alpha + \mathbf{e}_3 \bullet (m\mathbf{r}_{PC} \times \mathbf{a}_C) \quad \textbf{(8.40b)}$$

where I_C is the moment of inertia of the body about an axis through C perpendicular to the plane of motion. The two moments of inertia are related through the **parallel axis formula**

$$I_P = I_C + md^2$$

in which d is the distance between P and C.

Two special cases warrant attention. If P is chosen to be the mass center C, then $\mathbf{r}_{PC} = 0$, and the relationship is

$$M_{C3} = I_C\alpha$$

If the body is hinged about a fixed support and P is selected to be on the axis of the hinge, then

$\mathbf{a}_P = 0$, and the relationship is

$$M_{P3} = I_P\alpha$$

These last two relationships indicate the moment of inertia of the body is the property that provides resistance to changes in the angular velocity, much as mass provides resistance to changes in the velocity of a particle. For bodies of simple geometry, the integrals have been evaluated in terms of mass and the geometry, and results can be found in tabulated summaries. More complicated bodies can require tedious work to estimate the moment of inertia, or there are experiments based on the implications of Equation (8.40) that can be used to determine it. It is common to specify the moment of inertia by giving the mass of the body and its **radius of gyration**, k_p, defined by $I_P = mk_P^2$.

Example 8.26

A 23-kg rotor has a 127 mm radius of gyration about its axis of rotation. What average torque about its fixed axis of rotation is required to bring the rotor from rest to a speed of 200 rpm in 6 seconds?

Solution

The moment of inertia of the rotor is

$$I = (23\text{ kg})(0.127\text{ m})^2 = 0.371\text{ kg} \bullet \text{m}^2$$

and the average angular acceleration is

$$\alpha = \frac{(200\text{ rpm})}{6\text{ s}}\frac{(2\pi\text{ rad/r})}{(60\text{ s/min})} = 3.49\text{ rad/s}^2$$

For fixed-axis rotation,

$$M = I\alpha = (0.371\text{ kg} \bullet \text{m}^2)(3.49\text{ s}^{-2}) = 1.29\text{ N} \bullet \text{m}$$

Example 8.27

The car with rear-wheel drive in Exhibit 26 has sufficient power to cause the drive wheels to slip as it accelerates. The coefficient of friction between the drive wheels and roadway is μ. What is the acceleration in terms of g, μ, and the dimensions shown?

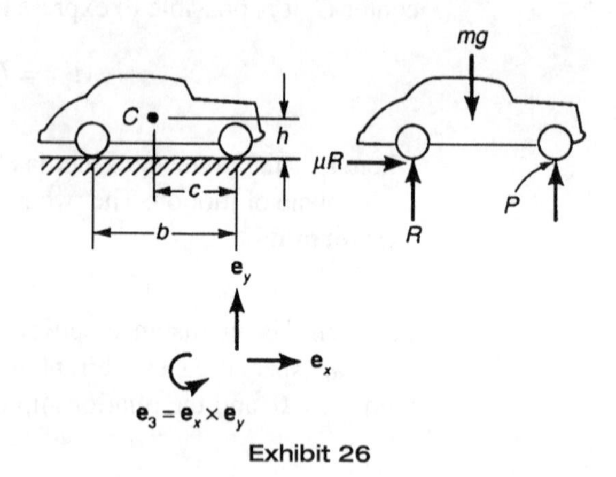

Exhibit 26

Solution

Assuming the car does not rotate, every point will have the same acceleration, $\mathbf{a} = a_x\mathbf{e}_x$. The free-body diagram shows the forces external to the car, with the label on the horizontal force at the drive wheels accounting for the fact that the friction limit has been reached there. Equation (8.28c) implies that $\mu R = ma_x$. Since the reaction R in this equation is unknown, another relationship must be introduced. Of several that could be written (for example, forces in another direction, moments about a selected point), it would be best if no additional unknowns are introduced. Thus, to avoid bringing the unknown reaction at the front wheels into the analysis, consider moments about point P, which are related by the moment law, Equation (8.40b):

$$-bR + c\,mg = I(0) + \mathbf{e}_3 \bullet [m(-c\mathbf{e}_x + h\mathbf{e}_y) \times a_x\mathbf{e}_x] = -mha_x$$

Eliminating R between this equation and the friction equation leads to

$$a_x = \frac{\mu c}{b - \mu h} g$$

Example 8.28

The uniform slender rod in Exhibit 27 slides along the wall and floor under the effects of gravity. If friction is negligible, what is the angular acceleration in terms of g, l, and the angle θ?

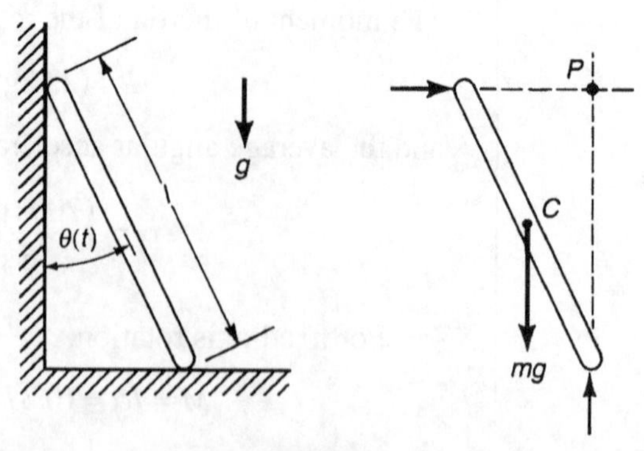

Exhibit 27

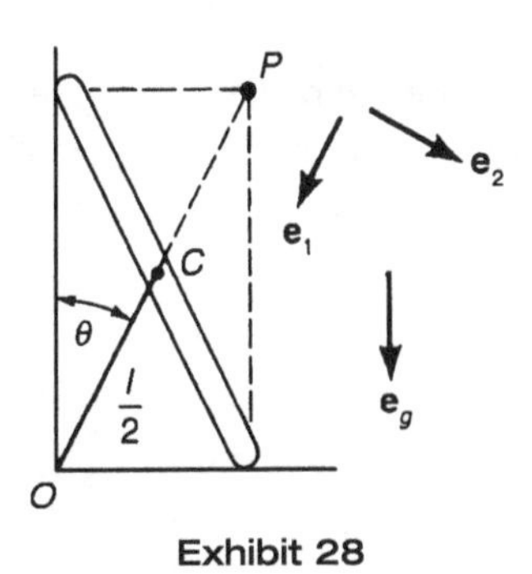

Exhibit 28

Solution

As indicated on the free-body diagram, the reactions from the wall and floor are horizontal and vertical since there is no friction. Because neither of these is known, a good strategy would be to avoid dealing with them; to this end, consider moments about point P, which will be related by Equation (8.40b). To evaluate the acceleration of the mass center C, observe that it follows a circular path with radius $l/2$ and center at O (see Exhibit 28). Thus, the acceleration of C is

$$\mathbf{a}_C = \frac{l}{2}(\dot{\theta}^2\mathbf{e}_1 + \ddot{\theta}\mathbf{e}_2)$$

Referring to the free-body diagram, we can write the moment of forces about P as

$$\mathbf{M}_P = \frac{l}{2}\mathbf{e}_1 \times mg\mathbf{e}_g = \frac{1}{2}mgl\sin\theta\mathbf{e}_3$$

Any of a number of references gives the moment of inertia of a slender, uniform rod about an axis through its center as

$$I_C = \frac{1}{12}ml^2$$

Substituting the above into the moment law, Equation (8.40b), yields

$$\frac{1}{2}mgl\sin\theta = \frac{1}{12}ml^2\ddot{\theta} + \mathbf{e}_3 \bullet \left[m\frac{l}{2}\mathbf{e}_1 \times \frac{l}{2}(\dot{\theta}^2\mathbf{e}_1 + \ddot{\theta}\mathbf{e}_2)\right] = \frac{ml^2}{3}\ddot{\theta}$$

which leads to the desired angular acceleration:

$$\ddot{\theta} = \frac{3g}{2l}\sin\theta$$

Example 8.29

The uniform, slender beam of length l is suspended by the two wires in the configuration shown in Exhibit 29 when the wire on the left is cut. Immediately after the wire is severed (that is, while the velocities of all points are still zero), what is the tension in the remaining wire?

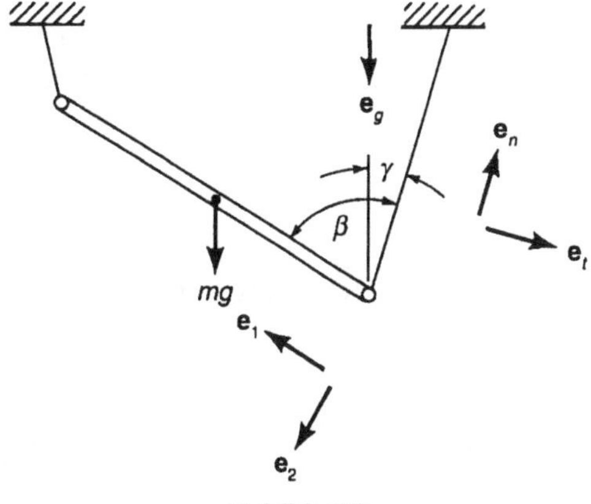

Exhibit 29

Solution

The free-body diagram (Exhibit 30) shows the desired force and the only other force acting on the beam, that of gravity. Summing moments about the center of mass gives a relationship between the desired tension and the angular acceleration of the bar:

$$\frac{l}{2}T\sin\beta = \frac{ml^2}{12}\alpha \qquad \textbf{(i)}$$

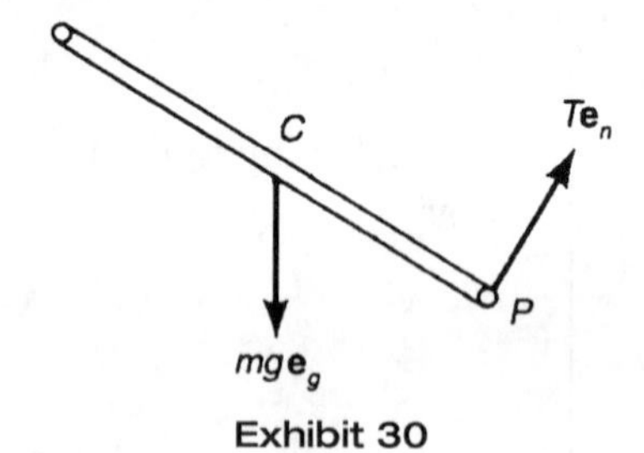

Exhibit 30

We also know that the sum of all forces is related to the acceleration of the mass center by

$$T\mathbf{e}_n + mg\mathbf{e}_g = m\mathbf{a}_C \qquad \textbf{(ii)}$$

Since the end P is constrained by the wire to follow a circular path, its acceleration may be expressed by

$$\mathbf{a}_P = a_t\mathbf{e}_t + \frac{v_P^2}{R}\mathbf{e}_n \qquad \textbf{(iii)}$$

in which a_t is another unknown quantity. Finally, this acceleration is related to that of the center of mass by

$$\mathbf{a}_C = \mathbf{a}_P + \frac{l}{2}\alpha\mathbf{e}_2 - \frac{l}{2}\omega^2\mathbf{e}_1 \qquad \textbf{(iv)}$$

Since velocities are still zero, the centripetal terms v_P^2/R and $\frac{l}{2}\omega^2$ are both zero. With this simplification, Equations (ii), (iii), and (iv) readily combine to give

$$T\mathbf{e}_n + mg\mathbf{e}_g = m\left(a_t\mathbf{e}_t + \frac{l}{2}\alpha\mathbf{e}_2\right)$$

To avoid dealing with the unknown a_t, we may dot-multiply each term in this equation by $\mathbf{e}_n$ with the result

$$T + mg\mathbf{e}_n \bullet \mathbf{e}_g = \frac{1}{2}ml\alpha\mathbf{e}_n \bullet \mathbf{e}_2$$

Referring to the specified geometry, the dot products are evaluated as

$$\mathbf{e}_n \bullet \mathbf{e}_g = \cos\,(180° - \gamma) = -\cos\,\gamma$$

$$\mathbf{e}_n \bullet \mathbf{e}_2 = \cos\,(90° + \beta) = -\sin\,\beta$$

and the equation can be written as

$$T+\frac{1}{2}ml\alpha\sin\beta = mg\cos\gamma \tag{v}$$

Now α is readily eliminated by substituting the expression for α obtained from Equation (i) into Equation (v), leading to the desired value of the tension:

$$T=\frac{mg\cos\gamma}{1+3\sin^2\beta}$$

Work and Kinetic Energy

If a rigid body has a number of forces $f_1, f_2, \ldots, f_n$ applied at points $P_1, P_2, \ldots, P_n$, the time rate at which these forces do work on the body (that is, the power transmitted to the body) can be evaluated as

$$\frac{dW}{dt}=\mathbf{f}_1\bullet\frac{d\mathbf{r}_1}{dt}+\mathbf{f}_2\bullet\frac{d\mathbf{r}_2}{dt}+\cdots+\mathbf{f}_n\bullet\frac{d\mathbf{r}_n}{dt}=\sum_i \mathbf{f}_i\bullet\mathbf{v}_i$$

But $\mathbf{v}_i$, the velocity of point P_i, can be related to the velocity of a selected point P of the body:

$$\mathbf{v}_i=\mathbf{v}_P+\omega\times\mathbf{r}_{Pi}$$

so that the power can also be expressed as

$$\frac{dW}{dt}=\sum_i \mathbf{f}_i\bullet(\mathbf{v}_P+\omega\times\mathbf{r}_{Pi})=\left(\sum_i \mathbf{f}_i\right)\bullet\mathbf{v}_P+\left(\sum_i \mathbf{r}_{Pi}\times\mathbf{f}_i\right)\bullet\omega$$

$$=\mathbf{f}\bullet\mathbf{v}_P+\mathbf{M}_P\bullet\omega \tag{8.41}$$

in which **f** is the resultant of all of the forces. P may be selected as any point of the body, and M_P is the resultant moment about P. For example, as a rotor turns about a fixed axis, there may be forces from the support bearings in addition to an accelerating torque about the axis of rotation. If the point P is selected to be on the axis of rotation, $\mathbf{v}_P$ will be zero, and the power transmitted to the rotor (which will, of course, induce a change in its kinetic energy) is simply the dot-product of the torque and the angular velocity. Negative or positive values are possible, depending on the angle between $\mathbf{M}_P$ and ω (that is, whether the moment component is in the same or the opposite direction as the rotation).

The kinetic energy of a rigid body is the sum of the kinetic energies of its individual elements, whose velocities can be related to the velocity of a selected point P and the angular velocity ω. Referring to Figure 8.13, we write this for plane motion as

$$T=\frac{1}{2}\int_m |\mathbf{v}|^2\,dm=\frac{1}{2}\int_m(\mathbf{v}_P+\omega\times\mathbf{r})\bullet(\mathbf{v}_P+\omega\times\mathbf{r})dm$$

$$=\frac{1}{2}\left(\int_m dm\right)v_P^2+\mathbf{v}_P\bullet\left[\omega\times\left(\int_m \mathbf{r}dm\right)\right]+\frac{1}{2}\int_m|\omega\times\mathbf{r}|^2dm$$

The integral in the first term is simply the mass m of the body. The integral in the second term is related to the mass and position of the center of mass by

$$\int_m \mathbf{r}dm = m\mathbf{r}_{PC}$$

For plane motion, $\omega \times \mathbf{r} = \omega r\mathbf{e}_2$, so the last term becomes

$$\int_m |\omega \times \mathbf{r}|^2 dm = \omega^2 \int_m r^2 dm = I_P \omega^2$$

The expression for kinetic energy for plane motion of a rigid body is then

$$T = \frac{1}{2}mv_P^2 + m\mathbf{v}_P \bullet (\omega \times \mathbf{r}_{PC}) + \frac{1}{2}I_P\omega^2 \quad \textbf{(8.42)}$$

Example 8.30

The wheel in Exhibit 31 is released from rest and rolls down the hill with sufficient friction to prevent slipping. Its mass is m, its radius is r, and its central radius of gyration is k. After the center of the wheel has dropped a vertical distance h, what is the speed of the center of mass of the wheel?

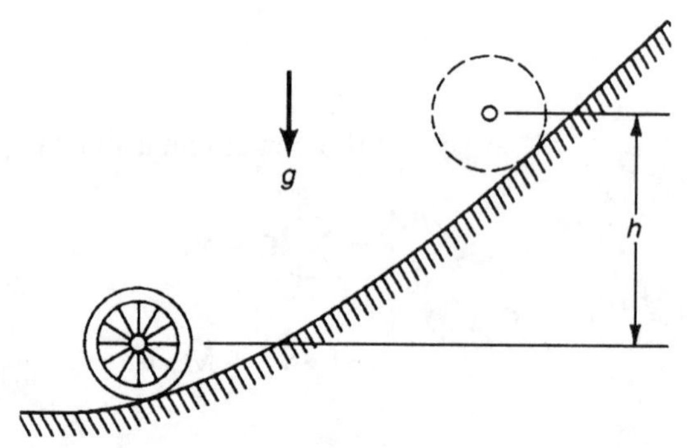

Exhibit 31

Solution

Since the velocity of the contact point is zero, the work of the force there is zero. The work of the force of gravity is then the total work done on the wheel and is simply $W = mgh$. Since the contact point is the instantaneous center, the speed of the center of the wheel is readily related to the angular velocity, $\omega = v_C/r$. The kinetic energy can be written from Equation (8.2), with P selected as any point on the wheel. If we choose P to be the center of mass,

$$T = \frac{1}{2}mv_C^2 + 0 + \frac{1}{2}mk^2\omega^2 = \frac{1}{2}\left(1 + \frac{k^2}{r^2}\right)mv_C^2$$

If, instead, we choose P to be the instantaneous center, the kinetic energy is

$$T = \frac{1}{2}m(0)^2 + 0 + \frac{1}{2}(mk^2 + mr^2)\omega^2 = \frac{1}{2}\left(1 + \frac{k^2}{r^2}\right)mv_C^2$$

Since the work must equal the change in kinetic energy,

$$mgh = \frac{1}{2}\left(1 + \frac{k^2}{r^2}\right) m v_C^2$$

and the speed of the center will be

$$v_C = \sqrt{\frac{2gh}{1 + \frac{k^2}{r^2}}}$$

SELECTED SYMBOLS AND ABBREVIATIONS

Symbol or Abbreviation	Description
$\mathbf{a}$	acceleration
$\mathbf{a}_t$	tangential component of acceleration
$\mathbf{e}_t$	unit vector tangent to path
$\mathbf{e}_n$	unit vector in principal normal direction
$\mathbf{e}_i$	unit vector in direction indicated by the specific value of i
$\mathbf{f}$	resultant force
g	gravitational field intensity
$\mathbf{g}$	impulse resultant force
$\mathbf{H}_o$	angular momentum about O
I_P	moment of inertia about P
κ	local curvature of path
k_P	radius of gyration about P
M_i	mass of ith particle
M	total mass
$\mathbf{M}_P$	moment of forces about P
N	coefficient of kinetic fraction
ρ	radius of curvature of path
$\mathbf{r}$	position vector
s	distance along path
t	time
$\boldsymbol{T}$	kinetic energy
$\mathbf{v}$	velocity
$\mathbf{W}$	work
α	angular acceleration
μ	coefficient of sliding friction
ω	angular velocity

PROBLEMS

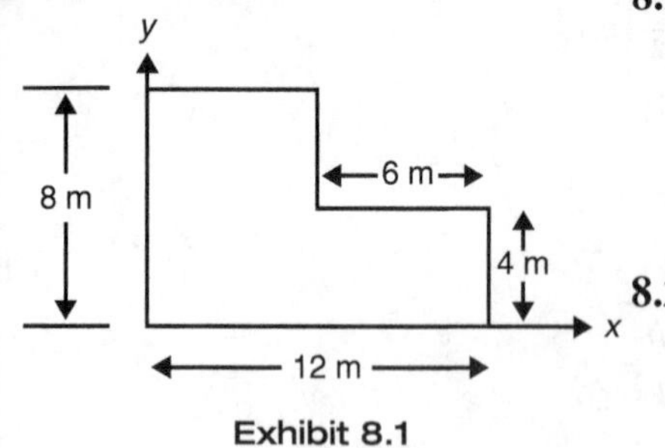

Exhibit 8.1

8.1 The uniform density flat plate shown in Exhibit 8.1 has a mass of 480 kg. The density of the body is most nearly:

a. 10 kg/m^2
b. 4.6 kg/m^2
c. 6.7 kg/m^2
d. 11.8 kg/m^2

8.2 Assuming the density of the plate in Exhibit 8.1 is 9 kg/m^2, the mass moment of inertia of the body is most nearly:

a. 10,400 kg-m^2
b. 18,300 kg-m^2
c. 9,600 kg-m^2
d. 7,400 kg-m^2

8.3 A particle is thrown vertically upward from the edge *A* of the ditch shown in Exhibit 8.3. If the initial velocity is 4 m/s, and the particle is known to hit the bottom, *B*, of the ditch exactly 6 seconds after it was released at *A*, determine the depth of this ditch. Neglect air resistance.

a. 24.0 m
b. 152.6 m
c. 200 m
d. 176.6 m

Exhibit 8.3

8.4 The slider *P* in Exhibit 8.4 is driven by a complex mechanism in such a way that (i) it remains on a straight path throughout; (ii) at the instant $t = 0$, the slider is located at A, (iii) at any general instant of time, the velocity of *P* is given by $v = (3t2 - t + 2)$ m/s. Determine the distance of *P* from point *O* when t = 2 s.

a. 26 m
b. 10 m
c. 6 m
d. 12 m

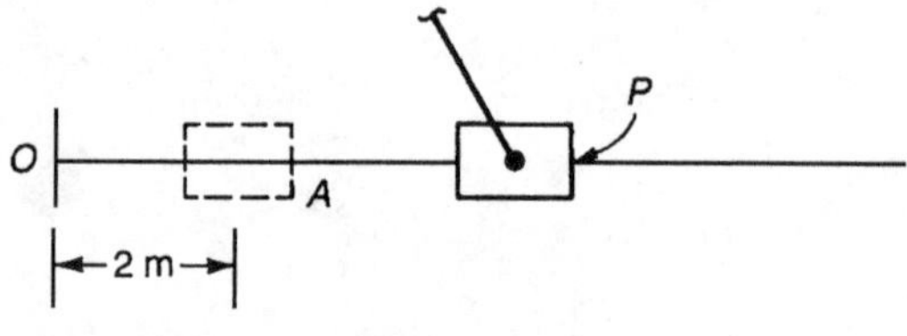

Exhibit 8.4

8.5 A particle in rectilinear motion starts from rest and maintains the acceleration profile shown in Exhibit 8.5. The displacement of the particle in the first 8 seconds is:

a. 4 m c. 24 m
b. 28 m d. 20 m

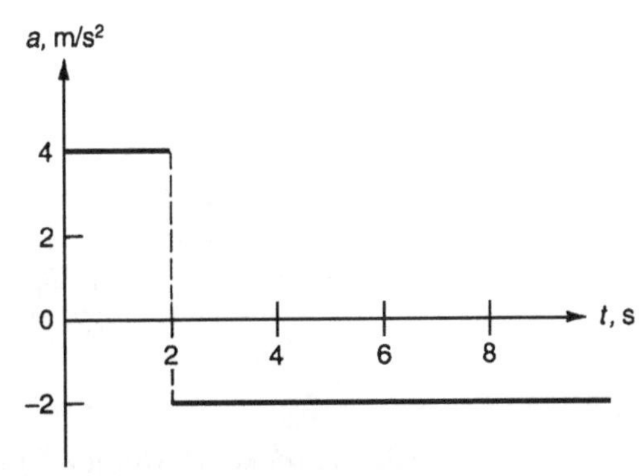

Exhibit 8.5

8.6 A ball is thrown by a player (Exhibit 8.6) from a position 2 m above the ground surface with a velocity of 40 m/s inclined at 60° to the horizontal. Determine the maximum height, H, the ball will attain.

a. 63.2 m c. 30.6 m
b. 61.2 m d. 31 m

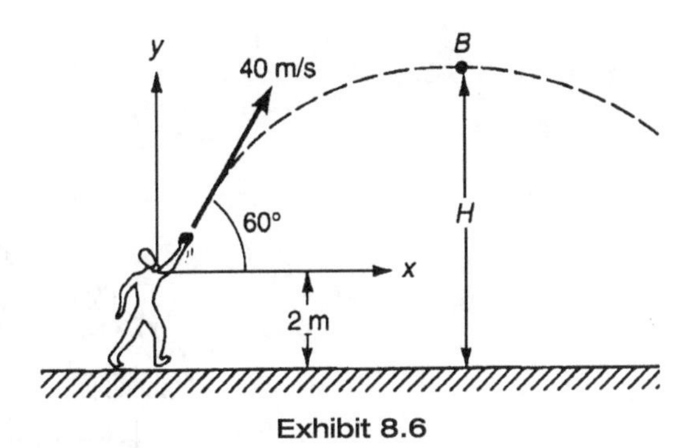

Exhibit 8.6

8.7 A golf ball (Exhibit 8.7) is struck horizontally from point A of an elevated fairway. Determine the initial speed that must be imparted to the ball if the ball is to strike the base of the flag stick on the green 140 meters away. Neglect air friction.

a. 34.3 m/s c. 90 m/s
b. 103 m/s d. 19.2 m/s

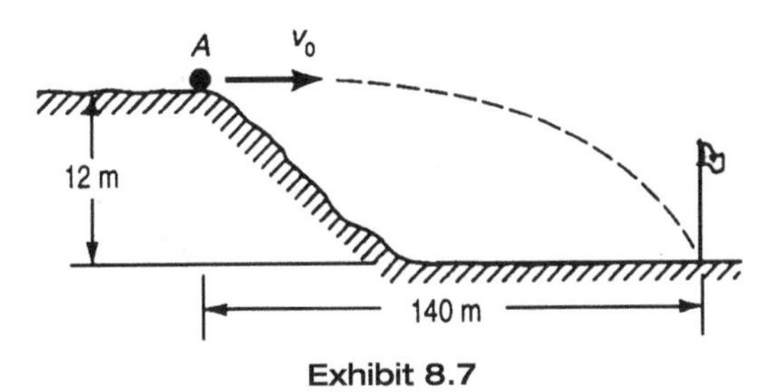

Exhibit 8.7

8.8 In Exhibit 8.8, the rod R rotates about a fixed axis at O. A small collar B is forced down the rod (toward O) at a constant speed of 3 m/s relative to the rod. If the value of θ at any given instant is $\theta = (t^2 + t - 2)$ rad, find the magnitude of the acceleration of B at time $t = 1$ second, when B is known the be 1 meter away from O.

a. 8.0 m/s^2 c. 18.4 m/s^2
b. 20.2 m/s^2 d. 3.0 m/s^2

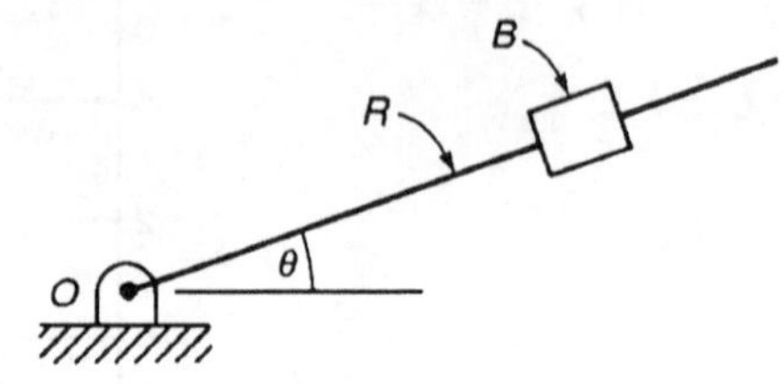

Exhibit 8.8

8.9 A rocket (Exhibit 8.9) is fired vertically upward from a launching pad at B, and its flight is tracked by radar from point A. Find the magnitude of the velocity of the rocket when $\theta = 45°$ if $\dot{\theta} = 0.1$ rad/s.

a. 36 m/s c. 90 m/s
b. 180 m/s d. 360 m/s

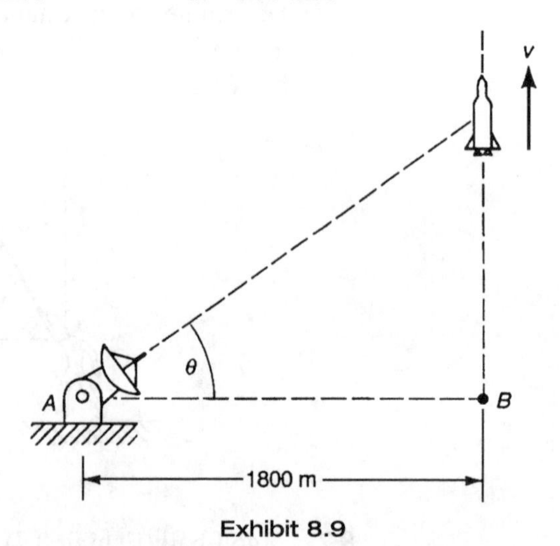

Exhibit 8.9

8.10 A particle is given an initial velocity of 50 m/s at an angle of 30° with the horizontal as shown in Exhibit 8.10. What is the radius of curvature of its path at the highest point, C?

a. 19.5 m c. 221 m
b. 255 m d. 191 m

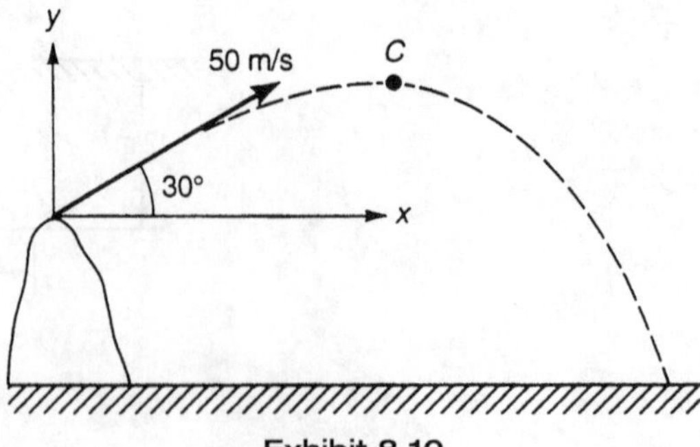

Exhibit 8.10

8.11 An automobile moves along a curved path that can be approximated by a circular arc of radius 110 meters. The driver keeps his foot on the accelerator pedal in such a way that the speed increases at the constant rate of 3 m/s^2. What is the total acceleration of the vehicle at the instant when its speed is 20 m/s?

a. 22.0 m/s^2
b. 3.6 m/s^2
c. 3.0 m/s^2
d. 4.7 m/s^2

8.12 A pilot testing an airplane at 800 kph wishes to subject the aircraft to a normal acceleration of 5 *gs* in order to fulfill the requirements of an on-board experiment. Find the radius of the circular path that would allow the pilot to do this.

a. 502 m
b. 3308 m
c. 1007 m
d. 1453 m

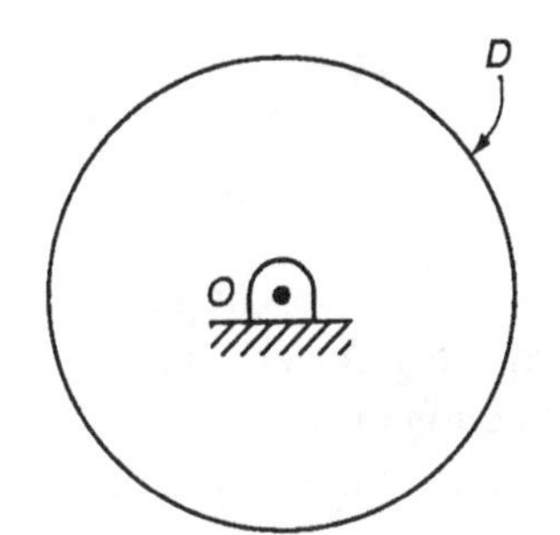

Exhibit 8.13

8.13 At the instant $t = 0$, the disk D in Exhibit 8.13 is spinning about a fixed axis through O at an angular speed of 300 rpm. Bearing friction and other effects are known to slow the disk at a rate that is k times its instantaneous angular speed, where k is a constant with the value $k = 1.2\ s^{-1}$. Determine when (from $t = 0$) the disk's spin rate is cut in half.

a. 6.5 s
b. 13.1 s
c. 0.8 s
d. 0.6 s

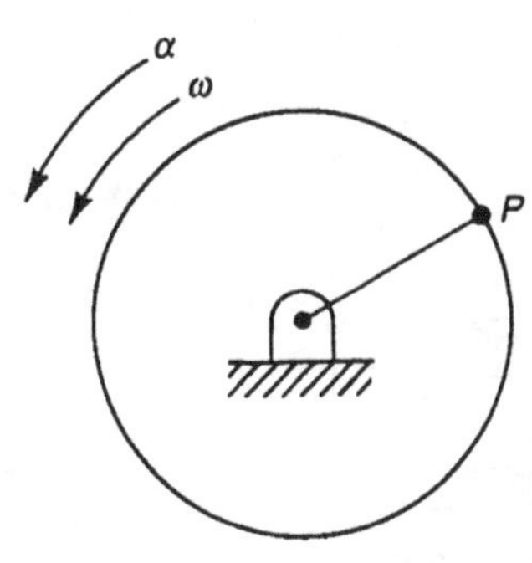

Exhibit 8.14

8.14 In Exhibit 8.14, a flywheel 2 m in radius is brought uniformly from rest up to an angular speed of 300 rpm in 30 s. Find the speed of a point P on the periphery 5 seconds after the wheel started from rest.

a. 10.5 m/s
b. 5.2 m/s
c. 62.8 m/s
d. 100.0 m/s

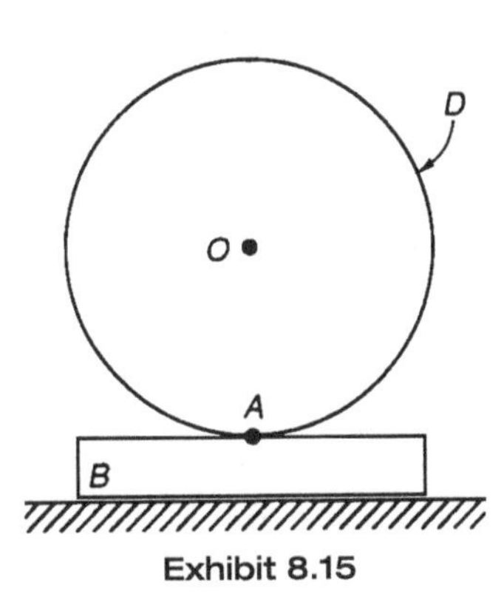

Exhibit 8.15

8.15 The block B (Exhibit 8.15) slides along a straight path on a horizontal floor with a constant velocity of 2 m/s to the right. At the same time, the disk, D, of 3-m diameter rolls without slip on the block. If the velocity of the center, O, of the disk is directed to the left and remains constant at 1 m/s, determine the angular velocity of the disk.

a. 0.3 rad/s counterclockwise
b. 2.0 rad/s counterclockwise
c. 0.7 rad/s counterclockwise
d. 0.7 rad/s clockwise

8.16 In Exhibit 8.16, the disk, D, rolls without slipping on a horizontal floor with a constant clockwise angular velocity of 3 rad/s. The rod, R, is hinged to D at A, and the end, B, of the rod touches the floor at all times. Determine the angular velocity of R when the line OA joining the center of the disk to the hinge at A is horizontal as shown.

a. 0.6 rad/s counterclockwise
b. 0.6 rad/s clockwise
c. 3.0 rad/s counterclockwise
d. 3.0 rad/s clockwise

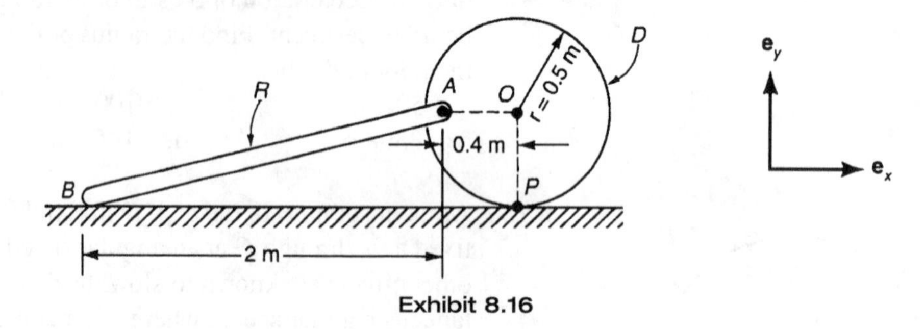

Exhibit 8.16

8.17 The fire truck in Exhibit 8.17 is moving forward along a straight path at the constant speed of 50 km/hr. At the same time, its 2-meter ladder OA is being raised so that the angle θ is given as a function of time by $\theta = (0.5t^2 - t)$ rad, where t is in seconds. The magnitude of the acceleration of the tip of the ladder when $t = 2$ seconds is:

a. 0
b. 4.0 m/s^2
c. 2.0 m/s^2
d. 2.8 m/s^2

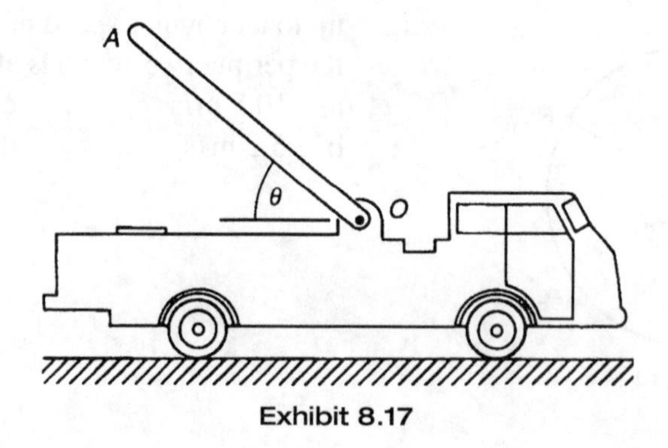

Exhibit 8.17

8.18 In Exhibit 8.18, the block B is constrained to move along a horizontal rectilinear path with a constant acceleration of 2 m/s^2 to the right. The slender rod, R, of length 2 m is pinned to B at O and can swing freely in the vertical plane. At the instant when $\theta = 0°$ (rod is vertical), the angular velocity of the rod is zero but its angular acceleration is 2.5 m/s^2 clockwise. Find the acceleration of the midpoint G of the rod at this instant ($\theta = 0°$).

a. 3.0 $m/s^2 \leftarrow$ c. 2.5 $m/s^2 \leftarrow$
b. 0.5 $m/s^2 \rightarrow$ d. 2.5 $m/s^2 \rightarrow$

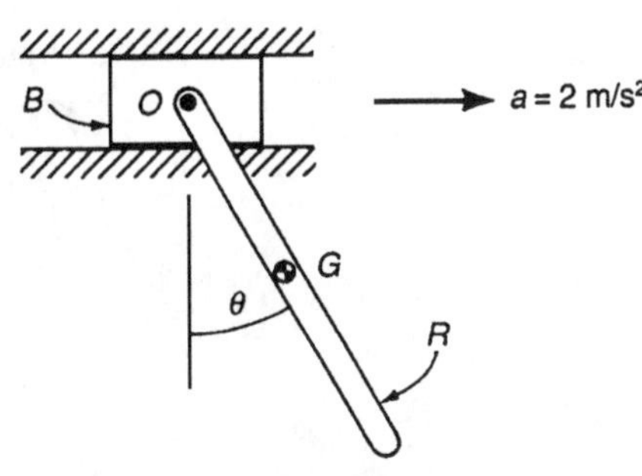

Exhibit 8.18

8.19 The block, B, in Exhibit 8.19, contains a square-cut circular groove. A particle, P, moves in this groove in the clockwise direction and maintains a constant speed of 6 m/s relative to the block. At the same time, the block slides to the right on a straight path at the constant speed of 10 m/s. Find the magnitude of the absolute velocity of P at the instant when $\theta = 30°$.

a. 8.7 m/s c. 4 m/s
b. 16 m/s d. 14 m/s

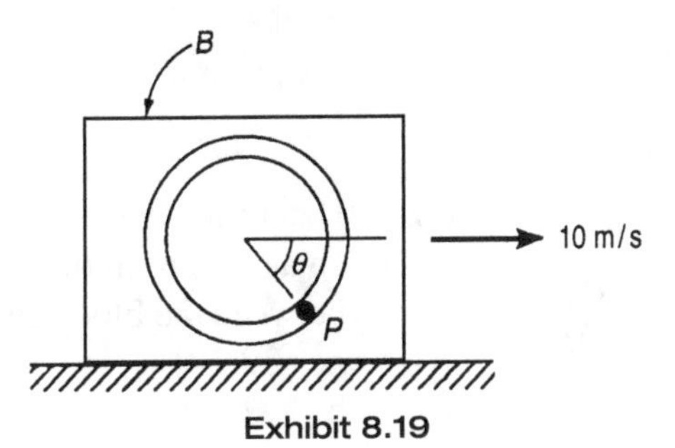

Exhibit 8.19

8.20 In Exhibit 8.20, a pin moves with a constant speed of 2 m/s along a slot in a disk that is rotating with a constant clockwise angular velocity of 5 rad/s. Calculate the absolute acceleration of this pin when it reaches the position C (directly above O). The unit vectors $\mathbf{e}_x$ and $\mathbf{e}_y$ are fixed to the disk.

a. $17.5\mathbf{e}_y$ m/s² c. $-2.5\mathbf{e}_y$ m/s²
b. $-17.5\mathbf{e}_y$ m/s² d. $-22.5\mathbf{e}_y$ m/s²

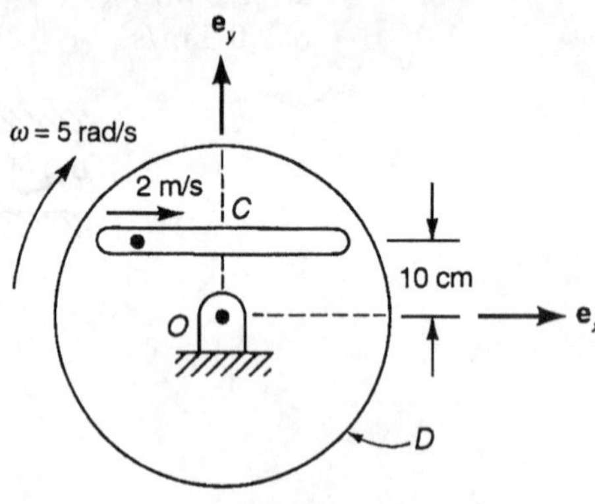

Exhibit 8.20

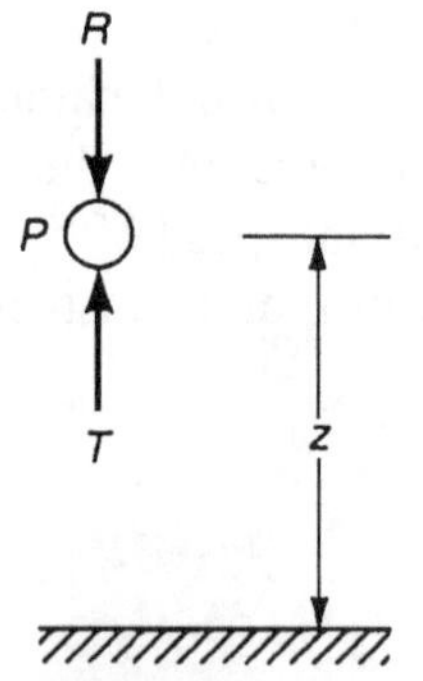

8.21 In Exhibit 8.21, a particle P of mass 5 kg is launched vertically upward from the ground with an initial velocity of 10 m/s. A constant upward thrust $T = 100$ newtons is applied continuously to P, and a downward resistive force $R = 2z$ newtons also acts on the particle, where z is the height of the particle above the ground. Determine the maximum height attained by P.

a. 6.0 m c. 15.8 m
b. 45.5 m d. 55.5 m

Exhibit 8.21

8.22 Determine the force P required to give the block shown in Exhibit 8.22 an acceleration of 2 m/s² up the incline. The coefficient of kinetic friction between the block and the incline is 0.2.

a. 39.2 N c. 44.6 N
b. 21.9 N d. 49.8 N

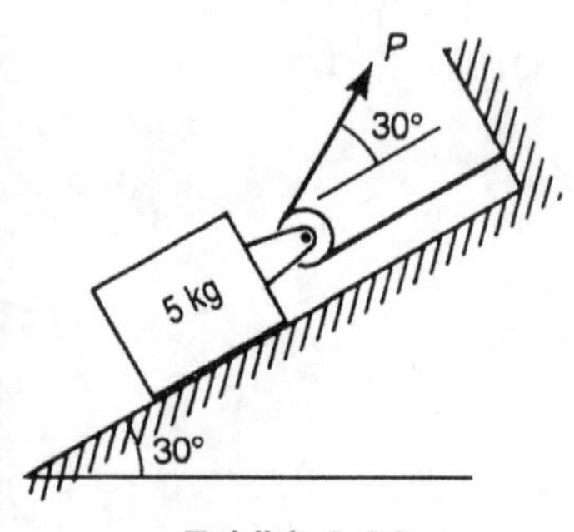

Exhibit 8.22

8.23 In Exhibit 8.23 the rod R rotates in the vertical plane about a fixed axis through the point O with a constant counterclockwise angular velocity of 5 rad/s. A collar B of mass 2 kg slides down the rod (toward O) so that the distance between B and O decreases at the constant rate of 1 m/s. At the instant when $\theta = 30°$ and $r = 400$ mm, determine the magnitude of the applied force P. The coefficient of kinetic friction between B and R is 0.1.

a. 9.9 N　　c. 10.5 N
b. 11.9 N　　d. 0.3 N

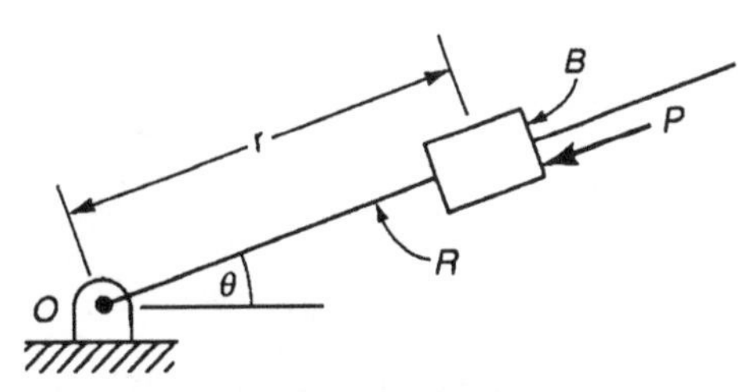

Exhibit 8.23

8.24 The 3-kg collar in Exhibit 8.24 slides down the smooth circular rod. In the position shown, its velocity is 1.5 m/s. Find the normal force (contact force) the rod exerts on the collar.

a. 12.2 N　　c. 19.2 N
b. 24.4 N　　d. 12.2 N

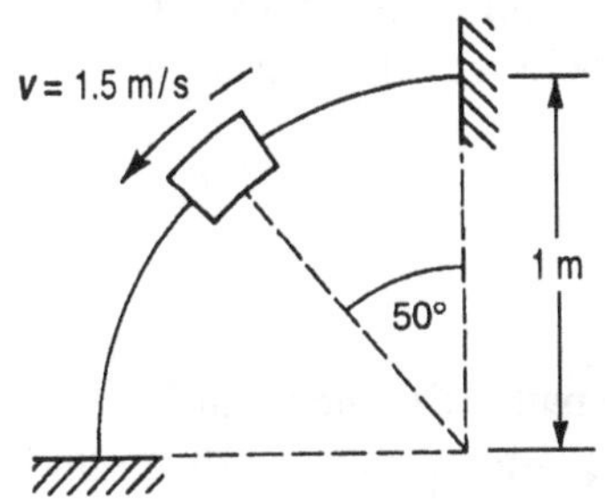

Exhibit 8.24

8.25 Forklift vehicles, Exhibit 8.25, tend to roll over if they are driven too fast while turning. For a vehicle of mass m with a mass center that describes a circle of radius R, find the relationship between the forward speed u and the vehicle dimensions and path radius at the onset of tipping.

a. $u = (Rgb/H)^{0.5}$　　c. $u = (gh)^{0.5}$
b. $u = (RgH/b)^{0.5}$　　d. $u = (bg)^{0.5}$

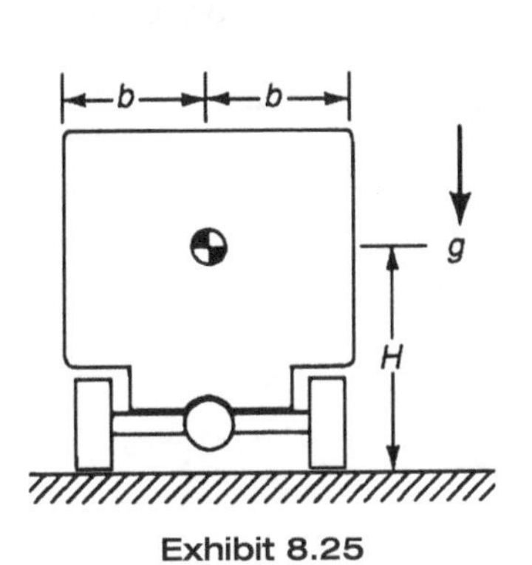

Exhibit 8.25

8.26 A toy rocket of mass 1 kg is placed on a horizontal surface, and the engine is ignited (Exhibit 8.26). The engine delivers a force equal to $(0.25 + 0.5t)$ N, where t is time in seconds, and the coefficient of friction between the rocket and the surface is 0.01. Determine the velocity of the rocket 7 seconds after ignition.

a. 14.0 m/s　　c. 13.3 m/s
b. 3.7 m/s　　d. 26.3 m/s

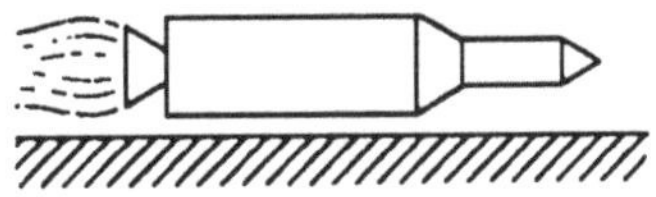

Exhibit 8.26

8.27 A 2000-kg pickup truck is traveling backward down a 10° incline at 80 km/hr when the driver notices through his rearview mirror an object on the roadway. He applies the brakes, and this results in a constant braking (retarding) force of 4000 N. How long does it take the truck to stop?

a. 11.1 s c. 2.3 s
b. 74.9 s d. 13.0 s

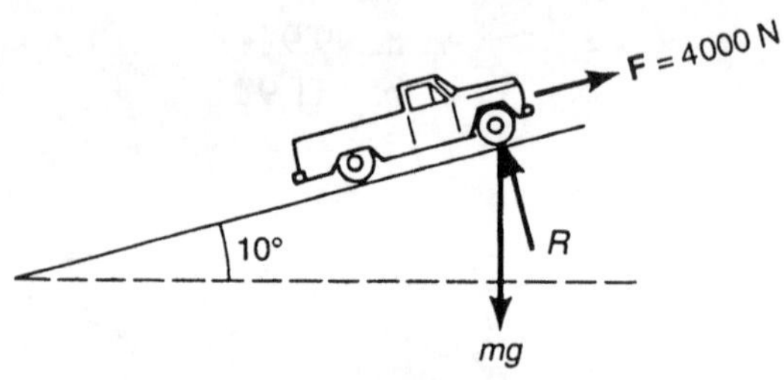

Exhibit 8.27

8.28 In Exhibit 8.28, a particle C of mass 2 kg is sliding down a smooth incline with a velocity of 3 m/s when a horizontal force $P = 15$ N is applied to it. What is the distance traveled by C between the instant when P is first applied and the instant when the velocity of C becomes zero?

a. 1.7 m c. 2.8 m
b. 5.6 m d. 0.9 m

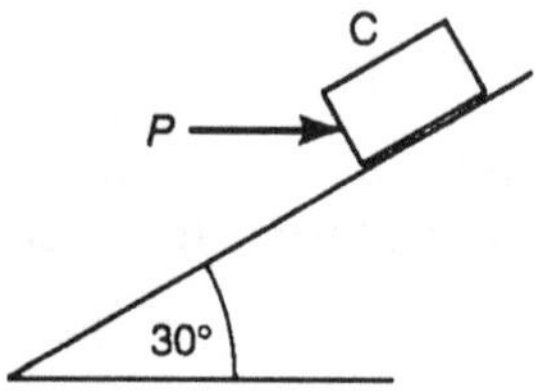

Exhibit 8.28

8.29 A particle moves in a vertical plane along the path ABC shown in Exhibit 8.29. The portion AB of the path is a quarter-circle of radius r and is smooth. The portion BC is horizontal and has a coefficient of friction μ. If the particle has mass m and is released from rest at A, determine the horizontal distance H that the particle will travel along BC before coming to rest.

a. μr c. r/μ
b. $2r$ d. $2r/\mu$

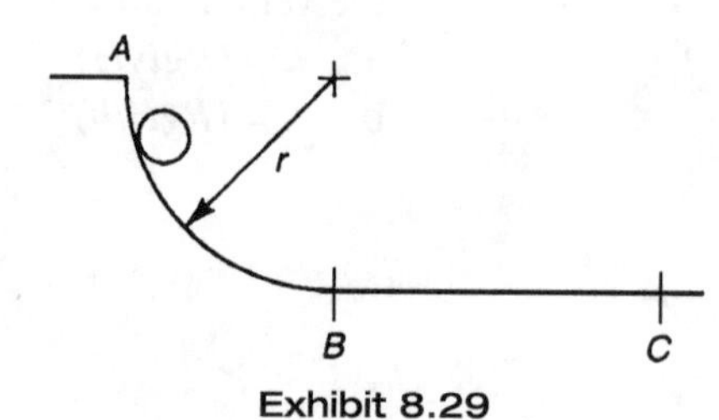

Exhibit 8.29

8.30 In Exhibit 8.30, a block of mass 2 kg is pressed against a linear spring of constant $k = 200$ N/m through a distance Δ on a horizontal surface. When the block is released at A, it travels along the straight horizontal path ADB and traverses point B with a velocity of 1 m/s. If the coefficient of kinetic friction between the block and the floor is 0.2, find Δ.

a. 0.22 m c. 0.26 m
b. 0.12 m d. 0.08 m

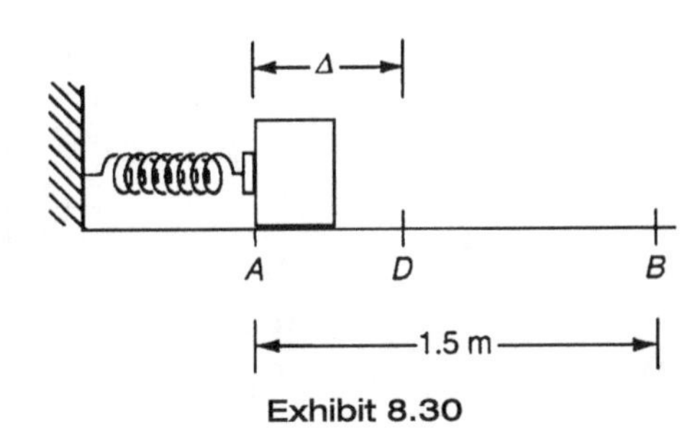

Exhibit 8.30

8.31 In Exhibit 8.31, a 6-kg block is released from rest on a smooth inclined plane as shown. If the spring constant $k = 1000$ N/m, determine how far the spring is compressed. Assume the acceleration of gravity, $g = 10$ m/s^2.

a. 0.40 m c. 0.83 m
b. 0.45 m d. 3.96 m

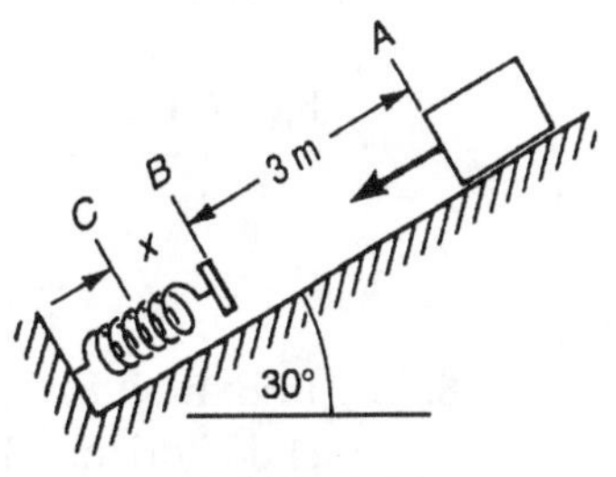

Exhibit 8.31

8.32 A train of joyride cars full of children in an amusement park is pulled by an engine along a straight-level track. It then begins to climb up a 5° slope. At a point B, 50 m up the grade when the velocity is 32 km/h, the last car uncouples without the driver noticing (Exhibit 8.32). If the total mass of the car with its passengers is 500 kg and the track resistance is 2% of the total vehicle weight, calculate the total distance up the grade where the car stops at point C.

a. 260 m c. 48.7 m
b. 37.6 m d. 87.6 m

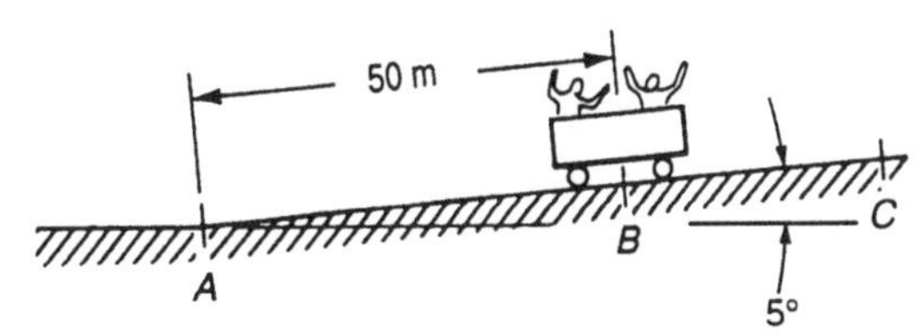

Exhibit 8.32

8.33 Two identical rods, each of mass 4 kg and length 3 m, are rigidly connected as shown in Exhibit 8.33. Determine the moment of inertia of the rigid assembly about an axis through the point *A* and perpendicular to the plane of the paper.

a. 19 kg-m^2 c. 18 kg-m^2
b. 23 kg-m^2 d. 15 kg-m^2

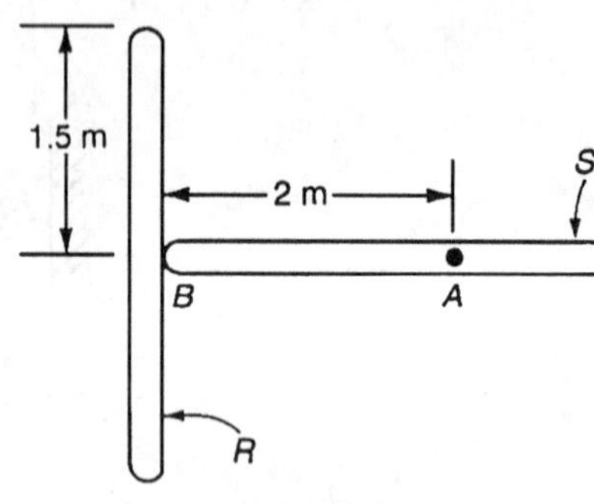

Exhibit 8.33

8.34 A torque motor, represented by the box in Exhibit 8.34, is to drive a thin steel disk of radius 2 m and mass 1.5 kg around its shaft axis. Ignoring the bearing friction about the shaft and the shaft mass, find the angular speed of the disk after applying a constant motor torque of 5 N-m for 5 seconds. The initial angular velocity of the shaft is 1 rad/s.

a. 8.3 rad/s c. 5.2 rad/s
b. 7.3 rad/s d. 9.3 rad/s

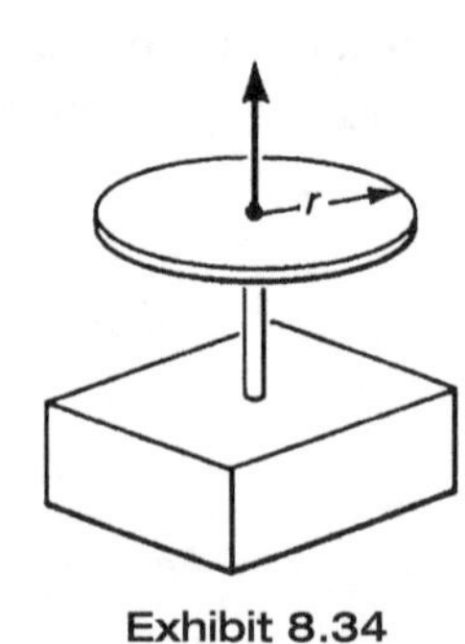

Exhibit 8.34

8.35 In Exhibit 8.35, the uniform slender rod *R* is hinged to a block *B* that can slide horizontally. Determine the horizontal acceleration α that must be given to *B* in order to keep the angle θ constant at 10°, balancing the rod in a tilted position.

a. 1.73 m/s^2 c. 9.81 m/s^2
b. 0 d. 9.66 m/s^2

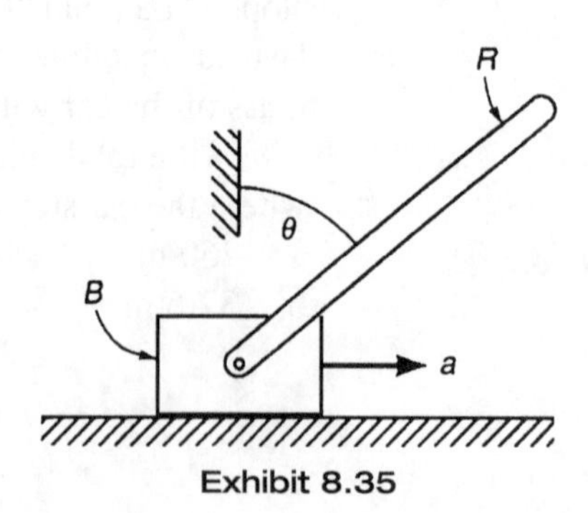

Exhibit 8.35

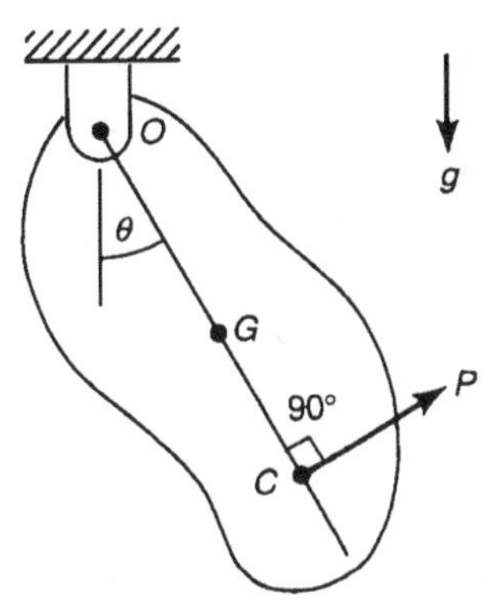

Exhibit 8.36

8.36 In Exhibit 8.36, a force P of constant magnitude is applied to the physical pendulum at point C and remains perpendicular to OC at all times. The pendulum moves in the vertical plane and has mass 3 kg; its mass center is located at G, and the distances are $OG = 1.5$ m, $OC = 2$ m. Also, $P = 10$ N and the radius of gyration of the pendulum about an axis through C and perpendicular to the plane of motion is 0.8 m. Determine the angular acceleration of the pendulum when $\theta = 30°$.

a. 5.31 rad/s^2 counterclockwise
b. 5.31 rad/s^2 clockwise
c. 0.26 rad/s^2 counterclockwise
d. 0.26 rad/s^2 clockwise

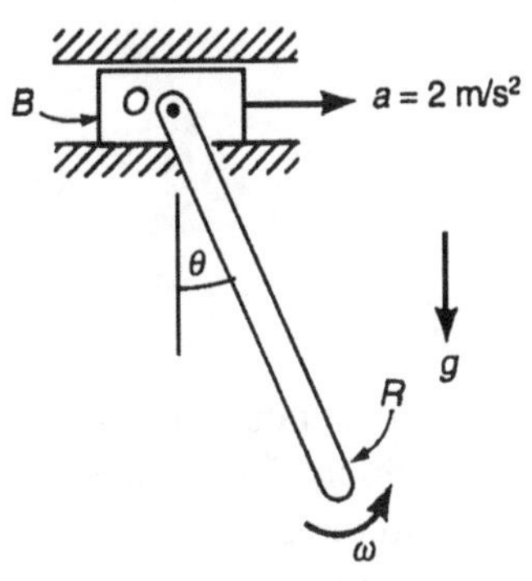

Exhibit 8.37

8.37 In Exhibit 8.37, the block B moves along a straight horizontal path with a constant acceleration of 2 m/s^2 to the right. The uniform slender rod R of mass 1 kg and length 2 m is connected to B through a frictionless hinge and swings freely about O as B moves. Determine the horizontal component of the reaction force at O on the rod when $\theta = 30°$ and $\omega = 2$ rad/s counterclockwise.

a. 4.31 N →
b. 4.31 N ←
c. 6.31 N →
d. 6.31 N ←

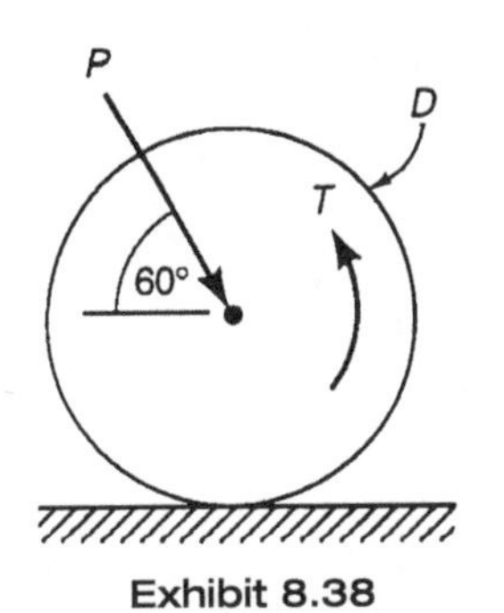

Exhibit 8.38

8.38 In Exhibit 8.38, a homogeneous cylinder rolls without slipping on a horizontal floor under the influence of a force $P = 6$ N and a torque $T = 0.5$ N-m. The cylinder has radius 1 m and mass 2 kg. If the cylinder started from rest, what is its angular velocity after 10 seconds?

a. 8.3 rad/s
b. 6.8 rad/s
c. 1.7 rad/s
d. 0.68 rad/s

Exhibit 8.39

8.39 A slender rod of length 2 m and mass 3 kg is released from rest in the horizontal position (Exhibit 8.39) and swings freely (no hinge friction). Find the angular velocity of the rod when it passes a vertical position.

a. 4.43 rad/s
b. 3.84 rad/s
c. 7.68 rad/s
d. 5.43 rad/s

8.40 The uniform slender bar of mass 2 kg and length 3 m is released from rest in the near-vertical position as shown in Exhibit 8.40, where the torsional spring is undeformed. The rod is to rotate clockwise about O and come gently to rest in the horizontal position. Determine the stiffness k of the torsional spring that would make this possible. The hinge is smooth.

a. 47.8 N-m/rad
b. 37.5 N-m/rad
c. 0.7 N-m/rad
d. 23.8 N-m/rad

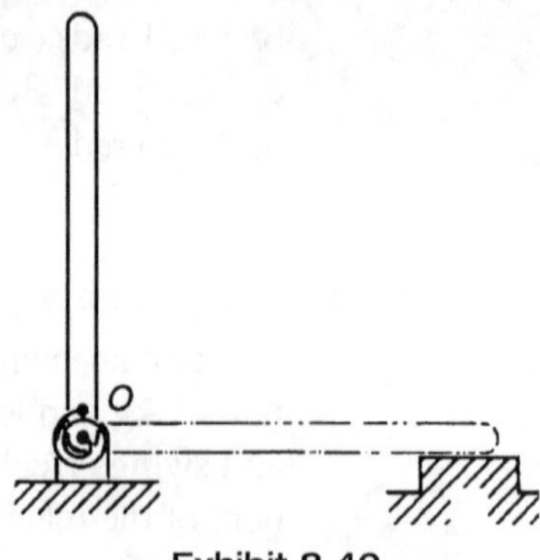

Exhibit 8.40

8.41 A solid homogeneous cylinder is released from rest in the position shown in Exhibit 8.41 and rolls without slip on a horizontal floor. The cylinder has a mass of 12 kg. The spring constant is 2 N/m, and the unstretched length of the spring is 3 m. What is the angular velocity of the cylinder when its center is directly below the point O?

a. 1.33 rad/s
b. 1.63 rad/s
c. 1.78 rad/s
d. 2.31 rad/s

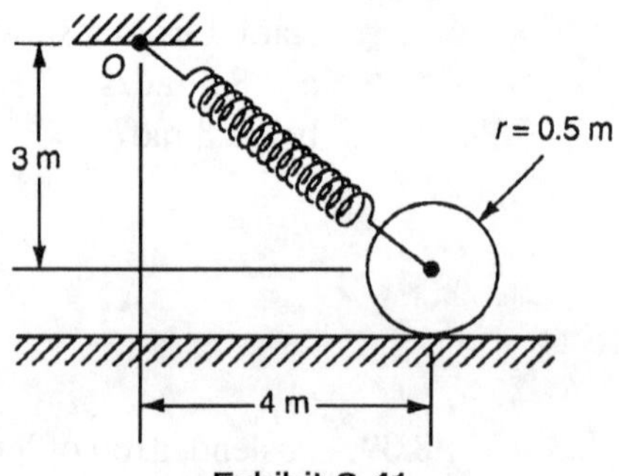

Exhibit 8.41

SOLUTIONS

8.1 c. The density of the plate is:

$$\rho = \frac{\text{mass}}{\text{area}} = \frac{480\ \text{kg}}{\left[(8\times6)+(4\times6)\right]\text{m}^2} = \frac{480\ \text{kg}}{72\ \text{m}^2} = 6.67\ \text{kg/m}^2$$

8.2 **a.** Assuming that the density of the plate is 9 kg/m^2, the mass moment of inertia of the flat plate may be found by:

$$I_x = \frac{1}{3}mh^2 = \frac{1}{3}(\rho A h^2) = \frac{1}{3}\rho(6)(8)^3 + \frac{1}{3}\rho(6)(4)^3$$

$$= \frac{1}{3}(9\ \text{kg/m}^2)(6\ \text{m})(8\ \text{m})^3 + \frac{1}{3}(9\ \text{kg/m}^2)(6\ \text{m})(4\ \text{m})^3$$

$$= 9216 + 1152 = 10{,}368\ \text{kg} - \text{m}^2$$

8.3 **b.** Let the depth of the ditch be h, and set up a vertical s-axis, positive upwards with the origin at A (Exhibit 8.3a). Then, for motion between A and B, $s_0 = 0$, $s = -h$, $v_0 = 4$ m/s, $a = -9.81$ m/s^2, and $t = 6$ s. Substituting these values in the relationship $s = s_0 + v_0 t + (at^2)/2$ yields $-h = 0 + 4(6) - 9.81(6)^2/2$ or $h = 152.6$ m.

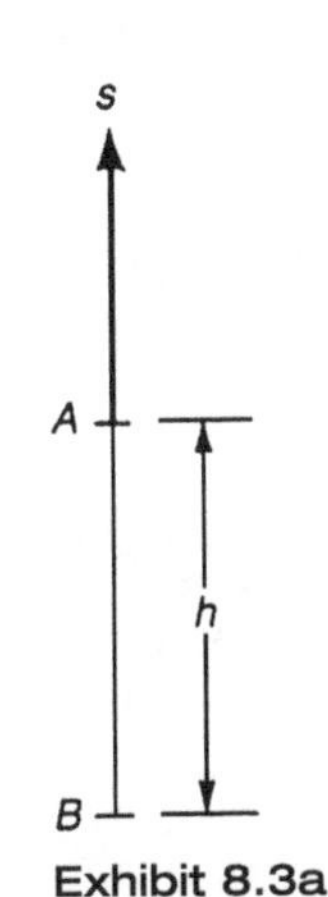

Exhibit 8.3a

8.4 **d.** Set up a horizontal s-axis with origin at O, and positive to the right (Exhibit 8.4a). Then, $sO = 2$ m, and $v = ds/dt$, so

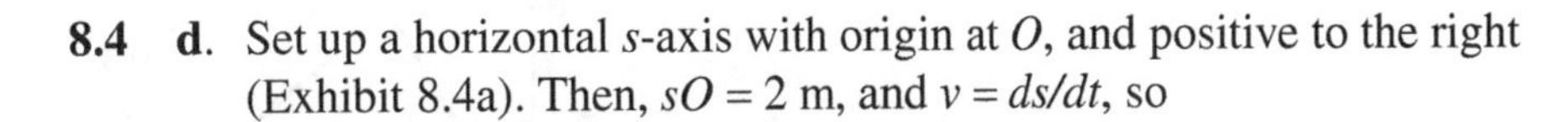

$$\int_{s_0}^{s} ds = \int_0^t v\,dt = \int_0^t (3t^2 - t + 2)\,dt$$

or $s = s_0 + t^3 - t^2/2 + 2t$. Substituting values into the equation yields $s(2\text{s}) = 12$ m.

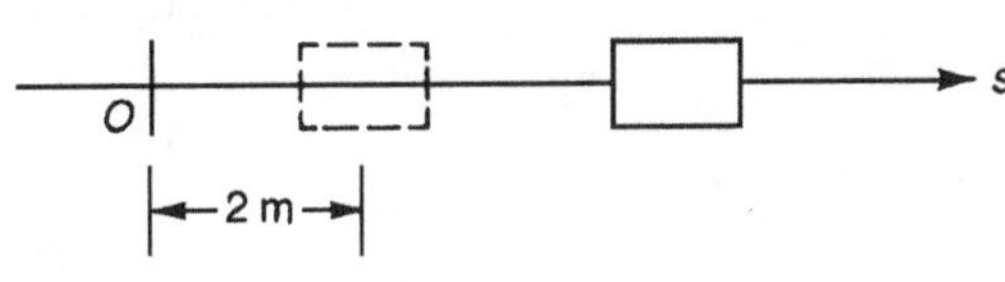

Exhibit 8.4a

8.5 d. The velocity-time curve for this particle is shown below the acceleration curve in Exhibit 8.5a. (Velocity at any given instant t_1 equals the area under the acceleration curve from time 0 to the time t_1, plus the initial velocity). Any area above the $a = 0$ line is counted as positive, and any area below is counted as negative.

The displacement D is the total area under the velocity curve between $t = 0$ and $t = 8$ s. The area of a triangle is $0.5(b)(h)$. Hence, $D = 0.5(6)(8) - 0.5(2)(4) = 20$ m.

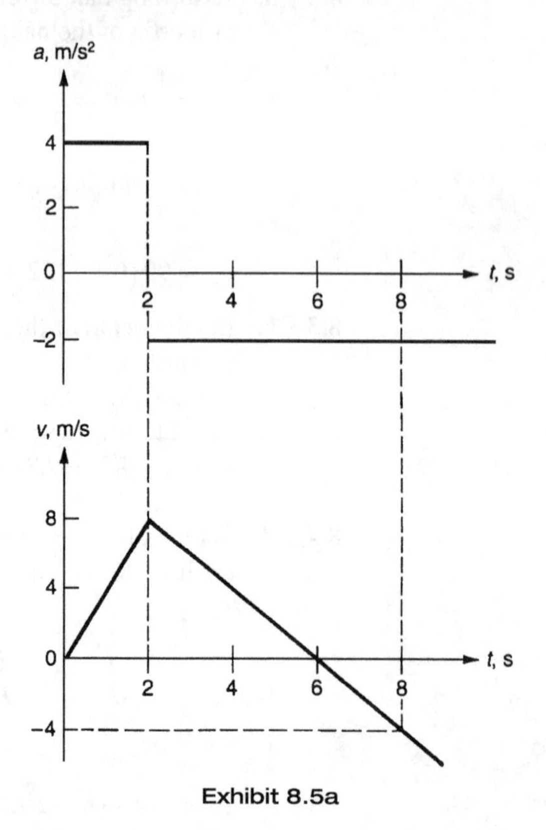

Exhibit 8.5a

8.6 a. When the ball reaches its highest position, B, the vertical component of the velocity of the ball is zero. Applying Equation (8.12) vertically,

$$v_y^2 = v_{yo}^2 + 2a_y(y - y_0)$$
$$0 = (40 \sin 60°)^2 - 2(9.81)(y - y_0)$$

Thus, $y - y_0 = 61.2$ m and $H = 63.2$ m.

8.7 **c.** Refer to Exhibit 8.7a.

Horizontal Motion:

$$x = x_0 + v_{x0}t + (a_x t^2)/2$$
$$140 = 0 + v_0 t + 0$$
$$t = 1.56 \text{ s}$$

Vertical Motion:

$$y = y_0 + y_{y0}t + (a_y t^2)/2$$
$$-12 = 0 + 0 - 0.5(9.81)t^2$$
$$v_o = 140/\text{t} = 140/1.56 = 89.7 \text{ m/s}$$

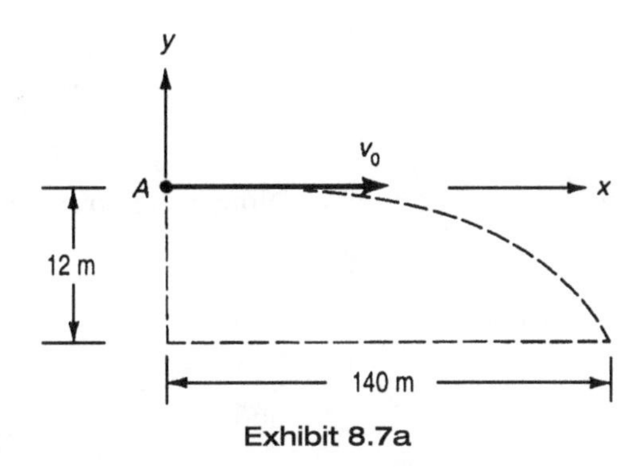

Exhibit 8.7a

8.8 **c.** In Exhibit 8.8a, we have

$$r = 1 \text{ m}; \quad \dot{r} = -3 \text{ m/s}; \quad \ddot{r} = 0$$

Since $\theta = (t^2 + t - 2)$ rad, differentiation gives

$$\dot{\theta} = (2t + 1) \text{ rad/s; at } t = 1 \text{ s, } \dot{\theta} = 2(1) + 1 = 3 \text{ rad/s}$$

Also, $\ddot{\theta} = 2 \text{ rad/s}^2$.

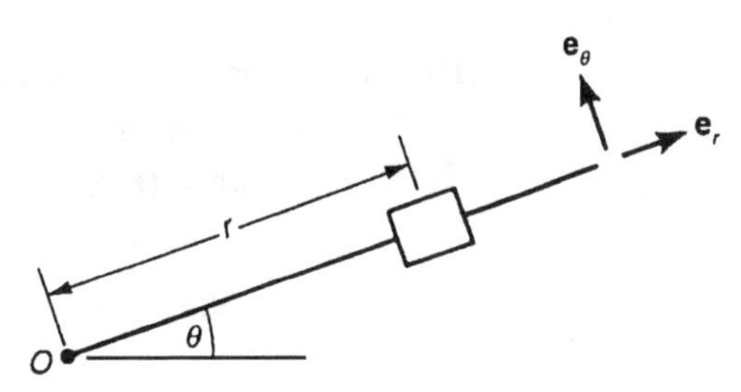

Exhibit 8.8a

Acceleration of the collar is therefore given by

$$\begin{aligned}\mathbf{a} &= (\ddot{r} - r\dot{\theta}^2)\mathbf{e}_r + (r\ddot{\theta} + 2\dot{r}\dot{\theta})\mathbf{e}_\theta \\ &= [0 - 1(3)^2]\mathbf{e}_r + [1(2) + 2(-3)(3)]\mathbf{e}_\theta \ \mathbf{m/s^2} \\ &= [-9\mathbf{e}_r - 16\mathbf{e}_\theta] \ \mathbf{m/s^2}\end{aligned}$$

Hence, $|a| = [(-9)^2 + (-16)^2]^{0.5} = 18.4 \text{ m/s}^2$.

8.9 d. Refer to Exhibit 8.9a. The velocity is

$$v = \left(v_r^2 + v_\theta^2\right)^{0.5} \quad \text{where } v_r = \dot{r} \text{ and } v_\theta = r\dot{\theta}$$

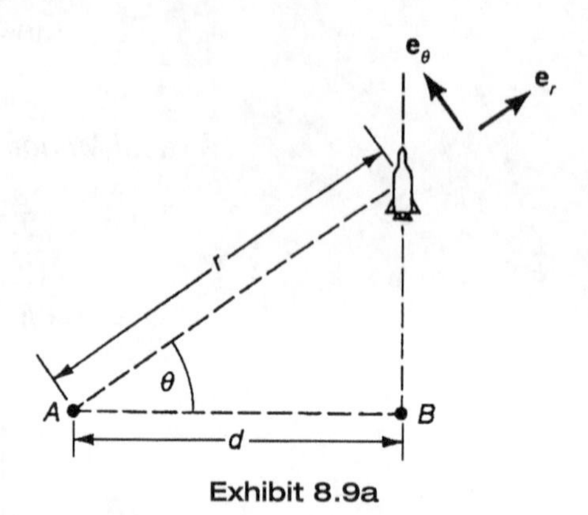

Exhibit 8.9a

Now, $r = d/\cos\theta$ so

$$v_r = \dot{r} = \frac{d\dot{\theta}\sin\theta}{\cos^2\theta} = \frac{d\dot{\theta}\tan\theta}{\cos\theta}$$

and

$$v_\theta r\dot{\theta} = \frac{d\dot{\theta}}{\cos\theta}$$

Thus

$$v^2 = v_r^2 + v_\theta^2 = \frac{d^2\dot{\theta}^2(\tan^2\theta + 1)}{\cos^2\theta} = \frac{d^2\dot{\theta}^2}{\cos^4\theta}; \quad \ddot{r} = 0$$

and

$$v = \frac{d\dot{\theta}}{\cos^2\theta} = \frac{1800(0.1)(2)}{1} \text{ m/s} = 360 \text{ m/s}$$

8.10 d. At the highest point, the vertical component of velocity is zero, and the acceleration is normal to the path. Thus, $v = v_x = v_{x0} + a_x t = 50$ cos 30° + 0. Also, the normal component of the acceleration is $a_n = -a_y = 9.81$ m/s^2. But $a_n = v^2/\rho$. Hence $\rho = v^2/a_n$ = (50 cos 30°)^2/9.81 = 191 m.

8.11 d. The tangential acceleration is given as $a_t = 3$ m/s^2; the normal acceleration is $a_n = v^2/\rho$ = [(20)^2/110] m/s^2 = 3.6 m/s^2. Hence, the total acceleration is a = [(3)^2 + (3.6)^2]$^{0.5}$ m/s^2 = 4.7 m/s^2.

8.12 c. The normal acceleration is given by $a_n = v^2/\rho$. So, $\rho = v^2/a_n$. Substituting values (converted to consistent units), we obtain

$$\rho = \frac{[800(1000)]^2}{[60(60)]^2(5)(9.81)} = 1007 \text{ m}$$

8.13 **d.** At any given instant, the angular acceleration of the disk is $\alpha = -k\omega = d\omega/dt$. So,

$$\int_{\omega_0}^{\omega} \frac{d\omega}{\omega} = -k \int_0^t dt$$

$$\ln \omega|_{\omega_0}^{\omega} = \ln(\omega/\omega_0) = -kt \text{ and } t = -(1/k)\ln(\omega/\omega_0) = -(1/1.2)\ln(0.5) = 0.6\,s$$

8.14 **a.** Initially, $\omega_0 = 0$. At $t = 30$ s, $\omega = 300$ rpm $= [300\,(2\pi)/60]$ rad/s $= 10\pi$ rad/s. For uniformly accelerated rotational motion,

$$\omega = \omega_0 + \alpha t$$
$$\alpha = (\omega - \omega_0)/t = (10\pi - 0)/30 \text{ rad/s}^2 = \pi/3 \text{ rad/s}^2$$
$$\omega(5) = \omega_0 + \alpha t = 0 + (\pi/3)(5) \text{ rad/s} = 5\pi/3 \text{ rad/s}$$
$$v_P(5) = \omega(5)r = (5\pi/3)\,(2) \text{ m/s} = 10.5 \text{ m/s}$$

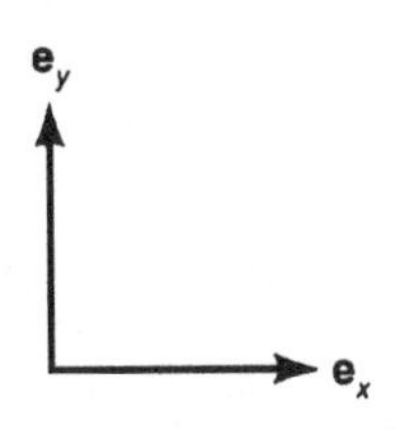

Exhibit 8.15a

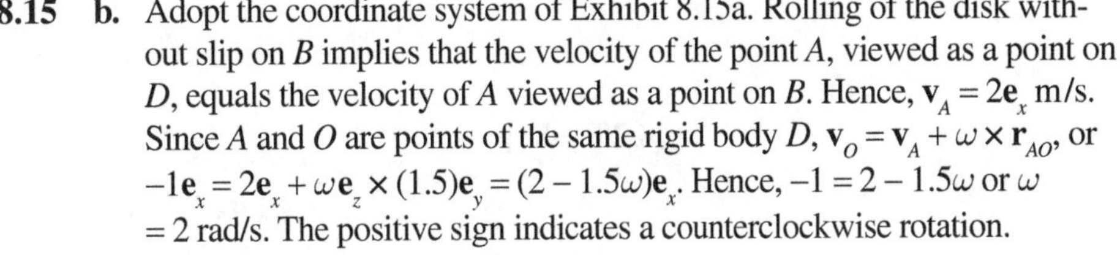

8.15 **b.** Adopt the coordinate system of Exhibit 8.15a. Rolling of the disk without slip on B implies that the velocity of the point A, viewed as a point on D, equals the velocity of A viewed as a point on B. Hence, $\mathbf{v}_A = 2\mathbf{e}_x$ m/s. Since A and O are points of the same rigid body D, $\mathbf{v}_O = \mathbf{v}_A + \omega \times \mathbf{r}_{AO}$, or $-1\mathbf{e}_x = 2\mathbf{e}_x + \omega\mathbf{e}_z \times (1.5)\mathbf{e}_y = (2 - 1.5\omega)\mathbf{e}_x$. Hence, $-1 = 2 - 1.5\omega$ or $\omega = 2$ rad/s. The positive sign indicates a counterclockwise rotation.

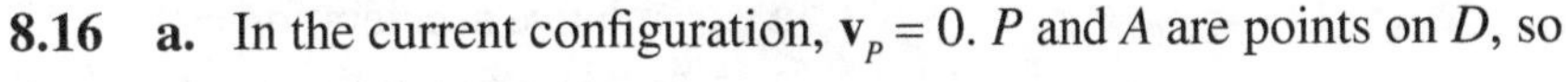

8.16 **a.** In the current configuration, $\mathbf{v}_P = 0$. P and A are points on D, so

$$\mathbf{v}_A = \mathbf{v}_P + \omega_D \times \mathbf{r}_{PA} = 0 - 3\mathbf{e}_z \times (0.5\mathbf{e}_y - 0.4\mathbf{e}_x)$$
$$= (1.5\mathbf{e}_x + 1.2\mathbf{e}_y) \text{ m/s}$$

Similarly, because B and A are points on R, $\mathbf{v}_B = \mathbf{v}_A + \omega_R \times \mathbf{r}_{AB}$. Therefore,

$$v_B\mathbf{e}_x = 1.5\mathbf{e}_x + 1.2\mathbf{e}_y + \omega_R\mathbf{e}_z \times (-2\mathbf{e}_x - 0.5\mathbf{e}_y)$$
$$= (1.5 + 0.5\omega_R)\mathbf{e}_x + (1.2 - 2\omega_R)\mathbf{e}_y$$

Equating the coefficients of $\mathbf{e}_y$ yields $0 = 1.2 - 2\omega_R$, or $\omega_R = 0.6$ rad/s. The positive sign indicates that ω_R is in the positive $\mathbf{e}_z$ direction, so the rotation is counterclockwise.

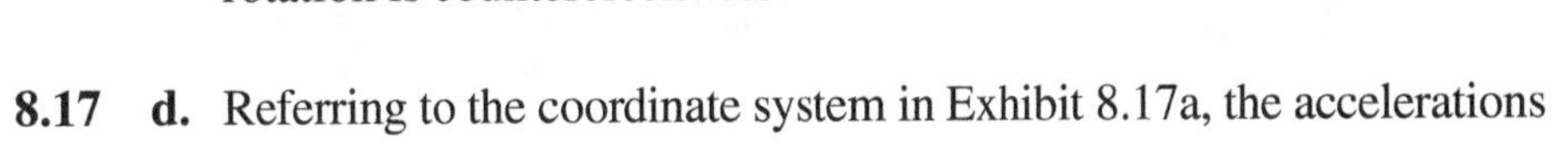

8.17 **d.** Referring to the coordinate system in Exhibit 8.17a, the accelerations are

$$\mathbf{a}_A = \mathbf{a}_O + \alpha \times \mathbf{r}_{OA} + \omega \times (\omega \times \mathbf{r}_{OA})$$
$$\mathbf{a}_O = 0 \text{ (constant velocity)}$$

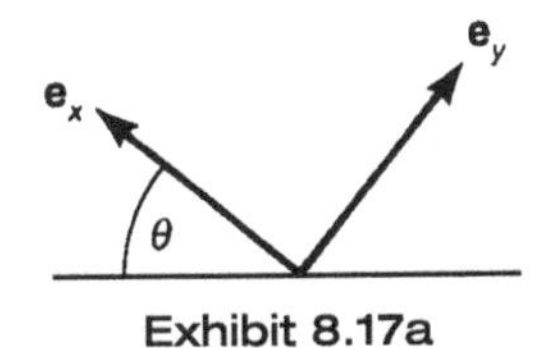

Exhibit 8.17a

Since $\theta = (0.5t^2 - t)$ rad, $\dot{\theta} = (t - 1)$rad/s and $\ddot{\theta} = 1$ rad/s^2.

So, at $t = 2$ s, $\omega = \dot{\theta}\mathbf{e}_z = 1\mathbf{e}_z$ rad/s. Since $\boldsymbol{a} = \ddot{\theta}\,\mathbf{e}_z = 1\mathbf{e}_z$ rad/s^2,

$$\mathbf{a}_A = 0 + \mathbf{e}_z \times (2\mathbf{e}_x) + \mathbf{e}_z \times [\mathbf{e}_z \times (2\mathbf{e}_x)] = (2\mathbf{e}_y - 2\mathbf{e}_x) \text{ m/s}^2$$

Hence, $|\mathbf{a}_A| = (2^2 + 2^2)^{0.5}$ m/s^2 = 2.8 m/s^2.

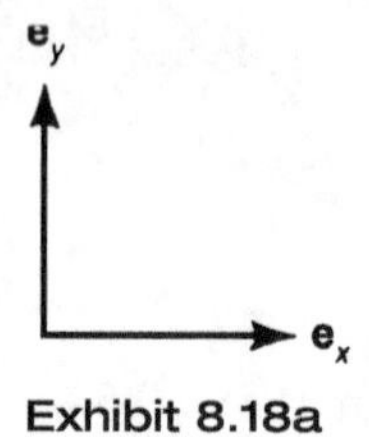

Exhibit 8.18a

8.18 b. With the coordinate system in Exhibit 8.18a, we can write

$$\mathbf{a}_O = 2\mathbf{e}_x \text{ m/s}^2;\ \omega = 0;\ \alpha = -2.5\mathbf{e}_z \text{ rad/s}^2$$

Now, $\mathbf{a}_G = \mathbf{a}_O + \mathbf{a} \times \mathbf{r}_{OG} + \omega \times (\omega \times \mathbf{r}_{OG})$ where $\mathbf{r}_{OG} = -1\mathbf{e}_y$ m. Hence

$$\mathbf{a}_G = [2\mathbf{e}_x - 2.5\mathbf{e}_z \times (-1)\mathbf{e}_y + 0] \text{ m/s}^2 = -0.5\mathbf{e}_x \text{ m/s}^2$$

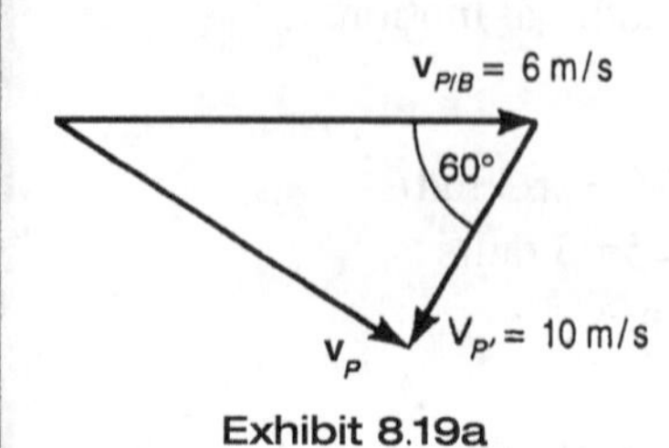

Exhibit 8.19a

8.19 a. We know that $\mathbf{v}_P = \mathbf{v}_{P/B} + \mathbf{v}_{P'}$ where $\mathbf{v}_{P/B}$ is the velocity of P relative to B and $\mathbf{v}_{P'}$ is the velocity of the point P' of the block that coincides with P at the instant under consideration (coincident point velocity). Here, $|\mathbf{v}_{P/B}| = 6$ m/s and $|\mathbf{v}_{P'}| = 10$ m/s (velocity of block). Because v_P is the vector sum of $\mathbf{v}_{P/B}$ and $\mathbf{v}_{P'}$, as shown in Exhibit 8.19a, we can use the law of cosines,

$$(\mathbf{v}_P)^2 = 10^2 + 6^2 - 2(10)(6)\cos 60° = 76$$
$$\mathbf{v}_P = 8.7 \text{ m/s}$$

8.20 d.

$$\mathbf{a}_P = \mathbf{a}_{P/D} + \mathbf{a}_{P¢}\rho' + \mathbf{a}_C$$

Here,

$\mathbf{a}_{P/D}$ = relative acceleration = 0

$\mathbf{a}_{P'}$ = acceleration of the point of D that is coincident with P at the instant under consideration:

$$\begin{aligned}\mathbf{a}_{P'} &= \mathbf{a}_O + \alpha \times \mathbf{r}_{OC} + \omega \times (\omega \times \mathbf{r}_{OC})\\ &= 0 + 0 + (-5\mathbf{e}_z) \times [(-5\mathbf{e}_z) \times (0.1\mathbf{e}_y)]\\ &= -2.5\mathbf{e}_y \text{ m/s}^2\end{aligned}$$

$\mathbf{a}_C$ = Coriolis acceleration = $2\omega \times \mathbf{v}_{P/D}$

$\mathbf{a}_C = 2(-5\mathbf{e}_z) \times 2\mathbf{e}_x = -20\mathbf{e}_y$ m/s^2

Finally,

$$\mathbf{a}_P = [-2.5\mathbf{e}_y - 20\mathbf{e}_y] \text{ m/s}^2 = -22.5\mathbf{e}_y \text{ m/s}^2$$

8.21 **d.** The free-body diagram is shown in Exhibit 8.21a. Apply Newton's second law:

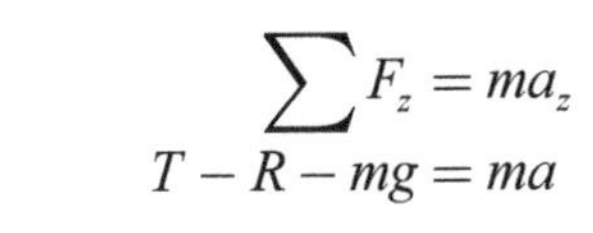

$$\sum F_z = ma_z$$
$$T - R - mg = ma$$

$$a = [(T - R)/m] - g = [(100 - 2z)N/5\text{ kg}] - 9.81\text{ m/s}^2$$

so

$$a = 10.2 - 0.4z = \frac{v\,dv}{dz}$$

and

$$\int_0^H (10.2 - 0.4z)\,dz = \int_{10}^0 v\,dv$$

where H is the highest height attained. Note also that $v = 0$ at this height. Integration yields,

$$10.2H - 0.2H^2 = [v^2/2]_{10}^0 = -50$$

or

$$0.2H^2 - 10.2H - 50 = 0$$

Solving this quadratic (and discarding the negative value) gives the maximum height,

$$H = 55.5\text{ m}$$

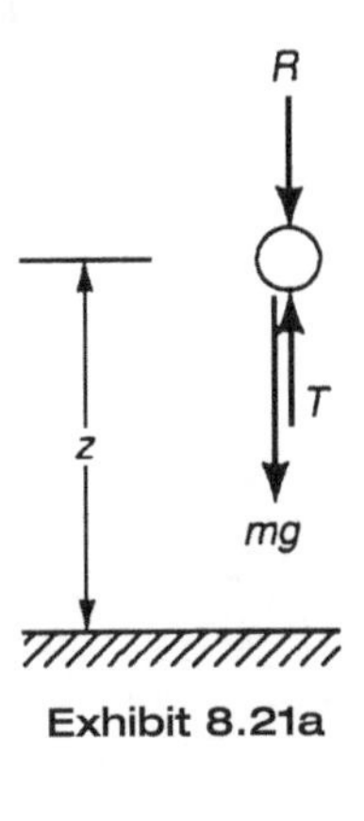

Exhibit 8.21a

8.22 **b.** Refer to Exhibit 8.22a. Apply Newton's second law in the x and y directions.

$$\sum F = ma_x \qquad \textbf{(i)}$$

$$P + P\cos 30° - 0.2N - 5(9.81)\sin 30° = 5(2)$$

$$\Sigma F_y = 0$$
$$N + P\sin 30° - 5(9.81)\cos 30° = 0 \qquad \textbf{(ii)}$$

Solving Equations (i) and (ii) simultaneously yields $P = 21.9$ N.

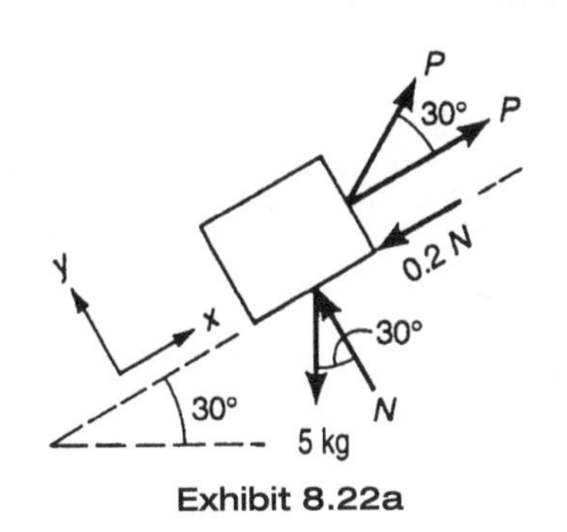

Exhibit 8.22a

8.23 **a.** In Exhibit 8.23a, apply Newton's second law in the radial and transverse directions.

$$\sum F_\theta = ma_\theta$$

$$N - mg\cos\theta = m(r\ddot{\theta} + 2\dot{r}\dot{\theta}) \qquad \textbf{(i)}$$

$$\sum F_r = ma_r$$

$$\mu N - P - mg\sin\theta = m(\ddot{r} - r\dot{\theta}^2) \qquad \textbf{(ii)}$$

Substitute values into Equations (i) and (ii):

$$N - 2(9.81)\ \cos 30° = 2[0 + 2(-1)(5)] \qquad \textbf{(iii)}$$

$$0.1N - P - 2(9.81)\sin 30° = 2[0 - 0.4(5)^2] \qquad \textbf{(iv)}$$

From Equation (iii), $N = -3.0$ newtons. Substituting this value into Equation (iv) gives $P = 9.9\text{N}$.

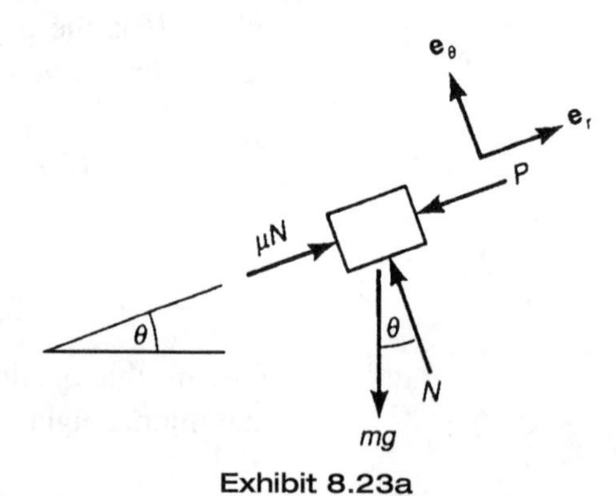

Exhibit 8.23a

8.24 **a.** Apply Newton's second law to the diagrams in Exhibit 8.24:

$$\sum F_n = ma_n$$

$$mg\cos\theta - N = mv^2/\rho$$

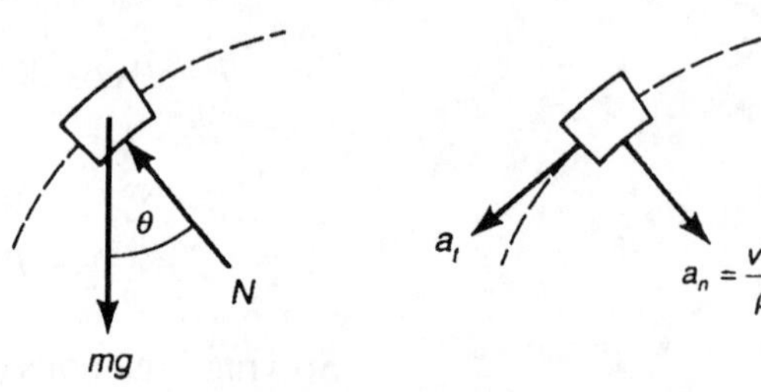

Exhibit 8.24a

or

$$\begin{aligned} N &= m[g\cos\theta - v^2/\rho] \\ &= (3)[9.81\cos 50° - 1.5^2/1] \\ &= 12.2 \text{ N} \end{aligned}$$

The positive sign indicates that N is directed as shown in the free-body diagram, Exhibit 8.24a.

8.25 **a.** At the onset of tipping, the free-body diagram and the inertia force diagram are as shown in Exhibit 8.25a. Take moments about point A:

$$mgb = m(u^2/R)H$$

Hence,

$$u = (Rgb/H)^{0.5}$$

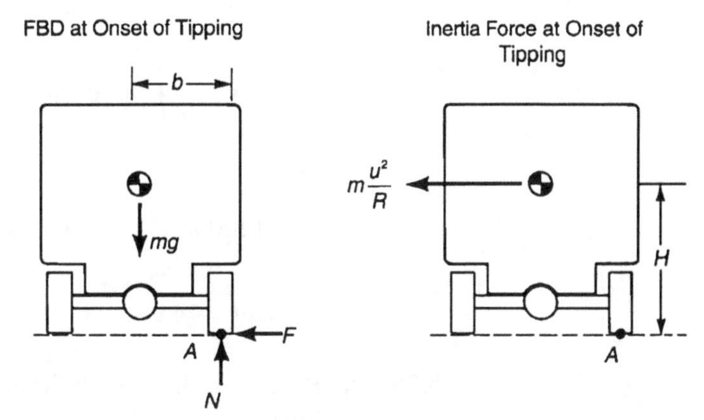

Exhibit 8.25a

8.26 **c.** In Exhibit 8.26a, the sum of the forces in the vertical direction yields

$$N = mg$$

Apply the impulse-momentum principle in the horizontal direction [Equation (8.32)]:

$$\int_{t_1}^{t_2} F_{\text{horiz}}\, dt = mv_2 - mv_1$$

$$F_{\text{horiz}} = (0.25 + 0.5t) - 0.1N$$

Thus

$$\int_0^7 [0.25 + 0.5t - (0.01)(9.81)]dt = (1)v_2 - 0$$

or

$$0.25t + 0.25t^2 - 0.098t\Big|_0^7 = v_2 = 13.3 \text{ m/s}$$

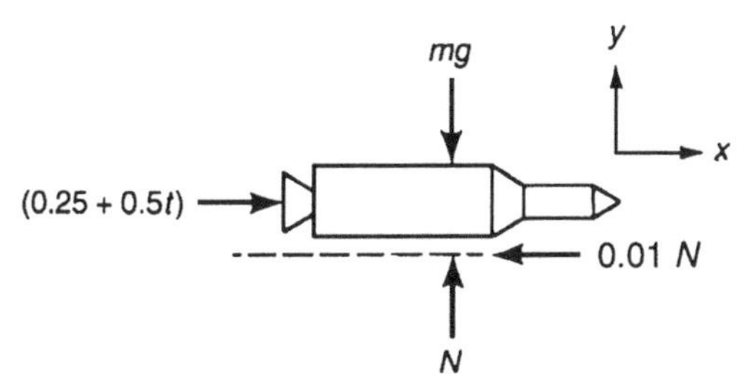

Exhibit 8.26a

8.27 **b.** Apply the impulse-momentum principle between the instant t_1 when the brakes are applied and the instant t_2 when the truck comes to a stop:

$$\int_{t_1}^{t_2} \mathbf{F}dt = m\mathbf{v}_2 - m\mathbf{v}_1$$

In the direction tangent to the road surface,

$$\int_0^t (mg \sin 10° - F)dt = 0 - mv_1$$

or

$$[2000(9.81) \sin 10° - 4000]t = -2000\frac{80(1000)}{60(60)}$$

so that $t = 74.9$ s.

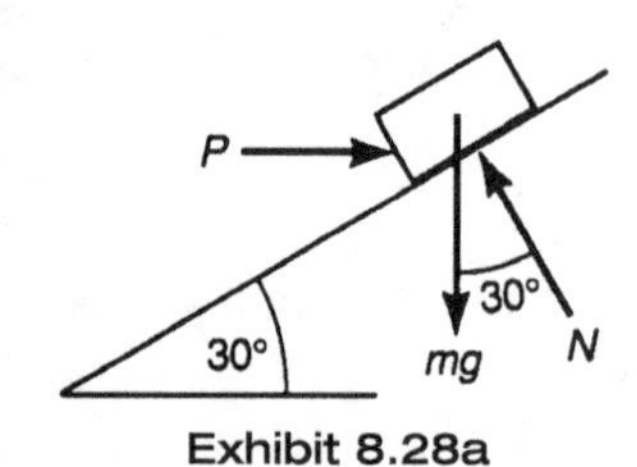

Exhibit 8.28a

8.28 **c.** Refer to Exhibit 8.28a. The subscript 1 is used for the instant when the force P is used first applied, and the subscript 2 is used for the instant when the block comes to rest. Apply the work-energy principle between 1 and 2:

$$W_{1-2} = T_2 - T_1$$
$$(mg \sin 30° - P\cos 30°)\Delta x = 0 - \tfrac{1}{2}mv_1^2$$

which yields

$$\Delta x = \frac{\frac{1}{2}mv_1^2}{P\cos 30° - mg \sin 30°} = \frac{0.5(2)3^2}{15\cos 30° - 2(9.81)\sin 30°} = 2.83 \text{ m}$$

8.29 **c.** Consult Exhibit 8.29a. Apply the work-energy principle between A and B, and then between B and C.

$$A \to B: \quad W_{A-B} = T_B - T_A$$

$$mgr = \frac{1}{2}mv_B^2 - 0 \qquad \textbf{(i)}$$

$$B \to C: \quad W_{B-C} = T_C - T_B$$

$$N = mg \qquad \textbf{(ii)}$$

$$-\mu NH = 0 - \frac{1}{2}mv_B^2 \qquad \textbf{(iii)}$$

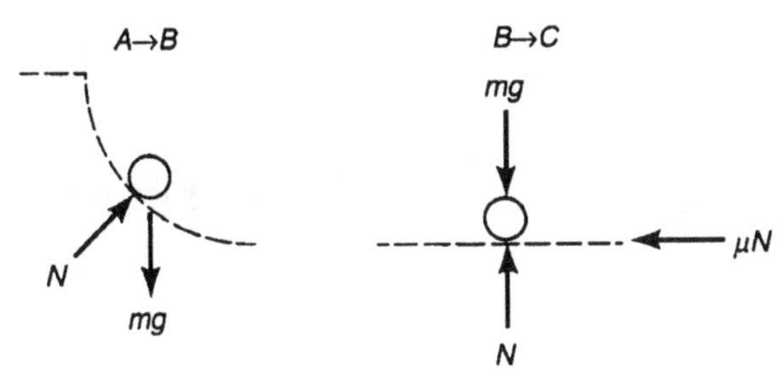

Exhibit 8.29a

Substituting Equations (i) and (ii) into Equation (iii),

$$-\mu mgH = -mgr, \quad \text{or} \quad H = r/\mu$$

8.30 **c.** Let D be the position at which the spring has its natural (unstretched) length. Apply the work-energy principle (Exhibit 8.30a) from A to D:

$$W_{A-D} = T_D - T_A$$

$$\frac{1}{2}k\Delta^2 - \mu N\Delta = \frac{1}{2}mv_D^2 - 0$$

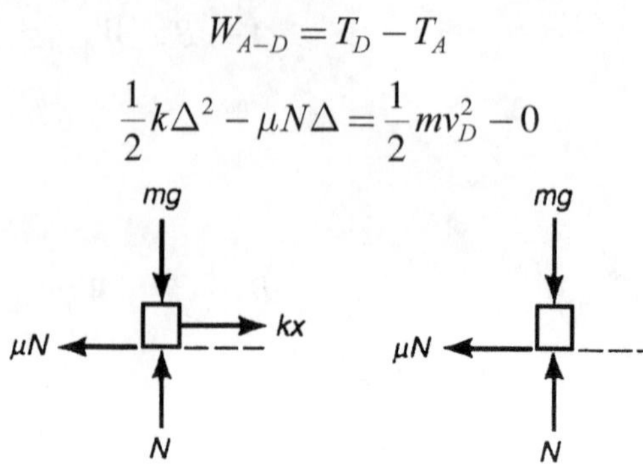

Exhibit 8.30a

Since $N = mg$, we have

$$\frac{1}{2}mv_D^2 = \frac{1}{2}k\Delta^2 - \mu mg\Delta \qquad \text{(i)}$$

Now apply the work-energy principle from D to B:

$$W_{D-B} = T_B - T_D$$

$$-\mu N(1.5 - \Delta) = \frac{1}{2}mv_B^2 - \frac{1}{2}mv_D^2$$

Again, since $N = mg$, and $\frac{1}{2}mv_D^2$ is given by Equation (i), we have

$$-\mu mg(1.5 - \Delta) = \frac{1}{2}mv_B^2 - \frac{1}{2}k\Delta^2 + \mu mg\Delta$$

or

$$\Delta = \sqrt{\frac{2\left(1.5\mu mg + 0.5mv_B^2\right)}{k}} = 0.26 \text{ m}$$

8.31 **b.** Apply the work-energy principle between A and B, and then between B and C.

$$W_{A-B} = T_B - T_A$$

With the forces shown in Exhibit 8.31a,

$$mg \sin 30°(3) = \frac{1}{2}mv_B^2 - 0$$

$$W_{B-C} = T_C - T_B$$

$$-\frac{1}{2}kx^2 + mg\sin 30°(x) = 0 - \frac{1}{2}mv_B^2 = -mg\sin 30°(3)$$

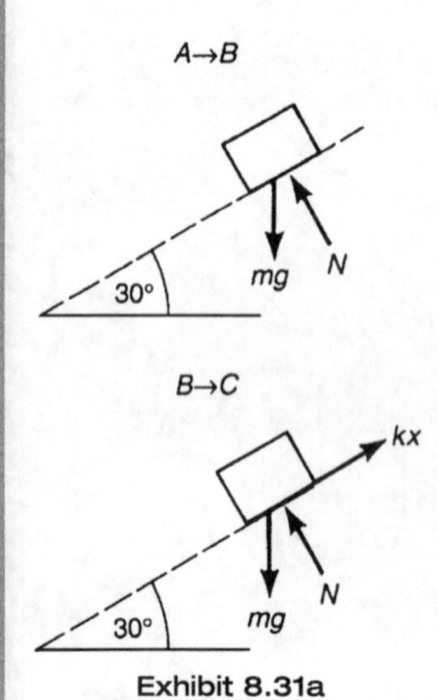

Exhibit 8.31a

Substituting values, we obtain the quadratic equation $500x^2 - 30x - 90 = 0$, which can be solved to yield $x = 0.45$ m.

8.32 d. Using the diagram in Exhibit 8.32a, apply the work-energy principle between B and C:

$$W_{B-C} = T_c - T_B$$

$$-mg \sin 5°(x) - 0.02\, mgx = 0 - \frac{1}{2} mv_B^2$$

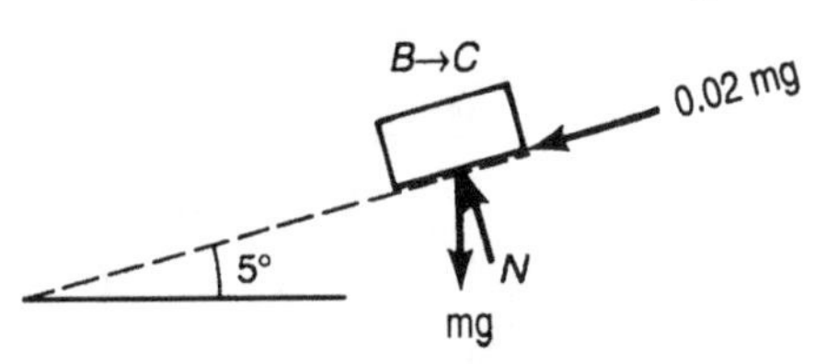

Exhibit 8.32a

or

$$x = \frac{\frac{1}{2} mv_B^2}{mg \sin 5° + 0.02\, mg} = \frac{0.5(500)\left[\frac{(32 \times 1000)}{60 \times 60}\right]^2}{500(9.81)\sin 5° + 0.02(500)9.81} = 37.6 \text{ m}$$

Total distance up the grade is (50 + 37.6) m = 87.6 m.

8.33 b. Consult Exhibit 8.33a. The moment of inertia of each rod about its mass center is

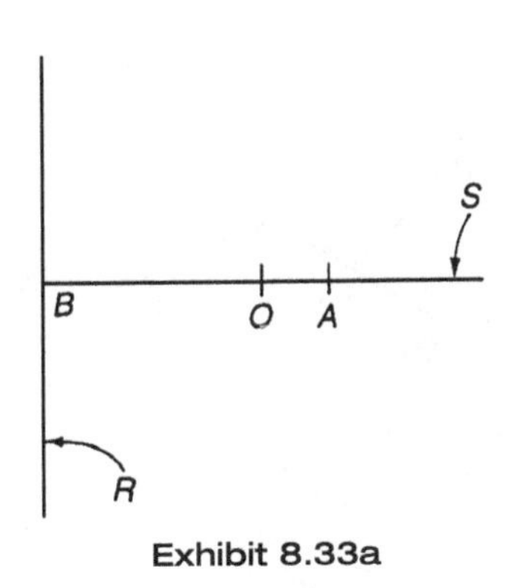

Exhibit 8.33a

$$I_{R/B} = I_{S/O} = \frac{1}{12} ml^2 = \frac{1}{12}(4)(3)^2 = 3 \text{ kg-m}^2$$

Here, O is the mass center of S, and B is the mass center of R. Apply the parallel axes theorem:

$$I_{R/A} = I_{R/B} + m(2)^2 = 3 + 4(2)^2 = 19 \text{ kg-m}^2$$

$$I_{S/A} = I_{S/O} + m(2 - 1.5)^2 = 3 + 4(0.5)^2 = 4 \text{ kg-m}^2$$

And, for the assemblage,

$$I_A = I_{R/A} + I_{S/A} = (19 + 4) \text{kg-m}^2 = 23 \text{ kg-m}^2$$

8.34 d. Since $M = I\alpha = (1/2)mr^2\alpha$, $\alpha = M/(0.5mr^2)$. Substituting values, we have $\alpha = 1.67$ rad/s^2. Because this angular acceleration is constant, the final angular velocity is given by

$$\omega = \omega_0 + \alpha t = 1 + 1.67(5) = 9.3 \text{ rad/s}$$

8.35 a. When the desired configuration is achieved, the rod is in translation. The free-body diagram and the inertia force diagram for the rod are shown in Exhibit 8.35a. Taking moments about point O,

$$mg(l/2)\sin\theta = ma(l/2)\cos\theta$$

where l is the length of the rod. Thus,

$$a = g\tan\theta = 9.81\tan 10° = 1.73\ \text{m/s}^2$$

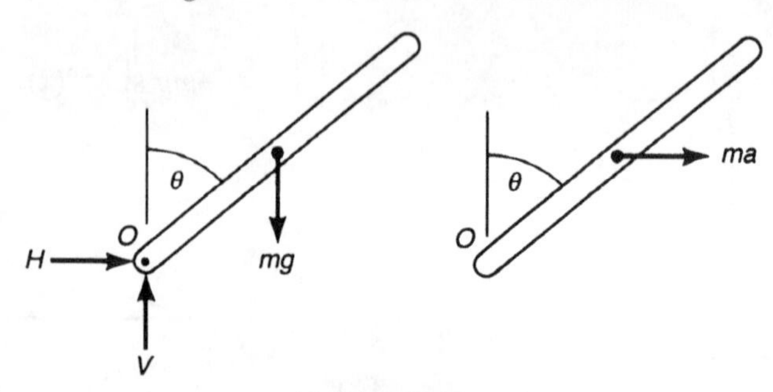

Exhibit 8.35a

8.36 d. From the diagrams in Exhibit 8.36a, and taking moments about O,

$$PH - mgl\ \sin\theta = I_G\alpha + ml^2\alpha$$

so that

$$\alpha = (PH - mgl\ \sin\theta)/(I_G + ml^2)$$

Now,

$$I_C = I_G + m(H - l)^2$$

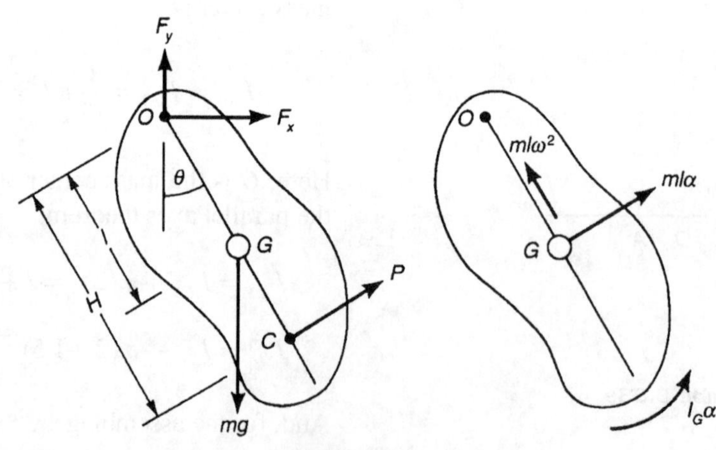

Exhibit 8.36a

from the parallel axis theorem. Thus,

$$I_G = I_C - m(H-l)^2 = mk^2 - m(H-l)^2$$

and

$$\alpha = \frac{PH - mgl\ \sin\theta}{m[k^2 - (H-l)^2] + ml^2}$$

Substitute values to get

$$\alpha = \frac{10(2) - 3(9.81)(1.5)(0.5)}{3[(0.8)^2 - (0.5)^2 + (1.5)^2]} = 0.26\ \text{rad/s}^2$$

8.37 **b.** C is the center of mass of the rod in Exhibit 8.37a. Summing moments about O gives

$$-mg(l/2)\sin\theta = (1/12)ml^2\alpha + m(l/2)^2\alpha + ma(l/2)\cos\theta \qquad \textbf{(i)}$$

Summing forces along the horizontal, gives

$$H = ma + m(l/2)\ \alpha\cos\theta - m\omega^2(l/2)\sin\theta \qquad \textbf{(ii)}$$

From Equation (i),

$$\alpha = -\frac{3}{2}\frac{(g\sin\theta\cos\theta)}{l}$$

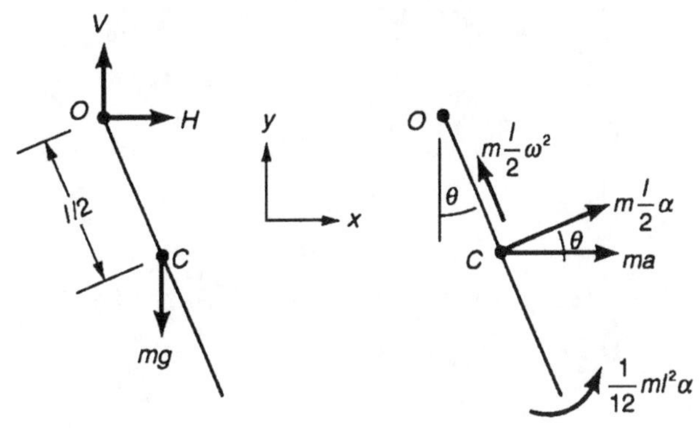

Exhibit 8.37a

Substituting values,

$$\alpha = -4.98\ \text{rad/s}^2 \qquad \textbf{(iii)}$$

Substituting Equation (iii) and the given values into Equation (ii) yields

$$H = (1)(2) + (1)(1)(-4.98)\cos 30° - (1)(2)^2(1)\sin 30° = -4.31\ \text{N}$$

8.38 a.

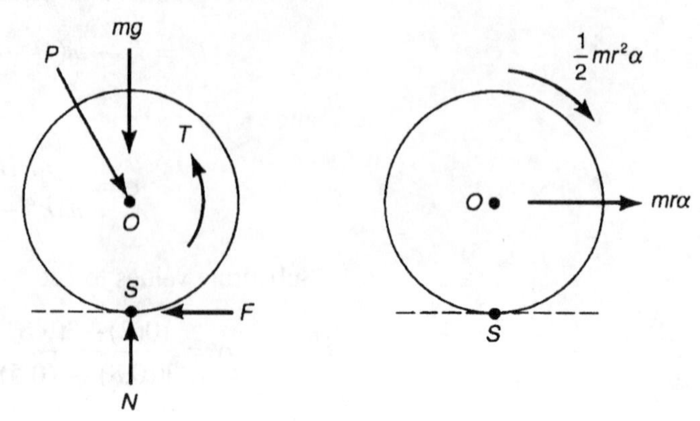

Exhibit 8.38a

Using the free-body diagram shown in Exhibit 8.38a, take moments about the contact point S:

$$rP\cos 60° - T = mr^2\alpha + (1/2)\,mr^2\alpha = (3/2)\,mr^2\alpha$$

or

$$\alpha = (rP\cos 60° - T)/(1.5mr^2) = \text{constant}$$

Substituting values, we find

$$\alpha = 0.83 \text{ rad/s}^2$$

With a constant angular acceleration, the angular velocity is

$$\omega = \omega_0 + \alpha t = 0 + 0.83(10) = 8.33 \text{ rad/s}$$

8.39 b. Exhibit 8.39a shows the forces acting on the rod as it swings from position 1 (horizontal) to position 2 (vertical). The work-energy principle gives

$$W_{1\rightarrow 2} = T_2 - T_1$$

That is,

$$mg\frac{l}{2} = \frac{1}{2}I_A\omega_2^2 - 0 = \frac{1}{2}\times\frac{1}{3}ml^2\omega_2^2$$

and

$$\omega^2 = \sqrt{\frac{3g}{l}} = \sqrt{\frac{(3)(9.81)}{2}} = 3.84 \text{ rad/s}$$

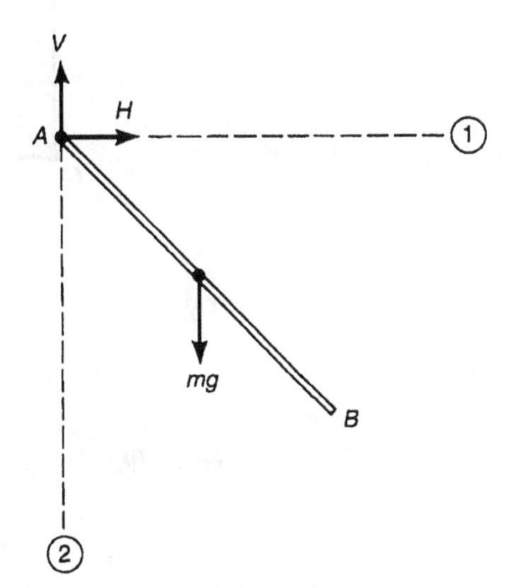

Exhibit 8.39a

8.40 **d.** Exhibit 8.40a shows the forces and torque acting on the rod as it rotates from position 1 (vertical) to position 2 (horizontal). The work-energy relation gives

$$W_{1-2} = T_2 - T_1$$

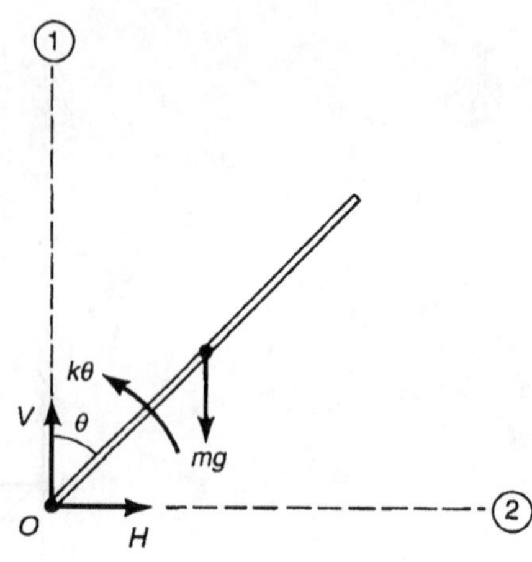

Exhibit 8.40a

That is,

$$mg\frac{l}{2} + \frac{l}{2}k(\theta_1^2 - \theta_2^2) = \frac{1}{2}I_0\omega_2^2 - \frac{1}{2}I_0\omega_1^2$$

Now, $\theta_1 = 0$, $\theta_2 = \pi/2\mathrm{k}$, and $\omega_2 = \omega_2 = 0$. Thus,

$$k = \frac{mgl}{\theta_2^2 - \theta_2^2} = \frac{2(9.8)(3)}{(\pi/2)^2} = 23.8 \text{ N-m/rad}$$

8.41 **a.** The forces acting on the cylinder during this motion are shown in Exhibit 8.41a. Applying the work-energy principle,

$$W_{1-2} = T_2 - T_1$$

F and *R* do no work because their point of application has zero velocity (rolling without slip); *mg* does no work because its point of application moves perpendicular to the force. Work done by the spring force is

$$W_{sp}\frac{1}{2}k\left(\Delta_1^2 - \Delta_2^2\right) = W_{1-2}$$

where

$$\Delta_1 = [(3^2 + 4^2)^{0.5} - 3]\,\text{m} = 2\text{ m}, \quad \text{and} \quad \Delta_2 = 0$$

$$T_1 = 0, \text{ and } T_2 = \frac{1}{2}m(v_P)^2 + \frac{1}{2}I_P\omega^2 = \frac{1}{2}m(\omega r)^2 + \frac{1}{2}\times\frac{1}{2}mr^2\omega^2 = \frac{3}{4}mr^2\omega^2$$

Substituting into the work-energy principle yields

$$W_{1-2} = \frac{1}{2}k\Delta_i^2 = \frac{3}{4}mr^2\omega^2$$

and

$$\omega = \left(\frac{2}{3}\frac{k}{m}\right)^{0.5}\frac{\Delta_1}{r} = \left(\frac{2(2)}{3(12)}\right)^{0.5} \times \frac{2}{0.5} = 1.33 \text{ rad/s}$$

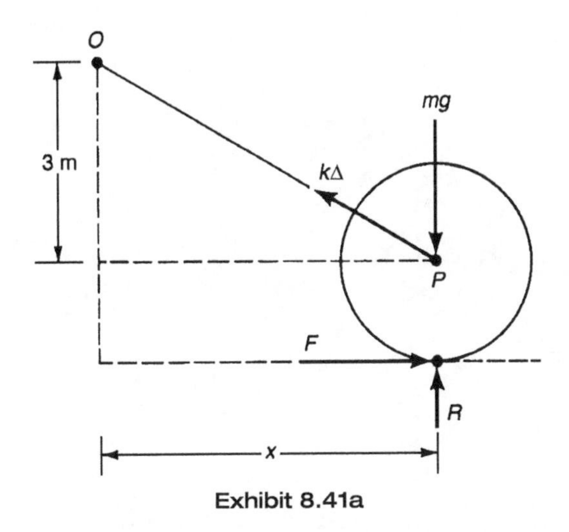

Exhibit 8.41a

CHAPTER 9

Mechanics of Materials

OUTLINE

Mechanics of materials deals with the determination of the internal forces (stresses) and the deformation of solids such as metals, wood, concrete, plastics, and composites. In mechanics of materials there are three main considerations in the solution of problems:

1. Static equilibrium
2. Force-deformation relations
3. Compatibility

Equilibrium refers to the equilibrium of forces. For the purposes of this chapter it is assumed that the system is in static equilibrium (i.e., not moving). The laws of statics must hold for the body and all parts of the body. *Force-deformation relations* refer to the relation of the applied forces to the deformation of the body. If certain forces are applied, then certain deformations will result. *Compatibility* refers to the compatibility of deformation. Upon loading, the parts of a body or structure must not come apart. These three principles will be emphasized throughout.

AXIALLY LOADED MEMBERS

If a force P is applied to a member as shown in Figure 9.1(a), then a short distance away from the point of application the force becomes uniformly distributed over the area as shown in Figure 9.1(b). The force per unit area is called the axial, or normal, stress and is given the symbol σ. Thus,

$$\sigma = \frac{P}{A} \tag{9.1}$$

The original length between two points A and B is L as shown in Figure 9.1(c). Upon application of the load P, the length L grows by an amount ΔL. The final length is $L + \Delta L$ as shown in Figure 9.1(d). A quantity measuring the intensity of deformation and being independent of the original length L is the strain ε, defined as

$$\varepsilon = \frac{\Delta L}{L} = \frac{\delta}{L} \tag{9.2}$$

where ΔL is denoted as δ.

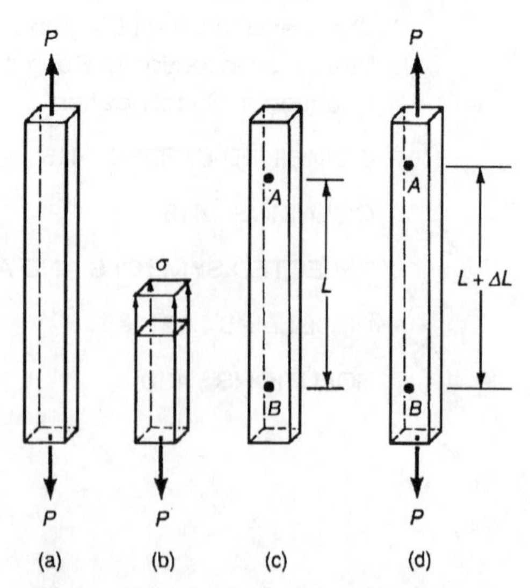

Figure 9.1 Axial member under force P

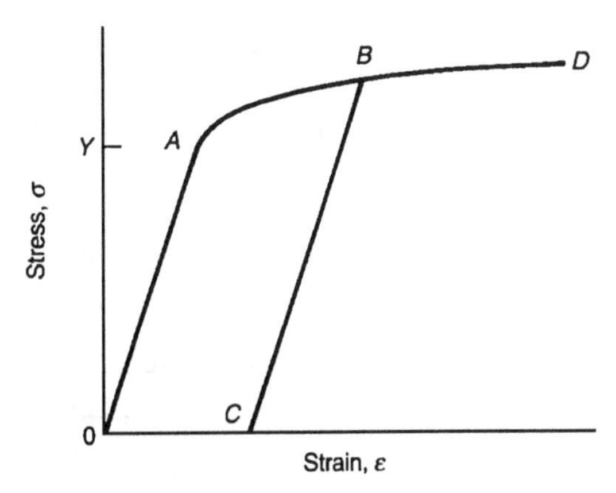

Figure 9.2 Stress-strain curve for a typical material

The relationship between stress and strain is determined experimentally. A typical plot of stress versus strain is shown in Figure 9.2. On initial loading, the plot is a straight line until the material reaches yield at a stress of Y. It is noted that in this example the yield strength point A is also the proportional limit (i.e., the maximum stress at which the stress-strain curve remains linear). If the stress remains less than yield, then subsequent loading and reloading continues along that same straight line. If the material is allowed to go beyond yield, then during an increase in the load the curve goes from A to D. If unloading occurs at some point B, for example, then the material unloads along the line BC, which has approximately the same slope as the original straight line from 0 to A. Reloading would occur along the line CB and then proceed along the line BD. It can be seen that if the material is allowed to go into the plastic region (A to D) it will have a permanent strain offset on unloading.

Modulus of Elasticity

The region of greatest concern is that below the yield point. The slope of the line between 0 and A is called the modulus of elasticity and is given the symbol E, so

$$\sigma = E\varepsilon \tag{9.3}$$

This is Hooke's law for axial loading; a more general form will be considered in a later section. The modulus of elasticity is a function of the material alone and not a function of the shape or size of the axial member.

The relation of the applied force in a member to its axial deformation can be found by inserting the definitions of the axial stress [Equation (9.1)] and the axial strain [Equation (9.2)] into Hooke's law [Equation (9.3)], which gives

$$\frac{P}{A} = E\frac{\delta}{L} \tag{9.4}$$

or

$$\delta = \frac{PL}{AE} \tag{9.5}$$

In the examples that follow, wherever it is appropriate, the three steps of (1) static equilibrium, (2) force-deformation, and (3) compatibility will be explicitly stated.

Example 9.1

The steel rod shown in Exhibit 1 is fixed to a wall at its left end. It has two applied forces. The 3 kN force is applied at the Point B and the 1 kN force is applied at the Point C. The area of the rod between A and B is $A_{AB} = 1000\ \text{mm}^2$, and the area of the rod between B and C is $A_{BC} = 500\ \text{mm}^2$. Take $E = 210$ GPa. Find (a) the stress in each section of the rod and (b) the horizontal displacement at the points B and C.

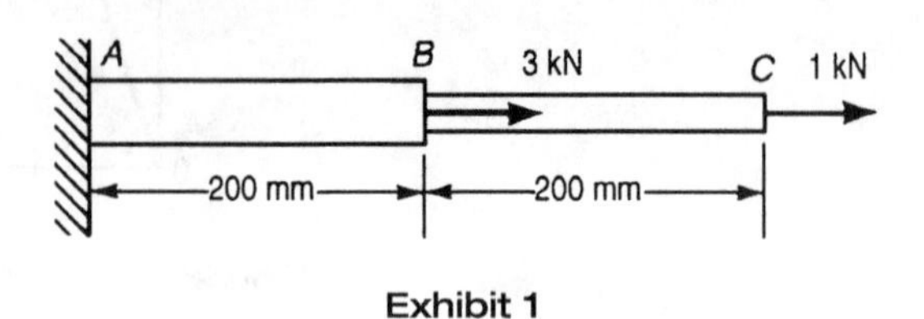

Exhibit 1

Solution—Static Equilibrium

Draw free-body diagrams for each section of the rod (Exhibit 2). From a summation of forces on the member BC, $F_{BC} = 1$ kN. Summing forces in the horizontal direction on the center free-body diagram, $F_{BA} = 3 + 1 = 4$ kN. Summing forces on the left free-body diagram gives $F_{AB} = F_{BA} = 4$ kN. The stresses then are

$$\sigma_{AB} = 4\ \text{kN}/1000\ \text{mm}^2 = 4\ \text{MPa}$$
$$\sigma_{BC} = 1\ \text{kN}/500\ \text{mm}^2 = 2\ \text{MPa}$$

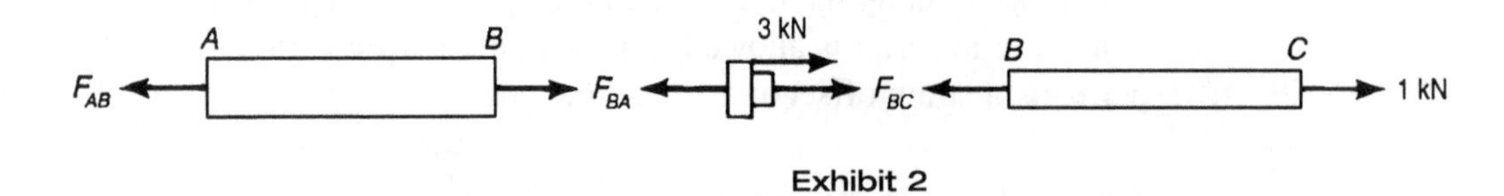

Exhibit 2

Solution—Force-Deformation

$$\delta_{AB} = \left(\frac{PL}{AE}\right)_{AB} = \frac{(4\ \text{kN})(200\ \text{mm})}{(1000\ \text{mm}^2)(210\ \text{GPa})} = 0.00381\ \text{mm}$$

$$\delta_{BC} = \left(\frac{PL}{AE}\right)_{BC} = \frac{(1\ \text{kN})(200\ \text{mm})}{(500\ \text{mm}^2)(210\ \text{GPa})} = 0.001905\ \text{mm}$$

Solution—Compatibility

Draw the body before loading and after loading (Exhibit 3).

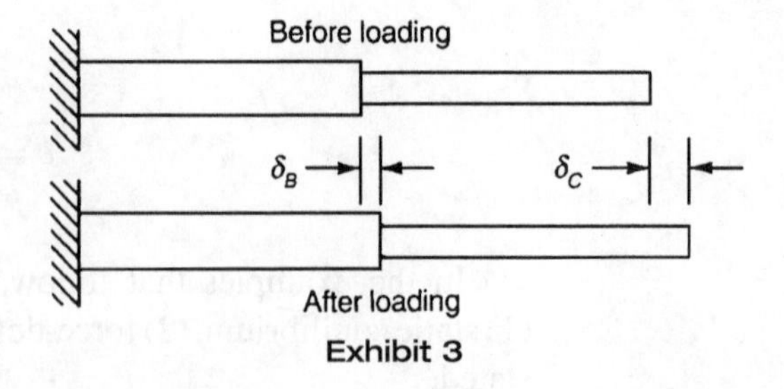

Exhibit 3

It is then obvious that

$$\delta_B = \delta_{AB} = 0.00381 \text{ mm}$$

$$\delta_C = \delta_{AB} + \delta_{BC} = 0.00381 + 0.001905 = 0.00571 \text{ mm}$$

In this first example the problem was statically determinate, and the three steps of static equilibrium, force-deformation, and compatibility were independent steps. The steps are not independent when the problem is statically indeterminate, as the next example will show.

Example 9.2

Consider the same steel rod as in Example 9.1 except that now the right end is fixed to a wall as well as the left (Exhibit 4). It is assumed that the rod is built into the walls before the load is applied. Find (a) the stress in each section of the rod, and (b) the horizontal displacement at the point B.

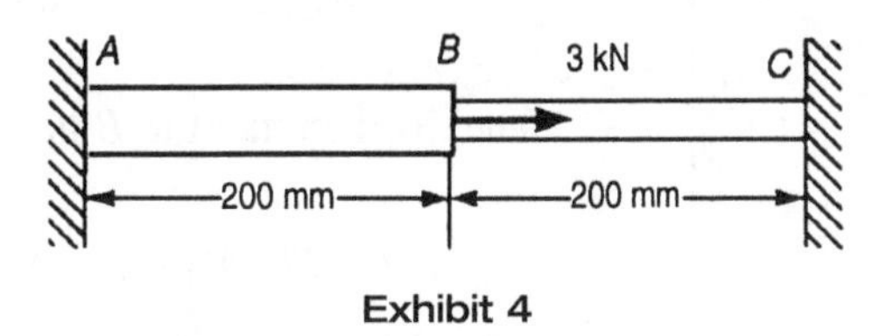

Exhibit 4

Solution—Equilibrium

Draw free-body diagrams for each section of the rod (Exhibit 5). Summing forces in the horizontal direction on the center free-body diagram:

$$-F_{AB} + F_{BC} + 3 = 0$$

It can be seen that the forces cannot be determined by statics alone so that the other steps must be completed before the stresses in the rods can be determined.

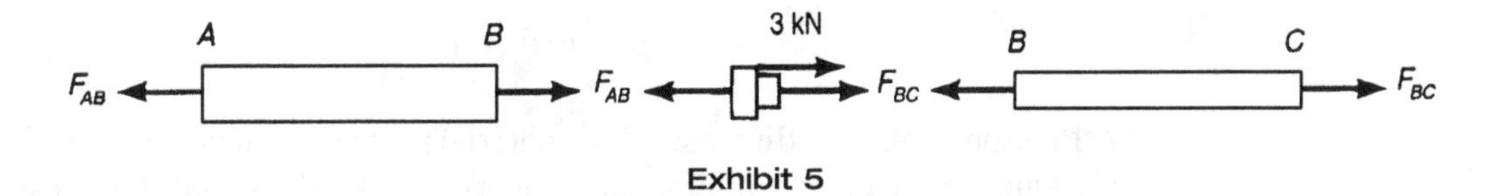

Exhibit 5

Solution—Force-Deformation

$$\delta_{AB} = \left(\frac{PL}{AE}\right)_{AB} = \frac{F_{AB}L}{A_{AB}E}$$

$$\delta_{BC} = \left(\frac{PL}{AE}\right)_{BC} = \frac{F_{BC}L}{A_{BC}}$$

The static equilibrium, force-deformation, and compatibility equations can now be solved as follows (see Exhibit 6). The force-deformation relations are put into the compatibility equations (recall: $A_{AB} = 2A_{BC}$):

$$\frac{F_{AB}L}{2A_{BC}E} = -\frac{F_{BC}L}{A_{BC}E}$$

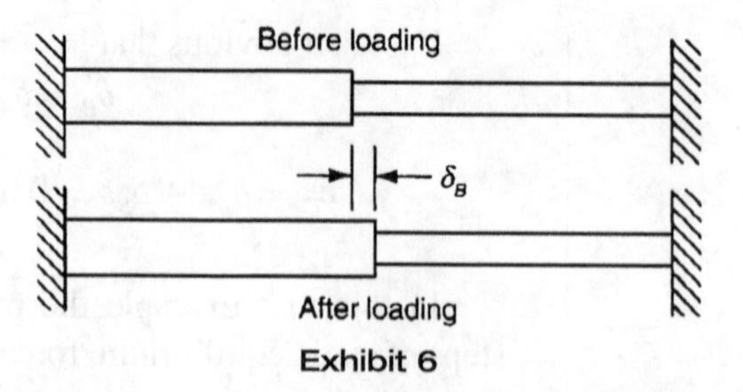

Exhibit 6

Then, $F_{AB} = -2F_{BC}$. Insert this relationship into the equilibrium equation:

$$-F_{AB} + F_{BC} + 3 = 0 = 2F_{BC} + F_{BC} + 3; \quad F_{BC} = -1 \text{ kN and } F_{AB} = 2 \text{ kN}$$

The stresses are

$$\sigma_{AB} = 2 \text{ kN/1000 mm}^2 = 2 \text{ MPa (tension)}$$
$$\sigma_{BC} = -1 \text{ kN/500 mm}^2 = -2 \text{ MPa (compression)}$$

The displacement at B is

$$\delta_A = \delta_{AB} = F_{AB}L/(AE) = (2 \text{ kN})(200 \text{ mm})/[(1000 \text{ mm}^2)(210 \text{ GPa})] = 0.001905 \text{ mm}$$

Poisson's Ratio

The axial member shown in Figure 9.1 also has a strain in the lateral direction. If the rod is in tension, then stretching takes place in the axial or longitudinal direction while contraction takes place in the lateral direction. The ratio of the magnitude of the lateral strain to the magnitude of the longitudinal strain is called Poisson's ratio ν.

$$\nu = -\frac{\text{Lateral strain}}{\text{Longitudinal strain}} \tag{9.6}$$

Poisson's ratio is a dimensionless material property that never exceeds 0.5. Typical values for steel, aluminum, and copper are 0.30, 0.33, and 0.34, respectively.

Example 9.3

A circular aluminum rod 10 mm in diameter is loaded with an axial force of 2 kN. What is the decrease in diameter of the rod? Take $E = 70$ GN/m² and $\nu = 0.33$.

Solution

The stress is $\sigma = P/A = 2 \text{ kN}/(\pi 5^2 \text{ mm}^2) = 0.0255 \text{ GN/m}^2 = 25.5 \text{ MN/m}^2$.

The longitudinal strain is $\varepsilon_{1on} = \sigma/\text{E} = (25.5 \text{ MN/m}^2)/(70 \text{ GN/m}) = 0.000364$.

The lateral strain is $\varepsilon_{1at} = -\nu\ \varepsilon_{1on} = -0.33(0.000364) = -0.000120$.

The decrease in diameter is then $-\text{D}\ \varepsilon_{1at} = -(10 \text{ mm})(-0.000120) = 0.00120$ mm.

Thermal Deformations

When a material is heated, thermal strains are created. If the material is unrestrained, the thermal strain is

$$\varepsilon_t = \alpha(t - t_0) \tag{9.7}$$

where α is the linear coefficient of thermal expansion, t is the final temperature, and t_0 is the initial temperature. Since strain is dimensionless, the units of α are $°F^{-1}$ or $°C^{-1}$ (sometimes the units are given as in/in/°F or m/m/°C, which amounts to the same thing). The total strain ε_T is equal to the strain from the applied loads plus the thermal strain. For problems where the load is purely axial, this becomes

$$\varepsilon_T = \frac{\sigma}{E} + \alpha(t - t_0) \tag{9.8}$$

The deformation δ is found by multiplying the strain by the length L:

$$\delta = \frac{PL}{AE} + \alpha L(t - t_0) \tag{9.9}$$

Example 9.4

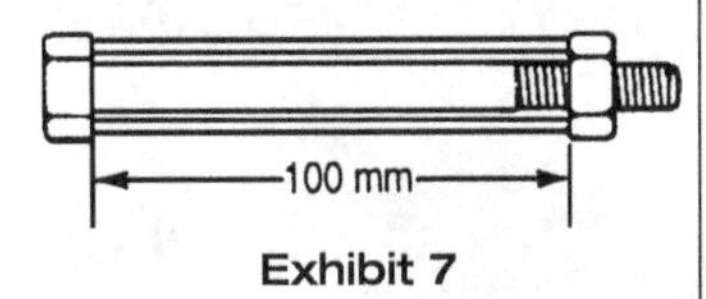

Exhibit 7

A steel bolt is put through an aluminum tube as shown in Exhibit 7. The nut is threaded on the bolt until it just makes contact with the tube (i.e., no torque). The temperature of the entire assembly is then raised by 60°C. Because the coefficient of thermal expansion of aluminum is greater than that of steel, the bolt will be put in tension and the tube in compression. Find the force in the bolt and the tube. For the steel bolt, take $E = 210$ GPa, $\alpha = 12 \times 10^{-6}\ °C^{-1}$ and $A = 32\ mm^2$. For the aluminum tube, take $E = 69$ GPa, $\alpha = 23 \times 10^{-6}\ °C^{-1}$ and $A = 64\ mm^2$.

Solution—Equilibrium

Draw free-body diagrams (Exhibit 8). From equilibrium of the bolt it can be seen that $P_B - P_T = 0$.

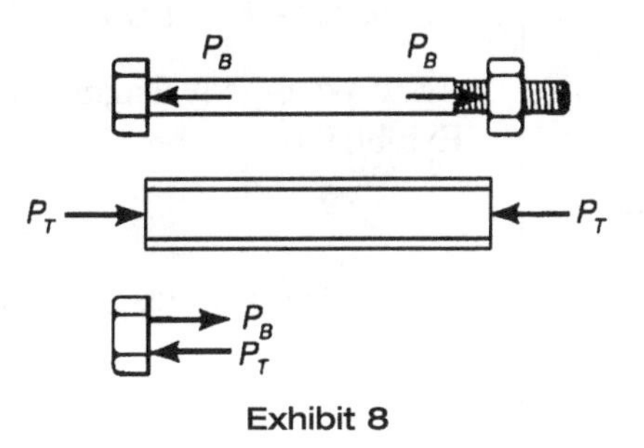

Exhibit 8

Solution—Force-Deformation

Note that both members have the same length and the same force magnitude, P.

$$\delta_B = \frac{PL}{A_B E_B} + \alpha_B L(t - t_0)$$

$$\delta_T = -\frac{PL}{A_T E_T} + \alpha_T L(t - t_0)$$

The minus sign in the second expression occurs because the tube is in compression.

Solution—Compatibility

The tube and bolt must both expand the same amount; therefore,

$$\delta_B = \delta_T$$

$$\delta_B = \delta_T = \frac{P \times (100 \text{ mm})}{32 \text{ mm}^2 \times 210 \text{ GPa}} + 12 \times 10^{-6} \frac{1}{°\text{C}} \times 100 \text{ mm} \times 60 \text{ °C}$$

$$= -\frac{P \times (100 \text{ mm})}{64 \text{ mm}^2 \times 69 \text{ GPa}} + 23 \times 10^{-6} \frac{1}{°\text{C}} \times 100 \text{ mm} \times 60 \text{ °C}$$

Solving for P gives $P = 1.759$ kN.

Variable Load

In certain cases the load in the member will not be constant but will be a continuous function of the length. These cases occur when there is a distributed load on the member. Such distributed loads most commonly occur when the member is subjected to gravitation, acceleration, or magnetic fields. In such cases, Equation (9.5) holds only over an infinitesimally small length $L = dx$. Equation (9.5) then becomes

$$d\delta = \frac{P(x)}{AE} dx \qquad \textbf{(9.10)}$$

or equivalently

$$\delta = \int_0^L \frac{P(x)}{AE} dx \qquad \textbf{(9.11)}$$

Example 9.5

An aluminum rod is hanging from one end. The rod is 1 m long and has a square cross-section 20 mm by 20 mm. Find the total extension of the rod resulting from its own weight. Take $E = 70$ GPa and the unit weight $\gamma = 27$ kN/m^3.

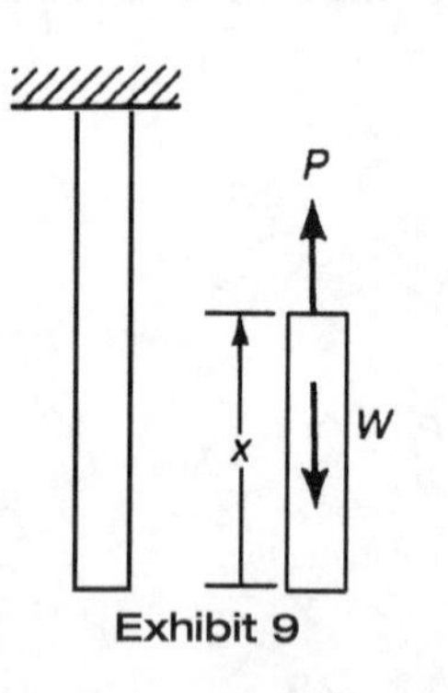

Exhibit 9

Solution—Equilibrium

Draw a free-body diagram (Exhibit 9). The weight of the section shown in Exhibit 9 is

$$W = \gamma V = \gamma A x = P$$

which clearly yields P as a function of x, and Equation (9.11) gives

$$\delta = \int_0^L \frac{\gamma A x}{AE} dx = \frac{\gamma}{E}\int_0^L x dx = \frac{\gamma L^2}{2E} \frac{\left(27 \frac{\text{kN}}{\text{m}^3}\right)(\text{lm})^2}{2\left(70 \frac{\text{GN}}{\text{m}^2}\right)} = 0.1929 \ \mu\text{m}$$

THIN-WALLED CYLINDER

Consider the thin-walled circular cylinder subjected to a uniform internal pressure q as shown in Figure 9.3. A section of length a is cut out of the vessel in (a). The cut-out portion is shown in (b). The pressure q can be considered as acting across

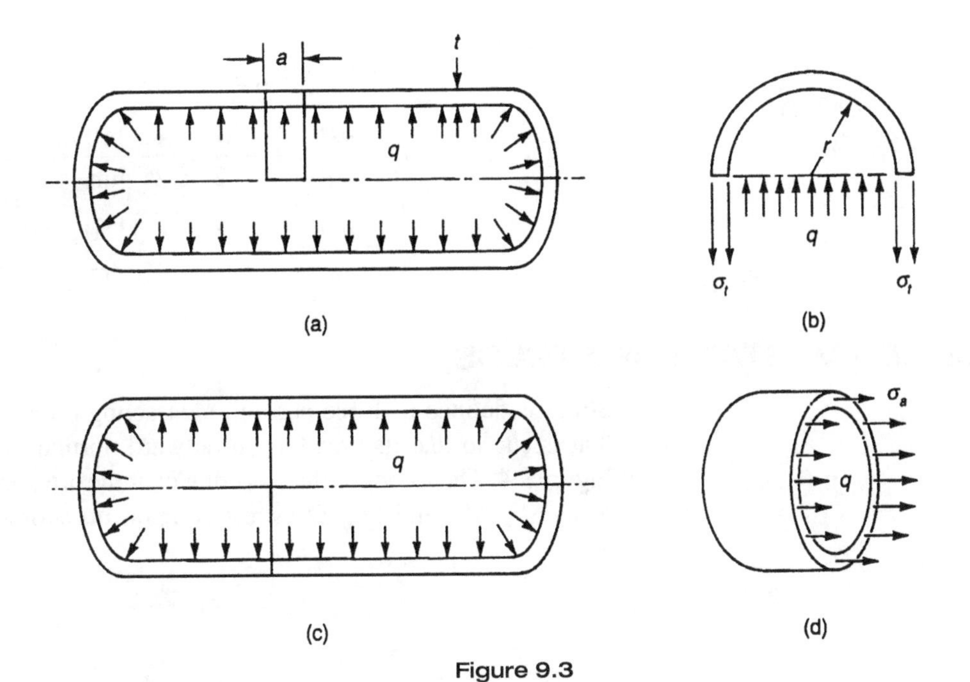

Figure 9.3

the diameter as shown. The tangential stress σ_t is assumed constant through the thickness. Summing forces in the vertical direction gives

$$qDa - 2\sigma_t\, ta = 0 \tag{9.12}$$

$$\sigma_t = \frac{qD}{2t} \tag{9.13}$$

where D is the inner diameter of the cylinder and t is the wall thickness. The axial stress σ_a is also assumed to be uniform over the wall thickness. The axial stress can be found by making a cut through the cylinder as shown in (c). Consider the horizontal equilibrium for the free-body diagram shown in (d). The pressure q acts over the area πr^2 and the stress σ_a acts over the area πDt, which gives

$$\sigma_a \pi Dt = q\pi\left(\frac{D}{2}\right)^2 \tag{9.14}$$

so

$$\sigma_a = \frac{qD}{4t} \tag{9.15}$$

Example 9.6

Consider a cylindrical pressure vessel with a wall thickness of 25 mm, an internal pressure of 1.4 MPa, and an outer diameter of 1.2 m. Find the axial and tangential stresses.

Solution

$$q = 1.4 \text{ MPa}; \; D = 1200 - 50 = 1150 \text{ mm}; \; t = 25 \text{ mm}$$

$$\sigma_t = \frac{qD}{2t} = \frac{1.4 \text{ MPa} \times 1150 \text{ mm}}{2 \times 25 \text{ mm}} = 32.2 \text{ MPa}$$

$$\sigma_a = \frac{qD}{4t} = \frac{1.4 \text{ MPa} \times 1150 \text{ mm}}{4 \times 25 \text{ mm}} = 16.1 \text{ MPa}$$

GENERAL STATE OF STRESS

Stress is defined as force per unit area acting on a certain area. Consider a body that is cut so that its area has an outward normal in the x direction as shown in Figure 9.4. The force ΔF that is acting over the area ΔA_x can be split into its components ΔF_x, ΔF_y, and ΔF_z. The stress components acting on this face are then defined as

$$\sigma_x = \lim_{\Delta A_x \to 0} \frac{\Delta F_x}{\Delta A_x} \tag{9.16}$$

$$\tau_{xy} = \lim_{\Delta A_x \to 0} \frac{\Delta F_x}{\Delta A_x} \tag{9.17}$$

$$\tau_{xz} = \lim_{\Delta A_x \to 0} \frac{\Delta F_z}{\Delta A_x} \tag{9.18}$$

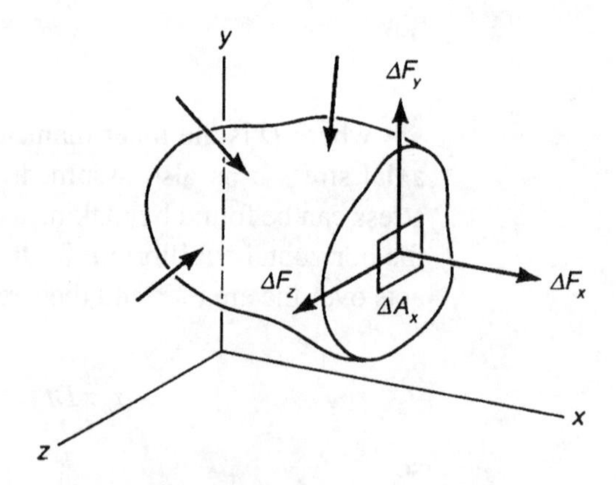

Figure 9.4 Stress on a face

The stress component σ_x is the normal stress. It acts normal to the x face in the x direction. The stress component τ_{xy} is a shear stress. It acts parallel to the x face in the y direction. The stress component τ_{xz} is also a shear stress and acts parallel to the x face in the z direction. For shear stress, the first subscript indicates the *face* on which it acts, and the second subscript indicates the *direction* in which it acts. For normal stress, the single subscript indicates both face and direction. In the general state of stress, there are normal and shear stresses on all faces of an element as shown in Figure 9.5.

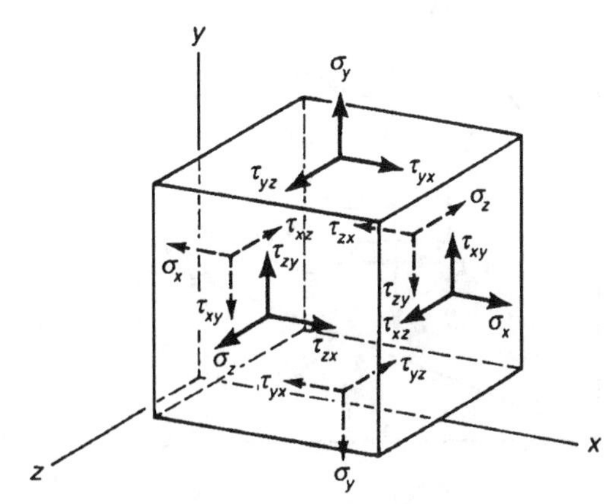

Figure 9.5 Stress at a point (shown in positive directions)

From equilibrium of moments around axes parallel to x, y, and z and passing through the center of the element in Figure 9.5, it can be shown that the following relations hold:

$$\tau_{xy} = \tau_{yx}; \qquad \tau_{yz} = \tau_{zy}; \qquad \tau_{zx} = \tau_{xz} \tag{9.19}$$

Thus, at any point in a body the state of stress is given by six components: $\sigma_x, \sigma_y, \sigma_z, \tau_{xy}, \tau_{yz}$, and τ_{zx}. The usual sign convention is to take the components shown in Figure 9.5 as positive. One way of saying this is that normal stresses are positive in tension. Shear stresses are positive on a positive face in the positive direction. A **positive face** is defined as a face with a positive outward normal.

PLANE STRESS

In elementary mechanics of materials, we usually deal with a state of plane stress in which only the stresses in the x-y plane are nonzero. The stress components σ_z, τ_{xz}, and τ_{yz} are taken as zero.

Mohr's Circle—Stress

In plane stress, the three components σ_x, σ_y, τ_{xy} define the state of stress at a point, but the components on any other face have different values. To find the components on an arbitrary face, consider equilibrium of the wedges shown in Figure 9.6.

Summation of forces in the x' and y' directions for the wedge shown in Figure 9.6(a) gives

$$\sum F_{x'} = 0 = \sigma_{x'}\Delta A - \sigma_x \Delta A(\cos\theta)^2 - \sigma_y \Delta A(\sin\theta)^2 - 2\tau_{xy}\Delta A \sin\theta\cos\theta \tag{9.20}$$

$$\sum F_{y'} = 0 = \tau_{x'y'}\Delta A + \left(\sigma_x - \sigma_y\right)\Delta A \sin\theta\cos\theta - \tau_{xy}\Delta A\left[\left(\cos\theta\right)^2 - \left(\sin\theta\right)^2\right] \tag{9.21}$$

Canceling ΔA from each of these expressions and using the double angle relations gives

$$\sigma_{x'} = \frac{\sigma_x + \sigma_y}{2} + \frac{\sigma_x - \sigma_y}{2}\cos 2\theta + \tau_{xy}\sin 2\theta \tag{9.22}$$

$$\tau_{x'y'} = -\frac{\sigma_x - \sigma_y}{2}\sin 2\theta + \tau_{xy}\cos 2\theta \tag{9.23}$$

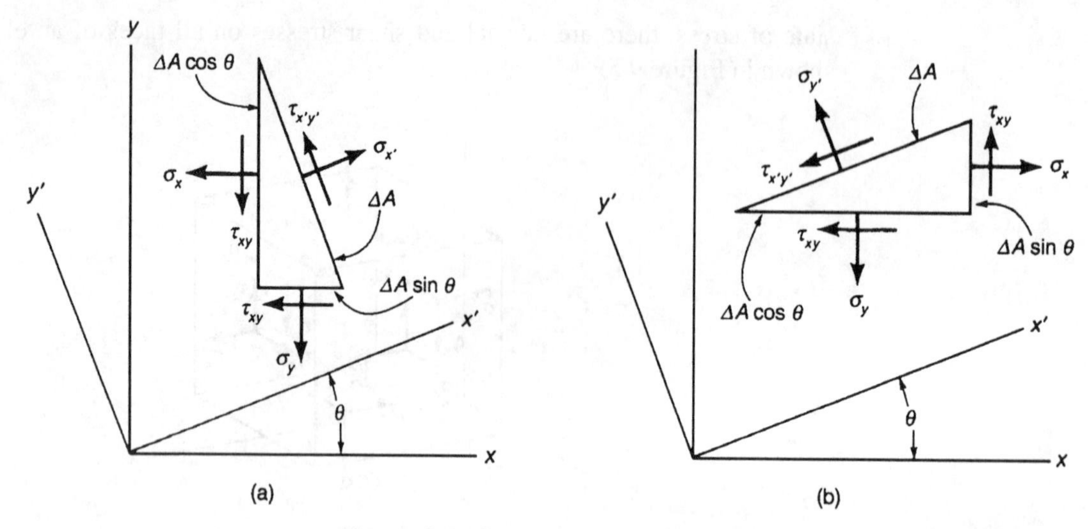

Figure 9.6 Stress on an arbitrary face

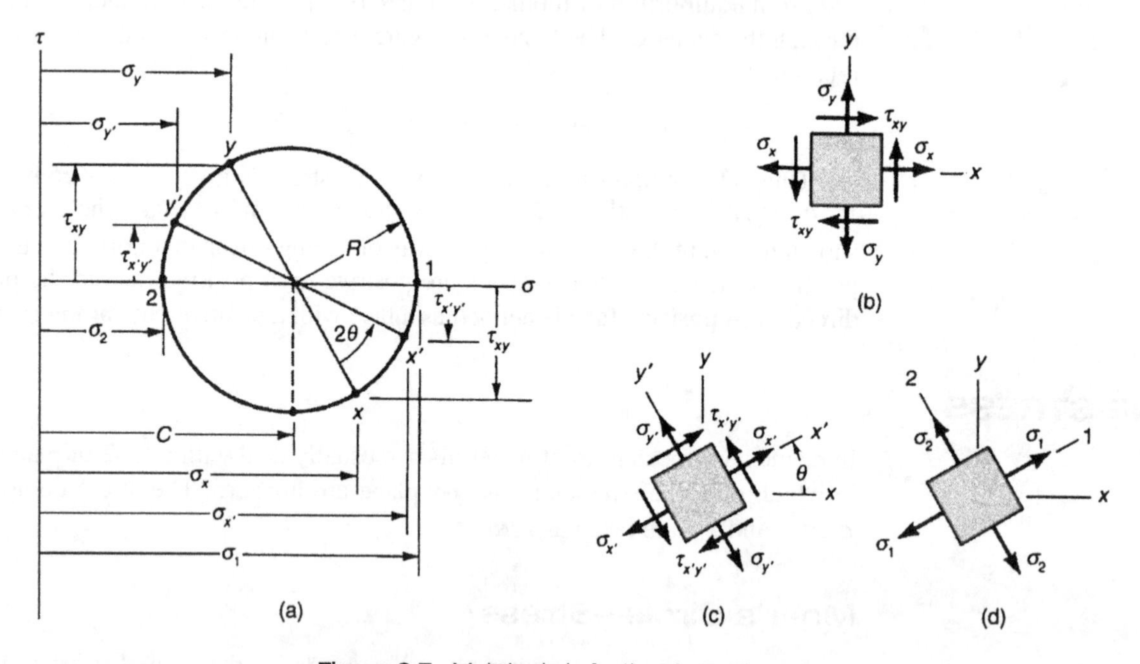

Figure 9.7 Mohr's circle for the stress at a point

Similarly, summation of forces in the y' direction for the wedge shown in Figure 9.6(b) gives

$$\sigma_{y'} = \frac{\sigma_x + \sigma_y}{2} - \frac{\sigma_x - \sigma_y}{2}\cos 2\theta - \tau_{xy}\sin 2\theta \qquad \textbf{(9.24)}$$

Equations (9.22), (9.23), and (9.24) are the parametric equations of Mohr's circle; Figure 9.7(a) shows the general Mohr's circle; Figure 9.7(b) shows the stress on the element in an x-y orientation; Figure 9.7(c) shows the stress in the same element in an x'-y' orientation; and Figure 9.7(d) shows the stress on the element in the 1-2 orientation. Notice that there is always an orientation (for example, a 1-2 orientation) for which the shear stress is zero. The normal stresses σ_1 and σ_2 on these 1-2 faces are the principal stresses, and the 1 and 2 axes are the principal axes of stress. In three-dimensional problems the same is true. There are always three mutually perpendicular faces on which there is no shear stress. Hence, there are always three principal stresses.

To draw Mohr's circle knowing σ_x, σ_y, and τ_{xy}:

1. Draw vertical lines corresponding to σ_x and σ_y as shown in Figure 9.8(a) according to the signs of σ_x and σ_y (to the right of the origin if positive and to the left if negative).
2. Put a point on the σ_x vertical line a distance τ_{xy} below the horizontal axis if τ_{xy} is positive (above if τ_{xy} is negative) as in Figure 9.8(a). Name this point x.
3. Put a point on the σ_y vertical line a distance τ_{xy} in the opposite direction as on the σ_x vertical line also as shown in Figure 9.8(a). Name this point y.
4. Connect the two points x and y, and draw the circle with diameter xy as shown in Figure 9.8(b).

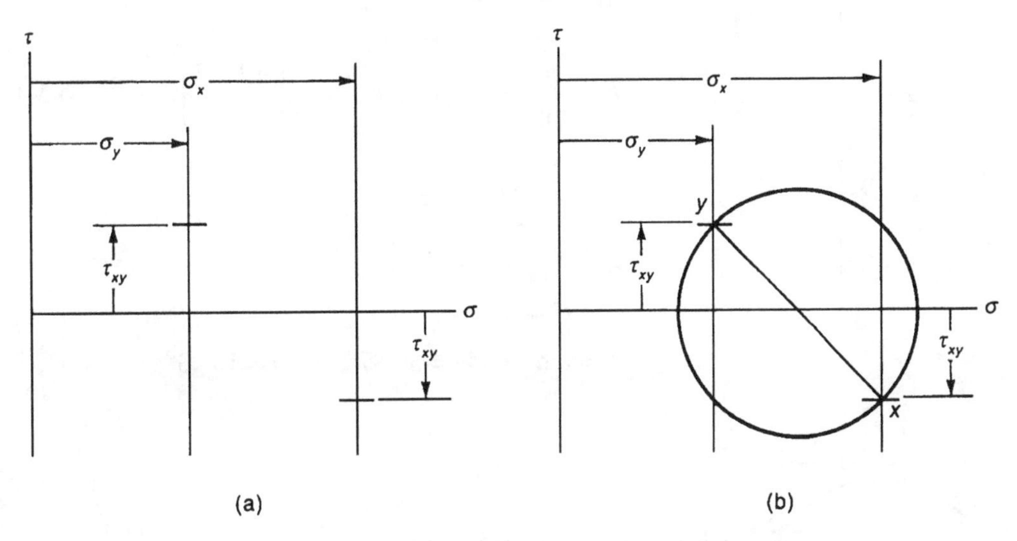

Figure 9.8 Constructing Mohr's circle

Upon constructing Mohr's circle you can now rotate the xy diameter through an angle of 2θ to a new position $x'y'$, which can determine the stress on any face at that point in the body as shown in Figure 9.7. Note that rotations of 2θ on Mohr's circle correspond to θ in the physical plane; also note that the direction of rotation is the same as in the physical plane (that is, if you go clockwise on Mohr's circle, the rotation is also clockwise in the physical plane). The construction can also be used to find the principal stresses and the orientation of the principal axes.

Problems involving stress transformations can be solved with Equations (9.22), (9.23), and (9.24), from construction of Mohr's circle, or from some combination. As an example of a combination, it can be seen that the center of Mohr's circle can be represented as

$$C = \frac{\sigma_x + \sigma_y}{2} \tag{9.25}$$

The radius of the circle is

$$R = \sqrt{\left(\frac{\sigma_x - \sigma_y}{2}\right)^2 + \tau_{xy}{}^2} \tag{9.26}$$

The principal stresses then are

$$\sigma_1 = C + R; \quad \sigma_2 = C - R \tag{9.27}$$

Example 9.7

Given $\sigma_x = -3$ MPa; $\sigma_y = 5$ MPa; $\tau_{xy} = 3$ MPa. Find the principal stresses and their orientation.

Solution

Mohr's circle is constructed as shown in Exhibit 10. The angle 2θ was chosen as the angle between the y axis and the 1 axis clockwise from y to 1 as shown in the circle. The angle θ in the physical plane is between the y axis and the 1 axis also clockwise from y to 1. The values of σ_1, σ_2, and 2θ can all be scaled from the circle. The values can also be calculated as follows:

$$R = \sqrt{\left(\frac{\sigma_x - \sigma_y}{2}\right)^2 + \tau_{xy}{}^2} = \sqrt{\left(\frac{-3-5}{2}\right)^2 + 3^2} = 5 \text{ MPa}$$

$$C = \frac{\sigma_x + \sigma_y}{2} = \frac{-3+5}{2} = 1 \text{ MPa}$$

$$\sigma_1 = C + R = 6 \text{ MPa}$$

$$\sigma_2 = C - R = -4 \text{ MPa}$$

$$2\theta = \tan^{-1}(3/4) = 36.87°; \qquad \theta = 18.43°$$

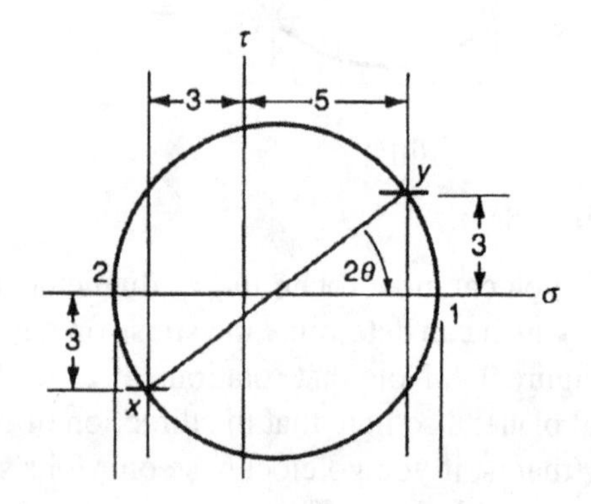

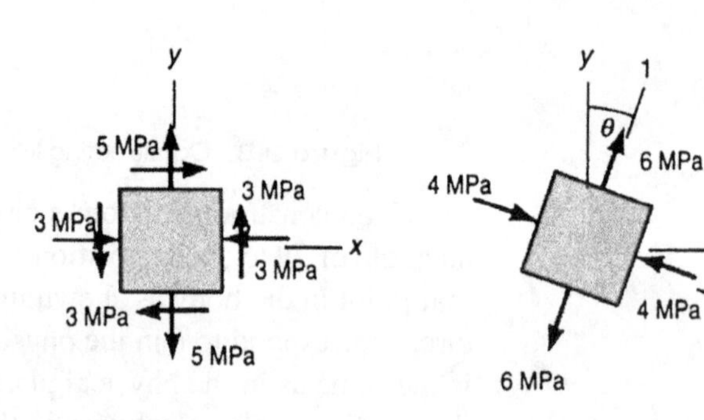

Exhibit 10

STRAIN

Axial strain was previously defined as

$$\varepsilon = \frac{\Delta L}{L} \tag{9.28}$$

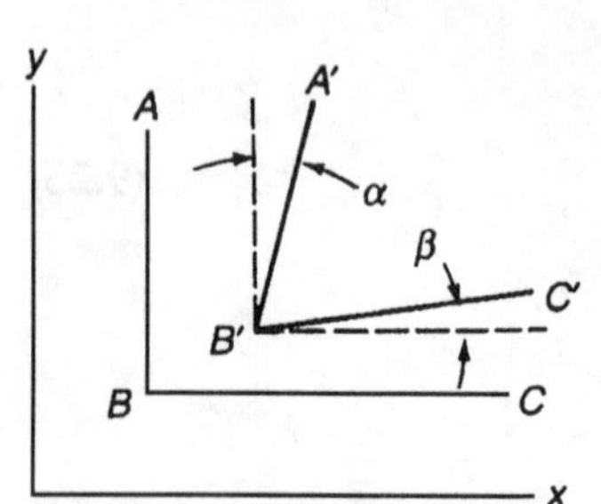

Figure 9.9 Definition of shear strain

In the general case, there are three components of axial strain, ε_x, ε_y and ε_z. Shear strain is defined as the decrease in angle of two originally perpendicular line segments passing through the point at which strain is defined. In Figure 9.9, AB is vertical and BC is horizontal. They represent line segments that are drawn before loading. After loading, points A, B, and C move to A', B', and C', respectively. The angle between $A'B'$ and the vertical is α, and the angle between B' and C' and the horizontal is β. The original right angle has been decreased by $\alpha + \beta$, and the shear strain is

$$\gamma_{xy} = \alpha + \beta \tag{9.29}$$

In the general case, there are three components of shear strain, γ_{xy}, γ_{yz}, and γ_{zx}.

Plane Strain

In two dimensions, strain undergoes a similar rotation transformation as stress. The transformation equations are

$$\varepsilon_{x'} = \frac{\varepsilon_x + \varepsilon_y}{2} + \frac{\varepsilon_x - \varepsilon_y}{2} \cos 2\theta + \frac{\gamma_{xy}}{2} \sin 2\theta \tag{9.30}$$

$$\frac{\gamma_{x'y'}}{2} = -\frac{\varepsilon_x - \varepsilon_y}{2} \sin 2\theta + \frac{\gamma_{xy}}{2} \cos 2\theta \tag{9.31}$$

$$\varepsilon_{y'} = \frac{\varepsilon_x + \varepsilon_y}{2} - \frac{\varepsilon_x - \varepsilon_y}{2} \cos 2\theta - \frac{\gamma_{xy}}{2} \sin 2\theta \tag{9.32}$$

These equations are the same as Equation (9.22), (9.23), and (9.24) for stress, except that the σ_x has been replaced with $\varepsilon_{x'}$, σ_y with $\varepsilon_{y'}$, and τ_{xy} with $\gamma_{xy}/2$. Therefore, Mohr's circle for strain is treated the same way as that for stress, except for the factor of two on the shear strain.

Example 9.8

Given that $\varepsilon_x = 600\ \mu$; $\varepsilon_y = -200\ \mu$; $\gamma_{xy} = +800\ \mu$, find the principal strains and their orientation. The symbol μ signifies 10^{-6}.

Solution

From the Mohr's circle shown in Exhibit 11, it is seen that $2\theta = 45°$; so, $\theta = 22.5°$ counterclockwise from x to 1. The principal strains are $\varepsilon_1 = 766\ \mu$ and $\varepsilon_2 = -366\ \mu$.

The principal strains can also be found by computation in the same way as principal stresses,

$$R = \sqrt{\left(\frac{\varepsilon_x - \varepsilon_y}{2}\right)^2 + \left(\frac{\gamma_{xy}}{2}\right)^2} = \sqrt{\left(\frac{600 + 200}{2}\right)^2 + \left(\frac{-800}{2}\right)^2} = 565.7\ \mu$$

$$C = \frac{\varepsilon_x + \varepsilon_y}{2} = \frac{600 - 200}{2} = 200\ \mu$$

$$\varepsilon_1 = C + R = 766\ \mu$$

$$\varepsilon_2 = C - R = -366\ \mu$$

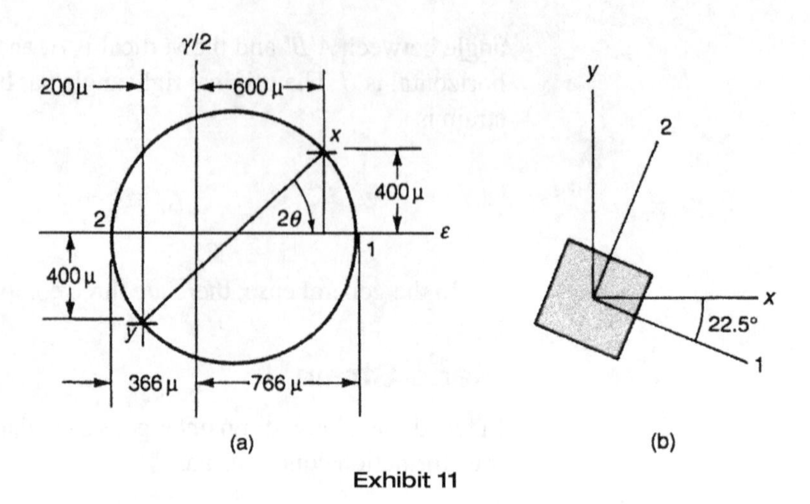

Exhibit 11

HOOKE'S LAW

The relationship between stress and strain is expressed by Hooke's law. For a linear elastic isotropic material it is

$$\varepsilon_x = \frac{1}{E}(\sigma_x - v\sigma_y - v\sigma_z) \tag{9.33}$$

$$\varepsilon_y = \frac{1}{E}(\sigma_y - v\sigma_z - v\sigma_x) \tag{9.34}$$

$$\varepsilon_z = \frac{1}{E}(\sigma_z - v\sigma_x - v\sigma_y) \tag{9.35}$$

$$\gamma_{xy} = \frac{1}{G}\tau_{xy} \tag{9.36}$$

$$\gamma_{xy} = \frac{1}{G}\tau_{xy} \tag{9.37}$$

$$\gamma_{zx} = \frac{1}{G}\tau_{zx} \tag{9.38}$$

Further, there is a relationship between E, G, and v, which is

$$G = \frac{E}{2(1+v)} \tag{9.39}$$

Thus, for an isotropic material there are only two independent elastic constants. An **isotropic material** is one that has the same material properties in all directions. Notable exceptions to isotropy are wood- and fiber-reinforced composites.

Example 9.9

A steel plate in a state of plane stress is known to have the following strains: $\varepsilon_x = 650\ \mu$, $\varepsilon_y = 250\ \mu$, and $\gamma_{xy} = 400\ \mu$. If $E = 210$ GPa and $v = 0.3$, what are the stress components, and what is the strain ε_z?

Solution

In a state of plane stress, the stresses $\sigma_z = 0$, $\tau_{xz} = 0$, and $\tau_{yz} = 0$. From Hooke's law,

$$\varepsilon_x = \frac{1}{E}(\sigma_x - \nu\sigma_y - 0)$$

$$\varepsilon_y = \frac{1}{E}(\sigma_y - \nu\sigma_x - 0$$

Using these relations to solve for stress gives

$$\sigma_x = \frac{E}{1-\nu^2}(\varepsilon_x + \nu\varepsilon_y) = \frac{210\text{ GPa}}{1-0.3^2}[650\,\mu + 0.3(250\,\mu)] = 167.3\text{ Mpa}$$

$$\sigma_y = \frac{E}{1-\nu^2}(\varepsilon_y + \nu\varepsilon_x) = \frac{210\text{ GPa}}{1-0.3^2}[250\,\mu + 0.3(650\,\mu)] = 102.7\text{ Mpa}$$

From Hooke's law, the strain γ_{xy} is

$$\gamma_{xy} = \frac{\tau_{xy}}{G};\ G = \frac{E}{2(1+v)};\ \tau_{xy} = \frac{E\gamma_{xy}}{2(1+v)} = \frac{(210\text{ GPa})\,(400\ \mu)}{2(1+0.3)} = 32.3\text{ Mpa}$$

The strain in the z direction is

$$\varepsilon_z = \frac{1}{E}(0 - v\sigma_x - v\sigma_y) = \frac{-v}{E}(\sigma_x + \sigma_y)$$
$$= \frac{-0.3}{210\text{ GPa}}(167.3\text{ MPa} + 102.7\text{ MPa}) = -386\,\mu$$

ELASTIC AND PLASTIC DEFORMATION

Figure 9.10 shows the conventional stress-strain diagram for a ductile material. The elastic region is the area in which the material will stretch when loaded and relax back to its original size and shape when unloaded. Hooke's law for axial loading ($\sigma = E\varepsilon$) refers to the relationship between stress and strain in the elastic region of the stress-strain diagram.

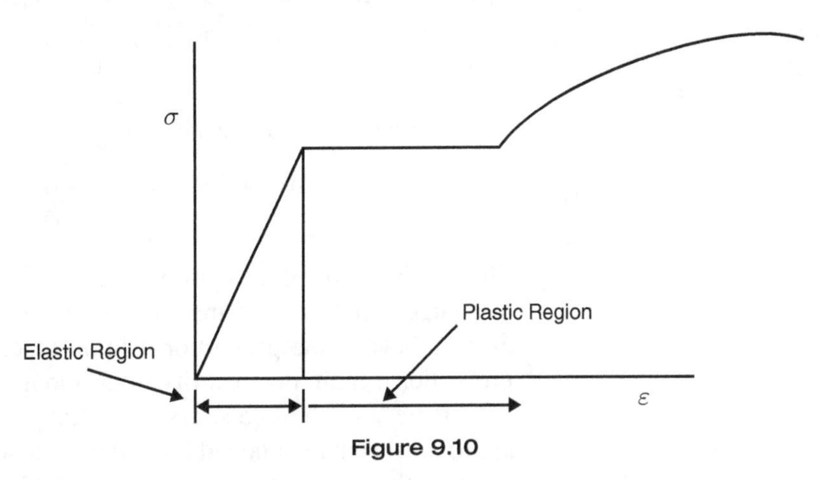

Figure 9.10

The constant of proportionality of Hooke's Law is obtained from the linear portion of the stress-strain diagram. The majority of our engineering work is in the area of elastic deformation, especially in the area of repetitive loads.

Example 9.10

A steel bar made up of two segments, as shown in Exhibit 12, is loaded with a 100 kN tension load. Determine the elongation of the bar assuming that it is stretched only in the elastic deformation region.

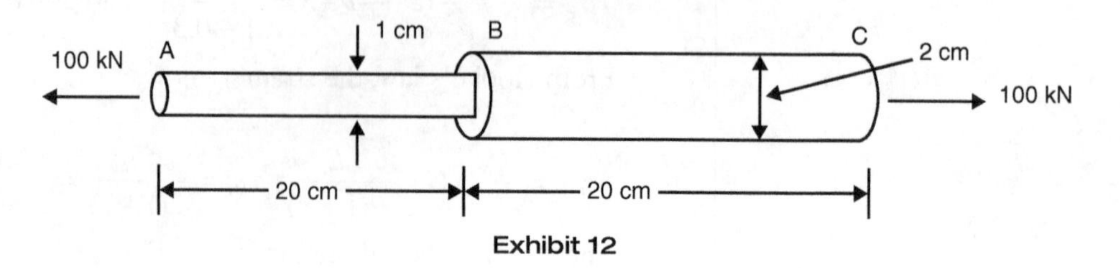

Exhibit 12

Solution

$$\delta_{AB} = \frac{PL}{AE} = \frac{(100 \text{ kN})(20 \text{ cm})}{\frac{\pi d^2}{4} \text{cm}^2 (220 \text{ GPa})}$$

$$\delta_{AB} = \frac{(100 \times 10^3 \text{ N})(0.20 \text{ m})}{\frac{\pi}{4} 0.01^2 \text{ m}^2 \left(220 \times 10^9\right) \frac{\text{N}}{\text{m}^2}} = 0.00116 \text{ m} = 0.116 \text{ cm}$$

$$\delta_{BC} = \frac{PL}{AE} = \frac{(100 \times 10^3 \text{ N})(0.20 \text{ m})}{\frac{\pi}{4} 0.02^2 \text{ m}^2 \left(220 \times 10^9\right) \frac{\text{N}}{\text{m}^2}} = 0.000289 \text{ m} = 0.0289 \text{ cm}$$

The total elastic deformation is

$$\delta_{AB} + \delta_{BC} = 0.116 + 0.0289 = 0.1449 \text{ cm}$$

The plastic region of the conventional stress-strain diagram in Figure 9.10 consists of a large region beyond the knee of the elastic region. That is, the plastic region extends beyond the proportional limit. Once the specimen is loaded beyond the proportional limit, the majority of the elongation cannot be relaxed.

Figure 9.11 shows a stress-strain diagram for an alloy. The upper curve is the approximate curve obtained by running a loading test on the material. The stress is given as the vertical line running from 50 MPa to 350 MPa, and the approximate strain ranges from 0 to 0.40 mm/mm. The diagram also has an inner curve, which is a magnification of the extremely left portion of the first curve. The strain scale of the magnified portion runs from 0 to 0.0040 mm/mm. This means the scale is magnified by 100.

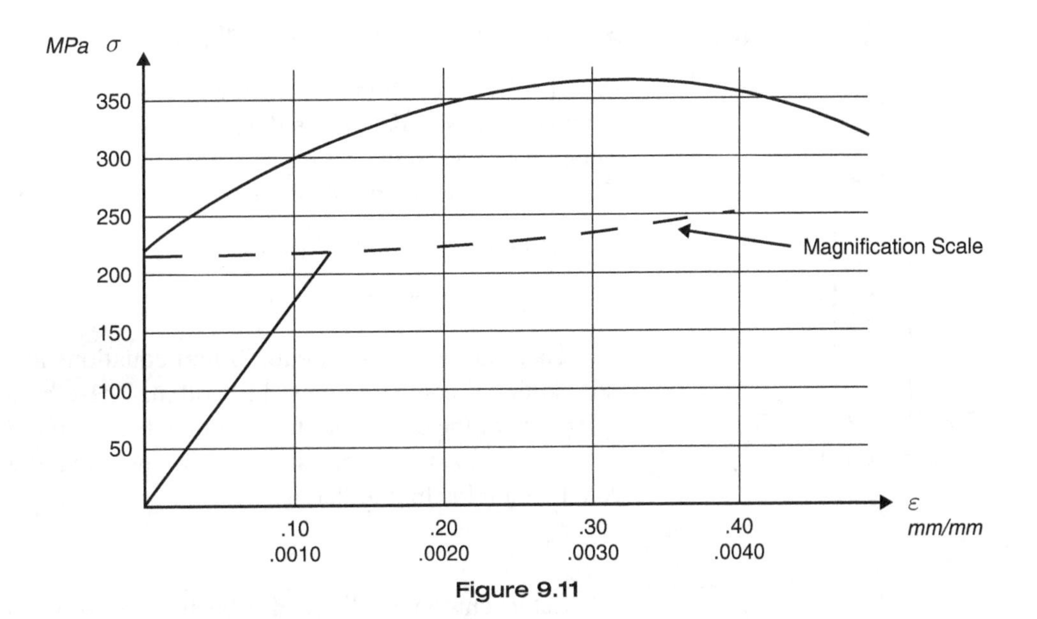

Figure 9.11

Example 9.11

Using the stress-strain diagram in Figure 9.11, determine (i) modulus of elasticity and (ii) the proportional limit.

Solution

(i) Modulus of elasticity: Using the straight-line portion of the magnified scale:

$$E = \frac{\sigma}{\varepsilon} = \frac{180 \text{ MPa}}{0.0010 \text{ mm/mm}} = 180 \text{ GPa}$$

(ii) Proportional limit: This point may be selected on the magnification scale where the knee occurs:

$$\sigma_{PL} = 220 \text{ MPa}$$

Example 9.12

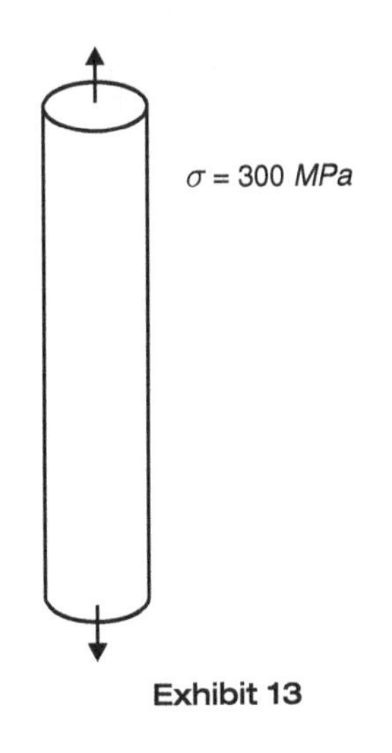

Exhibit 13

Consider that a bar made of the alloy in Example 9.11 is loaded to 300 MPa in tension, as shown in Exhibit 13. Determine (i) the modulus of resilience; (ii) the elastic recovery of the bar; and (iii) the permanent set.

Solution

(i) For modulus of resilience:

$$\mu_r = \frac{1}{2}\sigma_{PL}\varepsilon_{PL} = \frac{1}{2}(220 \text{ MPa})(0.0012 \text{ mm/mm}) = 0.132 \text{ MJ/m}^3$$

(ii) For the elastic recovery of the bar:

$$\text{Elastic recovery} = \frac{300 \times 10^6}{E} = \frac{300 \times 10^6}{180 \times 10^9} = 0.001667 \text{mm/mm}$$

(iii) Permanent set = 0.085 – 0.001667 = 0.083 mm/mm

STATICALLY INDETERMINATE STRUCTURES

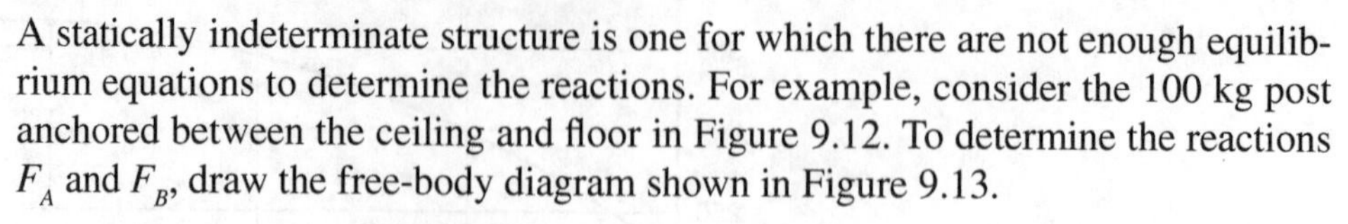

A statically indeterminate structure is one for which there are not enough equilibrium equations to determine the reactions. For example, consider the 100 kg post anchored between the ceiling and floor in Figure 9.12. To determine the reactions F_A and F_B, draw the free-body diagram shown in Figure 9.13.

Figure 9.12

Using the equation $\Sigma Fy = 0$,

$$F_B + F_A = W_t$$

There are not enough equilibrium equations available to solve this problem, so additional equations must be written. These additional equations involve the geometry of the deformation in the members of the structure.

For this axially loaded structure, we note that the structure does not change in length upon loading it; that is,

$$\delta_{AB} = 0$$

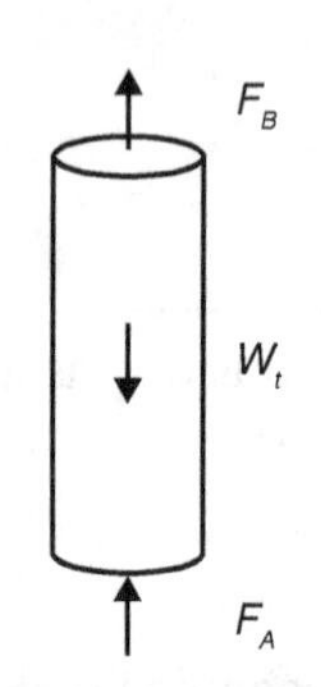

Figure 9.13

Let us cut the axially loaded beam into two parts (see Figure 9.14) and apply the loads.

Using the elongation equation:

$$\delta_B = \frac{PL}{AE} = \frac{F_B L_B}{AE} = 0$$

$$\delta_A = \frac{PL}{AE} = \frac{F_A L_A}{AE} = 0$$

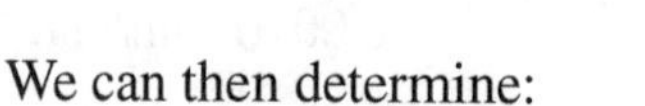

$$F_A + F_B = P = W_t$$

We can then determine:

$$F_A = P\left(\frac{L_B}{L}\right) \qquad F_B = P\left(\frac{L_A}{L}\right)$$

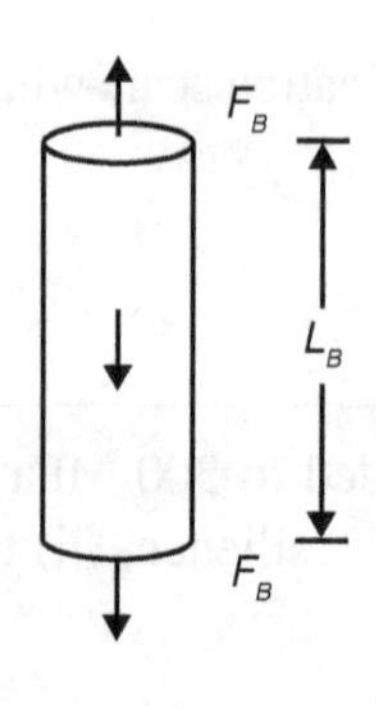

The post is cut into two pieces:

$$L_B = 1.5 \text{ m} \qquad L_A = 2.0 \text{ m}$$

Now determine F_A and F_B:

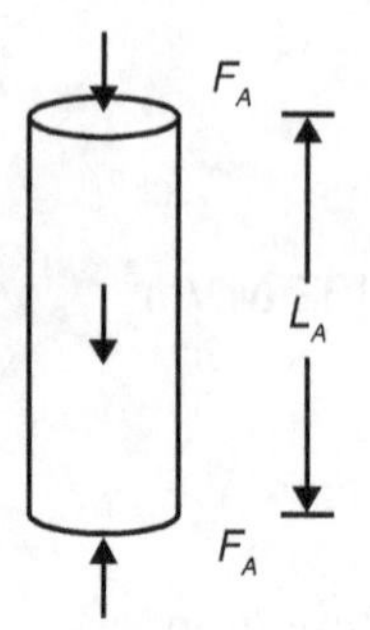

Figure 9.14

$$F_A = \frac{PL_B}{L} = \frac{(100 \text{ kg})\left(9.81\dfrac{\text{N}}{\text{kg}}\right)(1.5 \text{ m})}{3.5 \text{ m}} = 420 \text{ N}$$

$$F_B = \frac{PL_A}{L} = \frac{(100)(9.81)(2.0)}{3.5} = 560 \text{ N}$$

Example 9.13

A column constructed from concrete and one steel rod has an applied load of 400 kN. Assume that 80% of the load is carried by the concrete and 20% by the steel. See Exhibit 14. Determine the diameter of the rod.

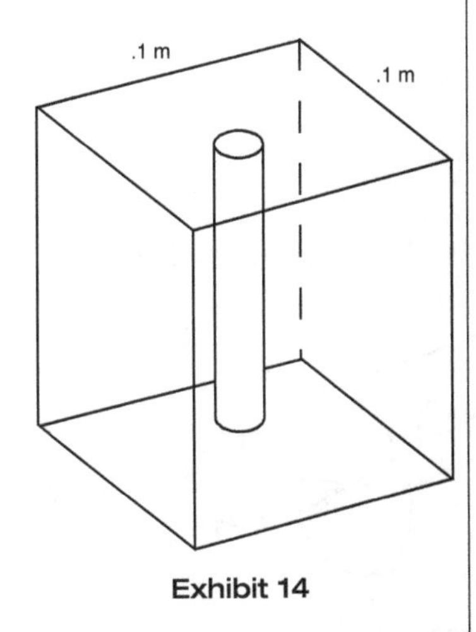

Exhibit 14

Solution

$$E_{\text{concrete}} = 25 \text{ GPa} \qquad E_{\text{steel}} = 180 \text{ GPa}$$

Knowing that $\delta_{\text{concrete}} = \delta_{\text{steel}}$

$$P_{\text{steel}} = 80 \text{ kN} \qquad P_{\text{concrete}} = 320 \text{ kN}$$

$$\left(\frac{P_{\text{concrete}}L}{AE}\right)_{\text{concrete}} = \left(\frac{P_{\text{steel}}L}{AE}\right)_{\text{steel}}$$

$$\frac{320\times10^3}{\left[.1^2 - \frac{\pi}{4}d^2\right]25\times10^9} = \frac{80\times10^3}{\left[\frac{\pi}{4}d^2\right]180\times10^9}$$

$$565.5\,d^2 = .25 \qquad d = 20.67 \text{ m}$$

TORSION

Torsion refers to the twisting of long members. Torsion can occur with members of any cross-sectional shape, but the most common is the circular shaft. Another fairly common shaft configuration, which has a simple solution, is the hollow, thin-walled shaft.

Circular Shafts

Figure 9.15(a) shows a circular shaft before loading; the r-θ-z cylindrical coordinate system is also shown. In addition to the outline of the shaft, two longitudinal lines, two circumferential lines, and two diametral lines are shown scribed on the shaft. These lines are drawn to show the deformed shape loading. Figure 9.15(b) shows the shaft after loading with a torque T. The **double arrow notation** on T indicates a moment about the z-axis in a right-handed direction. The effect of the torsion is that each cross-section remains plane and simply rotates with respect to other cross-sections. The angle ϕ is the twist of the shaft at any position z. The rotation $\phi(z)$ is in the θ direction.

The distance b shown in Figure 9.15(b) can be expressed as $b = \phi r$ or as $b = \gamma z$. The shear strain for this special case can be expressed as

$$\gamma_{\phi z} = r\frac{\phi}{z} \tag{9.40}$$

For the general case where ϕ is not a linear function of z the shear strain can be expressed as

$$\gamma_{\phi z} = r\frac{d\phi}{dz} \tag{9.41}$$

where $d\phi/dz$ is the twist per unit length or the rate of twist.

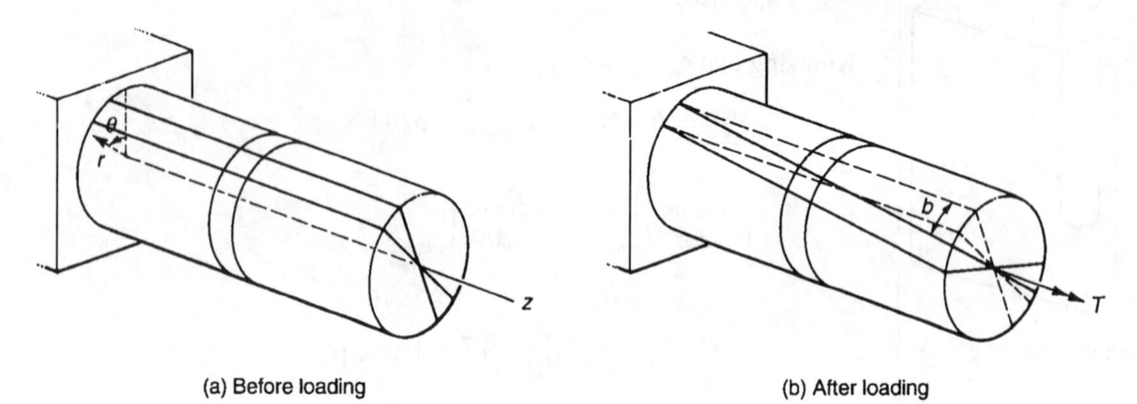

Figure 9.15 Torsion in a circular shaft

The application of Hooke's law gives

$$\tau_{\phi z} = G\gamma_{\phi z} = Gr\frac{d\phi}{dz} \tag{9.42}$$

The torque at the distance z along the shaft is found by summing the contributions of the shear stress at each point in the cross-section by means of an integration:

$$T = \int_A \tau_{\phi z} r\, dA = G\frac{d\phi}{dz}\int_A r^2 dA = GJ\frac{d\phi}{dz} \tag{9.43}$$

where $J = \int_A r^2 dA$ is the polar moment of inertia of the cross-sectional area of the shaft about the axis of the shaft. For a solid shaft with an outer radius of r_o the polar moment of inertia is

$$J = \frac{\pi r_o^4}{2} \tag{9.44}$$

For a hollow circular shaft with outer radius r_o and inner radius r_i, the polar moment of inertia is

$$J = \frac{\pi}{2}\left(r_o^4 - r_i^4\right) \tag{9.45}$$

Note that the J that appears in Equation (9.43) is the polar moment of inertia only for the special case of circular shafts (either solid or hollow). For any other cross-section shape, Equation (9.43) is valid only if J is redefined as a torsional constant *not equal* to the polar moment of inertia. Equation (9.42) can be combined with Equation (9.43) to give

$$\tau_{\phi z} = \frac{Tr}{J} \tag{9.46}$$

The maximum shear stress occurs at the outer radius of the shaft and at the location along the shaft where the torque is maximum.

$$\tau_{\phi z\,\max} = \frac{T_{\max} r_o}{J} \tag{9.47}$$

The angle of twist of the shaft can be found by integrating Equation (9.43)

$$\phi = \int_0^L \frac{T}{GJ} dz \tag{9.48}$$

For a uniform circular shaft with a constant torque along its length, this equation becomes

$$\phi = \frac{TL}{GJ} \tag{9.49}$$

Example 9.14

The hollow circular steel shaft shown in Exhibit 15 has an inner diameter of 25 mm, an outer diameter of 50 mm, and a length of 600 mm. It is fixed at the left end and subjected to a torque of 1400 N • m as shown in Exhibit 15. Find the maximum shear stress in the shaft and the angle of twist at the right end. Take $G = 84$ GPa.

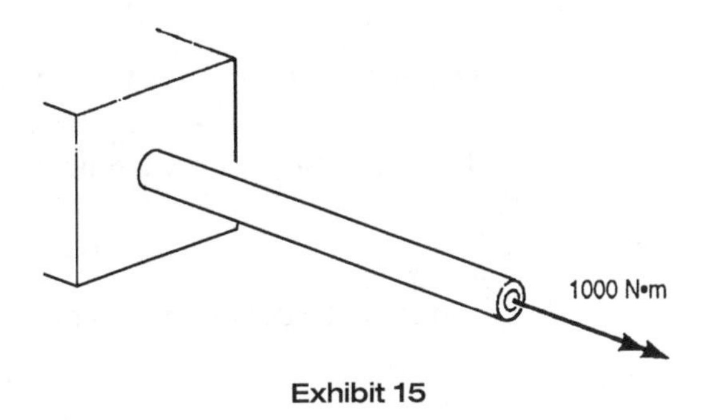

Exhibit 15

Solution

$$J = \frac{\pi}{2}\left(r_o^4 - r_i^4\right) = \frac{\pi}{2}\left[(25\text{ mm})^4 - (12.5\text{ mm})^4\right] = 575 \times 10^3\text{ mm}^4$$

$$\tau_{\theta z\,\max} = \frac{T_{\max} r_o}{J} = \frac{(1400\text{N} \bullet \text{m})(25\text{ mm})}{575 \times 10^3 \text{mm}^4} = 60.8\text{ MPa}$$

$$\phi = \frac{TL}{GJ} = \frac{(1400\text{N} \bullet \text{m})(600\text{ mm})}{(84\text{ GPa})(575 \times 10^3\text{ mm}^4)} = 0.01738\text{ rad}$$

Hollow, Thin-Walled Shafts

In hollow, thin-walled shafts, the assumption is made that the shear stress τ_{sz} is constant throughout the wall thickness t. The shear flow q is defined as the product of τ_{sz} and t. From a summation of forces in the z direction, it can be shown that q is constant—even with varying thickness. The torque is found by summing the contributions of the shear flow. Figure 9.16 shows the cross-section of the thin-walled tube of nonconstant thickness. The z coordinate is perpendicular to the plane of the paper. The shear flow q is taken in a counter-clockwise sense. The torque produced by q over the element ds is

$$dT = q\,r\,ds$$

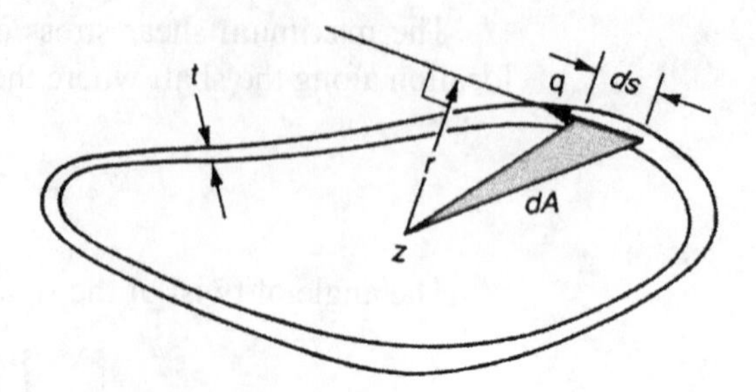

Figure 9.16 Cross-section of thin-walled tube

The total torque is, therefore,

$$T = \int qr\, ds = q \int r\, ds \tag{9.50}$$

When ds is small, the area dA is the area of the triangle of base ds and height r, which can be approximated by

$$dA = \tfrac{1}{2}(\text{base})(\text{height}) = \frac{r\, ds}{2} \tag{9.51}$$

so that

$$\int r\, ds = 2A_m \tag{9.52}$$

where A_m is the area enclosed by the wall (including the hole). It is best to use the centerline of the wall to define the boundary of the area; hence A_m is the mean area. The expression for the torque is

$$T = 2A_m q \tag{9.53}$$

and from the definition of q the shear stress can be expressed as

$$\tau_{sz} = \frac{T}{2A_m t} \tag{9.54}$$

Example 9.15

A torque of 10 kN • m is applied to a thin-walled rectangular steel shaft whose cross-section is shown in Exhibit 16. The shaft has wall thicknesses of 5 mm and 10 mm. Find the maximum shear stress in the shaft.

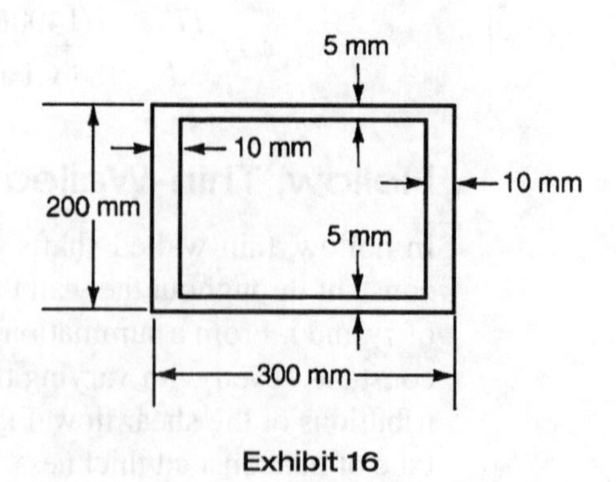

Exhibit 16

Solution

$$A_m = (200 - 5)(300 - 10) = 56{,}550 \text{ mm}^2$$

The maximum shear stress will occur in the thinnest section, so $t = 5$ mm.

$$\tau_{sz} = \frac{T}{2A_m t} = \frac{10 \text{ kN} \bullet \text{m}}{2(56{,}550 \text{ mm}^2)(5 \text{ mm})} = 17.68 \frac{\text{MN}}{\text{m}^2}$$

BEAMS

Shear and Moment Diagrams

Shear and moment diagrams are plots of the shear forces and bending moments, respectively, along the length of a beam. The purpose of these plots is to clearly show maximums of the shear force and bending moment, which are important in the design of beams. The most common sign convention for the shear force and bending moment in beams is shown in Figure 9.17. One method of determining the shear and moment diagrams is by the following steps:

1. Determine the reactions from equilibrium of the entire beam.
2. Cut the beam at an arbitrary point.

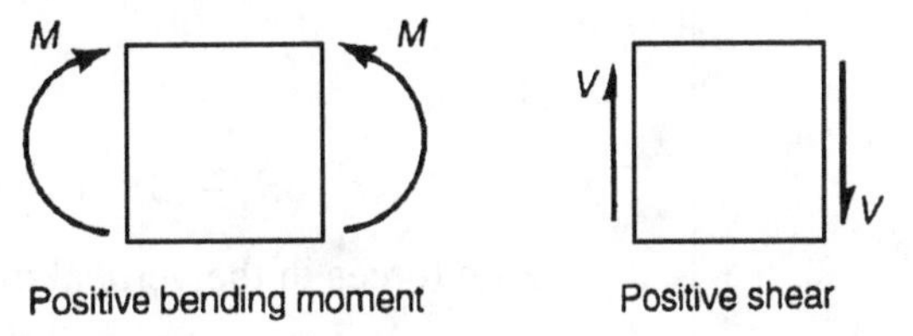

Figure 9.17 Sign convention for bending moment and shear

3. Show the unknown shear and moment on the cut using the positive sign convention shown in Figure 9.17.
4. Sum forces in the vertical direction to determine the unknown shear.
5. Sum moments about the cut to determine the unknown moment.

Example 9.16

For the beam shown in Exhibit 17, plot the shear and moment diagram.

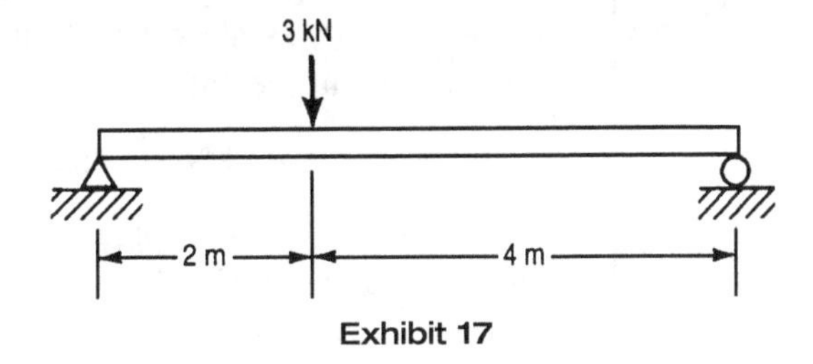

Exhibit 17

Solution

First, solve for the unknown reactions using the free-body diagram of the beam shown in Exhibit 18(a). To find the reactions, sum moments about the left end,

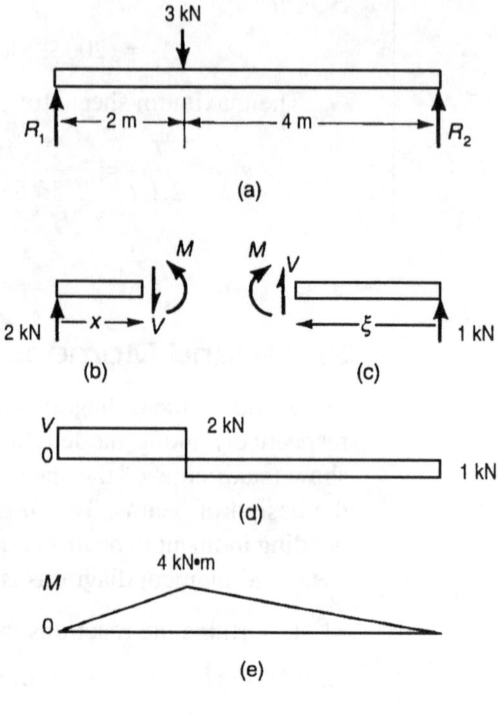

Exhibit 18

which gives

$$6R_2 - (3)(2) = 0 \quad \text{or} \quad R_2 = 6/6 = 1 \text{ kN}$$

Sum forces in the vertical direction to get

$$R_1 + R_2 = 3 = R_1 + 1 \quad \text{or} \quad R_1 = 2 \text{ kN}$$

Cut the beam between the left end and the load as shown in Exhibit 18(b). Show the unknown moment and shear on the cut using the positive sign convention shown in Figure 9.17. Sum the vertical forces to get

$$V = 2 \text{ kN (independent of } x)$$

Sum moments about the cut to get

$$M = R_1 x = 2x$$

Repeat the procedure by making a cut between the right end of the beam and the 3-kN load, as shown in Exhibit 18(c). Again, sum vertical forces and sum moments about the cut to get

$$V = 1 \text{ kN (independent of } \xi\text{), and } M = 1\xi$$

The plots of these expressions for shear and moment give the shear and moment diagrams shown in Exhibit 18(d) and 18(e).

It should be noted that the shear diagram in this example has a jump at the point of the load and that the jump is equal to the load. This is always the case. Similarly, a moment diagram will have a jump equal to an applied concentrated moment. In this example, there was no concentrated moment applied, so the moment was everywhere continuous.

Another useful way of determining the shear and moment diagram is by using differential relationships. These relationships are found by considering an element of length Δx of the beam with a distributed applied load q per unit length. The forces on that element are shown in Figure 9.18. Summation of forces in the y direction gives

$$q\Delta x + V - V - \frac{dV}{dx}\Delta x = 0 \qquad \textbf{(9.55)}$$

which gives

$$\frac{dV}{dx} = q \qquad \textbf{(9.56)}$$

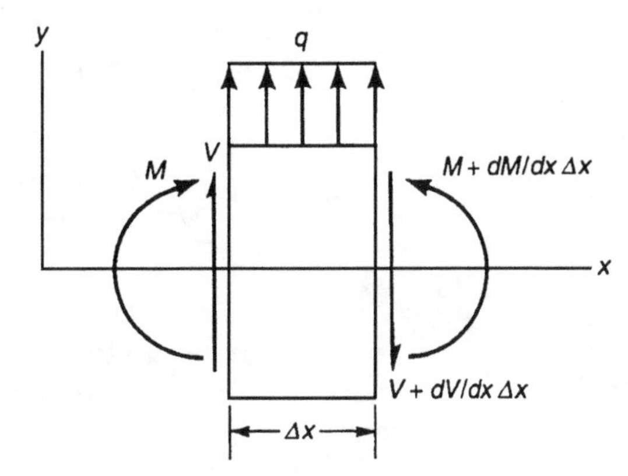

Figure 9.18

Summing moments and neglecting higher order terms gives

$$-\mathrm{M} + \mathrm{M} + \frac{dM}{dx}\Delta x - V\Delta x = 0 \qquad \textbf{(9.57)}$$

which gives

$$\frac{dM}{dx} = V \qquad \textbf{(9.58)}$$

Integral forms of these relationships are expressed as

$$V_2 - V_1 = \int_{x_1}^{x_2} q\,dx \qquad \textbf{(9.59)}$$

$$M_2 - M_1 = \int_{x_1}^{x_2} V\,dx \qquad \textbf{(9.60)}$$

Example **9.17**

The simply supported uniform beam shown in Exhibit 19 carries a uniform load of w_0. Plot the shear and moment diagrams for this beam.

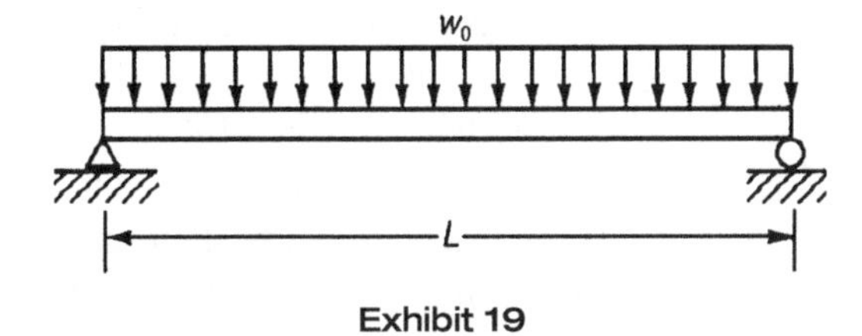

Exhibit 19

Solution

As before, the reactions can be found first from the free-body diagram of the beam shown in Exhibit 20(a). It can be seen that, from symmetry, $R_1 = R_2$. Summing vertical forces then gives

$$R = R_1 = R_2 = \frac{w_0 L}{2}$$

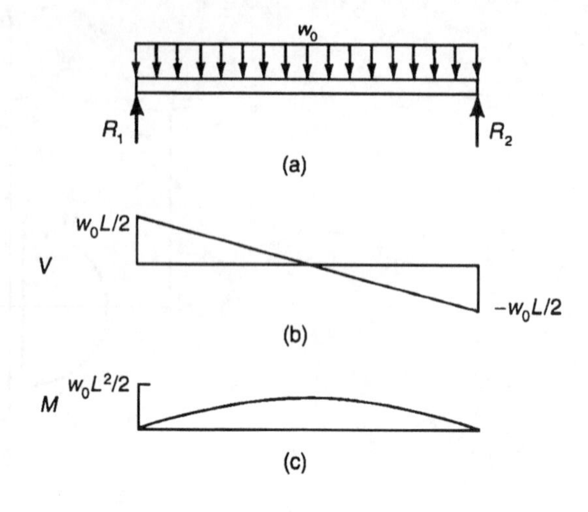

Exhibit 20

The load $q = -w_0$, so Equation (9.59) reads

$$V(x) - V_0 = \int_0^x q dx = qx$$

where V_0 is the shear force at $x = 0$.

$$V = V_0 - \int_0^x w_0 \, dx = \frac{w_0 L}{2} - w_0 x$$

Noting that the moment at $x = 0$ is zero, Equation (9.60) gives

$$M(x) = M_0 + \int_0^x \left(\frac{w_0 L}{2} - w_0 x \right) dx = 0 + \frac{w_0 Lx}{2} - \frac{w_0 x^2}{2} = \frac{w_0 x}{2}(L - x)$$

It can be seen that the shear diagram is a straight line, and the moment varies parabolically with x. Shear and moment diagrams are shown in Exhibit 20(b) and Exhibit 20(c). It can be seen that the maximum bending moment occurs at the center of the beam where the shear stress is zero. The maximum bending moment always has a relative maximum at the place where the shear is zero because the shear is the derivative of the moment, and relative maxima occur when the derivative is zero.

Often, it is helpful to use a combination of methods to find the shear and moment diagrams. For instance, if there is no load between two points, then the shear diagram is constant, and the moment diagram is a straight line. If there is a uniform load, then the shear diagram is a straight line, and the moment diagram is parabolic. The following example illustrates this method.

Example 9.18

Draw the shear and moment diagrams for the beam shown in Exhibit 21(a).

Solution

Draw the free-body diagram of the beam as shown in Exhibit 21(b). From a summation of the moments about the right end,

$$10\,R_1 = (4)(7) + (3)(2) = 34; \quad \text{so } R_1 = 3.4\,\text{kN}$$

From a summation of forces in the vertical direction,

$$R_2 = 7 - 3.4 = 3.6\,\text{kN}$$

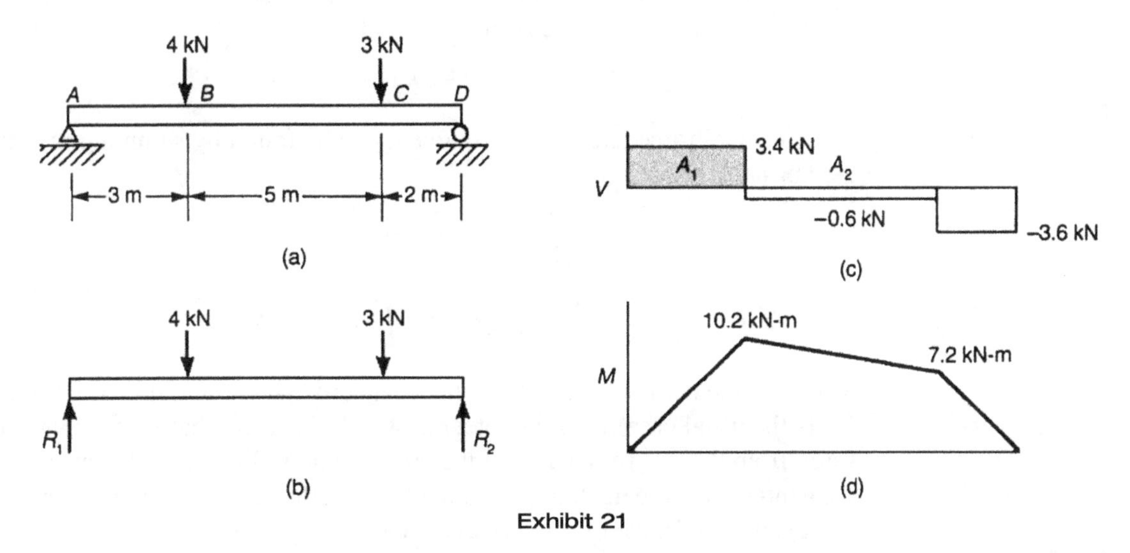

Exhibit 21

The shear in the left portion is 3.4 kN, the shear in the right portion is −3.6 kN, and the shear in the center portion is 3.4 − 4 = −0.6 kN. This is sufficient information to draw the shear diagram shown in Exhibit 21(c). The moment at *A* is zero, so the moment at *B* is the shaded area A_1 and the moment at *C* is $A_1 - A_2$.

$$M_B = A_1 = (3.4\,\text{kN})(3\,\text{m}) = 10.2\,\text{kN} \bullet \text{m}$$

$$M_C = A_1 - A_2 = (3.4\text{kN})(3\text{m}) - (0.6\text{kN})(5\text{m}) = 7.2\text{kN} \bullet \text{m}$$

The moments at *A* and *D* are zero, and the moment diagram consists of straight lines between the points *A*, *B*, *C*, and *D*. There is, therefore, enough information to plot the moment diagram shown in Exhibit 21(d).

Stresses in Beams

The basic assumption in elementary beam theory is that the beam cross-section remains plane and perpendicular to the neutral axis as shown in Figure 9.19 when the beam is loaded. This assumption is strictly true only for the case of pure bending (constant bending moment and no shear) but gives good results even when shear is present. Figure 9.19 shows a beam element before and after loading. It can be seen that there is a line of length *ds* that does not change length due to deformation. This line is called the neutral axis. The distance *y* is measured from this neutral axis. The strain in the *x* direction is $\Delta L/L$. The change in length $\Delta L = -y d\phi$ and the length is *ds*, so

$$\varepsilon_x = -y\frac{d\phi}{ds} = -\frac{y}{\rho} = -\kappa y \qquad \textbf{(9.61)}$$

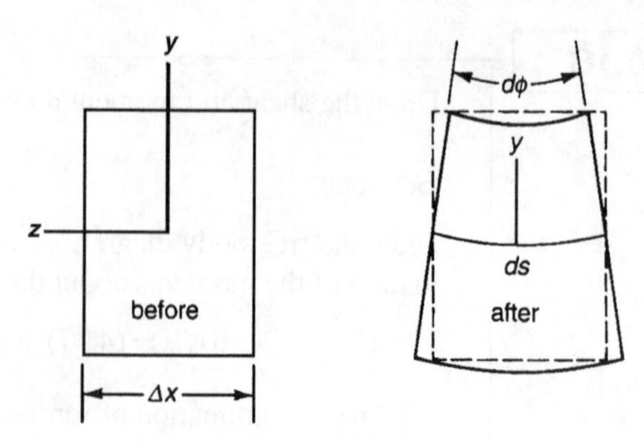

Figure 9.19

where ρ is the radius of curvature of the beam and κ is the curvature of the beam. Assuming that σ_y and σ_z are zero, Hooke's law yields

$$\sigma_x = -E\kappa y \tag{9.62}$$

The axial force and bending moment can be found by summing the effects of the normal stress σ_x,

$$P = \int_A \sigma_x \, dA = -E\kappa \int_A y \, dA \tag{9.63}$$

$$M = -\int_A y\sigma_x \, dA = E\kappa \int_A y^2 dA = EI\kappa \tag{9.64}$$

where I is the moment of inertia of the beam cross-section. If the axial force is zero (as is the usual case) then the integral of $y\,dA$ is zero. That means that y is measured from the centroidal axis of the cross-section. Since y is also measured from the neutral axis, the neutral axis coincides with the centroidal axis. From Equation (9.62) and (9.64), the bending stress σ_x can be expressed as

$$\sigma_x = -\frac{My}{I} \tag{9.65}$$

The maximum bending stress occurs where the magnitude of the bending moment is a maximum and at the maximum distance from the neutral axis. For symmetrical beam sections the value of $y_{max} = \pm C$, where C is the distance to the extreme fiber, so the maximum stress is

$$\sigma_x = \pm\frac{MC}{I} = \pm\frac{M}{S} \tag{9.66}$$

where S is the section modulus ($S = I/C$).

Example 9.19

A 100 mm × 150 mm wooden cantilever beam is 2 m long. It is loaded at its tip with a 4-kN load. Find the maximum bending stress in the beam shown in Exhibit 22. The maximum bending moment occurs at the wall and is M_{max} = 8 kN • m.

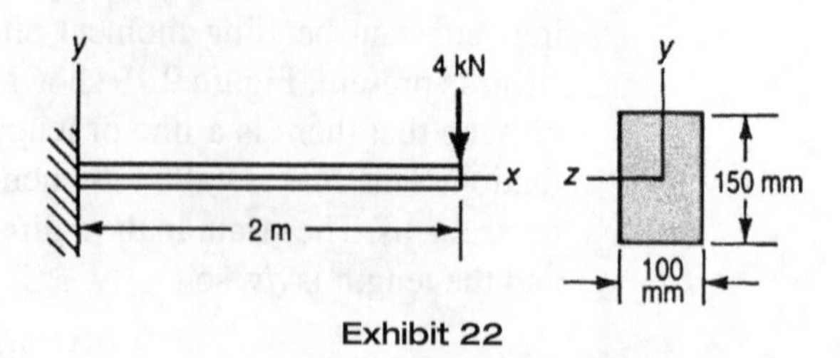

Exhibit 22

Solution

$$I = \frac{bh^3}{12} = \frac{100(150)^3}{12} = 28.1 \times 10^6 \text{ mm}^4$$

$$\sigma_{x\max} = \frac{|M|_{\max} c}{I} = \frac{(8 \text{ kN} \bullet \text{m})(75 \text{ mm})}{28.1 \times 10^6 \text{ mm}^4} = 21.3 \text{ MPa}$$

Shear Stress

To find the shear stress, consider the element of length Δx shown in Figure 9.20(a). A cut is made in the beam at $y = y_1$. At that point the beam has a thickness b. The shaded cross-sectional area above that cut is called A_1. The bending stresses acting on that element are shown in Figure 9.20(b). In general, the stresses are slightly larger at the right side than at the left side so that a force per unit length q is needed for equilibrium. Summation of forces in the x direction for the free-body diagram shown in Figure 9.20(b) gives

$$-F = q\Delta x = \int_{A_1} \sigma \, dA - \int_{A_1} \left(\sigma + \frac{d\sigma}{dx} \Delta x \right) dA = -\int_{A_1} \frac{d\sigma}{dx} \Delta x \, dA \quad \textbf{(9.67)}$$

From the expression for the bending stress ($\sigma = -My/I$) it follows that

$$\frac{d\sigma}{dx} = -\left(\frac{dM}{dx} \right) \frac{y}{I} = -V \frac{y}{I} \quad \textbf{(9.68)}$$

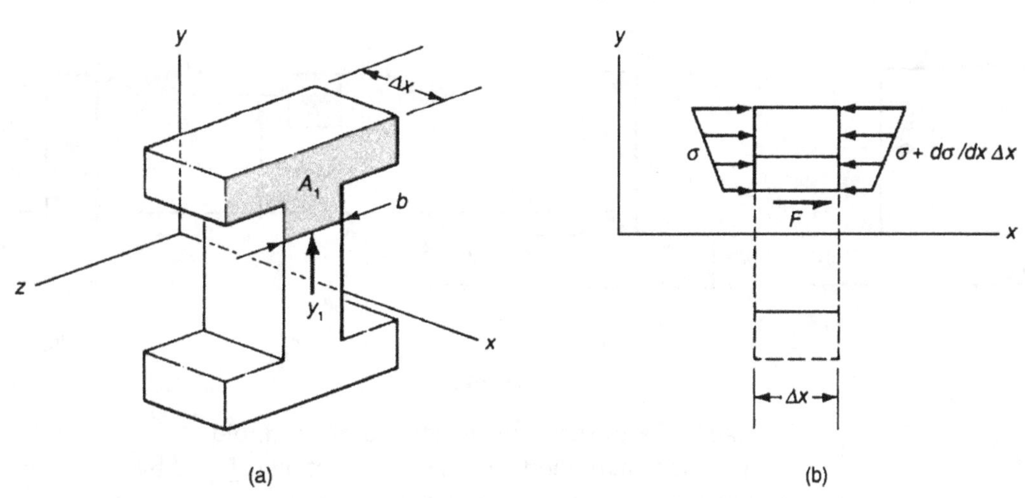

Figure 9.20 Shear stress in beams

Substituting Equation (9.68) into Equation (9.67) gives

$$q = \frac{V}{I} \int_{A_1} y \, dA = \frac{VQ}{I} \quad \textbf{(9.69)}$$

If the shear stress τ is assumed to be uniform over the thickness b then $\tau = q/b$ and the expression for shear stress is

$$\tau = \frac{VQ}{Ib} \quad \textbf{(9.70)}$$

where V is the shear in the beam, Q is the moment of area above (or below) the point in the beam at which the shear stress is sought ($y = y_1$), I is the moment of inertia of the entire beam cross-section, and b is the thickness of the beam cross-section at the point where the shear stress is sought ($y = y_1$). The definition of Q from Equation (9.69) is

$$Q = \int_{A_1} y\,dA = A_1 y \qquad \textbf{(9.71)}$$

Example 9.20

The cross-section of the beam shown in Exhibit 23 has an applied shear of 10 kN. Find (a) the shear stress at a point 20 mm below the top of the beam and (b) the maximum shear stress from the shear force.

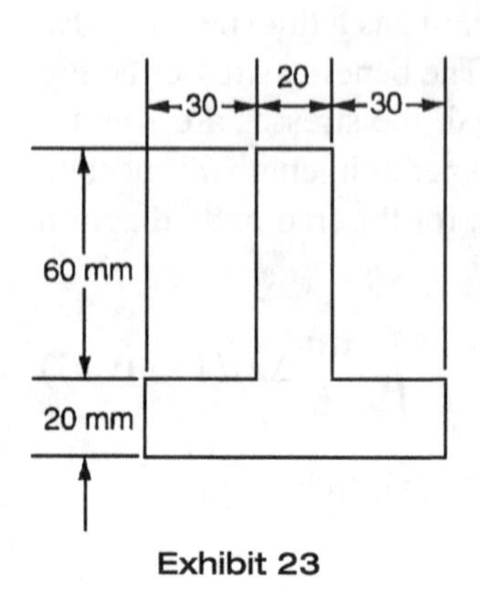

Exhibit 23

Solution

The section is divided into two parts by the dashed line shown in Exhibit 24(a). The centroids of each of the two sections are also shown in Exhibit 24(a). The centroid of the entire cross-section is found as follows:

$$\bar{y} = \frac{\sum_{n=1}^{N} \bar{y}_n A_n}{\sum_{n=1}^{N} A_n} = \frac{(60)(20)(30+20)+(80)(20)(10)}{(60)(20)+(80)(20)} = 27.14 \text{ mm (from bottom)}$$

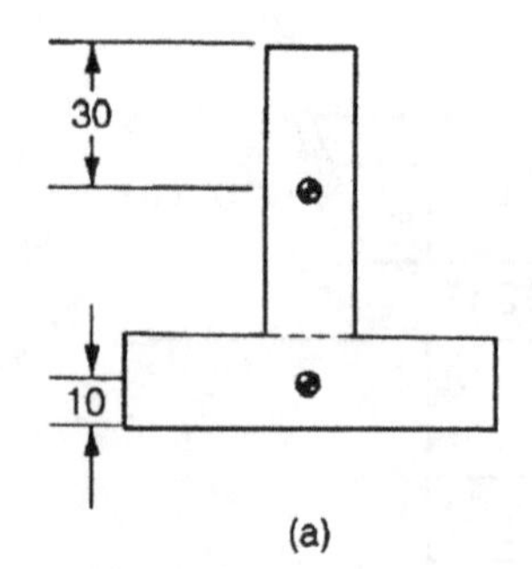

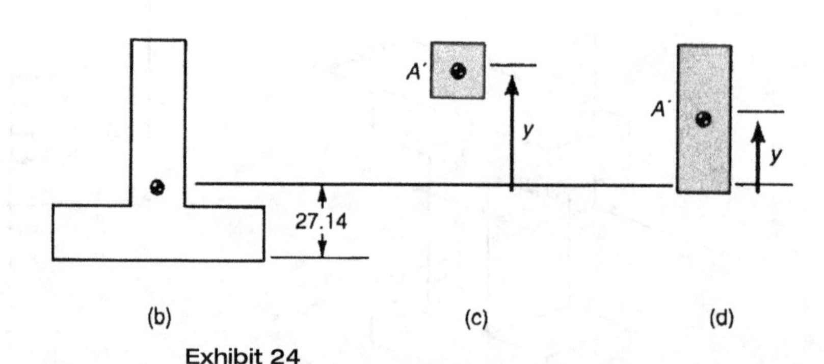

(a) (b) (c) (d)

Exhibit 24

Exhibit 24(b) shows the location of the centroid.

The moment of inertia of the cross-section is found by summing the moments of inertia of the two sections taken about the centroid of the entire section. The moment of inertia of each part is found about its own centroid; then the parallel axis theorem is used to transfer it to the centroid of the entire section.

$$I = \sum_{n=1}^{N} I_n + A_n \bar{y}_n$$

$$= \frac{(20)(60)^3}{12} + (20)(60)(50-27.14)^2 + \frac{(80)(20)^3}{12} - (20)(80)(27.14-10)^2$$

$$= 1.510 \times 10^6 \text{ mm}^4$$

For the point 20 mm below the top of the beam, the area A' and the distance y are shown in Exhibit 24(c). The distance y is from the neutral axis to the centroid of A'. The value of Q is then

$$Q = \int_{A'} y\,dA = A'y = (20)(20)(70 - 27.14) = 17{,}140\ \text{mm}^3$$

$$\tau = \frac{VQ}{Ib} = \frac{(10\text{kN})(17{,}140\ \text{mm}^3)}{(1.510 \times 10^6\ \text{mm}^4)(20\ \text{mm})} = 0.00568\frac{\text{kN}}{\text{mm}^2} = 5.68\ \text{MPa}$$

The maximum Q will be at the centroid of the cross-section. Since the thickness is the same everywhere, the maximum shear stress will appear at the centroid. The maximum moment of area Q_{max} is

$$Q = \int_{A'} y\,dA = A'\,y_1 = (20)(80 - 27.14)\frac{(80 + 27.14)}{2} = 56{,}600\ \text{mm}^3$$

$$\tau = \frac{VQ}{Ib} = \frac{(10\text{kN})(56{,}600\ \text{mm}^3)}{(1.510 \times 10^6\ \text{mm}^4)(20\ \text{mm})} = 0.01875\frac{\text{kN}}{\text{mm}^2} = 18.75\ \text{MPa}$$

Deflection of Beams

The beam deflection in the y direction will be denoted as y, while most modern texts use v for the deflection in the y direction. The *Fundamentals of Engineering Supplied-Reference Handbook* uses the older notation. The main assumption in the deflection of beams is that the slope of the beam is small. The slope of the beam is dy/dx. Since the slope is small, the slope is equal to the angle of rotation in radians.

$$\frac{dy}{dx} = \text{rotation in radians} \tag{9.72}$$

Because the slope is small it also follows that

$$\kappa = \frac{1}{\rho} \approx \frac{d^2y}{dx^2} \tag{9.73}$$

From Equation (9.62) this gives

$$\frac{d^2y}{dx^2} = \frac{M}{EI} \tag{9.74}$$

This equation, together with two boundary conditions, can be used to find the beam deflection. Integrating twice with respect to x gives

$$\frac{dy}{dx} = \int \frac{M}{EI}dx + C_1 \tag{9.75}$$

$$y = \iint \frac{M}{EI}dx + C_1x + C_2 \tag{9.76}$$

where the constants C_1 and C_2 are determined from the two boundary conditions. Appropriate boundary conditions are on the displacement y or on the slope dy/dx. In the common problems of uniform beams, the beam stiffness EI is a constant and can be removed from beneath the integral sign.

Example 9.21

The uniform cantilever beam shown in Exhibit 25 has a constant, uniform, downward load w_0 along its length. Find the deflection and slope of this beam.

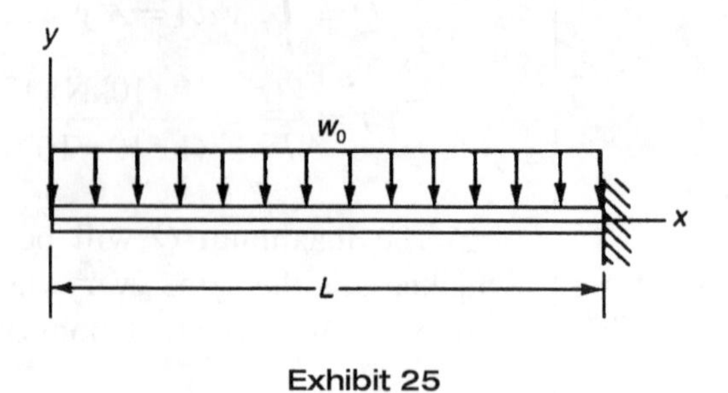

Exhibit 25

Solution

The moment is found by drawing the free-body diagram shown in Exhibit 26. The uniform load is replaced with the statically equivalent load w_0x at the position $x/2$. Moments are then summed about the cut giving

$$M = -w_0 \frac{x^2}{2}$$

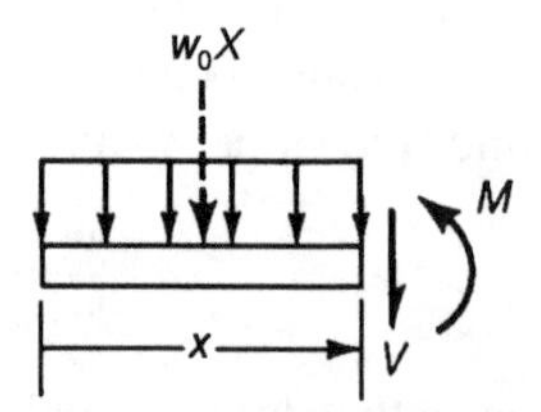

Exhibit 26

Integrating twice with respect to x,

$$\frac{dy}{dx} = \int \frac{M}{EI} dx + C_1 = \frac{1}{EI} \int \left(-w_0 \frac{x^2}{2} \right) dx + C_1 = -\frac{1}{6} \frac{w_0 x^3}{EI} + C_1$$

$$y = \int \left(-\frac{1}{6} \frac{w_0 x^3}{EI} \right) dx + C_1 x + C_2 = -\frac{1}{24} \frac{w_0 x^4}{EI} + C_1 x + C_2$$

At $x = L$ the displacement and slope must be zero so that

$$y(L) = 0 = -\frac{1}{24} \frac{w_0 L^4}{EI} + C_1 L + C_2$$

$$\frac{dy}{dx}(L) = 0 = -\frac{1}{6} \frac{w_0 L^3}{EI} + C_1$$

Therefore,

$$C_1 = \frac{1}{6}\frac{w_0 L^3}{EI}; \qquad C_2 = -\frac{1}{8}\frac{w_0 L^4}{EI}$$

Inserting C_1 and C_2 into the previous expressions gives

$$y = -\frac{w_0}{24EI}(x^4 - 4xL^3 + 3L^4)$$

$$\frac{dy}{dx} = \frac{w_0}{6EI}(L^3 - x^3)$$

Fourth-Order Beam Equation

The second-order beam Equation (9.74) can be combined with the differential relationships between the shear, moment, and distributed load. Differentiate Equation (9.74) with respect to x, and use Equation (9.58).

$$\frac{d}{dx}\left(EI\frac{d^2 y}{dx^2}\right) = \frac{dM}{dx} = V \tag{9.77}$$

Differentiate again with respect to x and use Equation (9.56).

$$\frac{d^2}{dx^2}\left(EI\frac{d^2 y}{dx^2}\right) = \frac{dV}{dx} = q \tag{9.78}$$

For a uniform beam with a constant Young's Modulus (that is, constant EI) the fourth-order beam equation becomes

$$EI\frac{d^4 y}{dx^4} = q \tag{9.79}$$

This equation can be integrated four times with respect to x. Four boundary conditions are required to solve for the four constants of integration. The boundary

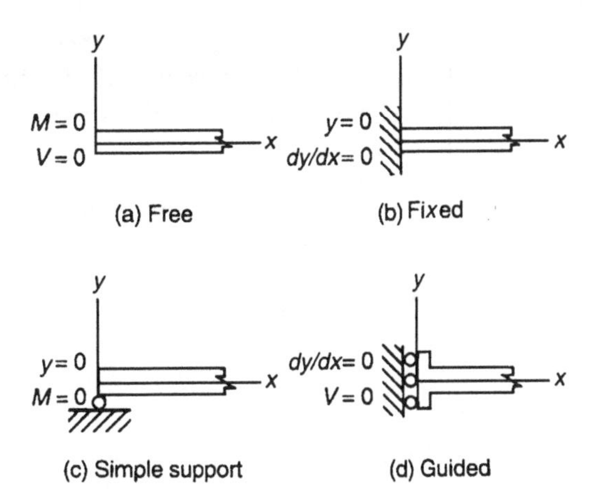

Figure 9.21 Boundary conditions for beams

conditions are on the displacement, slope, moment, and/or shear. Figure 9.21 shows the appropriate boundary conditions on the end of a beam, even with a distributed loading. If there is a concentrated force or moment applied at the end of a beam, that force or moment enters the boundary condition. For instance, an upward load of P at the left end for the free or guided beam would give $V(0) = P$ instead of $V(0) = 0$.

Example 9.22

Consider the uniformly loaded uniform beam shown in Exhibit 27. The beam is clamped at both ends. The uniform load w_0 is acting downward. Find an expression for the displacement as a function of x.

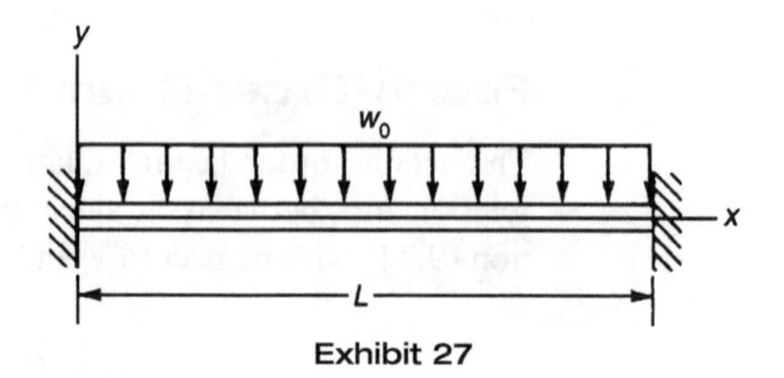

Exhibit 27

Solution

The differential equation is

$$EI\frac{d^4y}{dx^4} = q = -w_0 \,(\text{constant})$$

Integrate four times with respect to x.

$$V = EI\frac{d^3y}{dx^3} = -w_0 x + C_1$$

$$M = EI\frac{d^2y}{dx^2} = -w_0\frac{x^2}{2} + C_1 x + C_2$$

$$EI\frac{dy}{dx} = -w_0\frac{x^3}{6} + C_1\frac{x^2}{2} + C_2 x + C_3$$

$$EIy = -w_0\frac{x^4}{24} + C_1\frac{x^3}{6} + C_2\frac{x^2}{2} + C_3 x + C_4$$

The four constants of integration can be found from four boundary conditions. The boundary conditions are

$$y(0) = 0; \quad \frac{dy}{dx}(0) = 0; \quad y(L) = 0; \quad \frac{dy}{dx}(L) = 0$$

These lead to the following:

$$EIy(0) = 0 = C_4$$
$$EI\frac{dy}{dx}(0) = 0 = C_3$$
$$EIy(L) = 0 = -w_0\frac{L^4}{24} + C_1\frac{L^3}{6} + C_2\frac{L^2}{2}$$
$$EI\frac{dy}{dx}(L) = 0 = -w_0\frac{L^3}{6} + C_1\frac{L^2}{2} + C_2L$$

Solving the last two equations for C_1 and C_2 gives

$$C_1 = \frac{1}{2}w_0L; \qquad C_2 = -\frac{1}{12}w_0L^2$$

Inserting these values into the equation for y gives

$$y = -\frac{w_0x^2}{EI}\left(\frac{1}{24}x^2 - \frac{1}{12}xL + \frac{1}{24}L^2\right)$$

Some solutions for uniform beams with various loads and boundary conditions are shown in Table 9.1.

Table 9.1 Deflection and slope formulas for beams

	Beam	Deflection, v	Slope, v'
1.	y, P, x, a, b, L	For $0 \le x \le a$ $y = \frac{Px^2}{6EI}(3a - x)$ For $a \le x \le L$ $y = \frac{Pa^2}{6EI}(3x - a)$	For $0 \le x \le a$ $\frac{dy}{dx} = \frac{px}{2EI}(2a - x)$ For $a \le x \le L$ $\frac{dy}{dx} = \frac{Pa^2}{2EI}a$
2.	y, w_0, x, L	$y = -\frac{w_0x^2}{24EI}(x^2 - 4Lx + 6L^2)$	$\frac{dy}{dx} = -\frac{w_0x}{6EI}(x^2 - 12Lx + 12L^2)$
3.	y, P, x, a, b, L	For $0 \le x \le a$ $y = \frac{Pbx}{6LEI}(L^2 - b^2 - x^2)$ For $a \le x \le L$ $y = \frac{Pa(L - x)}{6LEI}(2Lx - a^2 - x^2)$	For $0 \le x \le a$ $\frac{dy}{dx} = \frac{Pb}{6LEI}(L^2 - b^2 - 3x^2)$ For $a \le x \le L$ $\frac{dy}{dx} = \frac{Pa}{6LEI}(2L^2 + a^2 - 6Lx + 3x^2)$

Table 9.1 Deflection and slope formulas for beams (Continued)

Beam	Deflection, v	Slope, v'
4.	$y=-\frac{w_0 x}{24EI}(L^3-2Lx^2+x^3)$	$\frac{dy}{dx}=-\frac{w_0}{24EI}(L^3-6Lx^2+4x^3)$
5.	$y=-\frac{M_0 x}{6EIL}(L^2-x^2)$	$\frac{dy}{dx}=-\frac{M_0}{6EIL}(L^2-3x^2)$

Superposition

In addition to the use of second-order and fourth-order differential equations, a very powerful technique for determining deflections is the use of superposition. Because all of the governing differential equations are linear, solutions can be directly superposed. Use can be made of tables of known solutions, such as those in Table 9.1, to form solutions to many other problems. Some examples of superposition follow.

Example 9.23

Find the maximum displacement for the simply supported uniform beam loaded by two equal loads placed at equal distances from the ends as shown in Exhibit 28.

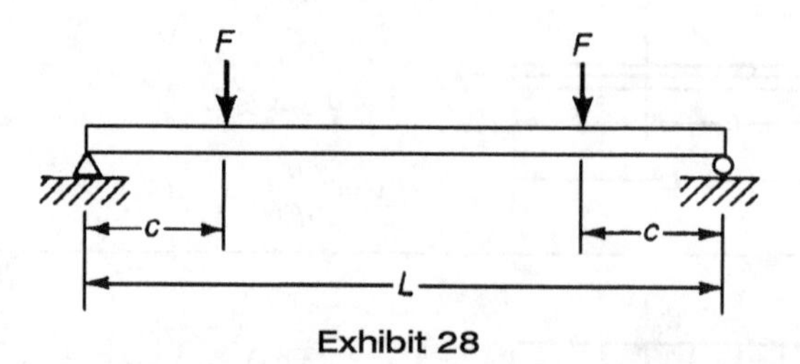

Exhibit 28

Solution

The solution can be found by superposition of the two problems shown in Exhibit 29. From the symmetry of this problem, it can be seen that the maximum deflection will be at the center of the span. The solution for the beam shown in Exhibit 29(a) is found as case 3 in Table 9.1. In Exhibit 29(a) the center of the span is to the left of the load F so that the formula from the table for $0 \le x \le a$ is chosen. In the formula, $x = L/2$, $c = b$, and $P = -F$ so that

$$y_a\left(\frac{L}{2}\right)=\frac{Pbx}{6LEI}\left(L^2-b^2-x^2\right)=-\frac{Fc\left(\frac{L}{2}\right)}{6LEI}\left[L^2-c^2-\left(\frac{L}{2}\right)^2\right]=-\frac{Fc}{48EI}\left(3L^2-4c^2\right)$$

The central deflection of the beam in Exhibit 29(b) will be the same, so the maximum downward deflection, Δ, will be

$$\delta=2y_a\left(\frac{L}{2}\right)=-\frac{Fc}{24EI}(3L^2-4c^2)$$

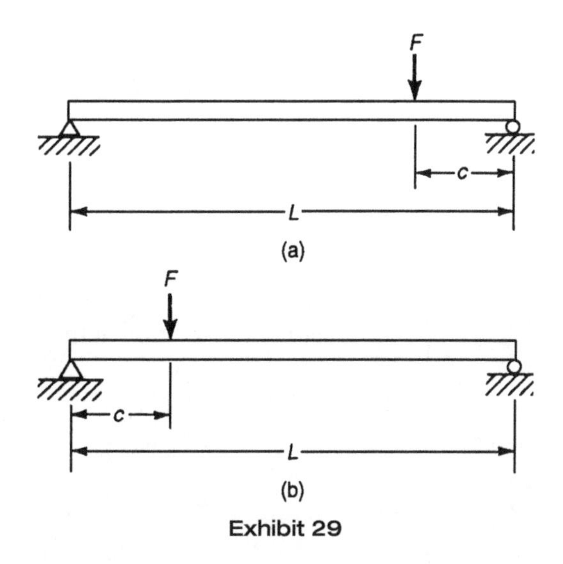

Exhibit 29

Example 9.24

Find an expression for the deflection of the uniformly loaded, supported, cantilever beam shown in Exhibit 30.

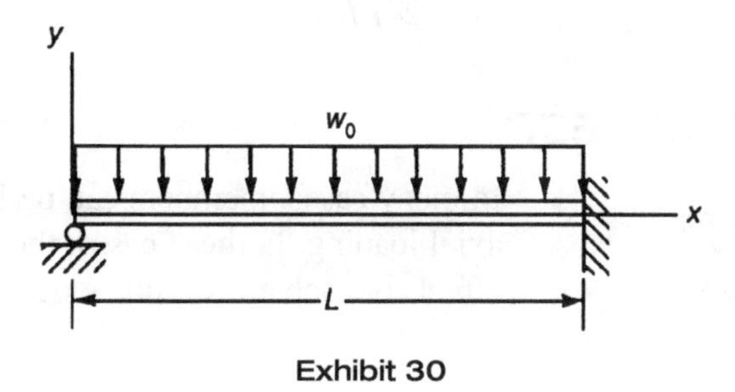

Exhibit 30

Solution

Superpose cases 4 and 5 as shown in Exhibit 31 so that the moment M_0 is of the right magnitude and direction to suppress the rotation at the right end. The rotation

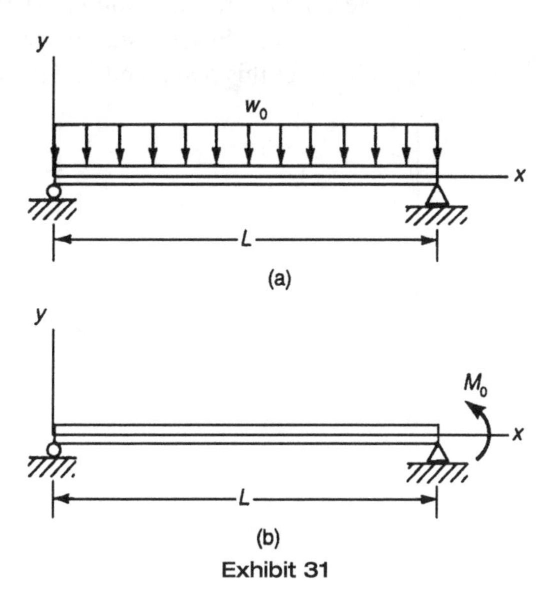

Exhibit 31

for each case from Table 9.1 is

$$\left.\left(\frac{dy}{dx}\right)_4\right|_{x=L} = -\frac{w_0}{24EI}(L^3 - 6L^3 + 4L^3) = \frac{w_0 L^3}{24EI}$$

$$\left.\left(\frac{dy}{dx}\right)_5\right|_{x=L} = -\frac{M_0}{6EIL}(L^2 - 3L^2) = \frac{M_0 L}{3EI}$$

Setting the rotation at the end equal to zero gives

$$\left.\left(\frac{dy}{dx}\right)_4\right|_{x=L} + \left.\left(\frac{dy}{dx}\right)_5\right|_{x=L} = 0 = \frac{w_0 L^3}{24EI} + \frac{M_0 L}{3EI}$$

$$M_0 = -\frac{w_0 L^2}{8}$$

Substituting this expression into the formulas in the table and adding gives

$$y = -\frac{w_0 x}{24EI}(L^3 - 2Lx^2 + x^3) + \frac{w_0 L^2}{8}\frac{x}{6EI}(L^2 - x^2) = -\frac{w_0 x}{48EI}(L^3 - 3Lx^2 + 2x^3)$$

COMBINED STRESS

In many cases, members can be loaded in a combination of bending, torsion, and axial loading. In these cases, the solution of each portion is exactly as before; the effects of each are simply added. This concept is best illustrated by an example.

Example 9.25

In Exhibit 32, there is a thin-walled, aluminum tube AB, which is attached to a wall at A. The tube has a rectangular cross-section member BC attached to it. A vertical load is placed on the member BC as shown. The aluminum tube has an outer diameter of 50 mm and a wall thickness of 3.25 mm. Take $P = 900$ N, $a = 450$ mm, and $b = 400$ mm. Find the state of stress at the top of the tube at the point D. Draw Mohr's circle for this point, and find the three principal stresses.

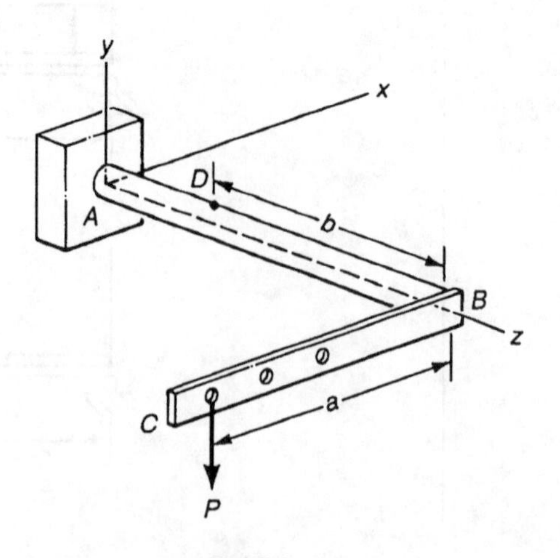

Exhibit 32

Solution

Cut the tube at the Point D. Draw the free-body diagram as in Exhibit 33(a). From that free-body diagram, a summation of moments at the cut about the z-axis gives

$$T = Pa = (900\ \text{N})(450\ \text{mm}) = 405\ \text{N} \bullet \text{m}$$

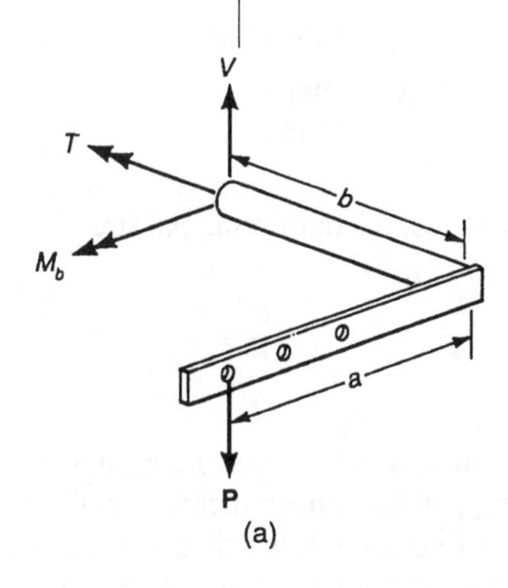

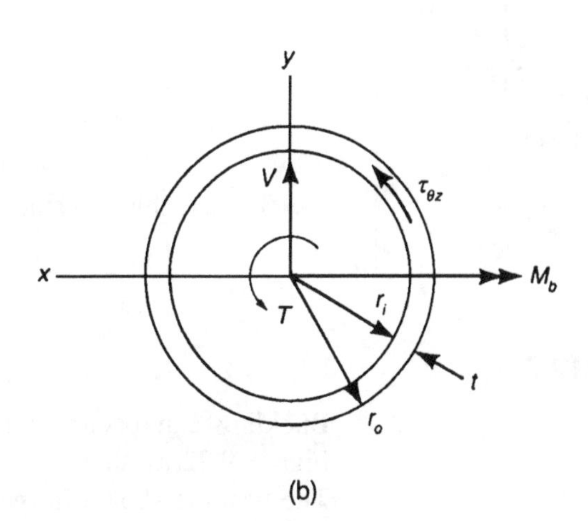

Exhibit 33

A summation of moments at the cut about an axis parallel with the x-axis gives

$$M_b = Pb = (900\ \text{N})(400\ \text{mm}) = 360\ \text{N} \bullet \text{m}$$

A summation of vertical forces gives

$$V = P$$

Exhibit 33(b) shows the force and moments acting on the cross-section. The bending and shearing stresses caused by these loads are

$$\sigma_z = \frac{M_b y}{I_{xx}} \qquad \text{(from } M_b\text{)}$$

$$\tau_{\theta z} = \frac{Tr}{I_z} \qquad \text{(from } T\text{)}$$

$$\tau_{zy} = \frac{VQ}{I_{xx} b} \qquad \text{(from } V\text{)}$$

The shearing stress attributed to V will be zero at the top of the beam and can be neglected. The moments of inertia are

$$I_{xx} = \frac{\pi\left(r_o^4 - r_i^4\right)}{4} = \frac{\pi(25^4 - 21.75^4)}{4} = 131 \times 10^3\ \text{mm}^4$$

$$I_z = \frac{\pi\left(r_o^4 - r_i^4\right)}{2} = 2I_{xx} = 262 \times 10^3\ \text{mm}^4$$

At the top of the tube r = 25 mm and y = 25 mm, so the stresses are

$$\sigma_z = \frac{M_b y}{I_{xx}} = \frac{(360\ \text{N} \bullet \text{m})(25\,\text{mm})}{131 \times 10^3\ \text{mm}^4} = 68.7\ \text{MPa}$$

$$\tau_{\theta z} = \frac{Tr}{I_z} = \frac{(405\ \text{N} \bullet \text{m})(25\,\text{mm})}{262 \times 10^3\ \text{mm}^4} = 38.6\ \text{MPa}$$

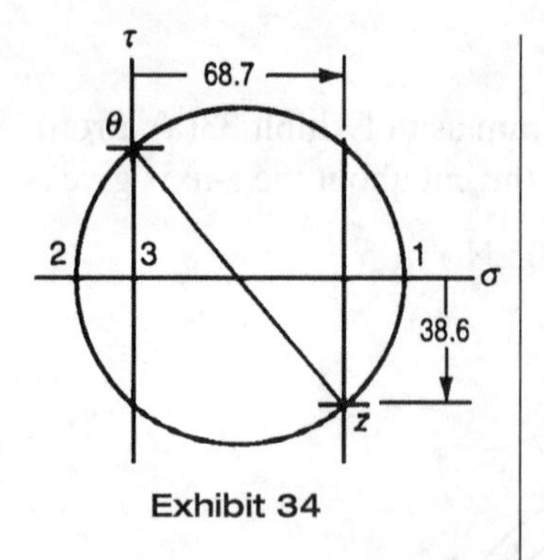

Exhibit 34

The Mohr's circle plot for this is shown in Exhibit 34.

$$R = \sqrt{\left(\frac{\sigma_z - \sigma_\theta}{2}\right)^2 + \tau_{\theta z}^2} = \sqrt{\left(\frac{68.7 - 0}{2}\right)^2 + 38.6^2} = 51.7\,\text{MPa}$$

$$C = \frac{\sigma_z \sigma_\theta}{2} = \frac{68.7}{2} = 34.4\,\text{MPa}$$

$$\sigma_1 = C + R = 86.1\,\text{MPa}$$

$$\sigma_2 = C - R = -17.3\,\text{MPa}$$

Because this is a state of plane stress, the third principal stress is

$$\sigma_3 = 0$$

COLUMNS

Buckling can occur in slender columns when they carry a high axial load. Figure 9.22(a) shows a simply supported slender member with an axial load. The beam is shown in the horizontal position rather than in the vertical position for convenience. It is assumed that the member will deflect from its normally straight configuration as shown. The free-body diagram of the beam is shown in Figure 9.22(b). Figure 9.22(c) shows the free-body diagram of a section of the beam. Summation of moments on the beam section in Figure 9.22(c) yields

$$M + Py = 0 \tag{9.80}$$

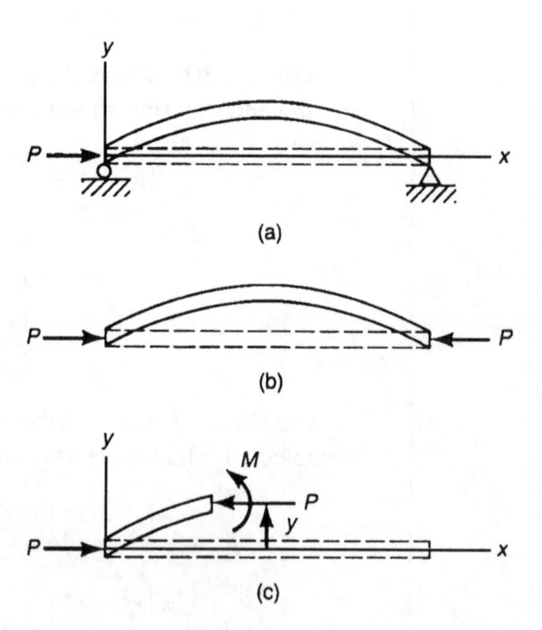

Figure 9.22 Buckling of simply supported column

Since M is equal to EI times the curvature, the equation for this beam can be expressed as

$$\frac{d^2 y}{dx^2} + \lambda y = 0 \tag{9.81}$$

where

$$\lambda^2 = \frac{P}{EI} \tag{9.82}$$

The solution satisfying the boundary conditions that the displacement is zero at either end is

$$v = \sin(\lambda x), \text{ where } \lambda = n\pi/L \; n = 1, 2, 3\ldots \tag{9.83}$$

The lowest value for the load P is the buckling load, so $n = 1$ and the critical buckling load, or Euler buckling load, is

$$P_{cr} = \frac{\pi^2 EI}{L^2} \tag{9.84}$$

For other than simply supported boundary conditions, the shape of the deflected curve will always be some portion of a sine curve. The simplest shape consistent with the boundary conditions will be the deflected shape. Figure 9.23 shows a sine curve and the beam lengths that can be selected from the sine curve. The critical buckling load can be redefined as

$$P_{cr} = \frac{\pi^2 EI}{L_e^2} = \frac{\pi^2 EI}{(kL)^2} = \frac{\pi^2 E}{(k\,L/r)^2} \tag{9.85}$$

where the radius of gyration r is defined as $\sqrt{I/A}$. The ratio L/r is called the slenderness ratio.

From Figure 9.23, it can be seen that the values for L_e and k are as follows:

For simple supports: $L = L_e$; $L_e = L$; $k = 1$

For a cantilever: $L = 0.5L_e$; $L_e = 2L$; $k = 2$

For both ends clamped: $L = 2L_e$; $L_e = 0.5L$; $k = 0.5$

For supported-clamped: $L = 1.43L_e$; $L_e = 0.7L$; $k = 0.7$

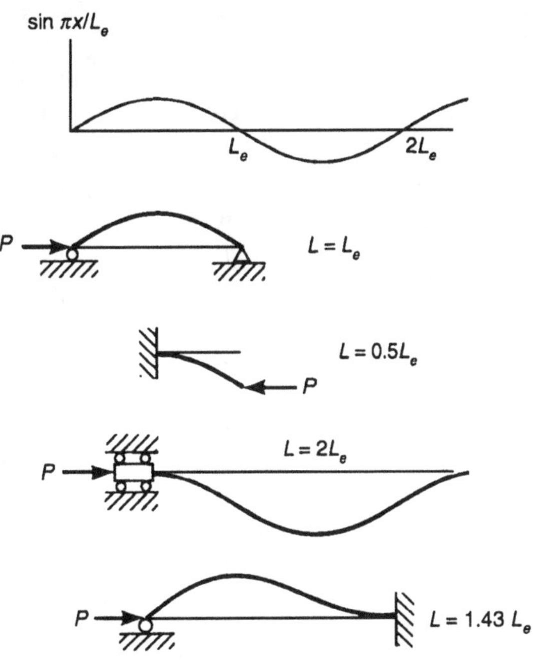

Figure 9.23 Buckling of columns with various boundary conditions

In dealing with buckling problems, keep in mind that the member must be slender before buckling is the mode of failure. If the beam is not slender, it will fail by yielding or crushing before buckling can take place.

Example 9.26

A steel pipe is to be used to support a weight of 130 kN as shown in Exhibit 35. The pipe has the following specifications: $OD = 100$ mm, $ID = 90$ mm, $A = 1500$ mm^2, and $I = 1.7 \times 10^6$ mm^4. Take $E = 210$ GPa and the yield stress $Y = 250$ MPa. Find the maximum length of the pipe.

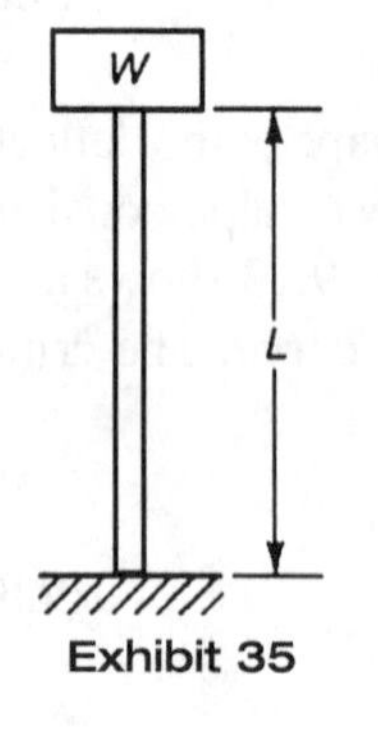

Exhibit 35

Solution

First, check to make sure that the pipe won't yield under the applied weight. The stress is

$$\sigma = \frac{P}{A} = \frac{130\,\text{kN}}{1500\,\text{mm}^2} = 86.7\,\text{MPa} < Y$$

This stress is well below the yield, so buckling will be the governing mode of failure. This is a cantilever column, so the constant k is 2. The critical load is

$$P_{cr} = \frac{\pi^2 EI}{(2L)^2}$$

Solving for L gives

$$L = \pi\sqrt{\frac{EI}{4P}} = \pi\sqrt{\frac{(210\,\text{GPa})(1.7\times10^6\,\text{mm}^4)}{4(130\,\text{kN})}} = 2.60\,\text{m}$$

The maximum length is 2.6 m.

SELECTED SYMBOLS AND ABBREVIATIONS

Symbol or Abbreviation	Description
σ	stress
ε	strain
ν	Poisson's ratio
kip	kilopound
E	modulus of elasticity
δ	deformation
W	weight
P	load
P, p	pressure
I	moment of inertia
τ	shear stress
T	torque
A	area
M	moment
V	shear
L	length
F	force

PROBLEMS

9.1 The stepped circular aluminum shaft in Exhibit 9.1 has two different diameters: 20 mm and 30 mm. Loads of 20 kN and 12 kN are applied at the end of the shaft and at the step. The maximum stress is most nearly:

a. 23.4 MPa c. 28.3 MPa
b. 26.2 MPa d. 30.1 MPa

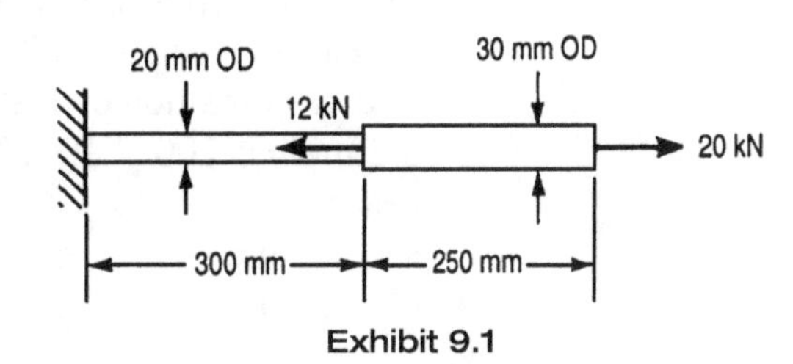

Exhibit 9.1

9.2 For the same shaft as in Problem 9.1 take $E = 69$ GPa. The end deflection is most nearly:

a. 0.18 mm c. 0.35 mm
b. 0.21 mm d. 0.72 mm

9.3 The shaft in Exhibit 9.3 is the same aluminum stepped shaft considered in problems 9.1 and 9.2, except now the right-hand end is also built into a wall. Assume that the member was built in before the load was applied. The maximum stress is most nearly:

a. 12.2 MPa c. 13.1 MPa
b. 12.7 MPa d. 15.2 MPa

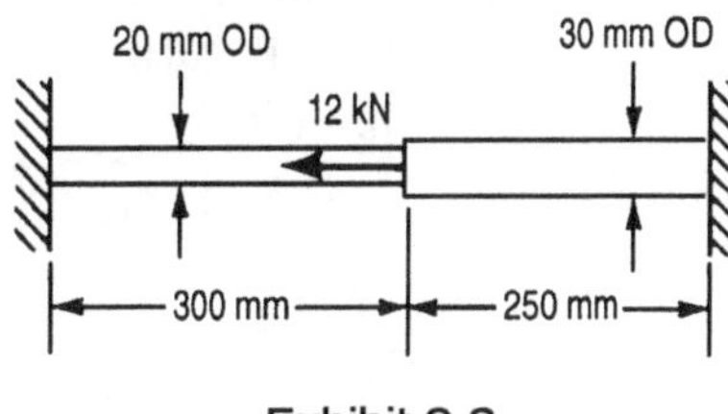
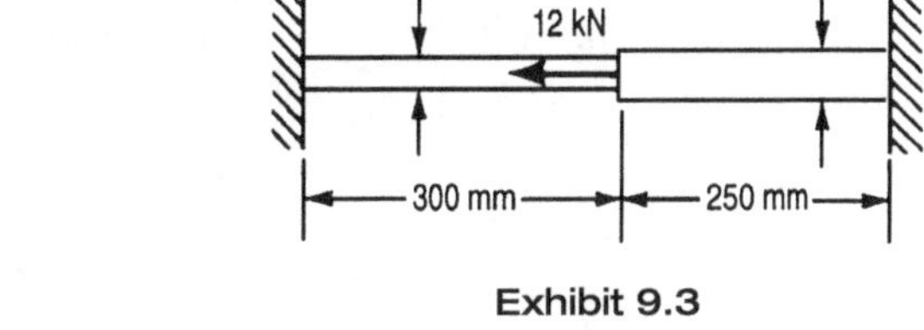

Exhibit 9.3

9.4 For the same shaft as in Problem 9.3 the deflection of the step is most nearly:

a. 0.038 mm c. 0.064 mm
b. 0.042 mm d. 0.086 mm

L

F

Exhibit 9.5

9.5 The uniform rod shown in Exhibit 9.5 has a force F at its end which is equal to the total weight of the rod. The rod has a unit weight γ. The total deflection of the rod is most nearly:

a. $1.00\,\gamma\,L^2/E$ c. $1.50\,\gamma\,L^2/E$
b. $1.25\,\gamma\,L^2/E$ d. $1.75\,\gamma\,L^2/E$

300 mm

Exhibit 9.6

9.6 At room temperature, 22°C, a 300-mm stainless steel rod (Exhibit 9.6) has a gap of 0.15 mm between its end and a rigid wall. The modulus of elasticity $E = 210$ GPa. The coefficient of thermal expansion $\alpha = 17 \times 10^{-6}$/°C. The area of the rod is 650 mm². When the temperature is raised to 100 °C, the stress in the rod is most nearly:

a. 175 MPa (tension) c. −17.5 MPa (compression)
b. 0 MPa d. −175 MPa (compression)

9.7 A steel cylindrical pressure vessel is subjected to a pressure of 21 MPa. Its outer diameter is 4.6 m, and its wall thickness is 200 mm. The maximum principal stress in this vessel is most nearly:

a. 183 MPa
b. 221 MPa
c. 362 MPa
d. 432 MPa

9.8 A pressure vessel shown in Exhibit 9.8 is known to have an internal pressure of 1.4 MPa. The outer diameter of the vessel is 300 mm. The vessel is made of steel; $v = 0.3$ and $E = 210$ GPa. A strain gage in the circumferential direction on the vessel indicates that, under the given pressure, the strain is 200×10^{-6}. The wall thickness of the pressure vessel is most nearly:

a. 3.2 mm
b. 4.3 mm
c. 6.4 mm
d. 7.8 mm

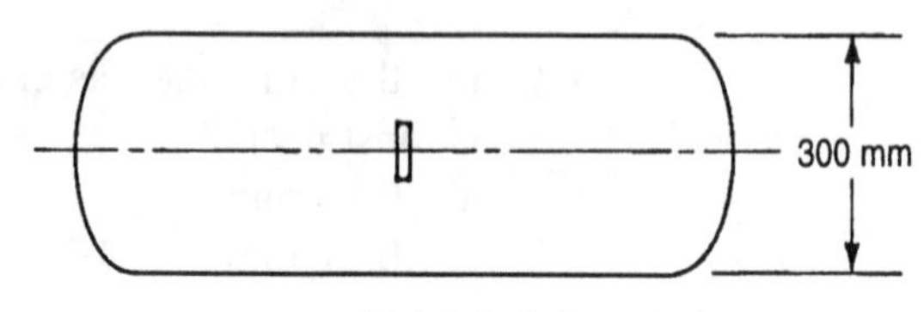

Exhibit 9.8

9.9 An aluminum pressure vessel has an internal pressure of 0.7 MPa. The vessel has an outer diameter of 200 mm and a wall thickness of 3 mm. Poisson's ratio is 0.33 and the modulus of elasticity is 69 GPa for this material. A strain gage is attached to the outside of the vessel at 45° to the longitudinal axis as shown in Exhibit 9.9. The strain on the gage would read most nearly:

a. 40×10^{-6}
b. 60×10^{-6}
c. 80×10^{-6}
d. 160×10^{-6}

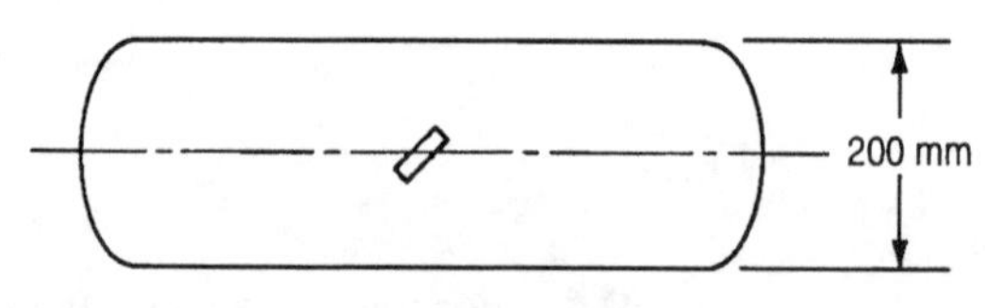

Exhibit 9.9

9.10 If $\sigma_x = -3$ MPa, $\sigma_y = 5$ MPa, and $\tau_{xy} = -3$ MPa, the maximum principal stress is most nearly:

a. 4 MPa
b. 5 MPa
c. 6 MPa
d. 7 MPa

9.11 Given that $\sigma_x = 5$ MPa, $\sigma_y = -1$ MPa, and the maximum principal stress is 7 MPa, the shear stress τ_{xy} is most nearly:

a. 1 MPa
b. 2 MPa
c. 3 MPa
d. 4 MPa

9.12 Given $\varepsilon_x = 800\,\mu$, $\varepsilon_y = 200\,\mu$, and $\gamma_{xy} = 400\,\mu$, the maximum principal strain is most nearly:

a. 840 μ
b. 860 μ
c. 900 μ
d. 960 μ

9.13 A steel plate in a state of plane stress has the same strains as in Problem 9.12: $\varepsilon_x = 800\ \mu$, $\varepsilon_y = 200\ \mu$, and $\gamma_{xy} = 400\ \mu$. Poisson's ratio $v = 0.3$ and the modulus of elasticity $E = 210$ GPa. The maximum principal stress in the plane is most nearly:

a. 109 MPa
b. 132 MPa
c. 173 MPa
d. 208 MPa

9.14 A stepped steel shaft shown in Exhibit 9.14 has torques of 10 kN • m applied at the end and at the step. The maximum shear stress in the shaft is most nearly:

a. 760 MPa
b. 810 MPa
c. 870 MPa
d. 930 MPa

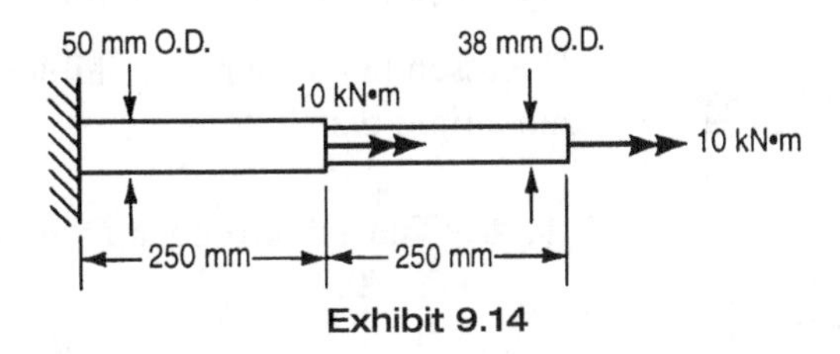

Exhibit 9.14

9.15 The shear modulus for steel is 83 MPa. For the same shaft as in Problem 9.14, the rotation at the end of the shaft is most nearly:

a. 0.014°
b. 0.14°
c. 1.4°
d. 14°

9.16 The same stepped shaft as in problems 9.14 and 9.15 is now built into a wall at its right end before the load is applied (Exhibit 9.16). The maximum stress in the shaft is most nearly:

a. 130 MPa
b. 200 MPa
c. 230 MPa
d. 300 MPa

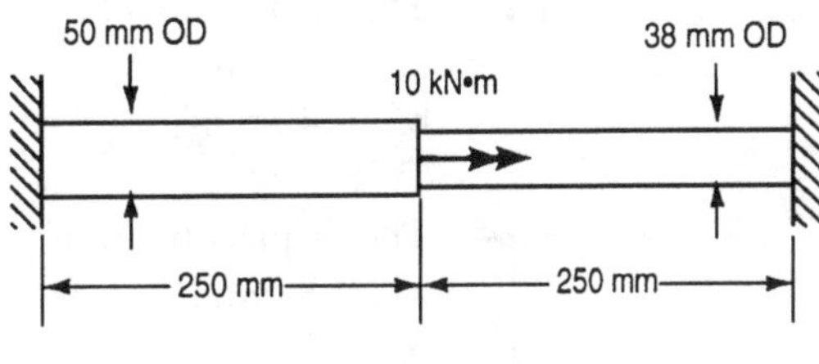

Exhibit 9.16

9.17 For the same shaft as in Problem 9.16, the rotation of the step is most nearly:

a. 0.2°
b. 1.1°
c. 1.8°
d. 2.1°

Use the following information for problems 9.18 and 9.19.

A steel rod with modulus of elasticity of 180 GPa is subjected to the loads as shown in Exhibit 9.18. The cross-sectional area of the rod is 1 cm².

9.18 The displacement of point B is most nearly:
a. 4.60 mm c. 2.85 mm
b. 7.30 mm d. 1.08 mm

9.19 The displacement of point A is most nearly:
a. 2.85 mm c. 5.61 mm
b. 8.20 mm d. 3.85 mm

Exhibit 9.18

Use the following information for problems 9.20 through 9.24.

The assembly shown in Exhibit 9.20 consists of three steel rods with E = 180 GPa and a rigid bar BCD.

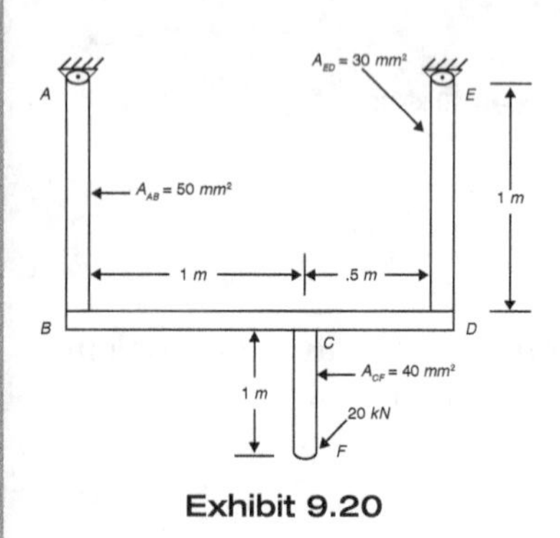

Exhibit 9.20

9.20 The axial force at D in kN is most nearly:
a. 25 ↓ c. 20 ↑
b. 13.3 ↓ d. 16.5 ↑

9.21 The axial force at B in kN is most nearly:
a. 6.7 ↓ c. 6.7 ↑
b. 8.5 ↑ d. 9.3 ↓

Use the free-body diagram in Exhibit 9.22 for problems 9.22 through 9.24.

9.22 The displacement of point D is most nearly:
a. 4.6 mm c. 6.2 mm
b. 2.5 mm d. 1.8 mm

9.23 The displacement of point B is most nearly:
a. 0.74 mm c. 0.55 mm
b. 0.40 mm d. 0.38 mm

Exhibit 9.22

9.24 The displacement of point F is most nearly:
a. 2.8 mm c. 6.2 mm
b. 3.5 mm d. 4.7 mm

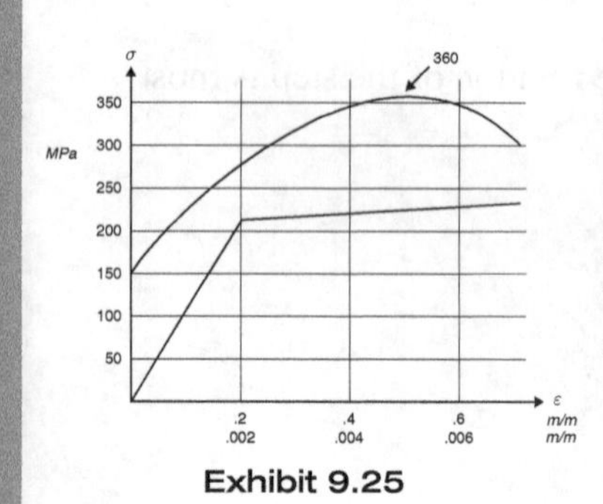

Exhibit 9.25

Use Exhibit 9.25 and the following information for the solution of problems 9.25 through 9.28.

Exhibit 9.25 shows a stress-strain diagram for a steel alloy rod, which is loaded in tension and is 1 cm in diameter. The inner curve shown is the magnified curve in which the strain scale is larger by a factor of 100 for the same stress scale. The outer curve is the typical approximate stress-strain curve.

9.25 The modulus of elasticity of the rod in GPa is most nearly:
a. 120 c. 113
b. 430 d. 265

9.26 The yield load is most nearly:
a. 18 kN c. 25 kN
b. 12 kN d. 30 kN

9.27 The ultimate load is most nearly:
a. 15 kN c. 32kN
b. 21.6 kN d. 28.3 kN

9.28 The modulus of resilience is most nearly:
a. 1.2 MJ/m^3 c. 2.4 MJ/m^3
b. 0.23 MJ/m^3 d. 6.8 MJ/m^3

Use Exhibit 9.29 and the following information for problems 9.29 through 9.31.

A truss is constructed of three members of a certain alloy that has a modulus of elasticity of 150 GPa. Each of the members has a cross-sectional area of 2 cm^2.

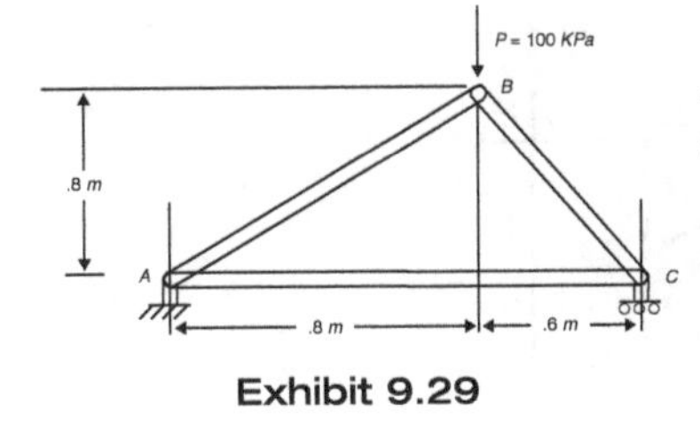

Exhibit 9.29

9.29 The load in bar *BC* in kN is most nearly:
a. 55 c. 71
b. 80 d. 93

9.30 The stress in bar *AB* in MPa is most nearly:
a. 300 c. 250
b. 350 d. 270

9.31 The elongation of bar *AC* in mm is most nearly:
a. 2.0 c. 2.8
b. 3.5 d. 1.5

50 mm O.D.
T

Exhibit 9.32

9.32 A strain gage shown in Exhibit 9.32 is placed on a circular steel shaft that is being twisted with a torque *T*. The gage is inclined 45° to the axis. If the strain reads $\varepsilon_{45} = 245\ \mu$, the torque is most nearly:
a. 1000 N • m c. 1570 N • m
b. 1230 N • m d. 2635 N • m

Exhibit 9.33

9.33 A shaft whose cross section is in the shape of a semicircle is shown in Exhibit 9.33 and has a constant wall thickness of 3 mm. The shaft carries a torque of 300 N • m. Neglecting any stress concentrations at the corners, the maximum shear stress in the shaft is most nearly:
a. 32 MPa c. 59 MPa
b. 48 MPa d. 66 MPa

9.34 The maximum magnitude of shear in the beam shown in Exhibit 9.34 is most nearly:
a. 40 kN c. 60 kN
b. 50 kN d. 75 kN

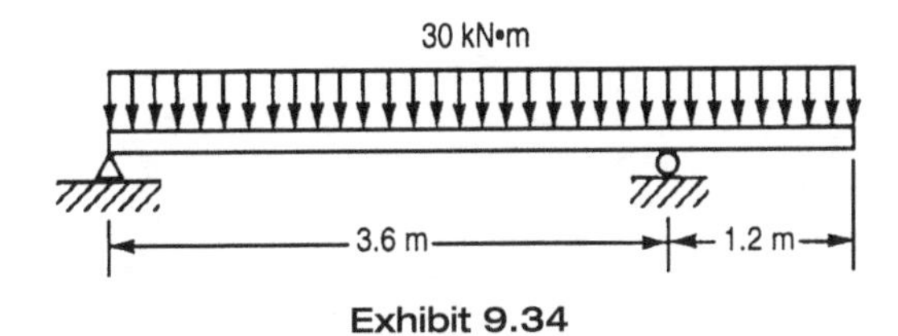

Exhibit 9.34

9.35 For the same beam as in Problem 9.34, the magnitude of the largest bending moment is most nearly:

a. 21.0 kN • m   c. 38.4 kN • m
b. 26.3 kN • m   d. 42.1 kN • m

9.36 The shear diagram shown in Exhibit 9.36 is for a beam that has zero moments at either end. The maximum concentrated force on the beam is most nearly:

a. 60 kN upward   c. 0
b. 30 kN upward   d. 30 kN downward

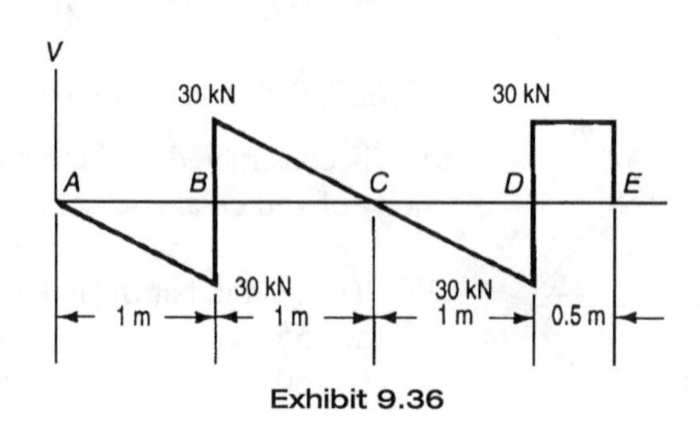

Exhibit 9.36

9.37 For the same beam as in Problem 9.36 the largest magnitude of the bending moment is most nearly:

a. 0   c. 12 kN • m
b. 8 kN • m   d. 15 kN • m

9.38 The 4-m long, simply supported beam shown in Exhibit 9.38 has a section modulus $Z = 1408 \times 10^3$ mm³. The allowable stress in the beam is not to exceed 100 MPa. The maximum load, w (including its own weight), that the beam can carry is most nearly:

a. 50 kN • m   c. 60 kN • m
b. 40 kN • m   d. 70 kN • m

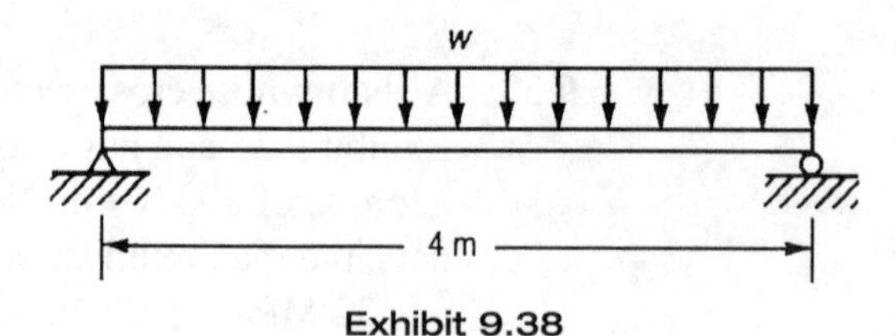

Exhibit 9.38

9.39 The standard wide flange beam shown in Exhibit 9.39 has a moment of inertia about the z-axis of $I = 365 \times 10^6\ \text{mm}^4$. The maximum bending stress is most nearly:

a. 4.5 MPa
b. 5.0 MPa
c. 6.5 MPa
d. 8 MPa

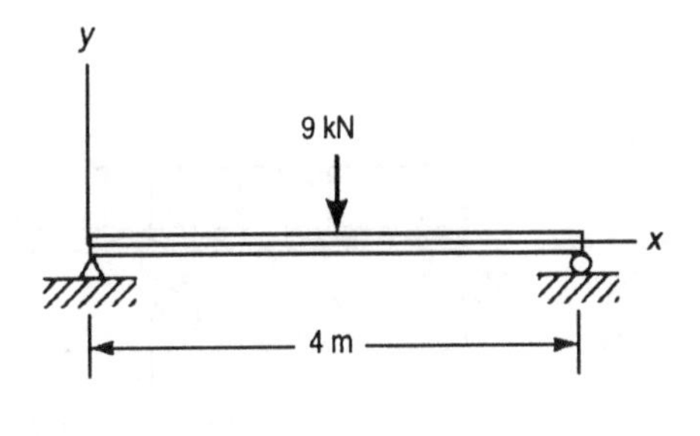

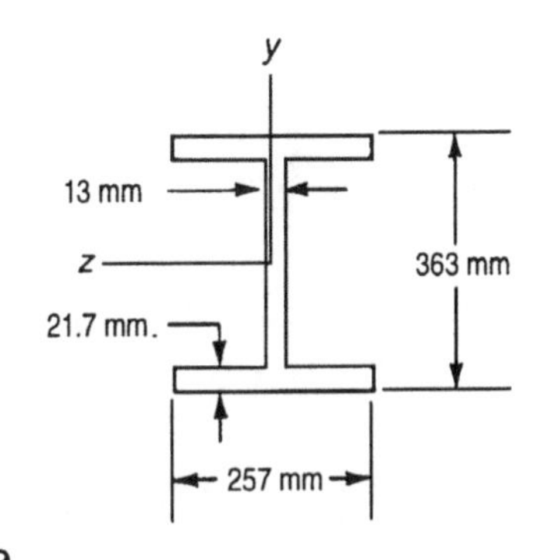

Exhibit 9.39

9.40 For the same beam as in Problem 9.39, the maximum shear stress τ_{xy} in the web is most nearly:

a. 1 MPa
b. 1.5 MPa
c. 2.0 MPa
d. 2.5 MPa

9.41 The deflection at the end of the beam shown in Exhibit 9.41 is most nearly:

a. 0.330 FL^3/EI (downward)
b. 0.380 FL^3/EI (downward)
c. 0.410 FL^3/EI (downward)
d. 0.440 FL^3/EI (downward)

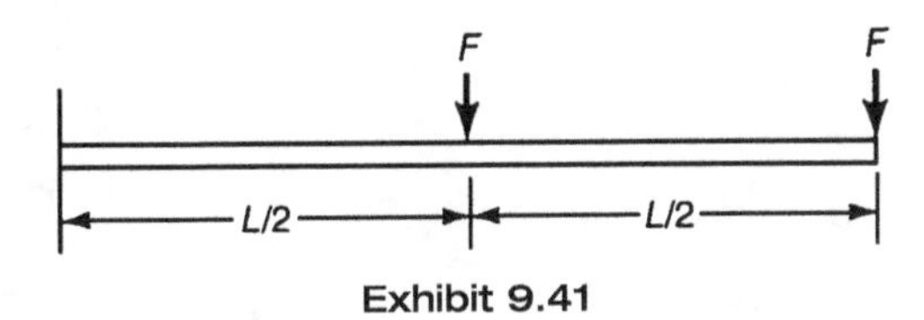

Exhibit 9.41

9.42 A uniformly loaded beam (Exhibit 9.42) has a concentrated load wL at its center that has the same magnitude as the total distributed load w. The maximum deflection of this beam is most nearly:

a. 0.029 wL^4/EI (downward)
b. 0.034 wL^4/EI (downward)
c. 0.043 wL^4/EI (downward)
d. 0.056 wL^4/EI (downward)

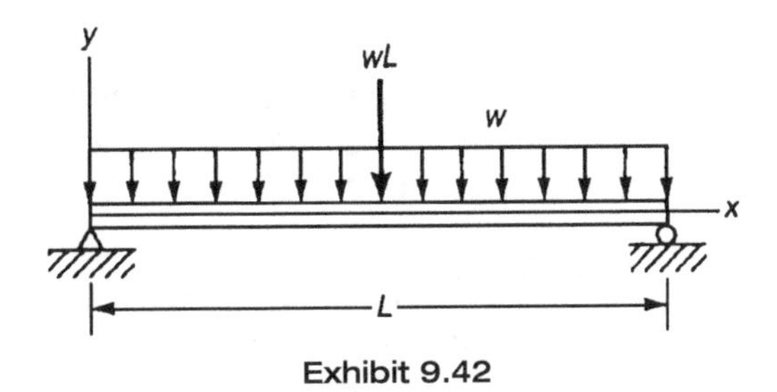

Exhibit 9.42

9.43 The reaction at the center support of the uniformly loaded beam shown in Exhibit 9.43 is most nearly:

a. $0.525\,wL$ c. $0.575\,wL$
b. $0.550\,wL$ d. $0.625\,wL$

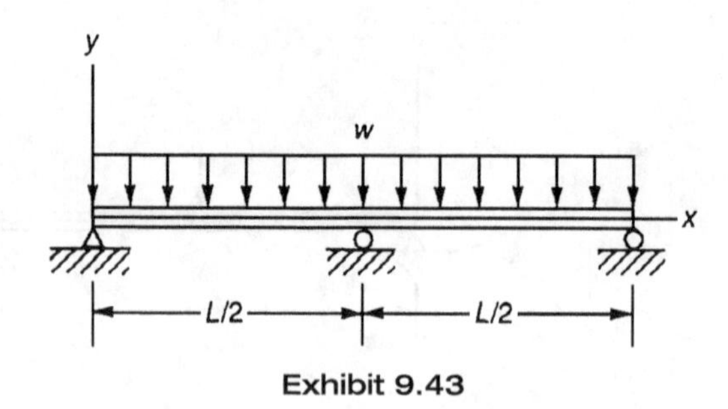

Exhibit 9.43

9.44 A solid circular rod has a diameter of 25 mm (Exhibit 9.44). It is fixed into a wall at *A* and bent 90° at *B*. The maximum bending stress in the section *BC* is most nearly:

a. 21.7 MPa c. 32.6 MPa
b. 29.3 MPa d. 45.7 MPa

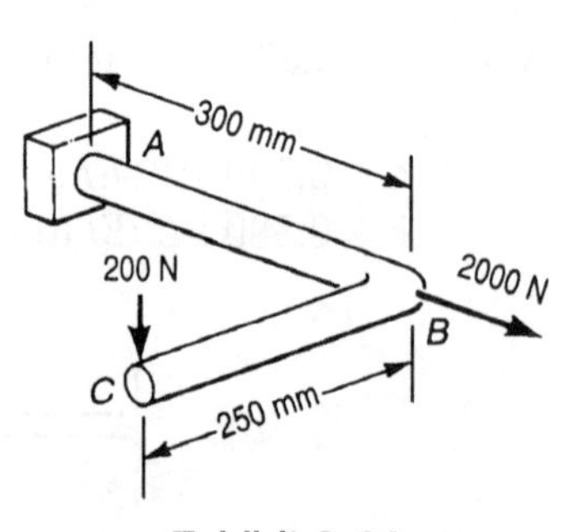

Exhibit 9.44

9.45 For the same member as in Problem 9.44, the maximum bending stress in the section *AB* is most nearly:

a. 21 MPa c. 31 MPa
b. 25 MPa d. 39 MPa

9.46 For the same member as in Problem 9.44, the maximum shear stress due to torsion in the section *AB* is most nearly:

a. 15.2 MPa c. 17.4 MPa
b. 16.3 MPa d. 18.5 MPa

9.47 For the same member as in Problem 9.44, the maximum stress due to the axial force in the section *AB* is most nearly:

a. 4 MPa c. 6 MPa
b. 5 MPa d. 8 MPa

9.48 For the same member as in Problem 9.44, the maximum principal stress in the section *AB* is most nearly:

a. 17 MPa
b. 27 MPa
c. 39 MPa
d. 44 MPa

9.49 A truss is supported so that it can't move out of the plane (Exhibit 9.49). All members are steel and have a square cross section 25 mm by 25 mm. The modulus of elasticity for steel is 210 GPa. The maximum load *P* that can be supported without any buckling is most nearly:

a. 14 kN
b. 25 kN
c. 34 kN
d. 51 kN

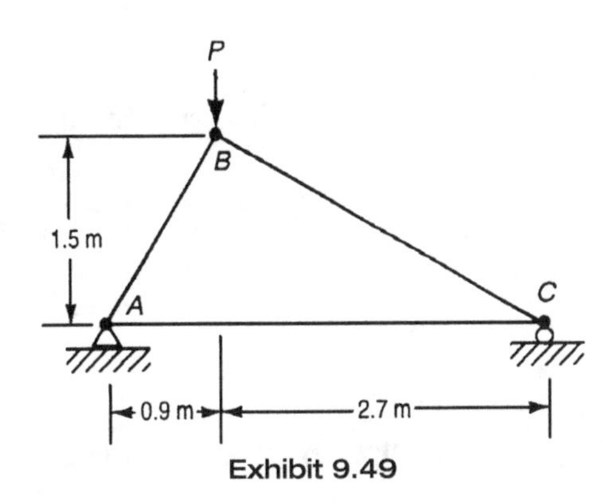

Exhibit 9.49

9.50 A beam is pinned at both ends (Exhibit 9.50). In the *x-y* plane it can rotate about the pins, but in the *x-z* plane the pins constrain the end rotation. In order to have buckling equally likely in each plane, the ratio *b*/*a* is most nearly:

a. 0.5
b. 1.0
c. 1.5
d. 2.0

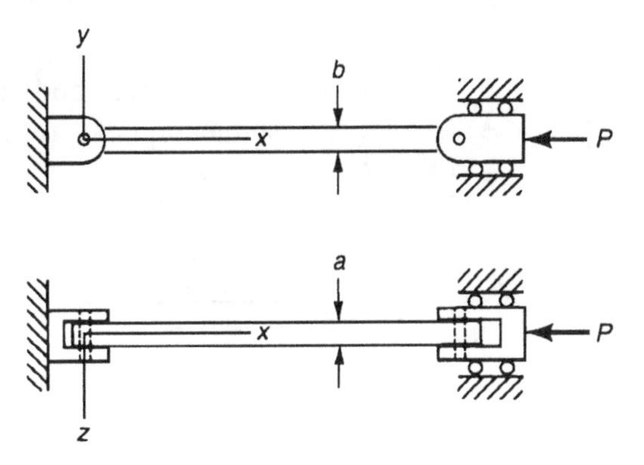

Exhibit 9.50

SOLUTIONS

9.1 **c.** Draw free-body diagrams. Equilibrium of the center free-body diagram gives

$$F_1 = 20 - 12 = 8 \text{ kN}$$

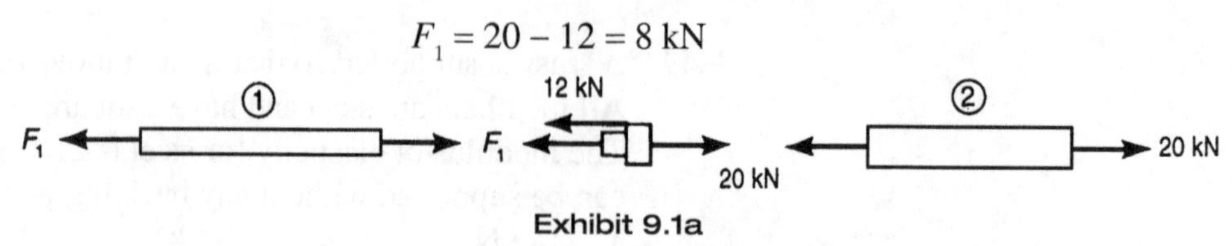

Exhibit 9.1a

The areas are

$$A_1 = \pi r^2 = \pi(10 \text{ mm})^2 = 314 \text{ mm}^2$$

$$A_2 = \pi r^2 = \pi(15 \text{ mm})^2 = 707 \text{ mm}^2$$

The stresses are

$$\sigma_1 = \frac{P}{A} = \frac{8\,\text{kN}}{314\,\text{mm}^2} = 25.5\,\text{MPa}$$

$$\sigma_2 = \frac{P}{A} = \frac{20\,\text{kN}}{707\,\text{mm}^2} = 28.3\,\text{MPa}$$

9.2 **b.** The force-deformation equations give

$$\delta_1 = \frac{P_1 L_1}{A_1 E_1} = \frac{(8\,\text{kN})(300\,\text{mm})}{(314\,\text{mm}^2)(69\,\text{GPa})} = 0.1107\,\text{mm}$$

$$\delta_2 = \frac{P_2 L_2}{A_2 E_2} = \frac{(20\,\text{kN})(250\,\text{mm})}{(707\,\text{mm}^2)(69\,\text{GPa})} = 0.1025\,\text{mm}$$

Compatibility of deformation gives

$$\delta_{\text{end}} = \delta_1 + \delta_2 = 0.1107 + 0.1025 = 0.213\,\text{mm}$$

9.3 **c.** Draw the free-body diagrams. From the center free-body diagram, summation of forces yields

$$F_2 = 12 \text{ kN} + F_1$$

Exhibit 9.3a

Force-deformation relations are

$$\delta_1 = \frac{P_1 L_1}{A_1 E_1} = \frac{(F_1\,\text{kN})(300\,\text{mm})}{(314\,\text{mm}^2)(69\,\text{GPa})} = 0.01384\,F_1$$

$$\delta_2 = \frac{P_2 L_2}{A_2 E_2} = \frac{(F_2\,\text{kN})(200\,\text{mm})}{(707\,\text{mm}^2)(69\,\text{GPa})} = 0.00410\,F_2$$

Compatibility gives

$$\delta_{end} = 0 = \delta_1 + \delta_2 = 0.01384\ F_1 + 0.00410\ F_2$$

Substitution of the equilibrium relation, $F_2 = 12 + F_1$, into the above equation gives

$$0 = 0.01384\ F_1 + 0.00410\ (12 + F_1)$$
$$F_1 = -2.74\ \text{kN},\ F_2 = 9.26\ \text{kN}$$

The stresses are

$$\sigma_1 = \frac{P}{A} = \frac{(-2.74\,\text{kN})}{(314\,\text{mm}^2)} = -8.73\text{MPa}; \qquad \sigma_2 = \frac{P}{A} = \frac{(9.26\,\text{kN})}{(707\,\text{mm}^2)} = 13.10\,\text{MPa}$$

9.4 a. The same three-step process as in Problem 9.3 must be carried out. Since this process has already been completed, the results can be used. The deflection can be expressed as

$$\delta = \delta_1 = -\delta_2 = 0.01384\ F_1 = 0.01384\ (-2.74) = -0.0380\ \text{mm}$$

9.5 c. Draw a free-body diagram. Summation of forces in the vertical direction gives

$$P = \gamma AL + \gamma Ax$$

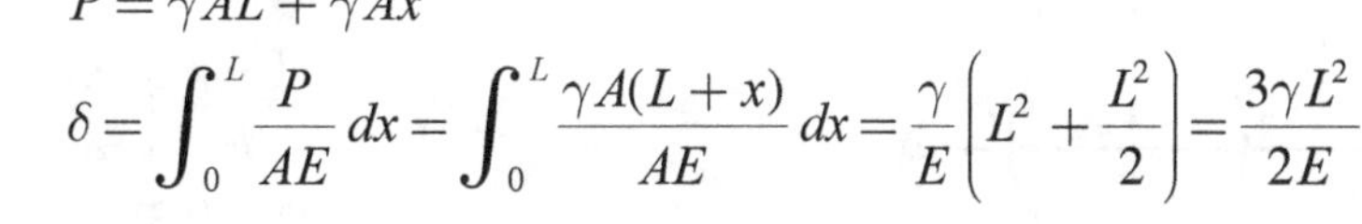

$$\delta = \int_0^L \frac{P}{AE}\,dx = \int_0^L \frac{\gamma A(L+x)}{AE}\,dx = \frac{\gamma}{E}\left(L^2 + \frac{L^2}{2}\right) = \frac{3\gamma L^2}{2E}$$

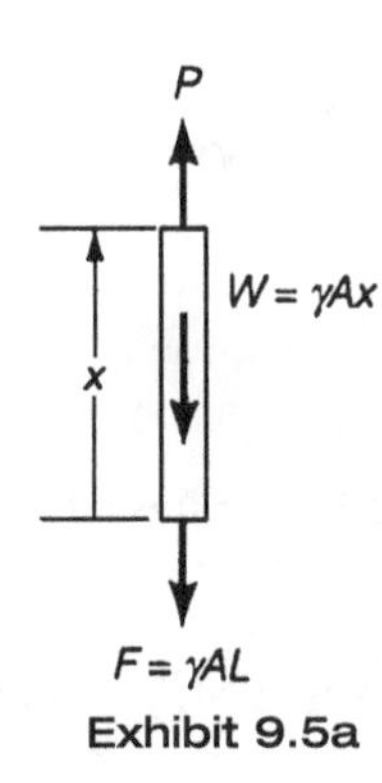

Exhibit 9.5a

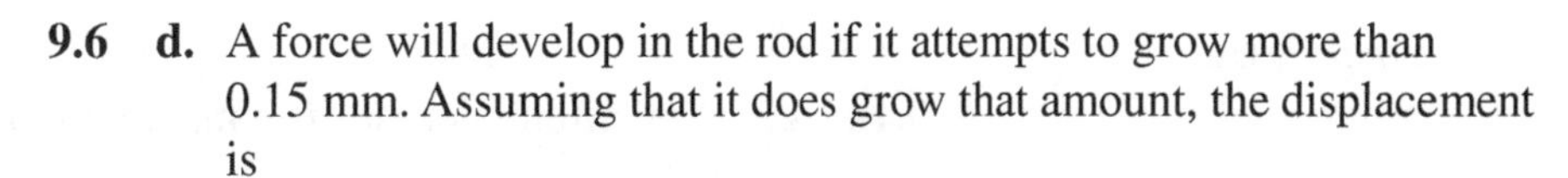

9.6 d. A force will develop in the rod if it attempts to grow more than 0.15 mm. Assuming that it does grow that amount, the displacement is

$$\delta = \frac{PL}{AE} + \alpha L(t - t_o) = 0.15\,\text{mm}$$
$$= \frac{P(300\,\text{mm})}{(650\,\text{mm})(210\,\text{GPa})} + \left(17\times10^{-6}\,\frac{1}{°\text{C}}\right)(300\,\text{mm})(100\,°\text{C} - 22\ °\text{C})$$
$$0.15\ \text{mm} = 0.00220\,P + 0.3978$$
$$P = -112.7\ \text{kN}$$
$$\sigma = \frac{P}{A} = \frac{-112.7\,\text{kN}}{650\,\text{mm}^2} = -173.5\,\text{MPa}$$

9.7 b. In a cylindrical pressure vessel the three principal stresses are

$$\sigma_t = qD/2t; \quad \sigma_a = qD/4t; \quad \sigma_r \approx 0$$

The maximum is σ_t, which gives

$$\sigma_t = \frac{qD}{2t} = \frac{(21\,\text{MPa})(4600\,\text{mm} - 400\,\text{mm})}{2(200\,\text{mm})} = 221\,\text{MPa}$$

9.8 b. The stresses in the pressure vessel are, as in the last problem,

$$\sigma_t = \frac{qD}{2t}; \qquad \sigma_a = \frac{qD}{4t}; \qquad \sigma_r \approx 0$$

The wall thickness is usually thin enough so that it can be assumed that

$$D_i \approx D_o$$

The tangential strain can be found from Hooke's law:

$$\varepsilon_t = \frac{1}{E}(\sigma_t - v\sigma_\theta - v\sigma_r) = \frac{1}{E}\left(\frac{qD}{2t} - v\frac{qD}{4t} - v0\right) = 0.425\frac{qD}{Et}$$

The thickness is then

$$t = 0.425\frac{pD}{E\varepsilon_t} = 0.425\frac{(1.4\,\text{MPa})(300\,\text{mm})}{(210\,\text{GPa})(200\times10^{-6})} = 4.25\,\text{mm}$$

9.9 d. Draw Mohr's circle for the stress. At 45° in the physical plane (90° on Mohr's circle) the two normal stresses are $3qD/8t$. Hooke's law gives

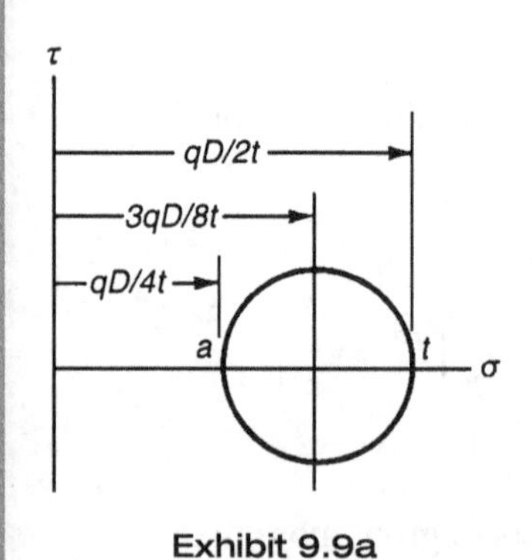

Exhibit 9.9a

$$\varepsilon_{45} = \frac{1}{E}(\sigma_{45} - v\sigma_{-45} - v\sigma_r) = \frac{1}{E}\left(\frac{3qD}{8t} - v\frac{3qD}{8t} - v0\right)$$

$$= \frac{(1-v)}{E}\left(\frac{3qD}{8t}\right)$$

$$\varepsilon_{45} = \frac{(1-0.33)}{(69\,\text{GPa})}\frac{3(0.7\,\text{MPa})(200\,\text{mm} - 6\,\text{mm})}{8(3\,\text{mm})} = 164.8\times10^{-6}$$

9.10 c. Draw Mohr's circle. The maximum principal stress is 6 MPa. As an alternative,

$$R = \sqrt{\left(\frac{\sigma_x - \sigma_y}{2}\right)^2 + \tau_{xy}^2} = \sqrt{\left(\frac{-3-5}{2}\right)^2 + (-3)^2} = 5$$

$$C = \frac{\sigma_x + \sigma_y}{2} = \frac{-3+5}{2} = 1$$

$$\sigma_1 = R + C = 6\,\text{MPa}$$

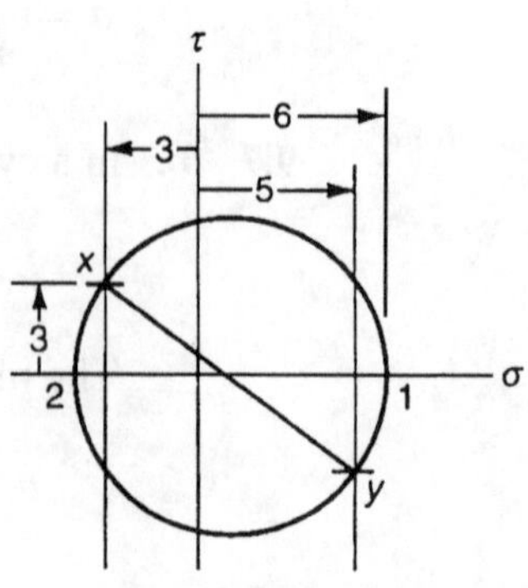

Exhibit 9.10a

9.11 d. Draw Mohr's circle. The center of the circle is

$$C = \frac{\sigma_x + \sigma_y}{2} = \frac{5-1}{2} = 2$$

The radius is then $R = 7 - 2 = 5$. The shear stress can be found from the Mohr's circle or from the expression

$$R = \sqrt{\left(\frac{\sigma_x - \sigma_y}{2}\right)^2 + \tau_{xy}^2}; \qquad \tau_{xy}^2 = R^2 - \left(\frac{\sigma_x - \sigma_y}{2}\right)^2$$

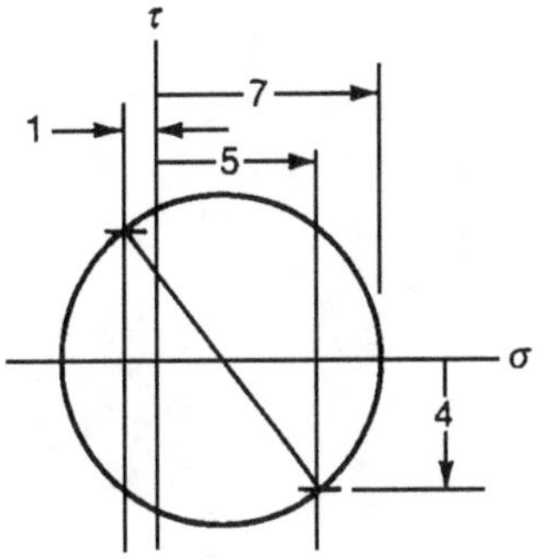

Exhibit 9.11a

In either case, the shear stress $\tau_{xy} = 4$ MPa.

9.12 b. Draw Mohr's circle. ε_1 can be scaled from the circle or computed as follows:

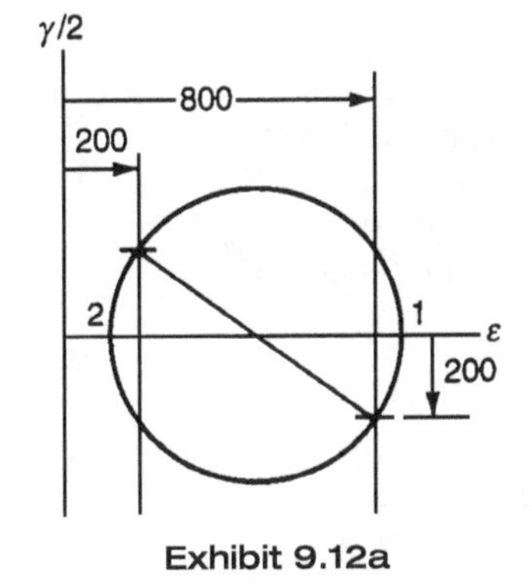

Exhibit 9.12a

$$R = \sqrt{\left(\frac{\varepsilon_x - \varepsilon_y}{2}\right)^2 + \left(\frac{\gamma_{xy}}{2}\right)^2} = \sqrt{\left(\frac{800-200}{2}\right)^2 + \left(\frac{400}{2}\right)^2} = 361\,\mu$$

$$C = \frac{\varepsilon_x + \varepsilon_y}{2} = \frac{800+200}{2} = 500\,\mu$$

$$\varepsilon_1 = C + R = 500\,\mu + 361\,\mu = 861\,\mu$$

$$\varepsilon_2 = C - R = 500\mu - 361\mu = 139\mu$$

ε_1 is the maximum.

9.13 d. Problems of this type can be done by using Hooke's law first and then Mohr's circle or by using Mohr's circle first and then applying Hooke's law. Since Mohr's circle was already drawn for this problem in the previous solution, the second approach will be followed. The principal strains were found to be the following: $\varepsilon_1 = 861\ \mu$; $\varepsilon_2 = 139\ \mu$. Hooke's law in plane stress is

$$\varepsilon_1 = \frac{1}{E}(\sigma_1 - \nu\sigma_2); \qquad \varepsilon_2 = \frac{1}{E}(\sigma_2 - \nu\sigma_1)$$

Inverting these relationships gives

$$\sigma_1 = \frac{E}{1-v^2}(\varepsilon_1 + v\varepsilon_2); \qquad \sigma_2 = \frac{E}{1-v^2}(\varepsilon_2 + v\varepsilon_1)$$

The maximum principal stress is σ_1, which is

$$\sigma_1 = \frac{E}{1-v^2}(\varepsilon_1 + v\varepsilon_2) = \frac{210\,\text{GPa}}{1-0.3^2}[861\times10^{-6} + 0.3(139\times10^{-6})] = 208\ \text{MPa}$$

9.14 d. Draw free-body diagrams. The torque in shaft 1 is $T_1 = 10 + 10 = 20$ kN • m. The torque in shaft 2 is $T_2 = 10$ kN • m.

$$\tau_1 = \frac{T_1 r_1}{J_1} = \frac{(20\,\text{kN} \bullet \text{m})(25\,\text{mm})}{0.5\pi(25\,\text{mm})^4} = 815\ \text{MPa}$$

$$\tau_2 = \frac{T_2 r_2}{J_2} = \frac{(10\,\text{kN} \bullet \text{m})(19\,\text{mm})}{0.5\pi(19\,\text{mm})^4} = 928\ \text{MPa}$$

The largest stress is 928 MPa.

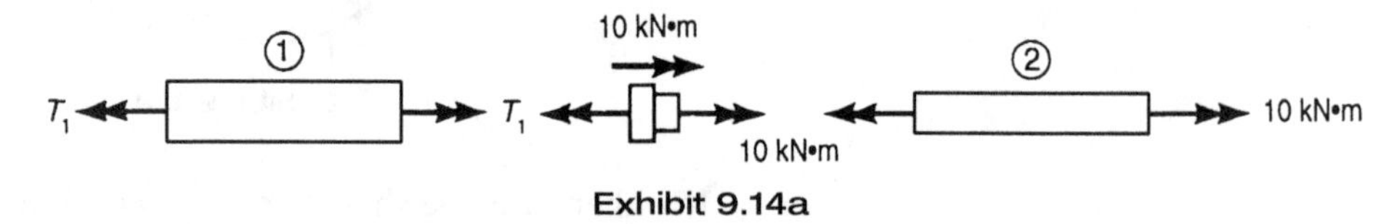

Exhibit 9.14a

9.15 d. From the force-deformation relations,

$$\phi_1 = \frac{T_1 L_1}{GJ_1} = \frac{(20\,\text{kN} \bullet \text{m})(250\,\text{mm})}{(83\,\text{GPa})0.5\pi(25\,\text{mm})^4} = 0.982\,\text{rad} = 5.63°$$

$$\phi_2 = \frac{T_2 L_2}{GJ_2} = \frac{(10\,\text{kN} \bullet \text{m})(250\,\text{mm})}{(83\,\text{GPa})0.5\pi(19\,\text{mm})^4} = 0.1471\,\text{rad} = 8.43°$$

From compatibility,

$$\phi = \phi_1 + \phi_2 = 5.63° + 8.43° = 14.06°$$

9.16 d. Draw the free-body diagrams. Equilibrium of the center free body gives

$$T_1 = T_2 + 10$$

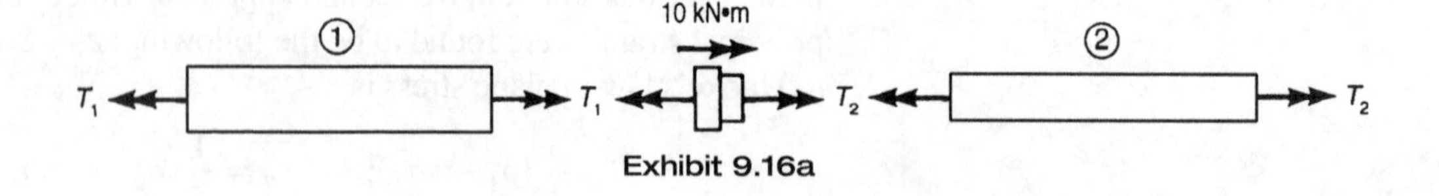

Exhibit 9.16a

The force-deformation relations are

$$\phi_1 = \frac{T_1 L_1}{GJ_1} = \frac{(10 + T_2)(250\,\text{mm})}{(83\,\text{GPa})0.5\pi(25\,\text{mm})^4} = 49.1\times10^{-3} + 4.91\times10^{-3}\,T_2$$

$$\phi_2 = \frac{T_2 r_2}{J_2} = \frac{T_2(250\,\text{mm})}{(83\,\text{GPa})\ 0.5\pi(19\,\text{mm})^4} = 14.7\times10^{-3}\,T_2$$

Compatibility requires that

$$\phi_1 + \phi_2 = 0 = 49.1 \times 10^{-3} + (4.91 \times 10^{-3} + 14.7 \times 10^{-3})T_2$$

Solving for the torques gives

$$T_2 = \frac{5.305}{2.207} = -2.50\ \text{kN} \bullet \text{m}$$
$$T_1 = T_2 + 10 = -2.50 + 10 = 7.50\ \text{kN} \bullet \text{m}$$

The stresses then are

$$\tau_1 = \frac{T_1 r_1}{J_1} = \frac{(7.50\ \text{kN} \bullet \text{m})(25\ \text{mm})}{0.5\pi(25\ \text{mm})^4} = 306\ \text{MPa}$$
$$\tau_2 = \frac{T_2 r_2}{J_2} = \frac{(-2.5\ \text{kN} \bullet \text{m})(19\ \text{mm})}{0.5\pi(19\ \text{mm})^4} = -232\ \text{MPa}$$

9.17 d. The same three-step process as in Problem 9.16 must be carried out. Since this process has already been completed, the results can be used. The rotation can be expressed as

$$\phi = \phi_1 = -\phi_2 = \frac{T_1 L_1}{GJ_1} = \frac{(7.50\text{kN} \bullet \text{m})(250\ \text{mm})}{(83\ \text{GPa})0.5\pi(25\ \text{mm})^4} = 0.0368\ \text{rad} = 2.11°$$

9.18 c. To determine displacement of B, calculate the load at B (Exhibit 9.18a).

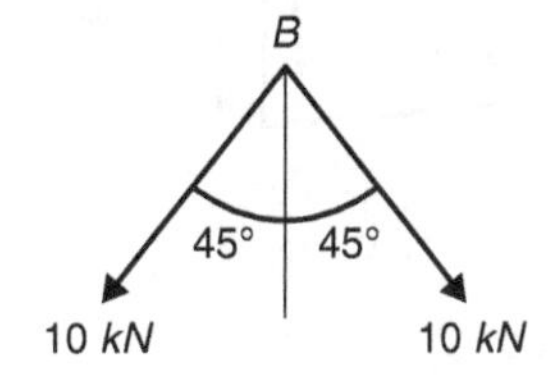

Exhibit 9.18a

(Load at B) = 2 (10 sin 45) = 14.14 kN

Therefore, the load at C = 14.14 + 20 = 34.14 kN

$$\delta_{CB} = \frac{PL_{CB}}{AE} = \frac{\left(34.14 \times 10^3\ \text{N}\right)\left(1.5\ \text{m}\right)}{\left(\frac{1}{100}\right)^2 \text{m}^2 \left(180 \times 10^9\right)\ \text{N/m}^2} = 0.002845\ \text{m} = 2.85\ \text{mm}$$

9.19 d. Determine displacement at A:

δ_A = 2.845 + 1 = 3.85 mm

$$\delta_A = \delta_B + \delta_{AB} = 2.845\ \text{mm} + \frac{\left(20 \times 10^3\ \text{N}\right)\left(0.5\ \text{m}\right)}{\left(\frac{1}{100}\right)^2 \text{m}^2 \left(180 \times 10^9\right)\text{N/m}^2}$$

9.20 b. Using the free-body diagram of $BCDF$ (Exhibit 9.20a):

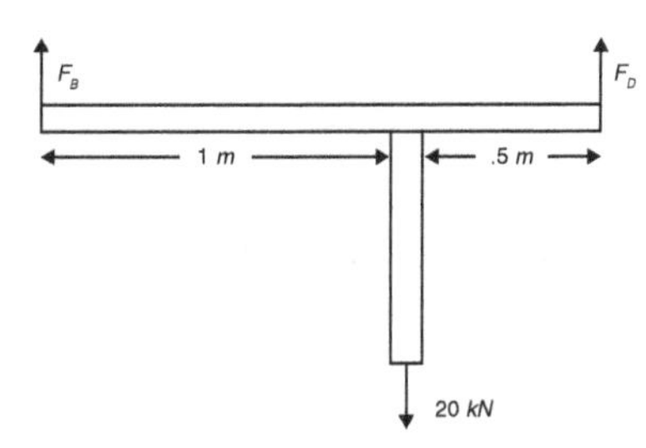

Exhibit 9.20a

$M_B = 0 = 20$ kN (1 m) – F_D (1.5 m)

$F_D = 13.3$ kN ↑

$\Sigma F_y = 0 = F_D + F_B - F_F$

9.21 c. To determine the load at D, we must write an equilibrium equation of forces:

$$13.3 \uparrow + F_B - 20 = 0 \qquad F_B = 6.7 \text{ kN} \uparrow$$

9.22 b. The displacement of point D is found:

$$\delta_D = \frac{PL}{AE} = \frac{(13.3 \times 10^3 \text{ kN})(1 \text{ m})}{(30 \times 10^{-6})\text{m}^2 (180 \times 10^9)\text{N/m}^2}$$

$$\delta_D = 0.00246 \text{ m} = 2.46 \text{ mm}$$

9.23 a. The displacement of point B is found:

$$\delta_B = \frac{PL}{AE} = \frac{(6.7 \times 10^3 \text{ kN})(1 \text{ m})}{(50 \times 10^{-6} \text{m}^2)(180 \times 10^9 \text{ N/m}^2)}$$

$$\delta_B = 0.000744 \text{ m} = 0.744 \text{ mm}$$

9.24 d. To determine the displacement of F, draw a displacement triangle from B to D, as shown in Exhibit 9.24.

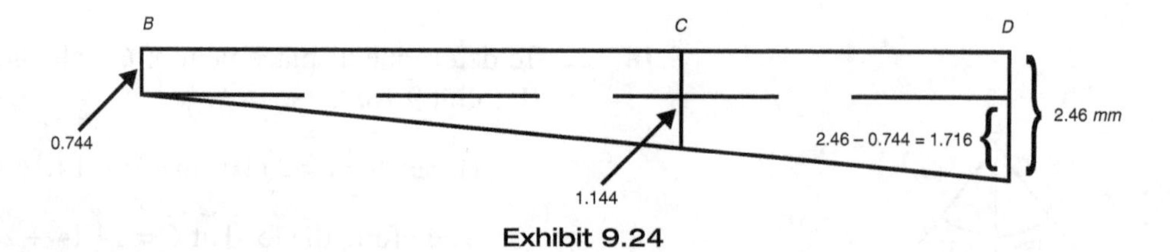

Exhibit 9.24

From the triangle, we see that point C is displaced:

$$\delta_C = 0.744 \text{ mm} + 1.144 \text{ mm} = 1.888 \text{ mm}$$

$$\delta_{F/C} = \frac{PL}{AE} = \frac{(20 \times 10^3 \text{ N})(1 \text{ m})}{(40 \times 10^{-6}) \text{ m}^2 \ (180 \times 10^9)\text{N/m}^2} = 0.00277 \text{ m}$$

$$\delta_F = 1.888 + 2.77 = 4.65 \text{ mm}$$

9.25 c. Find the modulus of elasticity by using the values from the magnified stress-strain portion of the curve.

$$E = \frac{\sigma}{\varepsilon} = \frac{225 \text{ MPa}}{0.002 \text{ mm}} = 112.5 \text{ GPa}$$

9.26 a. The yield load can be obtained by taking the yield stress off the curve and multiplying by the area of the bar.

$$P_y = \sigma_y A$$

$$P_y = (225 \times 10^6 \text{ N/m}^2)(\pi / 4)(0.01)^2 \text{ m}^2 = 17.7 \text{ kN}$$

9.27 **d.** Obtain the ultimate load by taking the ultimate stress off the curve and multiplying by the area of the bar.

$$P_{ult} = \sigma_{ult} A = (360 \times 10^6 \text{ Pa})(\pi / 4)(0.01)^2 = 28.3 \text{ kN}$$

9.28 **b.** The modulus of resilience may also be obtained using values from the stress-strain curve.

$$\mu_r = \frac{1}{2}\frac{\sigma_{PL}^2}{E} = \left(\frac{1}{2}\right)\frac{\left(225 \times 10^6\right)^2}{112.5 \times 10^9} = 0.225 \text{ MJ/m}^3$$

9.29 **c.** Sketch a free-body diagram (Exhibit 9.29a) and use the equilibrium equations to determine the load on BC.

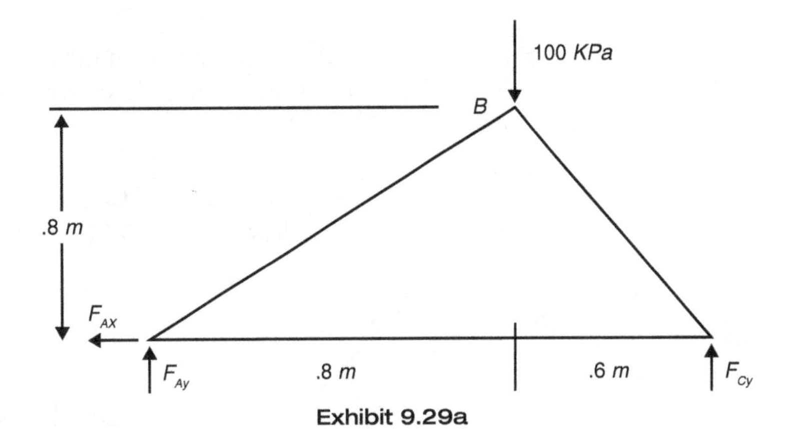

Exhibit 9.29a

$\Sigma F_X = 0 = -F_{AX} = 0; \quad F_{AX} = 0$

$M_A = 0 = 100 \text{ kPa } (0.8 \text{ m}) - F_{Cy} (1.4 \text{ m})$

Therefore:

$F_{Cy} = 57 \text{ kN}; \quad F_{Ay} = 43 \text{ kN}$

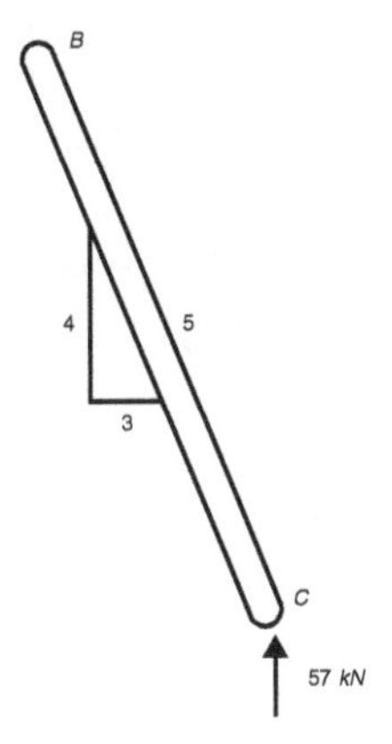

Exhibit 9.29b

Making a free-body diagram of bar BC (Exhibit 9.29b), calculate the axial load of BC.

$4/5\ F_{BC} = F_{Cy} = 57 \text{ kN, so } F_{BC} = 71.25 \text{ kN}$

9.30 **a.** Taking a free-body diagram of AB (Exhibit 9.30), we solve for the stress in AB

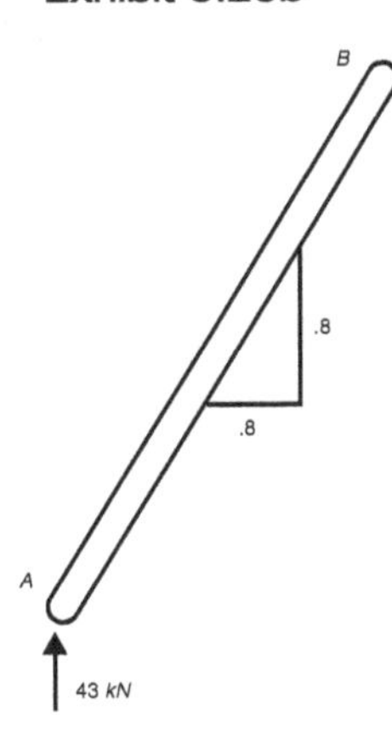

Exhibit 9.30

$$F_{AB} = \frac{F_{AY}}{\sin 45} = \frac{43}{.707} = 60.8 \text{ kN}$$

$$\sigma_{AB} = \frac{F_{AB}}{A} = \frac{60.8 \text{ kN}}{2 \text{ cm}^2} = \frac{60.8 \times 10^3 \text{ N}}{0.0002 \text{ m}^2} = 304 \text{ MPa}$$

9.31 **a.** To determine the elongation of bar AC, use equation:

$$\delta_{AC} = \frac{F_{AC} L_{AC}}{AE}$$

Knowing that $F_{AC} = 43$ kN,

$$\delta_{AC} = \frac{(43 \times 10^3 \text{ N})(1.4 \text{ m})(1000 \text{ mm/m})}{(0.0002 \text{ m}^2)(150 \times 10^9 \text{ N/m}^2)} = 2.0 \text{ mm}$$

9.32 **a.** For a torsion problem, the shear strain is

$$\gamma_{\phi z} = \frac{\tau_{\phi z}}{G} = \frac{Tr}{GJ}$$

Other shear strains in the $r - \phi$ orientation are zero. Mohr's circle for this state of strain is shown in Exhibit 9.32(a). From Mohr's circle,

$$\varepsilon_{45} = \frac{\gamma_{\phi z}}{2} = \frac{\tau_{\phi z}}{2G} = \frac{Tr}{2GJ}$$

$$T = \frac{2GJ\varepsilon_{45}}{r} = \frac{2(83\,\text{GPa})[0.5\pi(25\,\text{mm})^4](245 \times 10^{-6})}{25\,\text{mm}} = 998\,\text{N} \bullet \text{m}$$

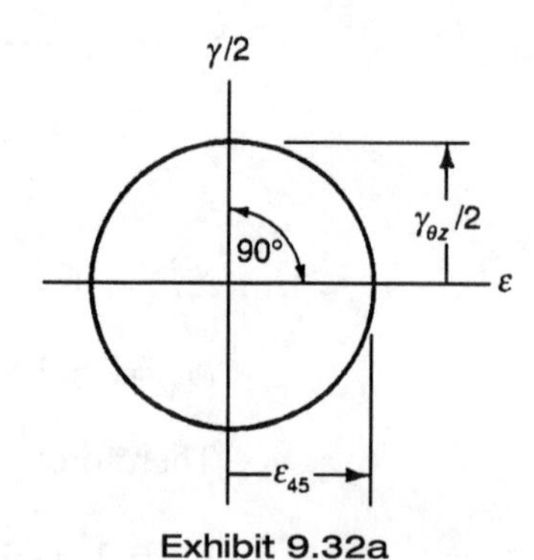

Exhibit 9.32a

9.33 **d.** In thin-walled shafts, the shear stress is

$$\tau_{sz} = \frac{T}{2At}$$

The area A is the cross-sectional area of the shaft including the hole, so

$$A = \frac{\pi r^2}{2} = \frac{\pi(25\,\text{mm} - 3\,\text{mm})^2}{2} = 760\,\text{mm}^2$$

$$\tau_{sz} = \frac{T}{2At} = \frac{300\,\text{N} \bullet \text{m}}{2(760\,\text{mm}^2)(3\,\text{mm})} = 65.7\,\text{MPa}$$

9.34 c. Draw the free-body diagram of the beam, replacing the distributed load with its statically equivalent loads. Summation of moments about the left end gives

$$0 = -3.6\,R_2 + (108)(1.8) + (36)(4.2)$$
$$R_2 = 96\text{ kN}$$

Summation of forces in the vertical direction gives

$$R_1 = 144 - 96 = 48\text{ kN}$$

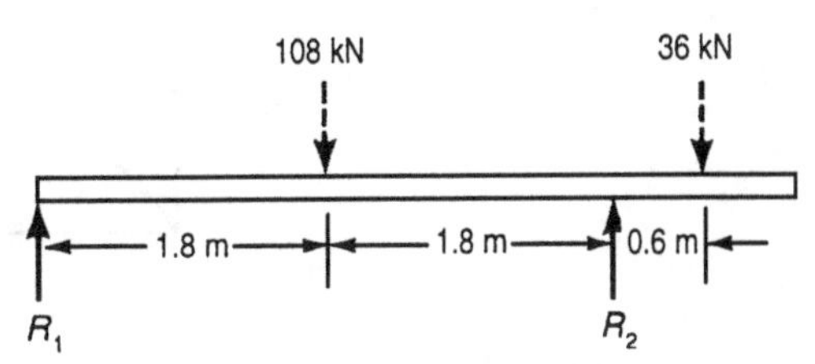

Exhibit 9.34a

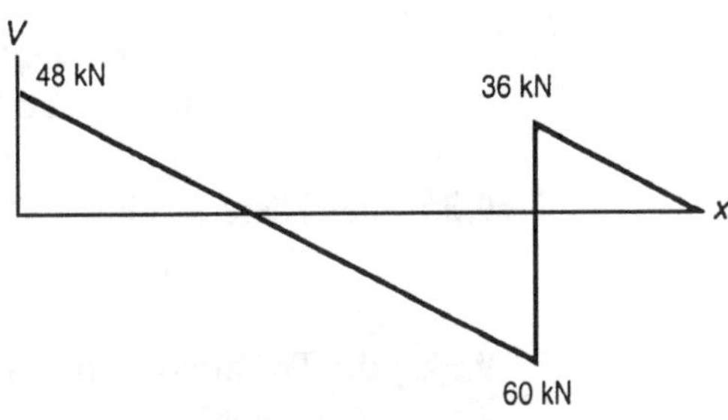

Exhibit 9.34b

This is enough information to plot the shear diagram (Exhibit 9.34b). The largest magnitude of shear is 60 kN.

9.35 c. The maximum bending moment occurs where the shear is zero. From the shear diagram, the distance to the zero from the left end can be found by similar triangles.

$$\frac{48}{x} = \frac{108}{3.6}; \qquad x = \frac{(48)(3.6)}{108} = 1.6\text{ m}$$

The areas of the shear diagrams are the changes in moment (Exhibit 9.35a).

$$A_1 = \frac{(48\text{ kN})(1.6\text{ m})}{2} = 38.4\text{ kN} \bullet \text{m}$$
$$A_2 = \frac{(36\text{ kN})(1.2\text{ m})}{2} = 21.6\text{ kN} \bullet \text{m}$$

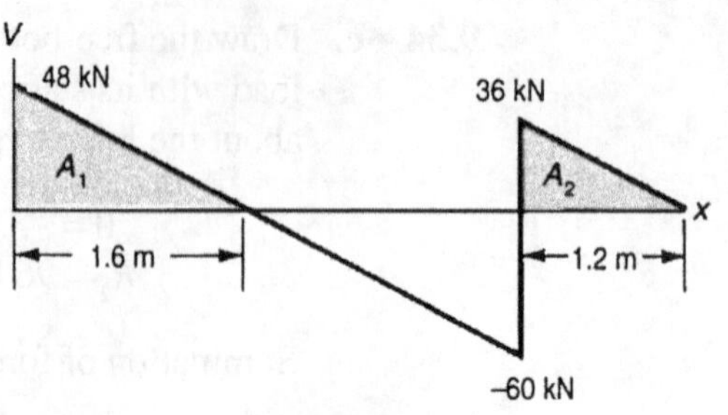

Exhibit 9.35a

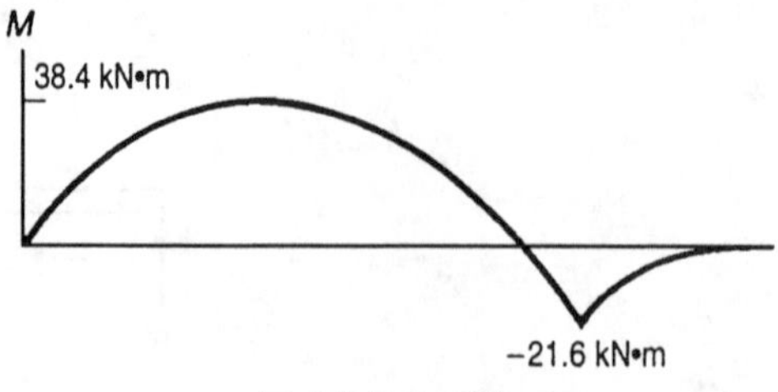

Exhibit 9.35b

The moment diagram is shown in Exhibit 9.35b. The maximum bending moment is 38.4 kN • m.

9.36 a. There is a jump at *B* and *D* of 60 kN upward and a downward jump of 30kN at *E*. These jumps correspond to concentrated forces.

9.37 d. The areas of the shear diagrams (Exhibit 9.37) are the changes in moment. Since the moments are zero on either end,

$$M_B = A_1 = \frac{(30\,\text{kN})(1\,\text{m})}{2} = 15\ \text{kN} \bullet \text{m}$$

$$M_C = A_1 + A_2 = \frac{(30\,\text{kN})(1\,\text{m})}{2} + \frac{(30\,\text{kN})(1\,\text{m})}{2} = 0$$

$$M_D = A_3 = (30\,\text{kN})(-0.5\,\text{m}) = -15\ \text{kN} \bullet \text{m}$$

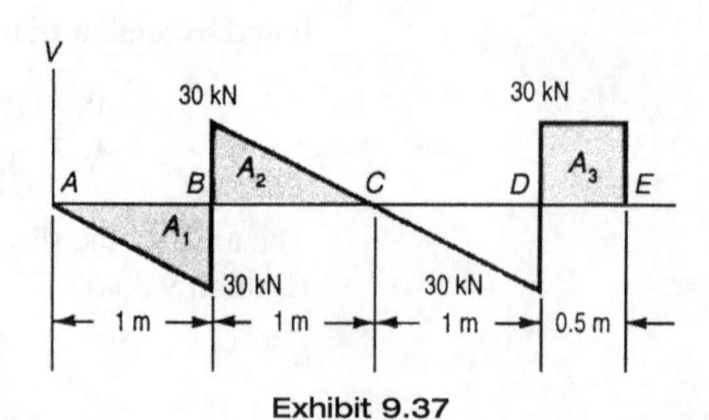

Exhibit 9.37

The largest magnitude of the bending moment is therefore 15 kN • m.

9.38 **d.** It is obvious that each support will carry half of the load, so the reactions are $wL/2$. The shear diagram is shown in Exhibit 9.38a. The maximum bending moment is

$$M = A_1 = \frac{1}{2}\left(\frac{wL}{2}\right)\left(\frac{L}{2}\right) = \frac{wL^2}{8}$$

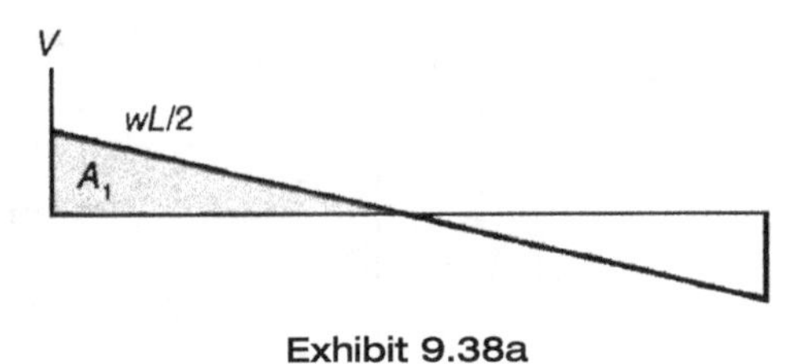

Exhibit 9.38a

The maximum bending stress is

$$\sigma_{\max} = \frac{M}{Z} = \frac{wL^2}{8Z} = 100\,\text{MPa}$$

$$w = \frac{(100\,\text{MPa})\,8Z}{L^2} = \frac{(100\,\text{MPa})(8)(1408 \times 10^3\ \text{mm}^3)}{(4\,\text{m})^2} = 70.4\,\text{kN/m}$$

9.39 **a.** Draw the free-body diagram and the shear and moment diagrams as shown in Exhibit 9.39a. The maximum bending stress is

$$\sigma_{\max} = \frac{M_{\max}c}{I} = \frac{(9\text{kN} \bullet \text{m})\left(\dfrac{363\ \text{mm}}{2}\right)}{\left(365 \times 10^6\ \text{mm}^4\right)} = 4.48\ \text{MPa}$$

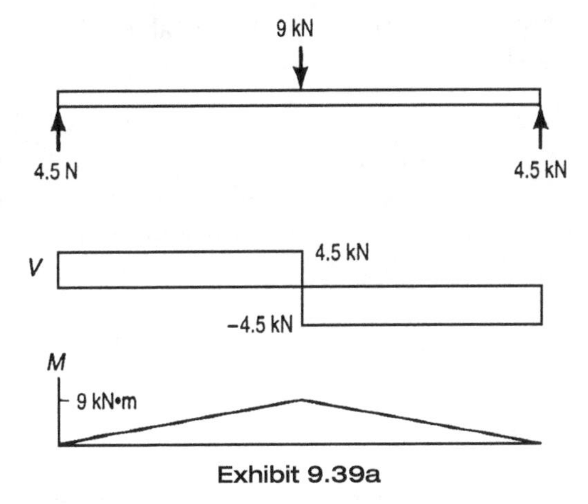

Exhibit 9.39a

9.40 a. From the previous problem, the maximum shear in the beam is 4.5 kN. The maximum shearing stress will take place at the centroid (Exhibit 9.40), so a cut must be made there in order to calculate Q. The moment of the area Q is, therefore,

$$Q = A_1\bar{y}_1 + A_2\bar{y}_2$$

$$Q = (257\text{ mm})(21.7\text{ mm})\left(\frac{363\text{ mm}}{2} - \frac{21.7\text{ mm}}{2}\right) + \ldots + \left(\frac{363\text{ mm}}{2} - 21.7\text{ mm}\right)$$

$$\times (13\text{ mm})\left(\frac{\dfrac{363\text{ mm}}{2} - 21.7\text{ mm}}{2}\right)$$

$$Q = 1.117 \times 10^6\text{ mm}^3$$

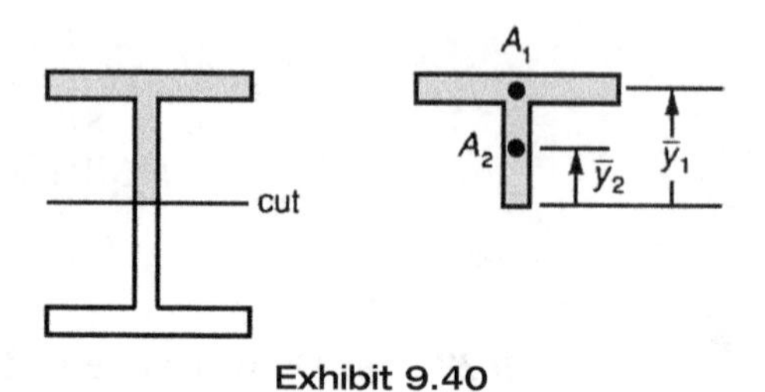

Exhibit 9.40

The maximum shear stress is then

$$\tau = \frac{VQ}{Ib} = \frac{(4.5\text{ kN})(1.117 \times 10^6\text{ mm}^3)}{(365 \times 10^6\text{mm}^4)(13\text{ mm})} = 1.060\text{ MPa}$$

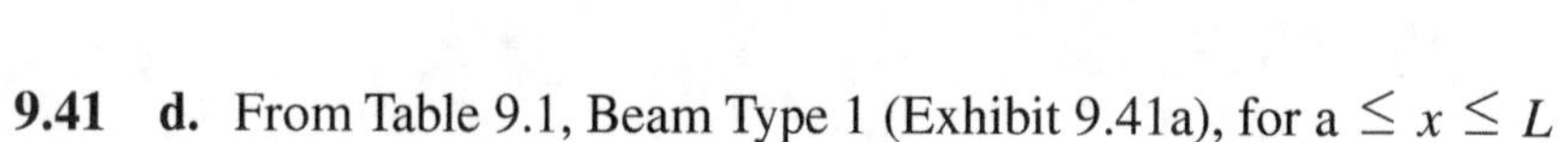

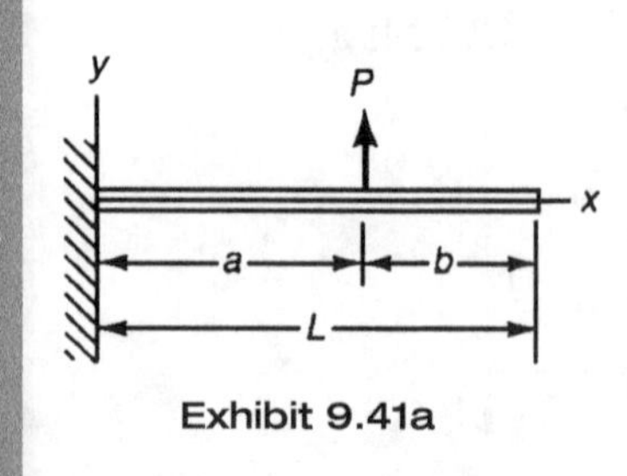

Exhibit 9.41a

9.41 d. From Table 9.1, Beam Type 1 (Exhibit 9.41a), for a $\leq x \leq L$

$$y = \frac{Pa^2}{6EI}(3x - \text{a})$$

For the load at the half-way point, $a = L/2$, $x = L$, and $P = -F$. For the load at the end, $a = L$, $x = L$, and $P = -F$. Therefore,

$$y = \frac{-F\left(\dfrac{L}{2}\right)^2}{6EI}\left[3L - \left(\frac{L}{2}\right)\right] + \frac{-FL^2}{6EI}(3L - L) = -0.4375\frac{FL^3}{EI}$$

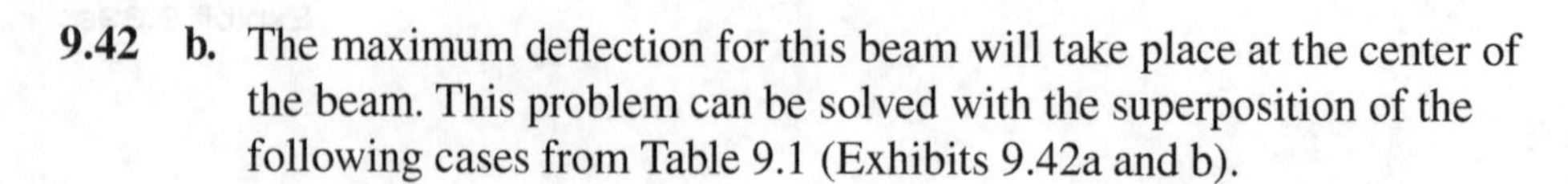

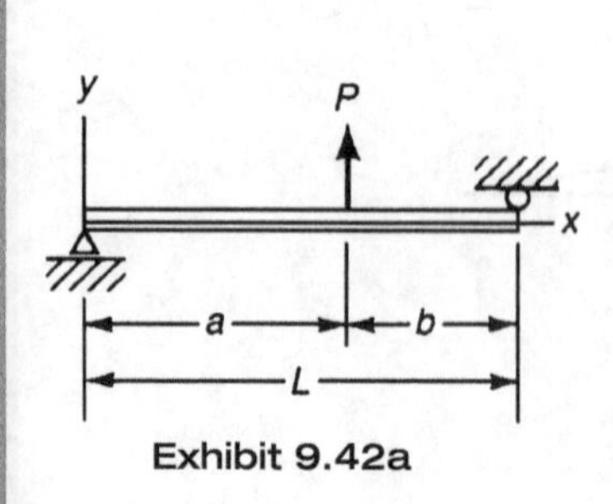

Exhibit 9.42a

9.42 b. The maximum deflection for this beam will take place at the center of the beam. This problem can be solved with the superposition of the following cases from Table 9.1 (Exhibits 9.42a and b).

For $0 \leq x \leq a$,

$$y = \frac{Pbx}{6LEI}(L^2 - b^2 - x^2)$$

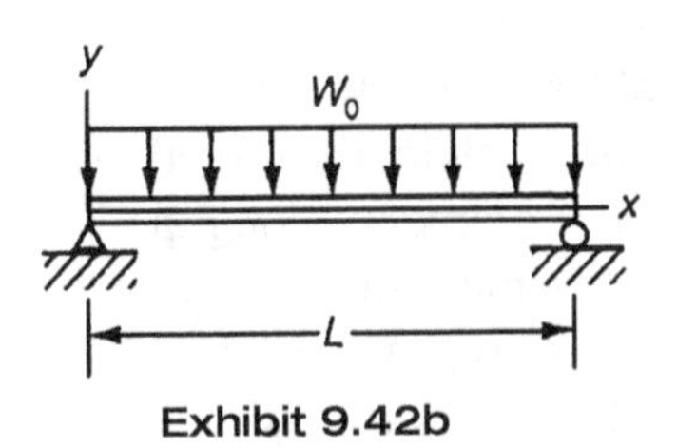

Exhibit 9.42b

For this problem, $P = -wL$, $a = b = L/2$, and $x = L/2$.

$$y = -\frac{wx}{24EI}(L^3 - 2Lx^2 + x^3)$$

For this problem, $x = L/2$. The total deflection is, therefore,

$$y = \frac{Pbx}{6LEI}\left(L^2 - b^2 - x^2\right) - \frac{wx}{24EI}\left(L^3 - 2Lx^2 + x^3\right)$$

$$y = \frac{(-wL)\left(\frac{L}{2}\right)\left(\frac{L}{2}\right)}{6LEI}\left[L^2 - \left(\frac{L}{2}\right)^2 - \left(\frac{L}{2}\right)^2\right] - \frac{w\left(\frac{L}{2}\right)}{24EI}\left[L^3 - 2L\left(\frac{L}{2}\right)^2 + \left(\frac{L}{2}\right)^3\right]$$

$$v = -0.0339\frac{wL^4}{EI} = -\frac{13wL^4}{384EI}$$

9.43 d. This problem can be solved from superposition of the same two cases as used in Problem 9.42. For the concentrated load solution, $b = L/2$ and P is left as an unknown. In both, $x = L/2$. The center support means the beam does not deflect in the center. Therefore,

$$y = 0\frac{P\left(\frac{L}{2}\right)\left(\frac{L}{2}\right)}{6LEI}\left[L^2 - \left(\frac{L}{2}\right)^2 - \left(\frac{L}{2}\right)^2\right] - \frac{w\left(\frac{L}{2}\right)}{24EI}\left[L^3 - 2L\left(\frac{L}{2}\right)^2 + \left(\frac{L}{2}\right)^3\right]$$

$$y = 0 = \frac{PL^3}{48\,EI} = -\frac{5wL^4}{384EI}$$

$$P = \frac{5}{8}wL$$

9.44 c. Draw the free-body diagram (Exhibit 9.44a). From a summation of moments about the cut at B, the maximum bending moment in BC is the moment $M = 200$ N × 250 mm or 50 kN • m. The maximum bending stress is

$$\sigma = \frac{Mc}{I} = \frac{(50\,\text{kN} \bullet \text{mm})(12.5\,\text{mm})}{0.25\,\pi(12.5\,\text{mm})^4} = 32.6\text{ MPa}$$

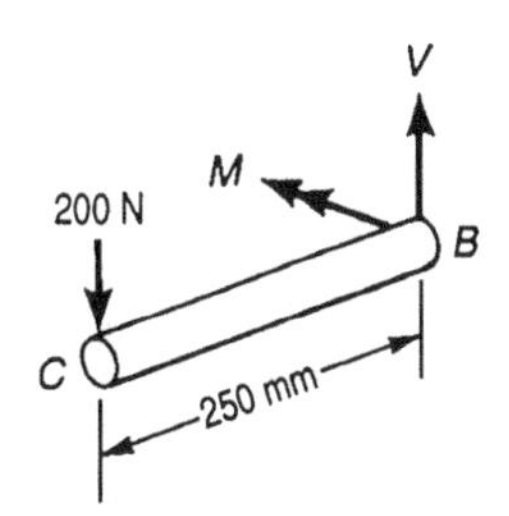

Exhibit 9.44a

9.45 d. Draw the free-body diagram (Exhibit 9.45). The maximum stresses in section *AB* will occur at *A*. Summation of forces in the vertical direction gives $V_A = 200$ N. Summation of forces along the direction of the rod *AB* gives $P = 2000$ N. Summation of moments along the rod *AB* gives $T_A = 200$ N × 250 mm or 50 kN • mm. Summation of moments at the cut perpendicular to the rod *AB* gives $M_A = 200$ N × 300 mm = 600 kN • mm. The maximum bending stress is

$$\sigma = \frac{Mc}{I} = \frac{(60\,\text{kN} \bullet \text{mm})(12.5\,\text{mm})}{0.25\pi(12.5\,\text{mm})^4} = 39.1\,\text{MPa}$$

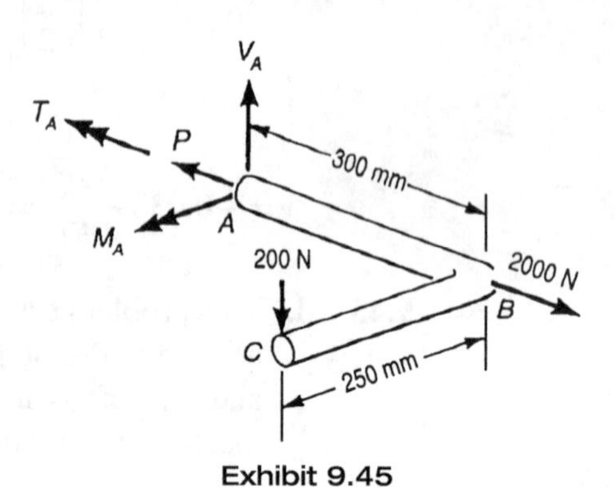

Exhibit 9.45

9.46 b. From the free-body diagram in Problem 9.45, the maximum torque is 50 kN • mm. The maximum shear stress is, therefore,

$$\tau_{max} = \frac{T_{max} r_0}{J} = \frac{(50\,\text{kN} \bullet \text{mm})(12.5\,\text{mm})}{0.5\pi(12.5\,\text{mm})^4} = 16.30\,\text{MPa}$$

9.47 a. From the free-body diagram in Problem 9.45, the maximum axial force is 2000 N. The maximum stress due to this force is, therefore,

$$\sigma = \frac{P}{A} = \frac{2000\,\text{N}}{\pi(12.5\,\text{mm})^2} = 4.07\,\text{MPa}$$

9.48 d. The stresses were found in the previous three problems. There is an axial stress due to both bending and axial loads. This stress is

$$\sigma = 39.1\text{ MPa} + 4.07\text{ MPa} = 43.2\text{ MPa}$$

The shear stress is 16.3 MPa. These are the only nonzero stresses. The maximum principal stress can be calculated as follows:

$$R = \sqrt{\left(\frac{\sigma_x - \sigma_y}{2}\right)^2 + \tau_{xy}^2} = \sqrt{\left(\frac{43.2 - 0}{2}\right)^2 + 16.3^2} = 22.0\text{ MPa}$$

$$C = \frac{\sigma_x + \sigma_y}{2} = \frac{43.2 + 0}{2} = 21.6\text{ MPa}$$

$$\sigma_1 = C + R = 22.0 + 21.6 = 43.6\text{ MPa}$$

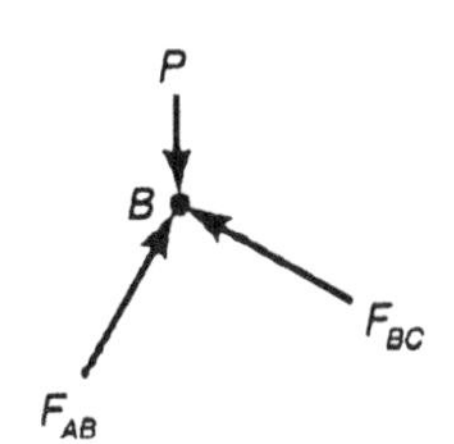

Exhibit 9.49a

9.49 **a.** Draw the free-body diagram of the joint B (Exhibit 9.49a). Summation of forces in the vertical direction gives

$$P = F_{AB}\frac{1.5}{\sqrt{1.5^2+0.9^2}} + F_{BC}\frac{1.5}{\sqrt{2.7^2+1.5^2}}$$

Summation of forces in the horizontal direction gives

$$F_{AB}\frac{0.9}{\sqrt{1.5^2+0.9^2}} = F_{BC}\frac{2.7}{\sqrt{2.7^2+1.5^2}}$$

Solving for F_{AB} and F_{BC} gives

$$F_{AB} = 0.875\,P; \quad F_{BC} = 0.515\,P$$

The member AC is in tension and does not need to be considered. The moment of inertia for both members is

$$I = \frac{bh^3}{12} = \frac{(25\,\text{mm})(25\,\text{mm})^3}{12} = 32,600\,\text{mm}^4$$

The critical buckling load for member AB is

$$P_{cr} = F_{AB} = \frac{\pi^2 EI}{L^2} = \frac{\pi^2(210\,\text{GPa})(32,600^4)}{(1.5\,\text{m})^2+(0.9\,\text{m})^2} = 22.0\,\text{kN}$$

The load P for buckling to occur in AB is

$$P = \frac{22\,\text{kN}}{0.875} = 25.2\,\text{kN}$$

The critical buckling load for member BC is

$$P_{cr} = F_{BC} = \frac{\pi^2 EI}{L^2} = \frac{\pi^2(210\,\text{GPa})(32,600\,\text{mm}^4)}{[(2.7\,\text{m})^2+(1.5\,\text{m})^2]} = 7.07\,\text{kN}$$

The load P for buckling to occur in BC is

$$P = \frac{7.07\,\text{kN}}{0.515} = 13.7\,\text{kN}$$

9.50 **d.** To buckle in the x-y plane the critical buckling load is

$$P_{cr} = \frac{\pi^2 EI}{L^2} \frac{\pi^2 E\left(\frac{ab^3}{12}\right)}{L^2}$$

To buckle in the x-z plane the critical buckling load is

$$P_{cr} = \frac{4\pi^2 EI}{L^2} \frac{4\pi^2 E\left(\frac{ba^3}{12}\right)}{L^2}$$

Equating these two representations of P_{cr} gives

$$ab^3 = 4ba^3; \quad b^2 = 4a^2; \quad \frac{b}{a} = 2$$

CHAPTER 10

Materials

OUTLINE

PAVEMENT DESIGN

Roadway pavements are divided into two general categories: rigid and flexible. Rigid pavements are usually constructed of portland cement concrete, which behaves much like a beam over any irregularities in the underlying supporting materials. Flexible pavements are usually constructed of bituminous materials that transfer vehicle loadings directly to the underlying support materials.

The damage to roadway pavements caused by passenger cars is very limited compared with that caused by trucks. Therefore, pavements are designed to support a specified number of heavy vehicle loadings over their design life. An equivalent single-axle load (ESAL) is a standard term used in pavement design to describe the damage caused by one pass of an 18,000-pound (18-kip) axle load over the pavement surface (in SI notation, 18,000 pounds = 80 kN). Consequently, in order to design a pavement, it is necessary to express all of the traffic that will use the pavement as an equivalent number of 18,000-pound axle loads. The use of an 18,000-pound standard stems from the maximum legal axle loadings in effect at the time many pavement design methods were developed. This conversion is accomplished by first determining an ESAL factor for each classification of vehicle using the pavement. An ESAL factor is a ratio relating the damage caused by a passing vehicle of specified weight relative to the damage caused by an 18-kip axle load, or

$$\text{ESAL factor} = D_i / D_{18} \tag{10.1}$$

where

D_i = damage caused by vehicle class i
D_{18} = damage caused by 18-kip axle load

Studies have shown that the ESAL factor for passenger cars is about 0.0008. The ESAL factors for heavy trucks, on the other hand, approach 2.4 when loaded to the legal limit and can be as high as 10 for overloaded trucks.

The truck factor is defined as the number of 18-kip single-load applications caused by a single passage of a vehicle. The load equivalency factor is the number of equivalent 18-kip single-axle load applications contributed by one passage of an axle. An average truck factor (f_i) can be calculated by multiplying the number of axles in each weight class (N_i) by the appropriate load equivalency factor (F_{Ei})

and dividing the sum of the products by the total number of vehicles involved (V_T). This relationship between truck factors and load equivalency factors is as follows:

$$f_i = \Sigma(N_i F_{Ei})/V_T \quad \textbf{(10.2)}$$

The accumulated annual design year ESAL for each category of axle load using *load equivalency factors* is calculated as follows (Garber and Hoel, 1997):

$$\text{ESAL}_i = \text{AADT}_i \times 365 \times F_{Ei} \times N_i \times g_{jt} \times f_d \quad \textbf{(10.3)}$$

where

ESAL_i = equivalent accumulated 18-kip single-axle load for axle category i
AADT_i = base-year annual average daily traffic for axle category i
F_{Ei} = load equivalency factor for axle category i
N_i = number of axles on each vehicle in axle category i
g_{jt} = growth factor for growth rate j and design period t
f_d = design lane factor (directional and lane distribution factor)

The accumulated annual design year ESAL using truck factors for each category of truck is

$$\text{ESAL}_i = \text{AADT}_i \times 365 \times f_i \times g_{jt} \times f_d \quad \textbf{(10.4)}$$

where

ESAL_i = equivalent accumulated 18-kip axle load for truck category i
AADT_i = base-year annual average daily traffic for vehicles in truck category i
f_i = truck factor for vehicles in truck category i

The accumulated ESAL for all categories of axle loads or vehicles is

$$\text{ESAL} = \Sigma\text{ESAL}_i \quad \textbf{(10.5)}$$

The three most commonly used pavement thickness design methods are the AASHTO (1993) method, the Asphalt Institute (AI) (1991) method, and the Portland Cement Association (PCA) (1984) method. The AASHTO method addresses both flexible and rigid pavements, while the Asphalt Institute method and the Portland Cement Association method are concerned with the design of flexible and rigid pavements, respectively. This section presents a review of these three pavement design methods as they are used to determine pavement thickness.

Because of the complexity of the procedures and their heavy reliance on nomographs and design charts, the reader should consult the individual design manuals for example problems. Garber and Hoel (1997) also provide an excellent overview of the three design methods reviewed in this section. Like much of the material in this chapter, it is not the intent of this section to relieve the reader of the need to consult these three design manuals directly for the details of the individual design methods.

AASHTO Design Method

The AASHTO pavement design method is documented in the AASHTO *Guide for Design of Pavement Structure* (AASHTO, 1993). The AASHTO method is based on results obtained from the "AASHTO Road Test" conducted in the late 1950s and early 1960s in northern Illinois. The method is an empirical method that relates pavement performance measurements (loss of serviceability) directly to traffic loading and volume characteristics, roadbed soil strength, pavement layer material characteristics, and environmental factors. The AASHTO Road Test resulted in many important concepts, including demonstration of the major influences of traffic loads and repetitions on design thickness.

Equally important was the development of the serviceability-performance method of analysis, which provided a quantifiable way of defining "failure" based on a user-defined definition rather than one based primarily on structural failure. This is achieved through an initial serviceability index, p_0, which is the serviceability index immediately after construction, and the terminal serviceability index, p_t, which is the minimum acceptable value before resurfacing or reconstruction is necessary. The AASHTO guide suggests the use of p_0 values of 4.2 for flexible pavements and 4.5 for rigid pavements. A terminal serviceability index (p_t) of 2.5 or higher is suggested for design of major highways, and an index of 2.0 is suggested for highways with lesser traffic volumes.

The AASHTO method also incorporates a reliability design factor (F_R) to account for uncertainty in traffic forecasts and pavement performance.

The 1986 AASHTO Guide considers the following factors in the design process: (1) pavement performance, (2) traffic, (3) roadbed soil, (4) materials of construction, (5) environment, (6) drainage, (7) reliability, (8) life-cycle costs, and (9) shoulder design.

The basic design equations used for flexible and rigid pavements in the 1993 AASHTO Design Guide are as follows:

Flexible pavements:

$$\begin{aligned}\log_{10}(W_{18}) = {} & Z_R \times S_0 + 9.36 \times \log_{10}(SN + 1) - 0.20 \\ & + \{\log_{10}(\Delta PSI/2.7)/[0.40 + 1094/(SN + 1)^{5.19}]\} \\ & + 2.32 \times \log_{10}(M_R) - 8.07 \end{aligned} \quad \textbf{(10.6)}$$

where

W_{18} = predicted number of 18-kip ESAL applications
Z_R = standard normal deviate
S_0 = combined standard error of the traffic forecast and performance prediction
ΔPSI = difference between the initial design serviceability index, p_0, and the design terminal serviceability index, p_t
M_R = resilient modulus (psi)
SN = structural number (indicative of the total pavement thickness required)

The structural number (SN) is computed as follows:

$$SN = a_1 D_1 + a_2 D_2 m_2 + a_3 D_3 m_3 \quad \textbf{(10.7)}$$

where

a_1, a_2, a_3 = layer coefficients of surface, base, and subbase courses, respectively
D_1, D_2, D_3 = thickness (inches) of surface, base, and subbase courses, respectively
m_i = drainage coefficient for layer i

Rigid pavements:

$$\log_{10}(W_{18}) = Z_R \times S_0 + 7.35 \times \log_{10}(D+1) - 0.06 + \frac{\log_{10}[\Delta PSI/3.0]}{1+\left[1.624\times 10^7/(D+1)^{8.46}\right]}$$
$$+(4.22-0.32p_t)\times \log_{10}\frac{S_c' \times C_d \times (D^{0.75}-1.132)}{215.63J\left[D^{0.75}-18.42/(E_c/k)\right]^{0.25}} \quad \textbf{(10.8)}$$

where

D = thickness of pavement slab (in.)
S_c' = modulus of rupture for portland cement concrete used on a specific project (psi)
J = load transfer coefficient used to adjust for load transfer characteristics of a specific design
C_d = drainage coefficient
E_c = modulus of elasticity for portland cement concrete (psi)
k = modulus of subgrade reaction (pci)

The AASHTO Guide provides a series of nomographs to solve these equations for the structural number (*SN*) for flexible pavements and the thickness of the pavement slab (*D*) for rigid pavements. The basic procedures are outlined below.

Flexible Pavements

The structural design of flexible pavements involves the determination of the thickness and vertical position of paving materials that can best be combined to provide a serviceable roadway for predicted traffic over the pavement's design life. The pavement structure is designed to use the most economical arrangement and minimum thickness of each material necessary to protect the underlying courses and the roadbed from stresses caused by traffic loads. The objective of the AASHTO design method is to determine a flexible pavement structural number (*SN*) adequate to carry the design ESAL. The basic method is applicable for 18-kip ESALs greater than 50,000 for the design period. Lower ESAL values are considered under the method described in the AASHTO Guide for low-volume roads. The flexible pavement design procedure is described in Section 3.1 (pp. II-31 to II-37) of the AASHTO Guide. Appendix H (pp. H-1 to H-7) of the AASHTO Guide presents a flexible pavement design example.

A summary of the basic steps in the AASHTO method for flexible pavement design is presented below. The section numbers, tables, and figures cited below refer to the AASHTO Guide (AASHTO, 1993).

Step 1: Determine the structural number. Figure 3.1 (p. II-32) of the AASHTO Guide presents the nomograph that can be used to solve Equation (10.207) for *SN* based on the following inputs.

1. Future traffic, W_{18}. Estimated as described in Section 2.1.2 (p. II-7). The basic procedure is similar to Equation (10.202), except the equivalence factors are based on the terminal serviceability index (p_t) and *SN*. The use of an assumed *SN* value of 5 will normally give results that are sufficiently accurate for design purposes. Appendix D of the AASHTO Guide presents tables of axle load equivalency factors for various axle configurations, and a range of values for p_t and *SN*. Appendix D also contains a worksheet to guide the analyst through the computations.

2. Reliability, *R*. Suggested levels of reliability for various functional classes of highways are presented in Table 2-2 (p. II-9) of the AASHTO Guide.

3. Overall standard deviation S_0. A value of 0.45 is recommended for flexible pavements.

4. Effective resilient modulus of roadbed material, M_R. The 1993 AASHTO Guide allows for the conversion of California Bearing Ratio (CBR) and stabilometer R-values to an equivalent M_R value through the following conversion factors:

$$M_R \text{ (psi)} = 1500 \times \text{CBR} \quad \textbf{(10.9)}$$

$$M_R \text{ (psi)} = 1000 + 555 \times (R\text{-value}) \quad \textbf{(10.10)}$$

5. Design serviceability loss, $\Delta PSI = p_0 - p_t$. Values of $p_0 = 4.2$ and $p_t = 2.0$ to 2.5 are suggested for flexible pavements.

Step 2: Evaluate need for staged construction (optional). If the analysis period (e.g., 20 years) is longer than the service life selected for the initial pavement structure, it will be necessary to consider staged construction and planned rehabilitation alternatives. In such cases, the analyst should consult Part III of the AASHTO Guide to develop design strategies that will last the entire analysis period. The design example in Appendix H of the AASHTO Guide provides an illustration of the application of staged construction alternatives. If staged construction is not a viable or necessary alternative, the analyst should proceed to step 3.

Step 3: Evaluate need to consider roadbed swelling and frost heave (optional). If the site of the highway construction project is in an area where the roadbed is subject to swelling and/or frost heave, the effects of these factors on the rate of serviceability loss must be taken into account. This is accomplished through the following iterative process. Table 3-1 (p. II-34) of the AASHTO Guide provides an example of the process.

Step 3.1: Select an appropriate *SN* for the initial pavement. The maximum initial *SN* recommended is that derived for conditions assuming no swelling or frost heave.

Step 3.2: Select a trial performance period that might be expected under the swelling/frost heave conditions anticipated. This number should be less than the maximum performance period corresponding to the pavement structural number selected in step 3.1.

Step 3.3: Use the nomograph in AASHTO Guide (Figure 2-2, p. II-11) to estimate the total serviceability loss due to swelling and frost heave ($\Delta PSI_{SW,\,FH}$) that can be expected for the trial performance period selected in step 3.2.

Step 3.4: Calculate the traffic serviceability loss (ΔPSI_{TR}).

$$\Delta PSI_{TR} = \Delta PSI - \Delta PSI_{SW,FH} \quad \textbf{(10.11)}$$

Step 3.5: Estimate the allowable cumulative 18-kip ESAL traffic corresponding to the traffic serviceability loss determined in step 3.4. Figure 3-1 (p. II-32) of the AASHTO Guide is used in this step.

Step 3.6: Estimate the year at which the cumulative 18-kip ESAL traffic from step 3.5 will be reached. This is accomplished with the aid of a cumulative-traffic-versus-time plot. See Figure 2.1 (p. II-8) of the AASHTO Guide for an example.

Step 3.7: Compare the trial performance period (step 3.2) with that calculated in step 3.6. If the difference is greater than one year, calculate the average of the two and use this average as the trial value for the start of the next iteration (return to step 3.2). If the difference is less than one year, convergence is achieved and the average is the predicted performance period of the initial pavement structure for the selected initial *SN*.

Step 4: Calculate layer thicknesses. In this step, a set of pavement layer thicknesses are determined that when combined will provide the load-carrying capacity corresponding to the design *SN*. Equation (10.208) is used to convert the *SN* to actual thicknesses of surfacing, base, and subbase courses. This equation does not have a unique solution; many combinations of layer thicknesses provide satisfactory solutions. The AASHTO Guide provides some guidance for determining a practical design. Sections 2.3.5 (p. II-17) and 2.4.1 (p. II-22) of the AASHTO Guide provide guidance concerning the selection of appropriate layer (a_i) and drainage (m_i) coefficients, respectively. Page II-35 of the AASHTO Guide suggests minimum practical thicknesses for each pavement course that should also be considered when determining layer thicknesses. Figure 3.2 (p. II-36) of the AASHTO Guide outlines a procedure for determining thicknesses of layers using a layered analysis approach.

The reader is referred to Appendix H of the AASHTO Guide for a comprehensive flexible pavement design example problem. Garber and Hoel (1997) also present an excellent overview of the AASHTO pavement design methods and present a number of example problems.

Rigid Pavements

Chapter 3, Section 3.2 (pp. II-37 to II-48) of the AASHTO Guide describes the rigid pavement design procedure. Sections 3.3 and 3.4 of the Guide address joint and reinforcement design, respectively. Appendix I of the AASHTO Guide presents a detailed rigid pavement design example problem. The following review of the basic steps in the AASHTO procedure is limited to the determination of the required slab thickness (*D*). As in the design of flexible pavements, the following procedure is applicable to pavements that are expected to carry traffic levels in excess of 50,000 18-kip ESALs over the design period.

Step 1: Determine effective modulus of subgrade reaction (k). The AASHTO Guide provides a worksheet to facilitate the determination of an appropriate value for *k* (see Table 3.2, p. II-38 of the AASHTO Guide). The determination of *k* involves eight steps and the use of a number of nomographs (see Figures 3-3 to 3-6, pp. II-39 to II-42 of the AASHTO Guide). These are outlined as follows:

Step 1.1: Determine levels of slab support to be considered in determining *k*. These include (1) subbase types, (2) subbase thicknesses, (3) loss of support due to erosion, and (4) depth to rigid foundation. A separate worksheet and corresponding *k*-value are prepared for each combination of these factors.

Step 1.2: Identify seasonal roadbed soil resilient modulus values (from Section 2.3.1, p. II-12 of the AASHTO Guide).

Step 1.3: Assign subbase elastic (resilient) modulus (E_{SB}) values for each season (from Section 2.3.3, p. II-16 of the AASHTO Guide).

Step 1.4: Estimate the composite *k*-value for each season. This is accomplished with the aid of Figure 3-3 (p. II-39) in the AASHTO Guide.

Step 1.5: Develop the k-value that includes effects of a rigid foundation near the surface (disregard if depth to rigid foundation is >10 ft). Figure 3-4 (p. II-40) in the Guide is used to estimate this modified k-value for each season.

Step 1.6: Estimate the required slab thickness and use Figure 3-5 (p. II-41) to determine the relative damage, u_r, in each season.

Step 1.7: Sum all the u_r values and divide by the number of seasonal increments (12 to 24) to determine the average relative damage. The effective modulus of subgrade reaction is the value corresponding to the average relative damage and projected slab thickness in Figure 3-5 (p. II-41).

Step 1.8: Adjust the effective modulus of subgrade reaction to account for potential loss of support from subbase erosion. This is accomplished with the aid of Figure 3-6 (p. II-42) in the AASHTO Guide.

Step 2: Determine required slab thickness. Figure 3-7 (pp. II-45 to II-46) is the nomograph used for determining slab thickness for each of the effective k-values determined in step 1. In addition to the design k-value, the nomograph requires the following inputs (section references refer to the AASHTO Guide):

1. Estimated future traffic, W_{18} (Section 2.1.2)
2. Reliability, R (Section 2.1.3)
3. Overall standard deviation, S_0 (Section 2.1.3)
4. Design serviceability loss, $\Delta PSI = p_0 - p_t$ (Section 2.2.1)
5. Concrete elastic modulus, E_c (Section 2.3.3)
6. Concrete modulus of rupture, S_c' (Section 2.3.4)
7. Load transfer coefficient, J (Section 2.4.2)
8. Drainage coefficient, C_d (Section 2.4.1)

If staged construction and roadbed swelling/frost heave are not factors in the analyses, then the thickness design procedure is considered complete at this point.

Step 3: Evaluate need for staged construction (optional). Same as step 2 for the flexible pavement design procedure.

Step 4: Evaluate the need to consider roadbed swelling and frost heave (optional). This step is almost identical to step 3 of the flexible pavement design procedure. For rigid pavements, the iterative process outlined in step 3 of the flexible pavement design procedure would begin with the selection of an appropriate slab thickness, D (instead of SN), and Figure 3-7 (pp. II-45 to II-46) is used to estimate the allowable cumulative 18-kip ESAL traffic corresponding to the computed traffic serviceability loss (step 4.5 of the process). As with the flexible pavement design procedure, the AASHTO Guide provides a worksheet to aid in the computations (see Table 3-4, p. II-48).

The reader is referred to Appendix I of the AASHTO Guide for a comprehensive example problem that illustrates the application of the rigid pavement design procedure. Garber and Hoel (1997) also present an excellent overview of the AASHTO (and other) pavement design methods and present a number of example problems.

Asphalt Institute Method

The Asphalt Institute (AI) method for thickness design of asphalt pavements is documented in *Thickness Design: Asphalt Pavements for Streets and Highways* (Manual Series No. 1 (MS-1), Asphalt Institute, Feb. 1991). The procedure is intended to allow the analyst to determine the minimum thickness of the asphalt layer to withstand the critical vertical compressive strains at the surface of the subgrade and the critical horizontal tensile strains at the bottom of the asphalt layer.

The procedures are applicable to the thickness design of pavement structures consisting of asphalt concrete surfaces, emulsified asphalt surfaces with surface treatment, asphalt concrete bases, emulsified asphalt bases, and untreated aggregate bases or subbases. The AI procedure makes extensive use of design charts in determining pavement thicknesses. The procedure consists of five main steps.

Step 1: Select or determine input data. The three basic design inputs are traffic (expressed as the total number of equivalent 18-kip single-axle load applications over the design period), the subgrade resilient modulus (M_r), and the surface and base types to be considered (asphalt concrete; emulsified asphalt mix Types I, II, or III; or untreated base or subbase). Determination of the design year axle loads consists of the following steps:

Step 1.1: Determine the average number of each type of vehicle expected on the design lane during the first year of traffic. In the absence of local traffic count and classification data, Table IV-1 (p. 13 of the AI Manual) can be used to estimate the distribution of trucks on different classes of highways. Table IV-2 (p. 14 of the AI Manual) can be used to estimate the relative proportion of trucks in the design lane, usually the outside lane for multilane roadways.

Step 1.2: Determine from local axle weight data, or select from Table IV-5 (p. 20 of the AI Manual) a truck factor for each vehicle type identified in step 1.1.

Step 1.3: Select a rate to forecast the design period traffic. This may be a single factor for all vehicles, or a separate factor for each vehicle type. Table IV-3 (p. 15 of the AI Manual) is a table of uniform series compound amount factors that can be used in this step.

Step 1.4: Multiply the number of vehicles of each type by the truck factor and the growth factor(s) from steps 1.2 and 1.3.

Step 1.5: Sum the values from step 1.4 to determine the design ESAL.

Step 1.6: The AI Manual suggests the following graphical procedure for determining the design subgrade resilient modulus, M_r. Test six to eight samples of the subgrade material and convert the CBR or R-values using the following relationships:

$$M_r \text{ (psi)} = 1500(\text{CBR}) \quad \textbf{(10.12)}$$

$$M_r \text{ (psi)} = 1155 + 555(R\text{-value}) \quad \textbf{(10.13)}$$

Step 1.7: Arrange the test values in descending numerical order.

Step 1.8: For each test value, compute the percentage of the total number of values that are equal to or greater than the test value.

Step 1.9: Plot the results from step 1.8.

Step 1.10: Fit a smooth curve through the plotted points. (The curve should be S-shaped with the 50th percentile close to the average of the sample data.)

Step 1.11: Read the design value of M_r from the curve at the percentile that corresponds to the anticipated traffic volume as indicated below.

ESAL	M_r
$<10^6$	60
10^4 to 10^6	75
10^6	87.5

Step 1.12: Determine the base types to be considered. These include asphalt concrete; emulsified asphalt Types I, II, or III; or untreated base or subbase. Guidelines for selecting the base type are discussed in Chapter II of the AI Manual.

Step 2: Determine design thickness. The design thickness is determined by entering the appropriate design chart in the AI Manual with the ESAL and M_r values previously selected. The design charts are stratified into three sets of temperature conditions (cold, warm, hot) that are typical of conditions throughout most of North America.

Step 3: Evaluate the need for staged construction (optional). The staged construction design method recommended in the AI Manual involves three steps: (1) first stage design, (2) preliminary design of second-stage overlay, and (3) final design of second stage overlay. Basically, the method involves reducing the design ESAL to account for the remaining life in preceding stages of the design. The AI Manual suggests that designs for equivalent axle loads in excess of 3×10^6 should be considered candidates for staged construction. The method is described in detail on pp. 41–43 of the AI Manual.

Step 4: Conduct an economic analysis of the design alternatives. The AI Manual recommends the use of present worth analysis to compare the design alternatives.

Step 5: Select final design. The selection of a base type or the decision to use staged construction is often based on the results of an economic analysis of the design alternatives.

Portland Cement Association Method

The Portland Cement Association (PCA) method for the thickness design of rigid pavements is documented in *Thickness Design for Concrete Highway and Street Pavements* (PCA, 1984). The PCA method consists of two parts: fatigue analysis and erosion analysis. The minimum thickness that satisfies both analyses is the design thickness.

The PCA Manual describes two design procedures: one procedure for use when axle load data are available, and a second, simplified procedure that can be used if axle load data are not available. The following review is limited to the first procedure.

Unlike the AASHTO and Asphalt Institute methods reviewed earlier in this section, the PCA method considers only trucks with six or more axles. Like the other two methods, application of the PCA method relies heavily upon design charts, tables, and nomographs. However, the required pavement thickness calculations can be completed on a single worksheet (see p. 47 of the PCA Manual for a blank worksheet).

The application of the PCA method is illustrated through numerous sample problems in the Manual. Garber and Hoel (1997) provide an excellent summary of the PCA method and also present numerous example applications.

The PCA method requires as inputs the following design factors: a trial pavement thickness; type of joint (doweled or undoweled) and shoulder (with or without concrete shoulder); concrete flexural strength (modulus of rupture, *MR*); *k*-value of the subgrade or subgrade and subbase combination; load safety factor (LSF); axle load distribution consisting of weights, frequencies, and types of truck axle loads that the pavement will carry; and the expected number of axle load repetitions during the design period (calculated by multiplying the axle load distribution by the LSF). These design factors are discussed in Chapter 2 of the PCA Manual.

The step-by-step procedure for the fatigue analysis phase of the PCA method is as follows:

Step 1: Determine the equivalent fatigue stress factors. These factors are determined by entering Table 6a (no concrete shoulders) or Table 6b (concrete shoulders) on page 14 of the PCA Manual with the appropriate trial thickness, *k*-value, and axle configuration (single or tandem).

Step 2: Calculate the fatigue stress ratio factors. The stress ratio factors are calculated for each axle configuration by dividing the equivalent stress factors from step 1 by the concrete modulus of rupture.

Step 3: Determine allowable axle load repetitions. The allowable repetitions are determined from Figure 5 (p. 15 of the PCA Manual) based on the stress ratio factor.

Step 4: Calculate total percent fatigue damage. The percent fatigue damage is computed by dividing the expected repetitions of each axle load class by the allowable repetitions, multiplying by 100%, and summing across all load and axle configuration classes.

Step 5: Determine the erosion factors. These factors are a function of shoulder and joint type, the trial thickness, and the *k*-value. Tables 7a, 7b, 8a, and 8b (pp. 16 and 18 in the PCA Manual) are used to determine the erosion factors.

Step 6: Determine allowable repetitions. Figure 6a (without shoulders) and 6b (with shoulders) on pp. 17 and 19 of the PCA Manual are used to determine allowable repetitions based on the erosion factors determined in step 5.

Step 7: Calculate total percent erosion damage. The percent erosion damage is computed by dividing the expected repetitions by the allowable repetitions, multiplying by 100%, and summing across all load and axle configuration classes.

Step 8: Assess adequacy of trial thickness. The trial thickness is not adequate if the total for either fatigue damage (step 4) or erosion damage (step 7) is greater than 100%. If either is greater than 100%, select a greater trial thickness and repeat the analysis. A lesser trial thickness should be selected if the totals are much lower than 100%. [The PCA Manual (p. 13) provides some guidance in reducing the number of interactions required to complete the analyses.]

CHAPTER 11

Fluid Mechanics

OUTLINE

Fluid mechanics is the study of fluids at rest or in motion. The topic is generally divided into two categories: *liquids* and *gases*. Liquids are considered to be incompressible, and gases are compressible. The treatment of incompressible fluids and compressible fluids each has its own group of equations. However, there are times when a gas may be treated as incompressible (or at least uncompressed). For example, the flow of air through a heating duct is one such case. This chapter will concentrate on incompressible fluids.

FLUID PROPERTIES

Thermodynamic properties are important in incompressible fluid mechanics. Those of particular importance are density, specific gravity, specific weight, viscosity, and pressure. Temperature is also important but is primarily used in finding other properties such as density and viscosity in tables or graphs.

Density

The **density**, ρ, is the mass per unit volume and is the reciprocal of the specific volume, a property used in thermodynamics:

$$\rho = \frac{m}{V} = \frac{1\,\text{kg}}{v\,\text{m}^3}$$

Specific Gravity

The **specific gravity,** SG, is defined by the following equation:

$$\text{SG} = \frac{\rho\left(\frac{\text{kg}}{\text{m}^3}\right)}{\frac{1000\ \text{kg}}{\text{m}^3}}$$

where 1000 kg/m3 is the density of water at 4°C.

In many cases the specific gravity of a liquid is known or found from tables and must be converted to density using this equation.

Specific Weight

The **specific weight**, γ, of a fluid is its weight per unit volume and is related to the density as follows:

$$\gamma = \frac{W}{V} = \rho\left(\frac{g}{g_c}\right)\frac{\text{N}}{\text{m}^3}$$

where g = local acceleration of gravity, $\frac{\text{m}}{\text{s}^2}$, and g_c = gravitational constant:

$$g_c = \frac{\text{kg}\bullet\text{m}}{\text{N}\bullet\text{s}^2}$$

The density of water at 4°C is 1000 kg/m^3. Its specific weight at sea level (g = 9.81 m/s^2) is calculated as follows:

$$\gamma = \rho\frac{g}{g_c} = 1000\frac{\text{kg}}{\text{m}^3} = \frac{9.81\frac{\text{m}}{\text{s}^2}}{\frac{\text{kg m}}{\text{N s}^2}} = 9810\frac{\text{N}}{\text{m}^3}$$

The density, specific gravity, and specific weight of a liquid are generally considered to be constant, with little variation, over a wide temperature range.

Viscosity

The **viscosity** of a fluid is a measure of its resistance to flow; the higher the viscosity the more resistance to flow. Water has a relatively low viscosity, and heavy fuel oils have a high viscosity. The **dynamic (absolute) viscosity,** μ, of a fluid is defined as the ratio of shearing stress to the rate of shearing strain. In equation form:

$$\mu = \frac{\tau}{\frac{dV}{dy}} \frac{\text{N} \bullet \text{s}}{\text{m}^2} \left(\frac{\text{kg}}{\text{m} \bullet \text{s}} \right)$$

where τ = shearing stress (force per unit area), N/m^2, and dV/dy = rate of shearing strain, $1/s$.

Fluids may be classified as Newtonian or non-Newtonian. Newtonian fluids are those in which dV/dy in the above equation can be considered to be constant for a given temperature. Thus the shearing stress, τ (horizontal force divided by the surface area), of a plate on a thin layer, δ, of a Newtonian fluid, as shown in Figure 11.1, may be found from

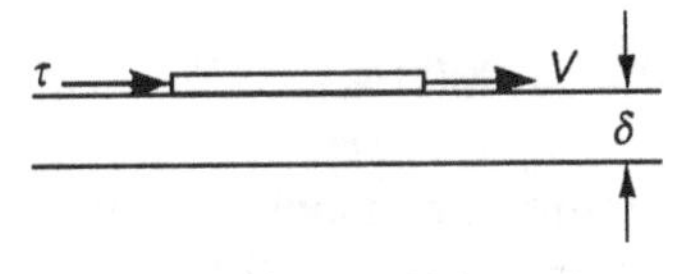

Figure 11.1

$$\tau = \mu \frac{dV}{dy} = \frac{\mu V}{\delta}$$

where V = velocity, m/s, and δ = thickness, m.

Most common fluids (liquids), such as water, oil, gasoline, and alcohol, are classified as Newtonian fluids.

The **kinematic viscosity** is defined by

$$\nu = \frac{\mu}{\rho} \frac{\text{m}^2}{\text{s}}$$

Both the dynamic and kinematic viscosities are highly dependent on temperature. The viscosity of most liquids decreases significantly (orders of magnitude) with increases in temperature, while the viscosity of gases increases mildly with increases in temperature. The viscosity of any gas is less than the viscosity of any liquid. Viscosities are generally found in tables and graphs.

The definition of viscosity assists in the development of the engineering definition of a fluid as follows:

A fluid is a substance that will deform readily and continuously when subjected to a shear force, no matter how small the force.

Pressure

Pressure, p, is the force per unit area of a fluid on its surroundings or vice versa. Pressure may be specified using two different datums. Absolute pressure, P_{abs}, is measured from absolute zero or a complete vacuum (void). At absolute zero there are no molecules and a negative absolute pressure does not exist. Absolute pressures are needed for ideal gas relations and in compressible fluid mechanics. Gage pressure, p_{gage}, on the other hand, uses local atmospheric pressure as its datum. Gage pressures may be positive (above atmospheric pressure) or negative (below atmospheric pressure). Negative gage pressure is also called vacuum. A complete vacuum occurs at a negative gage pressure that is equivalent to the atmospheric pressure or at absolute zero.

The relationship between absolute pressure and gage pressure is as follows:

$$P_{abs} = p_{gage} + p_{atm} \frac{N}{m^2} (Pa)$$

Actually, the pressure is usually expressed in kN/m^2 or kPa but should be converted to these units for use in most equations.

Example 11.1

A pressure gage measures 50 kPa vacuum in a system. What is the absolute pressure if the atmospheric pressure is 101 kPa?

Solution

Change vacuum to a negative gage pressure:

$$p_{abs} = -50 \text{ kPa} + 101 \text{ kPa} = 51 \text{ kPa} \bullet 1000 = 51{,}000 \text{ Pa}$$

Most pressure-measuring devices measure gage pressure. For incompressible fluid dynamics, gage pressure may be used in most equations. This capability simplifies equations significantly when one or more pressures in the system are atmospheric or $p_{gage} = 0$.

Surface Tension

Surface tension is another property of liquids. It is the force that holds a water droplet or mercury globule together, since the cohesive forces of the liquid are more than the adhesive forces of the surrounding air. The surface tension (or surface tension coefficient), σ, of liquids in air is available in tables and can be used to calculate the internal pressure, p, in a droplet from

$$p = \frac{4\sigma}{d}$$

where σ = surface tension of the liquid, kN/m, and d = droplet diameter, m. Values of surface tension for various liquids are found in tables as a function of the surrounding medium (air, etc.) and the temperature.

Surface tension is also the property that causes a liquid to rise (or fall) in a capillary tube. The amount of rise (or fall) depends on the liquid and the capillary tube material. When *adhesive* forces dominate, the liquid will rise—as with water. When cohesive forces dominate, it will fall—as with mercury. The capillary rise, h, can be calculated from the following equation:

$$h = \frac{4\sigma \cos\beta}{\gamma d}$$

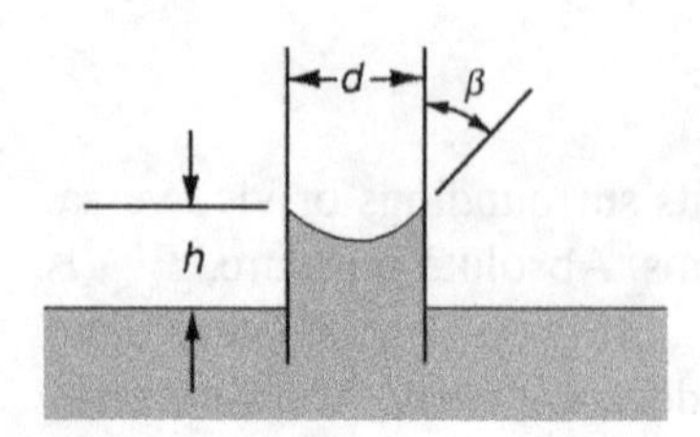

Figure 11.2

where β = angle made by the liquid with the tube wall, and d = diameter of capillary tube, as shown in Figure 11.2.

The angle, β, varies with different liquid/tube material combinations and is found in tables. β is within the range 0 to 180°. When $\beta > 90°$, h will be negative.

FLUID STATICS

Pressure-Height Relationship

For a static liquid, the pressure increases with depth (decreases with height) according to the following relationship

$$p_2 - p_1 = -\gamma(Z_2 - Z_1) = \gamma h$$

where h = depth from Point 1 to Point 2.

If p_1 is at the surface of a liquid that is open to the atmosphere, then the gage pressure at Point 2 is found from

$$p_2 = p = \gamma h$$

Example 11.2

Calculate the gage pressure at a depth of 100 meters in seawater, for which $\gamma = 10.1\ \text{kN/m}^3$.

Solution

$$p = \gamma h = \left(\frac{10.1\ \text{kN}}{\text{m}^3}\right)(100\ \text{m}) = 1010\frac{\text{kN}}{\text{m}^2} = 1010\ \text{kPa}$$

Manometers

A manometer is a device used to measure moderate pressure differences using the pressure-height relationship. The simplest manometer is the U-tube shown in Figure 11.3. The pressure difference between System 1 and System 2 is found from

$$p_1 - p_2 = \gamma_m h_m + \gamma_2 h_2 - \gamma_1 h_1$$

where γ_m, γ_1, and γ_2 = specific weight of manometer fluid, fluid in System 1, and fluid in System 2, respectively, and h_m, h_1, h_2 = depths as shown. If the fluids in both systems are gases and the manometer fluid is any liquid, then $\gamma_m >> \gamma_1$ or γ_2 and the equation simplifies to

$$p_1 - p_2 = \gamma_m h_m = \gamma h$$

If System 2 were the atmosphere ($p_2 = 0_{gage}$) then

$$p_1 = \gamma h = \text{gage pressure in System 1}$$

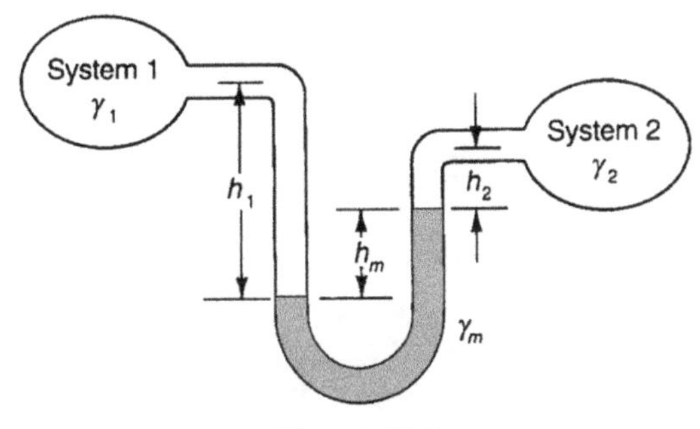

Figure 11.3

If System 1 were the atmosphere ($p_1 = 0$) then

$$p_2 = -\gamma h = \text{gage pressure in System 2}$$

The gage pressure in System 2 would be negative or a vacuum. Manometers are commonly used to measure system pressures between –101.3 kPa and +101.3 kPa. In many cases, particularly where the gage pressure is negative, the pressure may be given in millimeters of a fluid, and the equation above used to convert it to standard units.

Example **11.3**

A system gage pressure is given as 500 millimeters of mercury vacuum (mm Hg vac). What is the gage pressure in kPa? The specific gravity of mercury is 13.6.

Solution

The pressure is $p = p_2 = -\gamma h$ since vacuum is a negative gage pressure:

$$\gamma_m = (13.6)\left(9.81\frac{\text{kN}}{\text{m}^3}\right) = 133.4\frac{\text{kN}}{\text{m}^3}$$

$$p = -\gamma h = -133.4\frac{\text{kN}}{\text{m}^3} \times 0.5\,\text{m} = -66.7\text{ kPa}$$

The conversion factor from millimeters of mercury to N/m^2 (pascals) is 133.4.

A barometer is a special type of mercury manometer. In this case one leg of the U-tube is very wide. If we can adjust the scale on the narrow leg so that zero is at the level of the large leg, then the narrow leg will read the atmospheric pressure impinging on the wide leg corrected by the vapor pressure of the mercury. A barometer is shown in Figure 11.4.

There are several other types of manometers. A compound manometer consists of more than one U-tube in series between one system and another. The equation for $p_1 - p_2$ may be developed by starting at System 2 and adding γh's going downward and subtracting γh's going upward until System 1 is reached as follows:

$$p_2 + \sum \gamma h\text{ (downward)} - \sum \gamma h\text{ (upward)} = p_1$$

An inclined manometer is used to measure small pressure differentials. The measurement along the manometer must be multiplied by the sine of the angle of incline. An inclined manometer is generally "single leg," similar to the barometer previously described, and is shown in Figure 11.5. The pressure difference is found from $p_1 - p_2 = \gamma_m L \sin \alpha$, where L = length along manometer leg and α = angle of inclination.

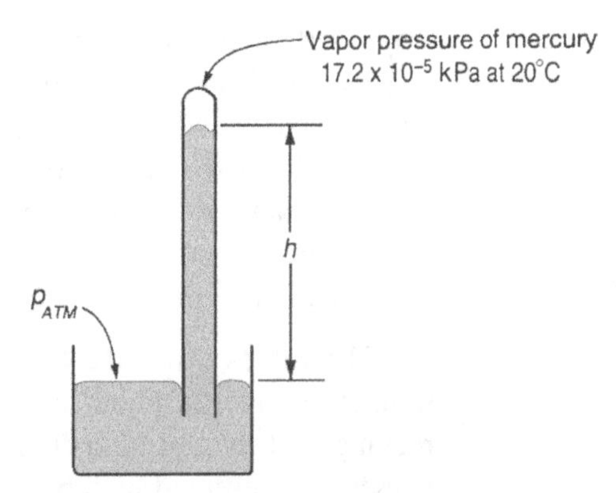

Figure 11.4

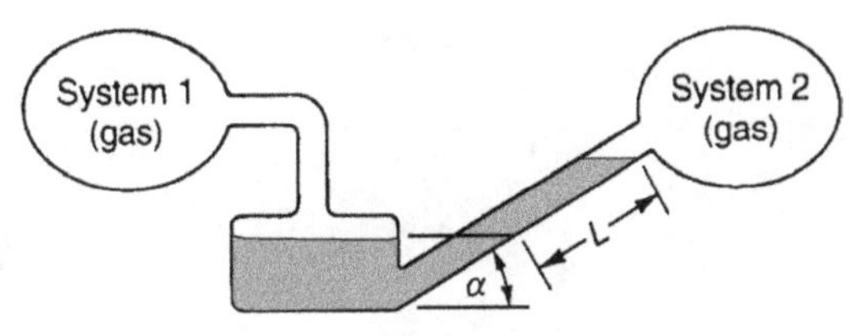

Figure 11.5

Forces on Flat Submerged Surfaces

A flat surface of arbitrary shape below a liquid surface is shown in Figure 11.6. Thew resultant force, F, on one side of the flat surface acts perpendicular to the surface. Its magnitude and location may be calculated from the following equations:

$$F = (p_0 + \gamma h_c)A$$

and

$$h_p = h_c + \frac{I_c \sin^2 \alpha}{\left(\frac{p_0}{\gamma} + h_c\right)A}$$

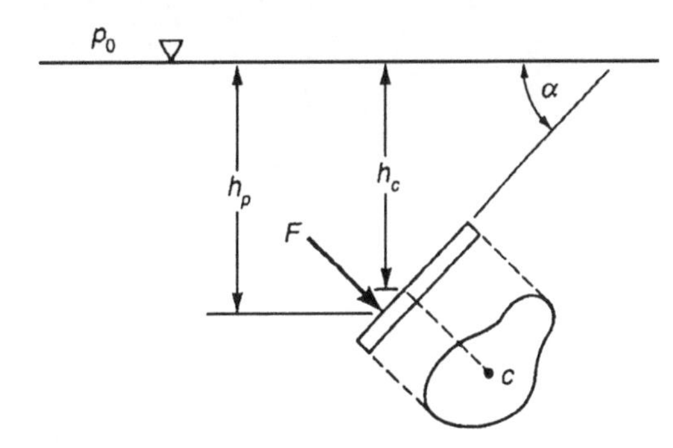

Figure 11.6

where

F = resultant force on the flat surface, N
p_0 = gage pressure on the surface, Pa
γ = specific weight of the fluid, N/m^3

h_c = vertical distance from fluid surface to the centroid of the flat surface area, m
A = area of flat surface, m^2
h_p = vertical distance from fluid surface to the center of pressure of the flat surface (where the equivalent, concentrated force acts), m
I_c = moment of inertia of the flat surface about a horizontal axis through its centroid, m^4
α = angle that the inclined flat surface makes with the horizontal surface

The values of h_c, the location of the centroid axis from the base $\bar{y}$, and the moment of inertia about the centroid I_c for common geometric shapes, such as rectangles, triangles, and circles, may be determined from existing tables. Typical values are presented in Table 11.1.

For the common case when p_0 is atmospheric pressure ($p_0 = 0$), the equations simplify to

$$F = \gamma h_c A$$

and

$$h_p = h_c + \frac{I_c \sin^2 \alpha}{h_c A}$$

From the above equations it is apparent that the center of pressure is always below the centroid except when the surface is horizontal ($\alpha = 0$). In that case the center of pressure is at the centroid. The deeper the flat surface is located below the fluid surface, the closer the center of pressure is to the centroid.

The pressure profile on the flat surface is generally trapezoidal (triangular, if the flat surface pierces the surface of a fluid exposed to atmospheric pressure). The slope of the pressure profile is equivalent to the specific weight of the fluid. Examples are shown in Figure 11.7.

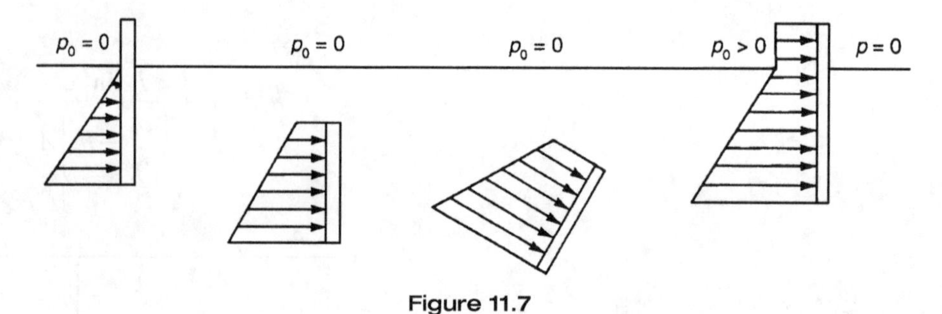

Figure 11.7

Table 11.1 Areas, centroids, and moments of inertia for selected areas

Section	Area of Section, A	Distance to Centroidal Axis, $\bar{y}$	Moment of Inertia about Centroidal Axis, I_c
Rectangle 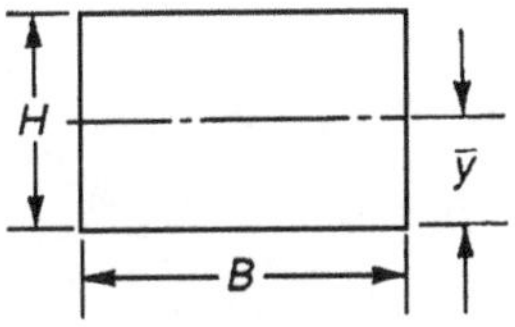	BH	$H/2$	$BH^3/12$
Triangle 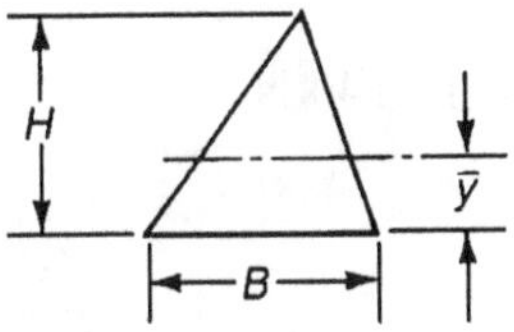	$BH/2$	$H/3$	$BH^3/36$
Circle 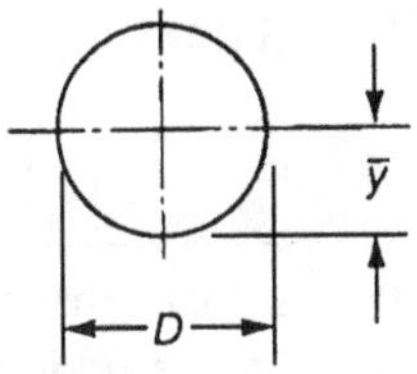	$\pi D^2/4$	$D/2$	$\pi D^4/64$
Ring 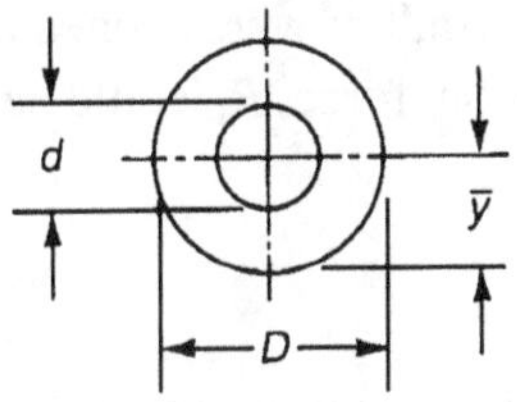	$\dfrac{\pi(D^2-d^2)}{4}$	$D/2$	$\dfrac{\pi(D^4-d^4)}{64}$
Semicircle 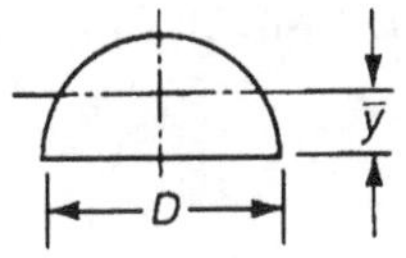	$\pi D^2/8$	$0.212D$	$(6.86 \times 10^{-3})\, D^4$
Quadrant 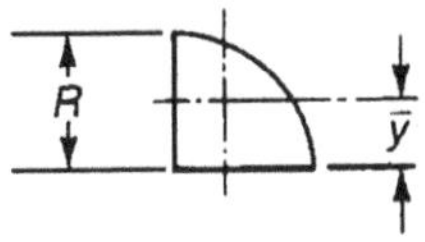	$\pi D^2/16$ $\pi R^2/4$	$0.212D$ $0.424R$	$(3.43 \times 10^{-3})D^4$ $(5.49 \times 10^{-2})R^4$
Trapezoid 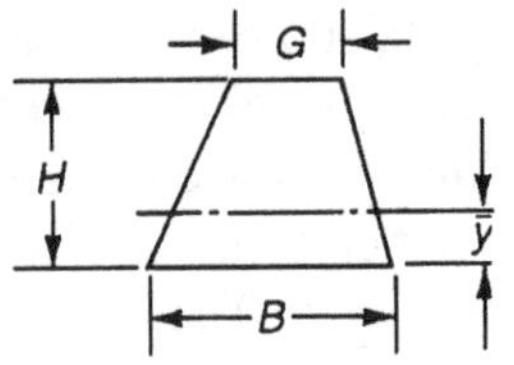	$\dfrac{H(G+B)}{2}$	$\dfrac{H(2G+B)}{3(G+B)}$	$\dfrac{H^3(G^2+4GB+B^2)}{36(G+B)}$

Example 11.4

A vertical side of a saltwater tank contains a round viewing window 60 cm in diameter with its center 5 meters below the liquid surface. If the specific weight of the saltwater is 10 kN/m³, find the force of the water on the window and where it acts.

Solution

Assume atmospheric pressure on the liquid surface, $p_0 = 0$.

$$d = 60\,\text{cm} = 0.6\text{ m}, \quad A = \frac{\pi(0.6)^2}{4} = 0.283\text{ m}^2$$

$$F = \gamma h_c A = 10\frac{\text{kN}}{\text{m}^3} \times 5\text{ m} \times 0.283\text{ m}^2 = 14.14\text{ kN}$$

$$I_c = \frac{\pi d^4}{64} = \frac{\pi(0.6\text{ m})^4}{64} = 0.00636\text{ m}^4$$

$$\alpha = 90°, \quad \sin\alpha = 1$$

$$h_p = h_c + \frac{I_c \sin^2\alpha}{h_c A} = 5\text{ m} + \frac{0.00636\text{ m}^4(1)^2}{5\text{ m}\bullet 0.283\text{ m}^2} = 5.0045\text{ m}$$

Example 11.5

In many cases problems involving fluid forces on flat surfaces are combined with a statics problem. The fluid force is just another force to be added into the statics equation.

In the previous example, suppose the circular window were hinged at the top with some sort of clamp at the bottom (Exhibit 1). What force, P, would be required of the clamp to keep the window closed?

Solution

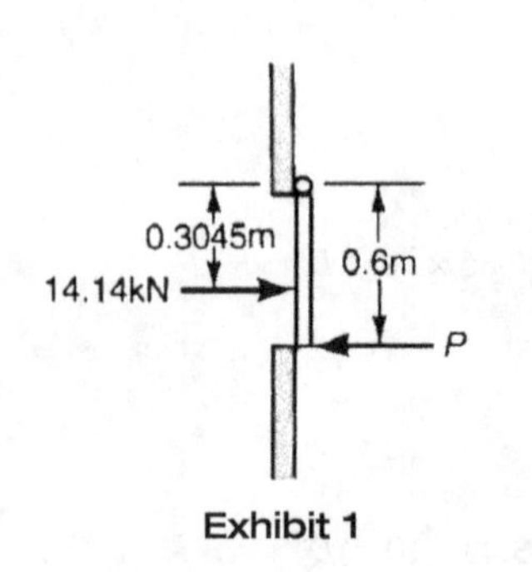

Exhibit 1

From Example 11.4 calculations, the force of the water is 14.14 kN located 5.0045 m below the fluid surface. The hinge is located 5 m – 0.6 m/2 = 4.7 m below the water surface. Thus the force location is 5.0045 m – 4.7 m = 0.3045 m below the hinge.

Summing moments about hinge,

$$\sum M_H = 0.6\text{ m}\bullet P - 0.3045\text{ m}\bullet 14.14\text{ kN} = 0$$

$$P = \frac{0.3045\text{ m}\bullet 14.14\text{ kN}}{0.6\text{ m}} = 7.18\text{ kN}$$

Buoyancy

In addition to the force of gravity, or weight, all objects submerged in a fluid are acted on by a buoyant force, F_B. The buoyant force acts upward and is equal to the weight of the fluid displaced by the object. This is known as Archimedes' principle. The upward buoyant force also acts through the center of gravity (or centroid) of the displaced volume, known as the center of buoyancy, B. Thus

$$F_B = \gamma_f V_D$$

where F_B = buoyant force, N; γ_f = specific weight of the fluid, N/m³; and V_D = volume displaced by the object, m³.

For a freely floating object (no external forces) the weight of the object (acting downward) is equal to the buoyant force on the object (acting upward) or

$$W = F_B = \gamma_f V_D$$

This equation is useful in determining what part of an object will float below the surface of a liquid. For objects partially submerged in a liquid and a gas, the buoyant force of the gas is usually neglected. However, the buoyant force on a totally submerged body in a gas is very important in the study of balloons, dirigibles, and so on.

Example 11.6

A wooden cube that is 15 centimeters on each side with a specific weight of 6300 N/m³ is floating in fresh water ($\gamma = 9{,}810$ N/m³) (Exhibit 2). What is the depth of the cube below the surface?

Solution

$$W = F_B = \gamma_f V_D$$

There are actually two buoyant forces on the cube, that of the water on the volume below the surface and that of the air on the volume above the surface. Neglecting the buoyant force of the air and rearranging the equation:

$$V_D = \frac{W}{\gamma_f} = \frac{\gamma_C V_C}{\gamma_f} = \frac{(6300\ \text{N/m}^3)(0.15\ \text{m})^3}{9810\ \text{N/m}^3} = 0.00217\ \text{m}^3$$

$$V_D = (0.15)^2 \bullet d = 0.00217\ \text{m}^3$$

$$d = \frac{0.00217\ \text{m}^3}{.0225\ \text{m}^2} = .0964\ \text{m} = 9.64\ \text{cm}$$

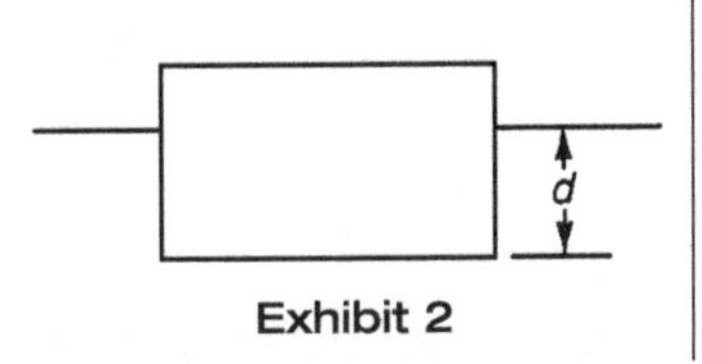

Exhibit 2

Neutral buoyancy exists when the buoyant force equals the weight when an object is completely submerged. The object will remain at whatever location it is placed below the fluid surface.

In the case of an object floating at the interface of two liquids, the total buoyant force is equal to the sum of the buoyant forces on the object created by each fluid on that part that is immersed. When external forces also act on a submerged or partially submerged object, they must be included in the force balance on the object. The force balance equation then becomes

$$W + \sum F_{ext}(\text{down}) = F_B + \sum F_{ext}(\text{up})$$

If weight is added to an object internally, or possibly on top of a partially submerged object, it will only affect the weight of the object. But if the weight is added externally, beneath the surface of the fluid, its buoyant force as well as its weight must be considered.

Example 11.7

If, in Example 11.6, a concrete weight (anchor) is added to the bottom of the cube externally, what anchor volume, V_A, would be required to make the cube float neutrally (below the surface). The specific weight of the concrete, γ_c, is 24 kN/m³.

Solution

Let the subscript C denote the properties of the cube and subscript A denote those of the anchor. Summing forces vertically,

$$W_C + W_A = F_{BC} + F_{BA}$$

$$\gamma_C V_C + \gamma_A V_A = \gamma_f V_D + \gamma_f V_A$$

Solving for V_A,

$$V_A = \frac{\gamma_f V_D - \gamma_C V_C}{\gamma_A - \gamma_f}$$

But for neutral buoyancy, the displaced volume, V_D, is equal to the total volume of the cube, V_C, and

$$V_A = \frac{(\gamma_f - \gamma_C) \bullet V_C}{\gamma_A - \gamma_f} = \frac{(9810 - 6300)\dfrac{\text{N}}{\text{m}^3} \bullet (0.15\text{ m})^3}{(24{,}000 - 9810)\dfrac{\text{N}}{\text{m}^3}} = 8.34 \times 10^{-4}\text{ m}^3$$

THE FLOW OF INCOMPRESSIBLE FLUIDS

The Continuity Equation

Most problems in fluid mechanics involve steady flow, meaning that the amount of mass in a system does not change with time. This is generally written as

$$\dot{m}_1 = \dot{m}_2 = \dot{m} = \text{mass rate}$$

where the subscript 1 denotes the entrance and the subscript 2 denotes the exit of the system. The mass rate may be written in terms of fluid properties:

$$\dot{m} = \rho A V$$

where ρ = fluid density, A = cross sectional area of flow, and V = average velocity of the fluid. Thus,

$$\rho_1 A_1 V_1 = \rho_2 A_2 V_2$$

and, since $\rho_1 = \rho_2 = \rho$ for an incompressible fluid, then

$$A_1 V_1 = A_2 V_2 = Q = \text{volume flow rate}$$

This equation is useful in determining one velocity when another velocity in the system is known.

Example **11.8**

Water is flowing in a 5 centimeter diameter pipe at a velocity of 5 m/s (Exhibit 3). The pipe expands to a 10-centimeter diameter pipe. Find the velocity in the 10-centimeter diameter pipe and the flow rate in liters (L) per minute.

Solution

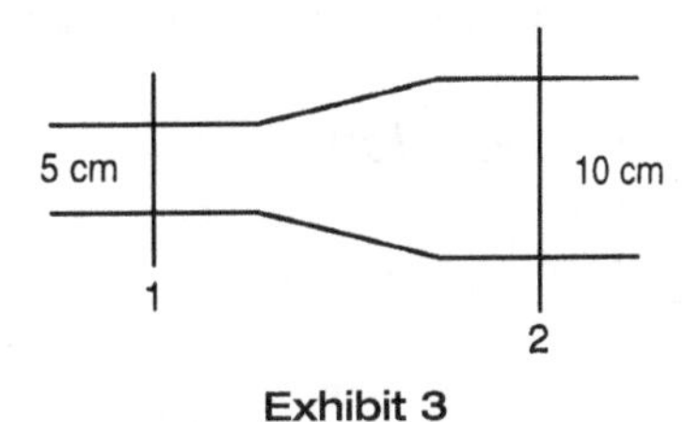

Exhibit 3

$$A_1V_1 = A_2V_2$$

$$V_2 = \frac{A_1}{A_2}V_1 = \frac{\pi d_1^2/4}{\pi d_2^2/4}V_1 = \left(\frac{d_1}{d_2}\right)^2 V_1 = \left(\frac{5\text{ cm}}{10\text{ cm}}\right)^2\left(5\frac{\text{m}}{\text{s}}\right) = 1.25\text{ m/s}$$

$$Q = A_1V_1 = \frac{\pi d_1^2}{4}V_1 = \frac{\pi(.05\text{ m})^2}{4}\bullet 5\frac{\text{m}}{\text{s}} = 0.00982\frac{\text{m}^3}{\text{s}}$$

$$Q = 0.00982\frac{\text{m}^3}{\text{s}}\frac{1000\text{ L}}{\text{m}^3}\bullet\frac{60\text{ s}}{\text{min}} = 589\text{ L/min}$$

In most cases the flow area may be calculated using the diameter. In some cases the nominal pipe size is known, such as a 4-inch Schedule 40 pipe. The exact inside dimensions of Schedule 40 and other pipes as well as dimensions for steel and copper tubing can be found in existing tables.

Reynolds Number

The **Reynolds number**, Re, is a dimensionless flow parameter that helps describe the nature of flow. It is sometimes defined as the ratio of dynamic forces to viscous forces. In terms of other fluid properties, it is defined as

$$\text{Re} = \frac{\rho V d}{\mu} = \frac{Vd}{v}$$

where V = fluid velocity, and d = characteristic length (diameter for pipes).

If the Reynolds number is below 2300, flow is laminar and occurs in layers with no mixing of adjacent fluid. Re = 2300 is known as the critical Reynolds number. Above the critical Reynolds number mixing begins to occur, and the flow becomes turbulent. As the Reynolds number increases, the flow becomes more turbulent.

For pipe flow with a circular cross section the Reynolds number may also be calculated from

$$\text{Re} = \frac{4\rho Q}{\pi d\mu} = \frac{4Q}{\pi d v}$$

The Reynolds number is a significant indicator of the influence of friction on the flow that occurs in pipes and other conduits as well as through flow meters. It is also important in the application of dynamic similarity to modeling and many other areas of fluid mechanics.

Example 11.9

For the pipe in Example 11.8, calculate the Reynolds number in the 5-centimeter diameter section of pipe. The kinematic viscosity of the water is 1.12×10^{-6} m²/s.

Solution

$$\mathrm{Re} = \frac{Vd}{\nu} = \frac{5\frac{\mathrm{m}}{\mathrm{s}} \bullet (0.05\ \mathrm{m})}{1.12 \times 10^{-6}\ \frac{\mathrm{m}^2}{\mathrm{s}}} = 2.2 \times 10^5$$

The flow is well into the turbulent regime.

The Energy Equation

The energy equation in fluid mechanics is similar to that used in thermodynamics. Instead of each energy term having the traditional units such as kJ/kg, energy is expressed in meters of head. For instance, kinetic energy is called velocity head. The general energy equation between two points in a system for incompressible steady flow (mass and energy in the system or at a point do not vary with time) is given by the following expression:

$$\frac{p_1}{\gamma} + \frac{V_1^2}{2g} + Z_1 + h_A - h_R = \frac{p_2}{\gamma} + \frac{V_2^2}{2g} + Z_2 + h_f$$

where

$\frac{p_1}{\gamma}, \frac{p_2}{\gamma}$ = pressure heads at Points 1 and 2

$\frac{V_1^2}{2g}, \frac{V_2^2}{2g}$ = velocity heads at Points 1 and 2

Z_1, Z_2 = potential or elevation heads at Points 1 and 2

h_A, h_R = the head added (pump) or removed (turbine) mechanically

h_f = head loss from friction in the pipe and fittings between Points 1 and 2

The energy equation in fluid mechanics assumes no heat transfer or changes in temperature. This equation, including its reduced forms, will solve most energy-related problems in fluid mechanics when used in conjunction with the continuity equation.

Bernoulli's Equation

Whereas Bernoulli's equation is usually derived from momentum principles using vector calculus, it can also be produced from the energy equation by introducing two additional restrictions to those of incompressible, steady flow and no heat transfer. If we restrict the energy equation to systems with no mechanical energy addition or removal (no pump or turbine) and with no (or negligible) friction losses, Bernoulli's equation is produced,

$$\frac{p_1}{\gamma}+\frac{V_1^2}{2g}+Z_1=\frac{p_2}{\gamma}+\frac{V_2^2}{2g}+Z_2$$

Bernoulli's equation can be used to solve a variety of problems.

Example **11.10**

Referring again to Example 11.8, calculate the pressure just after the expansion to the 10-centimeter diameter pipe if the pressure in the 5-centimeter pipe is 300 kPa. Friction is negligible. The specific weight of water is 9.81 kN/m³.

Solution

For the horizontal orientation $Z_1 = Z_2$ and Bernoulli's equation reduces to

$$\frac{p_1}{\gamma}+\frac{V_1^2}{2g}=\frac{p_2}{\gamma}+\frac{V_2^2}{2g}$$

$$p_2=p_1+\gamma\left(\frac{V_1^2-V_2^2}{2g}\right)$$

$$p_2=300\text{ kPa}+9.81\frac{\text{kN}}{\text{m}^3}\left[\frac{(5^2-1.25^2)\frac{\text{m}^2}{\text{s}^2}}{2\bullet 9.81\frac{\text{m}}{\text{s}^2}}\right]=311.7\text{ kPa}$$

Other Forms of the Energy Equation

An important rearrangement of the energy equation is to solve for the head added by a pump or removed by a turbine. For the head added by a pump:

$$h_A=\frac{p_2-p_1}{\gamma}+\frac{V_2^2-V_1^2}{2g}+Z_2-Z_1+h_f$$

Example **11.11**

A pump is being used to deliver 130 L/min of hot water from a tank through 15 meters of 2.5-cm diameter, smooth pipe, exiting through a 1.0 cm diameter nozzle 3 meters above the level of the tank as shown in Exhibit 4. The head loss from friction of the pipe is 8.33 m. The specific weight of the hot water is 9.53 kN/m³. Calculate the head delivered to the water by the pump.

Solution

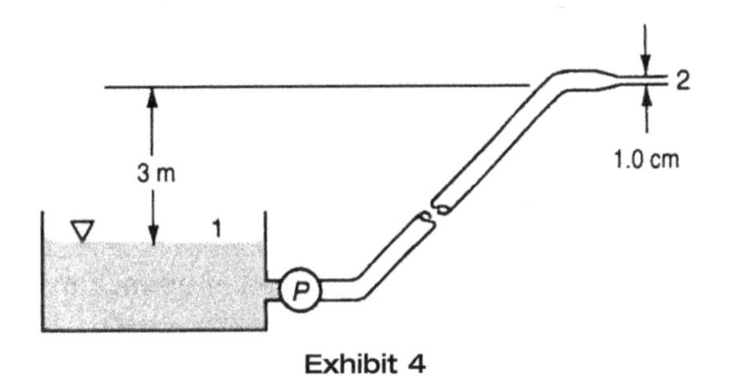

Exhibit 4

Select Points 1 and 2 as shown.

$$h_A = \frac{p_2 - p_1}{\gamma} + \frac{V_2^2 - V_1^2}{2g} + Z_2 - Z_1 + h_f$$

$$p_2 = p_1 = 0, \quad V_1 = 0, \quad Z_2 = 3\text{ m}, \quad Z_1 = 0, \quad h_f = 8.33\text{ m}$$

$$V_2 = \frac{Q}{A_2} = \frac{130\frac{\text{L}}{\text{min}} \bullet \frac{1\text{ m}^3}{1000\text{ L}} \bullet \frac{\text{min}}{60\text{ s}}}{\frac{\pi(.01)^2}{4}\text{m}^2} = 27.6\frac{\text{m}}{\text{s}}$$

$$h_A = \frac{\left(27.6\frac{\text{m}}{\text{s}}\right)^2}{2 \bullet 9.81\frac{\text{m}}{\text{s}^2}} + 3\text{ m} + 8.33\text{ m} = 50.2\text{ m}$$

In this problem the head of the pump serves three purposes: to increase the velocity of the water, raise its level, and overcome friction.

Pump and Turbine Power and Efficiency

The power delivered by a pump to a fluid or removed by a turbine from the fluid is given by the following:

$$P = \gamma Q h_A = \gamma Q h_R$$

The term γQ is the weight rate of flow. In the SI system, the units of power will usually be kN-m/s or kilowatts.

In selecting a pump or turbine, its efficiency is important. The efficiency may be calculated from

$$\eta_P = P/\dot{W} \bullet 100$$

$$\eta_T = \dot{W}/P \bullet 100$$

where η_p, η_T = pump and turbine efficiency, respectively, %, P = fluid power, and $\dot{W}$ = mechanical (or shaft) power actually delivered to the pump or by the turbine.

Example 11.12

From Example 11.11, calculate the power delivered to the water by the pump. If the efficiency of the pump is 60%, calculate the mechanical power delivered to the pump ($\dot{W}$).

Solution

$$P = \gamma Q h_A = 9.53\frac{\text{kN}}{\text{m}^3} \bullet 130\frac{\text{L}}{\text{min}} \bullet \frac{1\text{m}^3}{1000\text{L}} \bullet \frac{\text{min}}{60\text{ s}} \bullet 50.2\text{ m} = 1.04\text{ kW}$$

$$\eta = \frac{P}{\dot{W}} \bullet 100 \quad \text{or} \quad \dot{W} = \frac{P \bullet 100}{\eta} = \frac{1.04\text{ kW} \bullet 100}{60} = 1.73\text{ kW}$$

Head Loss from Friction in Pipes

Most of the terms in the energy equation will be known or calculated from the energy equation in conjunction with the continuity equation. Even the head loss from friction may be calculated if all other parameters are known. For example, if one wished to know the friction loss in a particular horizontal length of pipe, or in a fitting with equal entrance and exit areas, it could be calculated from the energy equation. Thus, $Z_1 = Z_2$, and since $A_1 = A_2$, then $V_1 = V_2$ and the energy equation becomes

$$\frac{p_1}{\gamma} = \frac{p_2}{\gamma} + h_f$$

If the pressure drop ($p_1 - p_2$) were known or measured, then

$$h_f = \frac{p_1 - p_2}{\gamma}$$

For most applications of the energy equation, the head loss from friction must be known and substituted into the energy equation to solve for an unknown pressure or height, pump or turbine head, or flow rate. For pipe flow, the head loss may be calculated from the Darcy equation:

$$h_f = f\frac{L}{d}\frac{V^2}{2g}$$

where h_f is the head loss from friction in a pipe of length L and diameter d, and f is the friction factor that is a function of the Reynolds number, Re, and the pipe relative roughness, ε/d.

The friction factor, f, can be found from the Moody diagram where f is plotted as a function of the Reynolds number and appears as a family of curves for different values of relative roughness, ε/d. *Relative roughness* is the roughness factor of the pipe, ε, divided by the pipe diameter, d. Typical roughness factors are shown in Table 11.2. Glass and plastic have the smallest roughness factors and are shown by the "smooth" curve on the Moody diagram. The Moody diagram is presented in Figure 11.8.

For Reynolds numbers below 2300 (laminar flow) the friction factor is independent of the relative roughness and may be calculated from the following relationship:

$$f = \frac{64}{\text{Re}}$$

Table 11.2 Moody pipe roughness

Material	Roughness, ε (m)
Glass, plastic	Smooth
Copper, brass, lead (tubing)	1.5×10^{-6}
Cast iron—uncoated	2.4×10^{-4}
Cast iron—asphalt coated	1.2×10^{-4}
Commercial steel or welded steel	4.6×10^{-5}
Wrought iron	4.6×10^{-5}
Riveted steel	1.8×10^{-3}
Concrete	1.2×10^{-3}

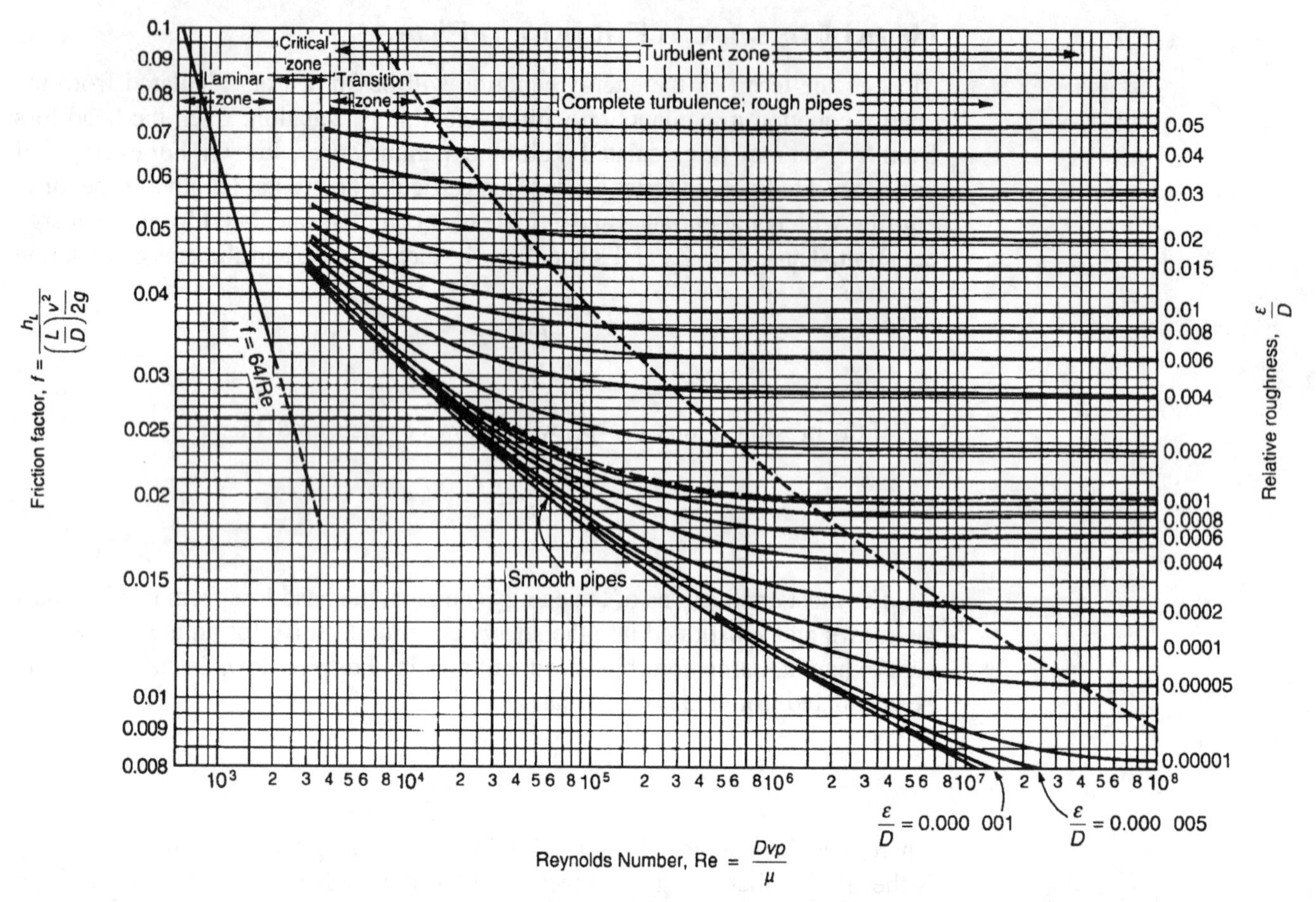

Figure 11.8 Moody diagram used with permission from *Transactions of the ASME*, 1944, vol. 66 by L. F. Moody

Substituting this value into the Darcy equation allows direct calculation of the head loss for laminar flow:

$$h_f = \frac{32\mu LV}{\gamma d^2}$$

This relation is known as the Hagen-Poiseuille equation. It may also be written in terms of flow rate, Q, where $V = Q/A = 4Q/\pi d^2$. Then

$$h_f = \frac{128\mu LQ}{\pi\gamma d^4}$$

Thus, the head loss from friction for laminar flow can be seen to be proportional to flow rate and inversely proportional to the diameter raised to the fourth power. If the diameter is doubled, and the flow remains laminar, the head loss will decrease by a factor of 16.

If the head loss from friction for turbulent flow is rewritten in terms of flow rate, Q, substitution of $4Q/\pi d^2$ for V yields

$$h_f = \frac{8fLQ^2}{\pi^2 gd^5}$$

Thus, the head loss from friction for turbulent flow is approximately proportional to the square of the flow rate and approximately inversely proportional to the diameter raised to the fifth power. If the diameter is doubled, the head loss will decrease by a factor of approximately 32. The proportionality is approximate because the friction factor, f, may change slightly with changes in flow rate and diameter.

Example 11.13

From Example 11.11 using the energy equation, the head loss from friction for 15 meters of 2.5-cm-diameter smooth pipe was given as 8.33 m. Show how this value was calculated. The dynamic viscosity, v, is 3.56×10^{-7} m²/s.

Solution

$$h_f = f\frac{L}{d}\frac{V^2}{2g}, \quad L = 15 \text{ m}, \ d = 2.5 \text{ cm} = .025 \text{ m}$$

$$V = \frac{Q}{A} = \frac{130\frac{\text{L}}{\text{min}} \bullet \frac{1 \text{ m}^3}{1000 \text{ L}} \bullet \frac{\text{min}}{60 \text{ s}}}{\frac{\pi(.025)^2}{4}\text{m}^2} = 4.41 \text{ m/s}$$

$$\text{Re} = \frac{Vd}{v}\frac{4.41\frac{m}{s} \bullet .025 \text{ m}}{3.56 \times 10^{-7}\frac{\text{m}^2}{\text{s}}} + 3.1 \times 10^5$$

Now enter the Moody diagram for this Reynolds number, and reflect off the ε/d = smooth line to read $f = 0.014$.

Hence,

$$h_f = 0.014 \bullet \frac{15 \text{ m}}{.025 \text{ m}} \bullet \frac{\left(4.41\frac{\text{m}}{\text{s}}\right)^2}{2 \bullet 9.81\frac{\text{m}}{\text{s}^2}} = 8.33 \text{ m}$$

Minor Losses

Flow losses from friction in pipe fittings, contractions, and enlargements are collectively known as minor losses. In problems where the pipe length is large, the minor losses in the system may be neglected. Minor losses are generally denoted by one of three methods:

(1) An equivalent length of pipe, L_e, is chosen, and the amount of the loss is calculated from

$$h_f = f\frac{L_e}{d}\frac{V^2}{2g}$$

(2) An equivalent length of pipe in diameters, $(L/d)_e$, is chosen, and the amount of loss is calculated from

$$h_f = f\left(\frac{L}{d}\right)_e\frac{V^2}{2g}$$

(3) A loss coefficient, C (or K), is chosen, and the amount of loss is calculated from

$$h_f = C\frac{V^2}{2g}$$

In more recent years the use of loss coefficients (C values) has become predominant and will be used in Fundamentals of Engineering exams. The C values may be a function of flow rate, fitting geometry, and/or diameter ratio (as in the case of contractions or enlargements) but in many cases may be considered constant over a wide range of conditions.

The equivalent lengths and C values for fittings and contractions are tabulated in catalogues, handbooks, and manuals. Representative values are listed in Table 10.3 and are illustrated in Figure 11.9.

The total friction loss in a piping system is the sum of the pipe losses and all minor losses.

Table 11.3 Resistance in valves and fittings expressed as equivalent length in pipe diameters $(L/d)_e$, and loss coefficient, C

Type	Equivalent Length in Pipe Diameters, $(L/d)_e$	Loss Coefficient, C
Globe valve—fully open	340	6.80
Angle valve—fully open	145	2.90
Gate valve—fully open	13	0.26
Check valve—swing type	135	2.70
Check valve—ball type	150	3.00
Butterfly valve—fully open	40	0.80
90° standard elbow	30	0.60
90° long-radius elbow	20	0.40
90° street elbow	50	1.00
45° standard elbow	16	0.32
45° street elbow	26	0.52
Close return bend	50	1.00
Standard tee with flow-through run	20	0.40
Standard tee with flow-through branch	60	1.20

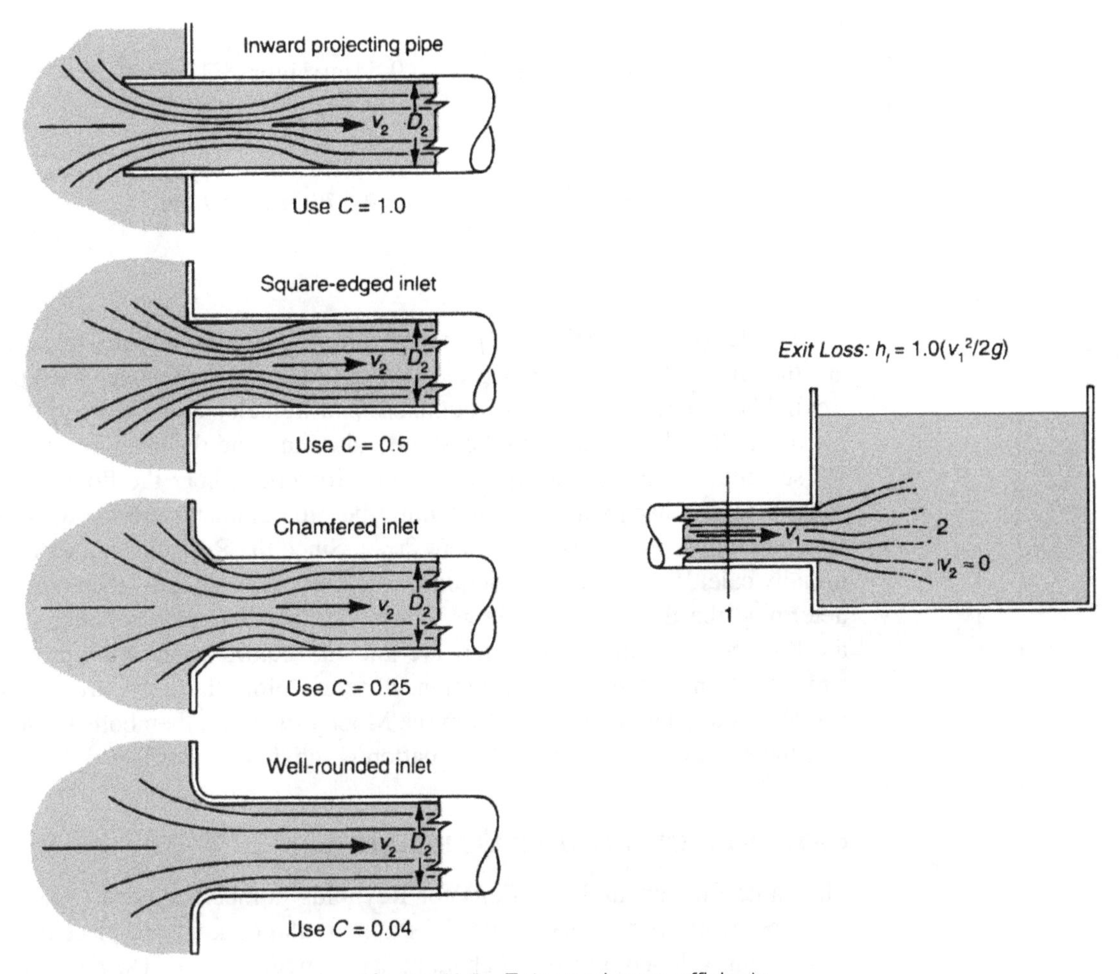

Figure 11.9 Entrance loss coefficients

Example 11.14

For Example 11.13, suppose there were three elbows ($C = 0.6$), two gate valves ($C = 0.26$), one globe valve ($C = 6.8$), and a square-edged entrance from the tank ($C = 0.5$). Calculate the head loss created by the minor losses for the existing flow conditions. Recalculate the head delivered by the pump considering both pipe and local losses.

Solution

Minor losses:

$$h_{fm} = \underset{\text{elbows}}{3C\frac{V^2}{2g}} + \underset{\text{gate vlvs.}}{2C\frac{V^2}{2g}} + \underset{\text{globe vlv.}}{C\frac{V^2}{2g}} + \underset{\text{entrance}}{C\frac{V^2}{2g}}$$

$$h_{fm} = [3(0.6) + 2(0.26) + 6.8 + 0.5]\frac{V^2}{2g}$$

$$h_{fm} = (9.62)\frac{\left(4.41\dfrac{\text{m}}{\text{s}}\right)^2}{2 \bullet 9.81\dfrac{\text{m}}{\text{s}^2}} = 9.54 \text{ m}$$

Total losses:

$$h_f = 8.33\text{m(pipe)} + 9.54\text{m(minor)} = 17.87\text{m}$$

$$h_A = \frac{\left(27.6\frac{\text{m}}{\text{s}}\right)^2}{2 \bullet 9.81\frac{\text{m}}{\text{s}^2}} = 3\text{m} + 17.87\text{m} = 59.7\text{m}$$

Problems to determine the pump or turbine head, a pressure, or an elevation are the simplest to solve. Since the flow rate and pipe diameter are known, the Reynolds number and relative roughness can be calculated directly; the friction factor can then be found from the Moody diagram, and the head loss is calculated for substitution into the energy equation. Problems where the flow rate is being sought—and other parameters including the pipe diameter are known—require a single iteration process (for turbulent flow). Since the Reynolds number cannot be initially calculated, an initial friction factor must be assumed—then corrected—to determine the flow rate. Problems where the pipe diameter is being sought, but the flow rate and other parameters are known, require an iteration process. After simplification of the energy equation, different pipe diameters are assumed and friction factors are determined from the Moody diagram, then both are substituted into the energy equation until the equation is satisfied.

Flow in Noncircular Conduits

The same fundamental equations for Reynolds number, relative roughness, and head loss from friction may be used for noncircular conduits. In place of the diameter, an equivalent diameter (or characteristic length) is used. The equivalent diameter is defined by

$$d_e = 4R_H = 4\frac{A}{WP}$$

where d_e = equivalent diameter, R_H = hydraulic radius = $\frac{A}{WP}$, A = cross-sectional area, and WP = wetted perimeter.

Example 11.15

Calculate the equivalent diameter of a rectangular conduit 0.6 meters wide and 0.3 meters high.

Solution

$$A = 0.6\,\text{m} \bullet 0.3\,\text{m} = 0.18\,\text{m}^2$$
$$WP = 2(0.6\,\text{m} + 0.3\,\text{m}) = 1.8\,\text{m}$$
$$d_e = 4\frac{A}{WP} = 4 \bullet \frac{0.18\,\text{m}^2}{1.8\,\text{m}} = 0.4\,\text{m}$$

Parallel Pipe Flow

The text to this point has addressed only flow in series piping systems; that is, all flow was considered to go through each pipe and fitting. If the flow divides

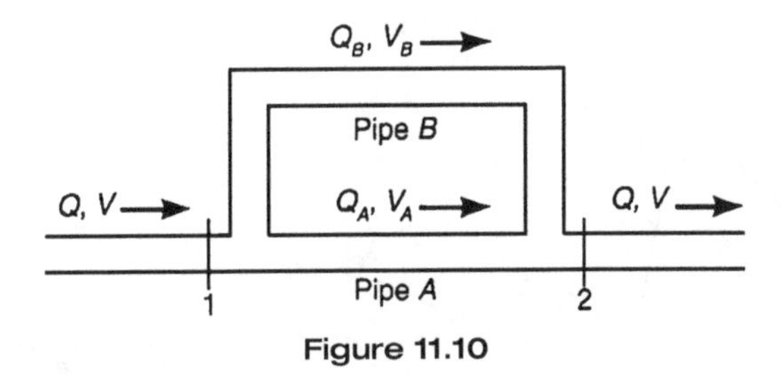

Figure 11.10

into two parallel branches and returns again to a single pipe as shown in Figure 11.10, the flow will divide so that the head loss in each branch is the same, or

$$h_{f1-2} = h_{fA} = h_{fB}$$

In the case when minor losses can be neglected, this relation becomes

$$h_{f1-2} = f_A \frac{L_A}{d_A} \frac{V_A^2}{2g} = f_B \frac{L_B}{d_B} \frac{V_B^2}{2g}$$

In addition, the continuity equation requires that

$$Q = QA + QB$$

Generally the flow rate, Q, or velocity, V, is known. Thus we have two equations and two unknowns, V_A and V_B. Values of f_A and f_B can be estimated and the two equations are then solved simultaneously for V_A and V_B. Using V_A and V_B to calculate the Reynolds numbers, corrected values of f_A and f_B can be found from the Moody diagram, and V_A and V_B are then recalculated. Then the head loss can be found from the above equation. For a flow that divides into three parallel branches, the same analysis can be used, where

$$h_{f1-2} = h_{fA} = h_{fB} = h_{fC}$$
$$Q = Q_A + Q_B + Q_C$$

In this case, we have three equations, and three unknowns.

FORCES ATTRIBUTABLE TO CHANGE IN MOMENTUM

The force created by the change in momentum of a fluid undergoing steady flow is given by the impulse momentum equation

$$F = \Delta(\dot{m}V) = \rho Q V_2 - \rho Q V_1$$

where F = resultant force on the fluid stream, ρQ = mass rate of flow, V_1, V_2 = inlet and exit velocities, respectively, $\rho Q V_1$ = momentum per second at the inlet, and $\rho Q V_2$ = momentum per second at the exit.

Forces on Bends

The magnitude and direction of the resultant force in flow through a bend will depend on the change in velocity magnitude and/or the change in direction of the flow. If the flow occurs in a two-dimensional bend, the equation can be rewritten in scalar form. Using Figure 11.11,

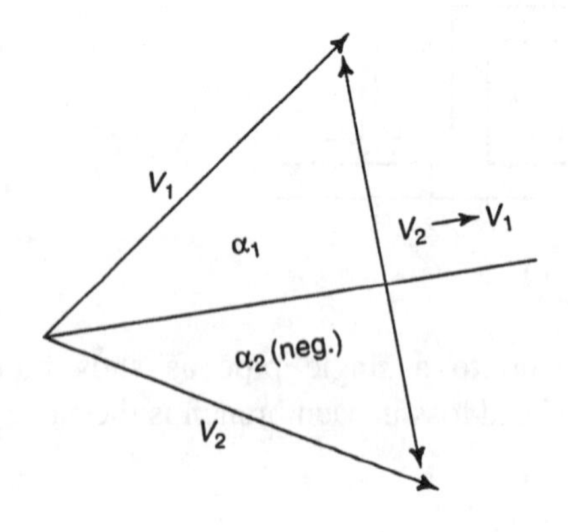

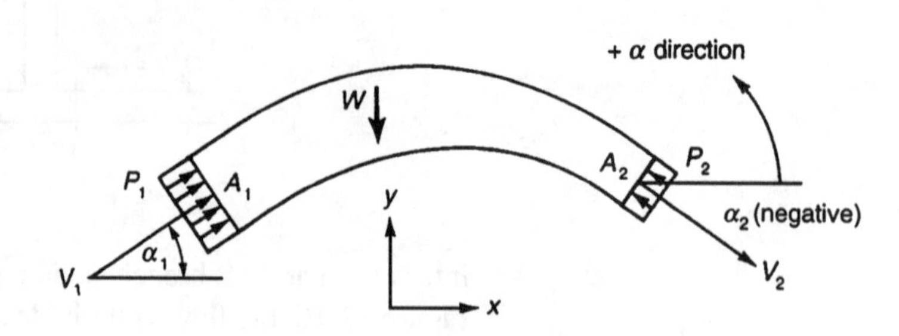

Figure 11.11

$$F_x = p_2 A_2 \cos\alpha_2 - p_1 A_1 \cos\alpha_1 + \rho Q(V_2 \cos\alpha_2 - V_1 \cos\alpha_1)$$

$$F_y - W = -p_2 A_2 \sin\alpha_2 - p_1 A_1 \sin\alpha_1 + \rho Q(V_2 \sin\alpha_2 - V_1 \sin\alpha_1)$$

$$F = \sqrt{F_x^2 + F_y^2}$$

where

F_x, F_y = resultant force component on the fluid by the bend in the x and y directions, respectively
p_1, p_2 = inlet and exit pressure on the fluid in the bend
A_1, A_2 = inlet and exit cross-sectional areas, respectively
α_1, α_2 = angles that the direction of flow makes with the positive x-direction at the entrance and exit, respectively
F = magnitude of resultant force
W = weight of the fluid in the bend

The equation may be simplified somewhat if the x-direction is chosen as the direction of the entering flow. The $\alpha_1 = 0$ and α_2 would be measured relative to that x-axis and

$$F_x = p_2 A_2 \cos\alpha_2 - p_1 A_1 + \rho Q(V_2 \cos\alpha_2 - V_1)$$

$$F_y - W = p_2 A_2 \sin\alpha_2 + \rho Q V_2 \sin\alpha_2$$

It is noted that the second equation, as written, requires that the y-axis remain parallel with gravity, since the coefficient of the weight W is unity. In many cases the weight of the fluid in the bend may not be significant. In addition to the impulse-momentum equations, it may be necessary simultaneously to utilize the energy and continuity equations in solving such problems.

Example 11.16

The 45° reducing bend discharges 0.008 m³/s of water into the atmosphere, as shown in Exhibit 5. The entrance diameter of the bend is 50 mm, and the exit diameter is 30 mm. Neglect the small elevation change, the weight of the fluid in the bend, and friction. Calculate the magnitude and force of the water on the bend. The density of the water is 1000 kg/m³, and its specific weight is 9810 N/m³.

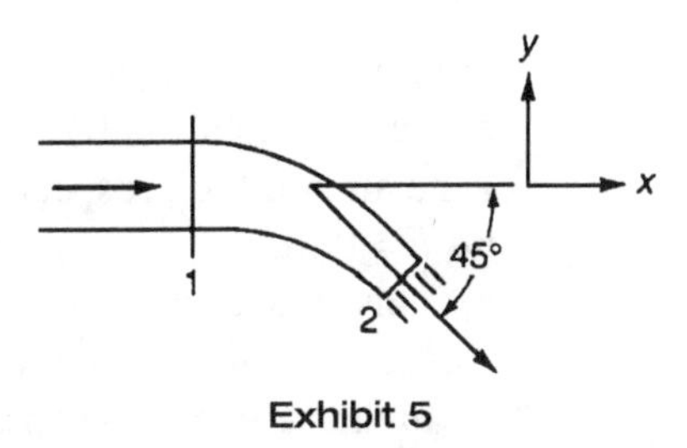

Exhibit 5

Solution

Select sections 1 and 2 as shown. Then $\alpha_1 = 0$, cos $\alpha_1 = 1$, sin $\alpha_1 = 0$, $\alpha_2 = -45°$, cos $\alpha_2 = 0.707$, sin $\alpha_2 = -0.707$, $p_2 = 0$, and the force-momentum equation becomes

$$F_x = -p_1 A_1 + \rho Q[V_2(0.707) - V_1]$$
$$F_y = \rho Q[V_2(-0.707)]$$

$$\rho Q = 1000\frac{\text{kg}}{\text{m}^3} \bullet 0.008\frac{\text{m}^3}{\text{s}} = 8.0\frac{\text{kg}}{\text{s}}$$

$$A_1 = \frac{\pi(0.05\,\text{m})^2}{4} = 0.00196\,\text{m}^2, \quad A_2 = \frac{\pi(0.03\,\text{m})^2}{4} = 0.00071\ \text{m}^2$$

$$V_1 = \frac{Q}{A_1} = \frac{0.008\ \text{m}^3/\text{s}}{0.00196\ \text{m}^2} = 4.08\,\text{m/s}, \quad V_2 = \frac{0.008\,\text{m}^3/\text{s}}{0.00071\text{m}^2} = 11.3\ \text{m/s}$$

The pressure p_1 may be found from the energy equation:

$$\frac{p_1}{\gamma} + \frac{V_1^2}{2g} + Z_1 = \frac{p_2}{\gamma} + \frac{V_2^2}{2g} + Z_2$$

$$p_1 = \gamma\left(\frac{V_2^2 - V_1^2}{2g}\right) = 9810\frac{\text{N}}{\text{m}^3} \bullet \frac{\left(11.3^2 - 4.08^2\right)\frac{\text{m}^2}{\text{s}^2}}{2 \bullet 9.81\frac{\text{m}}{\text{s}^2}} = 55{,}500\frac{\text{N}}{\text{m}^2}$$

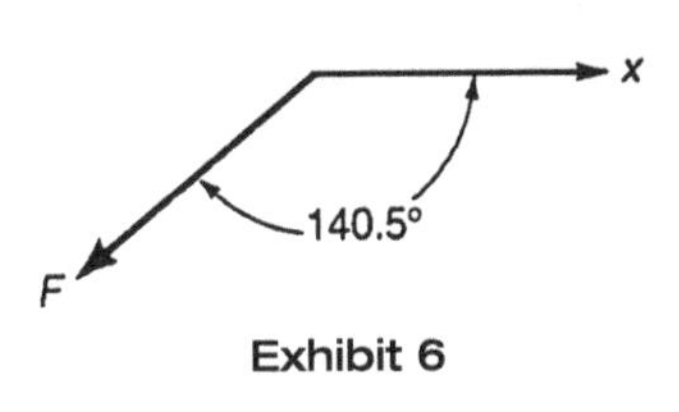

Exhibit 6

$$F_x = -55{,}500\frac{\text{N}}{\text{m}^2} \bullet 0.00196\text{m}^2 + 8.0\frac{\text{kg}}{\text{s}}\left(11.3\frac{\text{m}}{\text{s}} \bullet 0.707 - 4.08\frac{\text{m}}{\text{s}}\right)$$

$$F_x = -108.8\text{N} + 31.3\text{N} = -77.5\text{N}$$

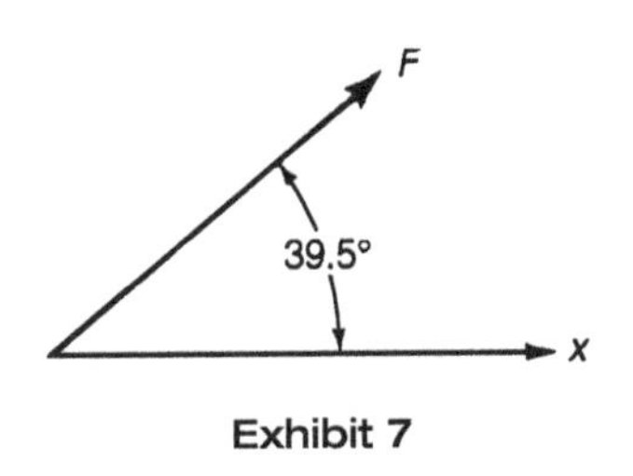

Exhibit 7

$$F_y = 8.0\frac{\text{kg}}{\text{s}} \bullet 11.3\frac{\text{m}}{\text{s}}(-0.707) = -63.9\text{N}$$

$$F = \sqrt{(-77.5)^2 + (-63.9)^2} = 100.4\text{N}$$

$$\text{at } \theta = -90.0° - \arctan\frac{-77.5}{-63.9} = -140.5°$$

This is the force of the bend on the fluid (Exhibit 6). The force of the fluid on the bend is equal and opposite (Exhibit 7).

Jet Engine and Rocket Thrust

The impulse-momentum relationship may be used to calculate the thrust created by a jet engine. Since the inlet and exit pressures are zero, the thrust may be calculated from a combination of equations for both the air and fuel:

$$F = \rho_a Q_a (V_2 - V_1) + \rho_f Q_f V_2$$

where V_1 = entering velocity relative to the engine (for engines in flight, this is the velocity of the aircraft), V_2 = exit velocity relative to the engine, ρ_a, ρ_f = density of the air and fuel, respectively, Q_a, Q_f = flow rate of the air and fuel, respectively, and F = thrust.

This equation assumes that the fuel enters the engine perpendicular to the thrust direction and leaves with the exhaust flow.

In the case of rocket propulsion, since the fuel and oxidizer initially are at rest, the equation for thrust becomes

$$F = \rho_m Q_m V_2$$

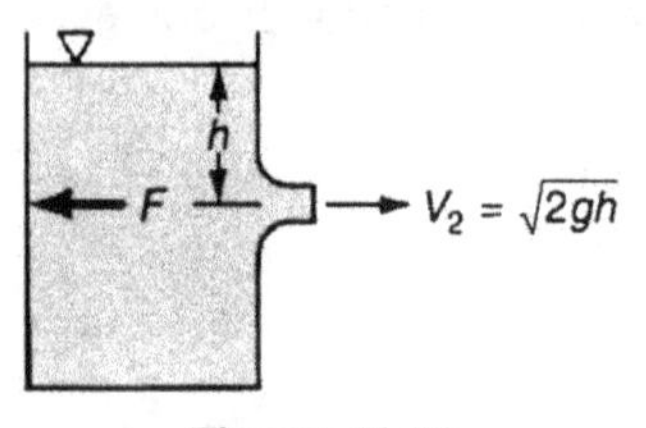

Figure 11.12

where ρ_m = density of the fuel-oxidizer mixture, and Q_m = flow rate of the fuel-oxidizer mixture.

A particular example of jet propulsion is shown Figure 11.12. The thrust is given by

$$F = \rho Q V_2 = \rho (A_2 V_2) V_2 = \rho A_2 V_2^2$$

but by Bernoulli's equation $V_2 = \sqrt{2gh}$. Substituting,

$$F = \rho A_2 (2gh) = 2\gamma A_2 h$$

where γ = specific weight of the liquid, A_2 = area of nozzle exit, h = height of fluid surface above the nozzle, and F = thrust or propulsion force.

Forces on Stationary Vanes

The impulse-momentum equation can also be used to determine forces on stationary vanes. Consider the stationary vane shown in Figure 11.13.

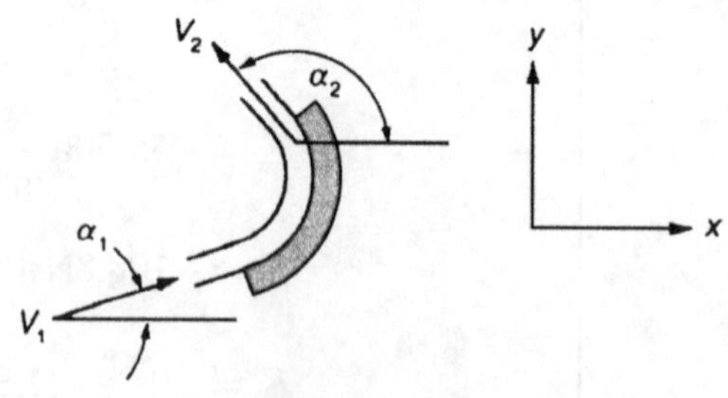

Figure 11.13

Since the fluid is open to the atmosphere, $p_1 = p_2 = 0$. If friction is neglected, it can be shown from Bernoulli's equation that $V_1 = V_2 = V$. The impulse-momentum equation reduces to

$$F_x = \rho Q(V_1 \cos \alpha_1 - V_2 \cos \alpha_2) = \rho QV(\cos \alpha_1 - \cos \alpha_2)$$
$$F_y = \rho QV(\sin \alpha_1 - \sin \alpha_2)$$

where F_x, F_y = *force of the fluid on the vane* in the x and y directions, respectively; α_1, α_2 = angle between the positive x-direction and the entrance and exit velocities, respectively; ρQ = mass flow rate; and V = fluid velocity.

If we reorient the vane so that the entrance velocity is in the x-direction, as shown in Figure 11.14, the equation can be rewritten as

$$F_x = \rho QV(1 - \cos \alpha_2)$$
$$F_y = -\rho QV \sin \alpha_2$$

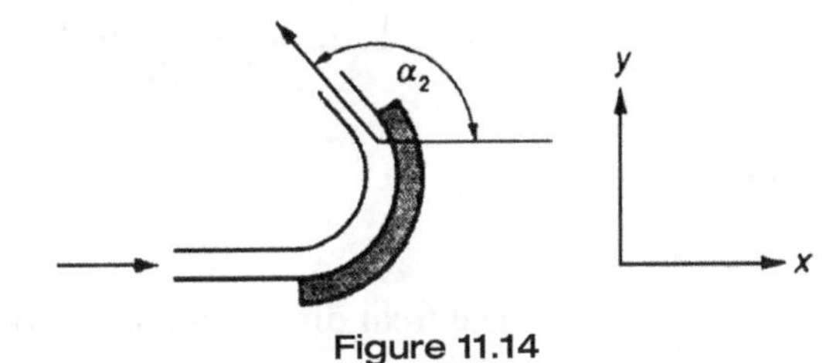

Figure 11.14

Example 11.17

Water impinges upon a stationary vane with a velocity of 15 m/s through a cross-sectional area of 6 square centimeters. The vane is oriented so that the fluid enters the vane cavity in the x-direction and leaves at an angle of 60° (Exhibit 8). The density of the water is 1000 kg/m³. Calculate the force on the vane.

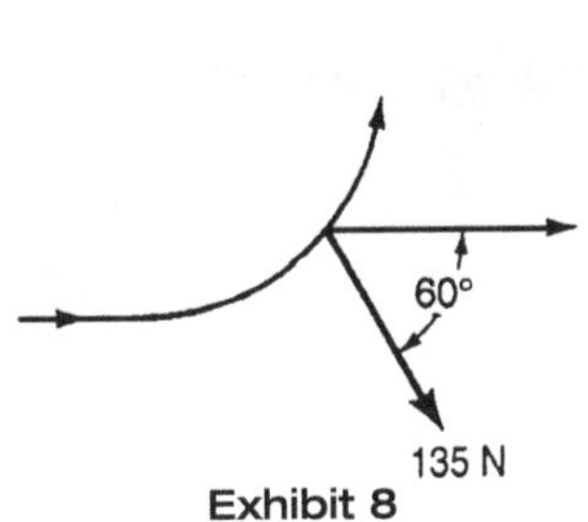

Exhibit 8

Solution

$$Q = AV = 6\,\text{cm}^2 \bullet \left(\frac{1\,\text{m}}{100\,\text{cm}}\right)^2 \bullet 15\,\text{m/s} = 0.009\frac{\text{m}^3}{\text{s}}$$

$$F_y = \rho QV(1 - \cos \alpha_2) = \frac{1000\,\text{kg}}{\text{m}^3} \bullet 0.009\frac{\text{m}^3}{\text{s}} \bullet 15\frac{\text{m}}{\text{s}} \bullet (1 - \cos 60°)$$

$$F_x = 67.5\frac{\text{kg} - \text{m}}{\text{s}^2} = 67.5\,\text{N}$$

$$F_y = -\rho QV \sin \alpha_2 = \frac{1000\,\text{kg}}{\text{m}^3} \bullet 0.009\frac{\text{m}^3}{\text{s}} \bullet 15\frac{\text{m}}{\text{s}} \bullet \sin 60°$$

$$F_y = -116.9\,\text{N}$$

$$F = \sqrt{(67.5)^2 + (-116.9)^2} = 135.0\,\text{N}$$

$$\text{at } \theta = \arctan\frac{-116.9}{67.5} = -60°$$

Forces on Moving Vanes

If the fluid enters the vane cavity in the x-direction and the vane is also moving in the x-direction with a velocity, v, as shown in Figure 11.15(a), then the force of the fluid on the moving vane in the direction of motion may be calculated from

$$F_x = \rho Q'(V - v)(1 - \cos \alpha_2)$$
$$F_y = -\rho Q'(V - v)\sin \alpha_2$$
$$Q' = A(V - v)$$

where v = velocity of the vane, and $V - v$ = fluid velocity relative to the vane.

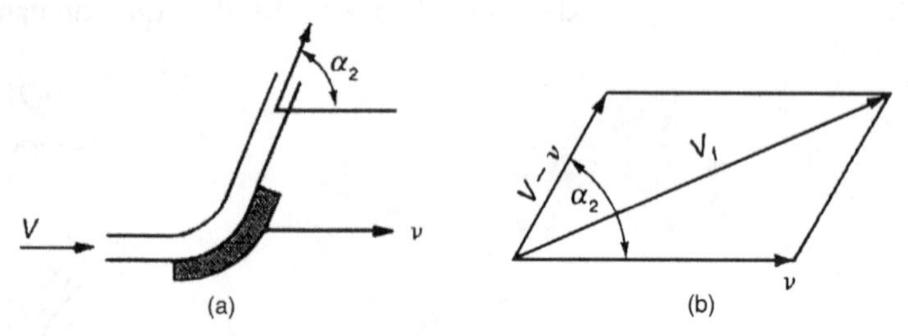

Figure 11.15

The final direction and magnitude of the jet is shown by V_f in the vector diagram [Figure 11.15(b)].

An impulse turbine contains a series of moving vanes as described above, one immediately replacing another as the turbine rotor rotates. See Figure 11.16.

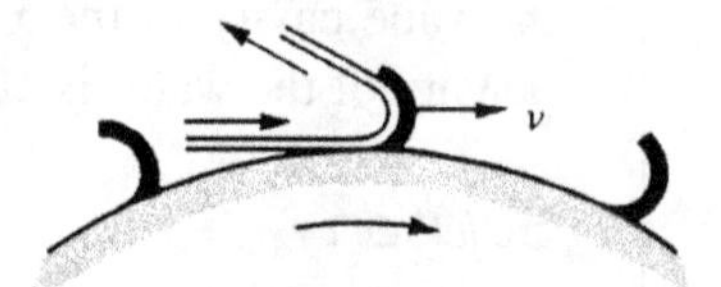

Figure 11.16

The x-direction force of the fluid on the series of vanes is given by

$$F = \rho Q(V - v)(1 - \cos \alpha_2)$$

The power delivered to the turbine, P, is given by

$$P = Fv = \rho Qv(V - v)(1 - \cos \alpha_2)$$

The maximum power for a given discharge angle, α_2, occurs when $V = 2v$ and is given by

$$P_{\max} = \frac{\rho Q V^2}{4}(1 - \cos \alpha_2)$$

The discharge angle that produces the maximum power possible is $\alpha_2 = 180°$. This maximum power is calculated from

$$P_{\max} = \frac{\rho Q V^2}{2} = g_c \gamma\, Q \frac{V^2}{2g}$$

VELOCITY AND FLOW MEASURING DEVICES

Pitot Tubes

For liquids flowing at relatively low pressures the mean static pressure may be measured using a piezometric tube indicated in Figure 11.17 by h_1. The stagnation pressure (i.e., the pressure at which the velocity is zero) is indicated by h_2.

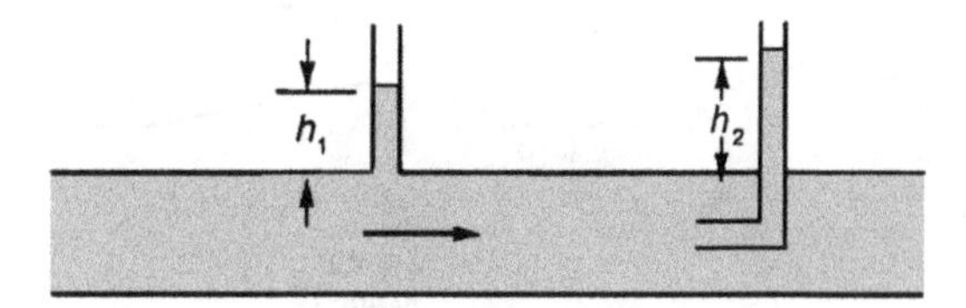

Figure 11.17

The relationship of each measurement to pressure, specific weight, and velocity is shown in the following two equations:

$$h_1 = \frac{p}{\gamma}, \quad h_2 = \frac{p_s}{\gamma} = \frac{p}{\gamma} + \frac{V^2}{2g}$$

Combining these relations, the velocity in the duct is

$$V = \sqrt{2g\left(\frac{p_s - p}{\gamma}\right)} = \sqrt{2g(h_2 - h_1)}$$

where V = velocity, γ = specific weight, p = static pressure, and p_s = stagnation pressure.

The combination of the two tubes as a single device is known as a pitot tube. If a manometer is connected between the static and stagnation pressure taps, velocities at moderate pressures may be calculated using the following equation:

$$V = \sqrt{2gh_m\left(\frac{\gamma_m}{\gamma} - 1\right)}$$

where h_m = height indicated by the manometer, and γ_m = specific weight of manometer fluid.

Example 11.18

A pitot tube is used to measure the mean velocity in a pipe where water is flowing. A manometer containing mercury is connected to the pitot tube and indicates a height of 150 mm. The specific weights of the water and mercury are 9810 N/m³ and 133,400 N/m³, respectively. Calculate the velocity of the water.

Solution

$$V = \sqrt{2gh_m\left(\frac{\gamma_m}{\gamma} - 1\right)}$$

$$V = \sqrt{2 \bullet 9.81\frac{\text{m}}{\text{s}^2} \bullet 0.15\text{m} \bullet \left(\frac{133{,}400\frac{\text{N}}{\text{m}^3}}{9810\frac{\text{N}}{\text{m}^3}} - 1\right)} = 6.09\frac{\text{m}}{\text{s}}$$

The pitot tube equation may also be used for compressible fluids with Mach numbers less than or equal to 0.3.

Flow Meters

There are three commonly used meters that measure flow rate in fluid systems: venturi meters, flow nozzles, and orifice meters. All three operate on the same basic principle, their equations being developed by combining the Bernoulli and continuity equations. The three meters are shown in Figure 11.18.

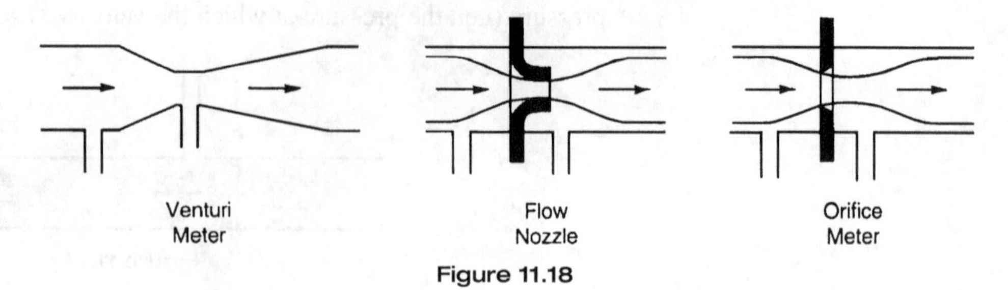

Figure 11.18

The equation for flow rate is given by

$$Q = \frac{c_v c_c}{\sqrt{1 - c_c^2 \left(\frac{A_2}{A_1}\right)^2}} \bullet A_2 \bullet \sqrt{2g\left(\frac{p_1 - p_2}{\gamma} + Z_1 - Z_2\right)}$$

or if a manometer is used between the pressure taps:

$$Q = \frac{c_v c_c}{\sqrt{1 - c_c^2 \left(\frac{A_2}{A_1}\right)^2}} \bullet A_2 \bullet \sqrt{2gh_m\left(\frac{\gamma_m}{\gamma} - 1\right)}$$

where

Q = flow rate

$p_1 - p_2$ = pressure difference between a point before the entrance to the meter, and the point of narrowest flow cross section in the meter

$Z_1 - Z_2$ = height difference between a point before the entrance to the meter and the point of narrowest flow cross section in the meter

A_1 = area of entrance

A_2 = area of narrowest flow cross section, except in the orifice, where it is the orifice area

h_m = height indicated by manometer

γ_m = specific weight of manometer fluid

γ = specific weight of fluid

c_v = coefficient of velocity

c_c = coefficient of contraction

For the orifice meter (and sometimes the flow nozzle) the coefficient terms in the equation are combined as follows:

$$c = \frac{c_v c_c}{\sqrt{1 - c_c^2\left(\frac{A_2}{A_1}\right)^2}}$$

and the flow rate equation is then written as

$$Q = cA_2\sqrt{2g\left(\frac{p_1 - p_2}{\gamma} + Z_1 - Z_2\right)} = cA_2\sqrt{2gh_m\left(\frac{\gamma_m}{\gamma} - 1\right)}$$

where c = orifice (or flow nozzle) coefficient.

The following values of c_c, c_v, and c are used for the various flow meters:

Venturi: $c_c = 1,\ 0.95 < c_v < 0.99$; c_v (nominal) = 0.984
Flow nozzle: $c_c = 1,\ 0.95 < c_v < 0.99$; c_v (nominal) = 0.98
Orifice: $c_c = 0.62,\ c_v = 0.98$; c (nominal) = 0.61

Actual values of c for orifice meters (and flow nozzles) vary with the diameter ratio, $d_0{:}d_1$, and the Reynolds number and are found on existing graphs. Curves for the values of c_v for venturi meters and flow nozzles are also available.

Flow from a Tank

The flow from a tank through various types of exit configurations can be calculated by using the energy equation and experimentally determined configuration coefficients. Consider a tank as shown in Figure 11.19. For frictionless flow, the flow-rate can be calculated using the energy and continuity equations, which reduce to

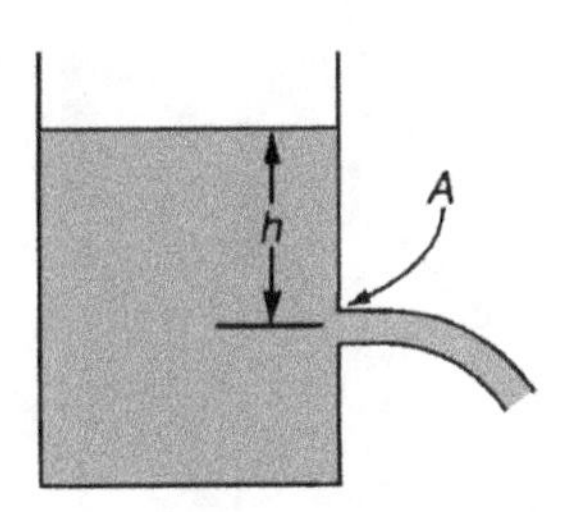

Figure 11.19

$$Q = AV = A\sqrt{2gh}$$

Considering friction, the flow rate may be calculated from

$$Q = cA\sqrt{2gh}, \quad c = c_v c_c$$

where Q = flow rate from the tank, h = height of water level above the exit, and c = coefficient of discharge for the exit.

If the friction in the exit is neglected, then $c = 1$.

For a sharp-edged orifice, $c_v = 0.98$, $c_c = 0.62$, and $c = 0.61$.

For a rounded exit, $c_v = 0.98$, $c_c = 1.00$, and $c = 0.98$.

For a short tube exiting from the tank, $c_v = 0.80$, $c_c = 1.00$, and $c = 0.80$.

For a re-entrant pipe, $c_v = 0.98$, $c_c = 0.52$, and $c = 0.51$.

For the special case when the flow from the tank discharges beneath the surface of the same fluid outside the tank, the flow rate from the tank is given by

$$Q = cA\sqrt{2gh(h_1 - h_2)}$$

where h_1 = height of fluid above exit in tank, and h_2 = height of fluid above exit outside tank.

SIMILARITY AND DIMENSIONLESS NUMBERS

The Reynolds number is a dimensionless number defined as the ratio of inertial forces to viscous forces. To test a model of some prototype, such as an air foil or length of pipe, the Reynolds number of the model must be equal to the Reynolds number of the prototype, or

$$(\mathrm{Re})_m = (\mathrm{Re})_p$$

$$\left(\frac{\rho V l}{\mu}\right)_m = \left(\frac{\rho V l}{\mu}\right)_p$$

where l = characteristic length.

A model is generally smaller geometrically than its prototype. Thus, if the characteristic length, l, of a model is to be one-tenth that of the prototype then one or more of the other terms in the Reynolds number must be adjusted to retain the Reynolds number the same for the model. For instance, the velocity, V, could be increased by a factor of ten using the same fluid or the fluid could be changed (liquid to gas) such that μ/ρ decreases by a factor of 10 with the same velocity, or a combination of the two. This condition is known as **dynamic similarity** of the prototype and model.

In fact, there are several independent force ratios that should be maintained in developing a model, depending on what forces are predominant in the situation. These force ratios involve pressure, inertia, viscosity, gravity, elasticity, and surface tension. The force ratios required for dynamic similarity are defined as follows, where the subscript m denotes model and p denotes prototype:

Inertia Force/Pressure Force Ratio

$$\left(\frac{\rho V^2}{p}\right)_m = \left(\frac{\rho V^2}{p}\right)_p$$

Inertia Force/Viscous Force Ratio

$$\left(\frac{\rho V l}{\mu}\right)_m = \left(\frac{\rho V l}{\mu}\right)_p = \text{Reynolds number, Re}$$

Inertia Force/Gravity Force Ratio

$$\left(\frac{V^2}{lg}\right)_m = \left(\frac{V^2}{lg}\right)_p = \text{Froude number, F}$$

Inertia Force/Elastic Force Ratio

$$\left(\frac{\rho V^2}{E}\right)_m = \left(\frac{\rho V^2}{E}\right)_p = \text{Cauchy number, } C_a$$

(E = modulus of elasticity of fluid)

Inertia Force/Surface Tension Force

$$\left(\frac{\rho l V^2}{\sigma}\right)_m = \left(\frac{\rho l V^2}{\sigma}\right)_p = \text{Weber number, } W_e$$

In many applications, one or more of the force ratios may be neglected because the forces are negligible.

INCOMPRESSIBLE FLOW OF GASES

The relationships thus far developed are primarily for the flow of incompressible fluids (liquids). Many are also applicable to the flow of compressible fluids (gases), for example, the continuity equation, equation for the Reynolds number, and so forth. The energy equation for incompressible flow also may be used under the following conditions:

1. The change in pressure in the pipe length is less than 10 percent of the inlet pressure. The density and specific weight at inlet conditions (pressure and temperature) should be used.
2. The change in pressure in the pipe length is between 10 and 40 percent of the inlet pressure. The density and specific weight at the average of the inlet and outlet conditions should be used. In some problems the outlet (or inlet) pressure is sought. In this case, the inlet (or outlet) conditions are used to find initial values of density and specific weight, and the approximate outlet (or inlet) pressure is then calculated. An iterative process ensues.

It may be necessary to utilize the perfect gas law from thermodynamics to calculate various properties, particularly the density of the gas given the pressure and temperature. The perfect gas law may be written in the following form

$$\rho = \frac{P}{RT}$$

where $R = \overline{R}/MW$ = gas constant
MW = gas molecular weight
$\overline{R}$ = universal gas constant

It should also be noted that the speed of sound, c, in a perfect gas is given by

$$c = \sqrt{kRT}$$

where k = ratio of specific heats, c_p/c_v
c_p = specific heat at constant pressure
c_v = specific heat at constant volume

It is apparent from the equation above that the speed of sound (acoustic velocity) in a gas depends only on its temperature.

The mach number Ma is the ratio of the actual fluid velocity to the speed of sound:

$$Ma = V/c$$

The accuracy of utilizing incompressible fluid flow equations for the flow of gases decreases with increasing velocities and their use is not recommended for mach numbers greater than 0.2.

SELECTED SYMBOLS AND ABBREVIATIONS

Symbol or Abbreviation	Description
α	angle with horizontal
β	angle with vertical
C	loss coefficient
C_a	Cauchy number
c	discharge coefficient, speed of sound
c_c	contraction coefficient
c_p	gas specific heat at constant pressure
c_v	velocity coefficient, gas specific heat at constant volume
δ	small thickness
d	diameter
d_e	equivalent diameter
ε	roughness factor
F_B	buoyant force
f	Darcy friction factor
F	Froude number
γ	specific weight
g_c	gravitational constant
h	depth of fluid
h_A	pump head
h_f	head loss from friction
h_R	turbine head
I	moment of inertia
k	ratio of specific heats
l	characteristic length
$\dot{m}$	mass flow rate
Ma	mach number
MW	molecular weight
μ	dynamic viscosity
η	efficiency
v	kinematic viscosity
P	power to or from fluid
p	pressure
Q	volume rate of flow
ρ	density
R	gas constant
$\overline{R}$	universal gas constant
Re	Reynolds number
R_H	hydraulic radius
σ	surface tension

(Continued)

(Continued)

Symbol or Abbreviation	Description
s	seconds
SG	specific gravity
τ	shear stress
T	absolute temperature, °R, °K
V	fluid velocity, volume
V_D	displaced volume
v	vane velocity, specific volume
$\dot{W}$	mechanical (shaft) power
W_c	Weber number
WP	Wetted perimeter
Z	elevation

PROBLEMS

11.1 Kinematic viscosity may be expressed in units of:
a. m^2/s
b. s^2/m
c. kg • s/m
d. kg/s

11.2 The absolute viscosity of a fluid varies with pressure and temperature and is defined as a function of:
a. density and angular deformation rate
b. density, shear stress, and angular deformation rate
c. density and shear stress
d. shear stress and angular deformation rate

11.3 An open chamber rests on the ocean floor in 50 m of sea water (SG = 1.03). The air pressure in kilopascals that must be maintained inside to exclude water is nearest to:
a. 318
b. 431
c. 505
d. 661

11.4 What is the static gage pressure in pascals in the air chamber of the container in Exhibit 11.4? The specific weight of the water is 9810 N/m^3.
a. −14,700 Pa
b. −4500 Pa
c. 0
d. +4500 kPa

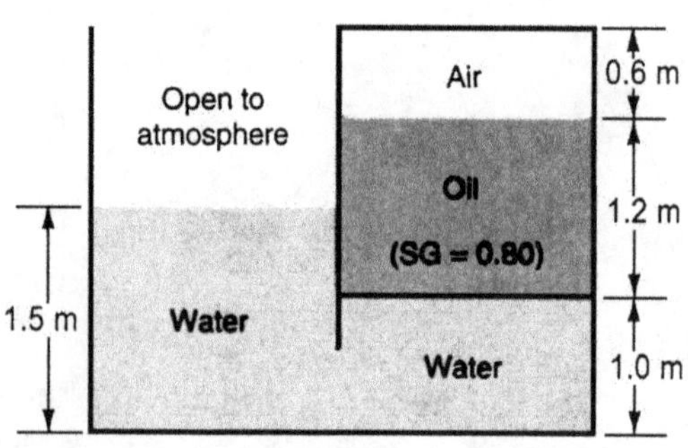

Exhibit 11.4

11.5 The pressure in kilopascals at a depth of 100 meters in fresh water is nearest to:
a. 268 kPa
b. 650 kPa
c. 981 kPa
d. 1620 kPa

11.6 What head, in meters of air, at ambient conditions of 100 kPa and 20 °C, is equivalent to 15 kPa?
a. 49
b. 131
c. 257
d. 1282

11.7 With a normal barometric pressure at sea level, the atmospheric pressure at an elevation of 1200 meters is nearest to:
a. 87.3 kPa
b. 83 kPa
c. 115.3 kPa
d. 101.3 kPa

11.8 The funnel in Exhibit 11.8 is full of water. The volume of the upper part is $0.165m^3$ and of the lower part is $0.057m^3$. The force tending to push the plug out is:

a. 1.00 kN
b. 1.47 kN
c. 1.63 kN
d. 2.00 kN

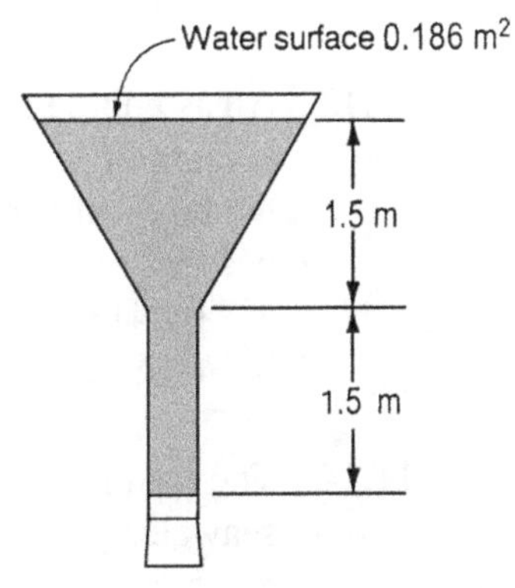

Exhibit 11.8

11.9 An open-topped cylindrical water tank has a horizontal circular base 3 meters in diameter. When it is filled to a height of 2.5 meters, the force in Newtons exerted on its base is nearest to:

a. 17,340
b. 34,680
c. 100,000
d. 170,000

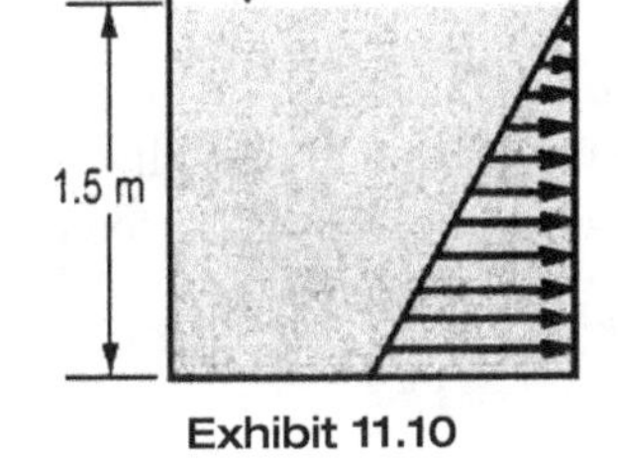

Exhibit 11.10

11.10 A cubical tank with 1.5 meter sides is filled with water (see Exhibit 11.10). The force, in kilonewtons, developed on one of the vertical sides is nearest to:

a. 4.1
b. 8.3
c. 16.5
d. 33.0

11.11 A conical reducing section (see Exhibit 11.11) connects an existing 10-centimeter-diameter pipeline with a new 5-centimeter-diameter line. At 700 kPa under no-flow conditions, what tensile force in kilonewtons is exerted on the reducing section?

a. 5.50
b. 2.07
c. 1.37
d. 4.13

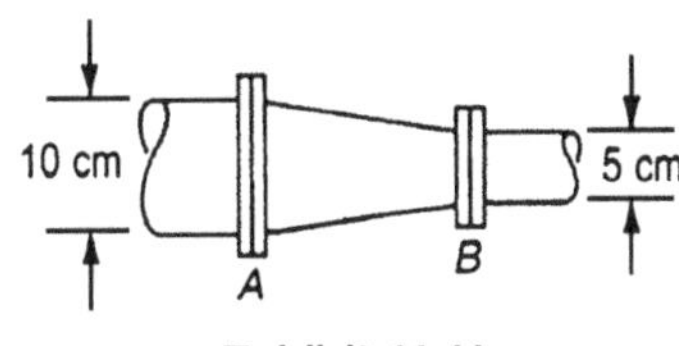

Exhibit 11.11

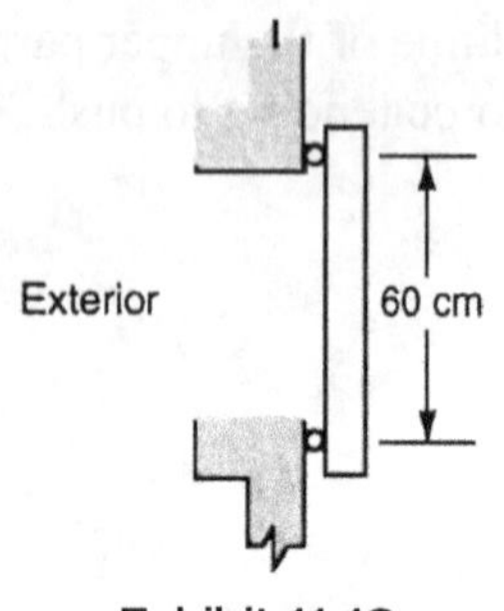

Exhibit 11.12

11.12 A circular access (see Exhibit 11.12) port 60 cm in diameter seals an environmental test chamber that is pressurized to 100 kPa above the external pressure. What force in newtons does the port exert upon its retaining structure?

a. 7100
b. 9500
c. 14,100
d. 28,300

11.13 A gas bubble rising from the ocean floor is 2.5 centimeters in diameter at a depth of 15 meters. Given that the specific gravity of seawater is 1.03, the buoyant force in newtons being exerted on the bubble at this instant is nearest to:

a. 0.0413
b. 0.0826
c. 0.164
d. 0.328

11.14 The ice in an iceberg has a specific gravity of 0.922. When floating in seawater (SG = 1.03), the percentage of its exposed volume is nearest to:

a. 5.6
b. 7.4
c. 8.9
d. 10.5

11.15 A cylinder of cork is floating upright in a container partially filled with water. A vacuum is applied to the container that partially removes the air within the vessel. The cork will:

a. rise somewhat in the water
b. sink somewhat in the water
c. remain stationary
d. turn over on its side

11.16 A floating cylinder 8 cm in diameter and weighing 9.32 newtons is placed in a cylindrical container that is 20 cm in diameter and partially full of water. The increase in the depth of water when the float is placed in it is:

a. 10 cm
b. 5 cm
c. 3 cm
d. 2 cm

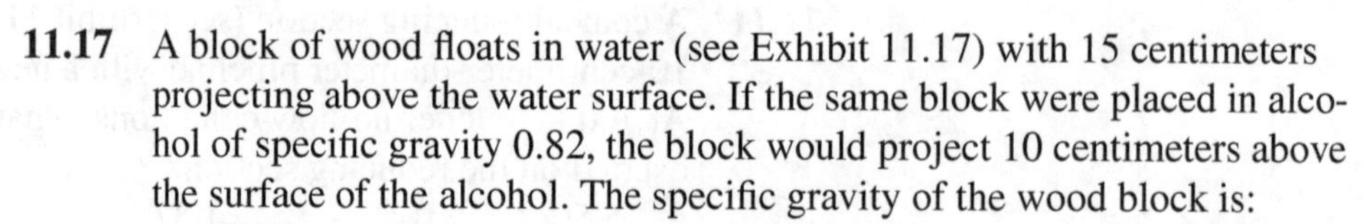

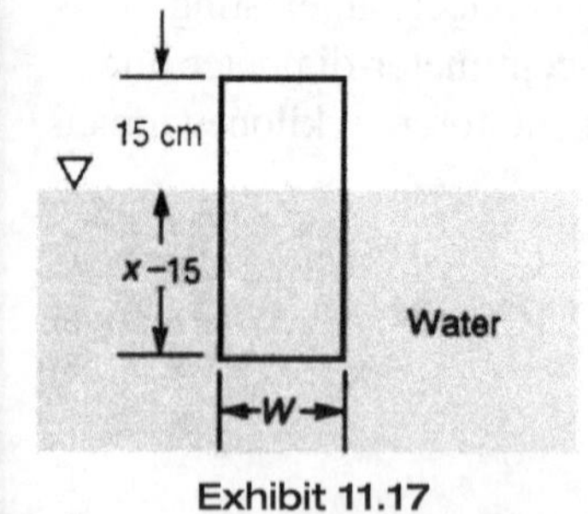

Exhibit 11.17

11.17 A block of wood floats in water (see Exhibit 11.17) with 15 centimeters projecting above the water surface. If the same block were placed in alcohol of specific gravity 0.82, the block would project 10 centimeters above the surface of the alcohol. The specific gravity of the wood block is:

a. 0.67
b. 3.00
c. 0.55
d. 0.60

11.18 The average velocity in a full pipe of incompressible fluid at Section 1 in Exhibit 11.18 is 3 m/s. After passing through a conical section that reduces the stream's cross-sectional area at Section 2 to one-fourth of its previous value, the velocity at Section 2, in m/s, is:

a. 1.0
b. 1.5
c. 3
d. 12

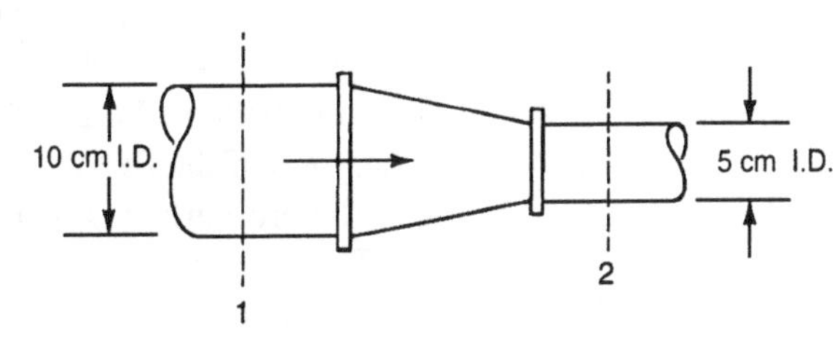

Exhibit 11.18

11.19 Refer to Exhibit 11.18. If the static pressure at Section 1 is 700 kPa and the 10-cm-diameter pipe is full of water undergoing steady flow at an average velocity of 10 m/s at A, the mass flow rate in kg/s at Section 2 is nearest to:

a. 10.0
b. 19.5
c. 78.5
d. 98.6

11.20 Air flows in a long length of 2.5-cm-diameter pipe. At one end the pressure is 200 kPa, the temperature is 150°C, and the velocity is 10 m/s. At the other end, the pressure has been reduced by friction and heat loss to 130 kPa. The mass flow rate in kg/s at any section along the pipe is nearest to:

a. 0.008
b. 0.042
c. 0.126
d. 0.5

11.21 Water flows through a long 1.0 cm I.D. hose at 10 liters per minute. The water velocity in m/s is nearest to:

a. 1
b. 2.12
c. 4.24
d. 21.2

11.22 Gasoline ($\rho = 800$ kg/m^3) enters and leaves a pump system with the energy in N • m/N of fluid that is shown in the following table:

	Entering	**Leaving**
Potential energy, Z meters above datum	1.5	4.5
Kinetic energy, $V^2/(2g_c)$	1.5	3.0
Flow energy, p/γ	9.0	45
Total energy	12.0	52.5

The pressure increase in kPa between the entering and leaving streams is nearest to:

a. 283
b. 566
c. 722
d. 803

11.23 Use the data of Problem 11.22. If the volume flow rate of the gasoline (800 kg/m³) is 55 liters per minute, the theoretical pumping power, in kW, is nearest to:

a. 0.3
b. 0.5
c. 3.2
d. 300

11.24 Water flowing in a pipe enters a horizontal venturi meter whose throat area at *B* is 1/4 that of the original and final cross sections at *A* and *C*, as shown in Exhibit 11.24. Continuity and energy conservation demand that which one of the following be *TRUE*?

a. The pressure at *B* is increased.
b. The velocity at *B* is decreased.
c. The potential energy at *C* is decreased.
d. The flow energy at *B* is decreased.

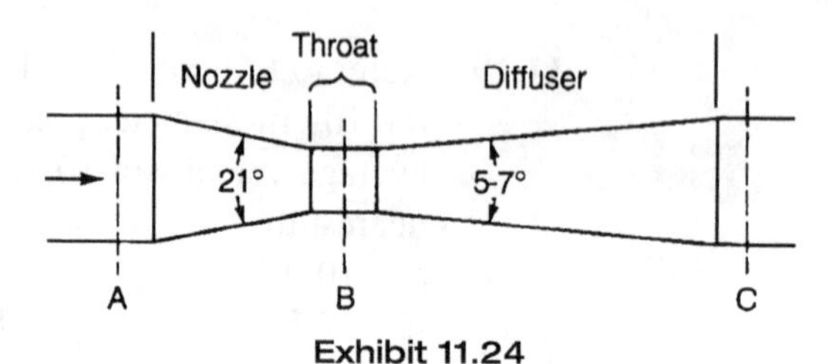

Exhibit 11.24

11.25 Given the energy data in N•m/N shown below existing at two sections across a pipe transporting water in steady flow, what frictional head loss in feet has occurred?

	Section A	Section B
Potential energy	20	40
Kinetic energy	15	15
Flow energy	100	75
Total	135	130

a. 0
b. 5
c. 130
d. 265

11.26 The power in kilowatts required in the absence of losses to pump water at 400 liters per minute from a large reservoir to another large reservoir 120 meters higher is nearest to:

a. 5.85
b. 7.85
c. 15.70
d. 30.00

11.27 The theoretical velocity generated by a 10-meter hydraulic head is:

a. 3 m/s
b. 10 m/s
c. 14 m/s
d. 16.4 m/s

11.28 What is the static head corresponding to a fluid velocity of 10 m/sec?

a. 5.1 m
b. 10.2 m
c. 16.4 m
d. 50 m

11.29 The elevation to which water will rise in a piezometer tube is termed the:

a. stagnation pressure
b. energy grade line
c. hydraulic grade line
d. friction head

11.30 A stream of fluid with a mass flow rate of 30 kg/s and a velocity of 6 m/s to the right has its direction reversed 180° in a "U" fitting. The net dynamic force in N exerted by the fluid on the fitting is nearest to:

a. 180
b. 360
c. 2030
d. 4300

11.31 The thrust in newtons generated by an aircraft jet engine on takeoff, for each 1 kg/s of exhaust products whose velocity has been increased from essentially 0 to 150 m/s, is nearest to:

a. 150
b. 1300
c. 3600
d. 7100

11.32 For the configuration in Exhibit 11.32, compute the velocity of the water in the 300-meter branch of the 15-cm-diameter pipe. Assume the friction factors in the two pipes are the same and that the incidental losses are equal in the two branches. The velocity in m/s is:

a. 10.0
b. 4.2
c. 1.8
d. 3.7

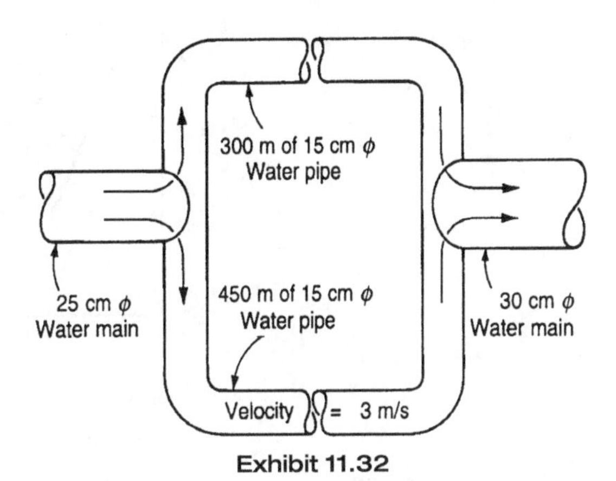

Exhibit 11.32

11.33 Which of the following statements most nearly approximates conditions in turbulent flow?

a. Fluid particles move along smooth, straight paths.
b. Energy loss varies linearly with velocity.
c. Energy loss varies as the square of the velocity.
d. Newton's law of viscosity governs the flow.

11.34 For turbulent flow of a fluid in a pipe, all of the following are true *EXCEPT*:

a. the average velocity will be nearly the same as at the pipe center.
b. the energy lost to turbulence and friction varies with kinetic energy.
c. pipe roughness affects the friction factor.
d. the Reynolds number will be less than 2300.

11.35 If the fluid flows in parallel, adjacent layers and the paths of individual particles do not cross, the flow is said to be:

a. laminar
b. turbulent
c. critical
d. dynamic

11.36 Which of the following constitutes a group of parameters with the dimensions of power?

a. ρAV
b. pAV
c. $\dfrac{DV\rho}{\mu}$
d. $\dfrac{\rho V^2}{P}$

11.37 At or below the critical velocity in small pipes or at very low velocities, the loss of head from friction:

a. varies linearly with the velocity
b. can be ignored
c. is infinitely large
d. varies as the velocity squared

11.38 The Moody diagram in Exhibit 11.38 is a log-log plot of friction factor vs. Reynolds number. Which of the lines *A*–*D* represents the friction factor to use for turbulent flow in a smooth pipe of low roughness ratio (ε/D)?

a. *A*
b. *B*
c. *C*
d. *D*

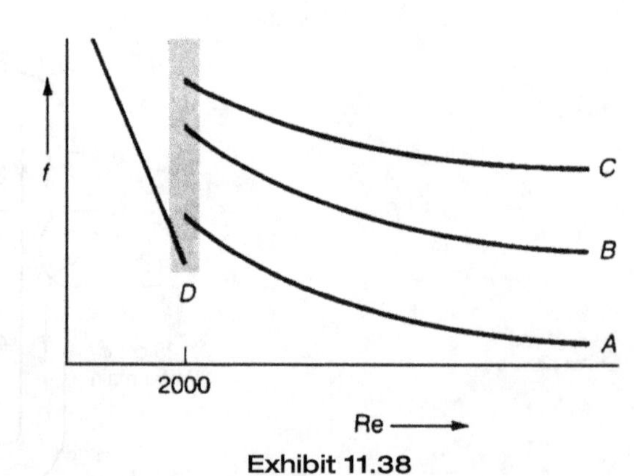

Exhibit 11.38

11.39 A 60-cm water pipe carries a flow of 0.1 m³/s. At Point *A* the elevation is 50 meters and the pressure is 200 kPa. At Point *B*, 1200 meters downstream from *A*, the elevation is 40 meters and the pressure is 230 kPa. The head loss, in feet, between *A* and *B* is:

a. 6.94
b. 15.08
c. 20.88
d. 100.2

11.40 Entrance losses between tank and pipe, or losses through elbows, fittings, and valves are generally expressed as functions of:

a. kinetic energy
b. pipe diameter
c. friction factor
d. volume flow rate

11.41 A 5-cm-diameter orifice discharges fluid from a tank with a head of 5 meters. The discharge rate, Q, is measured at 0.015 m^3/s. The actual velocity at the *vena contracta* (v.c.) is 9 m/s. The coefficient of discharge, C_D, is nearest to:

a. 0.62
b. 0.77
c. 0.99
d. 0.86

11.42 At normal atmospheric pressure, the maximum height in meters to which a nonvolatile fluid of specific gravity 0.80 may be siphoned is nearest to:

a. 4.0
b. 6.4
c. 10.3
d. 12.9

11.43 The water flow rate in a 15-centimeter-diameter pipe is measured with a differential pressure gage connected between a static pressure tap in the pipe wall and a pitot tube located at the pipe centerline. Which volume flow rate Q in cubic meters per second results in a differential pressure of 7 kPa?

a. 0.005
b. 0.066
c. 0.50
d. 1.00

11.44 The hydraulic formula $CA\sqrt{2gh}$ is used to find the:

a. discharge through an orifice
b. velocity of flow in a closed conduit
c. length of pipe in a closed network
d. friction factor of a pipe

11.45 The hydraulic radius of an open-channel section is defined as:

a. the wetted perimeter divided by the cross-sectional area
b. the cross-sectional area divided by the total perimeter
c. the cross-sectional area divided by the wetted perimeter
d. one-fourth the radius of a circle with the same area

11.46 To calculate a Reynolds number for flow in open channels and in cross-sections, one must utilize hydraulic radius, R, and modify the usual expression for circular cross sections, which is

$$\text{Re} = \frac{DV\rho}{\mu} = \frac{VD}{\nu}$$

where D = diameter, V = velocity, ρ = density, μ = absolute viscosity, and ν = kinematic viscosity.

Which of the following modified expressions for Re is applicable to flow in open or noncircular cross sections?

a. $\frac{RD}{\nu}$
b. $\frac{RV\rho}{\mu}$
c. $\frac{2RD}{\nu}$
d. $\frac{4RV}{\nu}$

SOLUTIONS

11.1 **a.** Kinematic viscosity $\nu = \dfrac{\text{Absolute viscosity}}{\text{Density}} = \dfrac{\mu}{\rho}$

Units of absolute viscosity: N• s/m² or kg/m-s

Units of density: kg/m³

The dimensions of kinematic viscosity would be

$$\nu = \frac{\mu \text{ (kg/m•s)}}{\rho \text{ (kg/m}^3\text{)}} = \frac{\text{m}^2}{\text{s}}$$

11.2 **d.** The absolute viscosity is proportional to the shear stress (τ) divided by the angular deformation rate. Density is not involved in the definition. The rate of angular deformation $\cong \dfrac{dV}{dy}$.

$$\text{Thus, } \tau = \mu \frac{dV}{dy}.$$

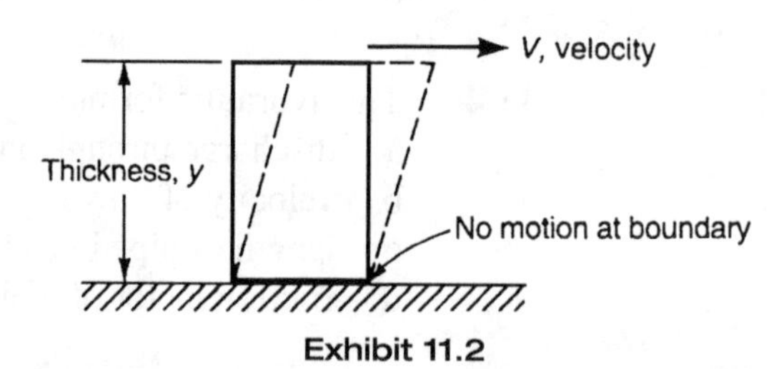

Exhibit 11.2

11.3 **c.** The internal pressure must equal the local external pressure. Externally, the pressure is

$$p = \gamma h = (\text{SG})(\gamma_{\text{water}})(h)$$

$$p = (1.03)\left(9.81 \frac{\text{kN}}{\text{m}^3}\right)(50 \text{ m}) = 505 \text{ kPa (gage)}$$

11.4 **b.** Since the situation is static, gage pressure at the base is 1.5 m of water. In the air chamber it is 1.5 m of water, less 1 m of water less 1.2 m of oil.

$$p = \gamma h$$

$$p = 1.5\,(9810) - 1\,(9810) - 1.2\,(0.8)\,(9810) = -4513 \text{ Pa}$$

11.5 **c.** Pressure $= \gamma h = 9810\,(100) = 981{,}000$ Pa $= 981$ kPa

11.6 **d.** The density of air can be calculated from the ideal gas law using

$$R = 0.286 \frac{\text{kN•m}}{\text{kg•K}}$$

$$\rho = \frac{p}{RT} = \frac{100 \text{ kN/m}^3}{0.286 \dfrac{\text{kN • m}}{\text{kg • K}}(20 + 273)\text{K}} = 1.19 \frac{\text{kg}}{\text{m}^3}$$

The specific weight of air $\gamma = 1.19\frac{\text{kg}}{\text{m}^3} \bullet \left(\frac{9.81\frac{\text{m}}{\text{s}^2}}{1\frac{\text{kg} \bullet \text{m}}{\text{N} \bullet \text{s}^2}} \right) = 11.7\frac{\text{N}}{\text{m}^3}$

$$p = \gamma h \text{ or } h = \frac{p}{\gamma} \text{ and } h = \frac{15{,}000\frac{\text{N}}{\text{m}^2}}{11.7 \text{ N/m}^3} = 1282 \text{ m}$$

11.7 **a.** Assuming atmospheric pressure at sea level at 101.3 kPa and a constant specific weight of air at 11.7 N/m³ (as previously calculated):

$$p = p_{\text{SL}} - \gamma h = 101.3 - \frac{11.7(1200)}{1000} = 87.26 \text{ kPa}$$

11.8 **b.**

$$\text{Force} = PA = \gamma h A = 9.81\frac{\text{kN}}{\text{m}^3} \times 3 \text{ m} \times 500 \text{ cm}^2 \times \left(\frac{1 \text{ m}}{100 \text{ cm}}\right)^2 = 1.47 \text{ kN}$$

11.9 **d.** The pressure at the tank base $= p = \gamma_W h = (9810)(2.5) = 24{,}325$ N/m³

Area of tank base, $A = \frac{\pi}{4}(3)^2 = 7.07\,\text{m}^2$

Force on tank base $= pA = 24{,}325\ (7.07) = 171{,}978$ N

11.10 **c.** The average pressure exerted on one side is the pressure that exists at the centroid of the side times the area of the side.

$$F = \gamma h_c A$$

where h_c = the depth in meters from the fluid-free surface to the centroid of the area, and A = area. Since the sides are square, $h_c = 1.5/2 = 0.75$ m.

$$F = 9.81(0.75)(1.5 \times 1.5) = 16.55 \text{ kN}$$

11.11 **d.** The static force at $A = \left(700\frac{\text{kN}}{\text{m}^2}\right)\left[\frac{\pi}{4}(0.1 \text{ m})^2\right] = 5.50$ kN tension on the bolts at A.

The static force at $B = \left(700\frac{\text{kN}}{\text{m}^2}\right)\left[\frac{\pi}{4}(0.05 \text{ m})^2\right] = 1.37$ kN tension on the bolts at B.

The end restraint by the pipes opposes a net force of 5.50 – 1.37 = 4.13 kN to the right on the reducing section.

11.12 **d.** Area of port $= \frac{\pi}{4}D^2 = \frac{\pi}{4}(0.6)^2 = 0.283\,\text{m}^2$

Pressure = 100 kPa = $100{,}000\frac{\text{N}}{\text{m}^2}$

$F = pA = 100{,}000\ (0.283) = 28{,}300$ N

11.13 b. The volume of the bubble equals the volume of the displaced seawater, which equals

$$V_D = \frac{4}{3}\pi r^3 = \frac{4}{3}\pi(0.0125)^3 = 8.18\times10^{-6}\,\text{m}^3$$

Since the specific weight of seawater is

$$(\text{SG})(\gamma_W) = 1.03\left(9810\,\frac{\text{N}}{\text{m}^2}\right) = 10{,}104\,\frac{\text{N}}{\text{m}^2}$$

The buoyant force, B, is

$$B = \gamma V_D (10{,}104)(8.18\times10^{-6}) = 0.0826\ \text{N}$$

11.14 d. A buoyant force is equal to the weight of fluid displaced. At equilibrium, or floating, the weight downward is equal to the buoyant force.

Let V_1 = total volume of the iceberg in m³. Its weight is $V_1(9810)(0.922) = 9045(V_1)$ N.

Let V_2 = immersed volume of the iceberg, which equals the volume of seawater displaced. The weight of seawater displaced is then $V_2(9810)(1.03) = 10{,}104(V_2)$ N.

Hence $\frac{V_2}{V_1} = \frac{9045}{10{,}104} = 0.895$ is the volume fraction of the iceberg immersed, and the volume fraction exposed is 1 – 0.895 = 0.105 = 10.5%.

11.15 b. Archimedes' principle applies equally well to gases. Thus a body located in any fluid, whether liquid or gaseous, is buoyed up by a force equal to the weight of the fluid displaced. A balloon filled with a gas lighter than air readily demonstrates the buoyant force.

Thus the weight of the cork is equal to the weight of water displaced plus the weight of air displaced. When the air within the vessel is removed, the cork is no longer provided a buoyant force equal to the weight of air displaced. For equilibrium, the cork will sink somewhat in the water.

11.16 c. $V_D = \frac{W}{\gamma} = \frac{9.32\text{N}}{9810\frac{\text{N}}{\text{m}^2}} = 0.00095\ \text{m}^3 = 950\ \text{cm}^3$

The change in total volume, ΔV, beneath the water surface equals the area of the cylindrical container, A, times the change in water level, dh, or $dV = A\,dh$. The depth of the water will increase

$$dh = \frac{dV}{A} = \frac{950\ \text{cm}^3}{\frac{\pi}{4}(20)^2} = 3.02\ \text{cm}$$

11.17 d. Let x = height of wood block, W = width of wood block, L = length of wood block, and γ = specific weight of water, N/m^3. The weight of the block is equal to the weight of the liquid displaced.

Weight of the block in water = (x – 15)WLγ(1.0)

Weight of the block in alcohol = (x – 10)WLγ(0.82)

Since the weight of the block is constant,

$$(x-15)WL\gamma = 0.82(x-10)WL\gamma$$
$$x-15 = 0.82x-8.2$$
$$x = \frac{6.8}{0.18} = 37.8\text{ cm}$$

The specific gravity of the wood block is, by definition,

$$\frac{\text{Volume of water displaced}}{\text{Total volume}} = \frac{(x-15)WL}{xWL} = \frac{37.8-15}{37.8} = 0.603$$

11.18 d. Continuity requires that

$$Q = A_1V_1 = A_2V_2 = A_1(3) = \frac{A_1}{4}V_2, \qquad V_2 = 12\text{ m/s}$$

11.19 c. Continuity requires that the mass flow rate be the same at all sections in steady flow. Calculate $\dot{m}$ at Section 1, where the velocity is given, using $\dot{m} = \rho AV$. This will also be the mass flow rate at Section 2.

Cross-sectional area at Section 1:

$$\frac{\pi}{4}(0.10)^2 = .00785\text{ m}^2$$

$$\dot{m} = \rho AV\left(1000\frac{\text{kg}}{\text{m}^3}\right)(.00785\text{m}^2)(10\text{m/s}) = 78.5\text{kg/s}$$

11.20 a. The mass flow rate $\dot{m} = \rho Q = \rho AV$. The density of air at 200 kPa and 150°C (423°K) is obtained from the ideal gas law:

$$\frac{p}{\rho} = RT$$

Use $R = 286\frac{\text{N}\bullet\text{m}}{\text{kg}\bullet°\text{K}}$ for air.

$$\rho = \frac{P}{RT} = \frac{200{,}000\frac{\text{N}}{\text{m}^2}}{286\frac{\text{N}\bullet\text{m}}{\text{kg}\bullet°\text{K}}\bullet 423°\text{K}} = 1.65\frac{\text{kg}}{\text{m}^3}$$

The cross-sectional area $= A = \frac{\pi}{4}D^2 = 0.785(.025\text{ m})^2$ = 490 × 10^{-6} m^2, and

$$\dot{m} = \rho Av = \left(1.65\frac{\text{kg}}{\text{m}^3}\right)(490\times10^{-6}\text{ m}^2)\left(10\frac{\text{m}}{\text{s}}\right) = 0.00809\frac{\text{kg}}{\text{s}}$$

11.21 b.

$$Q = 10\frac{\text{L}}{\text{min}} \times \frac{\text{m}^3}{1000\text{ L}} \times \frac{\text{min}}{60\text{ s}} = 167 \times 10^{-6}\frac{\text{m}^3}{\text{s}}$$

The cross-sectional area $A = \frac{\pi}{4}D^2 = 0.785(0.01)^2 = 78.5 \times 10^{-6}\text{ m}^2$

$$V = \frac{Q}{A} = \frac{167 \times 10^{-6}}{78.5 \times 10^{-6}} = 2.13\text{ m/s}$$

11.22 a. The pressure (flow) energy change is 45 – 9 = 36 N•m/N, or

$$\gamma = \frac{g}{g_c}\rho = \frac{9.81}{1.0}(800) = 7848\text{ N/m}^3$$

$$\frac{\Delta p}{\gamma} = 36, \qquad \Delta p = 36\gamma = 36(7848) = 282{,}500\text{ Pa} = 282.5\text{ kPa}$$

11.23 a. The volume flow rate is

$$55\frac{\text{L}}{\text{min}} \bullet \frac{\text{m}^3}{1000\text{ L}} \bullet \frac{\text{min}}{60\text{ s}} = 917 \times 10^{-6}\text{m}^3/\text{s}$$

Ignoring the head loss, h_L, from friction, the required energy input is $52.5 - 12 = 40.5\frac{\text{N}\bullet\text{m}}{\text{N}}$.

$$\text{Power} = \gamma Q h_A \left(7848\frac{\text{N}}{\text{m}^3}\right)(917 \times 10^{-6}\text{m}^3/\text{s})\left(40.5\frac{\text{N}-\text{m}}{\text{N}}\right)$$
$$= 291\text{ W} = 0.291\text{ kW}$$

11.24 d. In a venturi throat, the increased velocity required by continuity results in a *KE* (velocity) increase that occurs at the expense of pressure (flow) energy. Since the system is horizontal, no change in potential energy has occurred. At *B* the pressure (flow energy) decreases and *KE* increases. For a well-designed venturi, the conditions existent at *A* are essentially restored at *C*.

11.25 b. Apply an energy balance of the fluid flowing: Total energy in = Total energy out + Energy losses – Energy inputs. Thus, $135 = 130 + h_L - 0$. The head loss $h_L = 5$ N • m/m, or 5 meters.

11.26 b. Ignoring frictional losses, pump inefficiency, and noting that any changes in *KE* or pressure are essentially 0, pumping power is equal to the increase in potential energy between the reservoirs.

The potential energy increase per lb_m is Z or $h = 120$ meters or $\frac{\text{N}\bullet\text{m}}{\text{N}}$

The volume flow rate, Q, is

$$400\frac{\text{L}}{\text{min}} \bullet \frac{\text{m}^3}{1000\text{ L}} \bullet \frac{\text{min}}{60\text{ s}} = 6.67 \times 10^{-3}\text{m}^3/\text{s}$$

The power required is

$$P = \gamma Q h_a = 9.81\frac{\text{kN}}{\text{m}^3} \bullet 6.67 \times 10^{-3}/\text{s} \bullet 120\text{ m} = 7.85\text{ kW}$$

11.27 c.

$$h=\frac{V^2}{2g} \quad \text{or} \quad V=(2gh)^{1/2}=(2\times 9.81\times 10)^{1/2}=14\,\text{m/s}$$

11.28 a. The head is

$$h=\frac{V^2}{2g}=\frac{10^2}{2(9.81)}=5.10\ \text{m}$$

11.29 c. A **piezometer tube** indicates static pressure and is equivalent to a static pressure gage.

Stagnation pressure is an increased pressure developed at the entrance to a pitot tube when the velocity locally becomes zero.

The **hydraulic grade line** is a flow energy or pressure head in meters, which can be plotted vertically above the pipe centerline along the pipe.

The **energy grade line** is the total mechanical energy (flow energy or pressure head, plus kinetic energy or dynamic head, plus potential energy or height above datum) in meters, which may be plotted vertically above the datum along the pipe.

The **friction head** is the head loss h_f in meters caused by fluid friction.

The **critical depth** above the channel floor in open channels is the depth for minimum potential and kinetic energy for the given discharge. Tranquil flow (low *KE* and high *PE*) exists when the actual flow is above critical depth, and rapid flow (high *KE* and low *PE*) exists when the actual flow is below critical depth.

11.30 b. The steady impulse-momentum equation is $F_{\text{net}}=\dot{m}(\bar{v}_2-\bar{v}_1)$ if the pressure in the fluid stream is zero at each end of the "U." Then

$$F=30\ \text{kg/s}(-6-6)\,\text{m/s}=360\frac{\text{kg}\bullet\text{m}}{\text{s}^2}=360\ \text{N}$$

Since the original velocity was 6 m/s, the final reversed velocity is –6 m/s. This force from impulse-momentum is the force on the fluid to achieve the velocity change. In reaction, the fluid exerts an equal and opposite force, 360 N, to the right on the fitting.

11.31 a. The impulse-momentum equation is $F=pQ(V_2-V_1)$. Here $\rho Q=1$ kg/s, the final velocity of the exhaust is $V_2=150$ m/s, and $V_1=0$. Hence

$$F=\left(\frac{1\ \text{kg}}{\text{s}}\right)(150-0)\frac{\text{m}}{\text{s}}=150\frac{\text{kg}\bullet\text{m}}{\text{s}^2}=150\ \text{N}$$

11.32 d. There is a drop in the energy line from the 25-cm main to the 30-cm main. This head loss must be equal in both 15-cm lines, or

$$h_{f300} = h_{f450}$$

The Darcy equation is

$$h_f = f\frac{L}{d}\frac{V^2}{2g}$$

where h_f = head loss in meters, f = friction factor, L = length of pipeline in meters, d = diameter of pipe in meters, and g = 9.81 m/s^2. Thus, in this situation,

$$f\frac{300}{0.15}\frac{V_{300}^2}{2(9.81)} = f\frac{450}{0.15}\frac{V_{450}^2}{2(9.81)}$$

which reduces to

$$300V_{300}^2 = 450(3)^2$$

$$V_{300} = \sqrt{\frac{4050}{300}} = 3.67 \text{ m/s}$$

11.33 c. Laminar (streamline, viscous) flow is compared with turbulent flow in the following table:

	Laminar Flow	Turbulent Flow
Motion of fluid particles	Parallel to stream velocity. Paths of particles do not cross.	Particle paths cross and move in all directions.
Energy loss, h_f	$h_f = f\left(\frac{L}{D}\right)\left(\frac{V^2}{2g}\right)$ f is independent of surface roughness and decreases with Re. $f = \frac{64}{Re}$	$h_f = f\left(\frac{L}{D}\right)\left(\frac{V^2}{2g}\right)$ f varies with surface roughness, decreases with Re to a constant value. See Moody diagram.
Velocity distribution in pipe	Average is $^1/_2$ of maximum at centerline. parabolic distribution. Zero at wall.	Essentially same throughout, except for thin boundary layer at wall. Follows 1/7 power law.
Reynolds number $Re = DV\rho/\mu$	Less than 2300	Greater than 2300

Newton's law of viscosity defines μ on the basis of shear stress and the rate of fluid angular deformation. The Reynolds number contains μ as a contributing parameter. Very viscous liquids usually move in laminar flow.

On the basis of the above data, select (c). Do not confuse the energy loss, h_f, with the friction factor, f.

11.34 **d.** In turbulent flow the Reynolds number is *greater* than 2300.

11.35 **a.** Turbulent flow is highly agitated flow with individual particles crossing paths and colliding; critical flow is a point at which some property of the fluid—or some parameter related to it—changes; dynamic flow is redundant; uniform flow is of constant rate.

11.36 **b.** Choice (a) is mass flow rate, $\dot{m}$, in kg/s.

Choice (b) has these dimensions:

$$pAV = \left(\frac{\text{N}}{\text{m}^2}\right)(\text{m}^2)\left(\frac{\text{m}}{\text{s}}\right) = \frac{\text{N}\bullet\text{m}}{\text{s}} = \text{W}$$

Choice (c) is the Reynolds number, Re; it is the dimensionless ratio of inertial force to viscous force.

Choice (d) is the Euler number, Eu; it is the dimensionless ratio of inertial force to pressure force.

11.37 **a.** Below the critical velocity (Re < 2300) flow is laminar, and $f = 64/\text{Re}$. Substitution of this term into the Darcy equation for friction loss in pipes, $h_f = f\frac{L}{D}\frac{V^2}{2g}$, yields the Hagen-Poiseuille equation,

$$h_f = \frac{32\mu LV}{\gamma\, d^2}.$$

11.38 **a.** Line *D* applies to all roughness ratios in laminar flow (Re < 2300) because the boundary layer at the wall makes the friction factor independent of roughness ratio:

$$f = \frac{64}{\text{Re}}$$

In turbulent flow (Re > 2300), increasing roughness is represented by *A* for a smooth pipe to *C* for a very rough pipe; moreover, only the thinnest boundary layer exists in a turbulent flow, so the friction factor is very dependent on surface roughness.

11.39 **a.** Use an energy balance to determine h_f. Upon substituting the given data, the resulting equation is

$$Z_A + \frac{V_A^2}{2g} + \frac{P_A}{\gamma} = Z_B + \frac{V_B^2}{2g} + \frac{P_B}{\gamma} + h_f$$

Since the pipe diameter is unchanged, continuity requires that *V* be the same at both points. Thus the kinetic energy terms can be deleted from both sides of the equation.

$$Z_A + \frac{P_A}{\gamma} = Z_B + \frac{P_B}{\gamma} + h_f$$

$$h_f = \frac{200 - 230}{9.81} + 50 - 40 = 6.94\ \text{m}$$

11.40 a. Typical head losses for the above items are expressed as an empirical average constant, K or C times the kinetic energy, $V^2/2g$:

$$h_f = K\frac{V^2}{2g}$$

11.41 b. The discharge coefficient is $C_D = C_c C_v$, where C_c = coefficient of contraction = (area of v.c.)/(area of orifice) and C_v = coefficient of velocity = (actual velocity at v.c.)/(theoretical velocity at v.c.) ignoring losses.

The theoretical velocity at the v.c. is

$$V = \sqrt{2gh} = \sqrt{2(9.81)(5)} = 9.9 \text{ m/s}$$

$$C_v = \frac{9.0}{9.9} = 0.909$$

The area of the v.c. is

$$A = \frac{Q}{V} = \frac{.015}{9} = 0.00167 \text{ m}^2$$

The area of the orifice $= \frac{\pi}{4}D^2 = \frac{\pi}{4}(.05)^2 = 0.00196 \text{ m}^2$. Thus,

$$C_c = \frac{0.00167}{0.00196} = 0.852 \quad \text{and} \quad C_D = C_c C_v = (0.852)(0.909) = 0.774$$

11.42 d. The maximum height to which a fluid may be siphoned is determined when the pressure of the fluid column plus its vapor pressure equals the external pressure. The minimum pressure at the highest point is 0 kPa plus vapor pressure.

Ignoring the vapor pressure (small),

$$p = \gamma h \quad \text{or} \quad h = \frac{P}{\gamma}$$

$$h = \frac{-101.3 \text{ kN/m}^2}{(0.8)9.81 \text{ kN/m}^3} = 12.91 \text{ m in depth (or height)}$$

11.43 b. A pitot tube generates a stagnation pressure as fluid kinetic energy is converted to pressure head. Hence

$$V = \sqrt{2g\left(\frac{\Delta p}{\gamma}\right)}$$

$$V = \sqrt{2(9.81)\frac{\text{m}}{\text{s}^2}\left(\frac{7000 \text{ N/m}^3}{9810 \text{ N/m}^3}\right)} = 3.74 \text{ m/s}$$

$$Q = AV = \frac{\pi}{4}D^2V = \frac{\pi}{4}(0.15)^2(3.74) = 0.0661 \text{ m}^2/\text{s}$$

11.44 a. For a static head orifice discharging freely into the atmosphere

$$Q = CA\sqrt{2gh}$$

11.45 c. Hydraulic radius, R, is defined as cross-sectional area, divided by wetted perimeter.

11.46 d. Choices (a) and (c) are not dimensionless, as required for a Reynolds number. Since the hydraulic radius R = (cross sectional area)/(wetted perimeter), for a circular cross section,

$$R = \frac{\frac{\pi D^2}{4}}{\pi D} = \frac{D}{4} \quad \text{or} \quad D = 4R$$

Therefore,

$$\text{Re} = \frac{4RV\rho}{\mu} = \frac{4RV}{v}$$

CHAPTER 12

Hydraulics and Hydrologic Systems

OUTLINE

INTRODUCTION

This chapter will selectively review hydrostatics, the fundamental principles for conservation of fluid mass (continuity), energy and linear momentum, the selection and operation of centrifugal pumps and turbines, and elements of open channel flow. The review focuses primarily on water, assumes it is incompressible, and uses standard values for its properties. (Water density and viscosity depend somewhat on temperature, and these values can be obtained from tables in reference books.) Since entire books have been written on hydraulics, one can augment this review, if desired, by use of the supplementary references following the chapter.

Hydrostatics

The pressure distribution, p, in a motionless body of water is given by

$$\nabla p = \rho g \tag{12.1}$$

Using an (x, y, z) coordinate system, with x and y horizontal and z vertically upward, one obtains

$$\frac{dp}{dz} = -\rho g = -\gamma \tag{12.2}$$

in which ρ is the density (or mass density) of water (1.94 slugs/ft^3 or 1000 kg/m^3), g is the acceleration of gravity, and γ is the unit weight (or specific weight or weight density) for water (62.4 lb/ft^3 or 9800 N/m^3). The pressure in the horizontal (x, y) plane is uniform, that is, constant. Integration of Equation (12.2) between points 1 and 2 yields

$$p_2 - p_1 = -\gamma(z_2 - z_1) \tag{12.3}$$

Normally one can use either gage or absolute pressures in a problem so long as they are not mixed. **Gage pressure** registers zero at standard conditions (temperature = 273 K = 0°C; pressure = 1 atm = 760 torr = 760 mm Hg = 101.3 kPa = 14.7 lb/in.). The density or unit weight of other liquids is sometimes given in terms of the ratio $S = \rho/\rho_w = \gamma/\gamma_w$, where S is the specific gravity of the liquid.

Example 12.1

In Exhibit 1, reservoirs A and B are connected by a tube. The specific gravities of the liquids are $S_A = 0.8$, $S_1 = 1.3$, $S_2 = 1.6$, and $S_B = 1.0$. Determine the pressure difference between points A and B.

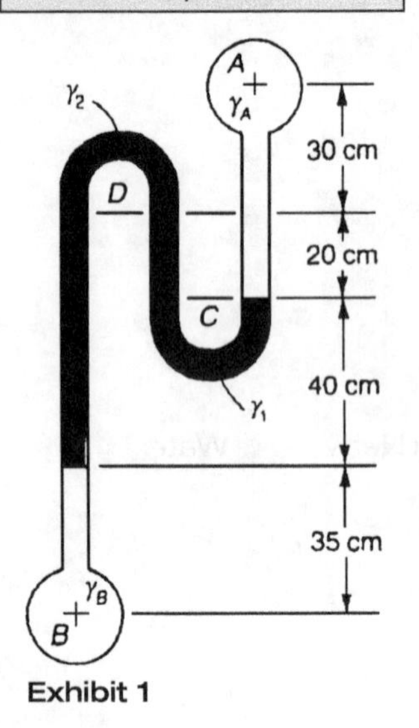

Exhibit 1

Solution

The pressures in opposing limbs of the tube at points C and D are equal on the left and right sides of each limb, that is, $p_{CL} = p_{CR}$ and $p_{DL} = p_{DR}$, where L and R indicate "left" and "right."

$$p_{CR} = p_A + (0.3 + 0.2)\,\gamma_A$$
$$p_{DR} = p_{CL} - 0.2\,\gamma_1$$
$$p_{DL} = p_B - 0.35\,\gamma_B - (0.4 + 0.2)\,\gamma_2.$$

Now algebra is used to eliminate the intermediate pressures:

$$p_B - p_A = 0.5\gamma_A - 0.2\gamma_1 + 0.6\gamma_2 + 0.35\gamma_B.$$

Inserting the specific gravity data, one obtains

$$p_B - p_A = [0.5(0.8) - 0.2(1.3) + 0.6(1.6) + 0.35(1.0)]\gamma_w$$
$$p_B - p_A = 1.45\gamma_w = (1.45\text{ m})(9800\text{ N/m}^3) = 14{,}200\text{ N/m}^2 = 14.2\text{ kPa}$$

The **buoyant force** on a floating or fully immersed object is

$$F_B = \gamma\,(\text{volume}) \qquad \textbf{(12.4)}$$

in which the volume is the amount of fluid of unit weight, γ, displaced.

The fluid pressure distribution on any submerged surface will develop a hydrostatic force on that surface. The overall magnitude and the direction of such a force on a surface of any shape can be computed by first determining the magnitude and direction of the force components in the horizontal and vertical directions; these component results are then combined by direct use of the principles of statics.

The magnitude of the force F on a submerged plane surface is

$$F = \int_A p\,dA = p_C A = \gamma h_C A \qquad \textbf{(12.5)}$$

in which A is the surface acted on, and p_C and h_C are the pressure and submerged depth of the centroid of this area.

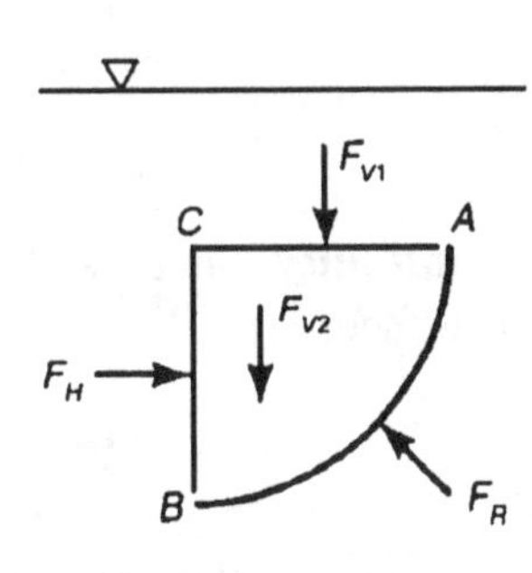

Figure 12.1

In Figure 12.1 the two-dimensional curved surface AB is acted upon by fluid above and to the left of it. A free-body diagram of the solid surface and a chunk of fluid above it is labeled ABC, and rectangular component forces F_H and F_{V1} act in the horizontal and vertical directions on the two faces CB and CA, which are each seen on edge. The line of the action of F_{V1} coincides with the centroid of area CA since this area is horizontal. The line of action of F_H lies below the centroid of area CB a distance $I/h_C A$, in which I is the moment of inertia of A about its centroid. The other vertical force component, F_{V2}, is equal to the weight of the fluid within the free body itself and acts through the centroid of this volume of fluid. By direct force summations, the components of the resultant force F_R are

$$(F_R)_h = F_H \text{ and } (F_R)_v = F_{V1} + F_{V2}$$

If surface AB is a plane inclined surface, Equation (12.5) may be used directly to find the magnitude of the force normal to it. The method described above for curved surfaces could still be used to find the location of this force, if needed.

Example 12.2

Exhibit 2 shows two vertical submerged plane areas. Determine the hydrostatic force on each area and the point of application of each equivalent concentrated force.

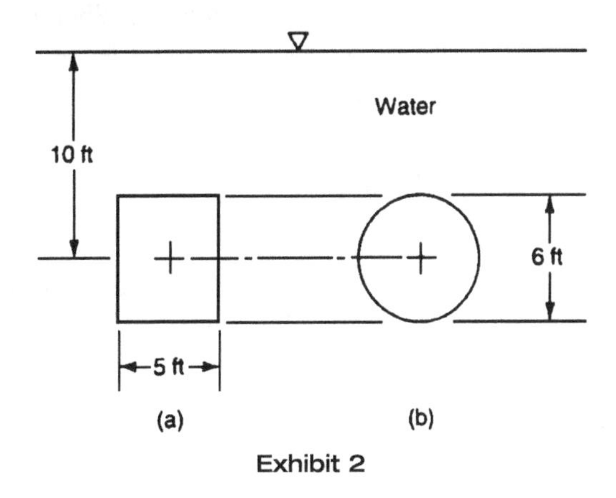

Exhibit 2

Solution

For the rectangle of area $A = (6)(5) = 30$ ft² in Exhibit 2(a),

$$F = \gamma h_C A = (62.4)(10)(30) = 18{,}700 \text{ lb}$$

For the circle of area $A = \pi(6)^2/4 = 28.3$ ft² in Exhibit 2(b),

$$F = \gamma h_C A = (62.4)(10)(28.3) = 17{,}700 \text{ lb}$$

The moments of inertia of these two shapes about their respective centroids are

$$I = \frac{bh^3}{12} = \frac{(5)(6)^3}{12} = 90.0 \text{ ft}^4 \text{ and } I = \frac{\pi}{4}R^4 = \frac{\pi}{4}(3)^4 = 63.6 \text{ ft}^4$$

Thus the distances down from the centroids to the actual point of application of these two forces are $I/(h_C A) = (90)/[(10)(30)] = 0.300$ ft and $I/(h_C A) = (63.6)/[(10)(28.3)] = 0.225$ ft, respectively.

CONSERVATION LAWS

Continuity

The principle of conservation of mass, often called simply **continuity**, can be written for a control volume of fixed volume, V, enclosed by a surface, S, as

$$\frac{\partial}{\partial t}\int_V \rho dV + \int_S \rho \mathbf{V} \cdot \mathbf{n}\, dS = 0 \qquad \textbf{(12.6)}$$

The two terms express, in turn, the accumulation of mass within the volume and the net outflow of fluid across the boundary, S. The dot product of the velocity vector $\mathbf{V}$ and the unit outer normal $\mathbf{n}$ gives the component of the velocity normal to dS. If the flow is steady, the first term is zero. Applied to a single streamtube in steady incompressible flow, the principle is often written as

$$Q = \int_A V dA = \text{ constant} \qquad \textbf{(12.7)}$$

or

$$Q = V_1 A_1 = V_2 A_2 \qquad \textbf{(12.8)}$$

in which Q is the discharge or volume rate of flow at any cross section of the streamtube, which is also expressible as the product of the mean velocity, V, and cross-sectional area, A. When more than one fluid stream enters or leaves the control volume, additional terms are needed in an overall continuity statement; they come from the evaluation of the second term in Equation (12.6).

Energy

A general statement of mechanical energy conservation between points 1 and 2 in a one-dimensional, steady incompressible flow from 1 to 2 may be written, per unit weight of fluid, as

$$\frac{V_1^2}{2g} + \frac{p_1}{\gamma} + z_1 = \frac{V_2^2}{2g} + \frac{p_2}{\gamma} + z_2 + h_L - E_m \qquad \textbf{(12.9)}$$

In Equation (12.9), $V^2/2g$ is the velocity head or kinetic energy per unit weight, p/γ is the pressure head or pressure energy per unit weight, and z is the elevation head or potential energy per unit weight. The head loss, h_L, is the accumulated loss in energy per unit weight occurring between points 1 and 2 caused by local and/or frictional effects. The last term, E_m, is the mechanical energy added to the flow between the two points by the action of hydraulic machinery. A pump adds energy to the flow, so E_m is positive; a turbine removes energy from the flow, and E_m is negative. In the absence of energy losses or gains (that is, h_L and E_m are both zero), one obtains the classic Bernoulli equation for ideal fluid flow.

Fluid power, P, is the product of the energy gained or lost per unit weight, E_m, and the weight rate of flow, $Q\gamma$, or $P = Q\gamma E_m$. If power is to be in horsepower, then one should divide this result by the factor 550 ft-lb/sec/horsepower. Depending on the specific application, power should sometimes be multiplied or divided by an efficiency factor η, as some of the problems in this chapter will demonstrate.

In practical pipeflow problems the most important cause of head loss is pipe friction. In a single pipe of length L and diameter D, the Darcy-Weisbach equation for frictional head loss is

$$h_L = h_f \frac{L}{D}\frac{V^2}{2g} \tag{12.10}$$

The friction factor is $f = f(\text{Re}, \varepsilon/D)$ in which the Reynolds number is $\text{Re} = VD\rho/\mu = VD/v$ and ε/D is called the relative roughness (μ = viscosity, v = kinematic viscosity). The Moody diagram, Figure 12.2, is a plot of the friction factor, f, as a function of Re and ε/D.

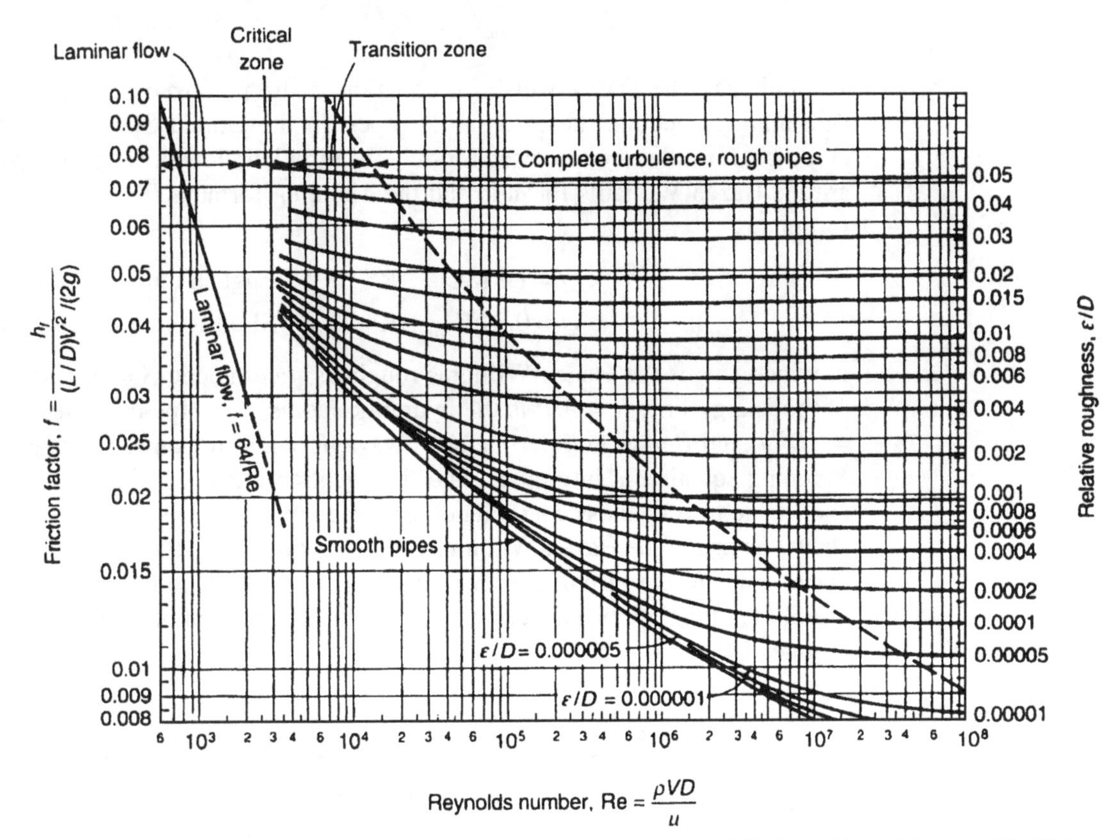

Figure 12.2 Moody Diagram (Source: L. F. Moody, *Transactions ASME,* 66, Nov. 1944, pp. 671–684.)

Below a transition or critical Reynolds number of approximately 2300, one has laminar flow in a pipe, and the unique relation $f = 64/\text{Re}$ applies. Above a Reynolds number of about 4000, turbulent flow normally occurs, with differing values of f in this regime depending also on ε/D; the lowest of these lines is for a smooth pipe. In the zone of wholly rough turbulent flow (above and to the right of the dashed line), the friction factor effectively depends only on ε/D and not on Re. Values of the absolute roughness, ε, for various pipe materials can be found in Table 12.1.

Table 12.1 Values of absolute roughness ε for new pipes

Material	ε	
	Inches	Millimeters
Riveted steel	0.036 to 0.36	0.91 to 9.1
Concrete	0.012 to 0.12	0.3 to 3.0
Wood stave	0.0072 to 0.036	0.18 to 0.9
Cast iron	0.0102	0.26
Galvanized iron	0.006	0.15
Asphalted cast iron	0.0048	0.12
Welded steel pipe	0.0018	0.046
Commercial steel or wrought iron	0.0018	0.046
PVC	0.000084	0.0021
Drawn tubing	0.00006	0.0015
Glass, brass, copper, lead	"Smooth"	"Smooth"

Source: L. F. Moody, *Transactions ASME*, 66, Nov. 1944, pp. 671–684.

Although the Darcy-Weisbach equation is the preferred equation to use in determining frictional head losses in pipes, in the United States two primarily empirical formulas are also used for only the turbulent flow of water in pipes. They are the Hazen-Williams formula and the Manning formula. The Hazen-Williams formula is

$$Q = 1.318 C_{HW} A R^{0.63} S^{0.54} \quad \text{English units}$$
$$Q = 0.849 C_{HW} A R^{0.63} S^{0.54} \quad \text{SI units} \qquad \textbf{(12.11)}$$

in which C_{HW} is the Hazen-Williams roughness coefficient, $S = h_L/L$ is the slope of the energy line, $R = A/P$ is the hydraulic radius, A = pipe cross-sectional area $= \pi D^2/4$, and P = wetted perimeter. Always, $R = D/4$ for pipes flowing full. The Manning equation is introduced later in Equation (12.29). Table 12.2 presents a short table of roughness coefficients for some pipe materials. Investigation would show that use of the Hazen-Williams equation with C_{HW} in the range of 130 to 150 corresponds to the turbulent transitional region of the Moody chart, whereas the use of the Manning formula corresponds to the assumption of a wholly rough flow regime.

Table 12.2 Roughness coefficients for pipe materials

Pipe Material	C_{HW}	n
PVC	150	0.009
Very smooth	140	0.010
Cement-lined ductile iron	140	0.012

(continued)

New cast iron, welded steel	130	0.014
Wood, concrete	120	0.016
Clay, new riveted steel	110	0.017
Old cast iron, brick	100	0.020
Badly corroded cast iron	80	0.035

Source: Larock et al., *Hydraulics of Pipeline Systems*, CRC Press, 2000.

All pipe systems are subject to local losses, sometimes called minor losses, of energy in addition to a frictional energy loss. The pipe entrance, exit, and any hardware fittings (valves, etc.) between these points that can change or disrupt the flow in the pipe all cause a local loss in energy. Each loss can be written as

$$h_L = K\frac{V^2}{2g} \quad \textbf{(12.12)}$$

and all such losses are summed as one traverses the pipe from end to end. At the pipe entrance the local loss coefficient depends on the local geometry. For a sharp-edged entrance $K = 0.5$; if the entrance is well rounded, then K is 0.04 to 0.10; if the entrance is reentrant, meaning that the pipe projects into the reservoir, then $K = 0.8$. The appropriate reference velocity, V, is usually, but not always, the mean velocity immediately downstream of the loss-causing device or location. For a sudden expansion from an upstream pipe of area A_1, to a downstream pipe of area A_2 and velocity V_2, the loss coefficient is $K = (A_2/A_1 - 1)^2$ and $V = V_2$. For a sudden contraction between sections 1 and 2, the loss coefficient K is given in Figure 12.3 as a function of A_2/A_1. One exception is the pipe exit to a reservoir, where V must be the upstream velocity and $K = 1.0$ for all exit shapes. A sampling of additional loss coefficients K is given in Table 12.3. More coefficients for other shapes and sizes may be found in textbooks and the references therein.

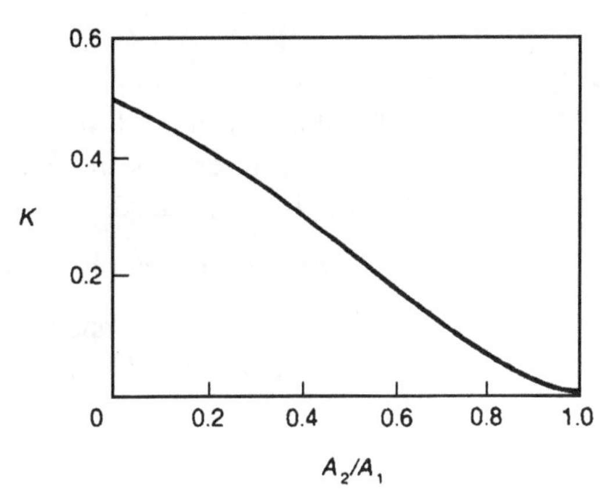

Figure 12.3

Table 12.3 Loss coefficients for a few standard (threaded) pipe fittings

Globe valve, wide open	10.0
Angle valve, wide open	5.0
Gate valve, wide open	0.2
Gate valve, half open	5.6
Return bend	2.2
Tee	1.8
90° elbow	0.9
45° elbow	0.4

Source: Streeter, *Handbook of Fluid Dynamics*, 1961, McGraw-Hill.

The energy equation, Equation (12.9), and the information that follows it, can be used to solve a variety of pipeflow problems. However, all of these problems will fit into one of three fundamental problem types:

1. Q and pipeline properties are known, and h_L is to be found.
2. The head difference, H, and the pipeline properties are known, and Q is to be found.
3. Q is prescribed, H is known, and the minimum required pipe diameter, D, is to be found.

The first two problem types are analysis problems, but the third type is typically encountered in a design context. In a Type-1 problem, one can immediately determine ε/D and Re and thus find f. Use of the Darcy-Weisbach expression, Equation (12.10), yields the final result directly. The other two problem types usually require iterative computations, as the next example shows.

Example **12.3**

In Exhibit 3 the head difference between the two reservoirs of water at 10°C is H = 15 m. The reservoirs are connected by L = 300 m of clean cast iron pipe. The pipe entrance is sharp edged.

a). If the pipe diameter is D = 0.3 m, find the discharge in the pipe.

b). Years have passed, and this pipe system is now required to deliver 0.5 m³/s. Is the original pipe diameter adequate, or is a larger pipe diameter now needed?

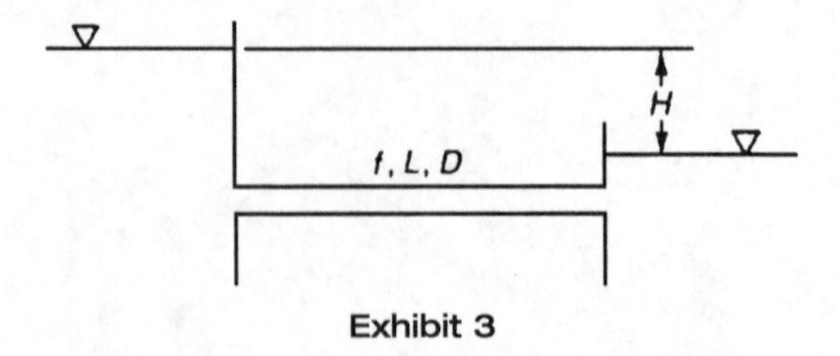

Exhibit 3

Solution

a). Between the two reservoirs the sum of the entrance, friction, and exit losses is equal to the head difference H:

$$K_{\text{ent}}\frac{V^2}{2g}+f\frac{L}{D}\frac{V^2}{2g}+K_{\text{exit}}\frac{V^2}{2g}=H$$

or

$$\left[K_{\text{ent}}+f\frac{L}{D}+K_{\text{exit}}\right]\frac{V^2}{2g}=H$$

For this pipe entrance K_{ent} = 0.5, and for all pipe exits K_{exit} = 1.0. Hence

$$\left[0.5+f\frac{300}{0.3}+1.0\right]\frac{V^2}{2(9.81)}=15.0 \tag{12.13}$$

Using Table 12.1, the relative roughness ε/D = 0.00026/0.3 = 0.00087. The Reynolds number is

$$\text{Re}=\frac{VD}{\nu}=\frac{V(0.3)}{1.3(10^{-6})}=2.3(10^5)V \tag{12.14}$$

which is not known initially. However, if one assumes that Re is large, then for the computed ε/D one can estimate directly from Figure 12.2 that f = 0.019 in the wholly rough flow region. Equation (12.13) can now be solved directly to obtain V = 3.79 m/s. Then Equation (12.14) yields Re = 8.7(10^5), and a check of Figure 12.2 suggests that f is between 0.019 and 0.020. If one re-solves Equation (12.13), V changes slightly; the discharge is $Q = VA$ = (3.79) $\times\ \pi(0.3)^2/4$ = 0.268 m^3/s.

b). Clearly the original pipe diameter is too small. Rewriting Equation (12.13) to display fully the role of diameter D,

$$\left[0.5+f\frac{300}{D}+1.0\right]\frac{1}{2g}\frac{Q^2}{(\pi/4)^2 D^4}=H \tag{12.15}$$

one sees that this equation is highly nonlinear in D. Moreover, the relative roughness and Re also depend on D. A trial solution is required with Q = 0.5 m^3/sec. One can begin by estimating that D should be approximately 0.4 m because Q has roughly doubled from part (a), and Q can be expected to vary with A or D^2. For a chosen value of D, ε/D and Re can be computed, yielding f = 0.018 from Figure 12.2. Hence H = 12.1 m for D = 0.4 m. As a next trial, D = 0.38 m gives H = 15.6 m. Interpolation suggests D = 0.383 m will give H = 15.0 m, which can be confirmed by calculations. One should select the next larger commercially available pipe size.

Momentum

In steady flow the conservation of linear momentum for a fixed control volume, V, enclosed by a surface S is

$$\mathbf{F}_S+\mathbf{F}_B=\int_S \mathbf{V}(\rho\mathbf{V}\bullet\mathbf{n})\,dS \tag{12.16}$$

In this vector equation the first two terms represent the surface and body forces, respectively, and the integral is the net outward flux of momentum through the surface of the control volume. The first term includes the overall effect of any distributed pressures and any viscous or shear stresses on the surface S. Usually the body force is just the weight of everything within the control volume; it acts in the direction of gravity. For flow in two dimensions with uniform flow across an entrance section 1 and an exit section 2, Equation (12.16) is equivalent to the two scalar component equations

$$\begin{aligned} \Sigma F_x &= \rho Q\left(V_{2x} - V_{1x}\right) \\ \Sigma F_y &= \rho Q\left(V_{2y} - V_{1y}\right) \end{aligned} \qquad \textbf{(12.17)}$$

in which the volume flow rate is Q.

Example 12.4

Determine the horizontal component of the force per unit width of the water on the radial gate shown in Exhibit 4.

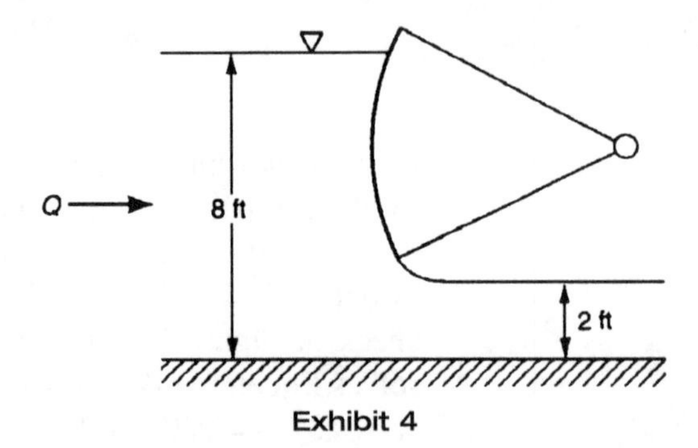

Exhibit 4

Solution

First find the discharge by applying the continuity and energy equations between sections 1 and 2, and then use the x-component of Equation (12.17) to find the force. Between the sections, the discharge per unit width $q = y_1V_1 = y_2V_2$ or $8V_1 = 2V_2$. Now assume as a reasonable approximation that

$$\frac{V_1^2}{2g} + y_1 = \frac{V_2^2}{2g} + y_2$$

or

$$\frac{1}{2g}\left(V_2^2 - V_1^2\right) = \frac{1}{2(32.2)}\left[1 - \left(\frac{2}{8}\right)^2\right]V_2^2 = 8 - 2 = 6$$

by using continuity, so that $V_2 = 20.3$ ft/s, $q = 2V_2 = 40.6$ ft²/s, and $V_1 = 5.08$ ft/s.

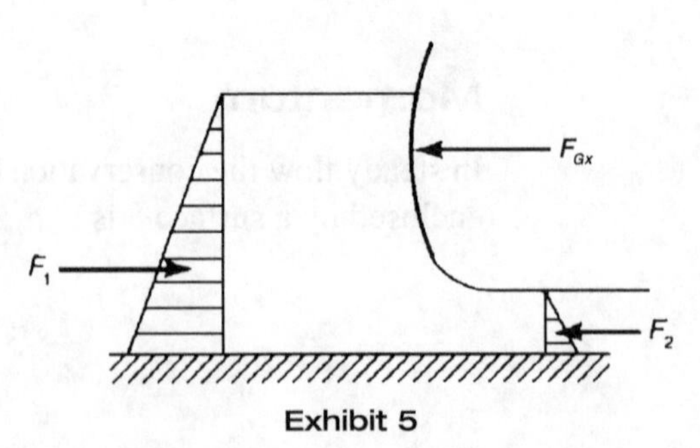

Exhibit 5

Now the horizontal component of Equation (12.16) is specialized to the control volume shown in Exhibit 5 to obtain

$$F_1 - F_2 - F_{Gx} = \rho Q(V_2 - V_1)$$

in which the first two terms are hydrostatic forces, and the third term is the integrated effect of the gate on the water in the x-direction. Thus

$$\frac{1}{2}\rho g\left(y_1^2 - y_2^2\right) - F_{Gx} = \rho q\left(V_2 - V_1\right)$$

$$\frac{1}{2}(62.4)\left(8^2 - 2^2\right) - F_{Gx} = 1.94(40.6)(20.3 - 5.08)$$

$$1870 - F_{Gx} = 1200$$

and the horizontal force of the water on the gate is equal and opposite this force, or 670 lb per unit width to the right.

PUMPS AND TURBINES

Turbomachinery, which incorporates a central rotating unit or impeller, is reviewed in this section. Positive-displacement pumps also play a significant role in engineering, but they usually operate at relatively lower discharges and will not be reviewed here. After introducing some basic terms, the review will look at nondimensional pump parameters and their relation to specific speed. The relation of the main centrifugal pump types to specific speed will be mentioned next, followed by a look at pump characteristic curves and their relation to hydraulic system requirements. A review of turbine types concludes the section.

Turbomachinery Similitude and Specific Speed

In the extended energy equation, Equation (12.9), E_m is the mechanical energy per unit weight added to or removed from the fluid stream between two points. For pumps this addition is the net pump head, H, which acts primarily to increase the pressure head. The power delivered to the fluid stream equals the product of H and the weight rate of flow $Q\gamma$ and is called the water power or horsepower, P_w. The mechanical power used in driving the pump is larger and is called brake horsepower, bhp = ωT, in which ω and T are the angular velocity and torque on the rotating shaft that drives the pump. The ratio P_w/bhp is the pump efficiency, η, which may be above 0.8 for large pumps operating near their best efficiency point, but which may be lower, even much lower, for pumps that are small or operating away from best conditions.

Similitude considerations for turbomachinery lead to three nondimensional parameters that can be used to summarize performance:

$$C_H = \frac{gH}{N^2 D^2}; \quad C_Q = \frac{Q}{ND^3}; \quad C_P = \frac{P}{\rho N^3 D^5} \qquad \textbf{(12.18)}$$

These coefficients are the head, discharge, and power coefficients, respectively. The length D is to be representative of the size of the unit; usually it is the impeller diameter. The rotative speed parameter, N, should have units of 1/time, but rev/s or rev/min is used conventionally. The coefficients in Equation (12.18) are, in principle, all dependent on Reynolds number and relative roughness, but at high Reynolds number this fact is often overlooked. Thus, if pumps 1 and 2 are of similar geometric shape and operate so that internal flow patterns are the same, then the units are called homologous and

$$\left(\frac{H}{N^2D^2}\right)_1 = \left(\frac{H}{N^2D^2}\right)_2; \quad \left(\frac{Q}{ND^3}\right)_1 = \left(\frac{Q}{ND^3}\right)_2; \quad \left(\frac{P}{\rho N^3 D^5}\right)_1 = \left(\frac{P}{\rho N^3 D^5}\right)_2 \tag{12.19}$$

These similarity laws are quite versatile. If pumps 1 and 2 have the same diameter, the rules show how H, Q, and P scale with N; or for fixed speed, N, the rules show how these quantities scale with size D. Note also that N may be given in any of the units rad/sec, rev/s, or rev/min in Equation (12.19). In Equation (12.18), however, it is important to know which of these units is being used.

All coefficients so far contain the size of the unit D; a grouping that displays the important variables, Q and H for pumps or P and H for turbines, without D appearing would be a valuable characterization of the performance of the unit. Specific speed is such a grouping; it is a shape parameter and takes on its greatest meaning when defined at the machine's best efficiency point (bep). In the United States, there are two forms that specific speed can take for pumps, and two others for turbines. In each pairing the first form is nondimensional and the second is the U.S. traditional form, which is commonly used but is far from being nondimensional.

$$\text{Pumps:} \quad n_s' = \frac{NQ^{1/2}}{g^{3/4}H^{3/4}} \qquad N_s' = \frac{(\text{rev/min})(\text{gal/min})^{1/2}}{[H(\text{ft})]^{3/4}} \tag{12.20}$$

$$\text{Turbines:} \quad n_s = \frac{NP^{1/2}}{\gamma^{1/2}g^{3/4}H^{5/4}} \qquad N_s = \frac{(\text{rev/min})(\text{horsepower})^{1/2}}{[H(\text{ft})]^{3/4}} \tag{12.21}$$

The first forms may be used in any dimensionally consistent set of units.

Pump Types

Specific speed, as a shape parameter, can be used to select the most appropriate kind of pump (or turbine) for a particular application. Pumps are normally classified according to the predominant direction of flow through the pump impeller. Experience has shown when the pump specific speed, N_s', is below roughly 4000 that the most efficient pump type uses a radial-flow impeller; that is, the primary direction of flow through the impeller is normal to the axis of rotation of the impeller. Between specific speeds of 4000 and 10,000, the most efficient pump type uses a mixed-flow impeller, and above a specific speed of 10,000 the efficient pump type uses an axial-flow impeller similar to a propeller. There is some overlap of intervals, however.

Information on pump performance is usually displayed in a set of characteristic curves. Most sets of curves are dimensional, but occasionally they are nondimensionalized. Figure 12.4 presents a set of characteristic curves for a pump operating at 900 rpm. In this case the best efficiency point (bep) is at 80% efficiency, corresponding to a head of 13.7 ft, a discharge of 2.5 ft³/s, and a required bhp of 3.5. Thus the computed pump specific speed N_s' is 4230, which is on the lower end of the mixed-flow range.

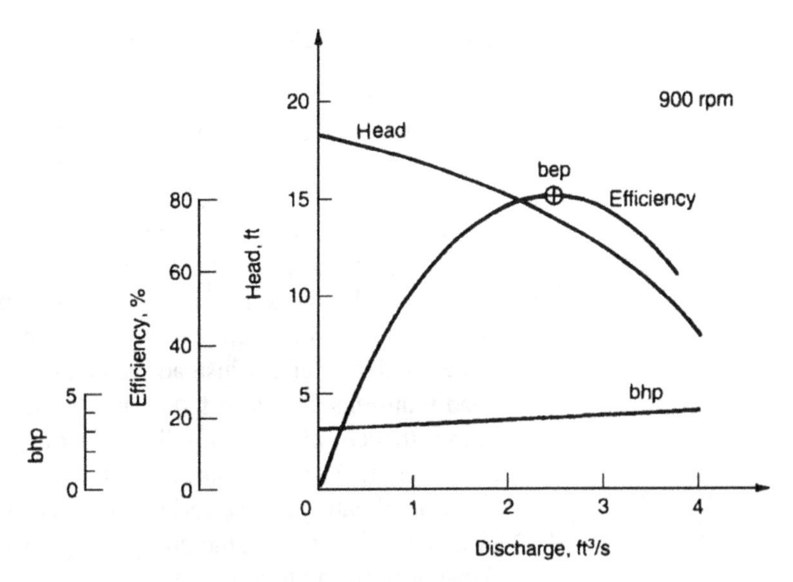

Figure 12.4

Net Positive Suction Head

The net positive suction head (NPSH) is the head difference between (1) the energy of the fluid per unit weight at the inlet side of the pump, measured above the inlet elevation, and (2) the vapor-pressure head of the fluid. That is,

$$\text{NPSH} = \frac{p_i}{\gamma} + \frac{V_i^2}{2g} - \frac{p_v}{\gamma} \tag{12.22}$$

In this expression all pressures must be computed as absolute, not gage, pressures. In order to avoid the onset of **cavitation**, which is the local conversion of liquid to vapor due to low pressure, each pump must be located so that the NPSH of the pump is exceeded. Referring to Figure 12.5, if Equation (12.9) is written between the reservoir and the pump inlet i, one obtains

$$R + \frac{p_{\text{atm}}}{\gamma} - h_L = \frac{p_i}{\gamma} + \frac{V_i^2}{2g} + R + z_i$$

The resulting relation for NPSH is therefore

$$\text{NPSH} = \frac{p_{\text{atm}}}{\gamma} - \frac{p_v}{\gamma} - h_L - z_i \tag{12.23}$$

This equation clearly shows that the distance z_i that the inlet can be placed above the reservoir surface is limited. Pump manufacturers routinely supply pump characteristic curves and NPSH data to users.

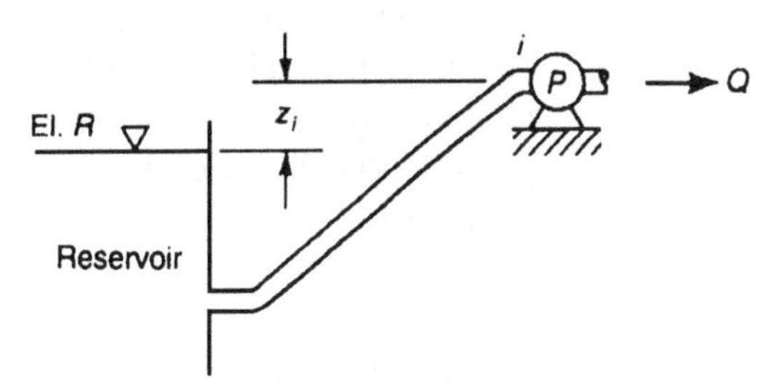

Figure 12.5

System Performance

Just as Figure 12.4 graphically presents a pump characteristic curve or relation between H and Q, Equation (12.9) describes the head requirement (H or E_m) of the hydraulic system between two points that must be satisfied if the system is to function properly. An inspection of Equation (12.9) shows that the system head requirement consists of two parts: a static lift that is the difference in elevation between points 1 and 2 and a frictional head component that grows in size with an increase in discharge (if f is constant, this part grows as Q^2). Matching the heads, either computationally or by use of a graph, determines the discharge at which the system will operate. If instead the desired discharge is prescribed, then the system head requirement is also fixed, and the problem is to select some pump or set of pumps that can efficiently deliver this discharge at this head. One then inspects the characteristic curves of various pumps until a suitable pump (or pumps) is found.

Complicating this selection process is the fact that, over time, pipes age and the system head increases; the pump ages as well and becomes less efficient. Together these factors lead to a decrease in system discharge. Sets of pumps are sometimes installed in series or in parallel to give system flexibility as the system ages and to aid in pump maintenance. Such arrangements also require additional piping and valving so individual units can be isolated (e.g., for servicing). The primary facts to recall in considering pump combinations are (1) the discharge is constant and heads add for pumps in series, and (2) discharges add whereas the head across a parallel combination is constant.

Turbines

Turbine specific speed, defined in Equation (12.21), can be used to classify the major types of hydropower turbines:

N_s	Type	ϕ
1–10	Impulse (Pelton)	~0.47
15–110	Francis	0.6–0.9
100–250	Propeller (Brightwood)	1.4–2.0

The quantity ϕ is the **peripheral speed factor,** which is the ratio of a typical runner speed to a typical fluid speed for the turbine. The specific speed allows one to identify clearly the different turbine types.

Impulse turbines are low-discharge, high-head devices in which one or more high-speed jets of water at atmospheric pressure act on carefully shaped vanes or "buckets" on the periphery of the rotating turbine wheel. Although its specific speed range is already narrow, it probably should be still narrower because efficiency drops rapidly at both ends of the range; for $N_s \leq 2$ the impulse wheel is relatively large and cumbersome with large electrical losses, and at $N_s > 8$ the fluid jet is not handled well by the buckets. The equations for force on a bucket F and for power P are

$$F = \rho Q v_r (1 - \cos\beta) \qquad P = Fu \tag{12.24}$$

in which the fluid speed relative to the runner is $v_r = V - u$, the speed of the fluid jet is V, the runner speed itself is $u = \omega r$, the angular velocity of the wheel is ω, the midbucket radius from the axis of rotation (the "pitch circle") is r, and β is the angle through which the fluid is turned on impact with the bucket. Normally β is 165° or a bit more.

Both Francis and propeller turbines are reaction turbines. The turbine runner is fully enclosed and acted upon by water under pressure.

Francis turbines are for moderate-head, moderate-discharge applications. At the lower values of N_s the flow through the rotating turbine runner is in the radial direction, whereas at the higher end of the N_s range the flow direction is mixed, being partly radial and partly axial.

For propeller turbines, the predominant flow direction through the runner is axial in order to handle a high discharge at low head. For all reaction turbines, the flow leaves the delivery pipe, called a penstock, and enters the turbine through a scroll case that wraps around the central turbine unit, gradually decreasing its cross section to force the flow toward the rotating impeller through fixed guide vanes and adjustable wicket gates.

The power, torque, and discharge for a reaction turbine are all related to the interaction of the flow with the turbine runner blades. Figure 12.6 depicts a pair of velocity diagrams for flow at the inlet, section 1, and the outlet, section 2, of a typical turbine blade. The runner angular velocity is ω. The absolute velocity, V, of the fluid can be viewed in two ways. It is composed of radial and tangential components V_r and V_t, respectively, but it is also composed of the blade velocity $u = \omega r$ and the fluid velocity w relative to the moving blade. Two angles help to define the velocity diagram: The angle at which the fluid enters the runner region is α, measured from the tangent to the circle surrounding the runner—also $\tan\alpha = V_r/V_t$; the second angle, β, is measured form u to w. The power and torque are

$$P = T\omega \qquad T = \rho Q\left[V_{t1}r_1 - V_{t2}r_2\right] \tag{12.25}$$

The discharge is the product of the radial velocity component and the area through which this velocity flows, or

$$Q = V_r\left(2\pi rb\right) \tag{12.26}$$

in which b is the thickness of the section. Useful velocity relations from Figure 12.6 are $V_r = w\sin\beta$ and $V_t = u + w\cos\beta = u + V_r\cot\beta$.

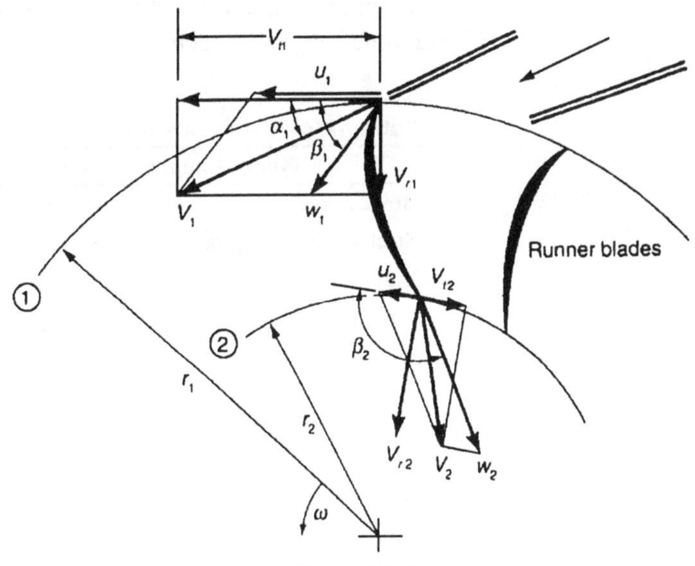

Figure 12.6

OPEN CHANNEL FLOW

Basic elements of open channel flow will be presented here, including descriptions of uniform flow and the use of the Manning equation for discharge determination, the role of the hydraulic jump, efficient section considerations, specific energy, and an introduction to gradually varied flow.

Uniform Flow and Manning Equation

The primary distinguishing feature of open channel flow is the presence of a constant-pressure free surface atop all such flows. If the flow properties at a point or section in a channel are unchanging with time, then the flow is steady; otherwise it is unsteady. If the flow properties in a steady flow are unchanging from one cross section to another in a channel, then the flow is uniform in space. In this case the slopes of the channel bottom, water surface, and energy grade line are all identical, and the shape of the channel section is unchanging with distance along the channel. Then one can select a longitudinal reach of a channel as a control volume and apply the steady linear momentum conservation principle, Equation (12.16), to the reach to find a balance between the force driving the flow, which is the component of the fluid weight down the slope S_o, and the frictional resisting force. The mean shear stress τ on the boundary of the channel cross section turns out to be

$$\tau = \gamma R S_o \qquad \textbf{(12.27)}$$

in which $R = A/P$ is called the hydraulic radius of the cross section, and A and P are, respectively, the area and the wetted perimeter of the section. With the aid of a bit of dimensional analysis, one can then develop the discharge equation originated by A. Chézy, which is

$$Q = VA = CA\sqrt{RS_o} \qquad \textbf{(12.28)}$$

The search for the best way to characterize or specify C has continued over many years. Currently the most popular approach is to adopt the **Manning** representation, which is $C = R^{1/6}/n$. Table 12.4 lists a variety of typical values for n, the Manning roughness factor.

Table 12.4 Values of roughness coefficient n

Type of Channel Surface	Minimum	Normal	Maximum
Brass, smooth	0.009	0.01	0.013
Steel, riveted and spiral	0.013	0.016	0.017
Cast iron, coated	0.01	0.013	0.014
Wrought iron, galvanized	0.013	0.016	0.017
Lucite	0.008	0.009	0.01
Glass	0.009	0.01	0.013
Cement, neat surface	0.01	0.011	0.013
Concrete, finished	0.011	0.012	0.014
Wood, stave	0.01	0.012	0.014
Clay, common drainage tile	0.011	0.013	0.017
Clay, vitrified sewer	0.011	0.014	0.017
Brick work, lined with cement mortar	0.012	0.015	0.017
Rubble masonry, cemented	0.018	0.025	0.03
Smooth steel surface, unpainted	0.011	0.012	0.014

Table 12.4 Values of roughness coefficient n (continued)

Type of Channel Surface	Minimum	Normal	Maximum
Cement, mortar	0.011	0.013	0.015
Wood, unplaned	0.011	0.013	0.015
Concrete, trowel finish	0.011	0.013	0.015
Concrete, gunite—good section	0.016	0.019	0.023
Gravel bottom with sides of formed concrete	0.017	0.02	0.025
Brick, glazed	0.011	0.013	0.015
Asphalt, smooth	0.013	0.013	—
Asphalt, rough	0.016	0.016	—
Vegetal lining	0.03	—	0.5
Excavated or dredged earth, straight and uniform:			
Clean, recently completed	0.016	0.018	0.02
Clean, after weathering	0.018	0.022	0.025
Gravel, uniform section, clean	0.022	0.025	0.03
With short grass, few weeds	0.022	0.027	0.033
Natural streams:			
Minor streams on plain, top width at flood stage <30 m, clean, straight, full stage, no rifts or deep pools	0.025	0.03	0.033
Mountain streams, no vegetation in channel, banks unusually steep, trees and brush along banks submerged at high stages, bottom with gravels, cobbles, and few boulders	0.03	0.04	0.05

Source: V. T. Chow, *Open-Channel Hydraulics* (table abridged), Copyright © Estate of V.T. Chow, and reproduced by permission.

The Manning equation for discharge determination in uniform, open channel flow is

$$Q = \frac{K}{n} A R^{2/3} S_o^{\,1/2} \tag{12.29}$$

For SI units use $K = 1.0$; for English units use $K = 1.49$. These choices for K allow one to retain one value of the Manning n for each physical roughness without any need for other unit conversions.

Example 12.5

A gunite concrete trapezoidal channel with 1:2 side slopes, which is shown in Exhibit 6, conveys 200 ft³/s on a slope $S_o = 0.0005$. Compute the depth of uniform flow.

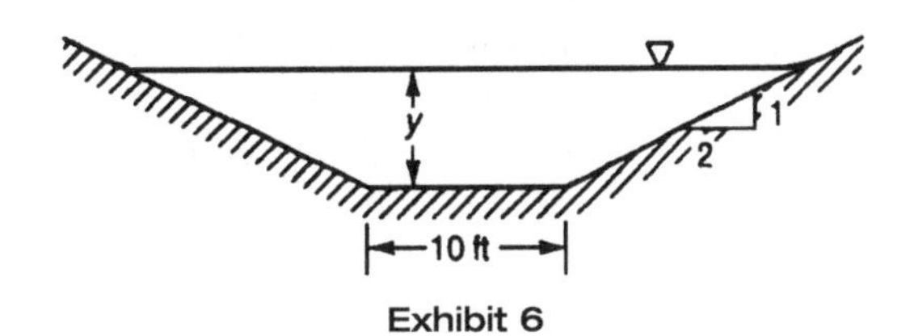

Exhibit 6

Solution

In English units the Manning equation is

$$Q = \frac{1.49}{n} AR^{2/3} S_o^{1/2}$$

From Table 12.4 the appropriate roughness coefficient is $n = 0.019$. Inserting the given information leads to

$$AR^{2/3} = 114 \tag{12.30}$$

in which

$$A = 10y + 2y^2$$
$$P = 10 + 2\sqrt{5}y \text{ and } R = A/P$$

Equation (12.30) now must be solved by successive trial. It is convenient to use a table in doing so:

Trial	*y*	*A*	*P*	*R*	$R^{2/3}$	$AR^{2/3} = 114?$
1	5.0	100.0	32.4	3.09	2.12	212
2	3.5	59.5	25.7	2.32	1.75	104
3	3.7	64.4	26.5	2.43	1.81	116
4	3.66	63.4	26.4	2.40	1.80	114

The normal depth, that is, the depth of uniform flow, is $y = 3.66$ ft.

Hydraulic Jump

When a change from high-speed to low-speed flow occurs in an open channel, a hydraulic jump is observed; over a short distance the depth of flow increases abruptly with a significant energy loss and a very turbulent appearance. The high-speed flow can be caused in a variety of ways, including flow under a gate or down any relatively steep slope. A schematic diagram of this phenomenon for flow on a horizontal channel bottom is given in Figure 12.7. To a good approximation, the flow through the jump can be analyzed directly by applying the linear momentum equation, Equation (12.17), for any shape of cross section. The result is

$$F_1 - F_2 = \rho Q(V_2 - V_1) \tag{12.31}$$

in which F and V represent here, respectively, the hydrostatic force and mean velocity at sections 1 and 2. This equation assumes that the overall effect of boundary shear is negligibly small. Specialization of Equation (12.31) to a particular cross-sectional shape will cause the effect of the depths y_1 and y_2 to appear in the equation. As an example, if this equation is specialized to a rectangular channel of width b and the discharge per unit width $q = Q/b$ is introduced, then one can solve Equation (12.31) directly for the depth ratios y_2/y_1 and y_1/y_2 in the forms

$$\frac{y_2}{y_1} = \frac{1}{2}\left[-1 + \left(1 + 8\text{Fr}_1^2\right)^{1/2}\right] \tag{12.32}$$

and

$$\frac{y_1}{y_2} = \frac{1}{2}\left[-1 + \left(1 + 8\text{Fr}_2^2\right)^{1/2}\right] \tag{12.33}$$

In these two equations, the square of the Froude number is $Fr^2 = q^2/(gy^3)$ and is evaluated at section 1 or 2, as the subscript indicates.

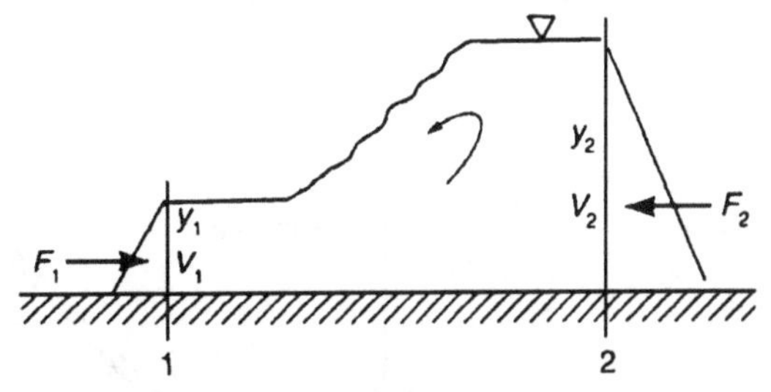

Figure 12.7

A useful relation for head loss across the jump can be obtained by inserting results from Equation (12.31) into the energy equation, Equation (12.9). If this is done for the rectangular channel, the head loss is found to be

$$h_L = \frac{(y_2 - y_1)^3}{4y_2y_1} \quad \textbf{(12.34)}$$

Efficient Section

When one uses the Manning equation, Equation (12.29), it is natural to want to use it efficiently, that is, to want to convey the largest discharge for a given cross-sectional area. So one seeks the largest hydraulic radius, R, for a given A, which means one must minimize the wetted perimeter, P. For both rectangular and trapezoidal cross sections, greatest efficiency is achieved when $R = y/2$. Moreover, the most efficient of all trapezoidal sections has sides that slope at a 60° angle, which means the most efficient trapezoid is a half hexagon. The most efficient section of all is the semicircle.

Specific Energy

Specific energy E is energy per unit weight above a channel bottom, so in general

$$E = y + \frac{V^2}{2g} \quad \textbf{(12.35)}$$

For a strictly two-dimensional channel (rectangular or extremely wide) this definition may be rewritten in terms of the discharge per unit width q to give

$$E = y + \frac{q^2}{2g}\frac{1}{y^2} \quad \textbf{(12.36)}$$

Figure 12.8 shows how this expression behaves as a function of y for two situations. In Figure 12.8(a) the specific energy, E, varies with y for a fixed constant value of discharge, and in Figure 12.8(b) the relation between q and y for fixed specific energy $E = E_o$ is displayed. In both cases an extremum is seen to occur at $y = y_c$, which is called the critical flow state. From these diagrams one can see that the **critical flow** state is the point at which minimum specific energy is required to pass a specified q through a channel section, and it is also the state at which the maximum discharge q is passed through a section for a fixed amount of energy E_o. Using differential calculus, one can find the critical depth and corresponding critical specific energy to be

$$y_c = \left(\frac{q^2}{g}\right)^{1/3} \qquad E_c = \frac{3}{2}y_c \quad \textbf{(12.37)}$$

for two-dimensional channels. For any channel of arbitrary cross section $A = A(y)$ with a width at the water surface that is $b = b(y)$, the critical flow state satisfies the relation

$$\frac{Q^2 b}{gA^3} = 1 \tag{12.38}$$

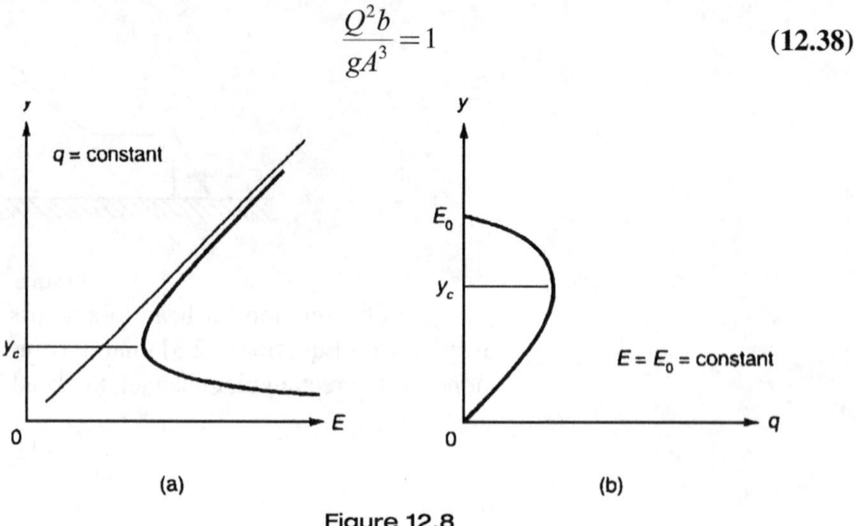

Figure 12.8

Example 12.6

Determine the critical depth for the data given in Example 12.5.

Solution

Equation (12.38) must be solved with $A(y)$ given in Example 12.5 and $b = 10 + 4y$. For the given data

$$\frac{A^3}{b} = \frac{Q^2}{g} = \frac{200^2}{32.2} = 1240$$

A small table will aid the search for the solution:

Trial	y	$A(y)$	$b(y)$	$A^3/b = 1240?$
1	3.0	48.0	22.0	5030
2	2.0	28.0	18.0	1220
3	2.1	29.8	18.4	1440
4	2.01	28.2	18.04	1240

Thus $y_c = 2.01$ ft in this case. The last two values of y in the table were simply chosen by estimating an interpolant between the previous values; this is easier than using a formal linear interpolation procedure, which is not strictly accurate when the basic equation is nonlinear.

Gradually Varied Flow

The governing equation for gradually varied flow can be developed by use of either the momentum or the energy principles, but the intermediate steps and the assumptions are not identical. Underlying either approach are the assumptions that the flow is one-dimensional and the pressure distribution is hydrostatic. The resulting equation can be written in either of two ways:

$$\frac{dy}{dx} = \frac{S_o - S}{1 - \text{Fr}^2} \quad \textbf{(12.39)}$$

$$\Delta x = \frac{E_2 - E_1}{S_o - S} \quad \textbf{(12.40)}$$

In Equation (12.39), which applies at one section in the flow, the rate of change of water depth y with respect to distance along the channel x is given as the difference between channel bottom slope S_o and the local slope of the energy line S, divided by 1 minus the square of the local Froude number. Hence $dy/dx = 0$ means the depth of flow is not changing (is uniform), not that the slope of the water surface is horizontal. The energy slope S is found by using the Manning equation, Equation (12.29), with S_o replaced by S. When Q is set, this equation will give $S = S(y)$ through the dependence of A and R on y. The Froude number is also dependent on depth y, since $\text{Fr}^2 = Q^2b/gA^3$.

Equation (12.40) is a discrete equation and applies between sections 1 and 2, which have different depths of flow. The specific energies at these two sections are E_1 and E_2, respectively. The average energy slope, S, can be determined in several ways. One effective way is to use Equation (12.29) with averaged values for the velocity and hydraulic radius, that is, $V_{\text{avg}} = (V_1 + V_2)/2$ and $R_{\text{avg}} = (R_1 + R_2)/2$.

From Equation (12.39) one can directly determine the qualitative behavior of the water surface profile. The quickest way to do this is to determine the depth of uniform flow (also called normal depth and denoted as y_o or y_n) by using Equation (12.29) with S_o. Next find the critical depth y_c from Equation (12.38), or Equation (12.37) if it is applicable. Finally consult Table 12.5, which graphically presents all the gradually varied flow profiles that are allowed by Equation (12.39). Use of this table requires only a knowledge of the local flow depth y in relation to y_o and y_c.

Table 12.5 Twelve surface profiles and associated data for gradually varied flow

Slope	Surface Profile	$\frac{dy}{dx}$	$\frac{d^2y}{dx^2}$	Fr	y_o	Graphic Profile
Mild	M_1	+	+	< 1	Exists	$y > y_o > y_c$
$S_o < S_c$	M_2	−	−	< 1		$y_o > y > y_c$
	M_3	+	+	> 1		$y_o > y_c > y$ (Mild Slope)
Steep	S_1	+	−	< 1	Exists	$y > y_c > y_o$
$S_o > S_c$	S_2	−	+	> 1		$y_c > y > y_o$
	S_3	+	−	> 1		$y_c > y_o > y$ (Steep Slope)
Critical	C_1	+	−	< 1	Exists	$y > (y_o = y_c)$
$S_o = S_c$	C_3	+	−	> 1		$y < (y_o = y_c)$ (Critical Slope)
Horizontal	H_2	−	−	< 1		$y > y_c$
$S_o = 0$	H_3	+	+	> 1		$y < y_c$ (Horizontal Slope)
Adverse	A_2	−	−	< 1	Imaginary	$y > y_c$
$S_o < 0$	A_3	+	+	> 1		$y < y_c$ (Adverse Slope)

Example 12.7

Two very long open channel reaches convey a discharge of 20 m³/s, as shown in Exhibit 7. Far upstream from the break in slope *B* the depth of uniform flow is 1.00 m; far downstream from *B* the uniform flow depth is 2.00 m. The rectangular channel cross section is 4 m wide and is lined with concrete ($n = 0.013$). Determine the flow profile between the regions of uniform flow, and compute approximately the length of the gradually varied flow region.

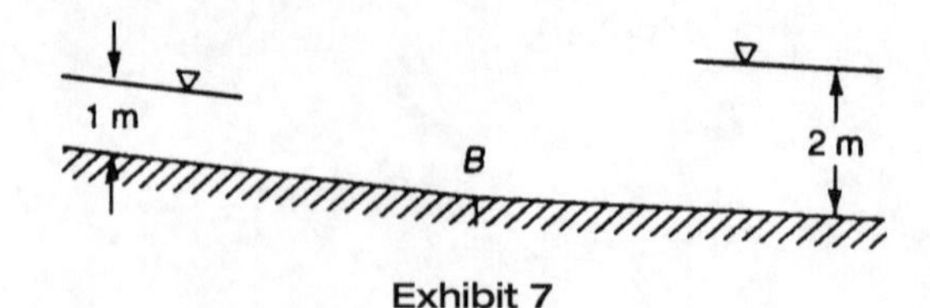

Exhibit 7

Solution

The discharge per unit width is $q = Q/b = 20/4 = 5.0$ m²/s. From Equation (12.37) the critical depth is $y_c = (q^2/g)^{1/3} = (5.0^2/9.81)^{1/3} = 1.37$ m. By direct comparison of y_c with the uniform flow depths, the flow far upstream is supercritical and far downstream it is subcritical. Since Table 12.5 shows that no gradually varied flow profile can cross critical depth, some other flow event is needed for the flow to cross y_c.

One begins by hypothesizing that the transition between uniform flows is some combination of a hydraulic jump and a gradually varied flow profile. If the jump occurs downstream from *B*, then an examination of the family of mild profiles in Table 12.5 shows that the jump must end at precisely the depth of downstream uniform flow, since a jump that ends on either side of y_o then moves away from this depth via an M_1 or M_2 profile. Thus, if a jump to $y_2 = 2.0$ m from some depth y_1 occurs, then Equation (12.33) with $\mathrm{Fr}_2^2 = q^2\left(gy_2^3\right) = (5.0)^2/[(9.81)(1.00)^3] = 0.319$ yields $y_1 = 0.88$ m. Since a review of both the steep and mild profiles in Table 12.5 shows that there is no way to move from $y = 1.00$ m to this new depth, the hypothesized jump cannot occur in this way.

The alternative is to assume that the jump is upstream of *B* and $y_1 = 1.00$ m. Using Equation (12.32) with $\mathrm{Fr}_1^2 = q^2\left(gy_1^3\right) = (5.0)^2/[(9.81)(1.00)^3] = 2.55$ yields $y_2 = 1.81$ m; then Table 12.5 shows that gradually varied flow to a depth of 2.00 m is possible via the S_1 profile. Hence all flow downstream of *B* is uniform at a depth of 2.00 m.

The length of the S_1 profile will be determined by using Equation (12.40). First the upstream channel-bottom slope, S_o, must be found from the Manning equation, Equation (12.29), with $K = 1$ and a depth of flow $y = 1.00$ m, which yields $S_o = 0.00725$. Now the approximate length of the reach of gradually varied flow will be found by the recommended procedure:

y, m	*V*, m/sec	*R*, m	*E*, m	*S*	Δ*x*, m
1.81	2.76	1.050	2.20		
				0.00113	8.2
1.90	2.63	1.080	2.25		
				0.000985	11.2
2.00	2.50	1.111	2.32		___
					19.4

In summary, a hydraulic jump from 1.00 m to 1.81 m occurs approximately 19.4 m upstream from *B* and is followed by an S_1 profile to a flow depth of 2.00 m at, and downstream of, *B*.

MANNING EQUATION

The Manning equation for the average velocity *V* in a steady open channel flow is

$$V = \frac{K}{n} R^{2/3} S^{1/2}$$

With SI units $K = 1.0$; with English units $K = 1.49$. The hydraulic radius is $R = A/P$, with A = flow cross-sectional area and P = wetted perimeter (i.e., the length of the interface along the fluid/solid boundary containing it). The slope S of the energy line is equal to the amount of head lost in a channel section of length L, or $S = h_L/L$. A short table of values for n, the Manning roughness factor, follows:

Channel Surface	Roughness Value, n
Concrete, finished	0.012
Concrete, gunite	0.019
Clay, vitrified sewer	0.014
Rubble masonry	0.025
Concrete, mortar	0.013
Concrete, troweled	0.013
Gravel, clean	0.025

Although the Manning equation is intended primarily for use in open channels, it is sometimes also used for pressurized flow in pipes.

Example 12.8

A rectangular open channel lined with rubble masonry is 5 m wide and laid on a slope of 0.0004. If the depth of uniform flow is 3 m, compute the discharge in m³/s.

Solution

From the table, the Manning roughness is approximately $n = 0.025$. The area and wetted perimeter are

$$A = (5)(3) = 15 \text{ m}^2$$
$$P = 5 + 2(3) = 11 \text{ m}$$
$$R = A/P$$

The Manning equation now gives

$$Q = AV = A\left(\frac{1}{n}\right)R^{2/3}S^{1/2}$$
$$= (15)\left(\frac{1}{0.025}\right)(15/11)^{2/3}(0.0004)^{1/2}$$
$$= 14.76 \text{ m}^3/\text{s}$$

Example 12.9

A gunite concrete trapezoidal channel with 1:2 side slopes, shown in Exhibit 8, conveys 60 m³/s on a slope $S_o = 0.0005$. Compute the depth of uniform flow.

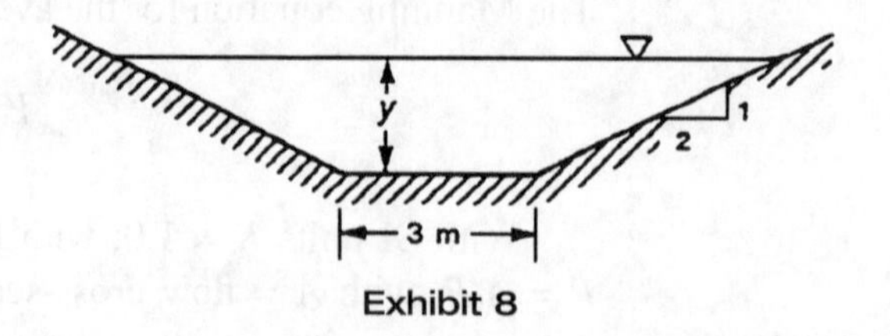

Exhibit 8

Solution

In SI units the Manning equation is

$$Q = (1/n)\, AR^{2/3}\, S_o^{2/3}$$

From the table the appropriate roughness coefficient is $n = 0.019$. Inserting the given information leads to

$$AR^{2/3} = 5.10 \qquad \textbf{(a)}$$

in which

$$A = 3y + 2y^2$$
$$P = 3 + 2\sqrt{5}\,y \text{ and } R = A/P$$

Equation (a) must now must be solved by successive trial. It is convenient to use a table in doing so:

Trial	y (m)	A (m²)	P (m)	R	$R^{2/3}$	$AR^{2/3}$ 5 5.10?
1	1.5	9.00	9.71	0.93	0.95	8.55
2	1.0	5.00	7.47	0.67	0.77	3.85
3	1.15	6.10	8.14	0.75	0.83	5.06
4	1.16	6.17	8.19	0.75	0.83	5.12

The normal depth, that is, the depth of uniform flow, is $y = 1.16$ m.

HAZEN-WILLIAMS EQUATION

Many empirical formulas for pipe friction have been developed over the past century. These formulas are usually based on tests involving the flow of water under fully turbulent conditions and are not normally reliable for use with other fluids. One relatively widely used such formula is the Hazen-Williams equation, which is

$$V = 0.849\, CR^{0.63}\, S^{0.54}$$

for SI units. For English units, replace 0.849 with 1.318. The Hazen-Williams coefficient C ranges from approximately 140 for very smooth and straight pipes to 120 for smooth masonry to 100 or less for old cast iron pipe. The other factors in the equation are defined as they were for the Manning equation.

Example 12.10

If 0.01 m³/s of water flows through a new 100 mm clean cast iron pipe ($C = 130$), determine the head loss in 1000 m of this pipe.

Solution

The discharge $Q = VA$ with

$$A = \frac{\pi}{4}(0.1)^2 = 0.00785\,\text{m}^2$$

Hence $V = 0.01/0.00785 = 1.273$ m/s

In this case the hydraulic radius is

$$R = \frac{\pi \frac{D^2}{4}}{\pi D} = \frac{D}{4} = \frac{0.1}{4} = 0.025 \text{ m}$$

The Hazen-Williams equation then yields

$$V = 1.273 = 0.849\,(130)(0.025)^{0.63}\,S^{0.54}$$
$$S = 0.0191 = h_L/L = h_L/1000 \text{ and } h_L = 19.1 \text{ m}$$

HYDROLOGIC ELEMENTS

Hydrology is in general a multidisciplinary subject that is the study of water movement and distribution on earth. This movement is a closed loop called the hydrologic cycle in which water is first evaporated primarily from the oceans, then transported as vapor by the atmosphere, and, under proper circumstances, precipitated to the earth's surface as rain or snow. The surface water may return to the atmosphere again as evaporation, it may infiltrate into the soil and reach the groundwater or be taken up by plants and transpired back into the air, or it may flow over the land surface and find its way into streams, rivers, or lakes, eventually flowing back into the oceans to complete the cycle.

Precipitation

The most common form of precipitation is liquid rain; when the amounts of other forms, such as snow, must be quantified, they are often melted first, and the amount is reported in terms of its liquid equivalent. The most common precipitation gage in the United States is the Weather Service 8-in.-diameter cylindrical gage, which directs captured rain into a measuring tube that is one tenth the cross-sectional area of the collector, and depths are then measured to 0.01 in. within it. Three types of recording gages are also in common use; they are the tipping-bucket gage, the weighing-type gage, and the float recording gage.

The average precipitation $\bar{p}$ over a region can be obtained from point data in one of three ways, all of which fit the formula

$$\bar{p} = \frac{1}{A_T} \Sigma A_i P_i \qquad A_T = \Sigma A_i \tag{12.41}$$

in which p_i is a point precipitation value, A_i is a weighting factor, and A_T is the sum of the weighting factors:

1. The simple arithmetic mean is appropriate when the individual values are all similar. In this case set each $A_i = 1$, and then A_T is just the number of points.
2. The widely used Thiessen average is a weighted average that in effect assumes that the value p_i best represents the true precipitation at all locations that are closer to gage i than to any other gage. Each A_i is the area surrounding gage i, and A_T is the total gaged area. The boundary of each A_i is formed by lines that are the perpendicular bisectors of lines drawn between the gages themselves.

3. The isohyetal method is the only method that allows a knowledge of basin topography to enter the calculation. One begins the computation by drawing contour lines of equal precipitation (isohyets) throughout the region. Then in Equation (12.41) A_i is the area between adjacent isohyets, p_i is the average precipitation between these adjacent isohyets, and A_T is the total gaged area.

Evapotranspiration

The quantification of evaporation or transpiration amounts (or of evapotranspiration, ET, the sum of the two) can become important to engineers who conduct water supply studies. There are several computational approaches and one primary experimental method of estimating evaporation; each approach has its problems and leads to imprecision in the result:

1. The water budget or mass conservation method attempts to account for all flows of water to and from the water body under study, including inflow, outflow, direct precipitation, and even seepage to the groundwater.
2. The energy budget is like the water budget, except the energy flows rather than mass flows are the basic accounting medium.
3. Direct empirical meteorological correlations are used to avoid the uncertainties of the first two methods, but attempts to avoid excessive complexity here usually lead to incomplete, and thus inaccurate, results.
4. A combination of the above methods has been relatively successful. For example, the Penman equation, when used with a set of charts, has become popular when all the required data can be obtained or estimated.
5. The National Weather Service Class A pan is 4 feet in diameter, 10 inches deep, made of unpainted galvanized iron, and used to measure evaporation by direct experiment. Multiplication of this result by a pan coefficient, typically about 0.7, then gives the evaporation from the adjacent larger water body. Difficult correlation studies are needed to ensure that the coefficient is appropriate to a particular application.

Infiltration

Infiltration of water into the soil is important in some studies. Horton's infiltration equation is widely used for this purpose, which is

$$f = f_c + (f_o - f_c)e^{-kt} \quad \textbf{(12.42)}$$

Here f is infiltration rate (in./hr); f_o and f_c are the initial and final infiltration rates, respectively; t is time (hr); and k (1/hr) is an empirically determined constant. Another common way of characterizing infiltration is via the ϕ index method. In this method one plots the overall precipitation rate versus time; a horizontal line called the ϕ index is drawn on the plot, such that the volume of rainfall excess above this line is equal to the actual volume of observed runoff. Thus the index indicates the average infiltration rate for the storm event.

Example 12.11

Exhibit 9 is a histogram that describes hourly rainfall for a 5-hour storm.

a). Previous experience has determined that the Horton infiltration parameters for the soil in this region are $f_o = 0.4$ in./hr, $f_c = 0.2$ in./hr, and $k = 0.5$/hr. Determine the volume of rainfall that infiltrates during the 5-hr period.

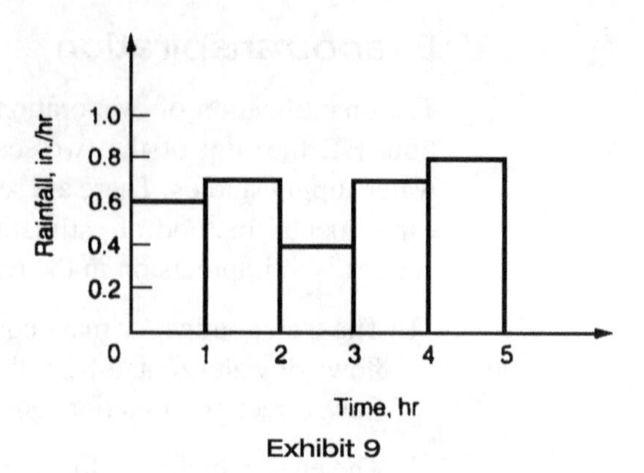

Exhibit 9

b). If the net runoff from this storm is known to be 1.8 in., compute the ϕ index for this event.

Solution

a). Using the Horton equation, the infiltrated volume V_I is

$$V_I = \int_0^5 f dt = 0.2\int_0^5 \left[1+e^{-0.5t}\right] dt$$

$$V_I = 0.2\left[t - \frac{1}{0.5}e^{-0.5t}\right]_0^5 = 0.2\left[5 - \frac{e^{-2.5}}{0.5} + \frac{1}{0.5}\right]$$

$$V_I = 1.37 \text{ in.}$$

b). The net runoff is the area in Exhibit 1 that lies above the ϕ index. Assuming $\phi < 0.4$ in, then one can write

$$1.8 = \sum_i \left[p_i - \phi\right] = (0.6-\phi)+(0.7-\phi)+(0.4-\phi)+(0.7-\phi)+(0.8-\phi)$$

$$\phi = 0.28 \text{ in.}$$

WATERSHED HYDROGRAPHS

A watershed or drainage basin is the region drained by a stream or river. When a precipitation event (a storm) occurs over the watershed, it causes several processes within the basin. First is the initial moistening of the land surface and the vegetation, followed by the local filling of small surface indentations (depression storage) and the buildup of some depth of water on the land surface (initial detention storage) before the flow of water over the land begins; at the same time infiltration begins. For the larger storms some, possibly even most, of the precipitation enters a stream and flows out of the basin. The discharge past this outflow point is a time-variant process. A plot of the outflow versus time is called a hydrograph; Figure 12.9 is a definition sketch of a hydrograph.

There are two components to any perennial streamflow, a relatively short-term component, which is the storm-induced surfaced water outflow, and a longer-term, slowly varying component called base flow, which is the contribution from the groundwater to the flow. In Figure 12.9 the storm hydrograph is caused by an effective storm precipitation of D hr, causing first the increasing discharge on the rising limb of the hydrograph from A to the crest C, and then the recessional limb from C to B when the storm-related discharge ceases. The base flow, below line AB in the figure, can be separated from the storm flow in any of several ways:

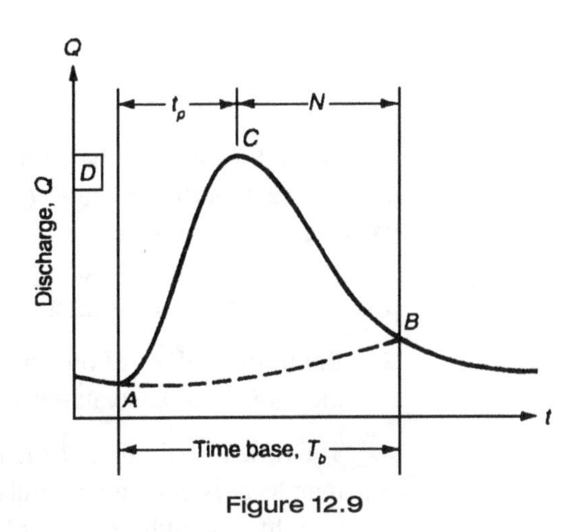

Figure 12.9

1. From the low point A before the storm outflow begins, simply draw a horizontal line.
2. From point A extend the upstream line to a point directly below the crest C. From that point draw a straight line to B, which is located a distance N (days) after point C. The value N is empirically found from

$$\begin{aligned} N &= A^{0.2} \qquad A \text{ in square miles} \\ &= 0.8A^{0.2} \qquad A \text{ in square kilometers} \end{aligned} \tag{12.43}$$

   The area of the drainage basin is A. Sometimes N is adjusted to the nearest full day.
3. Sometimes an attempt is made to mimic near B the character of the flow near A by patching the slope of the recession curve from A into the base-flow separation near B. This method leaves unanswered the choice of the remainder of the separation curve under the rising limb and crest. It must be drawn arbitrarily.

When the base flow has been removed from the original storm hydrograph, the remainder is direct storm runoff.

Unit Hydrograph

A unit hydrograph (unitgraph) has a volume of 1 in. (or 1 cm) of direct runoff over the drainage basin as a result of a storm of D hours' effective duration. Effective duration is the time interval when excess rainfall exists and direct runoff occurs. Any direct storm runoff has a volume

$$V = \int Q dt \tag{12.44}$$

which may also be written as $V = xA$, in which x is in inches (cm) and A is the basin area. Determination of the volume V, which is the area ABC in Figure 12.9, can be computed efficiently and accurately by use of the trapezoidal rule. Assume the time base T_b is divided into m intervals $\Delta t = T_b/m$ and the direct runoff ordinates Q_i, $i = 1$ to $m + 1$, are known with $Q_1 = Q_{m+1} = 0$. By the trapezoidal rule,

$$V = \int Qdt = (Q_1 + Q_2)\frac{\Delta t}{2} + (Q_2 + Q_3)\frac{\Delta t}{2} + \ldots + (Q_m + Q_{m+1})\frac{\Delta t}{2}$$

or

$$V = \Delta t \sum_{i=2}^{m} Q_i \qquad \textbf{(12.45)}$$

Normally x will not be 1.0 in. (cm). The unit hydrograph is simply obtained by dividing each of the ordinates Q_i of the direct storm runoff plot by x. The unitgraph has a variety of applications.

The suitability of the unitgraph for these uses, however, depends on the appropriateness of several assumptions, including these:

- Rainfall excesses of one duration D will always produce hydrographs with the same time base, independent of the intensity of the excess.
- The time distribution of the runoff does not change from storm to storm, so long as D is unchanged; thus an increase in runoff volume by $P\%$ increases each hydrograph ordinate Q_i by $P\%$. Moreover, the distribution is not affected by prior precipitation.

The development of a unit hydrograph that produces reliable results in applications will be enhanced if one follows some experience rules:

- Basin sizes should be between 1000 acres and 1000 square miles.
- The direct storm runoff should preferably be within a factor of 2 of 1.0 inch, and the storm structure should be relatively simple.
- The unitgraph should be derived from several storms of the same duration. In other words, compute several unit hydrographs, and then average them.

If one does not have sufficient storm data to derive a unitgraph, then theoretical or empirical methods may be used to develop a "synthetic" unitgraph based on information such as peak flow values and basin characteristics. Numerous such methods have been proposed. Two of the more commonly used synthetic methods are Snyder's method, originally developed for Appalachian watersheds, and the SCS method, developed by the Soil Conservation Service. They must be applied with care for best results; space does not permit an explanation here of these methods in the detail that is needed, so the reader may consult the chapter references for the complete methods.

Change of Unitgraph Duration

Each unit hydrograph is associated with an effective storm duration D. If one wants a unitgraph for some other effective storm duration without developing it directly from storm data, this can be done. (If the new storm duration differs from the existing one by no more than 25%, then normal practice is to use the existing one without alteration.) Two methods are used:

1. *Lagging.* This method can be used to construct a new unitgraph for a storm of effective duration nD, given the unitgraph for the storm having effective duration D, where n is an integer only. Simply add together n of the original

unit hydrographs, starting each successive unitgraph D hours after the beginning of the preceding one. This step produces a hydrograph associated with an effective duration of nD hours and having a runoff volume of n inches over the basin. Now divide all the hydrograph ordinates Q_i by n to obtain the new unitgraph. The method is easily set up in a table.

2. *S-curve*. This method is much more general and can be used to construct a unitgraph for either a shorter or longer effective storm duration than the original. Say the desired new effective storm duration is D_{new}. First one constructs the *S*-curve (it is a summation curve, that is, a sum of unitgraphs, and it also takes the general shape of an *S*) by successively lagging by D hours and summing (adding together) the ordinates of a total of T_b/D original unitgraphs. Next draw a second *S*-curve, lagged D_{new} hours after the first *S*-curve. The differences in ordinates of these two *S*-curves, each multiplied by the ratio D/D_{new}, will be the ordinates of the new unitgraph for the storm of effective duration D_{new}.

Example 12.12

Stream runoff from a 1500-acre watershed is plotted in Exhibit 10 for a storm having an effective duration D of 2 hours.

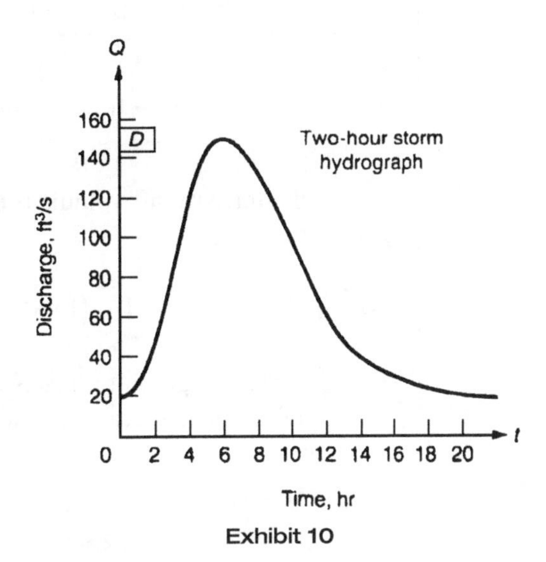

Exhibit 10

a). Compute the ordinates of, and plot, the 2-hour unit hydrograph.

b). Use the information for the 2-hour unit hydrograph to construct a 3-hour unit hydrograph.

c). Construct the composite storm hydrograph caused by 1.5 inches of excess precipitation falling in the first 2 hours, followed immediately by 0.7 inch of excess precipitation in the next 2 hours. Assume a base flow of 10 ft^3/s.

Solution

a). The computations are presented in Exhibit 11. First the amount of the base flow must be identified and separated from the overall runoff. Since little information is available in this problem and also because the runoff duration is relatively short, it is assumed that the base flow is a constant 20 ft^3/s.

Exhibit 11

Time, hr (1)	Stream Flow, ft³/s (2)	Storm Flow Q_i, ft³/s (3)	Unitgraph Ord. U_i, ft³/s (4)
0	20	0	0
2	60	40	58
4	113	93	135
6	150	130	188
8	127	107	155
10	96	76	110
12	65	45	65
14	43	23	33
16	27	7	10
18	20	0	0

The data in column 2 come directly from the hydrograph, Exhibit 10. The storm flow Q_i, column 3, is the stream flow minus the base flow. Selecting a time interval $\Delta t = 2$ hr for use in Equation (12.45), the storm runoff volume is

$$V = \Delta t \sum_{i=2}^{m} Q_i = (2\text{ hr})(521\text{ ft}^3/\text{s}) = 1042\frac{\text{ft}^3}{\text{s}}\text{-hr}$$

$$V = \left[1042\frac{\text{ft}^3}{\text{s}}\text{-hr}\right]\left[60^2\frac{\text{s}}{\text{hr}}\right] = 3.75\times10^6\text{ ft}^3$$

This storm runoff volume is equivalent to a depth x of water over the basin of

$$x = \frac{V}{A} = \frac{(3.75\times10^6\text{ ft}^3)(12\text{ in./ft})}{(1500\text{ acres})(43{,}560\text{ ft}^2/\text{acre})} = 0.69\text{ in.}$$

The unitgraph ordinates $U_i = Q_i/x$ are tabulated in column 4, and the unit hydrograph is plotted in Exhibit 12.

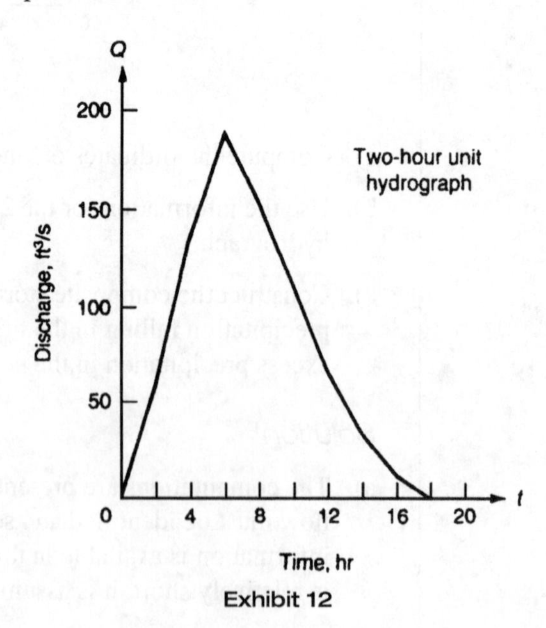

Exhibit 12

b). One constructs the S-curve by repeatedly lagging the 2-hour unitgraph, whose ordinates are listed in column 4 in Exhibit 11, and adding together all the values that are associated with each time instant. The individual ordinates S_i of the S-curve are

$$S_i = \sum_{n=1}^{i} U_n$$

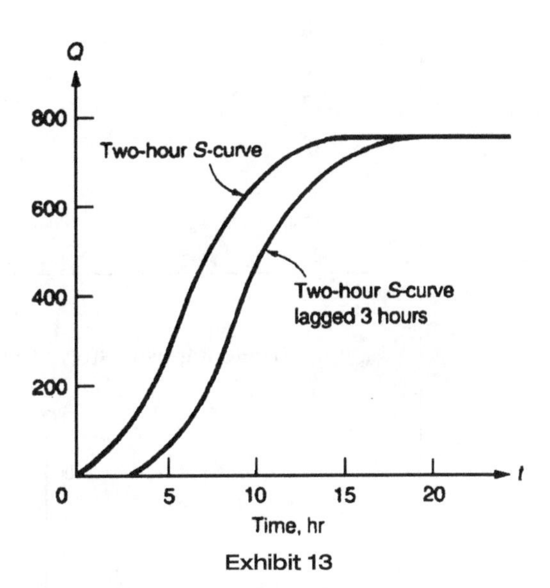

Exhibit 13

The 2-hr S-curve is plotted in Exhibit 13. Also shown is this same S-curve lagged 3 hours; the differences in ordinates of these two S-curves are then multiplied by the ratio $D/D_{new} = 2/3$ to scale the volume of the new hydrograph properly to end with the 3-hr unitgraph plotted in Exhibit 14.

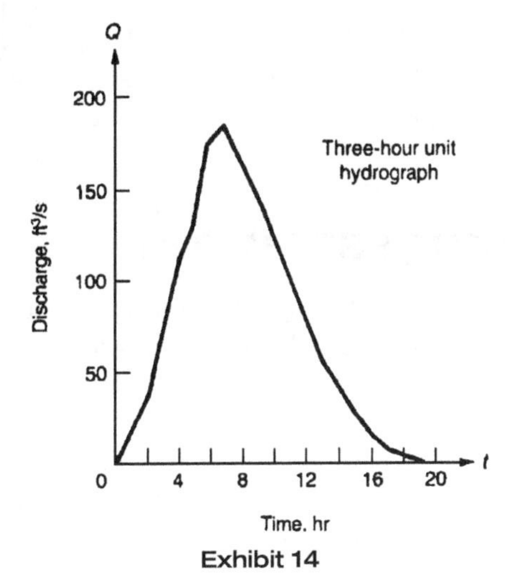

Exhibit 14

Scrutiny of this computational sequence shows that the peak discharge in the new unitgraph is slightly smaller than the peak of the 2-hr unitgraph, as one would expect.

c). Computations are tabulated in Exhibit 15. Time is measured from the start of the storm. The 2-hr unitgraph is multiplied by 1.5 for the first portion of the runoff, followed by a second unitgraph scaled by 0.7. Finally, the base flow is added.

Exhibit 15

Time, hr	Unitgraph Ord., U_i, ft³/s	1.5 × U_i, ft³/s	0.7 × U_i, lag 2 hr, ft³/s	Sum, with *BF*, ft³/s
0	0	0	—	10
2	58	87	0	97
4	135	203	41	254
6	188	282	95	387
8	155	233	132	375
10	110	165	109	284
12	65	98	77	185
14	33	50	46	106
16	10	15	23	48
18	0	0	7	17
20			0	10

The composite storm hydrograph is plotted in Exhibit 16.

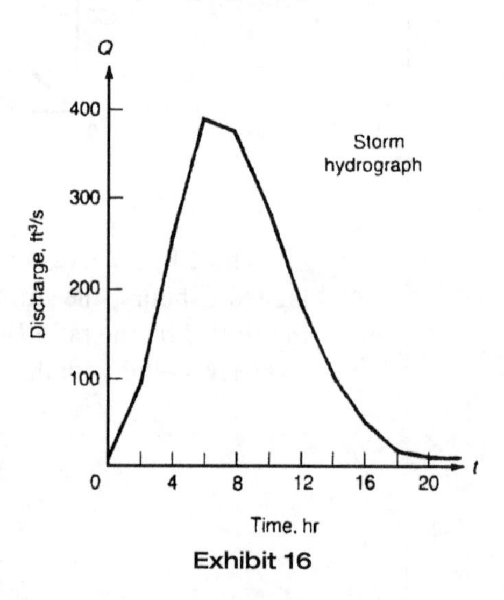

Exhibit 16

PEAK DISCHARGE ESTIMATION

Hydrographs convey a multitude of information. The volume of runoff over a time period is useful in water supply, flood control, and reservoir and detention basin studies. In other studies it is the peak discharge rate that is important—for example, in selecting pipe or culvert sizes or channel dimensions—and the other additional information is not needed.

The rational method is the most widely used method for the estimation of peak discharge Q_p (ft³/s) from runoff over small surface areas. In using it one assumes that a spatially and temporally uniform rainfall occurs for a time period that allows the entire catchment area to contribute simultaneously to the outflow. Clearly the satisfaction of these limitations becomes more difficult as the basin size increases, so this equation is normally limited to basins that are below 1 square mile (640 acres) in size. The equation is

$$Q_p = CiA \tag{12.46}$$

in which C is a nondimensional runoff coefficient that indicates the fraction of the incident rain that runs off the surface, i is the appropriate storm intensity (in./hr), and A is the watershed area (acres). Some add a dimensional conversion factor to this equation, but since 1 ft^3/s = 1.008 acre-in./hr, the conversion factor is usually ignored, as the other factors in the equation are not known with such accuracy. Table 12.6, adapted from Reference 1, gives reasonable ranges for C for various surfaces, as well as some guidance in selecting a value in the range.

Table 12.6 Runoff coefficients, C

Description of Area	Runoff Coefficients
Business	
Downtown	0.70 to 0.95
Neighborhood	0.50 to 0.70
Residential	
Single-family	0.30 to 0.50
Multiunits, detached	0.40 to 0.60
Multiunits, attached	0.60 to 0.75
Residential (suburban)	0.25 to 0.40
Apartment	0.50 to 0.70
Industrial	
Light	0.50 to 0.80
Heavy	0.60 to 0.90
Parks, cemeteries	0.10 to 0.25
Playgrounds	0.20 to 0.35
Railroad yard	0.20 to 0.35
Unimproved	0.10 to 0.30

It often is desirable to develop a composite runoff coefficient based on the percentage of different types of surface in the drainage area. This procedure often is applied to typical "sample" blocks as a guide to selection of reasonable values of the coefficient for an entire area. Coefficients with respect to surface type currently in use are:

Character of Surface	Runoff Coefficients
Pavement	
Asphaltic and concrete	0.70 to 0.95
Brick	0.70 to 0.85
Roofs	0.75 to 0.95
Lawns, sandy soil	
Flat, 2%	0.05 to 0.10
Average, 2 to 7%	0.10 to 0.15
Steep, 7%	0.15 to 0.20
Lawns, heavy soil	
Flat, 2%	0.13 to 0.17
Average, 2 to 7%	0.18 to 0.22
Steep, 7%	0.25 to 0.35

The coefficients in these two tabulations are applicable for storms of five- to ten-year frequencies. Less frequent, higher-intensity storms require the use of higher coefficients because infiltration and other losses have a proportionally smaller effect on runoff. The coefficients are based on the assumption that the design storm does not occur when the ground surface is frozen.

Source: *Design and Construction of Sanitary and Storm Sewers*, Manual No. 37, 1986; reproduced by permission of ASCE.

The intensity factor must also be chosen carefully. It is normally defined as the intensity of rainfall of a chosen frequency that lasts for a duration equal to the time of concentration t_c for the basin. Sometimes the frequency will be dictated by policy (one-year, five-year, or ten-year). After the frequency has been chosen, one usually consults an intensity-duration-frequency (IDF) plot to obtain i once the time of concentration has been picked. Conceptually this time is the time required for flow from the most remote point in the basin to reach the outlet, but in some cases it is simply estimated to be in the 5- to 15-minute range. Picking a shorter time usually leads to a higher-intensity i and a larger Q_p; in one sense this is conservative, but it may also be wasteful by causing one to design for an excessively large flow. The IDF plot, if developed properly, reports information that is the result of long-term statistical averages of many individual storms, not just the result of a compilation of relatively few data.

When the basin surface is not homogeneous, one should either subdivide the basin into smaller regions that are (nearly) uniform or compute a weighted average value for C, the weights being the areas.

Several other approaches to the estimation of peak discharge exist, the SCS methods being among the most prominent. If one wants to apply these methods properly, however, a lengthy description of the method and the supporting data and charts are required. One should consult the references at the end of this chapter for an adequate description of the procedures.

Example 12.13

A storm drain is to be extended to serve two developing areas in a suburb. Exhibit 17 presents the intensity-duration-frequency plot for this region as well as a schematic diagram of the developments. Area A consists of 40 acres of mostly single-family residential units, with some multiple-family units; the time of concentration is 15 min. Area B drains to inlet 2 and contains several small businesses. The transit time for storm water to move from inlet 1 to inlet 2 is $T = 5$ min. Assuming a five-year return period, estimate the peak discharges expected at the two inlets.

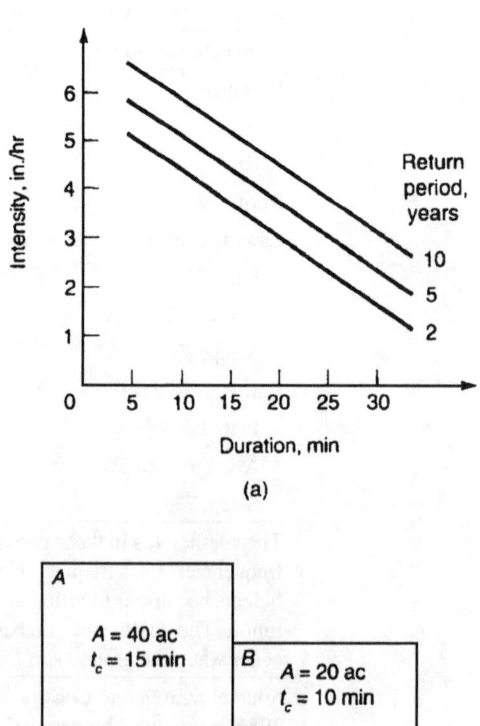

Exhibit 17

Solution

For area A one assumes a 15-min duration and finds $i = 4.50$ in./hr for a 5-yr return period from Exhibit 17(a). Referring to Table 12.6, it appears that $C = 0.45$ is reasonable for this residential area. For point 1 the peak discharge should be about

$$Q_p = CiA = (0.45)(4.50)(40) = 81 \text{ ft}^3/\text{s}$$

This peak is expected to appear at the second inlet location at 15 + 5 = 20 min after the storm begins.

If area B is considered separately, then a 10-min duration leads, via Exhibit 17(a), to $i = 5.17$ in./hr, the runoff coefficient may be nearly $C = 0.70$, and

$$Q_p = (0.70)(5.17)(20) = 72 \text{ ft}^3/\text{s}$$

at inlet 2 from area B. However, the two computed peak discharges do not both arrive at point 2 at the same instant. The peak flow from B arrives 10 min before the flow from A arrives.

To compensate for the fact that the two peak discharges do not coincide in time, the usual approximate procedure is to use an area-weighted coefficient C_w and a time of concentration that applies to the combination of the areas. Here the time of concentration is 20 min. Thus

$$C_w = \Sigma C_i A_i / \Sigma A_i = [0.45(40) + 0.70(20)]/[40 + 20] = 0.53$$

For the 5-yr return period and a 20-min duration Exhibit 17(a) gives $i = 3.83$ in./hr and

$$Q_p = (0.53)(3.83)(60) = 122 \text{ ft}^3/\text{s}$$

which is lower by some 30 ft³/s than the sum of the individual peak flows.

HYDROLOGIC ROUTING

Routing methods track water masses as a function of time as they course through streams, rivers, and reservoirs. Hydrologic routing is based on conservation of mass, supplemented by a relation between storage and discharge; it is an incomplete, approximate computation since it ignores momentum considerations, but it is often used because it can produce sufficiently accurate results with far less computational effort than is required in hydraulic routing, which does include the momentum equation. In this section the hydrologic routing of flows through reservoirs and rivers will be reviewed.

When the inflow hydrograph to either a reservoir or river reach is compared with the subsequent outflow hydrograph at the other end, two characteristic features are normally present: (1) the peak discharge of the inflow is attenuated—that is, reduced—in the outflow, and (2) the peak outflow occurs later than—that is, lags—the peak inflow. The difference between inflow I and outflow Q at any instant is equal to the rate of change of the storage S of water in the region between the inflow and outflow stations, or

$$I - Q = \frac{dS}{dt} \tag{12.47}$$

Usually this equation is integrated between two time instants t_n and t_{n+1} and the trapezoidal rule is applied over the interval $\Delta t = t_{n+1} - t_n$ to obtain

$$\left(I_{n+1} + I_n\right)\frac{\Delta t}{2} - \left(Q_{n+1} + Q_n\right)\frac{\Delta t}{2} = S_{n+1} - S_n \tag{12.48}$$

The typical routing problem begins with an inflow hydrograph given (a set of values I_n, $n = 1, N$). The value of the initial outflow must also be known. The remaining two unknowns in the equation are Q_{n+1} and S_{n+1}. Once the relation between storage and outflow is specified, the new outflow can be computed, and the computation can progress to the next time increment. This storage relation differs, however, depending on the application.

Reservoir Routing

Reservoir outflow either is controlled by gages and/or valves or is not controlled, owing to their absence. In uncontrolled reservoirs the storage relation is of the form $S = f(Q)$ when the reservoir water surface has no slope, as in short or deep reservoirs, or $S = f(Q, I)$ when the surface does slope, as in shallow reservoirs. For controlled reservoirs the storage representation may again be of either type, with the added problem that a separate storage relation must be determined for each combination of gate/valve settings. When $S = f(Q, I)$ the routing method is similar to river routing.

The storage indication, or Puls, method of hydrologic routing is commonly applied to reservoirs. When storage is assumed to be a function only of outflow, the method uses the following steps:

- Equation (12.48) is rearranged to give

$$I_n + I_{n+1} + \left(\frac{2}{\Delta t}S_n + Q_n\right) = \left(\frac{2}{\Delta t}S_{n+1} + Q_{n+1}\right) \qquad \textbf{(12.49)}$$

- From whatever data are given, a table or graph of $(2S/\Delta t + Q)$ versus Q is prepared; it is called a storage indication curve.
- The storage indication curve and inflow data are used in applying Equation (12.49) sequentially over time increments until the outflow has been computed as a function of time.

The Puls method is applied in Example 12.14.

Example 12.14

Some elevation-discharge and elevation-area data for a small reservoir with an ungated spillway are given below. An inflow sequence to the reservoir for part of a flood is given in a second table.

Elev., ft	0	1	2	3	4	5	6
Area, acres	1000	1020	1040	1050	1060	1080	1100
Outflow, ft³/s	0	525	1490	2730	4200	5880	7660

Date	Hour	Inflow, ft³/s
4/23	12 PM	1500
4/24	12 AM	1600
	12 PM	3100
4/25	12 AM	9600

Determine by routing the outflow discharge and reservoir water surface elevation at 12 AM on 25 April. Arrange the computations in a tabular form. Use a 12-hour routing period, and assume that the reservoir water level just reaches the spillway crest (elevation 0.0) at 12 PM on 23 April.

Solution

Since the outflow Q is given directly as a function of elevation, the first task is to determine the reservoir storage S as a function of elevation also. The given areas are the surface areas of the reservoir water surface; integrating these areas over the incremental elevation changes produces the incremental changes in storage. This computation will be tabulated along with the compilation of data points for the storage indication curve. Elevation values will also be used as the index n in the equations. The equations used in computing the table entries are

$$\bar{A}=\frac{1}{2}\left(A_n+A_{n+1}\right) \quad \Delta S=\bar{A}\Delta h \quad S_{AF}=\sum \Delta S$$

$$\frac{2}{\Delta t}S+Q=\frac{2S_{AF}\left(43{,}560\right)}{12\left(60^2\right)}+Q=2.02S_{AF}+Q, \text{ft}^3/\text{s}$$

Elev., n, ft	Area A, acres	Avg. Area, $\bar{A}$, acres	S_{AF}, acre-ft	$\frac{2}{\Delta t}S+Q$, ft^3/s
0	1000		0	0
		1010		
1	1020		1010	2560
		1030		
2	1040		2040	5600
		1045		
3	1050		3085	8950
		1055		
4	1060		4140	12,550
		1070		
5	1080		5210	16,400
		1090		
6	1100		6300	20,400

The resulting storage indication curve is plotted in Exhibit 18.

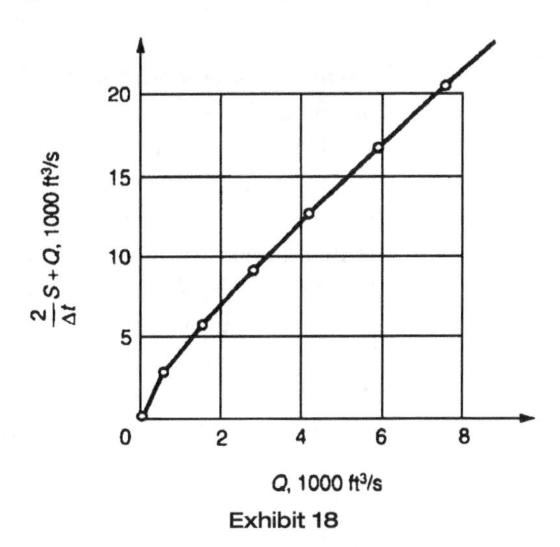

Exhibit 18

Now Equation (12.49) can be applied sequentially in the next table, with all flows in ft³/s:

n (a)	Time (b)	I (c)	$\frac{2}{\Delta t}S - Q$ (d)	$\frac{2}{\Delta t}S + Q$ (e)	Q (f)
1	4/23 12 PM	1500	0	—	0
2	4/24 12 AM	1600	1700	3100	700
3	4/24 12 PM	3100	2800	6400	1800
4	4/25 12 AM	9600		15,500	5500

All inflows were given data. Also $Q_1 = 0$ was given. Thus the value $(2S/\Delta t - Q)_1$ can be computed to be zero. Now all terms on the left side of Equation (12.49) are known for $n = 1$, and this equation gives $2S/\Delta t + Q = 3100$ for $n + 1 = 2$ in column (e). Entering the storage indication curve, Exhibit 10, with this value gives $Q = 700$ ft³/s [$n = 2$, column (f)]. Since $(2S/\Delta t + Q) - 2Q = 2S/\Delta t - Q$, column (d) with $n = 2$ is $3100 - 2(700) = 1700$. Applying Equation (12.49) with $n = 2$ then yields $1600 + 3100 + 1700 = 6400$ in column (e) for $n = 3$. And these operations are cyclically repeated until the solution is completed. Thus the outflow from the reservoir at 12 AM, 25 April, is $Q = 5500$ ft³/s. Using this discharge and the outflow-discharge data, the water surface elevation E at that time is, using interpolation,

$$E = 4.00 + \left(\frac{5500 - 4200}{5880 - 4200}\right) \times 1.00 = 4.77\text{ft}$$

above the spillway crest.

River Routing

All forms of hydrologic river routing begin with the assumption of some relation between storage in the river section and the inflow and outflow at the ends of the section. The most common of these methods is the Muskingum method, which assumes that this relation is a weighted linear relation between storage, inflow, and outflow taking the form

$$S = K[xI + (1 - x)Q] \qquad \textbf{(12.50)}$$

in which K is a proportionality factor with units of time, and x is the weighting factor giving the relative importance of the inflow and outflow contributions to storage. For example, for a simple reservoir one expects $S = f(Q)$ only so $x = 0$ could be chosen; if inflow and outflow are of equal importance, then $x = 0.5$ should be selected. For most streams x is between 0.2 and 0.3. The parameters K and x can be determined for a specific routing application if suitable data are available so that $[xI + (1 - x)Q]$ can be plotted versus storage S for several values of x between 0 and 0.5. The value of x that most nearly collapses the plotted data onto a single fitted straight line is used in the routing application, and $1/K$ is the slope of that fitted line.

The final form of the Muskingum routing equation is

$$Q_{n+1} = C_0 I_{n+1} + C_1 I_n + C_2 Q_n \qquad \textbf{(12.51)}$$

in which

$$C_0 = (\Delta t/2 - Kx)/D \quad (12.52)$$

$$C_1 = (\Delta t/2 - Kx)/D \quad (12.53)$$

$$C_2 = (K - Kx - \Delta t/2)/D \quad (12.54)$$

and

$$D = K - Kx + \Delta t/2 \quad (12.55)$$

Observe that one must always have $C_0 + C_1 + C_2 = 1$. These equations can be derived by using Equation (12.50) to express S_n and S_{n+1}, inserting the results in Equation (12.49), and rearranging the terms.

Example 12.15

Thirty-six hours of data for stream flow are given in the following table:

Time	6 AM	12 AM	6 PM	12 PM	6 AM	12 AM	6 PM
I, ft³/s	10	30	70	50	40	32	25

The Muskingum parameters have been determined to be $K = 10$ hr, $\Delta t = 6$ hr, and $x = 0.23$. The flow is steady in the reach at 6 AM on the first day. Determine the outflow hydrograph from this stream reach.

Solution

Direct computation using first Equation (12.55) and then Equations (12.52)–(12.54) will lead to

$$D = 10.70,\ C_0 = 0.065,\ C_1 = 0.495,\ \text{and}\ C_2 = 0.440$$

Use of a table is an aid in organizing the computations:

Time	I_n, ft³/s	C_0I_{n+1}	C_1I_n	C_2Q_n	Q_{n+1}, ft³/s
6 AM	10				10.0
12 AM	30	2.0	5.0	4.4	11.4
6 PM	70	4.6	14.9	5.0	24.5
12 PM	50	3.3	34.7	10.8	48.8
6 AM	40	2.6	24.8	21.5	48.9
12 AM	32	2.1	19.8	21.5	43.4
6 PM	25	1.6	15.8	19.1	36.5

The inflow data are reproduced in the first two columns. The next three columns contain the terms that appear on the right side of Equation (12.51); the last column is the sum of the three previous column entries, as Equation (12.51) indicates, and is the outflow hydrograph. According to these computations, the peak outflow is 48.9 ft³/s and occurs at 6 AM on the second day.

WELL HYDRAULICS

The basic equation describing local, steady groundwater movement is Darcy's law, which can be written

$$V = -Ki \tag{12.56}$$

In this equation V is the average velocity of a discharge Q that moves through a soil cross-sectional area A. Darcy's law indicates that V is the product of the local hydraulic conductivity K, which depends on the local soil or rock properties, and the local gradient i of the piezometric head $H = p/\gamma + z$, that is, $i = dH/dL$. This may also be interpreted as a difference in fluid energy between points, because the kinetic energy associated with groundwater flow is negligible. To obtain the actual fluid velocity, called the seepage velocity, in the subsurface saturated zone, one divides the average velocity by the local porosity. A variety of units are used in describing groundwater parameters, so one should take care to use consistent units in all computations.

Steady Flow

Equations for steady flow from a well in either an unconfined or a confined aquifer can be derived from Darcy's law. These simple equations have meaning and are accurate only when several simplifying assumptions are valid, including the following: (a) the aquifer, which is a geologic formation that contains enough saturated permeable material to yield significant quantities of water, must be large in extent and have uniform hydraulic properties, for example K is constant; (b) the pumping must occur at a constant rate for an extremely long time so that start-up transients no longer exist; (c) the well fully penetrates the aquifer; (d) the well depth is much larger than the drawdown near the well; and (e) the estimate of the gradient i is a good one.

An aquifer is called unconfined if the upper edge of the saturated zone (ignoring capillary effects) is at atmospheric pressure; this edge is called the water table. Figure 12.10 is a schematic cross section of a well in an unconfined, horizontal aquifer. A cylindrical coordinate system (x, y) is placed at the base of the well; the drawdown from the undisturbed water table is s; the gradient of the piezometric head is $i = dy/dx$ at the water table and is assumed to apply to the entire water column below it. The radius of the well is $x = r_w$. Applying Darcy's law gives

$$V = \frac{Q}{A} = \frac{Q}{2\pi xy} = -K\frac{dy}{dx} \tag{12.57}$$

Rearranging this expression and integrating it between points (r_1, h_1) and (r_2, h_2) along the water table yields

$$Q = \frac{\pi K\left[h_2^2 - h_1^2\right]}{\ln\left(r_2/r_1\right)} \tag{12.58}$$

as the expression for the steady pumping rate, or discharge, for this case.

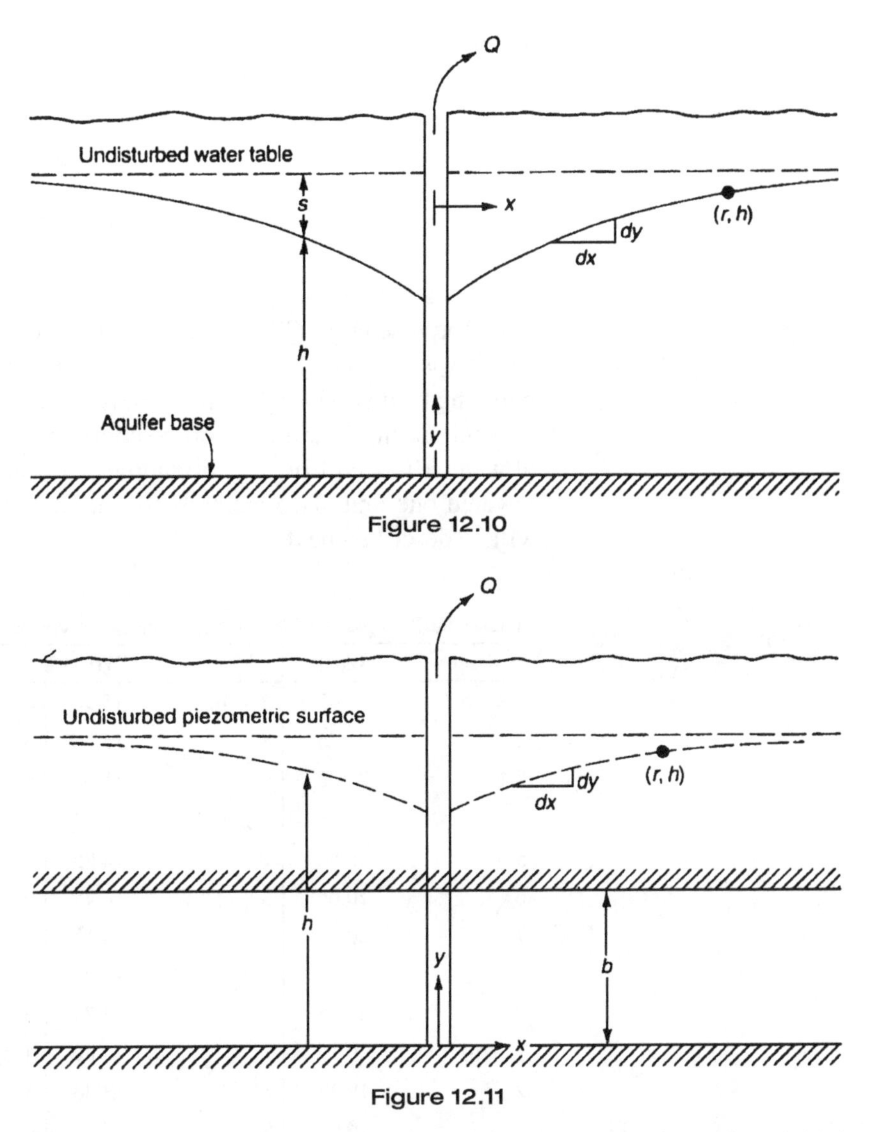

Figure 12.10

Figure 12.11

The case for steady pumping from a confined, horizontal aquifer of thickness m is similar to the first case, as shown in Figure 12.11. However, the gradient i is determined from the local slope of the piezometric head curve (shown dashed), which is no longer the same as the edge of the saturated zone. Equation (12.57) still applies to this case if the area through which flow occurs is corrected to $A = 2\pi xb$. Now the integration between points (r_1, h_1) and (r_2, h_2) on the piezometric surface results in

$$Q = \frac{2\pi Kb\left(h_2 - h_1\right)}{\ln\left(r_2/r_1\right)} \tag{12.59}$$

for the discharge. Sometimes the transmissibility $T = Kb$ is introduced into this equation.

Unsteady Flow

The first significant solution for unsteady flow to a well was originally developed by Theis for a confined aquifer. It expresses the drawdown s as

$$s = \frac{Q}{4\pi T} W\left(u\right) \tag{12.60}$$

in which

$$u = \frac{r^2 S}{4Tt} \tag{12.61}$$

and

$$W(u) = \int_u^{\infty} \frac{e^{-u} du}{u} = -0.5772 - \ln(u) + u - \frac{u^2}{2 \times 2!} + \ldots \tag{12.62}$$

is called the well function of u. Table 12.7 presents tabulated values for this function. The discharge Q is constant over the pumping period, and r is the radius at which s is computed (to find the drawdown at the well, use $r = r_w$) at time t after pumping begins. The solution depends on knowledge of two aquifer properties, the transmissibility T and the storage constant S. The storage constant is the amount of water removed from a unit volume of the aquifer when the piezometric head is lowered one unit. Two methods for the determination of these aquifer properties will be described next.

Table 12.7 Values of the function $W(u)$ for various values of u

u	$W(u)$	u	$W(u)$	u	$W(u)$	u	$W(u)$
1×10^{-10}	22.45	7×10^{-8}	15.90	4×10^{-5}	9.55	1×10^{-2}	4.04
2	21.76	8	15.76	5	9.33	2	3.35
3	21.35	9	15.65	6	9.14	3	2.96
4	21.06	1×10^{-7}	15.54	7	8.99	4	2.68
5	20.84	2	14.85	8	8.86	5	2.47
6	20.66	3	14.44	9	8.74	6	2.30
7	20.50	4	14.15	1×10^{-4}	8.63	7	2.15
8	20.37	5	13.93	2	7.94	8	2.03
9	20.25	6	13.75	3	7.53	9	1.92
1×10^{-9}	20.15	7	13.60	4	7.25	1×10^{-1}	1.823
2	19.45	8	13.46	5	7.02	2	1.223
3	19.05	9	13.34	6	6.84	3	0.906
4	18.76	1×10^{-6}	13.24	7	6.69	4	0.702
5	18.54	2	12.55	8	6.55	5	0.560
6	18.35	3	12.14	9	6.44	6	0.454
7	18.20	4	11.85	1×10^{-3}	6.33	7	0.374
8	18.07	5	11.63	2	5.64	8	0.311
9	17.95	6	11.45	3	5.23	9	0.260
1×10^{-8}	17.84	7	11.29	4	4.95	1×10^{0}	0.219
2	17.15	8	11.16	5	4.73	2	0.049
3	16.74	9	11.04	6	4.54	3	0.013
4	16.46	1×10^{-5}	10.94	7	4.39	4	0.004
5	16.23	2	10.24	8	4.26	5	0.001
6	16.05	3	9.84	9	4.14		

Source: Bedient and Huber, *Hydrology and Floodplain Analysis*, 3rd ed. ©2002 by Addison Wesley Publishing Co. Reprinted by permission.

The first method of determining T and S is by using the original Theis equations. An examination of these equations shows that a plot of $W(u)$ versus u, called a type curve, will have the same shape as a plot of s versus r^2/t on log-log graph paper. The two curves are plotted, and one graph is laid over the other so the curves lie on one another; then a so-called match point, which is a set of data for u, $W(u)$, s, and r^2/t, is taken from the plots, inserted in Equations (12.60) and (12.61), and the resulting relations are solved for S and T. Since the match point is used to establish a connection between one data plot and the other, the match point need not be on the curve itself, although most practitioners do choose the match point atop the superimposed curves.

The second method, called the Cooper-Jacob method, is appropriate when u is small (for example, $u \leq 0.01$ is a common rule). In this method s is plotted against pumping time on semilogarithmic paper; the curve eventually becomes linear. A fitted straight line is then extended to the point $s = 0$, where the value $t = t_0$ is noted. One then solves for the aquifer properties from

$$T = \frac{2.3Q}{4\pi(s_2 - s_1)} \log_{10}\left(\frac{t_2}{t_1}\right) \tag{12.63}$$

and

$$S = \frac{2.25Tt_0}{r^2} \tag{12.64}$$

Use of these equations is simplified if points 1 and 2 are chosen so that $t_2/t_1 = 10$; of course $s_2 - s_1$ is the difference in drawdown over this same time interval. The result in Equation (12.64) must be nondimensional.

Several approaches are possible for unsteady unconfined well flow, but the simplest is to use the Theis method, Equation (12.60), with modified definitions of T and S. Now $T = Kb$ is based on the saturated thickness when pumping commences, and S is the specific yield, the volume of water released when the water table drops one unit. This approach is accurate when the drawdown is small in comparison with the saturated thickness of the acquifer.

Example 12.16

A well has been pumped at a steady rate for a very long time. The well has a 12-inch diameter and fully penetrates an unconfined aquifer that is 150 feet thick. Two small observation wells are 70 and 150 feet from the well, and the corresponding observed drawdowns are 24 and 20 feet. If the estimated hydraulic conductivity is 10 ft /day (sandstone), what is the discharge?

Solution

The saturated aquifer thicknesses at the observation wells are $h_1 = 150 - 24 = 126$ ft and $h_2 = 150 - 20 = 130$ ft, and the use of Equation (12.58) leads directly to

$$Q = \frac{\pi K\left[h_2^2 - h_1^2\right]}{\ln(r_2/r_1)} = \frac{\pi(10)\left[(130)^2 - (126)^2\right]}{\ln(150/70)} = 42{,}200\,\text{ft}^3/\text{day}$$

This is equivalent to 0.49 ft³/s or 220 gal/min.

Example 12.17

Data on time t since pumping began versus drawdown s were collected from an observation well located 400 feet from a well that fully penetrated a confined aquifer that is 80 feet thick and is pumped at 200 gal/min. The data are presented in Exhibit 19.

Exhibit 19

t, min	s, ft	t, min	s, ft
35	2.82	103	4.43
41	3.12	131	4.60
48	3.25	148	5.00
60	3.60	205	5.35
80	3.98	267	5.80

Determine the aquifer properties T and S.

Solution

The data in Exhibit 19 have been used with $r = 400$ ft to compute s versus r^2/t, which have been plotted in Exhibit 20 on a sheet of log-log paper, and the plot has been placed on top of a log-log plot of $W(u)$ versus u (the type curve). The two plots have been moved around until the closest fit between the curves was found, taking care that the coordinate axes are parallel. If the match point is chosen as shown in the figure, then $s = 5$ ft, $r^2/t = 10^2$ ft³/min, $u = 0.0175$, and $W(u) = 3.50$. Rearranging Equation (12.60),

$$T = Q\frac{W(u)}{4\pi s} = \frac{200\,\text{gal/min}}{7.48\,\text{gal/ft}^3}\frac{(3.50)}{4\pi(5\text{ft})} = 1.49\,\text{ft}^2/\text{min}$$

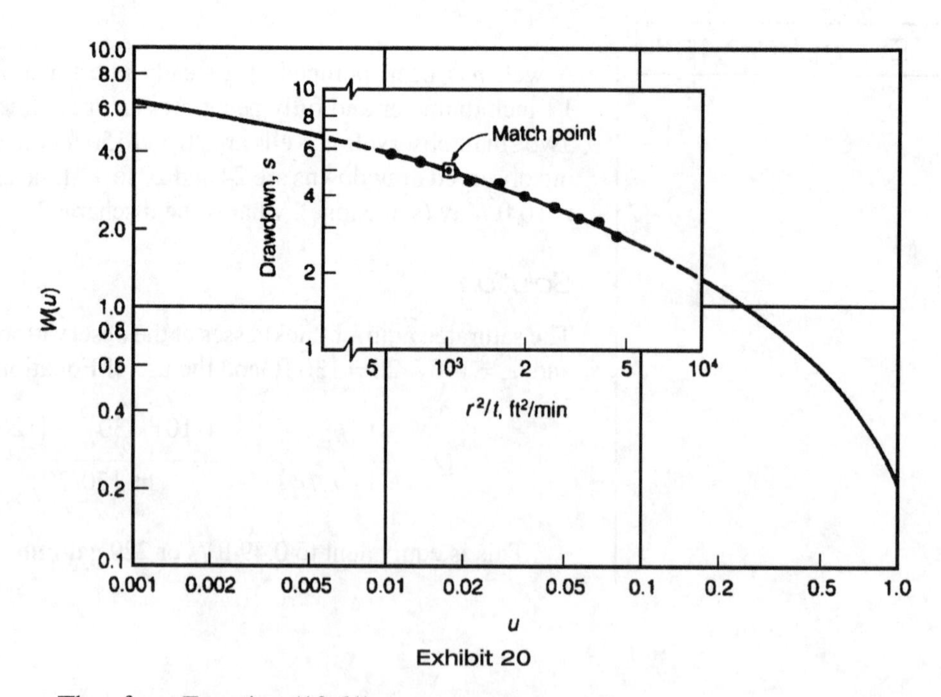

Exhibit 20

Then from Equation (12.61),

$$S = \frac{4Tu}{r^2/t} = \frac{4(1.49)(0.0175)}{10^3} = 1.04\times10^{-4}$$

Example 12.18

Last April pumping began at an 8-inch-diameter well at a steady rate of 300 gal/min while observations were made at a well 100 feet away. Values of elapsed time and drawdown were taken for 16 hours and plotted; see Exhibit 21. Use the plotted data to determine values for the transmissibility and storage constant of this aquifer. In addition, estimate the drawdown in the observation well after four months of steady pumping at 300 gal/min.

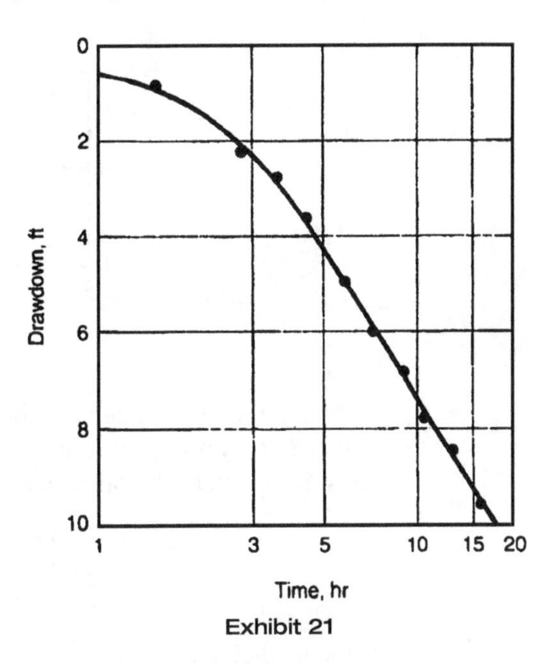

Exhibit 21

Solution

Exhibit 21 is a semilogarithmic plot of data in the form needed to apply the Cooper-Jacob (or modified Theis, as it is also called) method. If one extends the straight-line portion of the plot to $s = 0$, as shown in Exhibit 22, one can read from the plot the value $t_0 = 2$ hr. One also needs a pair of data points (s_1, t_1) and (s_2, t_2) for use in Equation (12.63). For example, at $t_1 = 4.0$ hr, $s_1 = 3.3$ ft, and at $t_2 = 10.0$ hr, $s_2 = 7.4$ ft. Then

$$T = \frac{2.3Q}{4\pi(s_2 - s_1)} \log_{10}\left(\frac{t_2}{t_1}\right) = \frac{2.3(300)}{4\pi(7.4 - 3.3)} \log_{10}\left(\frac{10.0}{4.0}\right) = 5.33\,\text{gal/min/ft}$$

$$T = \frac{5.33\,\frac{\text{gal/min}}{\text{ft}}}{7.48\,\frac{\text{gal}}{\text{ft}^3}}\left(60\,\frac{\text{min}}{\text{hr}}\right)\left(24\,\frac{\text{hr}}{\text{day}}\right) = 1026\,\text{ft}^2/\text{day}$$

The storage constant can then be found as

$$S = \frac{2.25Tt_0}{r^2} = \frac{2.25\left(1026\,\text{ft}^2/\text{day}\right)\left(\frac{2}{24}\,\text{day}\right)}{(100\,\text{ft})^2} = 0.0192$$

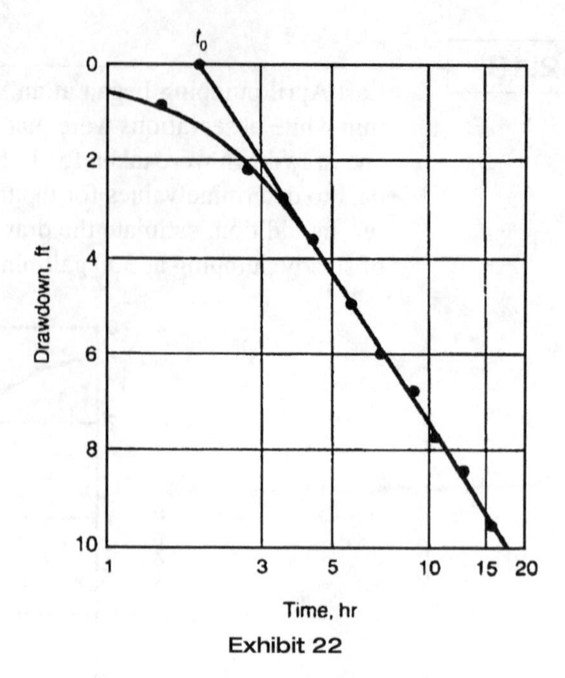

Exhibit 22

Equation (12.63) can also be used to find the drawdown after four months. If s_2 is the drawdown after four months, then t_2 is four months or approximately 122 days. If one picks the other point to be $s_1 = 0$ and $t_1 = t_0 = 2.0$ hr, then

$$s_2 = \frac{2.3Q}{4\pi T}\log_{10}\left(\frac{t_2}{t_1}\right) = \frac{2.3(300)}{4\pi(5.33)}\log_{10}\left(\frac{122(24)}{2}\right) = 32.6\text{ft}$$

is the predicted drawdown.

WATER DISTRIBUTION

Water distribution systems involve a **water source**, a **transmission line**, the **distribution network**, **pumping**, and **storage**.

Water Source

The engineering significance of the water source is primarily related to the water elevation and the water quality. The water elevation will determine the extent of pumping required to maintain adequate flows in the network. The range of elevations of the water source must be known as a function of flow rates and seasons of the year. Typical water sources include reservoirs, lakes, rivers, and underground aquifers.

Water quality is largely determined by the source. **Surface water** sources typically have low dissolved solids but high suspended solids. As a result, treatment of such sources requires coagulation followed by sedimentation and filtration to remove the suspended solids. **Groundwater** sources are low in suspended solids, but they may be high in dissolved solids and reduced compounds. Groundwater, if high in dissolved solids, requires chemical precipitation or ion exchange processes for the removal of dissolved ions.

Transmission Line

The transmission line is required to convey water from the sources and storage to the distribution of the system. Arterial main lines supply water to the various loops in the distribution system. The main lines should be arranged in loops or in parallel to allow for repairs. Such lines are designed using the Hazen-Williams formula for single pipes [Equation (12.65)], and the equivalent pipe method for single loops. Multiple loops are designed using **Hardy Cross methodology**:

$$v = kCr^{0.63}s^{0.54} \quad \textbf{(12.65)}$$

where v = pipe velocity (fps), k = constant (1.318 fps), C = roughness coefficient (100 for cast iron pipe), r = hydraulic radius (ft), and s = slope of hydraulic grade line (ft/ft).

The **equivalent pipe method** involves two procedures: (1) conversion of pipes of unequal diameters in series into a single pipe, and (2) conversion of pipes in a parallel flow from the same nodes into a single pipe.

The Distribution Network

The distribution network comprises the arterial mains, distribution mains, and smaller distribution piping. The mains are spaced in a series of loops to supply redundancy in case of failure. Diameters of mains should be larger than 6 inches.

Pressure Requirements

The controlling design variable for water distribution networks is the amount of pressure under maximum flow. Typically, a minimum value of 20 to 40 psi is required if pumper fire trucks are used, and 40 to 75 psi if fire flows are taken directly from hydrants. Static pressure should range from 60 to 80 psi. If pressures greater than 100 psi occur in the system, then the system should be separated into levels with separate storage elevations for each level.

Required Flows

Flow from the water distribution network is calculated as the sum of domestic, irrigation, industrial, commercial, and fire requirements. The design flow is the largest of either

$$Q^{\text{design}} = Q^{\text{hourly}} \quad \textbf{(12.66)}$$

where Q^{design} = design flow (cfs) and Q^{hourly} = maximum hourly (cfs), or,

$$Q^{\text{design}} = Q^{\text{daily}} + Q^{\text{fire}} \quad \textbf{(12.67)}$$

where Q^{daily} = maximum daily flow over the past three years (cfs) and Q^{fire} = fire flow (cfs).

The fire flow is recommended by the National Board of Fire Underwriters as

$$Q^{\text{fire}} = 1020\sqrt{P}\left(1 - 0.01\sqrt{P}\right) \quad \textbf{(12.68)}$$

where Q = demand in gpm and P = population in thousands (up to 200,000 persons). Above 200,000 persons, an additional 2000 to 5000 gpm need to be added to accommodate an additional fire.

Fire flows are dependent on the type of construction, floor area, height of buildings, fire hazards, and local codes. Fire flows in gpm can also be estimated for specific buildings from Equation (12.69) as

$$Q^{\text{fire}} = 18C(A)^{0.5} \quad \textbf{(12.69)}$$

where C = construction material constant (1.5 for combustible materials; 0.6 for fire resistant materials), and A = floor area (ft^2).

The various flows for a given community are typically obtained from historical data of water use. Without such data, estimates can be made from the average values listed in Table 12.8. The design flow calculated by Equation (12.66) or Equation (12.67) should be increased by 10% to allow for system losses.

Table 12.8 Water flow rates

Flow	Method of Estimation
Average domestic flow	Population served, 100 gal/capita-d, 3000 to 10,000 persons/km^2 (7765 to 25,889 persons/mi^2)
Irrigation flow	Maximum of 0.75 times average daily flow for arid climates
Industrial/commercial flow	Computed from known industries and area of commercial districts
Q^{daily}	2.5 times the sum of yearly average domestic, industrial, and irrigation
Q^{hourly}	1.5 times Q^{daily}
Q^{fire}	Equation (8.5)

Water Storage

Water storage is required to maintain pressure in the system, minimize pumping costs, and meet emergency demands. Typically the storage is placed so that the load center or distribution network is between the water source and the storage, as shown in Figure 12.12.

Storage requirements are calculated based on variations in hourly flow and fire requirements. If the pumping capacity is set at some value less than the maximum hourly flow, then all flow requirements above that value will have to be provided by storage. Such storage quantities can be estimated by integration of the area in Figure 12.13 above the pumping rate. If pumping rates are set at the average daily flow, the storage requirements can be estimated as about 30% of the total daily water requirements.

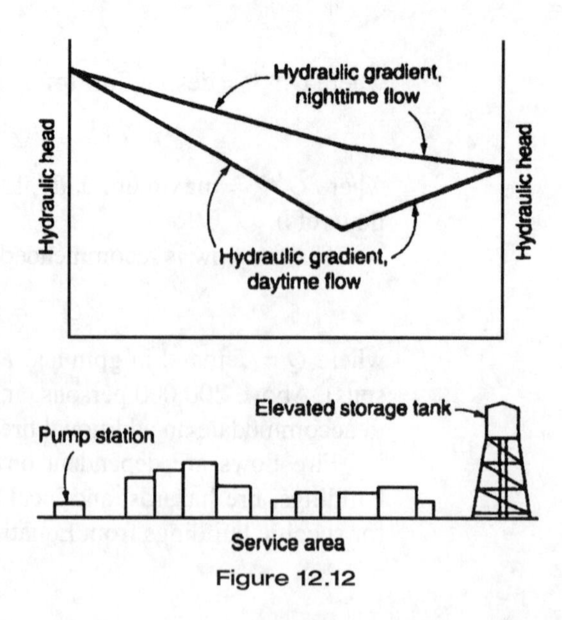

Figure 12.12

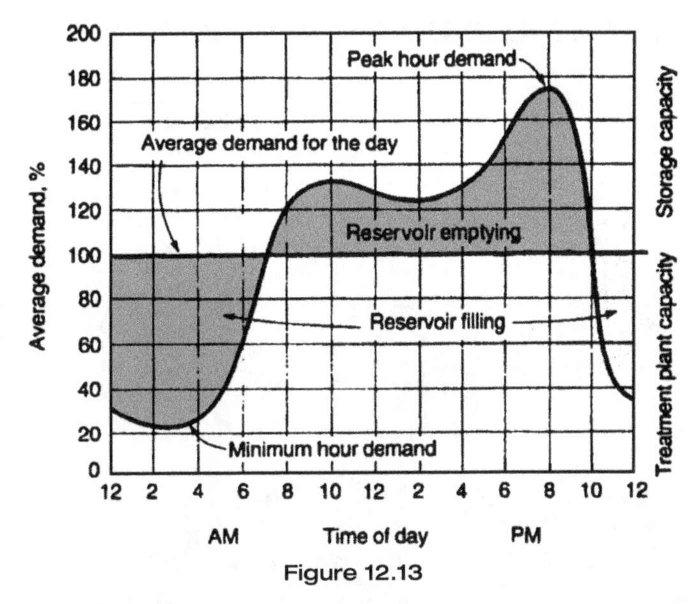

Figure 12.13

Storage requirements for fire flows depend on the required duration. Duration can be estimated as approximately one hour for each 1000 gpm of fire flow. The storage requirements for fire are equal to the duration times Q^{fire}. Storage must be supplied to meet the maximum daily requirement, which is equal to the average daily requirement plus the fire flow for the duration of the fire.

Pumping Requirements

The pumping system needs adequate capacity for design flows, taking into account the supply from storage. The required head must be adequate to maintain 20 psi at the load center during maximum flow. Excess pumping capacity is required to account for the possible failure of pumps and storage facilities.

Pressure Tests

The adequacy of the distribution system to maintain required pressure under fire-fighting conditions is tested by measuring pressure drops after opening the valves at a hydrant. For such tests, the fire flow that can be delivered is given by Equation (12.70):

$$\frac{Q^{\text{fire}}}{Q^{\text{test}}} = \frac{\left(P - P^{\text{fire}}\right)^{0.54}}{\left(P - P^{\text{test}}\right)^{0.54}} \qquad \textbf{(12.70)}$$

where Q^{fire}, Q^{test} = fire flow and test flow, respectively, and P, P^{fire}, P^{test} = static main pressure, minimum required pressure under fire flow (20 psi), and pressure under hydrant test, respectively.

SELECTED SYMBOLS AND ABBREVIATIONS

Symbol or Abbreviation	Description
bep	best efficiency point
E	specific energy
η	efficiency factor
Fr	Froude number
f	Darcy friction factor
ϕ	peripheral speed factor
g	gravitational acceleration
γ	unit weight
H	head difference
I	moment of inertia
K	local loss coefficient
μ	viscosity
N	rotative speed
NPSH	net positive suction head
P	power
Q	discharge
Re	Reynolds number
S	specific gravity
τ	mean shear stress
u	runner speed or blade velocity
ν	kinematic viscosity
V	velocity
ω	angular velocity

REFERENCES

ASCE. *Design and Construction of Sanitary and Storm Sewers.* Manual No. 37. American Society of Civil Engineers, New York, 1986.

Bedient, P. B., and W. C. Huber. *Hydrology and Foodplain Analysis*, 3rd ed. Prentice Hall, Upper Saddle River, NJ, 2002.

Chow, V. T. *Open-Channel Hydraulics*. McGraw-Hill, New York, 1959.

Chow, V. T., D. R. Maidment, and L. R. Mays. *Applied Hydrology*. McGraw-Hill, New York, 1988.

Henderson, F. M. *Open Channel Flow*. Macmillan, New York, 1966.

Larock, B. E., R. W. Jeppson, and G. Z. Watters. *Hydraulics of Pipeline Systems.* CRC Press, New York, 2000.

Linsley, R. K., M. A. Kohler, and J. L. H. Paulhus. *Hydrology for Engineers*, 2nd ed. McGraw-Hill, New York, 1975.

McCuen, R. H. *Hydrologic Analysis and Design*, 2nd ed. Prentice Hall, Englewood Cliffs, NJ, 1998.

Munson, B. R., D. F. Young, and T. H. Okiishi. *Fundamentals of Fluid Mechanics*, 3rd ed. Wiley, New York, 1998.

Sanks, R. L. *Pumping Station Design*. Butterworths, Boston, 1989.

Street, R. L., G. Z. Watters, and J. K. Vennard. *Elementary Fluid Mechanics*, 7th ed. Wiley, New York, 1996.

Sturm, T. W. *Open Channel Hydraulics*. McGraw-Hill, New York, 2001.

White, F. M. *Fluid Mechanics*, 3rd Ed. McGraw-Hill, New York, 1994.

White, F. M. *Fluid Mechanics*, 5th ed. McGraw-Hill, New York, 2003.

Viessman, W., Jr., and G. L. Lewis. *Introduction to Hydrology*, 4th ed. HarperCollins, New York, 1996.

CHAPTER 13

Structural Analysis

OUTLINE

The outline notes that constitute this chapter are a summary of what can be found in all textbooks upon this subject. Since no one person's choice of summary can hope to match the needs of all students simultaneously, it is recommended that a suitable textbook be reread in conjunction with these notes and that personal notes be appended to the text in this chapter.

Determinate structural analysis, often termed **statics**, deals with structures that do not move. There are only two types of motion: translation and rotation. If the structure (and all parts of it) neither translates nor rotates, it is said to be in **static equilibrium**. (It is true, of course, that any material under stress will undergo some change of size and/or shape because of those stresses, but these movements are considered negligible in the present context.)

In statics, translation is caused by **forces**, and rotations are caused by **moments**. These are vector quantities. To define a vector requires three characteristics, usually (a) a line of action, (b) a direction, and (c) a magnitude. In some contexts, moments whose vectors are out of the plane are termed **torques** or **twists**. The student is expected to be familiar with simple vectors and their manipulation.

Actions is a general term that includes both forces and moments.

NEWTON'S LAWS

The study of statics is based upon two of Newton's three laws of motion. These are (1) a body will remain in its state of rest (relative to some chosen reference point) or of motion in a straight line, unless acted upon by a force, and (2) to every action there is an equal and opposite reaction. The third law (a body under the action of a single force will translate with acceleration along the line of the force and in the same direction as the force) is used in the study of *dynamics*.

Law 2 is often restated as "Forces (actions) can exist only in equal and opposite pairs." The combination of laws 1 and 2 thus requires that the net action on a body at rest be zero. This requirement, in turn, defines the two vector equations

$$\text{Net force} = 0$$
$$\text{Net moment} = 0$$

This is the central principle of statics.

More generally, determinate structural analysis can be defined as the process of calculating (for a body in static equilibrium)

a). Any one, or all, of the actions acting upon the body

b). The movement (deflection) of any point (in any direction) within the body

c). The rotation of any line in the body

The process of structural analysis has just two steps. *First*, an idealized model of the structure (including the forces acting upon it) is agreed upon. In all but the most complex or large structures, it is adequate to use two-dimensional representations of real structures. When modeling the structure, it is also common practice to idealize the supports. Such idealizations as frictionless rollers, fully fixed supports, and frictionless pins are assumed to be familiar to the reader.

One other assumption is that structures are composed of rigid members. This does not mean that each member is infinitely rigid (i.e., does not change length or shape when loaded) but rather that these deformations are small relative to the original dimensions. Expressed another way, the deformations induced by loading do not change the geometry of the structure significantly. Luckily, this assumption is a good one for the vast majority of real engineering structures. The discerning reader may be about to object that many textbook examples (and reality) contain structures with ropes or wires, which are not stable if placed in compression. This objection is a good one but is countered by the observation (often implicit) that in all such cases the rope or wire must be in tension and hence cannot "buckle" or otherwise move.

Most engineers' drawings (and most examination problems) already include the simplifications just described.

The *second* step of the analysis process is the application of Newton's laws to the model.

The application of these equations of statics (often termed the equations of equilibrium) to a three-dimensional body involves just two vector equations (translation = 0 and rotation = 0). However, it is frequently more convenient to use the equivalent six independent scalar equations (three for translation and three for rotations). For two-dimensional models these reduce to three (two translations and one rotation). Thus, if a two-dimensional structural problem involves more than three unknowns, it is statically indeterminate and requires further knowledge for its solution. In what follows, the discussion is limited to two dimensions and stable static structures.

FREE BODY DIAGRAMS

The single most important concept in the *solution* of statics problems is that of the **free body diagram** (FBD). This concept is founded on the observation that if a (rigid) structure is in equilibrium, then every part of it must also be in equilibrium. This method has proven to minimize the likelihood of errors in solving structural problems.

In this approach an imaginary closed line is drawn through the structure. This line isolates a part of the structure, called the **free body**. The isolated part may be drawn by itself, but to maintain consistency with the original structure and loading, the engineer must examine the complete length of the closed line and insert all possible actions in their correct form and at their correct locations. Of course, some of these actions will be known and some unknown. All known actions must be included with their correct locations, magnitudes, and orientations. All unknown actions are given names, and their directions may be assumed (the mathematics will provide both the correct magnitude and the correct direction for each unknown).

It should be obvious that if the FBD is incorrect in *any way*, then the solution will be incorrect. A solution may be obtained, but if so, it is the solution to some *other* problem.

It is recommended that for each FBD used, a clear choice of axes be indicated and a positive direction also be chosen. These choices are, of course, arbitrary. It is then recommended that when applying the equations of equilibrium, all terms be written on one side of the equation and the result set equal to zero. The consistent use of this technique will reduce errors. By contrast, the practice, for example, of setting "all up forces equal to all down forces," which may seem attractive at first, is fraught with peril.

Facility in the analysis of structures is best obtained by solving multiple problems of as wide a variety as can be found. An excellent learning tool is to outline the steps of the solution without necessarily performing the mathematics. This permits a more rapid acquisition of the skills needed to understand how new problems are solved.

Note that in the following examples, units are not specified. The reader may select any units desired (e.g., metric versus "English" and small versus large).

In the following two problems the entire structure is used as the FBD.

Example 13.1

Find the reactions necessary to keep the structure shown in Exhibit 1 in equilibrium. Assume the axes, names, and directions shown in the figure.

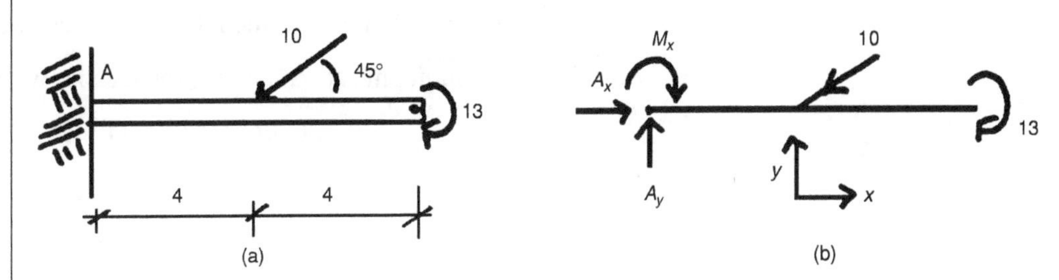

Exhibit 1 (a) Structure; (b) its FBD

Solution

Step 1. Sum forces in x direction:

$$A_x + (-).707 \times 10 = 0$$
$$A_x = +7.07 \text{ (assumed direction correct)}$$

Step 2. Sum forces in y direction:

$$A_y + (-).707 \times 10 = 0$$
$$A_y = + 7.07 \text{ (assumed direction correct)}$$

Step 3. Sum moments about an axis through A (assume clockwise positive):

$$(+)M_x + (+)10 \times 4 \times .707 + (+)13 = 0$$
$$M_x = -41.3 \text{ (assumed direction was incorrect)}$$

This completes the solution requested.

Example 13.2

Find the reactions necessary to keep the structure shown in Exhibit 2 in equilibrium.

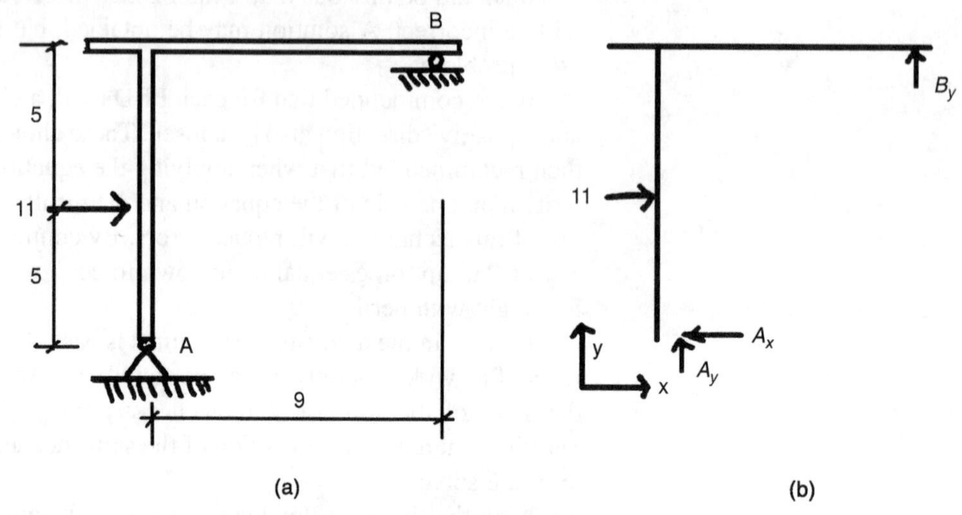

Exhibit 2 (a) Structure; (b) FBD of entire structure

Assume the names, axes, and directions shown.

Solution

Step 1. Summing forces in the positive x direction gives

$$+11 + (-)(A_x) = 0$$

Hence $(A_x) = +11$ (i.e., assumed direction is correct)

Step 2. Taking moments about an axis through A (with clockwise positive) gives

$$(-)B_y \times 9 + (+)11 \times 5 = 0$$

Hence

$$B_y = +\frac{55}{9} = +6.1$$

Step 3. Summing forces in the y direction gives

$$A_y + 6.1 = 0$$
$$A_y = -6.1 \text{ (i.e., assumed direction was incorrect)}$$

This completes the solution requested.

TRUSSES AND FRAMES

It should already be obvious to the reader that there is no single best way to solve structural analysis problems. It is true that a judicious choice of FBD and of the order in which equations are solved can shorten the calculation process, but all solution approaches will produce the correct answers if applied without error. However, it is convenient (but nothing more) to classify certain groups of structures, because they share characteristics and thus can be handled by a more "systematic" solution process. Two such groups are trusses and frames.

Trusses

Trusses must meet three criteria:

- All members (often termed **bars**) of the structure must be connected at only two points (these are called **joints**).
- All joints are frictionless pins.
- Loads are applied only at the joints.

The FBD of a bar of a truss can thus have only two forces acting upon it: one at each frictionless pin. Note that the bars need not be straight (although they almost always are). Application of the equations of statics shows that the line of action of both these forces must be directed along the line joining the two joints and that they must be equal in magnitude but oppositely directed. Hence a member of a truss is often described as a "two-force member." Only two possibilities then exist: Either the two forces are directed toward each other (in which case the bar is in compression) or they are directed away from each other (in which case the bar is in tension).

This feature of trusses simplifies their analysis, because if an FBD is drawn that cuts a member, it is known that only one action can exist at the cut, and it must be a force directed along the line that connects the joints (the sense and magnitude to be found as part of the solution).

There are two "classes" of analysis for trusses.

- **Method of Joints**: In this approach all FBDs are joints (with the exception that the entire body FBD is often used to find the reaction).
- **Method of Sections**: In this method the FBDs can include portions of the structure that are larger than a single joint. It should be noted that most designers use a combination of these methods when solving structures.

Example 13.3

Find the bar force in member CB in Exhibit 3(a) using the method of joints.

Use an FBD of the entire structure, shown in Exhibit 3(a), to obtain reactions. Assume the axes, names, and directions shown.

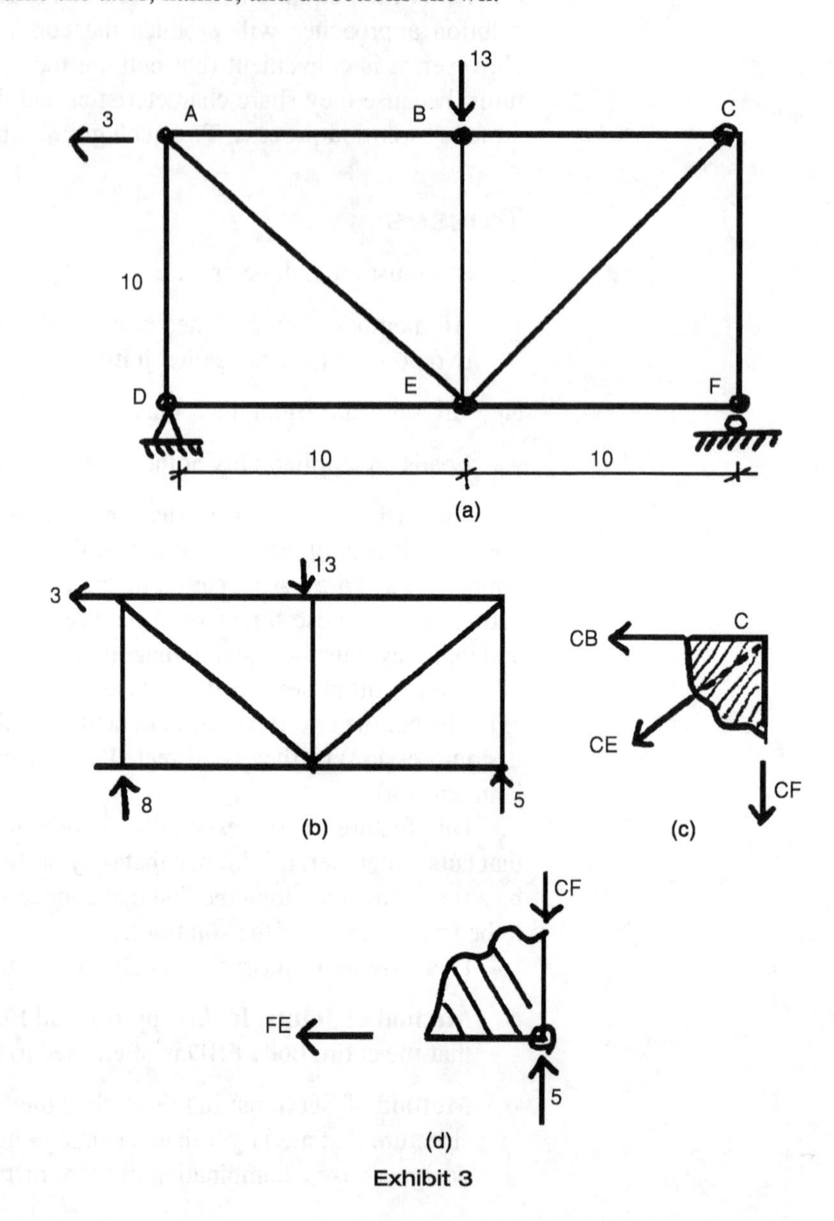

Exhibit 3

Solution

Solving by the methods shown above, the reader should obtain

$$D_y = 8.0 \text{ up}$$
$$F_y = 5.0 \text{ up}$$
$$D_x = 3 \text{ to right}$$

This completes the reactions, as shown in Exhibit 3(b).

Note that if we wish to attempt to find CB using an FBD of joint C, there will be three unknowns, which are not solvable; see Exhibit 3(c). We must find one of the other bar forces, CE or CF, in order to obtain a solution for CB. One possibility is to solve joint F first. This will give us CF and then permit a solution at joint C.

With Exhibit 3(d) and the assumed directions, joint F is solved to give CF = 5 (with bar in compression).

Now we move back to joint C in Exhibit 3(c) but with CF known in magnitude and direction—hence the arrow for CF in Exhibit 3(c) must be reversed. Taking moments about E with clockwise assumed positive,

$$CB = 5 \text{ and CB is in compression}$$

Example 13.4

Find the bar force in member DH in Exhibit 4(a) using the method of sections.

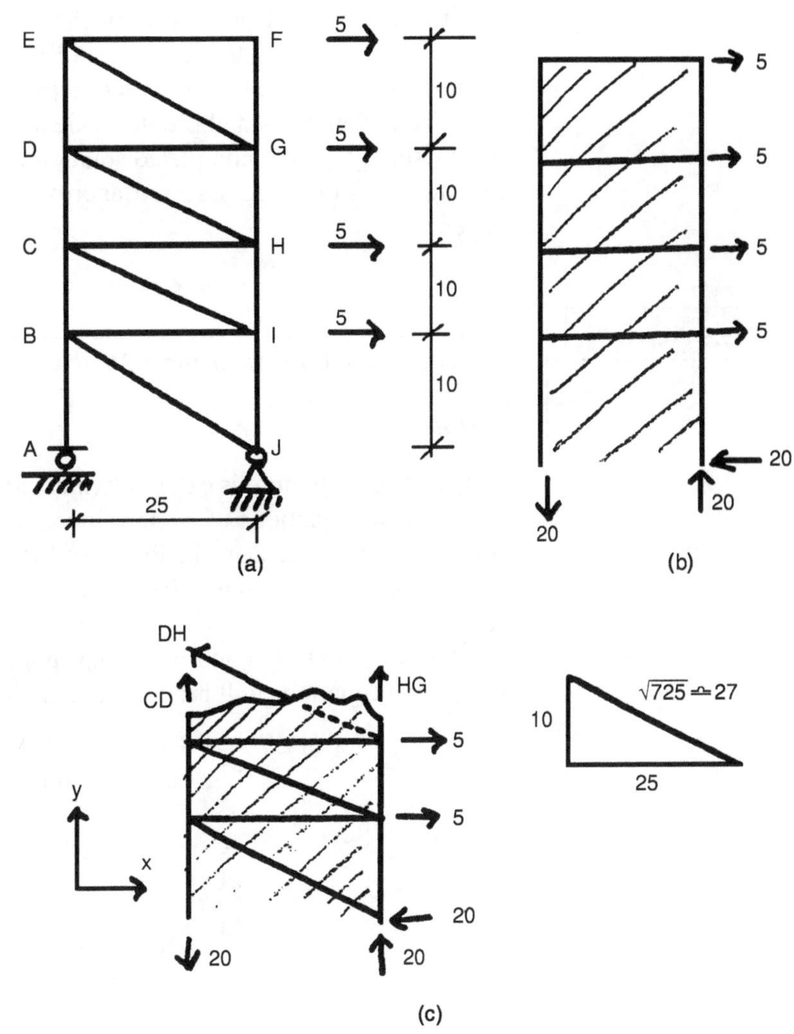

Exhibit 4 (a) Structure; (b) FBD entire structure; (c) FBD with horizontal cut through DC and GH

Solve to get the reactions shown in Exhibit 4(b).

Solution

Note that a horizontal cut can be used to give the FBD shown in Exhibit 4(c). Sum forces in x direction:

$$+5 + 5 - 20 - \text{DH} \times \frac{25}{27} = 0$$

DH = +10.8 (assumed direction correct, bar DH in tension)

Frames

Frames are multimember structures that do not meet the criteria for trusses (although individual members within a structure may be two-force members and thus help to simplify their solution). Thus, when an FBD cuts a member, there are (in two dimensions) three possible actions across the cut, and these must be included in the FBD.

Frames are clearly more complicated to solve than trusses, and there are no simple rules or guidelines for solution other than (a) the general rule that several FBDs will be needed for a solution and (b) the designer must experiment with several possible FBDs to find a solution.

It is an excellent practice (though very infrequently followed) to outline the steps of a solution (i.e., the sequence of FBDs to be used and the identification of the unknowns that can be found at each step) prior to performing any calculations.

When relatively small problems are solved, it is always a good idea to make a quick "survey" of the completed solution to check for gross errors. The equations of statics can often be assessed approximately without the need for pencil and paper.

Example 13.5

Determine the force in member AC in Exhibit 5(a).

Solution

Note that this structure is externally indeterminate; that is, the reactions cannot be found by the equations of statics alone. We note, however that bar AC is a two-force member, so we can take BCD and FBD and solve for the force AC. (Note that we cannot solve the second FBD completely, but we can get the bar force asked for.)

From Exhibit 5(b) and by taking moments about B (note that neither of the translation equations will produce a numerical result), with clockwise positive:

$$(+)17 \times 17 + (-).707 \times CA \times 11 = 0$$

$$CA = 37.2 \text{ (assumed direction correct, CA in tension)}$$

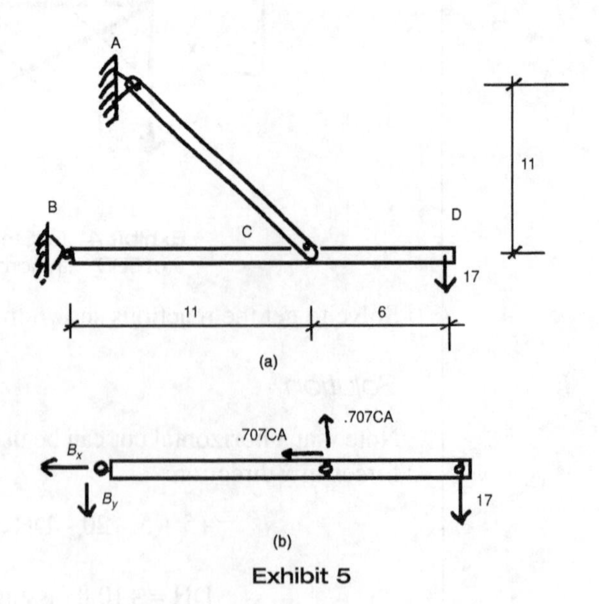

Exhibit 5

PROBLEMS

The reader is encouraged to attempt these problems alone prior to reviewing the sample solutions and to gain insight by comparing solutions where these differ.

In all cases the self-weight of the structure may be ignored unless instructions indicate otherwise.

Indicate the best answer from those offered (roundoff errors may lead to small discrepancies).

13.1 For the structure shown in Exhibit 13.1 the reactions are (in kN)

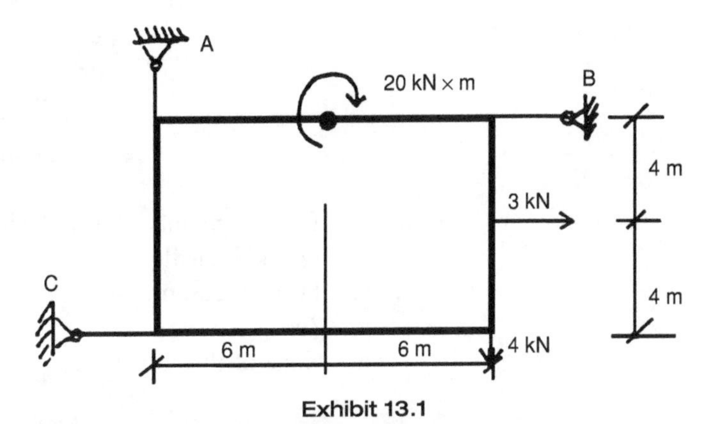

Exhibit 13.1

a. A = 4 kN down, B = 10 kN to left, C = 11.5 kN to right
b. A = 3 kN down, B = 10 kN to right, C = 10.0 kN to right
c. A = 4 kN up, B = 10 kN to left, C = 7.0 kN to right
d. A = 4 kN up, B = 8.5 kN to left, C = 11.5 kN to left

13.2 The truss shown in Exhibit 13.2 is used as a weighing device. If the gauge at A reads 9 kN, what is the weight of mass P?

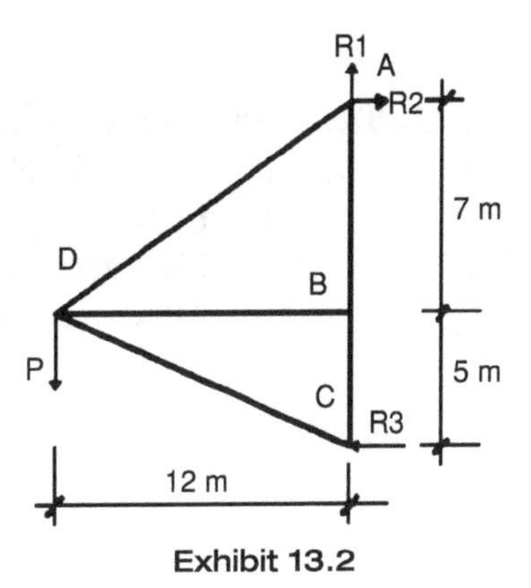

Exhibit 13.2

a. 9 kN
b. 12 kN
c. 4 kN
d. 8 kN

13.3 The force in member CB in Exhibit 13.3 is

a. 4.65 kN tensile
b. 4.65 kN compressive
c. 47.5 kN tensile
d. 8.00 kN compressive

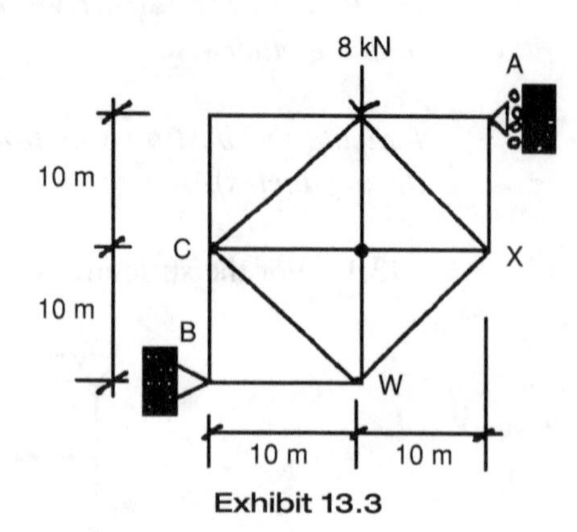

Exhibit 13.3

13.4 The force in member DE in Exhibit 13.4 is

a. 11.2 kN tensile
b. 11.3 kN compressive
c. Zero
d. None of the above

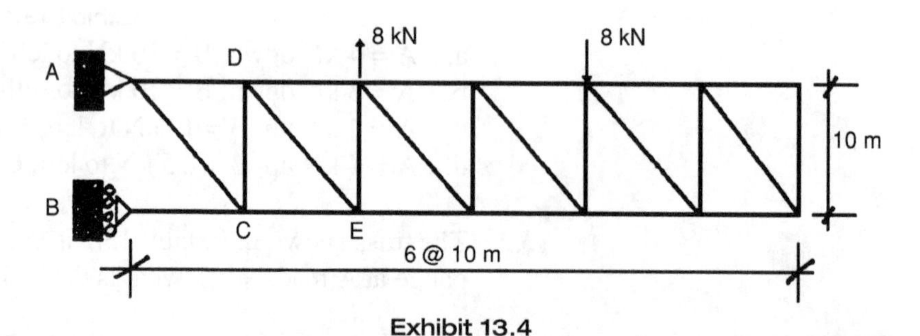

Exhibit 13.4

13.5 The forces in members FB and GH in Exhibit 13.5 are

a. FB = 2.5 kN tensile, GH = 11.66 kN tensile
b. FB = zero kN tensile, GH = 11.66 kN tensile
c. FB = zero kN tensile, GH = 11.66 kN compressive
d. FB = zero kN tensile, GH = 10.00 kN tensile

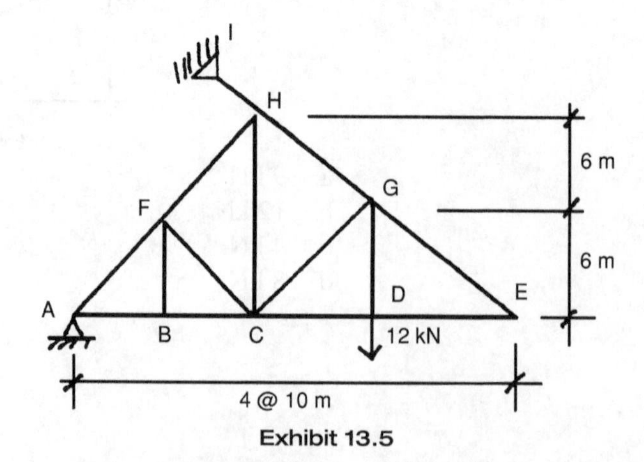

Exhibit 13.5

SOLUTIONS

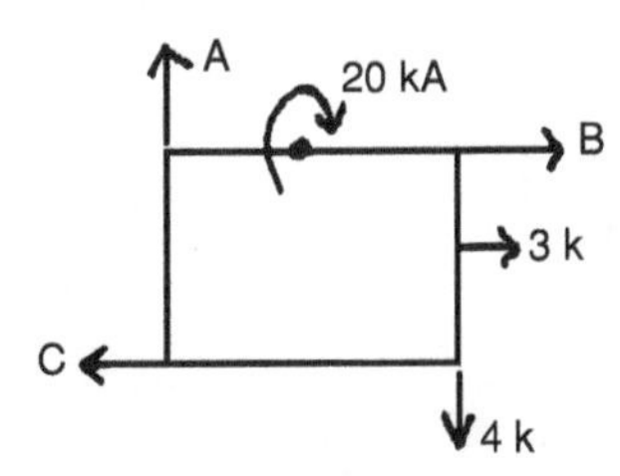

Exhibit 13.1a FBD whole structure

13.1 c.

$$\Sigma V = 0 \text{ (positive upward)}$$
$$+\text{A} - 4 = 0$$
$$A = 4 \text{ upward}$$

$$\Sigma M_c = 0 \text{ (positive clockwise)}$$
$$+20 + B(8) + 3(4) + 4(12) = 0$$
$$B = -10$$
$$B = 10 \text{ leftward}$$

$$\Sigma H = 0 \text{ (positive rightward)}$$
$$-C(-)10 + 3 = 0$$
$$C = -7$$
$$C = 7 \text{ rightward}$$

13.2 a.

$$\Sigma M_c = 0 \text{ (positive clockwise)}$$
$$+9 \times 12 - P(12) = 0$$
$$P = 9k$$

13.3 d.

Note: XY = 0 = ZA = AB.

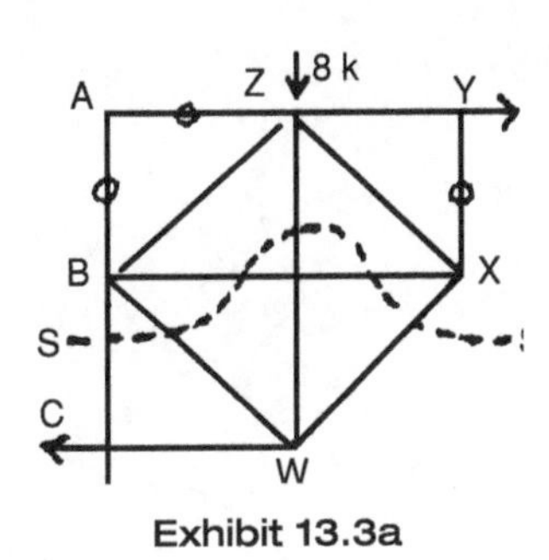

Exhibit 13.3a

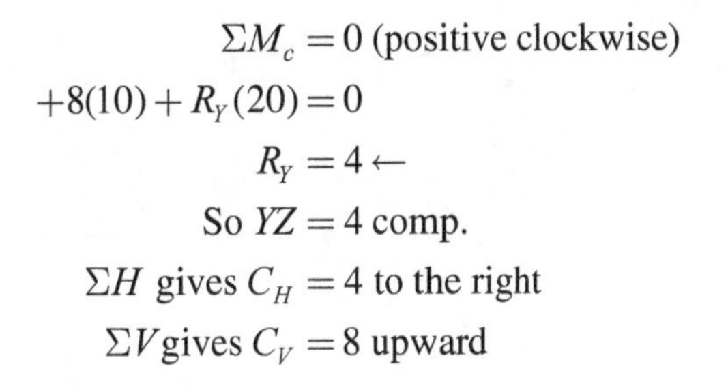

$$\Sigma M_c = 0 \text{ (positive clockwise)}$$
$$+8(10) + R_Y(20) = 0$$
$$R_Y = 4 \leftarrow$$
$$\text{So } YZ = 4 \text{ comp.}$$
$$\Sigma H \text{ gives } C_H = 4 \text{ to the right}$$
$$\Sigma V \text{ gives } C_V = 8 \text{ upward}$$

So

$$CB = 8c$$
$$CW = 4c$$

13.4 c.

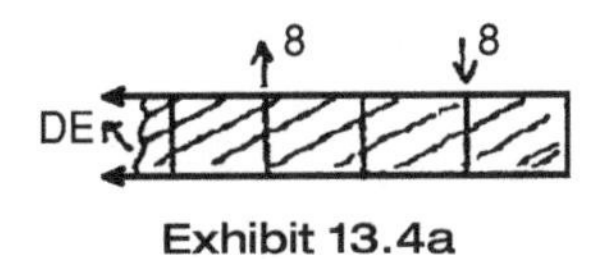

Exhibit 13.4a

FBD shown:

$$\Sigma V = 0$$
$$+8 - 8 + DE(0.7) = 0$$
$$DE = 0$$

13.5 b.

Joint B gives FB = 0

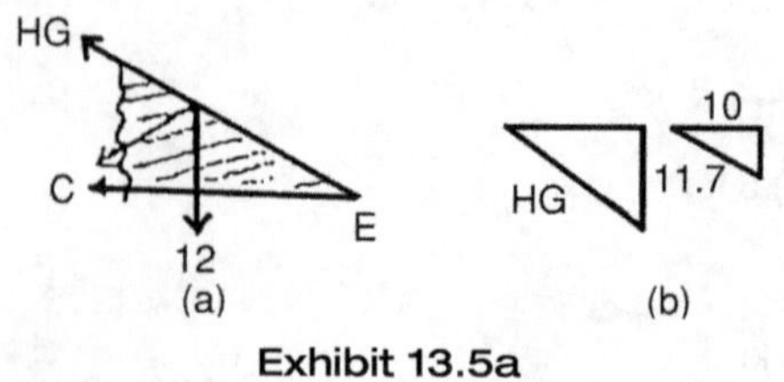

Exhibit 13.5a

FBD shown in Exhibit 13.5a:

$$\Sigma M_c = 0 \text{ (positive clockwise)}$$

$$+12\times 10-\left(\frac{10}{11.7}\text{HG}\right)\times 6$$

$$-\left(\frac{6}{11.7}\text{HG}\right)\times 10 = 0$$

$$120 - 5.13(\text{HG}) - 5.13(\text{HG}) = 0$$

$$\text{HG} = +\frac{120}{10.26} = 11.7$$

CHAPTER 14

Structural Design

Alan Williams

OUTLINE

DESIGN OF STEEL COMPONENTS

Introduction

Terminology

The notation used in this section is that adopted in the NCEES Handbook. NCEES specifies the use of the 13th edition of the AISC Manual and the use of English units of measurement. Both of these are adopted in this section. Appropriate design tables from the AISC Manual are reproduced in the NCEES Handbook.

Load and Resistance Factor Design

In accordance with AISC 360 Section B3, steel structures may be designed using the provisions for Load and Resistance Factor Design (LRFD) or using the provisions for Allowable Strength Design (ASD). NCEES uses the LRFD provisions in both the NCEES Handbook and the NCEES Sample Questions book. The LRFD provisions are adopted in this section. The applicable LRFD load combinations and resistance factors are reproduced in the NCEES Handbook. A load combination consists of a summation of applied loads each multiplied by an appropriate load factor, λ. A load combination constitutes a required strength level for a structure or member. The load factors account for the variability of the applied loads and the unlikely event of several transient loads occurring simultaneously.

Example 14.1

The gravity loads and seismic loads acting on a two-story steel frame are shown in Exhibit 1. The frame forms part of a wholesale store and the gravity loads consist of:

Roof dead load, $W_{Dr} = 50$ kips
Roof live load, $W_{Lr} = 10$ kips
Floor dead load, $W_D = 50$ kips
Floor live load, $W_L = 40$ kips

Using the load combinations given in the NCEES Handbook, the maximum design load on a column is most nearly:

a). 100 kips

b). 105 kips

c). 110 kips

d). 115 kips

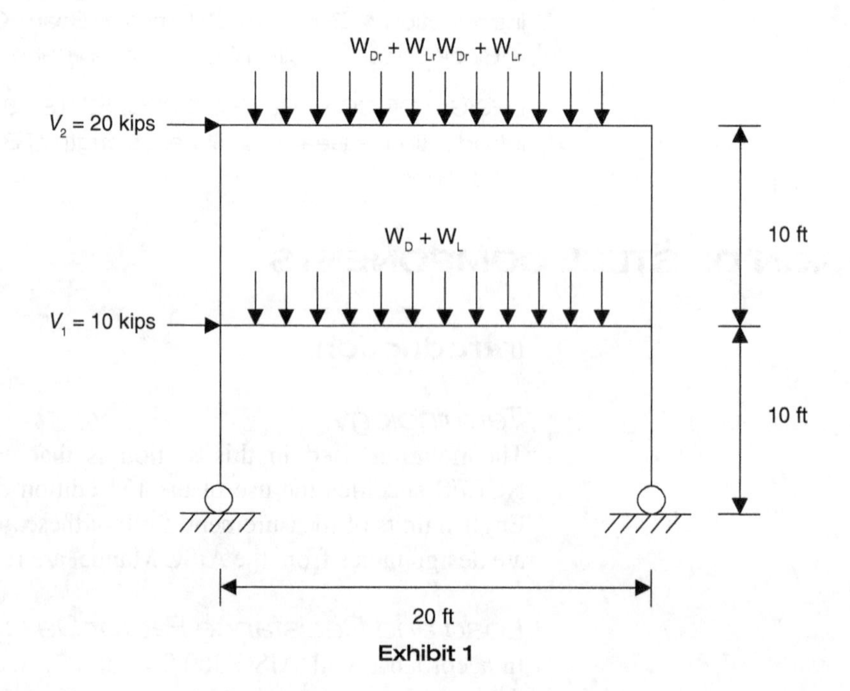

Exhibit 1

Solution

The axial load on one column due to the dead load is

$$D = (W_{Dr} + W_D)/2 = (50 + 50)/2 = 50 \text{ kips}$$

The axial load on one column due to the roof live load is

$$L_r = W_{Lr}/2 = 10/2 = 5 \text{ kips}$$

The axial load on one column due to the floor live load is

$$L = W_L/2 = 40/2 = 20 \text{ kips}$$

The axial load on the right hand column due to the seismic loads is

$$E = (20V_2 + 10V_1)/20 = (400 + 100)/20 = 25 \text{ kips}$$

The applicable load combinations are given in the NCEES Handbook. Rain, snow, and wind load are not applicable and the design load on a column considering dead load plus floor and roof live load is

$$\begin{aligned}\Sigma(\lambda Q_n) &= 1.2D + 1.6L + 0.5L_r \\ &= 1.2 \times 50 + 1.6 \times 20 + 0.5 \times 5 \\ &= 94.5 \text{ kips}\end{aligned}$$

The design load on a column considering dead load plus floor live load plus seismic load is

$$\begin{aligned}\Sigma(\lambda Q_n) &= 1.2D + 1.0E + 1.0L \\ &= 1.2 \times 50 + 1.0 \times 25 + 1.0 \times 20 \\ &= 105 \text{ kips} \ldots \text{ governs}\end{aligned}$$

The correct answer is b).

Beams

Nominal Strength of Beams

The nominal flexural strength of a member, M_n, is defined as the theoretical maximum member flexural strength as defined in AISC 360 Section F. The design strength, or available strength, of a member, $\phi_b M_n$, is the product of the resistance factor for flexure and the nominal strength. The resistance factor accounts for variations in materials, design equations, fabrication, and erection and, for flexure, is given by

$$\phi_b = 0.9$$

To ensure structural safety, AISC 360 Section B3 specifies that the design strength of a member must equal or exceed the required strength. Thus,

$$\phi_b M_n \geq \Sigma(\lambda Q_n)$$

Example 14.2

A beam of Grade 50 steel is adequately braced and has a lateral-torsional buckling modification factor of $C_b = 1.0$. The maximum factored moment applied to the component is $M_u = 300$ kip-ft. The lightest suitable W-shape required is most nearly:

a). W14 × 43

b). W14 × 48

c). W14 × 53

d). W14 × 61

Solution

Using the load table given in the NCEES Handbook, a W14 × 53 has a design strength of

$$\phi_b M_n = 327 \text{ kip-ft} > 300 \text{ kip-ft} \ldots \text{ satisfactory}$$

The correct answer is c).

The nominal strength of a flexural member depends on the following factors:

- The slenderness parameters of its compression flange and web. The more slender the elements, the more prone to local buckling of the member and the lower the nominal strength.
- The magnitude of the lateral-torsional buckling modification factor, C_b. A beam subjected to a uniform bending moment has a lateral-torsional buckling modification factor of 1.0. For other moment gradients, C_b exceeds 1.0 and the lateral-buckling strength is obtained by multiplying the basic strength by C_b, with a maximum permitted value of M_p.
- The onset of lateral-torsional buckling. As the effective length of a beam is increased, the compression flange tends to buckle laterally and the beam fails by elastic instability. Lateral bracing applied to the compression flange restrains the flange and prevents buckling.

Compact, Noncompact, and Slender Sections

Steel beams are classified as compact, noncompact, and slender in accordance with the slenderness parameters of its compression flange and web.

For a compact section, the criteria for determining the compactness of the flange are given in AISC 360 Table B4.1 as

$$b_f/2t_f < \lambda_{pf} = 0.38(E/F_y)^{0.5}$$

where

b_f = flange width
t_f = flange thickness
λ_{pf} = limiting slenderness parameter for compact flange
$\lambda = b_f/2t_f$
= beam flange slenderness parameter

A compact section is one that can develop the full plastic moment of resistance, M_p, before the onset of local buckling. The nominal flexural strength of a section with compact web and compact flange with adequate lateral bracing, is then

$$\begin{aligned} M_n &= M_p \\ &= F_y Z_x \end{aligned}$$

The plastic section modulus, Z_x, is tabulated in the NCEES Handbook for a selected number of W-sections. Most rolled W-shapes, with a yield stress not exceeding $F_y = 50$ ksi, qualify as compact flexural sections and all of the W-shapes tabulated in the NCEES Handbook, with the exception of W12 × 65, are compact flexural sections.

A noncompact section is one that can develop the yield stress in its compressive elements before the onset of local buckling but cannot develop a plastic hinge. The limiting buckling moment is given by

$$M_r = 0.75F_y S_x$$

and

$$M_r \le M_n < M_p$$

For a noncompact section the criteria for determining the compactness of the flange are given in AISC 360 Table B4.1 as

$$\lambda_{pf} = 0.38(E/F_y)^{0.5} < b_f/2t_f < \lambda_{rf} = 1.0(E/F_y)^{0.5}$$

where

λ_{rf} = limiting slenderness parameter for a noncompact flange

The nominal flexural strength of a section with compact web and noncompact flange with adequate lateral bracing, is given by AISC 360 Equation (F3-1) as

$$M_n = M_p - (M_p - 0.7F_yS_x)(\lambda - \lambda_{pf})/(\lambda_{rf} - \lambda_{pf})$$

Tabulated values of $\phi_b M_p$ in AISC Manual Table 3-2 allow for any reduction in M_n because of noncompactness.

A slender section is one that cannot develop the yield stress in its compressive elements before the onset of local buckling. There are no slender rolled I-shapes. In accordance with AISC 360 Section B4, a section is classified as slender when the slenderness ratio of the flange or web exceeds the limiting slenderness parameters for a noncompact section. For flange local buckling, a slender section is defined as

$$b_f/2t_f > \lambda_{rf} = 1.0(E/F_y)^{0.5}$$

Values of λ are tabulated in the NCEES Handbook Table of W-shape properties. The limiting slenderness parameter for a compact element, λ_p, and for a noncompact element, λ_r are given in AISC 360 Table B4.1 and the influence of λ on the nominal moment is shown in Figure 14.1.

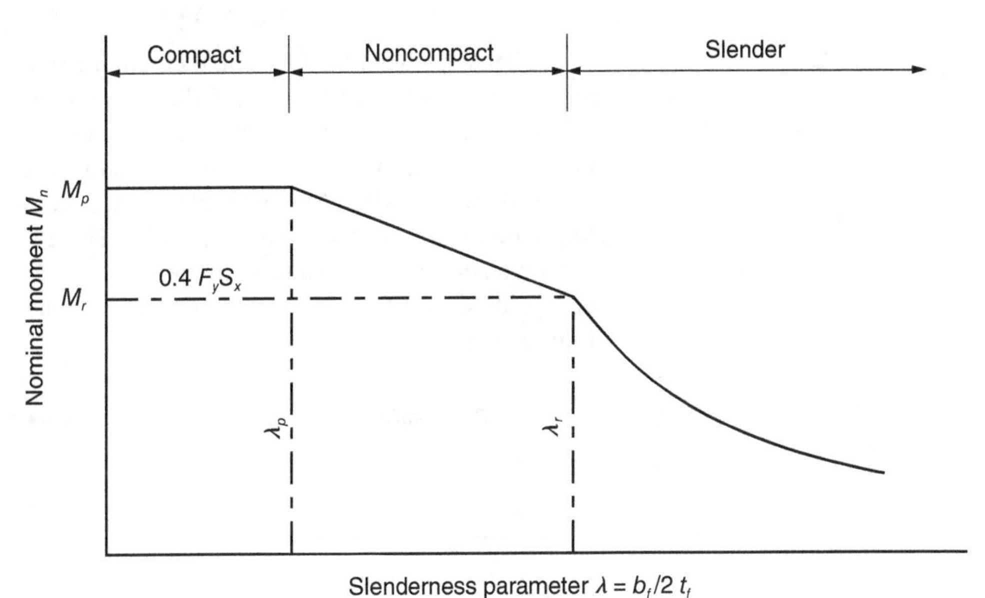

Figure 14.1 Nominal moment and flange slenderness

Lateral-Torsional Buckling Modification Factor

The lateral-torsional buckling modification factor, C_b, allows for the effect that applied bending moment has on the lateral-torsional buckling of a beam. The derivation of this factor, for a doubly symmetric shape such as a W-shape, is given by AISC 360 Equation (F1-1) as:

$$C_b = 12.5M_{max}/(2.5M_{max} + 3M_A + 4M_B + 3M_C)$$

where

M_A = absolute value of the bending moment at the quarter point of an unbraced segment

M_B = absolute value of the bending moment at the centerline of an unbraced segment

M_C = absolute value of the bending moment at the three-quarter point of an unbraced segment

M_{max} = absolute value of the maximum bending moment in the unbraced segment

The terms used in the derivation of C_b are illustrated in Fig. 14.2.

Load tables in the AISC Manual and in the NCEES Handbook are based on an assumed value of $C_b = 1.0$. When the value of C_b exceeds 1.0, tabulated values of the design moment may be multiplied by C_b with a maximum permitted value of the design moment of $\phi_b M_p$.

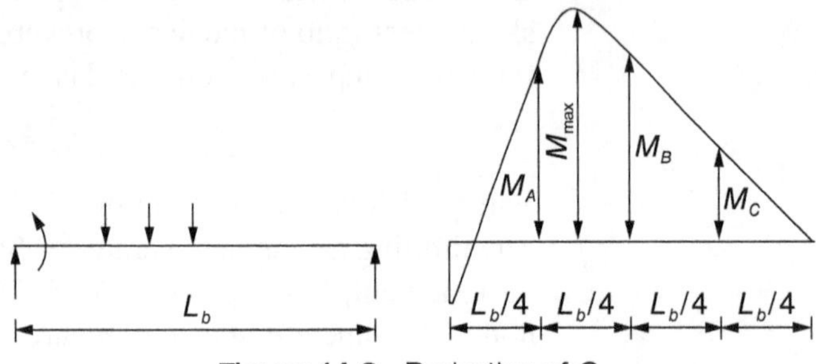

Figure 14.2 Derivation of C_b

A beam segment bent in single curvature and subjected to a uniform bending moment has a value of $C_b = 1.0$. Other moment gradients increase the value of C_b and increase the resistance of the beam to lateral torsional buckling. For any loading condition, the value of C_b may conservatively be taken as 1.0 and this is the value adopted in AISC Manual tables. For a cantilever, the lateral-torsional buckling modification factor is taken as unity. For a beam with a compression flange continuously braced along its entire length, such as a beam supporting a composite deck slab, $C_b = 1.0$. Values of C_b for various loading conditions are illustrated in Figure 14.3.

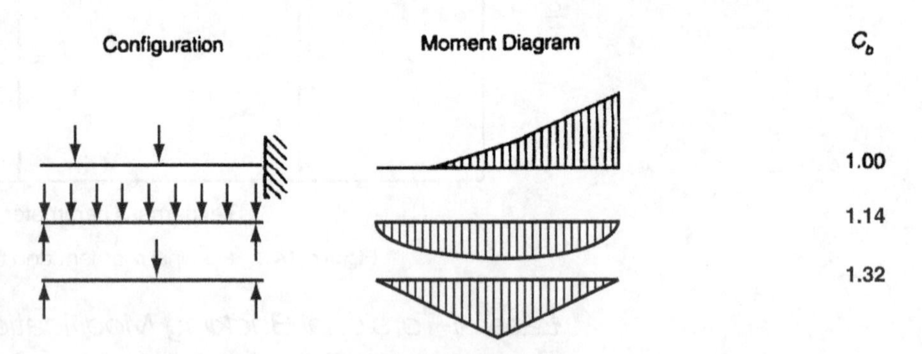

Figure 14.3 Typical values of C_b

Example 14.3

The simply supported beam shown in Exhibit 2 is laterally braced at the supports and at the location of the central point load. The self weight of the beam may be neglected. The value of C_b for each segment of the beam is most nearly:

a). 1.00

b). 1.14

c). 1.32

d). 1.67

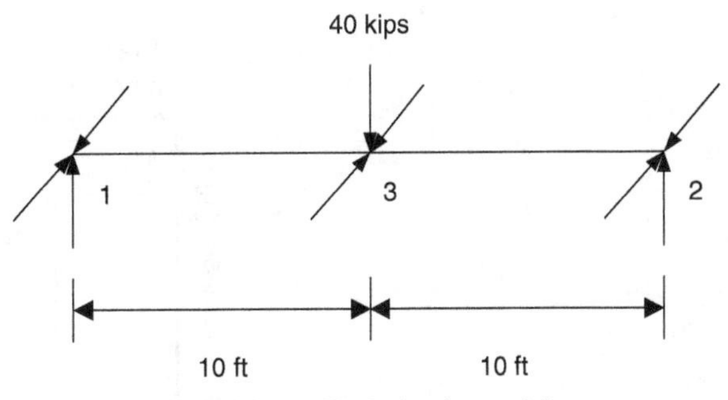

Exhibit 2 Typical values of C_b

Solution

The moment diagram is shown in Exhibit 2a and for segment 13

$$M_A = 20 \times 2.5 = 50 \text{ kip-ft}$$
$$M_B = 20 \times 5 = 100 \text{ kip-ft}$$
$$M_C = 20 \times 7.5 = 150 \text{ kip-ft}$$
$$M_{max} = 20 \times 10 = 200 \text{ kip-ft}$$

The modification factor C_b is given by AISC 360 Equation (F1-1) as

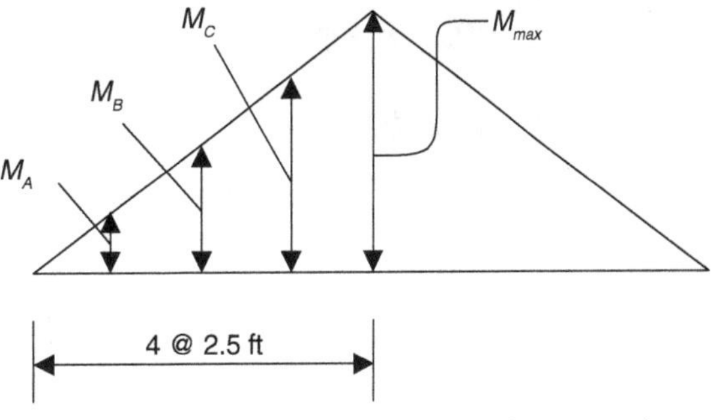

Exhibit 2a Welded connections for an angle

$$C_b = 12.5M_{max}/(2.5M_{max} + 3M_A + 4M_B + 3M_C)$$
$$= 12.5 \times 200/(2.5 \times 200 + 3 \times 50 + 4 \times 100 + 3 \times 150)$$
$$= 1.67$$

The correct answer is d).

Lateral Bracing of Beams, $C_b = 1.0$

As shown in Figure 4.4, when the length, L_b between points of lateral support on a compact beam with $C_b = 1.0$ is increased, the nominal flexural strength of the beam decreases as the beam passes through the following phases:

- *Plastic Phase* $L_b \leq L_p$. The plastic mode occurs when the beam is adequately braced to prevent lateral-torsional buckling. When L_b does not exceed the limiting unbraced length for the limit state of yielding, L_p, a compact section can develop the full plastic moment of resistance and the nominal flexural strength of the beam is

$$M_n = M_p$$

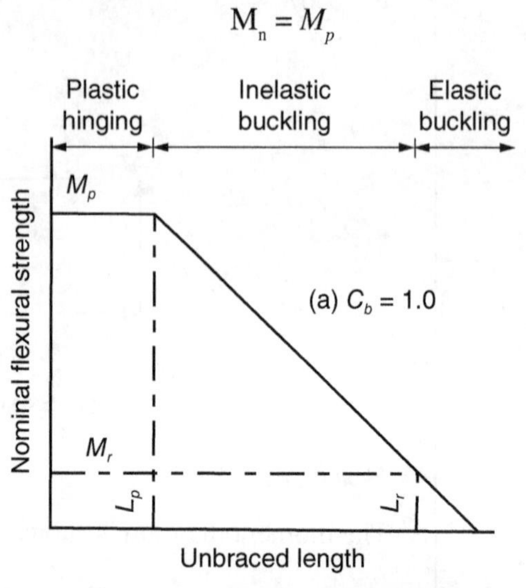

Figure 14.4 Nominal moment and unbraced length with $C_b = 1.0$

Example **14.4**

A W18 × 65 Grade 50 beam has an unbraced segment length of $L_b = 5$ ft with $C_b =$ 1.0. The design flexural strength of the segment is most nearly:

a). 297 kip-ft

b). 394 kip-ft

c). 432 kip-ft

d). 499 kip-ft

Solution

The unbraced segment length is

$$L_b = 5 \text{ ft}$$

AISC Manual Table 3-2 and NCEES Handbook indicate that

$$\phi_b M_p = 499 \text{ kip-ft}$$
$$L_p = 5.97 \text{ ft}$$
$$> L_b$$

Hence, the segment can achieve the full plastic moment of resistance and the design flexural strength is

$$\phi_b M_n = \phi_b M_p$$
$$= 499 \text{ kip-ft}$$

The correct answer is d).

- *Inelastic Phase* $L_p < L_b \leq L_r$. When L_b exceeds L_p and does not exceed L_r, the limiting unbraced length for the limit state of inelastic lateral-torsional buckling, collapse occurs prior to the development of the full plastic moment. When $L_b = L_r$ the nominal flexural strength of the beam is

$$M_n = M_r \text{ ... limiting buckling moment}$$
$$= 0.7F_yS_x$$

For a value of L_b between L_p and L_r, the nominal flexural strength is obtained by linear interpolation between M_r and M_p and the design flexural strength is given by

$$\phi_bM_n = [\phi_bM_p - (\text{BF})(L_b - L_p)]$$
$$\leq \phi_bM_p$$

where

$M_p = F_yZ_x$

$M_r = 0.7F_yS_x$

(BF) = flexural strength factor tabulated in AISC Manual Table 3-2 and NCEES Handbook

$= \phi_b(M_p - M_r)/(L_r - L_p)$

Example 14.5

A W18 × 65 Grade 50 beam has an unbraced segment length of L_b = 14 ft with C_b = 1.0. The design flexural strength of the segment is most nearly:

a). 297 kip-ft

b). 394 kip-ft

c). 432 kip-ft

d). 499 kip-ft

Solution

The unbraced segment length is

$$L_b = 13 \text{ ft}$$

AISC Manual Table 3-2 and NCEES Handbook indicate that

$$(\text{BF}) = 14.9 \text{ kips}$$
$$\phi_bM_p = 499 \text{ kip-ft}$$
$$\phi_bM_r = 307 \text{ kip-ft}$$
$$L_p = 5.97 \text{ ft}$$
$$< L_b$$
$$L_r = 18.8 \text{ ft}$$
$$> L_b$$

Hence, the segment cannot achieve the full plastic moment of resistance and the design flexural strength is

$$\phi_bM_n = \phi_bM_p - (\text{BF})(L_b - L_p)$$
$$= 499 - (14.9)(13 - 5.97)$$
$$= 394 \text{ kip-ft}$$

Alternatively, this value may be obtained directly from the design moment curves given in AISC Manual Table 3-10 and NCEES Handbook as

$$\phi_bM_n = 394 \text{ kip-ft}$$

The correct answer is b).

- *Elastic Phase* $L_b > L_r$. When the distance between braces exceeds L_r collapse occurs by elastic lateral-torsional buckling. The nominal flexural strength of the beam is

$$M_n = M_c \text{ ... elastic buckling moment}$$

Values of M_c may be obtained from the design moment curves given in AISC Manual Table 3-10 and NCEES Handbook.

Example 14.6

A W18 × 65 Grade 50 beam has an unbraced segment length of L_b = 19.3 ft with C_b = 1.0. The design flexural strength of the segment is most nearly:

a). 297 kip-ft

b). 394 kip-ft

c). 432 kip-ft

d). 499 kip-ft

Solution

The unbraced segment length is

$$L_b = 19.3 \text{ ft}$$

AISC Manual Table 3-2 and NCEES Handbook indicate that

$$L_r = 18.8 \text{ ft}$$
$$< L_b$$

Hence, the collapse occurs by elastic lateral-torsional buckling and the design flexural strength is obtained from the design moment curves given in AISC Manual Table 3-10 and NCEES Handbook as

$$\phi_b M_n = 297 \text{ kip-ft}$$

The correct answer is a).

Lateral Bracing of Beams, $C_b > 1.0$

Figure 14.5, shows the effect of $C_b > 1.0$ on the relationship between L_b and M_n. The nominal moment in the elastic and inelastic regions is obtained by multiplying the tabulated nominal strength values in AISC Manual Table 3-2 and NCEES Handbook by C_b. The maximum permitted value of the nominal flexural capacity is limited to M_p.

- *Plastic Phase* $L_b \leq L_m$. For a beam with a value of C_b greater than 1.0, the unbraced length for full plastic moment of resistance is extended beyond L_p to L_m which is given by

$$L_m = L_p + \phi_b M_p (C_b - 1.0)/C_b(\text{BF}) \text{ ... for } L_m \leq L_r$$

and the nominal flexural strength of the beam is

$$M_n = M_p$$

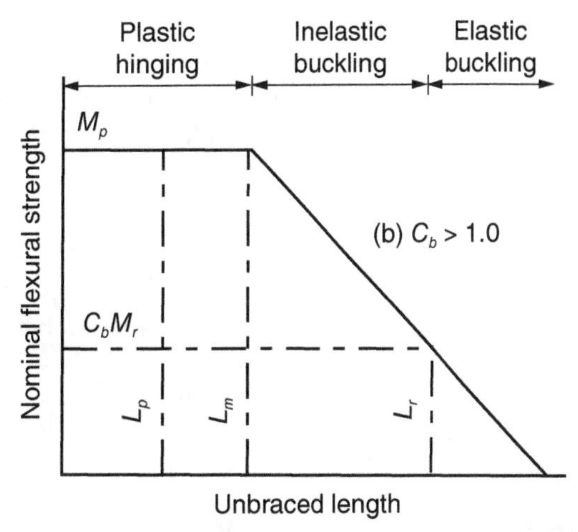

Figure 14.5 Nominal moment and unbraced length with $C_b = 1.0$

Example 14.7

A W18 × 65 Grade 50 beam has an unbraced segment length of L_b = 13 ft with Cb = 1.3. The design flexural strength of the segment is most nearly:

a). 386 kip-ft

b). 394 kip-ft

c). 432 kip-ft

d). 499 kip-ft

Solution

The unbraced segment length is

$$L_b = 13 \text{ ft}$$

AISC Manual Table 3-2 and NCEES Handbook indicate that

$$\begin{aligned} \phi_b M_p &= 499 \text{ kip-ft} \\ L_p &= 5.97 \text{ ft} \\ (\text{BF}) &= 14.9 \text{ kips} \end{aligned}$$

The extended length for full plastic moment of resistance is

$$\begin{aligned} L_m &= L_p + \phi_b M_p (C_b - 1.0)/C_b(\text{BF}) \\ &= 5.97 + 499(1.3 - 1)/(1.3 \times 149) \\ &= 13.70 \text{ ft} \\ &> L_b \end{aligned}$$

Hence, the segment can achieve the full plastic moment of resistance and the design flexural strength is

$$\begin{aligned} \phi_b M_n &= \phi_b M_p \\ &= 499 \text{ kip-ft} \end{aligned}$$

The correct answer is d).

- *Inelastic Phase* $L_m < L_b \le L_r$. When L_b exceeds L_m and does not exceed L_r, the limiting unbraced length for the limit state of inelastic lateral-torsional buckling, collapse occurs prior to the development of the full plastic moment. When $L_b = L_r$ the nominal flexural strength of the beam is

$$M_n = C_b M_r$$
$$= 0.7 C_b F_y S_x$$

For a value of L_b between L_m and L_r, the nominal flexural strength is obtained by linear interpolation between M_r and M_p and the design flexural strength is given by

$$\phi_b M_n = C_b[\phi_b M_p - (\text{BF})(L_b - L_p)]$$
$$\le \phi_b M_p$$

Example 14.8

A W18 × 65 Grade 50 beam has an unbraced segment length of L_b = 14 ft with C_b = 1.3. The design flexural strength of the segment is most nearly:

a). 386 kip-ft

b). 394 kip-ft

c). 493 kip-ft

d). 499 kip-ft

Solution

The unbraced segment length is

$$L_b = 14 \text{ ft}$$

AISC Manual Table 3-2 and NCEES Handbook indicate that

$$(\text{BF}) = 14.9 \text{ kips}$$
$$\phi_b M_p = 499 \text{ kip-ft}$$
$$L_p = 5.97 \text{ ft}$$
$$< L_b$$
$$L_r = 18.8 \text{ ft}$$
$$> L_b$$

Hence, the design flexural strength is

$$\phi_b M_n = C_b[\phi_b M_p - (\text{BF})(L_b - L_p)]$$
$$= 1.3[499 - (14.9)(14 - 5.97)]$$
$$= 493 \text{ kip-ft} \ldots \text{satisfactory}$$
$$< M_p$$

Alternatively, this value may be obtained directly from the design moment curves given in AISC Manual Table 3-10 and NCEES Handbook as

$$\phi_b M_n = 1.3 \times 379 \text{ kip-ft}$$
$$= 493 \text{ kip-ft}$$

The correct answer is c).

- *Elastic Phase* $L_b > L_r$. When the distance between braces exceeds L_r collapse occurs by elastic lateral-torsional buckling. The nominal flexural strength of the beam is

$$M_n = C_b M_c$$

Values of M_c may be obtained from the design moment curves given in AISC Manual Table 3-10 and NCEES Handbook.

Example 14.9

A W18 × 65 Grade 50 beam has an unbraced segment length of $L_b = 19.3$ ft with $C_b = 1.3$. The design flexural strength of the segment is most nearly:

a). 386 kip-ft

b). 394 kip-ft

c). 432 kip-ft

d). 499 kip-ft

Solution

The unbraced segment length is

$$L_b = 19.3 \text{ ft}$$

AISC Manual Table 3-2 and NCEES Handbook indicate that

$$L_r = 18.8 \text{ ft}$$
$$< L_b$$

Hence, the collapse occurs by elastic lateral-torsional buckling and the design flexural strength is obtained from the design moment curves given in AISC Manual Table 3-10 and NCEES Handbook as

$$\phi_b M_n = 1.3 \times 297$$
$$= 386 \text{ kip-ft}$$

The correct answer is a).

Shear Strength of Beams

The shear in rolled W-shape beams is considered to be resisted by the area of the web which is defined as

$$A_w = dt_w$$

where

d = overall depth of the beam
t_w = web thickness

It is assumed that the shear stress is uniformly distributed over this area and, for most W-shapes, the nominal shear strength is governed by yielding of the web. The nominal shear strength is given by AISC 360 Equation (G2-1) as

$$V_n = V_p$$
$$= 0.6F_y A_w$$

The design shear strength is

$$\phi_v V_n = 1.0 \times 0.6 F_y A_w$$
$$= 0.6 F_y A_w$$

Values of $\phi_v V_n$ are tabulated in AISC 360 Table 3-2.

Example 14.10

The simply supported beam shown in Exhibit 3 is a W18 × 65 Grade 50 beam. The design shear strength of the beam is most nearly:

a). 230 kips

b). 240 kips

c). 250 kips

d). 260 kips

Solution

AISC Manual Table 1-1 and NCEES Handbook indicate that

$$t_w = 0.45 \text{ in.}$$
$$d = 18.4 \text{ in.}$$

The design shear strength is

$$\phi_v V_n = 0.6 F_y A_w$$
$$= 0.6 \times 50 \times 18.4 \times 0.45$$
$$= 248 \text{ kips}$$

The correct answer is c).

Columns

Effective Length of Columns

The allowable stress in an axially loaded compression member is dependent on the slenderness ratio which is defined in AISC 360 Section E2 as

$$KL/r$$

where

r = the governing radius of gyration
KL = effective length of the member
K = effective length factor
L = unbraced length of the member

The effective length factor converts the actual column length L to an equivalent pin-ended column of length KL. The effective length factor accounts for the influence of restraint conditions on the behavior of the column and is determined in accordance with AISC 360 Commentary Section C2 and two methods are presented:

- Tabulated factors for single columns with well defined support conditions.
- Alignment charts for columns in a rigid framed structure.

Effective Length by Tabulated Factors

The value of the effective-length factor depends on the restraint conditions at each end of the column. AISC 360 Commentary Table C-C2.2 specifies effective-length factors for well defined, standard conditions of restraint and these are reproduced in NCEES Handbook and are illustrated in Figure 14.6. Values are indicated for ideal and practical end conditions, allowing for the fact that full fixity may not be realized. These values may only be used in simple cases when the tabulated end conditions are closely approximated in practice.

End Restraints	Ideal K	Practical K	Shape
Fixed at both ends	0.5	0.65	
Fixed at one end, pinned at the other end	0.7	0.8	
Pinned at both ends	1.0	1.0	
Sidesway prevented / Sidesway not prevented			
Fixed at one end with the other end fixed in direction but not held in position	1.0	1.2	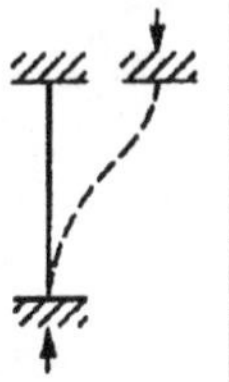
Pinned at one end with the other end fixed in direction but not held in position	2.0	2.0*	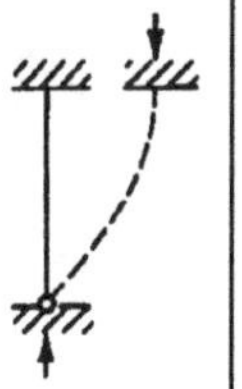
Fixed at one end with the other end free	2.0	2.1	

Figure 14.6 Effective-length factors for columns

For compression members in a plane truss, an effective-length factor of 1.0 may be used. In accordance with AISC 360 Section C3a, in structures where lateral stability is provided by diagonal bracing, shear walls, or equivalent means, the effective length factor for columns may be taken as 1.0. Braced-frame structures may be designed as vertically cantilevered pin-connected truss system with a value of K = 1.0 for all members.

For the braced frame shown in Figure 14.7, which has an infinitely rigid girder, the effective length factors may be obtained from Figure 14.6. For column 12, which is fixed at both ends, $K = 0.65$. For column 34 which is fixed at one end and pinned at the other, $K = 0.80$.

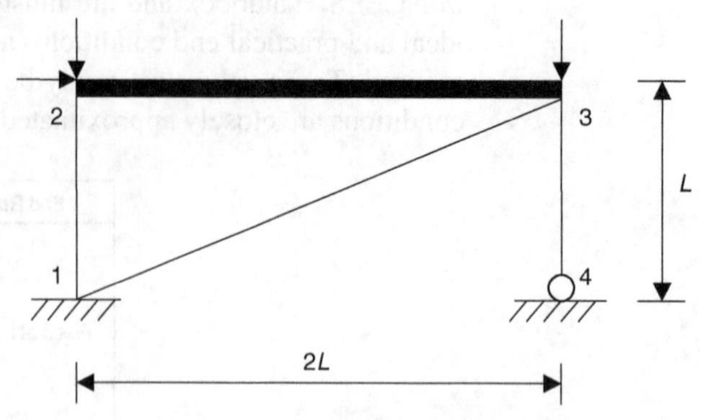

Figure 14.7 Braced frame

For the sway frame shown in Figure 14.8, which has an infinitely rigid girder, the effective length factors may be obtained from Figure 14.6. For column 12, which is fixed at both ends, $K = 1.2$. For column 34, which is fixed at one end and pinned at the other, $K = 2.0$.

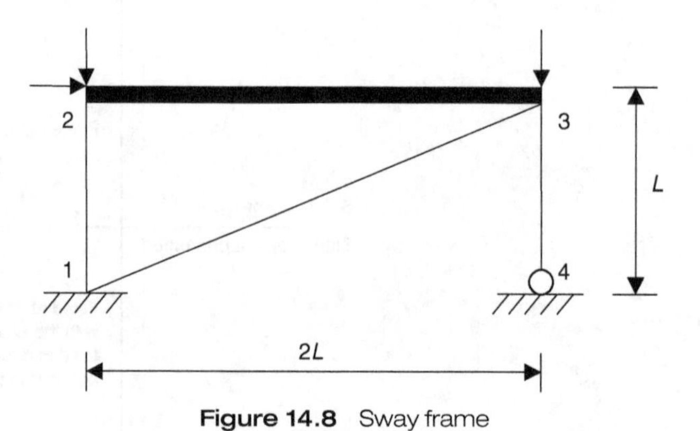

Figure 14.8 Sway frame

Effective Length by Alignment Charts

For compression members forming part of a frame with rigid joints, AISC 360 Commentary Tables C-C2.3 and C-C2.4 presents alignment charts for determining effective-length for the two conditions of sidesway prevented and sidesway permitted. These are reproduced in NCEES Handbook and illustrated in Figure 14.9. To utilize the alignment charts, the stiffness ratio at each end of the column under consideration must be determined and this is defined by

$$G = \Sigma(E_c I_c/L_c)/\Sigma(E_g I_g/L_g)$$

where

$\Sigma(E_c I_c/L_c)$ = the sum of the EI/L values for all columns meeting at the joint
$\Sigma(E_g I_g/L_g)$ = the sum of the EI/L values for all girders meeting at the joint

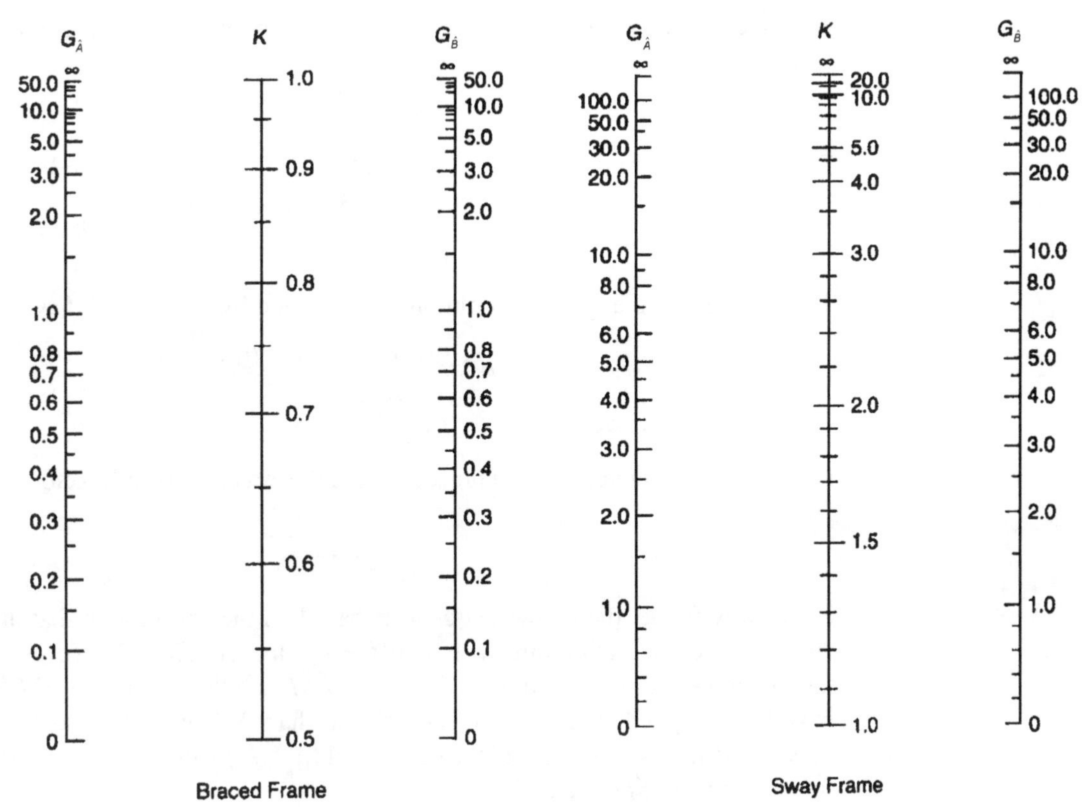

Figure 14.9 Alignment charts for effective length factor

For a column end pinned to a foundation, the stiffness ratio is theoretically infinity and AISC 360 Commentary Section C-C2 recommends a practical value of

$$G = 10$$

For a column end rigidly attached to a foundation, the stiffness ratio is theoretically zero and AISC 360 Commentary Section C-C2 recommends a practical value of

$$G = 1.0$$

For a braced frame, the alignment chart is based on the assumption that the girders are bent in single curvature with a stiffness value of $2EI/L$. However, when the far end of a girder is pinned, its stiffness is $3EI/L$ and the calculated (E_gI_g/L_g) value must be multiplied by 1.5 before entering the chart. When the far end of a girder is fixed, its stiffness is $4EI/L$ and the calculated (E_gI_g/L_g) value must be multiplied by 2.0 before entering the chart.

Referring to the braced frame in Figure 14.7 where the girder has 20 times the moment of inertia of each column and the brace may be considered pinned at each end, the column stiffness may be determined from the alignment chart for a braced frame. For the fixed connection to the foundation at end 1 of column 12, AISC 360 Commentary Section C-C2 recommends a stiffness ratio of

$$G_A = 1.0$$

The stiffness ratio at end 2 of column 12 is defined by

$$\begin{aligned} G_B &= \Sigma(E_cI_c/L_c)/\Sigma(E_gI_g/L_g) \\ &= (1/1)/(20/2) \\ &= 0.1 \end{aligned}$$

From the alignment chart for a braced frame, the effective-length factor for column 12 is

$$K_{12} = 0.65$$

For the pinned connection to the foundation at end 4 of column 34, AISC 360 Commentary Section C-C2 recommends a stiffness ratio of

$$G_A = 10$$

The stiffness ratio at end 3 of column 34 is defined by

$$\begin{aligned} G_B &= \Sigma(E_c I_c/L_c)/\Sigma(E_g I_g/L_g) \\ &= (1/1)/(20/2) \\ &= 0.1 \end{aligned}$$

From the alignment chart for a braced frame, the effective-length factor for column 34 is

$$K_{12} = 0.72$$

For a sway frame, the alignment chart is based on the assumption that the girders are bent in double curvature with a stiffness value of $6EI/L$. However, when the far end of a girder is pinned, its stiffness is $3EI/L$ and the calculated $(E_g I_g/L_g)$ value must be multiplied by 0.5 before entering the chart. When the far end of a girder is fixed, its stiffness is $4EI/L$ and the calculated $(E_g I_g/L_g)$ value must be multiplied by 0.67 before entering the chart.

Referring to the sway frame in Figure 14.8 where the girder has 20 times the moment of inertia of each column, the column stiffness may be determined from the alignment chart for a sway frame. For the fixed connection to the foundation at end 1 of column 12, AISC 360 Commentary Section C-C2 recommends a stiffness ratio of

$$G_A = 1.0$$

The stiffness ratio at end 2 of column 12 is defined by

$$\begin{aligned} G_B &= \Sigma(E_c I_c/L_c)/\Sigma(E_g I_g/L_g) \\ &= (1/1)/(20/2) \\ &= 0.1 \end{aligned}$$

From the alignment chart for a sway frame, the effective-length factor for column 12 is

$$K_{12} = 1.19$$

For the pinned connection to the foundation at end 4 of column 34, AISC 360 Commentary Section C-C2 recommends a stiffness ratio of

$$G_A = 10$$

The stiffness ratio at end 3 of column 34 is defined by

$$\begin{aligned} G_B &= \Sigma(E_c I_c/L_c)/\Sigma(E_g I_g/L_g) \\ &= (1/1)/(20/2) \\ &= 0.1 \end{aligned}$$

From the alignment chart for a sway frame, the effective-length factor for column 34 is

$$K_{12} = 1.70$$

Axially Loaded Members

The nominal compressive strength of a compact member is defined by AISC 360 Equation (E3-1) as

$$P_n = A_g F_{cr}$$

where

A_g = gross area of member
F_{cr} = critical stress

The design strength in compression is given by

$$\phi_c P_n = 0.9 A_g F_{cr}$$

where

ϕ_c = resistance factor for compression

For a short column with $KL/r \leq 4.71(E/F_y)^{0.5}$ and $F_y/F_e \leq 2.25$ inelastic buckling governs. Then AISC 360 Equation (E3-2) defines the critical stress as

$$F_{cr} = (0.658^{\kappa})F_y$$

where

$\kappa = F_y/F_e$
F_e = elastic critical buckling stress given by AISC 360 Equation (E3-4)
$= \pi^2 E/(KL/r)^2$
$\geq F_y/2.25$

For a long column with $KL/r > 4.71(E/F_y)^{0.5}$ and $F_y/F_e > 2.25$ elastic buckling governs. Then AISC 360 Equation (E3-3) defines the critical stress as

$$F_{cr} = 0.877 F_e$$

For a member with a yield stress of $F_y = 50$ ksi, the transition between AISC 360 Equation (E3-2) and Equation (E3-3) occurs at a slenderness ratio of $KL/r = 113$ with a value for the elastic critical buckling stress of $F_e = 22.2$ ksi. In accordance with AISC 360 Section E2, the slenderness ratio of a compression member should preferably not exceed 200.

Once the governing slenderness ratio of a column is established, the available critical stress may be obtained directly from AISC Manual Table 4-22 or from NCEES Handbook for steel members with a yield stress of $F_y = 50$ ksi.

Buckling About the Minor Axis

Values of the design axial strength for W-shapes are tabulated in AISC 360 Table 4-1 and NCEES Handbook for varying values of effective length KL. These tabulated values may be used directly when the slenderness ratio about the minor axis, $(KL/r)_y$, exceeds the slenderness ratio about the major axis, $(KL/r)_x$.

Example 14.11

A W12 × 50 Grade 50 column has a length of 9 feet and is pinned at each end and has no intermediate bracing. The maximum available design strength in compression of the column is most nearly:

a). 525 kips

b). 550 kips

c). 575 kips

d). 600 kips

Solution

The column is pinned at the ends and the effective length factor is

$$K = 1.0$$

The effective length is

$$\begin{aligned} KL &= 1.0 \times 9 \\ &= 9 \text{ ft} \end{aligned}$$

Since the column has no intermediate bracing, buckling about the minor axis governs. AISC Manual Table 4-1 and NCEES Handbook indicate that the available design strength in compression is

$$\phi_c P_n = 526 \text{ kips}$$

The correct answer is a).

Buckling About the Major Axis

When the minor axis of a W-shape is braced at closer intervals than the major axis, the slenderness ratio about both axes must be investigated to determine which governs. The larger of the two values will control the design. If the slenderness ratio about the minor axis, $(KL/r)_y$, governs, the design axial strength values tabulated in AISC 360 Table 4-1 and NCEES Handbook may be utilized directly. If the slenderness ratio about the major axis, $(KL/r)_x$, governs, the design axial strength may be determined from values of $\phi_c F_{cr}$ tabulated in AISC 360 Table 4-22 and NCEES Handbook.

Example 14.12

A W12 × 50 Grade 50 column has a length of 9 feet and is pinned at each end and braced at third points about the minor axis. The maximum available design strength in compression of the column is most nearly:

a). 520 kips

b). 560 kips

c). 600 kips

d). 640 kips

Solution

From AISC Manual Table 1-1 and NCEES Handbook the radii of a W12 × 50 column are

$$r_y = 1.98$$
$$r_x = 5.16$$

The gross area of the column is

$$A_g = 14.6 \text{ in.}^2$$

The effective length of the column about the minor axis is

$$KL = 1.0 \times 3 \times 12$$
$$= 36 \text{ in.}$$

The slenderness ratio about the minor axis is

$$(KL/r)_y = 36/1.98$$
$$= 18.18$$

The effective length of the column about the major axis is

$$KL = 1.0 \times 9 \times 12$$
$$= 108 \text{ ft}$$

The slenderness ratio about the major axis is

$$(KL/r)_x = 108/5.16$$
$$= 20.93 \ldots \text{ governs}$$
$$> (KL/r)_y$$

From AISC 360 Table 4-22 and NCEES Handbook the design critical stress is

$$\phi_c F_{cr} = 43.61 \text{ ksi}$$

The available design strength is

$$\phi_c P_n A_g = 43.61 \times 14.6$$
$$= 637 \text{ kips}$$

The correct answer is d).

Alternatively, when values of r_x/r_y are available for a column, the effective length about the major axis is divided by r_x/r_y to give an equivalent effective length about the minor axis, which has the same load carrying capacity as the actual effective length about the major axis. AISC Manual Table 4-1 and NCEES Handbook may then be used to obtain the available design strength in compression.

Example 14.13

A W12 × 50 Grade 50 column has a length of 9 feet and is pinned at each end and braced at third points about the minor axis. The maximum available design strength in compression of the column is most nearly:

a). 580 kips

b). 600 kips

c). 620 kips

d). 640 kips

Solution

From Example 14.12 a W12 × 50 column has a value of

$$r_x/r_y = 5.18/1.96$$
$$= 2.64$$

The equivalent effective length about the major axis with respect to the y-axis is

$$(KL_y)_{equiv} = (KL_x)/(r_x/r_y)$$
$$= 9/2.64$$
$$= 3.41 \text{ ft ... governs}$$
$$> KL_y$$

AISC Manual Table 4-1 and NCEES Handbook indicate that the available design strength in compression is

$$\phi_c P_n = 622 \text{ kips}$$

The correct answer is c).

Second-Order Effects

Second-order effects are produced in a member by the P-δ and P-Δ effects. The P-δ or B_1 effect is produced by the eccentricity of the axial force with respect to the displaced center line of the member and B_1 is known as the member amplification factor. The P-Δ or B_2 effect is produced by the eccentricity caused by the drift in a sway frame and B_2 is known as the sidesway amplification factor. The P-delta effects are shown in Figure 14.10.

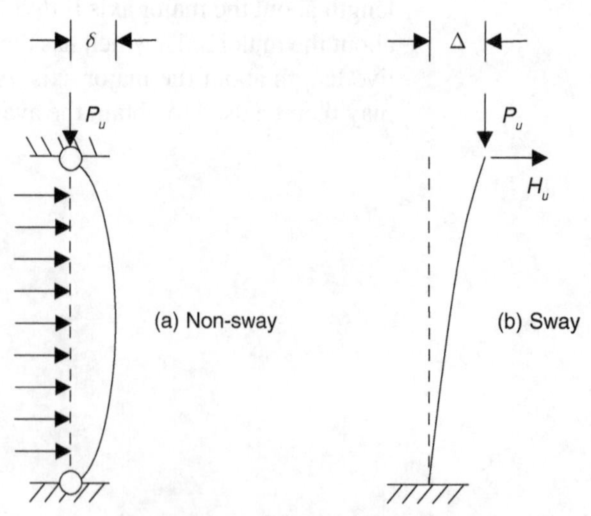

Figure 14.10 Second-order effects

The resultant second-order forces are given by AISC 360 Equations (C2-1a) and (C2-1b) as

$$M_r = B_1 M_{nt} + B_2 M_{lt}$$
$$P_r = P_{nt} + B_2 P_{lt}$$

where

M_r = required second-order flexural strength
P_r = required second-order axial strength
M_{nt} = first-order moment assuming no lateral translation of the member
P_{nt} = first-order axial force assuming no lateral translation of the member
B_1 = multiplier to account for the P-δ
M_{lt} = first-order moment in a member, due only to lateral translation
P_{lt} = first-order axial force in a member, due only to lateral translation
B_2 = multiplier to account for the P-Δ effect

For the special case when sidesway is inhibited $B_2 = 0$ and these equations reduce to

$$M_r = B_1 M_{nt}$$
$$P_r = P_{nt}$$

When the column is transversely loaded between supports and LRFD load combinations are used in design, the member magnification factor is given by

$$B_1 = 1/(1 - P_r/P_{e1})$$

where

P_{e1} = Euler buckling strength of the member in the plane of bending
$= \pi^2 EI/(K_1 L)^2$
L = length of column
K_1 = effective-length factor in the plane of bending
$= 1.0$ ... for lateral translation inhibited
EI = flexural rigidity of the column

Example 14.14

The W12 × 50 Grade 50 column, shown in Figure 14.10(a) has a length of 9 feet and is pinned at each end and has no intermediate bracing. The column has a factored axial load of $P_u = 300$ kips and a factored moment, due to the transverse load wind load, of $M_u = 100$ kip-ft. Allowing for second-order effects, determine the required axial and flexural strengths.

Solution

Since sidesway is inhibited, the first-order applied forces are

$$M_{nt} = M_u = 100 \text{ kip-ft}$$
$$P_{nt} = P_u = 300 \text{ kips}$$

and

$$B_2 = M_{lt} = P_{lt} = 0$$

The required second-order axial strength is

$$P_r = P_{nt} = P_u = 300 \text{ kips}$$

AISC Manual Table 1-1 and NCEES Handbook indicate that

$$I = 391 \text{ in.}^4$$
$$E = 29{,}000 \text{ ksi}$$
$$K_1 = 1.0 \ldots \text{ for lateral translation inhibited}$$

Euler buckling strength of the member in the plane of bending is

$$\begin{aligned} P_{e1} &= \pi^2 EI/(K_1 L)^2 \\ &= \pi^2 \times 29{,}000 \times 391/(1.0 \times 108)^2 \\ &= 9585 \text{ kips} \end{aligned}$$

The member magnification factor for a column transversely loaded between supports is given by

$$\begin{aligned} B_1 &= 1/(1 - P_r/P_{e1}) \\ &= 1/(1 - 300/9585) \\ &= 1.03 \end{aligned}$$

The required second-order flexural strength is

$$\begin{aligned} M_r &= B_1 M_{nt} \\ &= 1.03 \times 100 \\ &= 103 \text{ kip-ft} \end{aligned}$$

Beam-Columns

The design of symmetric members subjected to combined compression and bending is given in AISC Section H1.1.

For values of $P_r/P_c \geq 0.2$, AISC Equation (H1-1a) applies and for the special case of bending about the major axis only

$$P_r/P_c + (8/9)(M_{rx}/M_{cx}) \leq 1.0$$

For values of $P_r/P_c < 0.2$, AISC Equation (H1-1b) applies and for the special case of bending about the major axis only

$$P_r/2P_c + (M_{rx}/M_{cx}) \leq 1.0$$

where

P_r = required axial strength in compression, including second-order effects
M_r = required flexural strength, including second-order effects
P_c = allowable compressive strength in the absence of bending moment
$= \phi_c P_n$
M_c = allowable flexural strength in the absence of axial load
$= \phi_b M_n$
P_n = nominal compressive strength in the absence of bending moment
M_n = nominal flexural strength in the absence of axial load
ϕ_c = resistance factor for compression
$= 0.9$
ϕ_b = resistance factor for flexure
$= 0.9$

Example 14.15

The W12 × 50 Grade 50 column, shown in Figure 14.10(a) has a length of 9 feet and is pinned at each end and has no intermediate bracing. The column has a factored axial load of $P_u = 300$ kips and a factored moment, due to the transverse load wind load, of $M_u = 100$ kip-ft. Allowing for second-order effects, determine if the column is adequate.

Solution

The unbraced segment length is

$$L_b = 9 \text{ ft}$$

AISC Manual Table 3-2 and NCEES Handbook indicate that for the W12 × 50 column

$$\begin{aligned}(\text{BF}) &= 5.97 \text{ kips}\\ \phi_b M_p &= 270 \text{ kip-ft}\\ \phi_b M_r &= 169 \text{ kip-ft}\\ L_p &= 6.92 \text{ ft}\\ &< L_b\\ L_r &= 23.9 \text{ ft}\\ &> L_b\end{aligned}$$

Hence, the segment cannot achieve the full plastic moment of resistance and the design flexural strength in the absence of axial force is

$$\begin{aligned}M_c = \phi_b M_n &= \phi_b M_p - (\text{BF})(L_b - L_p)\\ &= 270 - (5.97)(9 - 6.92)\\ &= 258 \text{ kip-ft}\end{aligned}$$

From Example 14.14, the required second-order axial and flexural strengths are

$$\begin{aligned}P_r &= 300 \text{ kips}\\ M_r &= 103 \text{ kip-ft}\end{aligned}$$

From Example 14.11, the available design strength in compression in the absence of bending moment is

$$P_c = \phi_c P_n = 526 \text{ kips}$$

Hence,

$$\begin{aligned}P_r/P_c &= 300/526\\ &= 0.57\\ &> 0.2\end{aligned}$$

Hence, AISC Equation (H1-1a) applies and, for the special case of bending about the major axis only, the requirement is

$$\begin{aligned}P_r/P_c + (8/9)(M_{rx}/M_{cx}) &= 0.57 + (8/9)(103/258)\\ &= 0.93 \ldots \text{ satisfactory}\\ &< 1.0\end{aligned}$$

Tension Members

Tensile Strength

Two limit states govern the design of tension members. These are the limit state of yielding and the limit state of rupture. The limit state of yielding applies to failure in the gross cross-sectional area of the member. Yielding of a member may result in excessive elongation and lead to instability of the whole structure. The limit state of rupture applies to failure in the net cross-sectional area of the member. A reduction in area of a tension member occurs at connections where holes are punched in the member to accommodate bolts. Failure by rupture at a connection may be sudden and catastrophic. The available strength of a tensile member is the lower calculated value for the limit state of yielding or the limit state of rupture.

For tensile yielding in the gross section, the nominal strength is given by AISC 360 Eq. (D2-1) and NCEES Handbook as

$$P_n = F_y A_g$$

where

F_y = yield stress of the member
A_g = gross area of the member
= wt
w = width of plate
t = thickness of plate

The design tensile strength for tensile yielding is

$$\phi_t P_n = 0.9 F_y A_g$$

where

ϕ_t = resistance factor for tensile yielding
= 0.9

For tensile rupture in the weakest effective net area, the nominal strength is given by AISC 360 Eq. (D2-2) and NCEES Handbook as

$$P_n = F_u A_e$$

where

F_u = tensile strength of the member
A_e = effective net area of the member

The design tensile strength for tensile rupture is

$$\phi_t P_n = 0.75 F_u A_e$$

where

ϕ_t = resistance factor for tensile rupture
= 0.75

Effective Net Area, Flat Plate with Bolted Connection

To account for the effects of eccentricity and shear lag in tension members connected through only part of their cross-sectional elements, the effective net area is given by AISC 360 Equation (D3-1) and NCEES Handbook as

$$A_e = UA_n$$

where

A_n = net area of the member
U = shear lag factor given in AISC 360 Table D3.1

In the case of a flat plate with a bolted connection, the total cross-section is assumed to transfer the load without shear lag and AISC 360 Table D3.1 indicates that U = 1.0. Hence, for a flat plate with bolted connection

$$A_e = A_n$$

For a flat plate with bolted connection, the effective net area is illustrated in Figure 14.11 and defined in AISC 360 Section D3.2 and NCEES Handbook. For the straight perpendicular fracture 1-1, the effective net area of the plate is

$$A_e = t(w - 2d_h)$$

where

d_h = diameter of hole as defined in AISC 360 Section D3.2 and AISC 360 Table J3.3
= (d_n + 1/16 in.) ... from AISC 360 Section D3.2
= [(d_b + 1/16 in.) + 1/16 in.]
= (d_b + 1/8 in.)
d_b = bolt diameter
d_n = nominal hole diameter
= (d_b + 1/16 in.) ... from AISC 360 Table J3.3

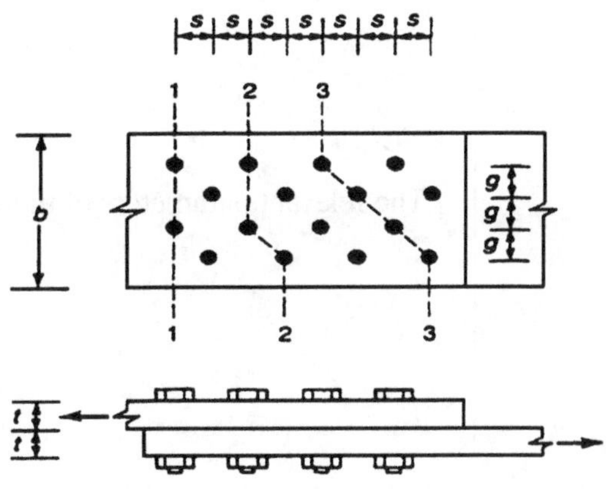

Figure 14.11 Net area calculation

For a staggered fracture, the effective net width is obtained, as specified in AISC 360 Section D3.2 and NCEES Handbook, by deducting from the gross plate width the sum of the bolt holes in the failure path and adding, for each gage space traversed by a diagonal portion of the failure path, the quantity $s^2/4g$ where

g = transverse center-to-center spacing between fasteners (gage)
s = bolt spacing in direction of load (pitch)

For the staggered fracture 2-2, the effective net area of the plate is

$$A_e = t(b - 3d_h + s^2/4g)$$

For the staggered fracture 3-3, the effective net area of the plate is

$$A_e = t(b - 4d_h + 3s^2/4g)$$

For bolted splice plates, the length of the connection is small compared with the total length of the member, and inelastic deformation at the connection is limited. Hence, AISC 360 Section J4.1 limits the effective net area of the connection to a value of

$$A_e = A_n \leq 0.85A_g$$

Example 14.16

The spliced joint shown in Exhibit 3 consists of A36 steel plates connected with ¾-inch diameter bolts in standard holes. The bolt pitch is 1½ inches and the gage 3 inches. Assuming that the bolts are satisfactory and that block shear does not govern, the design tensile strength of the plates is most nearly:

a). 195 kips

b). 205 kips

c). 215 kips

d). 225 kips

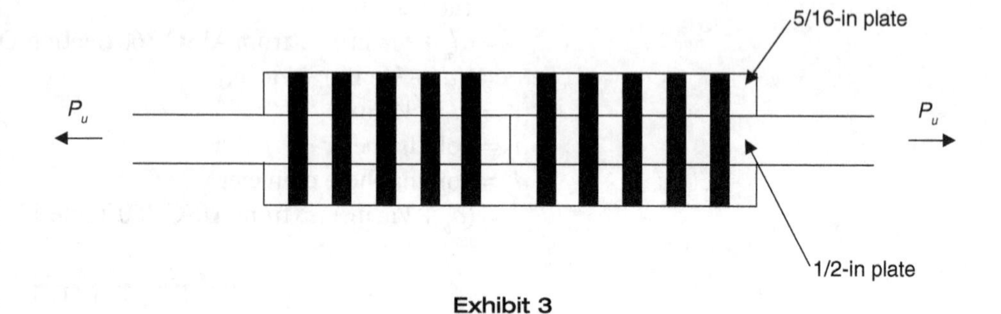

Exhibit 3

Solution

The relevant parameters of the ½-inch plate are

$$F_y = 36 \text{ ksi}$$
$$F_u = 58 \text{ ksi}$$
$$w = 12 \text{ in.}$$
$$t = 0.5 \text{ in.}$$
$$g = 3 \text{ in.}$$
$$s = 1.5 \text{ in.}$$
$$\begin{aligned} d_h &= (d_b + 1/8 \text{ in.}) \\ &= 0.75 + 0.125 \\ &= 0.875 \text{ in.} \end{aligned}$$

The gross area of the ½-inch plate is

$$\begin{aligned} A_g &= wt \\ &= 12 \times 0.5 \\ &= 6.0 \text{ in.}^2 \end{aligned}$$

and

$$\begin{aligned} 0.85A_g &= A_{e(max)} \\ &= 0.85 \times 6 \\ &= 5.10 \text{ in.}^2 \end{aligned}$$

For the straight perpendicular fracture 1-1, the effective net area of the ½-inch plate is

$$\begin{aligned} A_e &= t(w - 2d_h) \\ &= 0.5(12 - 2 \times 0.875) \\ &= 5.125 \text{ in.}^2 \\ &> 0.85A_g \text{ ... unsatisfactory} \end{aligned}$$

Use

$$A_{e(max)} = 5.10 \text{ in.}^2$$

For the staggered fracture 2-2, the effective net area of the ½-inch plate is

$$\begin{aligned} A_e &= t(w - 3d_h + s^2/4g) \\ &= 0.5[12 - 3 \times 0.875 + 1.5^2/(4 \times 3)] \\ &= 4.78 \text{ in.}^2 \end{aligned}$$

The connection consists of ten bolts, each of which may be considered to support 10 percent of the applied tensile force. Since bolt A is in front of fracture plane 2-2, the fracture plane is required to support only 90 percent of the applied force and the equivalent effective net area is

$$\begin{aligned} A_{e(equiv)} &= A_e/0.9 \\ &= 4.78/0.9 \\ &= 5.31 \text{ in.}^2 \end{aligned}$$

For the staggered fracture 3-3, the effective net area of the ½-inch plate is

$$\begin{aligned} A_e &= t(w - 4d_h + 3s^2/4g) \\ &= 0.5[12 - 4 \times 0.875 + 3 \times 1.5^2/(4 \times 3)] \\ &= 4.53 \text{ in.}^2 \ldots \text{ governs} \\ &< 0.85A_g \ldots \text{ satisfactory} \end{aligned}$$

The design tensile rupture capacity for this fracture condition is

$$\begin{aligned} \phi_t P_n &= 0.75 F_u A_e \\ &= 0.75 \times 58 \times 4.53 \\ &= 197.06 \text{ kips} \end{aligned}$$

The design tensile yield strength condition for the ½-inch plate is

$$\begin{aligned} \phi_t P_n &= 0.9 F_y A_g \\ &= 0.9 \times 36 \times 6.0 \\ &= 194.40 \text{ kips} \ldots \text{ governs} \end{aligned}$$

The answer is a).

Effective Net Area, Angle Member with Bolted Connection

For bolted connections, when an angle shape is connected through only one leg of its cross-section, the effective net area is given by AISC 360 Eq. (D3-1) and NCEES Handbook as

$$A_e = A_n U$$

The shear lag factor U allows for the effects of eccentricity and shear lag at the ends of the angle and is defined in AISC 360 Table D3.1 Case 2 and NCEES Handbook as

$$U = 1.0 - \bar{x}/l$$

where

$\bar{x}$ = eccentricity of connection as shown in Figure 14.12
= distance from the connection plane to the centroid of the angle

l = length of the connection as shown in Figure 14.12
= distance, parallel to the line of force, between the centers of the first and last fasteners in a line.

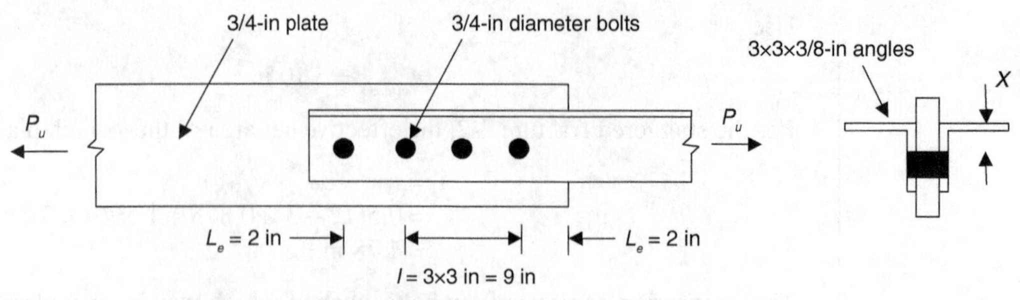

Figure 14.12 Calculation of shear lag factor

In lieu of applying this expression, it is permitted to adopt the following values for the shear lag factor, as given in AISC 360 Table D3.1 Case 8 as

- U = 0.60 when the angle is connected by two or three bolts in line in the direction of stress.
- U = 0.80 when the angle is connected by not less than four bolts in line in the direction of stress.

It is permitted to adopt the larger value for U given by Case 2 and Case 8.

Example 14.17

Figure 14.12 shows a bolted connection with two 3 × 3 × ⅜ inch angles bolted to a ¾-inch gusset plate with four ¾-inch diameter bolts in standard holes. All components are Grade 36 steel and the gross area of the double angles is $A_g = 4.22$ in.^{2} Assuming that bolt strength and block shear do not govern, the design tensile strength of the double angles is most nearly:

a). 105 kips

b). 115 kips

c). 125 kips

d). 135 kips

Solution

The relevant parameters of the ⅜-inch angles are

$$F_y = 36 \text{ ksi}$$
$$F_u = 58 \text{ ksi}$$

The hole diameter is defined in AISC 360 Section D3.2 and AISC 360 Table J3.3 as

$$\begin{aligned} d_h &= (d_b + 1/8 \text{ in.}) \\ &= 0.75 + 0.125 \\ &= 0.875 \text{ in.} \end{aligned}$$

The net area of the two angles is

$$\begin{aligned} A_n &= A_g - 2td_h \\ &= 4.22 - 2 \times 0.375 \times 0.875 \\ &= 3.56 \text{ in.}^2 \end{aligned}$$

The effective net area is given by AISC 360 Table D3.1 Case 8 for four bolts in line as

$$A_e = UA_n$$
$$= 0.80 \times 3.56$$
$$= 2.85 \text{ in.}^2$$

The design tensile rupture capacity for the fracture condition for the double angles is

$$\phi_t P_n = 0.75 F_u A_e$$
$$= 0.75 \times 58 \times 2.85$$
$$= 124 \text{ kips} \ldots \text{governs}$$

The design tensile yield strength condition for the double angles is

$$\phi_t P_n = 0.9 F_y A_g$$
$$= 0.9 \times 36 \times 4.22$$
$$= 137 \text{ kips}$$
$$> 124 \text{ kips}$$

The correct answer is c).

Effective Net Area, Flat Plate with Welded Connection

To account for the effects of eccentricity and shear lag in a flat plate member connected by welds, the effective net area is given by AISC 360 Equation (D3-1) and NCEES Handbook as

$$A_e = UA_n = UA_g$$

where

A_n = net area of the member
$= A_g$
U = shear lag factor given in AISC 360 Table D3.1 and NCEES Handbook

For the transverse fillet welded connection shown in Figure 14.13a, the total cross-section is assumed to transfer the load without shear lag and AISC 360 Table D3.1 Case 3 and NCEES Handbook indicates that $U = 1.0$. Hence, for a flat plate with transverse weld

$$A_e = A_g$$

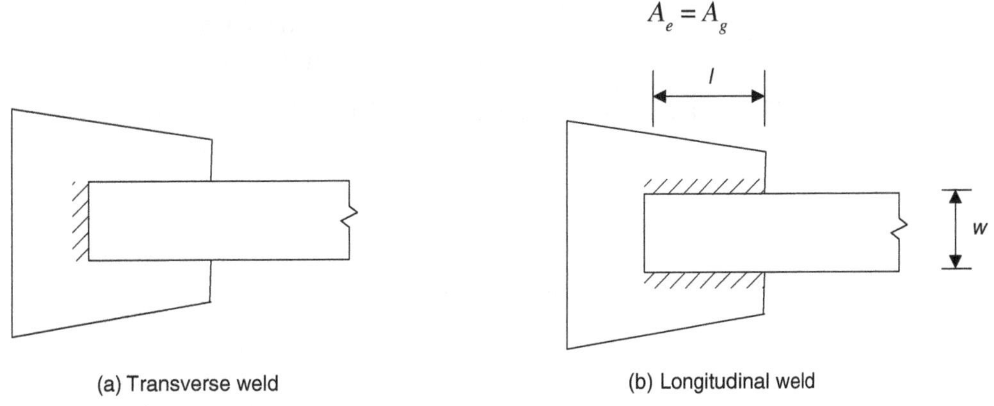

Figure 14.13 Welded connections for a plate

For welded connections, the effective net area is defined in AISC 360 Table D3.1 Case 3 and 4 and NCEES Handbook. For the transverse fillet welded connection shown in Figure 14.13a

$$A_e = A_g$$

For the longitudinal fillet welded connection shown in Figure 14.13b, shear lag occurs at the ends of the plate and the shear lag factor is defined in AISC 360 Table D3.1 Case 4 and NCEES Handbook as follows:

- For the condition $l \geq 2w$; U = 1.00
- For the condition $2w > l > 1.5w$; $U = 0.87$
- For the condition $1.5w > l > w$; $U = 0.75$

Example 14.18

Figure 14.13b shows a ½-inch plate connected to a gusset plate with longitudinal fillet welds as indicated with $l = 5$ in. and $w = 3$ in. The plate material is A36 steel. Assuming that the welds are satisfactory and that block shear does not govern, the design tensile strength of the plate is most nearly:

a). 49 kips

b). 57 kips

c). 65 kips

d). 73 kips

Solution

$$l/w = 5/3 = 1.67$$

From AISC 360 Table D3.1 Case 4 and NCEES Handbook, the shear lag factor is

$$U = 0.87$$

The gross area of the ½-inch plate is

$$A_g = wt = 3 \times 0.5 = 1.5 \text{ in.}^2$$

AISC 360 Equation (D3-1) and NCEES Handbook as

$$A_e = UA_g = 0.87 \times 1.5 = 1.31 \text{ in.}^2$$

The design tensile rupture capacity for the fracture condition is

$$\phi_t P_n = 0.75 F_u A_e = 0.75 \times 58 \times 1.31 = 57 \text{ kips}$$

The design tensile yield strength capacity is

$$\phi_t P_n = 0.9 F_y A_g = 0.9 \times 36 \times 1.5 = 49 \text{ kips} \ldots \text{governs} < 57 \text{ kips}$$

The correct answer is a).

Effective Net Area, Angle Member with Welded Connection

For an angle, when the axial force is transmitted only by a transverse fillet weld to one leg of the angle, as shown in Figure 14.14a the shear lag factor is given by AISC 360 Table D3.1 Case 3 and NCEES Handbook as

$$U = 1.0$$

The effective net area is given by AISC 360 Equation (D3-1) and NCEES Handbook as

$$\begin{aligned} A_e &= UA_g \\ &= A_g \\ &= \text{gross area of the directly connected leg} \end{aligned}$$

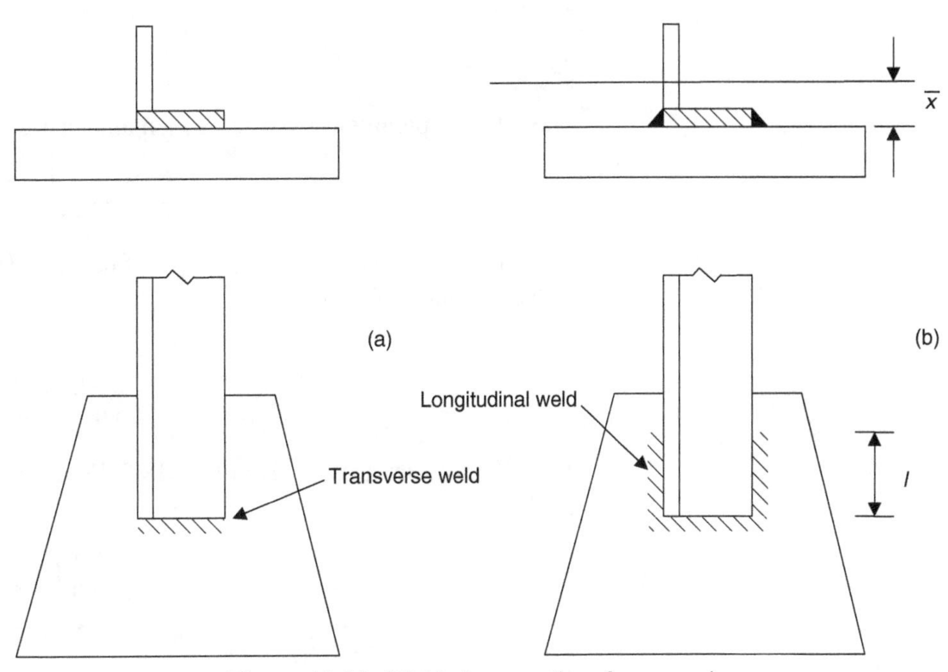

Figure 14.14 Welded connections for an angle

For a welded angle, when the axial force is transmitted only by longitudinal fillet welds or by longitudinal fillet welds in combination with a transverse fillet weld, as shown in Figure 14.14b, the shear lag factor U is defined by AISC 360 Table D3.1 Case 2 and NCEES Handbook as

$$\begin{aligned} U &= 1 - \bar{x}/l \\ &\leq 0.9 \end{aligned}$$

The length of the connection l is shown in Figure 14.14b and defined in AISC 360 Commentary Section D3.3 as the length of the weld, parallel to the line of force. The connection eccentricity $\bar{x}$ is shown in Figure 14.14b 5-7 and defined in AISC 360 Commentary Section D3.3 as the distance from the connection plane to the centroid of the angle resisting the connection force.

The effective net area is given by AISC 360 Equation (D3-1) and NCEES Handbook as

$$A_e = UA_g$$

where

A_g = gross area of the directly connected leg

Example 14.19

Figure 14.14b shows a welded connection with two 3 × 3 × ⅜ inch angles welded to a ¾-inch gusset plate with transverse and longitudinal welds. The length of each longitudinal weld is $l = 12$ in. All components are Grade 36 steel and the gross area of the double angles is $A_g = 4.22$ in.^{2} The connection eccentricity for each angle is $\bar{x} = 0.884$ in. Assuming that weld strength and block shear do not govern, the design tensile strength of the double angles is most nearly:

a). 105 kips

b). 115 kips

c). 125 kips

d). 135 kips

Solution

The relevant parameters of the ⅜-inch angles are

$$F_y = 36 \text{ ksi}$$
$$F_u = 58 \text{ ksi}$$

The value of the shear lag factor is defined in AISC 360 Table D3.1 Case 2 and NCEES Handbook as

$$\begin{aligned} U &= 1 - \bar{x}/l \\ &= 1 - 0.884/12 \\ &= 0.9 \text{ ... maximum} \end{aligned}$$

The effective net area is given by AISC 360 Equation (D3-1) and NCEES Handbook as

$$\begin{aligned} A_e &= UA_g \\ &= 0.90 \times 4.22 \\ &= 3.80 \text{ in.}^2 \end{aligned}$$

The design tensile rupture capacity for the fracture condition is

$$\begin{aligned} \phi_t P_n &= 0.75 F_u A_e \\ &= 0.75 \times 58 \times 3.80 \\ &= 165 \text{ kips} \end{aligned}$$

The design tensile yield strength capacity is

$$\begin{aligned} \phi_t P_n &= 0.9 F_y A_g \\ &= 0.9 \times 36 \times 4.22 \\ &= 137 \text{ kips ... governs} \end{aligned}$$

The correct answer is d).

Block Shear

Block Shear Strength for Plates and Angles

As shown in Figure 14.15, failure may occur by block shear at the ends of plates and angles in either bolted or welded connections. Block shear is a combination of shear along the length of the connected member and tension along the end of the connected member, and the block shear strength is the sum of the strengths of the shear area and tension area. Tension failure occurs by rupture in the net tension area. Shear failure occurs either by rupture in the net shear area or by shear yielding in the gross shear area and the minimum value governs.

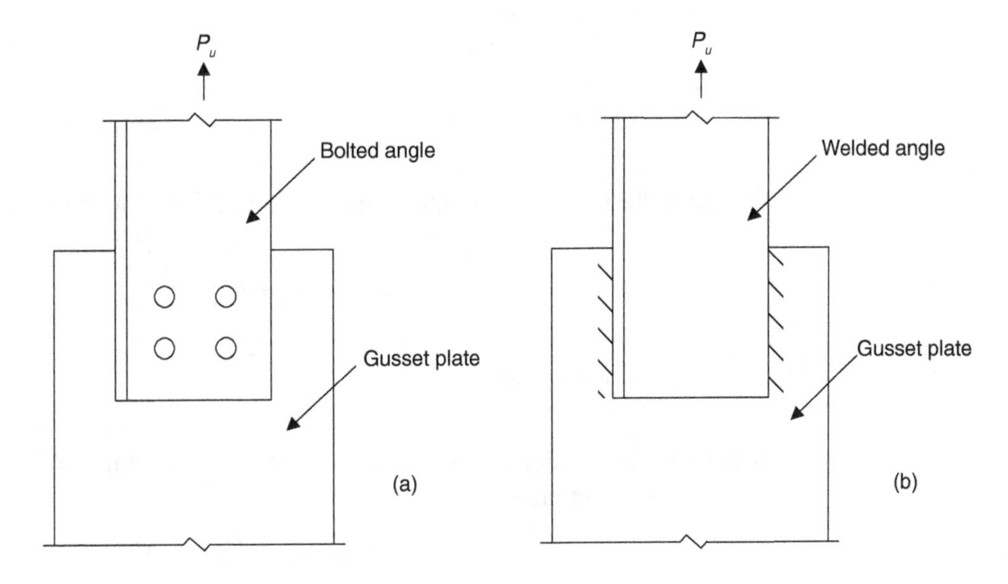

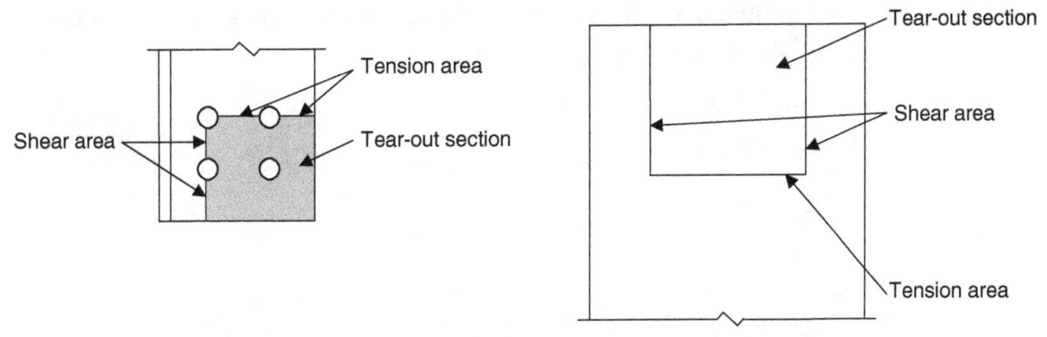

Figure 14.15 Block shear in bolted and welded connections

The nominal strengths are given in AISC 360 Section J4 and NCEES Handbook as the following:

- The nominal rupture strength in shear is

$$V_n = 0.6F_u A_{nv}$$

- The nominal yield strength in shear is

$$V_n = 0.6F_y A_{gv}$$

- The nominal rupture strength in tension is

$$P_n = U_{bs} F_u A_{nt}$$

where

A_{nv} = net shear area
A_{nt} = net tension area
A_{gv} = gross shear area
F_u = minimum tensile strength
F_y = minimum yield stress
U_{bs} = reduction coefficient
= 1.0 for uniform stress
= 0.5 for non-uniform stress

The block shear design strength is given by AISC 360 Equation (J4-5) as

$$\phi R_n = \phi(0.6F_yA_{gv} + U_{bs}F_uA_{nt})$$
$$\leq \phi(0.6F_uA_{nv} + U_{bs}F_uA_{nt})$$

where

ϕ = resistance factor
= 0.75

In the case of welded connections, net areas are not applicable and AISC 360 Equation (J4-5) reduces to

$$\phi R_n = \phi(0.6F_yA_{gv} + U_{bs}F_uA_{gt})$$

Example 14.20

Figure 14.15b shows a welded connection with two 3 × 3 × ⅜ inch angles welded to a ⅝-inch gusset plate with transverse and longitudinal welds. The length of each longitudinal weld is $l = 6$ in. The design block shear strength of the ⅝-inch gusset plate is most nearly:

a). 205 kips

b). 215 kips

c). 225 kips

d). 235 kips

Solution

The relevant parameters of the ⅝-inch gusset plate are

$$F_y = 36 \text{ ksi}$$
$$F_u = 58 \text{ ksi}$$

The tensile stress is uniform across the connection and

$$U_{bs} = 1$$

From Figure 14.15b, the gross shear area is

$$A_{gv} = 2 \times 6 \times 0.625$$
$$= 7.50 \text{ in.}^2$$

From Figure 14.15b, the gross tension area is

$$A_{gt} = 3 \times 0.625$$
$$= 1.875 \text{ in.}^2$$

The rupture strength in tension is

$$F_uA_{gt} = 58 \times 1.875$$
$$= 108.75 \text{ kips}$$

The yield strength in shear

$$0.6F_yA_{gv} = 0.6 \times 36 \times 7.50$$
$$= 162.00 \text{ kips}$$

The block shear design strength is given by the modified AISC 360 Equation (J4-5) as

$$\phi R_n = \phi(0.6F_yA_{gv} + U_{bs}F_uA_{gt})$$
$$= 0.75(162.00 + 108.75)$$
$$= 203 \text{ kips}$$

The correct answer is a).

Bolted Connections

Types of Bolts

The bolts used in steel structures are classified into the following two categories:

- *Common bolts* which are of Grade A307 and have a minimum tensile strength of 60 ksi.
- *High-strength bolts* which are of Grade A325 with a minimum tensile strength of 120 ksi, or Grade A490 with a minimum tensile strength of 150 ksi.

Bolts may be used in the following three types of connections:

- *Snug-tight connections* require the bolts to be tightened sufficiently to bring the plies into firm contact. No specific level of installed tension is specified to achieve this condition. Common bolts are always used in snug-tight connections. High-strength bolts may be used in snug-tight connections in order to take advantage of their higher shear strength. Transfer of the load from one connected part to another depends on the contact or bearing of the bolts against the side of the holes. Snug-tight connections may be used when pretensioned or slip-critical connections are not required.
- *Pretensioned connections* require the bolts to be pretensioned to a minimum value of 70 percent of the tensile strength of the bolt and the faying surfaces may be uncoated, coated, or galvanized without regard to the slip coefficient obtained. Transfer of the load from one connected part to another depends on the contact of the bolts against the side of the holes. Only high-strength bolts may be used in pretensioned connections. Pretensioned connections are required in the following situations:
  - Column splices in buildings with high ratios of height to width
  - Connections of members that provide bracing to columns in tall structures
  - Connections in structures carrying cranes of over five ton capacity
  - Connections for supports of running machinery and other sources of impact or stress reversal
- *Slip-critical connections* require the bolts to be pretensioned to a minimum value of 70 percent of the tensile strength of the bolt and the faying surfaces

must be prepared to produce a specific value of the slip coefficient. Transfer of the load from one connected part to another depends on the friction induced between the parts. Only high-strength bolts may be used in slip-critical connections. Slip-critical connections are required in the following situations:

- Where fatigue load with reversal of the loading direction occurs
- Where bolts are used in oversize holes or slotted holes parallel to the direction of load
- Where slip at the faying surfaces would be detrimental to the performance of the structure
- Where bolts are used in conjunction with welds

Snug-Tight Bolts in Shear

The minimum permissible distance between the centers of standard, oversized, or slotted holes is given by AISC 360 Sec. J3.3 as:

$$s_{min} = 2.67d$$

The preferred distance between the centers of standard, oversized, or slotted holes is:

$$s_{pref} = 3.0d$$

Design shear capacity is based on the nominal unthreaded cross-sectional area of the bolt A_b and is given by AISC 360 Section J3.6 as

$$\begin{aligned}\phi R_n &= \phi F_{nv} A_b \\ &= 0.75 F_{nv} A_b\end{aligned}$$

where

F_{nv} = nominal shear strength of bolt
R_n = nominal shear capacity,
ϕ = resistance factor
= 0.75

Values of the design shear capacity ϕR_n are given in AISC Manual Table 7-1 and NCEES Handbook. For high-strength bolts, a reduced design strength is applicable when threads are not excluded from the shear planes. High-strength bolts in slip-critical connections must also be checked for shear strength since the connection, at the strength limit state, may slip sufficiently to place the bolts in shear.

Snug-Tight Bolts in Bearing

Bolt bearing capacity is dependent on the diameter, spacing, and edge distance of a bolt in addition to the material of the connected parts and the acceptable deformation at the bolt hole. High-strength bolts in slip-critical connections must also be checked for bearing strength since the connection, at the strength limit state, may slip sufficiently to place the bolts in bearing.

The nominal bearing strength of connected parts is specified in AISC 360 Section J3.10 and, when deformation at the hole is a design consideration, is given by AISC 360 Equation (J3-6b) as

$$\begin{aligned}R_n &= 1.2 L_c F_u \text{ ... when tear out strength governs} \\ &\leq 2.4 d_b t F_u \text{ ... when bearing strength governs}\end{aligned}$$

The design bearing capacity is

$$\phi R_n = 0.75R_n$$

where

F_u = tensile strength of the critical connected part
L_c = clear distance, in the direction of force, between the edge of the hole and the edge of the adjacent hole or edge of the connected part
t = thickness of the connected part
d_b = nominal bolt diameter
R_n = nominal bearing capacity
ϕ = resistance factor
= 0.75

To ensure that tear out does not occur, the clear distance between adjacent holes or between the edge of a hole and the edge of the connected part is obtained from AISC 360 Equation (J3-6a) as

$$L_c \geq 2d_b$$

When deformation of the hole is not a design consideration, the nominal bearing capacity of each bolt is given by AISC 360 Equation (J3-6b) as

$$R_n = 1.5L_c tF_u \text{ ... when tear out strength governs}$$
$$\leq 3d_b F_u \text{ ... when bearing strength governs}$$

To ensure that tear out does not occur, the clear distance between adjacent holes or between the edge of a hole and the edge of the connected part is obtained from AISC 360 Equation (J3-6b) as

$$L_c \geq 2d_b$$

The clear distance between adjacent holes is

$$L_c = s - (d_b + 1/16)$$

The clear distance between the edge of a hole and the edge of the connected part is

$$L_c = L_e - (d_b + 1/16)/2$$

where

$(d_b + 1/16)$ = nominal hole diameter given in AISC 360 Table J3.3
s = bolt center-to-center spacing
L_e = edge distance, in the direction of force, between the bolt center and the edge of the connected part

Example 14.21

The connection shown in Figure 14.12 consists of four, Grade A325, ¾-inch diameter bolts. The bolts are snug-tight and threads are excluded from the shear planes. Deformation around the bolt holes is a design consideration and the bolt spacing is as indicated. The angles and gusset plate are fabricated from A36 steel. Assuming that the angles and gusset plate are satisfactory, determine the shear force that may be applied to the bolts in the connection.

Solution

The clear distance between adjacent holes is

$$\begin{aligned} L_c &= s - (d_b + 1/16) \\ &= 3 - (3/4 + 1/16) \\ &= 2.19 \text{ in.} \\ &> 2d_b \text{ ... full bearing capacity possible on interior bolts} \end{aligned}$$

The clear distance, in the direction of force, between the edge of the end hole and the edge of the connected parts is

$$\begin{aligned} L_c &= L_e - (d_b + 1/16)/2 \\ &= 2 - (3/4 + 1/16)/2 \\ &= 1.59 \text{ in.} \\ &> 2d_b \text{ ... full bearing capacity possible on end bolt} \end{aligned}$$

Hence, the design bearing capacity of each bolt on the ½-inch gusset plate is

$$\begin{aligned} \phi R_n &= \phi \times 2.4 d_b t F_u \\ &= 0.75 \times 2.4 \times 0.75 \times 0.5 \times 58 \\ &= 39\ 15 \text{ kips} \end{aligned}$$

Alternatively, from AISC Manual Tables 7-5 and 7-6 and from NCEES Handbook the design bearing capacity of each bolt on the ½-inch gusset plate is

$$\begin{aligned} \phi R_n &= 78.3/2 \\ &= 39.15 \text{ kips} \end{aligned}$$

The design bearing capacity of the four bolts is

$$\begin{aligned} \phi R_n &= 4 \times 39.15 \\ &= 157 \text{ kips ... governs} \end{aligned}$$

The double shear capacity of the four, Grade A325, ¾-inch diameter bolts with threads excluded from the shear planes is:

$$\begin{aligned} \phi_v R_{nv} &= 2n\phi_v F_{nv} A_b \\ &= 2 \times 4 \times 0.75 \times 60 \times 0.442 \\ &= 159 \text{ kips} \end{aligned}$$

Alternatively, from AISC Manual Table 7-1 and from NCEES Handbook the design shear capacity of each bolt on the ½-inch gusset plate is

$$\begin{aligned} \phi R_n &= 8 \times 19.9 \\ &= 159 \text{ kips} \\ &> 157 \text{ kips} \end{aligned}$$

The maximum shear force that may be applied to the bolts in the connection is

$$T_u = 157 \text{ kips}$$

Snug-Tight Bolts in Tension and Combined Shear and Tension

Design tensile capacity is based on the nominal unthreaded cross-sectional area of the bolt A_b and is given by AISC 360 Section J3.6 as

$$\begin{aligned} \phi R_n &= \phi F_{nt} A_b \\ &= 0.75 F_{nv} A_b \end{aligned}$$

where

F_{nv} = nominal tensile strength of bolt
R_n = nominal tensile capacity
ϕ = resistance factor
= 0.75

Values of the design tensile capacity ϕR_n are given in AISC Manual Table 7-2 and NCEES Handbook.

When a bearing-type connector is subjected to combined shear and tension, AISC 360 Section J3.7 specifies that the nominal tensile stress is reduced while the nominal shear strength is unaffected. The value of the reduced design tensile stress is

$$\phi R_n = \phi F'_{nt} A_b$$

where

F'_{nt} = modified nominal tensile stress
$= 1.3F_{nt} - f_v F_{nt}/\phi F_{nv}$
A_b = nominal area of bolt
R_n = nominal tensile capacity
ϕ = resistance factor
= 0.75
f_v = required shear stress using LRFD load combinations
F_{nt} = nominal tensile stress from AISC 360 Table J3.2
F_{nv} = nominal shear stress from AISC 360 Table J3.2

The required shear stress f_v must not exceed the design shear stress ϕF_{nv}. When either $f_v \le 0.2\phi F_{nv}$ or $f_t \le 0.2\phi F_{nv}$ the effects of combined stress need not be investigated.

Slip-Critical Bolts in Shear

The minimum pre-tension force T_b in a bolt is specified in AISC 360 Table J3.1 as

$$T_b = 0.70F_u$$

Slip-critical bolts in standard holes or slots transverse to the direction of load are designed for slip at the serviceability limit state. Slip-critical bolts in oversized holes or slots parallel to the direction of load are designed for slip at the strength level limit state. For both cases, the required strength is determined using LRFD load combinations, and the available strength is determined using the resistance factor appropriate to each case.

The frictional resistance developed in a slip-critical connection is dependent on the condition of the faying surfaces. Values of the mean slip coefficient μ for two types of surface condition are given in AISC 360 Sec. J3.8. The two conditions are the following:

- Class A surface conditions—unpainted clean mill scale surfaces or blast-cleaned surfaces with class A coatings with $\mu = 0.35$
- Class B surface conditions—unpainted blast-cleaned surfaces or blast-cleaned surfaces with class B coatings with $\mu = 0.50$

The design slip resistance, for both cases, is given by AISC 360 Eq. J3-4 as

$$\phi R_n = \phi \mu D_u h_{sc} T_b N_s$$

where

D_u = a multiplier that reflects the ratio of the mean installed bolt tension to the specified minimum bolt pretension
= 1.13
h_{sc} = modification factor for type of hole
= 1.00 ... for standard size holes
= 0.85 ... for oversized and short-slotted holes
= 0.70 ... for long-slotted holes
T_b = minimum bolt pretension given in AISC 360 Table J3.1
N = number of slip planes
ϕ = resistance factor
= 1.00 … for the serviceability limit state
= 0.85 … for the strength level limit state

Values for the design slip-critical shear resistance ϕR_n are given in AISC 360 Tables 7-3 and 7-4 and NCEES Handbook. These values are based on a Class A faying surface and are multiplied by 1.43 for a Class B faying surface. For slip-critical connections, the slip-critical shear resistance is identical for the cases of threads included or excluded from the shear plane.

Slip-Critical Bolts in Tension and Combined Shear and Tension

Design tensile capacity is independent of the pretension in the bolt and is based on the nominal unthreaded cross-sectional area of the bolt A_b and is given by AISC 360 Section J3.6 as

$$\phi R_n = \phi F_{nt} A_b$$
$$= 0.75 F_{nv} A_b$$

where

F_{nv} = nominal tensile strength of bolt
R_n = nominal tensile capacity
ϕ = resistance factor
= 0.75

Values of the design tensile capacity ϕR_n are given in AISC Manual Table 7-2 and NCEES Handbook.

When a slip-critical connection is subjected to combined shear and direct tension, as shown in Figure 14.16, the design tensile force in a connector is unaffected. However, the clamping force on the faying surface is reduced by the applied tensile force, and the design shear capacity must be reduced in proportion to the loss of pretension. In accordance with AISC 360 Section J3.9, the design shear capacity must be multiplied by the reduction factor

$$k_s = 1 - T_u / D_u T_b N_b$$

where

T_u = strength level tensile force on the bolt due to LRFD load combinations
T_b = specified pretension force on the bolt
N_b = number of bolts carrying the applied tension

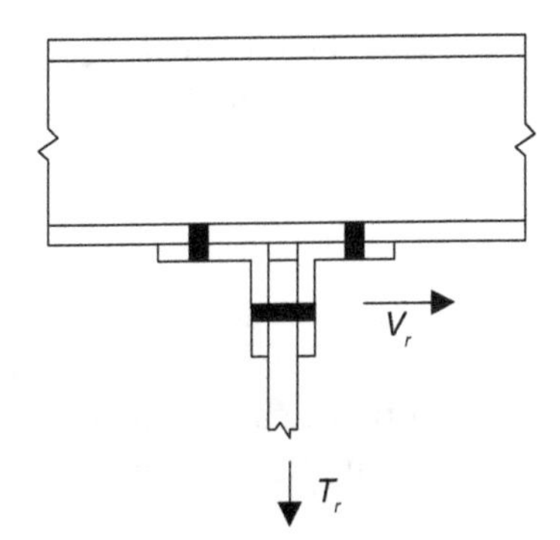

Figure 14.16 Shear and tension on a slip-critical connection

However, when the tensile force in a bolt is due to a bending moment applied in a plane perpendicular to the faying surface, as shown in Figure 14.17, no reduction in the design shear capacity is necessary since the increase in compressive force at the bottom of the connection compensates for the tension produced in the bolts at the top of the connection.

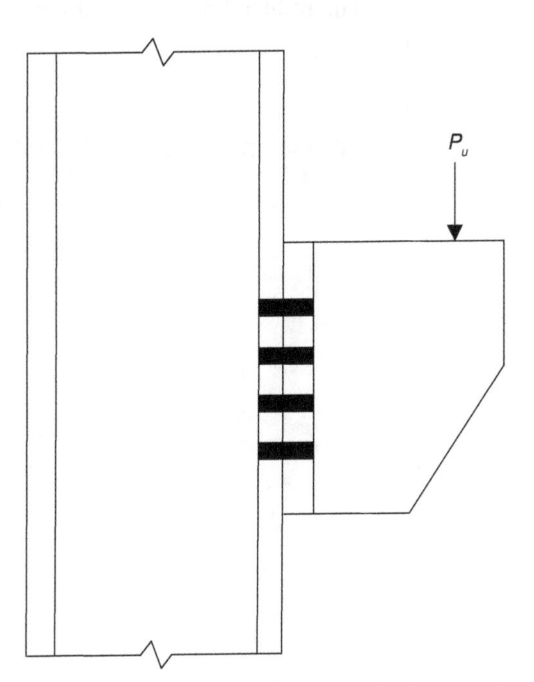

Figure 14.17 Tension due to applied moment

Example 14.22

The four bolts in the connection of the double angle bracket to the girder flange shown in Figure 14.16 are ¾-inch diameter Grade A325 slip-critical bolts in standard holes with a Class A faying surface. The pretension force in each bolt is 28 kips. The strength level tensile force applied to the connection is T_r = 40 kips. Assuming that the angles and the girder flange are satisfactory, determine the maximum strength level shear force that may be applied to the connection. Prying action may be neglected.

Solution

The strength level tensile force applied to each bolt in the connection is

$$\begin{aligned} T_u &= T_r/4 \\ &= 40/4 \\ &= 10 \text{ kips} \end{aligned}$$

The design tensile capacity of each bolt is obtained from AISC Manual Table 7-2 and NCEES Handbook as

$$\phi R_n = 29.8 \text{ kips}$$
$$> 10 \text{ kips} \ldots \text{ satisfactory}$$

The bolts are in standard holes and the serviceability limit state applies to slip at the connection. The design shear capacity for a ¾-inch diameter Grade A325 slip-critical bolt in a standard hole with a Class A faying surface, in the absence of tensile force, is given by AISC Manual Table 7-3 and NCEES Handbook as

$$\phi R_n = 11.1 \text{ kips}$$

The shear capacity reduction factor is

$$k_s = 1 - T_u/D_u T_b N_b$$
$$= 1 - 10/(1.13 \times 28 \times 1.0)$$
$$= 0.68$$

The reduced design shear capacity for each bolt is

$$k_s \phi R_n = 0.68 \times 11.1$$
$$= 7.6 \text{ kips}$$

The maximum strength level shear force that may be applied to the connection is

$$4 \times 7.6 = 30.4 \text{ kips}$$

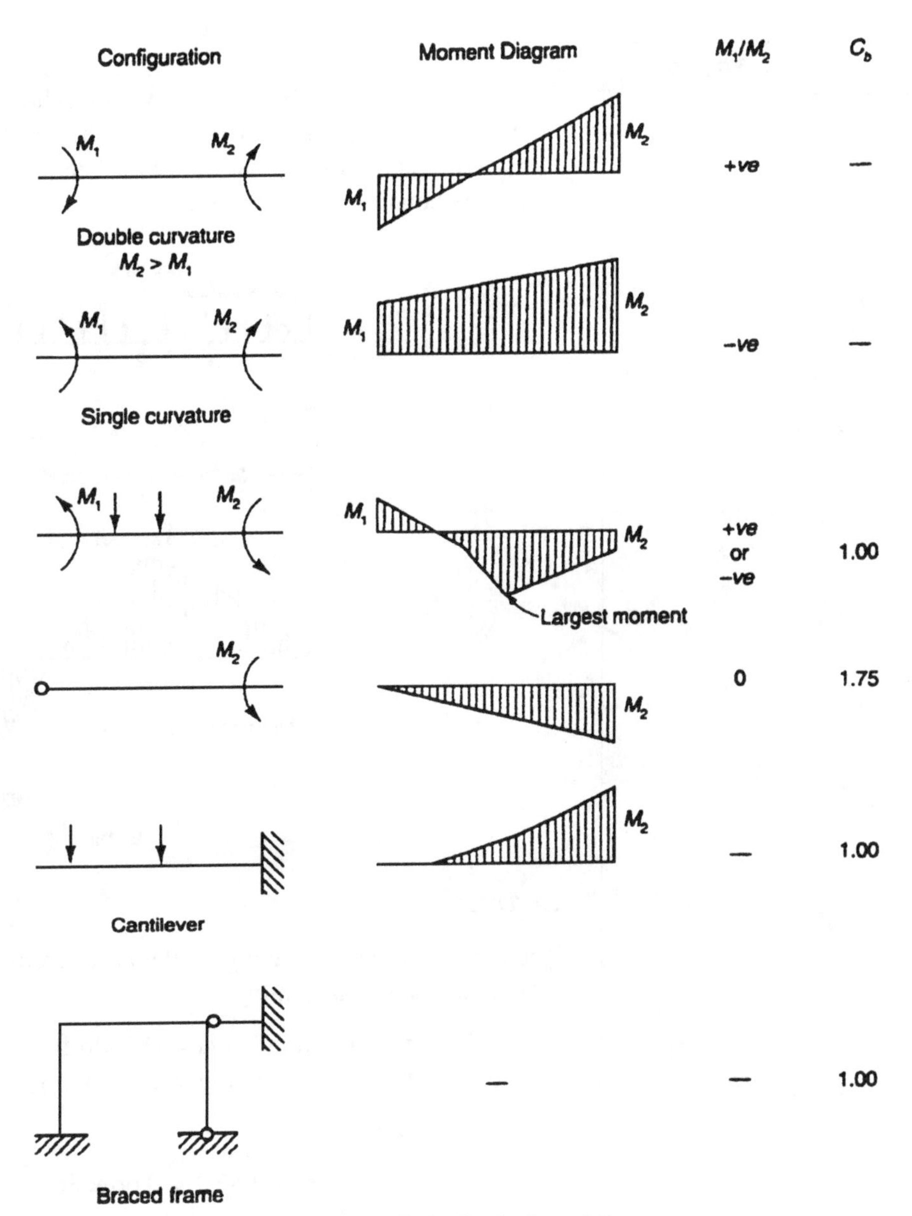

Figure 14.18 Derivation of C_b

Equation (F1-8) is independent of l/r_T and gives an allowable stress of

$$F_b = \frac{12{,}000 C_b A_f}{ld}$$

These expressions may be solved by calculator, or the allowable beam resisting moment may be obtained from AISC Moment Charts, which are based on a value of unity for C_b and are conservative for larger values of C_b.

Compact sections, solid rectangular sections bent about their weak axes, and solid round or square bars have an allowable stress given by AISC Equation (F2-1) of

$$F_b = 0.75F_y$$

Noncompact sections bent about their weak axes have an allowable stress given by AISC Equation (F2-2) of

$$F_b = 0.60F_y$$

Example 14.23

The W18 × 60 grade A36 beam shown in Exhibit 4 is laterally supported throughout its length. Determine whether the beam is adequate to support the applied loads indicated. The relevant properties of the beam are $S_x = 108$ in³, $F_y = 36$ kips/in², and allowable bending stress $F_b = 0.66F_y$.

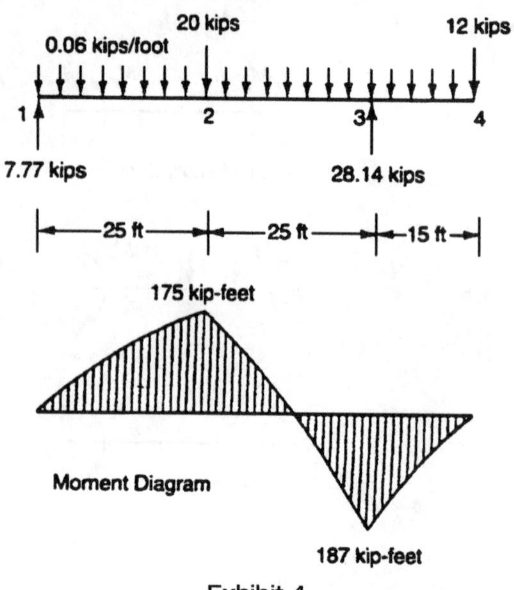

Exhibit 4

Solution

The bending moments acting on the beam from the applied loads and beam self-weight are shown in Exhibit 4.

$$M_x = \text{maximum moment} = 187 \text{ kip-ft}$$
$$f_b = \text{maximum bending stress} = M_x/S_x = 187 \times 12/108 = 20.78 \text{ kips/in}^2$$

The allowable stress is

$$F_b = 0.66F_y = 0.66 \times 36 = 23.76 \text{ kips/in}^2$$

Hence, the W18 × 60 is adequate.

Shear Stress

The allowable shear stress, based on the overall beam depth, is given by AISC Equation (F4-1) as

$$F_v = 0.40F_y$$

provided

$$h/t_w \le 380/\sqrt{F_y}$$

DESIGN OF CONCRETE COMPONENTS

Introduction

The basic requirement of designing for strength is to ensure that the design strength of a member is not less than the required ultimate strength. The latter consists of the service-level loads multiplied by appropriate load factors, and this is defined in ACI Section 9.2.

The design strength of a member consists of the theoretical ultimate strength of the member—the **nominal strength**—multiplied by the appropriate strength reduction factor, ϕ. Thus

$$\phi\ (\text{nominal strength}) \geq U$$

ACI Section 9.3 defines the reduction factor as $\phi = 0.90$ for flexure of tension-controlled sections, $\phi = 0.75$ for shear and torsion, $\phi = 0.75$ for compression members with spiral reinforcement, $\phi = 0.65$ for compression members with lateral ties, and $\phi = 0.65$ for bearing on concrete.

Beams

Nominal Strength of Beams

The nominal strength of a rectangular beam, with tension reinforcement only, is derived from the assumed ultimate conditions shown in Figure 14.5. ACI Section 10.2 specifies an equivalent rectangular stress block in the concrete of $0.85\,f_c'$, with a depth of

$$a = A_s f_y / 0.85\,f_c'\,b = \beta_1 c$$

where c = depth to neutral axis and β_1 = compression zone factor, given in ACI Section 10.2.7.3.

From Figure 14.19, the nominal strength of the member is derived as

$$M_n = A_s f_y d(1 - 0.59\rho f_y / f_c')$$

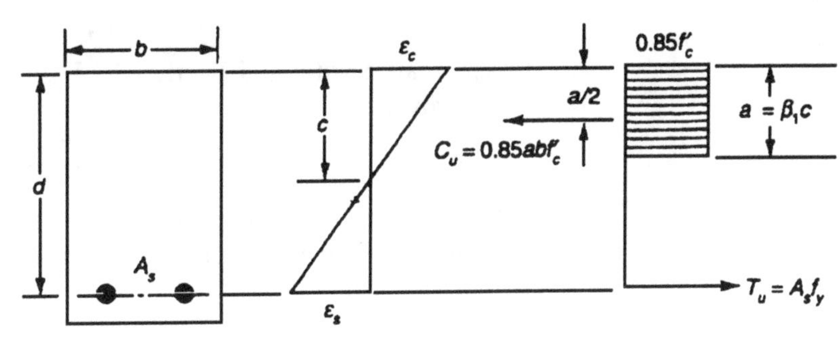

Figure 14.19 Member with tension reinforcement only

where $\rho = A_s / b_d$ = reinforcement ratio. For a tension-controlled section, $\phi M_n = 0.9M_n$ = design strength, and $\phi M_n \geq M_u$ = applied factored moment. This expression may also be rearranged to give the reinforcement ratio required to provide a given factored moment, M_u, as

$$\rho = 0.85 f_c' \left[1 - \sqrt{1 - K/0.383 f_c'}\right] / f_y$$

where $K = M_u / bd^2$.

These expressions may be readily applied using standard calculator programs and tables. For a balanced strain condition, the maximum strain in the concrete—and in the tension reinforcement—must simultaneously reach the values specified in ACI Section 10.3.2 as

$$\varepsilon_c = 0.003 = \text{concrete strain}$$
$$\varepsilon_s = f_y/E_s = \text{steel strain}$$
$$= 0.002 \text{ ... for Grad GO reinforcement}$$

The balanced reinforcement ratio is given as $\rho_b = 0.85 \times 87{,}000\,\beta_1 f_c'/f_y(87{,}000 + f_y)$. When $\varepsilon_s \le 0.002$, the section is compression-controlled and

$$c/d = 0.6$$
$$\phi = 0.65 \text{ ... section with rectangular stirrups}$$

When $\varepsilon_s \ge 0.005$, the section is tension-controlled and

$$c_t/d = 0.375$$
$$a_t = \beta_1 c_t$$
$$\phi = 0.90$$
$$\rho_t \le 0.319\beta_1 f_c'/f_y$$
$$A_t = \rho_t bd$$

For intermediate values of ε_s

$$\phi = 0.65 + (\varepsilon_s - 0.002)(250/3)$$
$$= 0.48 + 83\varepsilon_s$$

To ensure adequate ductility, the minimum allowable strain in tension reinforcement is

$$\varepsilon_{min} = 0.004$$

and

$$c/d = 0.429$$
$$\phi = 0.812$$
$$\rho_{max} \le 0.364\beta_1 f_c'/f_y$$

In accordance with ACI Section 10.5.1., the minimum allowable reinforcement ratio is given by

$$\rho_{\min} = 3\sqrt{f_c'}/f_y \ge 200/f_y$$

with the exception that the minimum reinforcement provided need not exceed one-third more than required by analysis. The expression $200/f_y$ governs when $f_c' < 4444$ lb/in.^{2}

Example 14.24

A reinforced concrete beam, with an overall depth of 16 inches, an effective depth of 14 inches, and a width of 12 inches, is reinforced with Grade 60 bars and has a concrete cylinder strength of 3000 pounds per square inch. Determine the area of tension reinforcement required for the beam to support a superimposed live load of one kip per foot run over an effective span of 20 feet.

Solution

The weight of the beam is

$$w_D = 16 \times 12 \times 0.150/144 = 0.20 \text{ kips per ft}$$

The dead load moment is

$$M_D = w_D\ell^2/8 = 0.2 \times 20^2/8 = 10 \text{ kip ft}$$

The live load moment is

$$M_L = w_L\ell^2/8 = 1.0 \times 20^2/8 = 50 \text{ kip ft}$$

The factored moment is

$$M_u = 1.2M_D + 1.6M_L = 1.2 \times 10 + 1.6 \times 50 = 92 \text{ kip ft}$$

The moment factor is

$$K = M_u/bd^2 = 92 \times 12/(12 \times 14^2) = 0.469 \text{ kips per square in.}$$

Assuming a tension-controlled section, the reinforcement ratio required to provide a given factored moment M_u is

$$\rho = 0.85f_c'[1 - (1 - K/0.383f_c')^{0.5}]/f_y = 0.85 \times 3\{1 - [1 - 0.469/(0.383 \times 3)]^{0.5}\}/60$$
$$= 0.010$$

The limiting reinforcement ratio for a tension-controlled section is

$$\rho_t \leq 0.319\beta_1 f_c'/f_y = 0.319 \times 0.85 \times 3/60 = 0.014 > \rho$$

Hence, the section is tension-controlled.

The minimum allowable reinforcement ratio is

$$\rho_{min} = 200/f_y = 200/60{,}000 = 0.0033 < \rho \text{ ... satisfactory}$$

The reinforcement area required is

$$A_s = \rho bd = 0.01 \times 12 \times 14 = 1.68 \text{ square in.}$$

When the applied factored moment exceeds the maximum design strength of a singly reinforced member that has the maximum allowable reinforcement ratio, compression reinforcement and additional tensile reinforcement must be provided, as shown in Figure 14.20. The difference between the applied factored moment and the maximum design moment strength of a singly reinforced section is

$$M_r = M_u - \phi M_t = \text{residual moment}$$

where

M_t = maximum design moment for a tension-controlled section.

The additional area of tensile reinforcement required is

$$A_T = M_r/\phi f_y(d - d') = A_s' f_s'/f_y$$

The depth of the stress block is

$$a = f_y A_{max} / 0.85 f'_c b$$

The depth of the neutral axis is

$$c = a/\beta_1$$

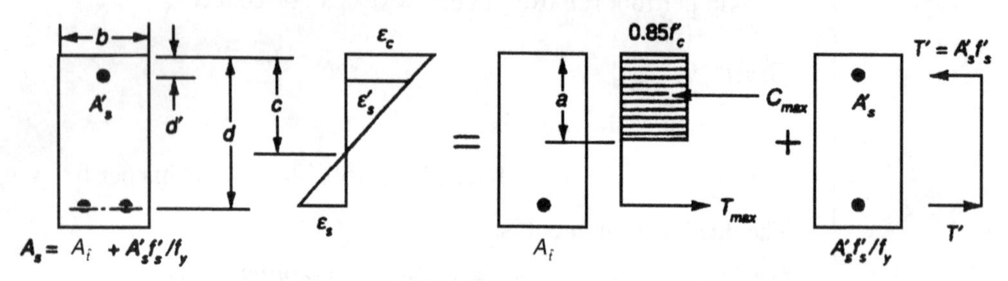

Figure 14.20 Member with compression reinforcement

The stress in the compression reinforcement is given by

$$f'_s = E_s E_u (1 - d'/c) \leq f_y$$

The required area of compression reinforcement is given by

$$A'_s = M_r / \phi f'_s (d - d')$$

The total required area of tension reinforcement is

$$A_s = A_i + A'_s f_s / f_y$$

In order to analyze a given member with compression reinforcement, an initial estimate of the neutral axis depth is required. The total compressive force in the concrete and compression reinforcement is then compared with the tensile force in the tension reinforcement. The initial estimate of the neutral axis depth is then adjusted until these two values are equal.

The conditions at ultimate load in a flanged member, when the depth of the equivalent rectangular stress block exceeds the flange thickness, are shown in Figure 14.21. The area of reinforcement required to balance the compressive force in the flange is given by

$$A_{sf} = C_f / f_y = 0.85 f'_c h_f (b - b_w) / f_y$$

The corresponding design moment strength is

$$M_f = \phi A_{sf} f_y (d - h_f / 2)$$

The residual moment to be developed by the web is $M_r = M_u - M_f$. The required reinforcement ratio to provide the residual moment is

$$\rho_w = 0.85 f'_c \left[1 - \sqrt{1 - 2K_w / 0.9 \times 0.85 f'_c}\right] / f_y$$

where $K_w = M_r / b_w d^2$.

The corresponding reinforcement area is

$$A_{sw} = b_u d \rho_w$$

The total reinforcement area required is

$$A_s = A_{sf} + A_{sw}$$

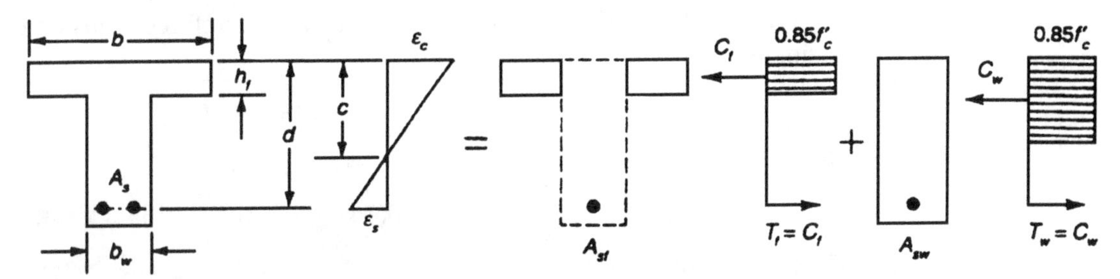

Figure 14.21 Flanged member with tension reinforcement

Example 14.25

The reinforced concrete beam shown in Exhibit 5 is reinforced with Grade 60 bars at the positions indicated and has a concrete cylinder strength of 3000 psi. The beam carries a superimposed load of 2 kips/ft run over an effective span of 6 m. Determine the areas of tension and compression steel required.

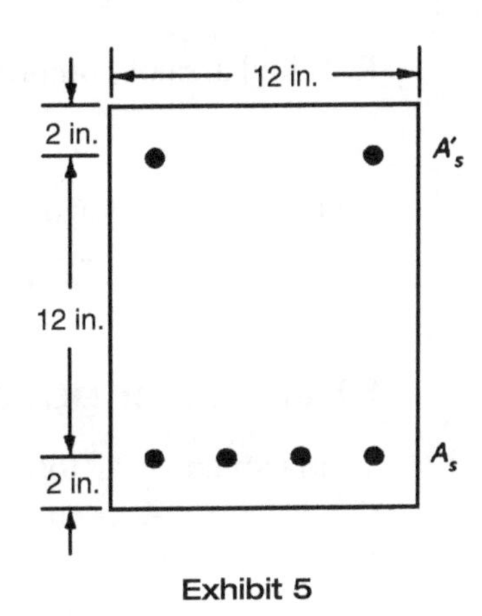

Exhibit 5

Solution

The weight of the beam is

$$w_D = 16 \times 12 \times 0.150/144 = 0.20 \text{ kips per ft}$$

The dead load moment is given by

$$M_D = w_D \ell^2/8 = 0.2 \times 20^2/8 = 10 \text{ kip ft}$$

The live load moment is given by

$$M_L = w_L \ell^2/8 = 2.0 \times 20^2/8 = 50 \text{ kip ft}$$

The factored moment is

$$M_u = 1.2M_D + 1.6M_L = 1.2 \times 10 + 1.6 \times 100 = 172 \text{ kip ft}$$

The compression zone factor is given by ACI Section 10.2.7 as

$$\beta_1 = 0.85$$

The limiting reinforcement ratio for a singly reinforced tension-controlled section is

$$\rho_t = 0.319\beta_1 f_c'/f_y = 0.319 \times 0.85 \times 3/60 = 0.0136$$

The limiting reinforcement area for a singly reinforced tension-controlled section is

$$A_t = bd\rho_t = 12 \times 14 \times 0.0136 = 2.29 \text{ in.}^2$$

The limiting depth of the stress block for a singly reinforced tension-controlled section is

$$a_t = \beta_1 c_t = 0.85 \times 0.375 \times 14 = 4.46 \text{ in.}$$

The corresponding moment is

$$M_t = A_t f_y (d - a_t/2) = 2.29 \times 60(14 - 4.46/2)/12 = 134.8 \text{ kip-ft}$$

The residual moment is

$$M_r = M_u - \phi M_t = 172 - 0.9 \times 134.8 = 50.7 \text{ kip-ft}$$

Stress in the compression steel is

$$f_s' = E_s \varepsilon_c (c_t - d')/c_t = 29{,}000 \times 0.003(5.25 - 2)/5.25 = 53.86 \text{ ksi}$$

Force in the compression steel is

$$C_s' = M_r/\phi(d - d') = 50.7 \times 12/0.9(14 - 2) = 56.33 \text{ kips}$$

Required area of compression steel is

$$A_s' = C_s'/f_s' = 56.33/53.86 = 1.05 \text{ in.}^2$$

Required area of tension steel is

$$A_s = A_t + A_s' f_s'/f_y = 2.29 + 1.05 \times 53.86/60 = 3.23 \text{ in.}^2$$

Shear Strength of Beams

The factored shear force acting on a member, in accordance with ACI Equations (11-1) and (11-2), is resisted by the combined design shear strength of the concrete and shear reinforcement. Thus, the factored applied shear is given by

$$V_u = \phi V_c + \phi V_s$$

where $\phi = 0.75$ = strength reduction factor for shear given by ACI Section 9.3.2.3; $V_c = 2\lambda\sqrt{f_c'}\,b_w d$ = nominal shear strength of concrete from ACI (11-3); $V_s = A_v f_y d/s$ = nominal shear strength of shear reinforcement from ACI Equation (11-15); A_v = area of shear reinforcement; and s = spacing of shear reinforcement, specified in ACI Section 11.4.5 as

$$s \leq d/2 \text{ or } 24 \text{ in. when } V_s \leq 4\sqrt{f_c'}\,b_w d,$$

$$s \leq d/4 \text{ or } 12 \text{ in. when } 4\sqrt{f_c'}\,b_w d < V_s \leq 8\sqrt{f_c'}\,b_w d$$

λ = 1.0 for normal weight concrete
= 0.85 for sand-lightweight concrete
= 0.75 for all-lightweight concrete

The dimensions of the section or the strength of the concrete must be increased, in accordance with ACI Section 11.4.7.9, to ensure that $V_s \leq 8\sqrt{f_c'}\,b_w d$.

As specified in ACI Section 11.4.6, a minimum area of shear reinforcement is required when

$$V_u > \phi V_c/2$$

and the minimum area of shear reinforcement is given by ACI Equation (11-13) as

$$A_{v(\min)} = 0.34 b_w s/f_y$$

but not less than

$$A_{v(\min)} = 0.45\sqrt{f_c'}b_w s/f_y$$

$$A_{v(\min)} = 50b_w s/f_y$$

When the support reaction produces a compressive stress in the member, ACI Section 11.1.3.1 specifies that the critical factored shear force is the force that is acting at a distance equal to the effective depth from the face of the support. Figure 14.22 summarizes the shear provisions of the Code, and Figure 14.23 illustrates the design principles involved.

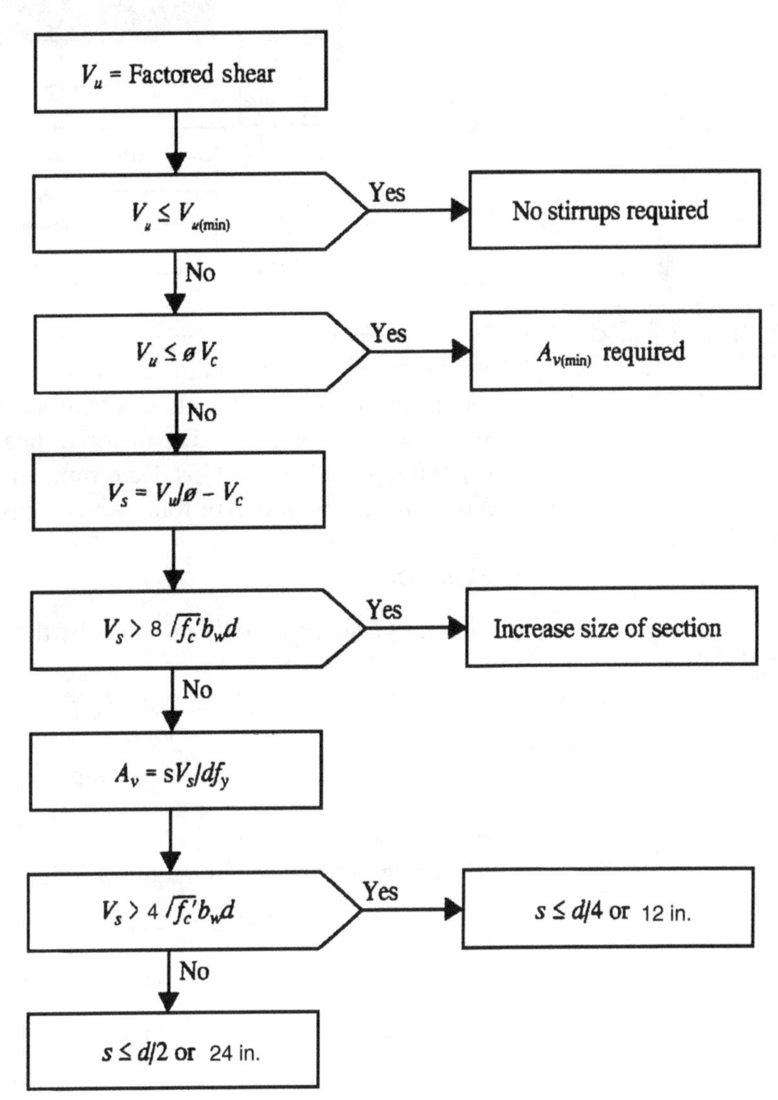

Figure 14.22 Flow chart for design for shear

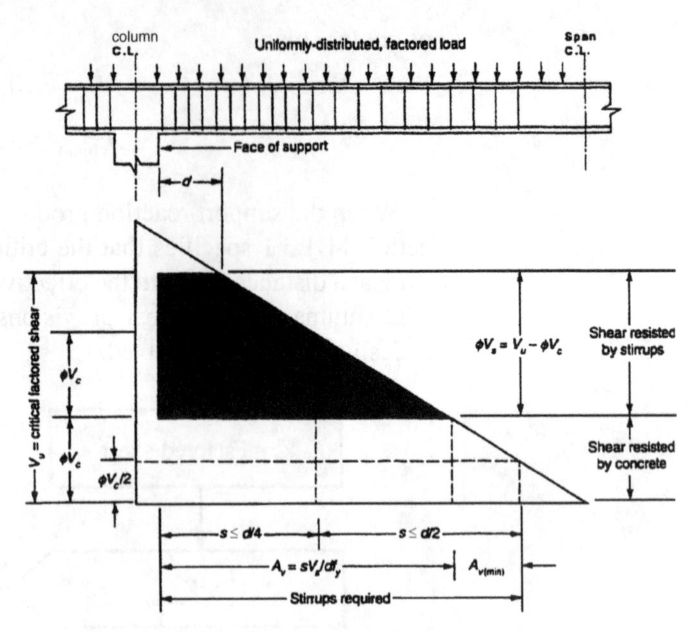

Figure 14.23 Shear in a reinforced concrete beam

Example 14.26

A reinforced normal weight concrete beam, which has an effective depth of 24 inches and a width of 20 inches, is reinforced with Grade 60 bars and has a compressive strength of 3000 psi. The factored shear force, at two locations on the beam, is (a) 180 kips and (b) 90 kips. Determine the required spacing, at each location, for No. 4 stirrups with two or four vertical legs as suitable.

Solution

The design shear strength provided by the concrete is

$$\begin{aligned}\phi V_c &= 2\phi b_w d\sqrt{f_c'}\\ &= 2\times 0.75\times 20\times 24(3000)^{0.5}/1000\\ &= 39.44 \text{ kips}\end{aligned}$$

(a)

The design shear strength required from the shear reinforcement is

$$\phi V_s = V_u - \phi V_c = 180 - 39.44 = 140.56 \text{ kips}$$

Since $\phi V_s < 4 \times \phi V_c$ the concrete section is adequate.

Since $\phi V_s > 2 \times \phi V_c$ the maximum stirrup spacing is given by

$$s = d/4 = 24/4 = 6 \text{ in.}$$

The area of shear reinforcement required is

$$\begin{aligned}A_v/s &= \phi V_s/\phi df_y = 140.56 \times 12(0.75 \times 24 \times 60)\\ &= 1.56 \text{ in.}^2/\text{ft}\end{aligned}$$

The required spacing of stirrups with four No. 4 vertical legs is

$$s = 4 \times 0.2 \times 12/1.50 = 6.4 \text{ in.}$$
$$s = 6 \text{ in.}$$

(b)
The design shear strength required from the shear reinforcement is

$$\phi V_s = V_u - \phi V_c = 90 - 39.44 = 50.56 \text{ kips}$$

Since $\phi V_s < 2 \times \phi V_c$, the maximum stirrup spacing is given by

$$s = d/2 = 24/2 = 12 \text{ in.}$$

$$A_v/s = 0.56$$

Using two No. 4 vertical legs,

$$s = 2 \times 0.02 \times 12/0.56$$
$$= 8.6 \text{ in.}$$

Slabs

Approximate Design Coefficients

In designing continuous beams and one-way slabs, approximate coefficients may be used instead of an exact analysis provided that the following conditions are observed:

- The loads are uniformly distributed.
- The ratio of live to dead loads does not exceed 3.
- Adjacent spans do not differ in length by more than 20% of the shorter span.
- Members are prismatic.

The bending moment coefficients are specified in ACI Section 8.3.3 and are shown in Figure 14.24. The bending moments are obtained from the expression

$$M_u = \tau w_u \ell_n^2$$

where

τ = approximate design coefficient
w_u = factored load per unit length
ℓ_n = length of clear span

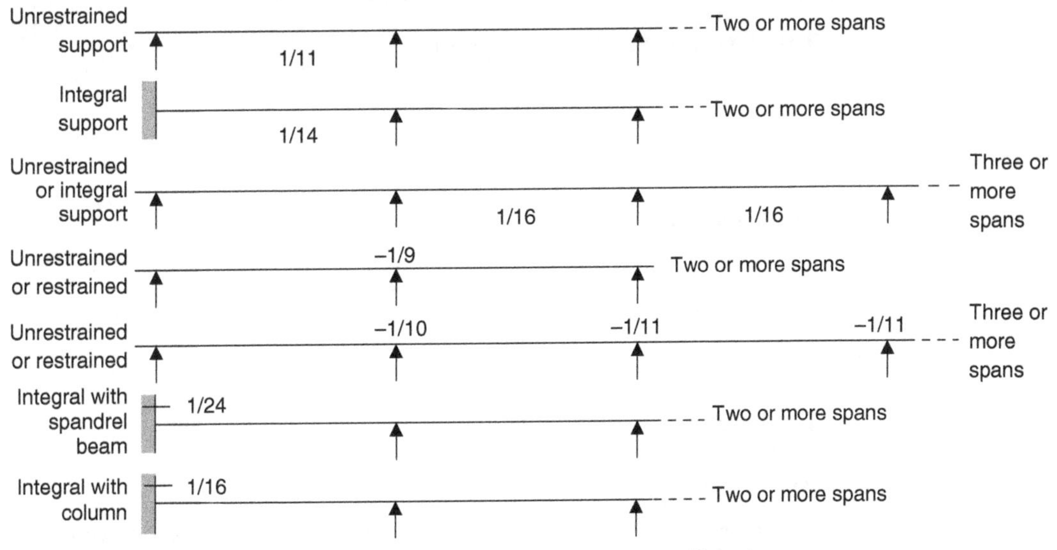

Figure 14.24 Bending moment coefficients

Shear force coefficients are also specified in ACI Section 8.3.3 and are shown in Figure 14.25. The shear force is obtained from the expression

$$V_u = \tau w_u \ell_n$$

where

τ = approximate design coefficient
w_u = factored load per unit length
ℓ_n = length of clear span

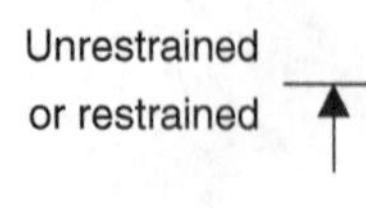

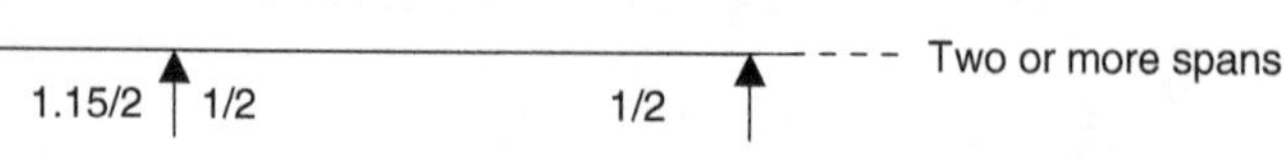

Figure 14.25 Shear force coefficients

Example 14.27

A reinforced concrete slab, continuous over two spans, supports a factored dead load of 3 kips per foot run and a factored live load of 2 kips per foot run. The clear distance between support faces is 15 feet. Determine the design bending moments and shear forces.

Solution

The total factored load is

$$\begin{aligned} w_u &= w_{uD} + w_{uL} \\ &= 3 + 2 \\ &= 5 \text{ kips per ft} \end{aligned}$$

For an unrestrained end support, the mid-span moment is

$$\begin{aligned} M_u &= \tau w_u \ell_n^2 \\ &= 5 \times 15^2 \times 1/11 \\ &= 102.27 \text{ kip-ft} \end{aligned}$$

For a two span beam, the bending moment at the center support is

$$\begin{aligned} M_u &= \tau w_u \ell_n^2 \\ &= 5 \times 15^2 \times 1/9 \\ &= 125 \text{ kip-ft} \end{aligned}$$

The shear force at the end support is

$$\begin{aligned} V_u &= \tau w_u \ell_n \\ &= 5 \times 15 \times 1/2 \\ &= 37.5 \text{ kips} \end{aligned}$$

The shear force at the center support is

$$\begin{aligned} V_u &= \tau w_u \ell_n \\ &= 5 \times 15 \times 1.15/2 \\ &= 43.13 \text{ kips} \end{aligned}$$

Columns

Short Columns

Reinforced concrete columns may be classified as either short columns or long columns. In the design of long columns, slenderness effects must be considered and the secondary moments, due to the $P - \Delta$ effects, added to the primary moments. Classification as a short column depends on the slenderness ratio of the column which is given by

$$kl_u/r$$

where

k = effective length factor determined from the alignment charts in ACI Section R10.10.1 and NCEES Handbook
l_u = unsupported column length
r = radius of gyration of column given in ACI Section 10.10.1 as
$r = 0.3 \times$ rectangular column width
$r = 0.25 \times$ circular column diameter

In an unbraced frame, a column is defined as a short column, in accordance with ACI Section 10.10.1, when the slenderness ratio is given by

$$kl_u/r < 22$$

For a braced frame, a column is defined as a short column, accordance with ACI Section 10.10.2 when the slenderness ratio is given by

$$kl_u/r \leq 34 - 12M_{1b}/M_{2b}$$

where

M_{2b} = larger factored moment at end of column
M_{1b} = smaller factored moment at end of column, negative if the column is bent in double curvature

Short Column with Axial Load Only

For a short column, the axial load carrying capacity may be illustrated by reference to the column shown in Figure 14.26.

The theoretical design axial load strength at zero eccentricity is

$$\phi P_0 = \phi[0.85 f_c'(A_g - A_{st}) + A_{st} f_y]$$

where

ϕ = strength reduction factor specified in Section 9.3
= 0.70 for compression members with spiral reinforcement and
= 0.65 for compression members with lateral ties
f_c' = concrete cylinder strength = 3 ksi
f_y = reinforcement yield strength = 60 ksi
A_g = gross area of the section = 100 in.^{2}
A_{st} = reinforcement area = 4 in.^{2}

Then

$$\begin{aligned}\phi P_0 &= 0.65[0.85 \times 3(100 - 4) + 4 \times 60] \\ &= 315 \text{ kips}\end{aligned}$$

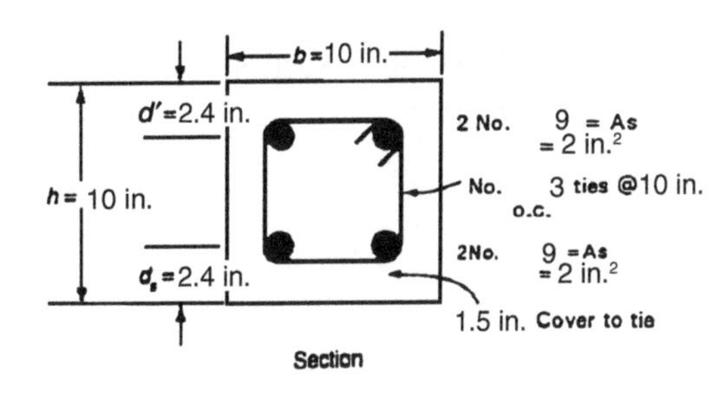

Figure 14.26 Compression in a short tied column

To account for accidental eccentricity, ACI Section 10.3.6 requires a spirally reinforced column to be designed for a minimum eccentricity of approximately $0.05h$, which gives a maximum design axial load strength at zero eccentricity, in accordance with ACI Equation (10-1), of

$$\phi P_{n\max} = 0.85\phi P_0$$

In the case of a column with lateral ties, a minimum eccentricity of approximately $0.10h$ is specified, which gives a maximum design axial load strength at zero eccentricity, in accordance with ACI Equation (10-2), of

$$\phi P_{n\max} = 0.80\phi P_0 = 0.80 \times 315 = 252 \text{ kips}$$

Short Column with Applied Moment

For a short column, the axial load carrying capacity reduces as the moment applied to the column increases and causes additional stresses in the column. Design may be facilitated by means of the interaction diagrams published in NCEES Handbook.

Example 14.28

A 24-inch diameter spirally reinforced column with 1½-inch cover to the ⅜-inch spiral is reinforced with 10 No. 9 bars. The concrete strength is 4000 pounds per square inch and the tensile strength of the reinforcement is 60,000 pounds per square inch. The unsupported length of the column is 9 feet, it is braced against side sway, and is bent in single curvature with factored end moments of $M_1 = M_2 =$ 250 kip-foot. Determine the design axial load strength of the column.

Solution

From ACI Section 10.10.1, the effective length factor of a column braced against sidesway may conservatively be taken as

$$k = 1.0$$

The radius of gyration, in accordance with ACI Section 10.10.1 is

$$r = 0.25h$$

where

$$h = \text{diameter of column} = 24 \text{ in.}$$

and

$$r = 0.25 \times 24 = 6 \text{ in.}$$

The slenderness ratio is

$$k\ell_u/r = 1 \times 9 \times 12/6 = 18$$

From ACI Section 10.10.1, slenderness effects in the non-sway column may be neglected when

$$k\ell_u/r \le 34 - 12M_1/M_2 \le 40$$

where

M_2 = larger factored moment at end of column
= 250 kip-ft

M_1 = smaller factored moment at end of column, positive if the column is bent in single curvature.
= 250 kip-ft

Then

$$M_1/M_2 = 1.0 \text{ ... for single curvature}$$

and

$$\begin{aligned} 34 - 12M_1/M_2 &= 34 - 12 \\ &= 22 \\ &> k\ell_u/r \end{aligned}$$

Hence, the column may be considered a short column and slenderness effects need not be considered.

The ratio of the distance between centroids of longitudinal steel to the overall diameter of the column is

$$\begin{aligned} \gamma &= (24 - 4.88)/24 \\ &= 0.80 \text{ ... the NCEES Handbook interaction diagram is applicable} \end{aligned}$$

The reinforcement ratio is

$$\begin{aligned} \rho_g &= 10/452 \\ &= 0.022 \end{aligned}$$

and

$$\begin{aligned} M_u/\phi A_g h f_c' &= 250 \times 12/(0.7 \times 452 \times 24 \times 4) \\ &= 0.099 \end{aligned}$$

Utilizing the interaction diagram in NCEES Handbook

$$P_u/\phi A_g f_c' = 0.82$$

The allowable design axial load is

$$\begin{aligned} P_u &= 0.82\phi A_g f_c' \\ &= 0.82 \times 0.7 \times 452 \times 4 \\ &= 1038 \text{ kips} \end{aligned}$$

Long Columns

Long columns are columns that do not conform to the requirements for a short column given in ACI Section 10.10.1. For long columns, it is necessary to consider the secondary bending stresses caused by the *P*-delta effect. The secondary stresses are produced by the eccentricity of the axial force about the displaced center line of the member. The P-delta effects are automatically determined in a second-order frame analysis. Alternatively, provided the slenderness ratio of a long column does not exceed 100, slenderness effects to be allowed for by means of the approximate moment magnifier method. A conventional first-order frame analysis is utilized and the P-delta effects estimated by amplifying the primary bending moments by a moment magnification factor. The column cross section may then be designed for the calculated axial force and the magnified bending moment using the short column design procedure. Separate procedures are used for sway and non-sway situations.

Long Columns, Non-Sway

For the non-sway case, the magnified factored moment given by ACI Equation (10–11) is

$$M_c = \delta M_2$$

where

M_2 = larger factored moment due to gravity or transverse loads
δ = moment magnification factor for non-sway frames
$= C_m/(1 - P_u/0.75P_c) \geq 1.0$
C_m = equivalent moment correction factor
$= 0.6 + 0.4M_1/M_2 \geq 0.4$ ... for non-sway frames without transverse loads
= 1.0 ... for members with transverse loads between supports
= 1.0 ... for a column bent in single curvature with equal end moments
= 1.0 ... for a column with pinned ends
= 0.6 ... for a column bent in single curvature with one end pinned
= 0.4 ... for a column with one end fixed
M_1 = smaller factored moment at end of column, negative if the column is bent in double curvature, positive if the column is bent in single curvature
0.75 = stiffness reduction factor
P_u = factored axial load at a given eccentricity
$\leq \Phi P_n$
P_n = nominal axial load capacity at a given eccentricity
P_c = Euler critical load from ACI Equation (10–13)
$= \pi^2 EI/(k\ell_u)^2$
$I \approx 0.25I_g$ ... from ACI Section R10.10.6.2

When the calculated moments at the ends of a non-sway column are very small or zero, ACI Section 10.10.6.5 specifies that $C_m = 1.0$ and the column design is based on a minimum factored moment of

$$M_2 = P_u(0.6 + 0.03h)$$

where

h = width or diameter of column

Long Columns with Sway

For the sway case, the magnified factored moments given by ACI Section 10.10.7 are

$$M_1 = M_{1ns} + \delta_s M_{1s}$$
$$M_2 = M_{2ns} + \delta_s M_{2s}$$

where

M_1 = smaller factored moment at end of column, negative if the column is bent in double curvature, positive if the column is bent in single curvature
M_2 = larger factored moment at end of column
M_{1ns} = factored moment at the end of the column at which M_1 acts, due to loads which cause no appreciable side sway, calculated using first-order frame analysis

M_{2ns} = factored moment at the end of the column at which M_2 acts, due to loads which cause no appreciable side sway, calculated using first-order frame analysis

M_{1s} = factored moment at the end of the column at which M_1 acts, due to loads which cause appreciable side sway, calculated using first-order frame analysis

M_{2s} = factored moment at the end of the column at which M_2 acts, due to loads which cause appreciable side sway, calculated using first-order frame analysis

δ_s = moment magnification factor for sway frames
= $1/[1 - \Sigma P_u/(0.75\Sigma P_c)]$ ... where the summations extend over all the columns in a story.
≥ 1.0

Walls

Wall Details

Structural walls are designed for axial force and bending moment in the same manner as columns. Walls may be utilized as exterior walls, interior walls, basement walls, and shear walls in a building and as retaining walls.

The minimum reinforcement required in the stem wall is specified by ACI Section 14.3. For Grade 60 bars larger than No.5, the reinforcement ratios, based on the gross concrete area, for vertical and horizontal reinforcement are given by

$$\rho_{vert} = 0.15 \text{ percent}$$
$$\rho_{hor} = 0.25 \text{ percent}$$

For bars not larger than No.5, the corresponding ratios are

$$\rho_{vert} = 0.12 \text{ percent}$$
$$\rho_{hor} = 0.20 \text{ percent}$$

For walls exceeding 10-inch thickness, with the exception of basement walls, two layers of reinforcement are required in both the vertical and horizontal directions. One layer consisting of not less than one-half and not more than two-thirds of the total reinforcement required for each direction is placed not less than 2 inches nor more than one-third the thickness of the wall from the exterior surface. The other layer, consisting of the balance of the required reinforcement in that direction, is placed not less than ¾-inch nor more than one-third the thickness of the wall from the interior surface.

Cantilever Retaining Wall

Active earth pressure behind the wall and passive pressure in front of the wall may conveniently be obtained using Rankine's theory, as detailed in NCEES Handbook. Alternatively, walls retaining drained earth may be designed for a pressure equivalent to that exerted by a fluid weighing not less than 30 pounds per cubic foot and having a depth equal to that of the retained earth. A live load surcharge behind the wall may be represented by an additional height of fill which produces the same uniform pressure behind the wall. Adequate drainage is necessary behind a retaining wall to prevent a build up of hydrostatic pressure.

Example 14.29

The fill behind the retaining wall in Exhibit 6 has a unit weight of 110 pounds per cubic foot with an equivalent fluid pressure of p_A =30 pounds per square foot per foot. All concrete has a compressive strength of 3000 pounds per square inch and reinforcement consists of Grade 60 deformed bars. Determine the vertical reinforcement required in the stem if a 2-inch cover is provided.

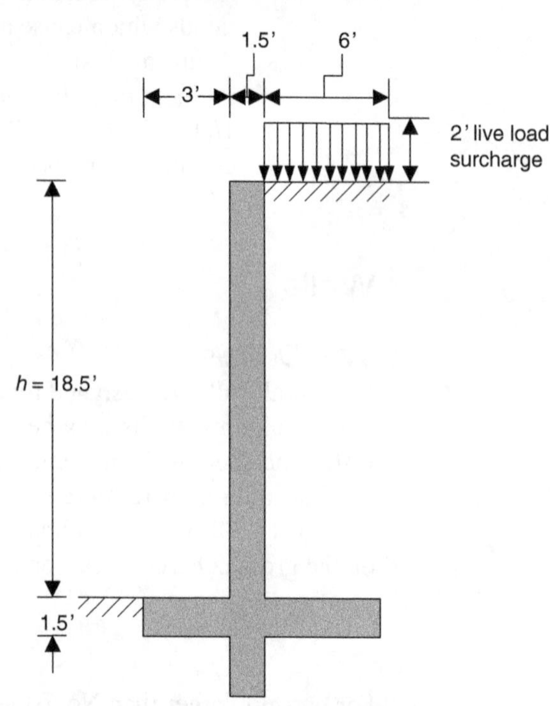

Exhibit 6 Retaining wall details

Solution

The horizontal loads acting on the stem are given by

$$H_A = \text{lateral pressure from backfill}$$
$$= p_A h^2/2$$
$$= 30 \times 18.5^2/2$$
$$= 5134 \text{ pounds}$$
$$H_L = \text{lateral pressure from surcharge}$$
$$= wp_A h$$
$$= 2 \times 30 \times 18.5$$
$$= 1110 \text{ pounds}$$

The maximum factored bending moment at the base of the stem is

$$M_u = 1.6(H_A \times h/3 + H_L \times h/2)$$
$$= 1.6(5134 \times 18.5/3 + 1110 \times 18.5/2)$$
$$= 67{,}083 \text{ pounds-feet}$$

Assuming No. 7 bars, the effective depth is

$$d = 18 - 2 - 0.875/2$$
$$= 15.56 \text{ in.}$$

The moment factor for a 12-inch width is $K = M_u/bd^2 = 67.083 \times 12/(12 \times 15.56^2) = 0.277$ ksi

Assuming a tension-controlled section, the reinforcement ratio required in the earth face to provide a given factored moment M_u is

$\rho = 0.85f_c'[1 - (1 - K/0.383f_c')^{0.5}]/f_y = 0.85 \times 3\{1 - [1 - 0.277/(0.383 \times 3)]^{0.5}\}/60 = 0.0055$

The limiting reinforcement ratio for a tension-controlled section is

$$\rho_t \leq 0.319\beta_1 f_c'/f_y = 0.319 \times 0.85 \times 3/60 = 0.014 > \rho$$

Hence, the section is tension-controlled.

The minimum allowable ratio is

$$\rho_{min} = 200/f_y = 200/60{,}000 = 0.0033 < \rho \text{ ... satisfactory}$$

The reinforcement area required is

$$A_s = \rho bd = 0.0055 \times 12 \times 15.56 = 1.03 \text{ in}^2/\text{ft}$$

Shear is not critical and No.7 bars at a spacing of 7 inches provides a reinforcement area of

$$A' = 0.60 \times 12/7$$
$$= 1.03 \text{ square in. per ft}$$
$$= A_s \text{ ... satisfactory}$$

Based on the gross concrete area, the reinforcement ratio required for the vertical reinforcement in the air face of the wall is given by ACI Section 14.3 as

$$\rho = 0.0012/2 \text{ ... bars not larger than No.5}$$

The required vertical reinforcement area is

$$A_s = 0.0012 \times 12 \times 18/2$$
$$= 0.13 \text{ square in. per ft}$$

Providing No.3 bars at a spacing of 10 inches gives

$$A'' = 0.11 \times 12/10$$
$$= 0.13 \text{ square in. per ft}$$
$$= A_s \text{ ... satisfactory}$$

Footings

Soil Bearing Pressure

An applied load with an eccentricity less than $L/6$ produces the pressure distribution shown in Figure 14.27. The maximum and minimum bearing pressure under the footing is given by

$$q = P/A \pm Pe/S$$
$$= P(1 \pm 6e/L)/BL$$

where

e = eccentricity of the applied load P
B = width of footing
L = length of footing
S = footing section modulus
$= BL^2/6$
A = footing area
$= BL$

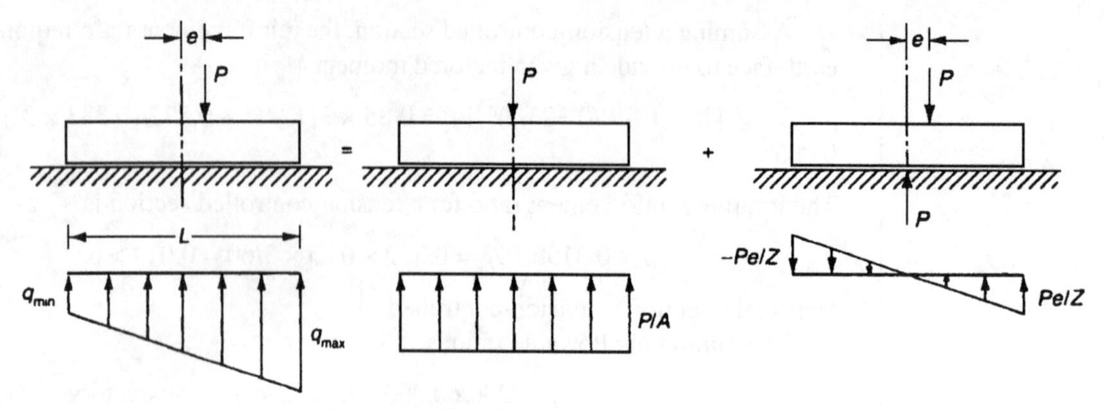

Figure 14.27 Footing with Eccentric Load

An axial load P plus a bending moment M applied to a footing produces an equivalent eccentricity of

$$e = M/P$$

and the bearing pressure under the footing is similarly obtained using this value of e.

When the value of the eccentricity is

$$e = L/6$$

the bearing pressure under the footing is

$$q_{max} = 2P/BL$$
$$q_{min} = 0$$

and the pressure distribution is triangular.

When the eccentricity exceeds $L/6$, as shown in Figure 14.28, no tension is possible between the soil and the footing, and the bearing pressure under the footing is given by

$$q_{max} = 2P/3Be'$$

where

$$e' = L/2 - e$$

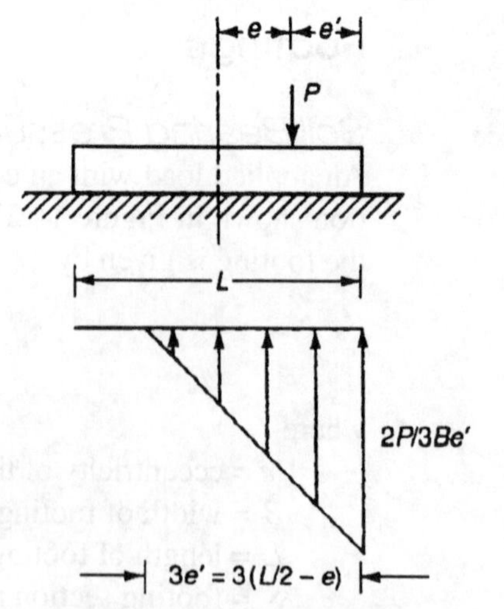

Figure 14.28 Footing with $e > L/6$

Reinforcement Details

The maximum spacing of the principal reinforcement in footings is limited by ACI Sections 7.6.5 and 10.5.4 to

$$s = 3h$$
$$\leq 18 \text{ in.}$$

where

h = footing thickness

The maximum spacing of distribution reinforcement, which is required to resist shrinkage and temperature stresses, is limited by ACI Section 7.12.2.2 to

$$s = 5h$$
$$\leq 18 \text{ in.}$$

The minimum spacing of reinforcement must be adequate to allow full consolidation of the concrete around the bars. The minimum clear spacing between parallel bars in a layer is specified by ACI Sections 7.6.1 and 3.3.2 as

$$s_{(min\ clear)} = d_b$$
$$\geq 1 \text{ in.}$$
$$\geq 1.33 \times \text{maximum aggregate size}$$

where

d_b = diameter of bar

The minimum reinforcement ratio specified for distribution steel in ACI Section 7.12.2.1 is

$$A_s/A_g = 0.0018 \text{ ... for Grade 60 reinforcement}$$

where

A_s = area of distribution reinforcement
A_g = gross area of footing cross section

The minimum cover to reinforcement when the concrete is cast against earth, as is the situation in the soffit of footings, is specified in ACI Section 7.7.1 as 3 inches. For concrete exposed to earth or weather, the minimum cover for No. 5 bars and smaller is specified as 1½ inches, and for larger bars as 2 inches.

Isolated Column Footing

An isolated column footing transfers the loads from a single column to the supporting soil. The size of the footing is determined by the allowable soil bearing pressure. The footing is designed for the following:

- Flexure
- Punching, or two-way, shear
- Flexural, or one-way, shear

The depth of the footing is generally governed by punching shear. The ACI Code, in Sections 15.4 and 15.5, specifies the critical sections in the footing for flexure and shear. For a reinforced concrete column, the critical section for flexure is defined in ACI Section 15.4.2 as being located at the face of the column. The location of the critical sections for flexural shear and punching shear are specified in ACI Sections 11.1.3.1, 11.11.1.2, and 15.5.2 and are illustrated in Figure 14.29.

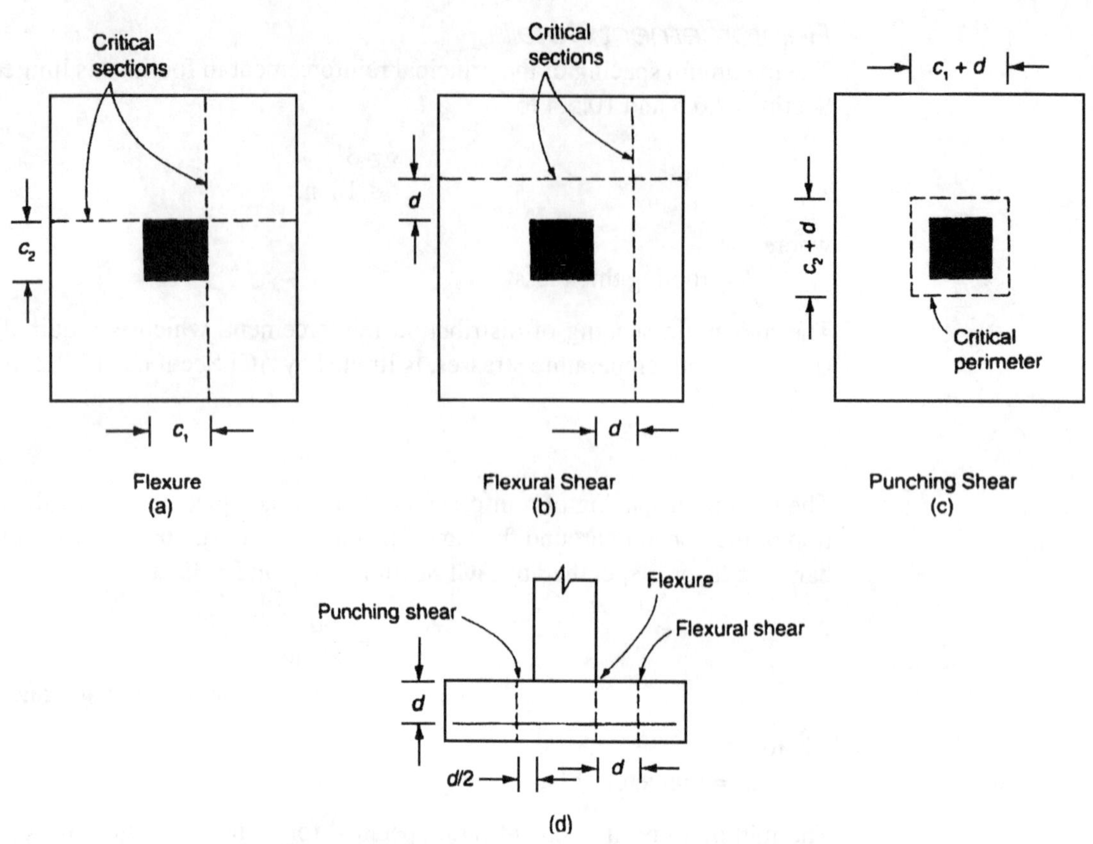

Figure 14.29 Critical sections for shear and flexure

The length of the critical perimeter for punching shear is given by

$$b_o = 2(c_1 + c_2) + 4d$$

where

c_1 = short side of column
c_2 = long side of column
d = effective depth of footing reinforcement

Reinforcement is designed for the maximum moment at the critical section and is distributed uniformly across the base in the case of a square footing. For a rectangular footing, reinforcement parallel to the shorter side should be concentrated in a central band width equal to the length of the shorter side. The area of reinforcement required in the central band is given by ACI Section 15.4.4.2 as

$$A_b = 2A_s/(\beta + 1)$$

where

A_s = total required reinforcement area
β = ratio of the long side to the short side of the footing
$= \ell_2/\ell_1$

The capacity of a footing for flexural shear is given by ACI Equation (11–3) as

$$\phi V_c = 2\phi bd\lambda(f_c')^{0.5}$$

where

ϕ = strength reduction factor
= 0.75 from ACI Section 9.3
b = width of footing
d = effective depth
f_c' = concrete strength
λ = lightweight concrete modification factor

The capacity of a footing for punching shear is given by ACI Equation (11–31) as

$$\phi V_c = (2 + 4/\beta_c)\phi b_o d\lambda(f_c')^{0.5}$$
$$\leq 4\phi b_o d\lambda(f_c')^{0.5} \text{ ... ACI Equation (11–33)}$$

where

β_c = ratio of long side to short side of column
$= c_2/c_1$
b_o = length of critical perimeter for punching shear
$= 2(c_1 + c_2) + 4d$

Load transfer between a reinforced concrete column and the footing may be provided by the bearing capacity of the column and the footing. The bearing capacity of the column concrete at the interface is given by ACI Section 10.14.1 as

$$\phi P_n = 0.85\phi f_c' A_1$$

where

ϕ = strength reduction factor
= 0.65 from ACI Section 9.3
A_1 = area of column
f_c' = strength of column concrete

The bearing capacity of the footing concrete at the interface is given by ACI Section 10.14.1 as

$$\phi P_n = 0.85\phi f_c' A_1 (A_2/A_1)^{0.5}$$
$$\leq 0.85\phi f_c' A_1 \times 2$$

where

f_c' = strength of footing concrete
ϕ = strength reduction factor
A_2 = area of the base of the pyramid, with side slopes of 1:2, formed within the footing by the column base.

In accordance with ACI Section 15.8.1.2, when the bearing strength of the concrete is exceeded, reinforcement must be provided at the interface to transfer the excess load and the capacity of this reinforcement is

$$\phi P_s = \phi A_s f_y$$

where

ϕ = strength reduction factor specified in ACI Section 9.3.2.2 for axial compression in a column
f_y = reinforcement yield strength
A_s = reinforcement area

In addition, in accordance with ACI Section 15.8.2.1, a minimum area of reinforcement must be provided across the interface given by

$$A_{s(min)} = 0.005A_1$$

Example 14.30

For each of the following problems, the dead load D = 350 kips and live load L = 175 kips acting on a normal weight concrete column, and footing is shown in Exhibit 7. Reinforcement consists of Grade 60 bars, and the concrete strength is 4000 pounds per square inch.

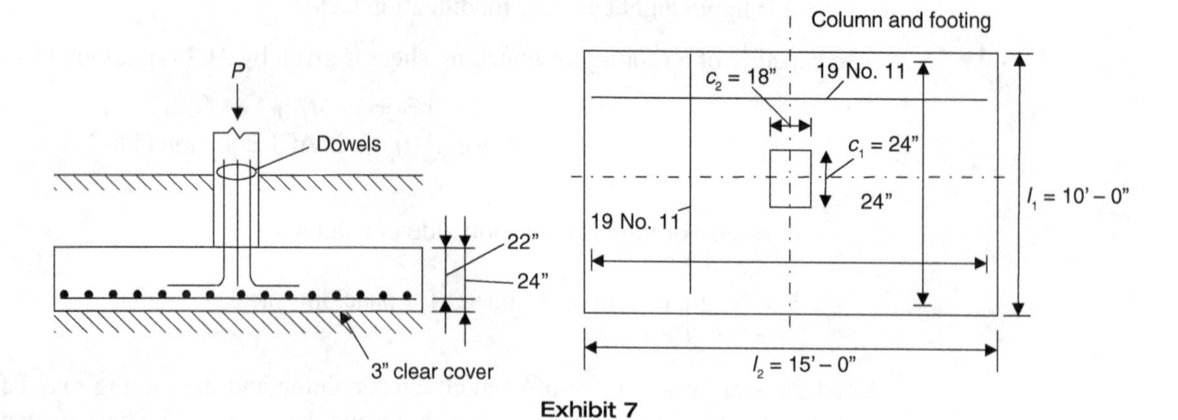

Exhibit 7

1. The factored pressure on the footing is most nearly:

   a). 4.0 kips/ft²

   b). 4.5 kips/ft²

   c). 5.0 kips/ft²

   d). 5.5 kips/ft²

2. The flexural shear acting on the footing, in the longitudinal direction, at the critical section is most nearly:

   a). 230 kips

   b). 240 kips

   c). 250 kips

   d). 260 kips

3. The design capacity of the footing for flexural shear, in the longitudinal direction, is most nearly:

   a). 230 kips

   b). 240 kips

   c). 250 kips

   d). 260 kips

4. The punching shear acting on the footing at the critical perimeter is most nearly:

   a). 640 kips

   b). 665 kips

   c). 690 kips

   d). 715 kips

5. The design capacity for punching shear at the critical perimeter is most nearly:

   a). 640 kips

   b). 665 kips

   c). 690 kips

   d). 715 kips

6. The bending moment acting on the footing, in the longitudinal direction, at the critical section is most nearly:

   a). 1000 kip-ft

   b). 1050 kip-ft

   c). 1100 kip-ft

   d). 1150 kip-ft

7. The design capacity for flexure at the critical section is most nearly:

   a). 2500 kip-ft

   b). 2600 kip-ft

   c). 2700 kip-ft

   d). 2800 kip-ft

8. Neglecting the vertical column reinforcement, the bearing capacity of the column concrete at the interface is most nearly:

   a). 800 kips

   b). 850 kips

   c). 900 kips

   d). 950 kips

9. Neglecting the vertical column reinforcement, the bearing capacity of the footing concrete at the interface is most nearly:

   a). 1800 kips

   b). 1850 kips

   c). 1900 kips

   d). 1950 kips

Solutions

1. The factored load on the footing is

$$\begin{aligned} P_u &= 1.2D + 1.6L \\ &= 1.2 \times 350 + 1.6 \times 175 \\ &= 700 \text{ kips} \end{aligned}$$

   The factored pressure on the footing is

$$\begin{aligned} q_u &= P_u/A_f \\ &= 700/(15 \times 10) \\ &= 4.67 \text{ kips per square ft} \end{aligned}$$

   The correct answer is b.)

2. The critical section for flexural shear in a footing, as defined by ACI Sections 15.5.2 and 11.1.3.1, is located a distance from the long side of the column equal to the effective depth. The distance of the critical section from the short edge of the footing is, then,

$$\begin{aligned} x &= \ell_2/2 - c_2/2 - d \\ &= 15/2 - 18/24 - 20/12 \\ &= 5.083 \text{ ft} \end{aligned}$$

The factored applied shear at the critical section is

$$\begin{aligned} V_u &= q_u \ell_1 x \\ &= 4.67 \times 10 \times 5.083 \\ &= 237 \text{ kips} \end{aligned}$$

The correct answer is b.)

3. The shear capacity of the normal weight concrete footing for beam action is given by ACI Equation (11-3) as

$$\begin{aligned} \phi V_c &= 2\phi \ell_1 d \lambda (f_c')^{0.5} \\ &= 2 \times 0.75 \times 10 \times 12 \times 22 \times 1.0(4000)^{0.5}/1000 \\ &= 250 \text{ kips} \end{aligned}$$

The correct answer is c).

4. The critical section for punching shear in a footing, as defined by ACI Sections 15.5.2 and 11.11.1.2, is located on the perimeter of a rectangle with sides of length

$$\begin{aligned} b_1 &= c_1 + d \\ &= 24 + 22 \\ &= 46 \text{ in.} \end{aligned}$$

and

$$\begin{aligned} b_2 &= c_2 + d \\ &= 18 + 22 \\ &= 40 \text{ in.} \end{aligned}$$

The factored applied shear at the critical perimeter is

$$\begin{aligned} V_u &= q_u(\ell_1 \ell_2 - b_1 b_2) \\ &= 4.67(10 \times 15 - 46 \times 40/144) \\ &= 641 \text{ kips} \end{aligned}$$

The correct answer is a).

5. The length of the critical perimeter is

$$\begin{aligned} b_o &= 2(b_1 + b_2) \\ &= 2(46 + 40) \\ &= 172 \text{ in.} \end{aligned}$$

The ratio of the long side to the short side of the column is

$$\begin{aligned} \beta &= c_1/c_2 \\ &= 24/18 \\ &= 1.33 \\ &< 2 \end{aligned}$$

Hence, the punching shear capacity of the footing is given by ACI Equation (11-33) as

$$\phi V_c = 4\phi b_o d\lambda(f_c')^{0.5}$$
$$= 4 \times 0.75 \times 172 \times 22 \times 1.0(4000)^{0.5}/1000$$
$$= 718 \text{ kips}$$

The correct answer is d).

6. The critical section for flexure in a footing, as defined by ACI Section 15.4.2, is located along the long side of the column. The factored applied moment at this section, for the longitudinal direction, is

$$M_u = q_u \ell_1(\ell_2/2 - c_2/2)^2/2$$
$$= 4.67 \times 10(15/2 - 18/24)^2/2$$
$$= 1064 \text{ kip-ft}$$

The correct answer is b).

7. The maximum reinforcement ratio for a tension-controlled section is

$$\rho_t = 0.319\beta_1 f_c'/f_y$$
$$= 0.319 \times 0.85 \times 4/60$$
$$= 0.0181$$

The reinforcement area provided in the footing in the longitudinal direction is

$$A_s = 19 \times 1.56$$
$$= 29.64 \text{ in.}^2$$

The reinforcement ratio provided is

$$\rho = A_s/(\ell_1 \times d)$$
$$= 29.64/(120 \times 22)$$
$$= 0.0112$$
$$< \rho_t \text{ ... the section is tension-controlled and } \phi = 0.9$$

The design flexural strength of the footing is

$$\phi M_n = \phi A_s f_y d(1 - 0.59\rho f_y/f_c')$$
$$= 0.9 \times 29.64 \times 60 \times 22(1 - 0.59 \times 0.0112 \times 60/4)/12$$
$$= 2644 \text{ kip-ft}$$

The correct answer is b).

8. Neglecting vertical reinforcement, the column load must be transferred to the footing by bearing on the concrete. The bearing capacity of the column concrete at the interface is given by ACI Section 10.14.1 as

$$\phi P_n = 0.85\phi f_c' A_1$$

where

ϕ = strength reduction factor
= 0.65 from ACI Section 9.3

A_1 = loaded area
= 24 × 18
= 432 square in.

Hence,

$$\phi P_n = 0.85 \times 0.65 \times 4 \times 432$$
$$= 955 \text{ kips}$$

The correct answer is d.

9. The area of the base of the pyramid, with side slopes of 1:2, formed within the footing by the loaded area is

$$A_2 = (c_1 + 4d)(c_2 + 4d)$$
$$= (24 + 4 \times 22)(18 + 4 \times 22)$$
$$= 11{,}872 \text{ square in.}$$

Hence,

$$(A_2/A_1)^{0.5} = (11{,}872/432)^{0.5}$$
$$= 5.2$$
$$> 2$$

Neglecting vertical reinforcement, the column load must be transferred to the footing by bearing on the concrete in the footing. The bearing capacity of the footing concrete is given by ACI Section 10.14.1 as

$$\phi P_n = 2 \times 0.85\phi f_c'A_1$$
$$= 2 \times 955$$
$$= 1910 \text{ kips}$$

The correct answer is c).

CHAPTER 15

Geotechnical Engineering

OUTLINE

PARTICLE SIZE DISTRIBUTION

The particle size distribution in a given soil is determined in the laboratory by sieve analysis and hydrometer analysis. For classification purposes, in coarse-grained soils the following two parameters can be obtained from a particle size distribution curve:

$$\text{Uniformity coefficient, } c_u = \frac{D_{60}}{D_{10}} \quad \textbf{(15.1)}$$

$$\text{Coefficient of gradation, } c_c = \frac{D_{30}^2}{D_{60} \times D_{10}} \quad \textbf{(15.2)}$$

where D_{10}, D_{30}, D_{60} = diameters through which, respectively, 10 percent, 30 percent, and 60 percent of the soil pass.

PARTICLE SIZE

Soils are assemblages of particles of various sizes and shapes with void spaces in between. They are formed primarily from decomposition of rocks. Based on the size of the particles present, soils can be described as gravel, sand, silt, or clay. Following are two grain size classification systems generally used by geotechnical engineers:

- System of the American Association of State Highway and Transportation Officials (AASHTO)
  Gravel: 75 mm to 2 mm
  Sand: 2 mm to 0.075 mm
  Silt and clay: Less than 0.075 mm
- Unified System
  Gravel: 76.2 mm to 4.75 mm
  Sand: 4.75 mm to 0.075 mm
  Silt and clay: Less than 0.075 mm

The particle size distribution in a given soil is determined in the laboratory by sieve analysis and hydrometer analysis.

A typical particle size distribution curve is shown in Figure 15.1. For classification purposes in coarse-grained soils, the following two parameters can be obtained from a particle size distribution curve:

- Uniformity coefficient:

$$C_c = \frac{D_{60}}{D_{10}} \quad \textbf{(15.3)}$$

- Coefficient of gradation:

$$C_z = \frac{D_{30}^2}{D_{10} \times D_{60}} \quad \textbf{(15.4)}$$

The coefficient of gradation is also sometimes referred to as the coefficient of curvature. The definitions of D_{10}, D_{30}, and D_{60} are shown in Figure 15.1.

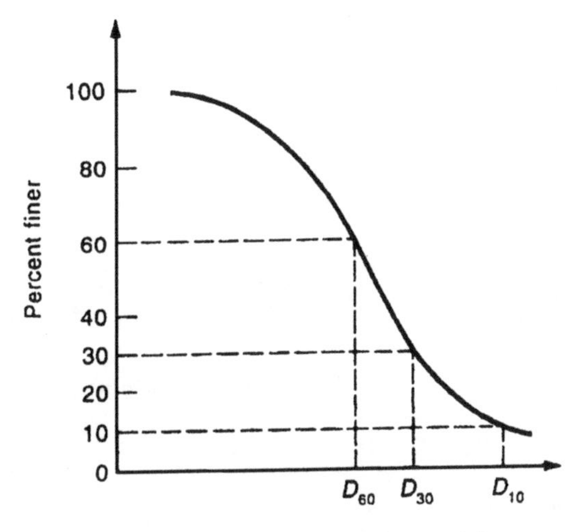

Figure 15.1

SPECIFIC GRAVITY OF SOIL SOLIDS, G_s

The specific gravity of soil solids is defined as

$$G_s = \frac{\text{unit weight of soil solids only}}{\text{unit weight of water}} \tag{15.5}$$

The general range of G_s for various soils is given in Table 15.1.

Table 15.1 General range of G_s for various soils

Soil Type	Range of G_s
Sand	2.63–2.67
Silt	2.65–2.7
Clay and silty clay	2.67–2.8
Organic soil	Less than 2

WEIGHT-VOLUME RELATIONSHIPS

Soils are three-phase systems containing soil solids, water, and air (Figure 15.2). Referring to Figure 15.2,

$$W = W_s + W_w \tag{15.6}$$

$$V = V_s + V_v = V_s + V_w + V_a \tag{15.7}$$

where W = total weight of the soil specimen, W_s = weight of the solids, W_w = weight of water, V = total volume of the soil, V_s = volume of soil solids, V_v = volume of voids, V_w = volume of water, and V_a = volume of air.

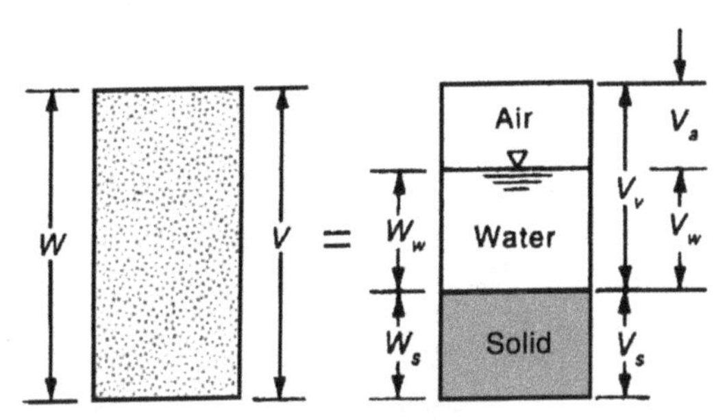

Figure 15.2

The *volume relationships* can then be given as follows:

$$\text{Void ratio} = e = \frac{V_v}{V_s} \tag{15.8}$$

$$\text{Porosity} = n = \frac{V_v}{V} \tag{15.9}$$

$$\text{Degree of saturation} = S = \frac{V_w}{V_v} \tag{15.10}$$

Similarly, the *weight relationships* are

$$\text{Moisture content} = w = \frac{W_w}{W_s} \tag{15.11}$$

$$\text{Moist unit weight} = \gamma = \frac{W}{V} \tag{15.12}$$

$$\text{Dry unit weight} = \gamma_d = \frac{W_s}{V} \tag{15.13}$$

Consider a soil sample with a unit volume of soil solids (Figure 15.3) to derive the following relationships:

1. $e = \frac{n}{1-n}$
2. $n = \frac{e}{1+e}$
3. $\gamma = \frac{G_s\gamma_w + wG_s\gamma_w}{1+e} = \frac{G_s\gamma_w(1+w)}{1+e}$
4. $\gamma_d = \frac{G_s\gamma_w}{1+e}$
5. $S = \frac{wG_s}{e}$

For *saturated soils*, $V_a = 0$ and $V_v = V_w$. Hence,

1. $S = 100\%$
2. $\gamma = \gamma_{sat} = \frac{\gamma_w(G_s + e)}{1+e} = \frac{\gamma_w(G_s + wG_s)}{1+wG_s}$
3. $\gamma_d = \frac{G_s\gamma_w}{1+e}$

In the preceding relationships, γ_w = unit weight of water = 62.4 lb/ft³ (or 9.81 KN/m³).

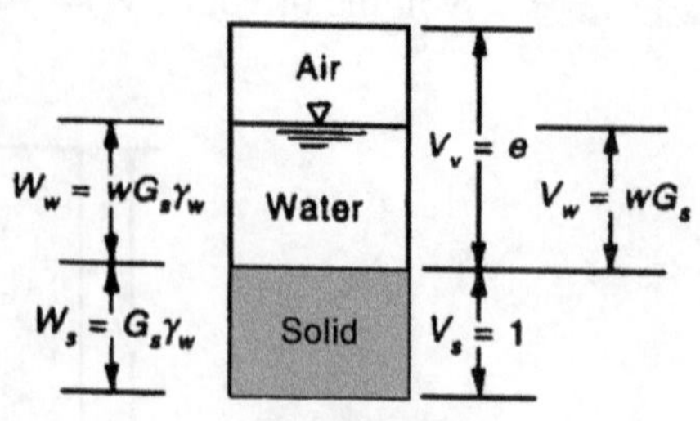

Figure 15.3

Example 15.1

A soil has a volume of 0.3 ft^3 and weighs 36 lb. Given $G_s = 2.67$ and moisture content $(w) = 18\%$, determine (a) moist unit weight (γ), (b) dry unit weight (γ_d), (c) void ratio (e), (d) porosity (n), and (e) degree of saturation (S).

Solution

a). $\gamma = \dfrac{W}{V} = \dfrac{36}{0.3} = 120 \text{ lb/ft}^3$

b). $\gamma_d = \dfrac{W_s}{V} = \dfrac{W}{(1+w)V} = \dfrac{36}{\left(1+\dfrac{18}{100}\right)0.3} = 101.7 \text{ lb/ft}^3$

c). $\gamma_d = \dfrac{G_s\gamma_w}{1+e}$

$$e = \frac{G_s\gamma_w}{\gamma_d} - 1 = \frac{(2.67)(62.4)}{101.7} - 1 = 0.64$$

d). $n = \dfrac{e}{1+e} = \dfrac{0.64}{1+0.64} = 0.39$

e). $S = \dfrac{wG_s}{e} = \dfrac{(0.18)(2.67)}{0.64} = 0.75(75\%)$

RELATIVE DENSITY

In granular soils, the degree of compaction is generally expressed by a nondimensional parameter called *relative density*, D_r, or

$$D_r = \frac{e_{\max} - e}{e_{\max} - e_{\min}} \quad \textbf{(15.14)}$$

where

e = actual void ratio in the field
$e_{\max}$ = void ratio in the loosest state
$e_{\min}$ = void ratio in the densest state

In many practical cases, a granular soil is qualitatively described by its relative density in the following manner.

D_r (%)	**Description**
0–15	Very loose
15–50	Loose
50–70	Medium
70–85	Dense
85–100	Very dense

Example 15.2

A sand has the following maximum and minimum dry unit weights:

$$\gamma_{d(\max)} = 17.29 \text{ kN/m}^3$$
$$\gamma_{d(\min)} = 15.41 \text{ kN/m}^3$$

Given $G_s = 2.66$ and dry unit weight in the field = 16.51 kN/m³, determine the relative density in the field.

Solution

$$\gamma_{d(\max)} = \frac{G_s \gamma_w}{1 + e_{\min}}$$

$$e_{\min} = \frac{G_s \gamma_w}{\gamma_{d(\max)}} - 1 = \frac{(2.66)(9.81)}{17.29} - 1 = 0.51$$

Similarly,

$$\gamma_{d(\min)} = \frac{G_s \gamma_w}{1 + e_{\max}}$$

$$e_{\min} = \frac{G_s \gamma_w}{\gamma_{d(\min)}} - 1 = \frac{(2.66)(9.81)}{15.41} - 1 = 0.69$$

Also,

$$e = \frac{G_s \gamma_w}{\gamma_{d(\text{field})}} - 1 = \frac{(2.66)(9.81)}{16.51} - 1 = 0.58$$

$$D_r = \frac{e_{\max} - e}{e_{\max} - e_{\min}} = \frac{0.69 - 0.58}{0.69 - 0.51} = 0.61 = 61\%$$

CONSISTENCY OF CLAYEY SOILS

When a cohesive soil is mixed with an excessive amount of water, it will be in a somewhat liquid state and flow like a viscous liquid. However, when this viscous liquid is gradually dried, with the loss of moisture it will pass into a plastic state. With further reduction of moisture, the soil will pass into a semisolid and then into a solid state. This is shown in Figure 15.4. The moisture content, in percent, at which the cohesive soil will pass from a liquid state to a plastic state is called the *liquid limit*. Similarly, the moisture contents at which the soil changes from a plastic state to a semisolid state and from a semisolid state to a solid state are referred to as the *plastic limit* and the *shrinkage limit*, respectively. These limits are referred to as the *Atterberg limits.*

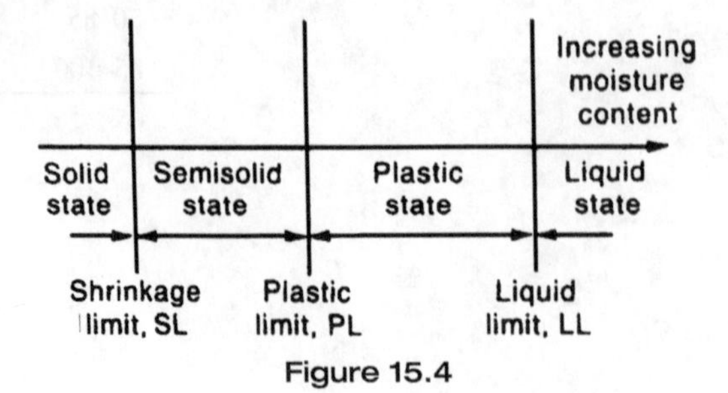

Figure 15.4

The liquid limit (LL) is the moisture content, in percent, at which the groove in a Casagrande liquid limit device closes for a distance of 0.5 in. after 25 blows. The plastic limit (PL) is the moisture content, in percent, at which the soil, when rolled into a thread of ⅛-in. diameter, crumbles. The *plasticity index* (PI) is defined as

$$PI = LL - PL \quad (15.15)$$

The shrinkage limit (SL) is the moisture content, in percent, at or below which the volume of the soil mass no longer changes from drying.

PERMEABILITY

The rate of flow of water through a soil of gross cross-sectional area A can be given by the following relationships (Figure 15.5);

$$v = ki \quad (15.16)$$

where
v = discharge velocity
k = coefficient of permeability
i = hydraulic gradient = h/L (see Figure 15.5)

$$q = vA = kiA \quad (15.17)$$

where
q = flow through soil in unit time
A = area of cross section of the soil at a right angle to the direction of flow

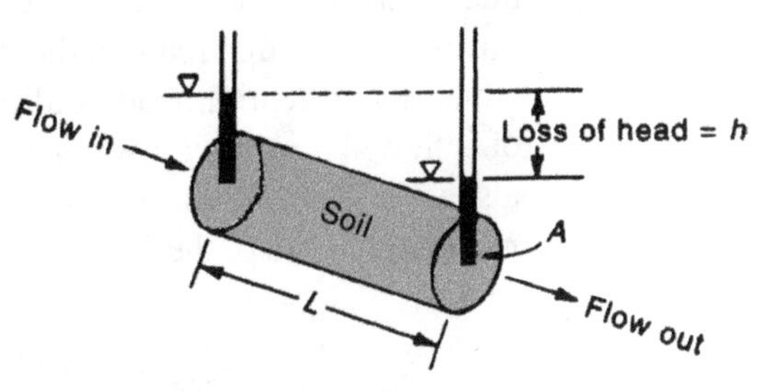

Figure 15.5

Equation (15.16) is known as *Darcy's law*. For granular soils, the coefficient of permeability can be estimated as

$$k \propto e^2 \quad (15.18)$$

and

$$k \propto \frac{e^3}{1+e} \quad (15.19)$$

where e is the void ratio. The range of the coefficient of permeability in various types of soil is given in Table 15.2.

Table 15.2 Range of k in various soils

Soil Type	Range of k (cm/sec)
Coarse sand	$1–10^{-2}$
Fine sand	$10^{-2}–10^{-3}$
Silt	$10^{-3}–10^{-5}$
Clay	$<10^{-6}$

Example 15.3

A sandy soil has a coefficient of permeability of 0.006 cm/sec at a void ratio of 0.5. Estimate k at a void ratio of 0.7.

Solution

From Equation (15.18),

$$\frac{k_1}{k_2} = \frac{e_1^2}{e_2^2}$$

so

$$\frac{0.006}{k_2} = \frac{(0.5)^2}{(0.7)^2}$$

$$k_2 = \frac{(0.006)(0.7)^2}{(0.5)^2} = 0.0118 \text{ cm/s}$$

FLOW NETS

In many cases, flow of water through soil varies in direction and in magnitude over the cross section. In those cases, calculation of rate of flow of water can be made by using a graph called a *flow net*. A flow net is a combination of a number of flow lines and equipotential lines. A flow line is a line along which a water particle will travel from the upstream to the downstream side. An equipotential line is one along which the potential head at all points is the same. Figure 15.6 shows an example of a flow net in which water flows from the upstream to the downstream around a sheet pile. Note that in a flow net the flow lines and equipotential lines cross at right angles. Also, the flow elements constructed are approximately square.

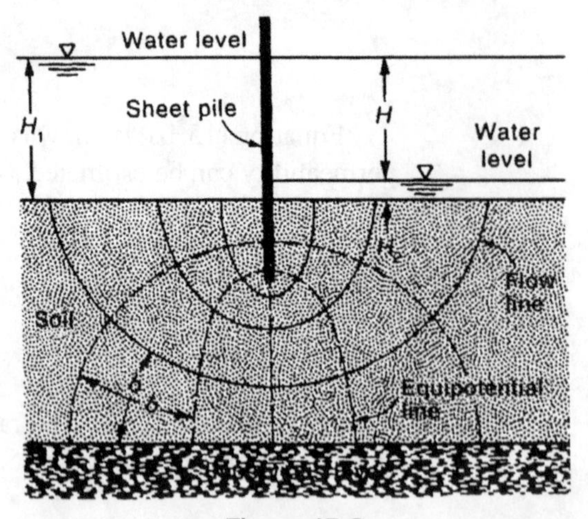

Figure 15.6

Referring to Figure 15.6, the flow in unit time (q) per unit length normal to the cross section shown is

$$q = k\frac{N_f}{N_d}H \tag{15.20}$$

where

N_f = number of flow channels
N_d = number of drops
H = head difference between the upstream and downstream side

(*Note:* In Figure 15.6, $N_f = 4$ and $N_d = 6$.)

Example 15.4

Refer to the flow net shown in Figure 15.6. Given $k = 0.001$ ft/min, $H_1 = 30$ ft, and $H_2 = 5$ ft, determine the seepage loss per day per foot under the sheet pile structure.

Solution

$$q = k\frac{N_f}{N_d}H = (0.001 \times 60 \times 24 \text{ ft/day})\left(\frac{4}{6}\right)(30 - 5) = 24 \text{ ft}^3\text{/day/ft}$$

EFFECTIVE STRESS

The total stress, σ, at a point in a soil mass is the sum of two components: (1) pore water pressure, u, and (2) effective stress, σ'. Thus

$$\sigma = \sigma' + u \qquad \textbf{(15.21)}$$

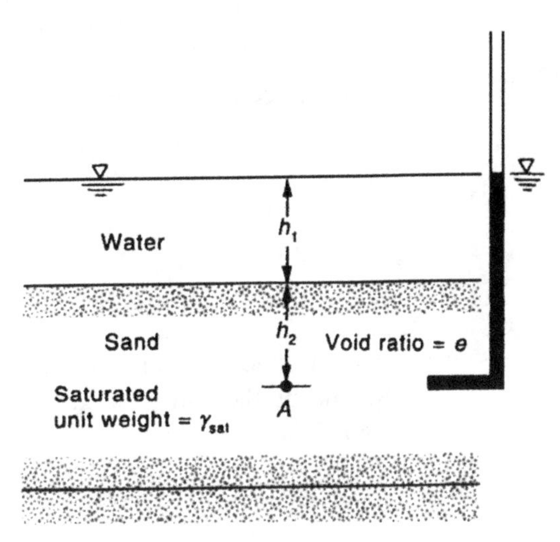

Figure 15.7

The effective stress is the sum of the vertical components of the forces developed at the points of contact of the solid particles per unit cross section of the soil mass. Referring to Figure 15.7, at point A

$$\sigma = h_1\gamma_w + h_2\gamma_{sat}$$
$$u = \gamma_w(h_1 + h_2)$$

so

$$\sigma' = \sigma - u = (h_1\gamma_w + h_2\gamma_{sat}) - \gamma_w(h_1 + h_2) = h_2(\gamma_{sat} - \gamma_w) = h_2\gamma'$$

In the preceding relationships, γ_w = unit weight of water, γ_{sat} = saturated unit weight of soil, and $\gamma' = \gamma_{sat} - \gamma_w$ = effective unit weight of soil. From the section on weight-volume relationships,

$$\gamma_{sat} = \frac{\gamma_w (G_s + e)}{1+e}$$

so

$$\gamma' = \gamma_{sat} - \gamma_w = \frac{\gamma_w (G_s + e)}{1+e} - \gamma_w = \frac{\gamma_w (G_s - 1)}{1+e} \quad \textbf{(15.22)}$$

For a quicksand condition, for example, when the effective stress $\sigma' = 0$, the hydraulic gradient is given as

$$i = i_{cr} = \frac{\gamma'}{\gamma_w} = \frac{G_s - 1}{1+e} \quad \textbf{(15.23)}$$

Example 15.5

Refer to Figure 15.7. For the soil: void ratio $e = 0.5$, $G_s = 2.67$, $h_1 = 1.5$ m, $h_2 =$ 3.05 m, determine the effective stress at A.

Solution

$$\gamma' = \frac{\gamma_w (G_s - 1)}{1+e} = \frac{9.81(2.67-1)}{1+0.5} = 10.92 \text{ kN/m}^3$$

So the effective stress is

$$\sigma' = h_2 \gamma' = (3.05)(10.92) = 33.31 \text{ kN/m}^2$$

Example 15.6

For the sandy soil shown in Figure 15.7, if there is an upward flow of water, what should be the hydraulic gradient for the quicksand condition? Given: $G_s = 2.65$ and $e = 0.7$.

Solution

For the quicksand condition

$$i_{cr} = \frac{\gamma'}{\gamma_w} = \frac{G_s - 1}{1+e} = \frac{2.65-1}{1+0.7} = 0.97$$

VERTICAL STRESS UNDER A FOUNDATION

Boussinesq (1883) proposed an equation to determine the increase of vertical stress (Δp) at a point (A) in a soil mass due to a point load (Q) on the surface, which can be expressed as (Figure 15.8)

$$\Delta p = \frac{3Q}{2\pi} \frac{z^3}{\left(r^2 + z^2\right)^{5/2}} \quad \textbf{(15.24)}$$

where

$$r = \sqrt{x^2 + y^2}\,.$$

Equation (15.24) can be used to determine the stress increase below the center of a *flexible* rectangular foundation. In Figure 15.9(a), the plan of a rectangular flexible foundation is shown. The length and the width of the foundation are L and B, respectively. The uniformly distributed load on the foundation is q. Figure 15.9(b) shows the increase of stress (Δp) *below the center of the foundation* [point A in Figure 15.9(a)] due to the distributed load q. In Figure 15.9(b), z is the vertical distance below the foundation.

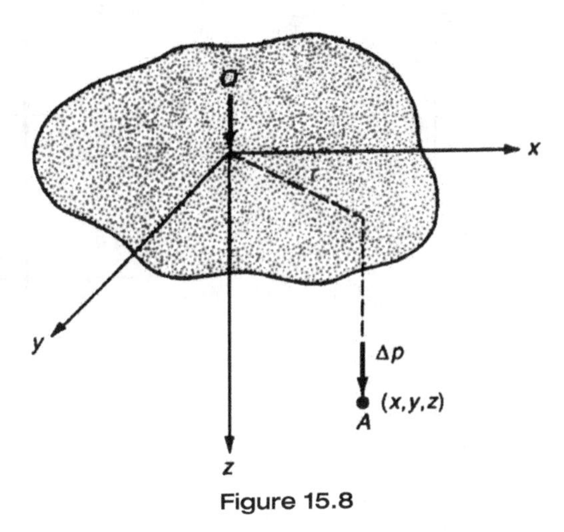

Figure 15.8

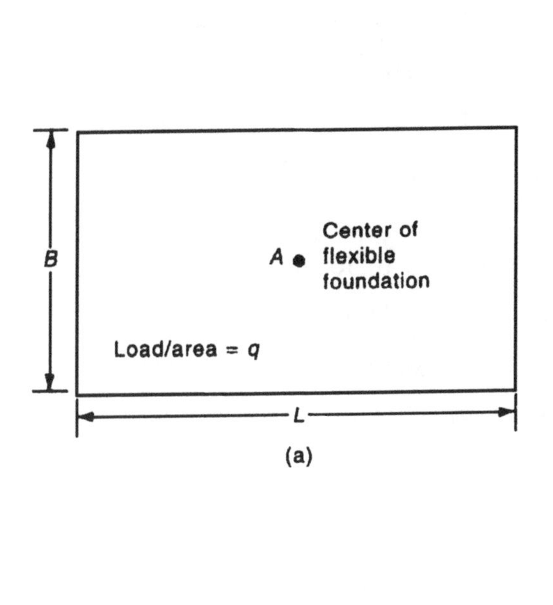

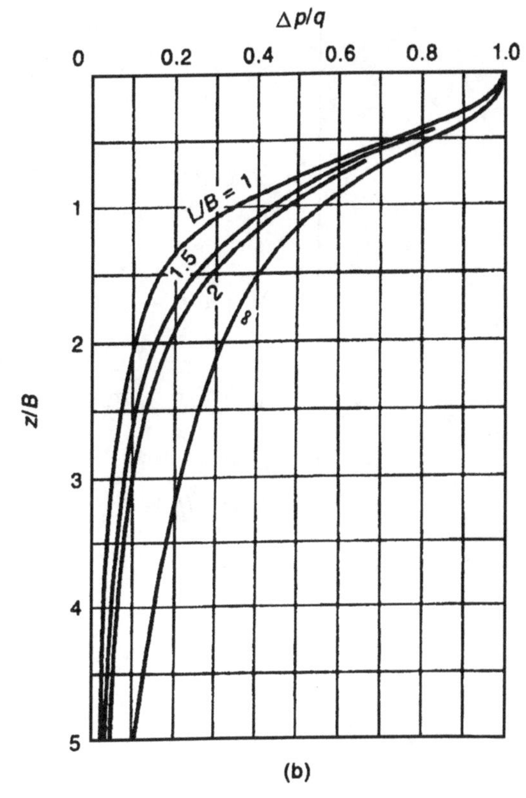

Figure 15.9

Example 15.7

Exhibit 1 shows a square foundation. The distributed load on the foundation, q, is 4000 lb/ft^2. Determine the average increase of stress in the clay layer, which has a thickness of 10 ft.

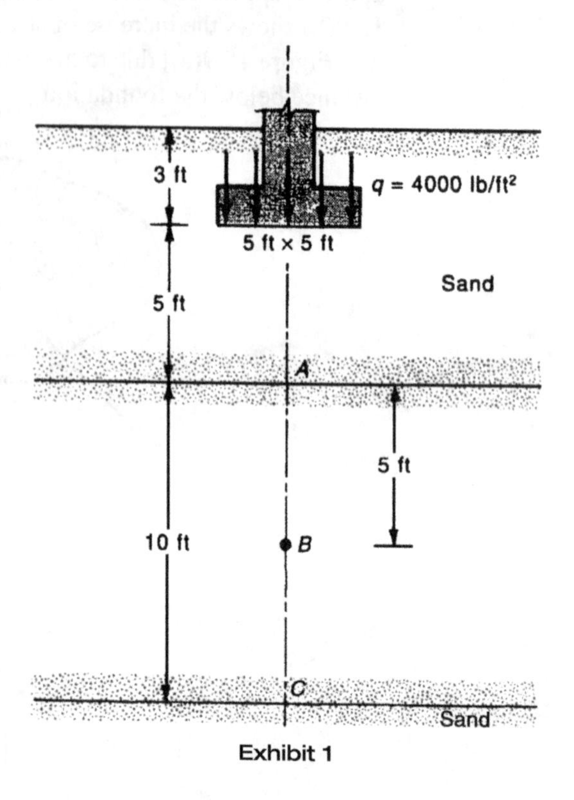

Exhibit 1

Solution

$$\Delta p_{\text{av}} = \frac{1}{6}(\Delta p_A + 4\Delta p_B + \Delta p_C)$$

At A, $z/B = 5/5 = 1$. From Figure 15.9(b), for $z/B = 1$, $\Delta p/q \approx 0.325$.

$$\Delta p_A = (0.325)(4000) = 1300 \text{ lb/ft}^2$$

At B, $z/B = 10/5 = 2$. For $z/B = 2$, $\Delta p/q \approx 0.1$.

$$\Delta p_B = (0.1)(4000) = 400 \text{ lb/ft}^2$$

At C, $z/B = 15/5 = 3$. For $z/B = 3$, $\Delta p/q \approx 0.04$.

$$\Delta p_C = (0.04)(4000) = 160 \text{ lb/ft}^2$$

$$\Delta p_{\text{av}} = \frac{1}{6}\left[1300 + (4)(400) + 160\right] = 510 \text{ lb/ft}^2$$

CONSOLIDATION

Consolidation settlement is the result of volume change in saturated clayey soils due to the expulsion of water occupied in the void spaces. In soft clays, the major portion of the settlement of a foundation may be due to consolidation. Based on the theory of consolidation, a soil can be divided into two major categories:

- *Normally consolidated clay.* In this case, the *present effective overburden pressure* is the maximum pressure to which the soil has been subjected in the recent geologic past.

- *Overconsolidated or preconsolidated clay.* In this case, the present effective overburden pressure is less than what the soil has encountered in the past. The past maximum effective overburden pressure is referred to as the *preconsolidation pressure* (p_c). The *overconsolidation ratio* (*OCR*) is defined as

$$OCR = \frac{p_c}{p} \tag{15.25}$$

where p_c is the preconsolidation pressure and p is the effective overburden pressure.

If a normally consolidated soil specimen is collected at a point A as shown in Figure 15.10(a), the nature of variation of void ratio (e) with the effective pressure (p) in the field will be as shown by the curve in Figure 15.10(b). If the soil at A [Figure 15.10(a)] is preconsolidated, the nature of e versus p in the field will be as shown by the curve in Figure 15.10(c). Note the slopes of the lines of the e versus log p plots. The following empirical equations may be used to estimate the slopes C_c and C_s:

$$C_c = \text{compression index} = 0.009(\text{LL} - 10) \tag{15.26}$$

where LL = liquid limit, in percent.

$$C_s = \text{swell index} = \frac{1}{5} \text{ to } \frac{1}{6} C_c \tag{15.27}$$

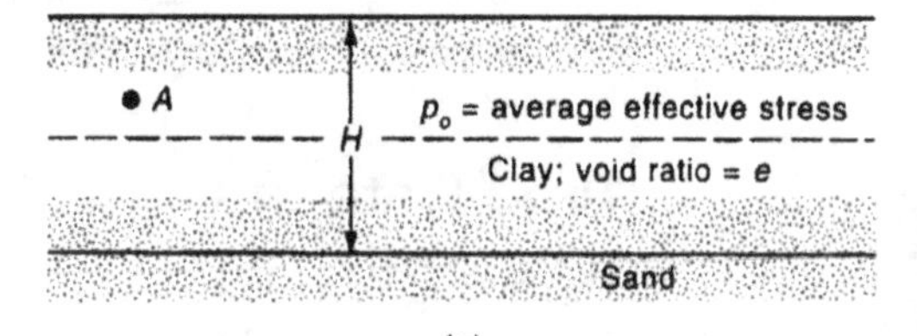

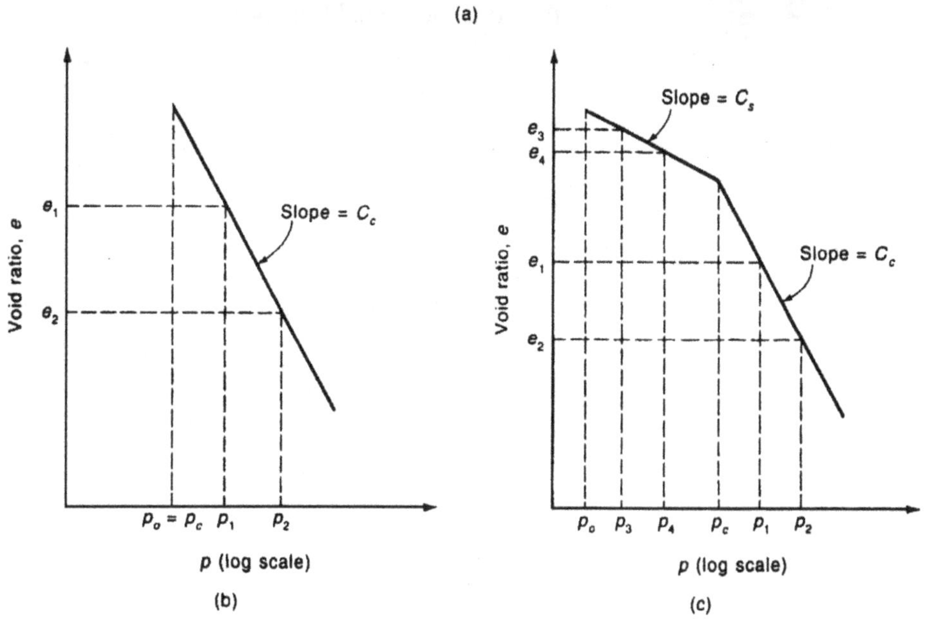

Figure 15.10

Calculation of Consolidation Settlement

Consolidation settlement of a saturated clay layer of thickness H [Figure 15.10(a)] can be calculated by the following procedure:

1. Calculate the *effective* overburden pressure, p_0, at the middle of the clay layer.
2. Determine the preconsolidation pressure (p_c) of the clay layer.
3. Estimate C_c and C_s by using Equations (15.26) and (15.27), or calculate [Figure 15.10(b) and (c)].

$$C_c = \frac{e_1 - e_2}{\log(p_2/p_1)} \tag{15.28}$$

$$C_s = \frac{e_3 - e_4}{\log(p_3/p_4)} \tag{15.29}$$

4. Calculate settlement, S, by using one of the following equations:

$$S = \frac{C_c H}{1+e_o}\log\left(\frac{p_o + \Delta p}{p_o}\right) \tag{15.30}$$

(for $p_o = p_c$ for normally consolidated clay)

$$S = \frac{C_s H}{1+e_o}\log\left(\frac{p_o + \Delta p}{p_o}\right) \tag{15.31}$$

(for $p_o + \Delta p \le p_c$)

$$S = \frac{C_s H}{1+e_o}\log\left(\frac{p_c}{p_o}\right) + \frac{C_c H}{1+e_o}\log\left(\frac{p_o + \Delta p}{p_c}\right) \tag{15.32}$$

(for $p_o < p_c < p_o + \Delta p$) where e_o = initial void ratio of the clay and Δp = average increase of pressure in the clay layer.

Time Rate of Consolidation

The average degree of consolidation, U, of a saturated clay layer is a function of the nondimensional time factor, T_v, or

$$U = f(T_v) \tag{15.33}$$

$$T_v = \frac{c_v t}{H_d^2} \tag{15.34}$$

where

c_v = coefficient of consolidation
t = time
H_d = length of the drainage path

For two-way drainage $H_d = H/2$, and for one-way drainage $H_d = H$ (see Figure 15.11).

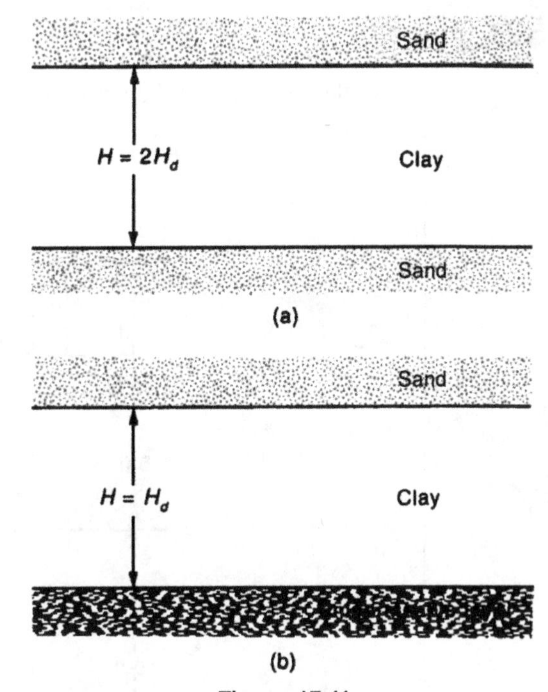

Figure 15.11

The coefficient of consolidation is defined as follows:

$$c_v = \frac{k}{\gamma_w \left(\frac{\Delta e / \Delta p}{1 + e_o} \right)} \quad \textbf{(15.35)}$$

where k = coefficient of permeability of the clay layer
Δe = change in void ratio due to an average change of pressure, Δp

The variation of U with T_v is given in Table 15.3.

Table 15.3 Variation of U with T_v

U (%)	T_v
0	0
10	0.008
20	0.031
30	0.071
40	0.126
50	0.197
60	0.287
70	0.403
80	0.567
90	0.848
100	∞

Example **15.8**

For the shallow foundation shown in Exhibit 2, determine (a) the consolidation settlement and (b) the time for 50% consolidation.

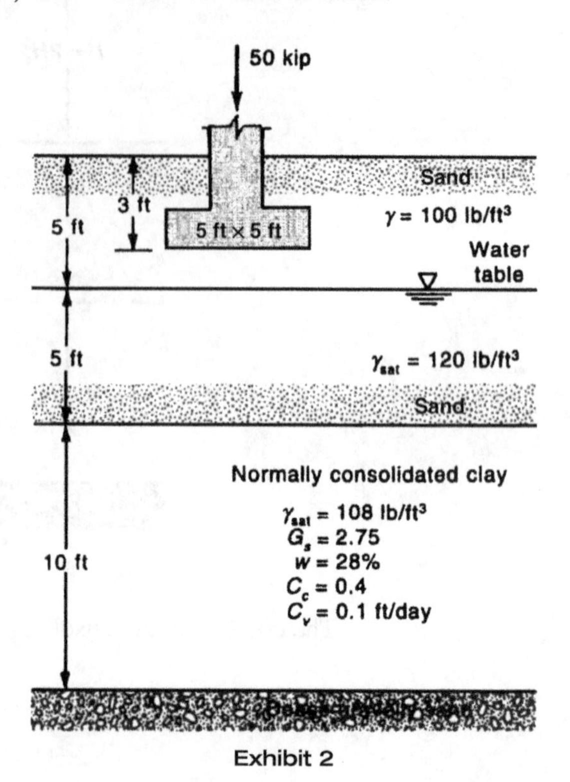

Exhibit 2

Solution

a). For normally consolidated clay

$$S = \frac{C_c H}{1 + e_o} \log \frac{p_o + \Delta p}{p_o}$$

At the middle of the clay layer,

$$\begin{aligned} p_o &= (100)(5) + (120 - 62.4)(5) + (108 - 62.4)(10/2) \\ &= 500 + 288 + 228 = 1016 \text{ lb/ft}^2 \\ e_o &= wG_s = (0.28)(2.75) = 0.77 \end{aligned}$$

Referring to Figure 15.10(b), at the top of the clay layer, $z/B = (10 - 3)/5 = 1.4$.

$$\Delta p_{\text{top}} = 0.16q = (0.16)(50{,}000/25) = 320 \text{ lb/ft}^2$$

At the middle of the clay layer, $z/B = (7 + 5)/5 = 2.4$.

$$\Delta p_{\text{middle}} = 0.07q = (0.07)(2000) = 140 \text{ lb/ft}^2$$

At the bottom of the clay layer, $z/B = (7 + 10)/5 = 3.4$.

$$\Delta p_{\text{bottom}} \approx 0.03q = (0.03)(2000) = 60 \text{ lb/ft}^2$$

$$\Delta p = \Delta p_{\text{av}} = (1/6)[320 + 4(140) + 60] = 156.6 \text{ lb/ft}^2$$

Hence

$$S = \frac{(0.4)(10)}{1 + 0.77} \log\left(\frac{1016 + 156.5}{1016}\right) = 0.14 \text{ ft} = 1.69 \text{ in.}$$

b). $T_v = \frac{c_v t}{H_d^2}$

For 50% consolidation, $T_v = 0.197$.

$$0.197 = \frac{\left(0.1\ \text{ft}^2/\text{day}\right)t}{\left(\frac{10}{2}\right)^2}$$

$$t = \frac{(0.197)\left(\frac{10}{2}\right)^2}{0.1} = 49.25\ \text{days}$$

SHEAR STRENGTH

The shear strength of a soil (s), in general, is given by the *Mohr-Coulomb failure criteria*, or

$$s = c + \sigma' \tan\phi \qquad \textbf{(15.36)}$$

where

c = cohesion
σ' = effective normal stress
ϕ = drained friction angle

For sands, $c = 0$, and the magnitude of ϕ varies with the relative density, size, and shape of the soil particles. A general range of the magnitude of ϕ is given in Table 15.4.

Table 15.4 General range of soil friction angle ϕ for sand

Type	Nature of Compaction	ϕ (deg)
Round grained	Loose	28–32
	Medium	30–36
	Dense	35–40
Angular grained	Loose	30–35
	Medium	35–40
	Dense	40–45

For normally consolidated clays, $c = 0$. So $s = \sigma' \tan\phi$. However, for overconsolidated clays, $c \neq 0$; thus $s = c + \sigma' \tan\phi$. An important concept for the shear strength of cohesive soils is the so-called $\phi = 0$ concept. This is the condition where drainage from the soil does not take place during loading. For such a case

$$s = c_u \qquad \textbf{(15.37)}$$

where c_u is the undrained shear strength.

The unconfined compression strength, q_u, of a cohesive soil is (Figure 15.12)

$$q_u = 2c_u \qquad \textbf{(15.38)}$$

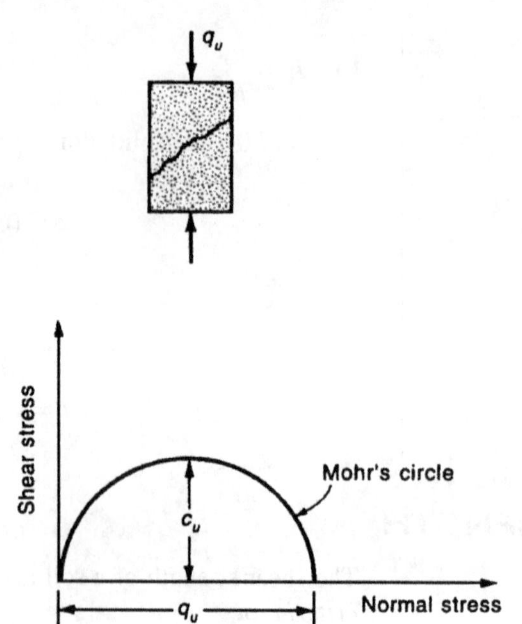

Figure 15.12

A general range of q_u with the consistency of the soil is given in Table 15.5.

Table 15.5 Range of unconfined compression strength of cohesive soils

Consistency	q_u (lb/ft²)	q_u (kN/m²)
Very soft	0–500	0–24
Soft	500–1000	24–48
Medium	1000–2000	48–96
Stiff	2000–4000	96–192
Very stiff	4000–8000	192–384

LATERAL EARTH PRESSURE

The lateral earth pressure behind a retaining wall can be one of the following three types:

- At-rest
- Active
- Passive

At-Rest Type

In the at-rest type of lateral earth pressure the retaining wall does not move away from or toward the backfill soil (Figure 15.13). Thus

$$\sigma_h' = K_o \sigma_v' \tag{15.39}$$

where

σ_h' = effective horizontal pressure on the wall
σ_v' = effective vertical pressure at a given depth
K_o = at-rest earth pressure coefficient

$$K_o = 1 - \sin\phi \text{ (for granular soil)} \tag{15.40}$$

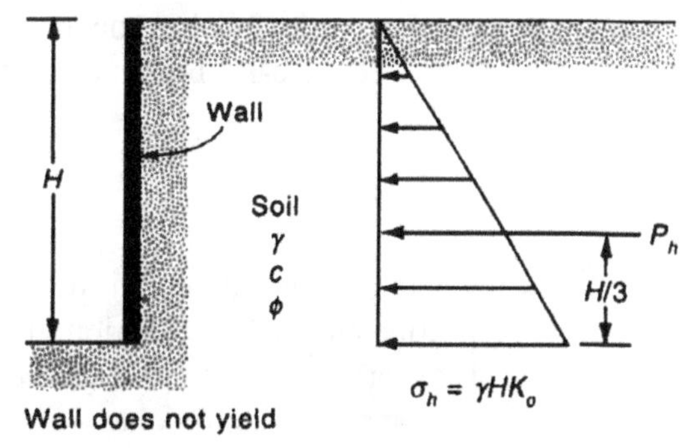

Figure 15.13 At-rest earth pressure

Thus, the total lateral force, P_h, per unit length of the wall (with dry granular soil as backfill) can be given as

$$P_h = \frac{1}{2} K_o \gamma_d H^2 \tag{15.41}$$

where H is the height of the wall
γ_d = dry unit weight of the soil

Active Type

Figure 15.14 shows the condition where the retaining wall moves away from the soil mass. For a frictionless retaining wall, the *Rankine active pressure* (σ'_a) is given as

$$\sigma'_a = K_a \sigma'_v - 2c\sqrt{K_a} \tag{15.42}$$

where c = cohesion, ϕ = soil friction angle, and

$$K_a = \tan^2(45 - \phi/2) \tag{15.43}$$

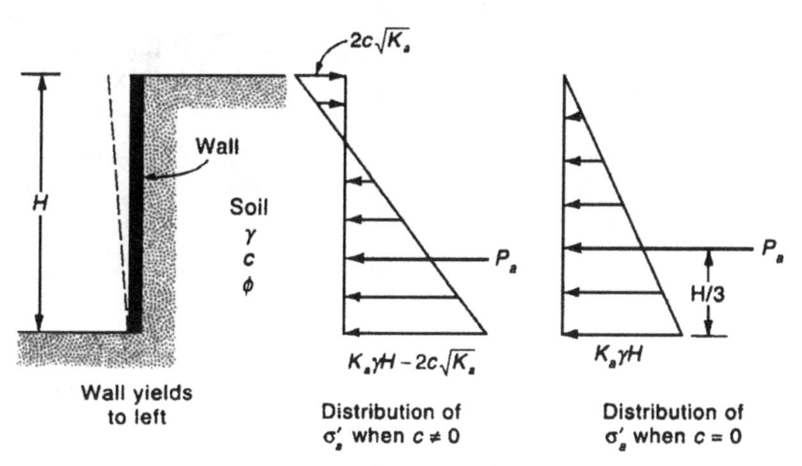

Figure 15.14 Active earth pressure

For most retaining wall construction, a granular backfill is used for drainage. In that case, $c = 0$. Hence

$$\sigma'_a = K_a \sigma'_v = \tan^2(45 - \phi/2)\sigma'_v$$

Referring to Figure 15.14, for a *dry granular backfill*

$$P_a = \frac{1}{2} \gamma H^2 \tan^2(45 - \phi/2) \tag{15.44}$$

For an inclined granular backfill and *frictionless wall* (Figure 15.15), Rankine's active earth pressure coefficient is

$$K_a = \cos\alpha \left(\frac{\cos\alpha - \sqrt{\cos^2\alpha - \cos^2\phi}}{\cos\alpha + \sqrt{\cos^2\alpha - \cos^2\phi}} \right) \quad \textbf{(15.45)}$$

where α is the angle that the backfill makes with the horizontal, and ϕ is the soil friction angle. The variation of K_a as expressed by Equation (15.45) with α and ϕ is given in Table 15.6. This table is important in the calculation of active earth pressure and the stability check of the retaining wall.

Table 15.6 Variation of K_a [Equation (15.45)]

	$\phi°$				
$a°$	30	32	34	36	40
0	0.361	0.307	0.283	0.260	0.217
5	0.366	0.311	0.286	0.262	0.219
10	0.380	0.321	0.294	0.270	0.225
15	0.409	0.341	0.311	0.283	0.235
20	0.461	0.374	0.338	0.306	0.250
25	0.573	0.434	0.385	0.343	0.275

The active force (with dry granular backfill) per unit length of the wall is given as

$$P_a = \frac{1}{2} K_a \gamma H^2 \quad \textbf{(15.46)}$$

In Figure 15.15, note the direction of the resultant active force.

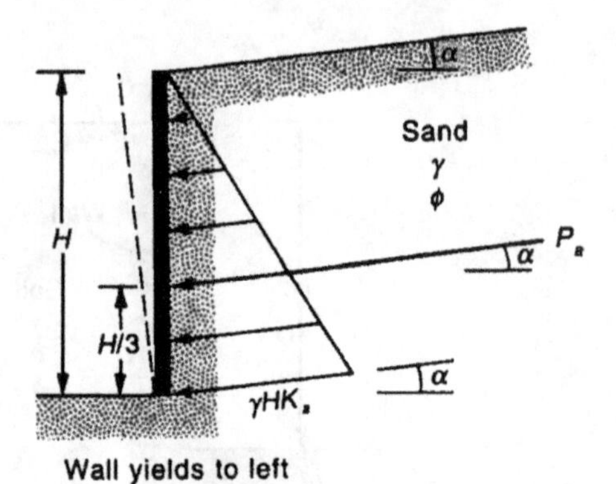

Figure 15.15

Passive Type

Figure 15.16 illustrates the case where the wall moves into the soil. If the wall is frictionless, the Rankine passive earth pressure at a given depth can be given as

$$\sigma'_p = K_p \sigma'_v + 2c\sqrt{K_p} \quad \textbf{(15.47)}$$

where σ'_p = passive earth pressure, σ'_v = effective vertical pressure at a certain depth, c = cohesion, and

K_p = Rankine passive earth pressure coefficient = $\tan^2(45 + \phi/2)$ **(15.48)**

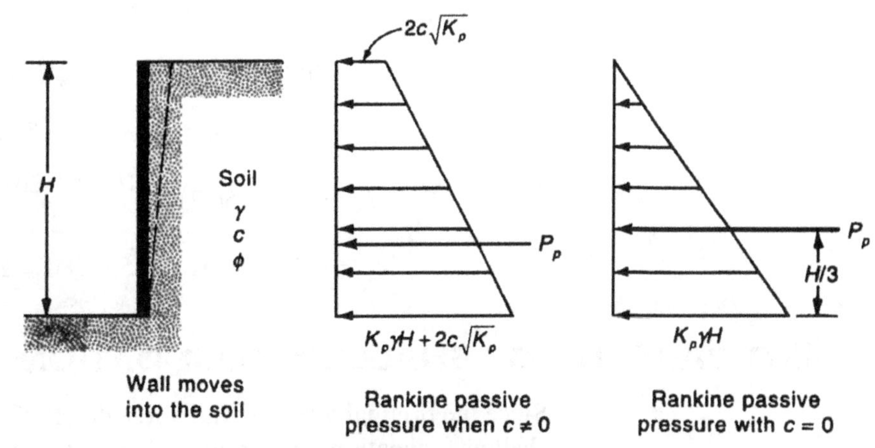

Figure 15.16 Passive earth pressure

Referring to Figure 15.16, for a dry granular backfill, $c = 0$, and

$$P_p = \frac{1}{2}\gamma H^2 \tan^2(45 + \phi/2) \tag{15.49}$$

where P_p = Rankine passive force per unit length of the wall.

Example 15.9

A retaining wall with a granular backfill is shown in Exhibit 3(a). Determine the Rankine active force per unit length of the wall P_a.

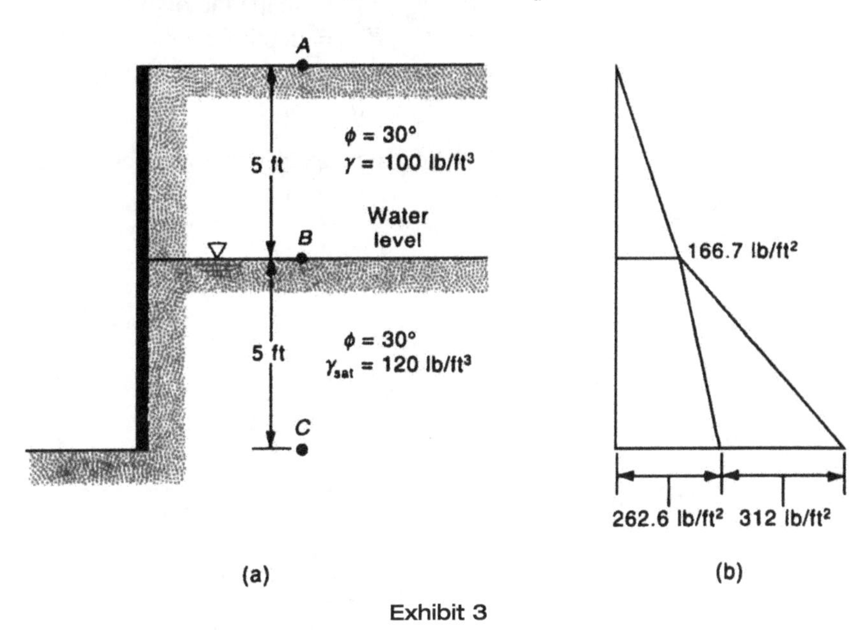

Exhibit 3

Solution

$$K_a = \tan^2(45 - \phi/2) = \tan^2(45 - 30/2) = 1/3$$

At level A:

$$\sigma'_a = 0$$

At level B:

$$\sigma'_a = (100)(5)(1/3) = 166.7 \text{ lb/ft}^2$$

$$\sigma_w = \text{ hydrostatic pressure} = 0$$

At level C:

$$\sigma'_a = [(100)(5)+(120-62.4)(5)](1/3) = 262.6 \text{ lb/ft}^2$$

$$\sigma_w = \text{ hydrostatic pressure} = (5)(62.4) = 312 \text{ lb/ft}^2$$

$$P_a = \frac{1}{2}(5)(166.7) + \left(\frac{166.7+262.6}{2}\right)(5) + \frac{1}{2}(5)(312)$$

$$= 416.75 + 1073.25 + 780 = 2270 \text{ lb/ft}$$

BEARING CAPACITY OF SHALLOW FOUNDATIONS

Since the original work of Karl Terzaghi (1943), the bearing capacity theories for shallow foundations have gone through extensive study. Figure 15.17 shows a shallow foundation that has a width B and length L. The depth of the foundation is D_f. The ultimate bearing capacity of the foundation can be estimated by the following equation:

$$q_u = cF_{cs}F_{cd}N_c + qF_{qs}F_{qd}N_q + \frac{1}{2}\gamma BF_{\gamma s}F_{\gamma d}N_\gamma \tag{15.50}$$

where

c = cohesion
γ = unit weight of soil
B = width of the foundation
F_{cs}, F_{qs}, $F_{\gamma s}$ = shape factors
F_{cd}, F_{qd}, $F_{\gamma d}$ = depth factors
N_c, N_q, N_γ = bearing capacity factors
$q = \gamma D_f$

The bearing capacity factors (Vesic, 1973) are

$$N_q = e^{\pi \tan\phi} \tan^2(45 + \phi/2) \tag{15.51}$$

$$N_c = (N_q - 1)\cot\phi \tag{15.52}$$

$$N_\gamma = 2(N_q + 1)\tan\phi \tag{15.53}$$

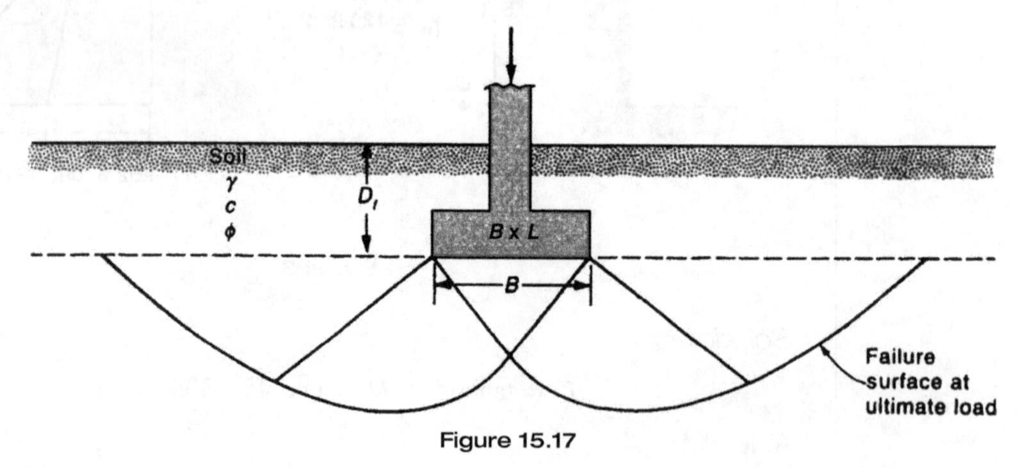

Figure 15.17

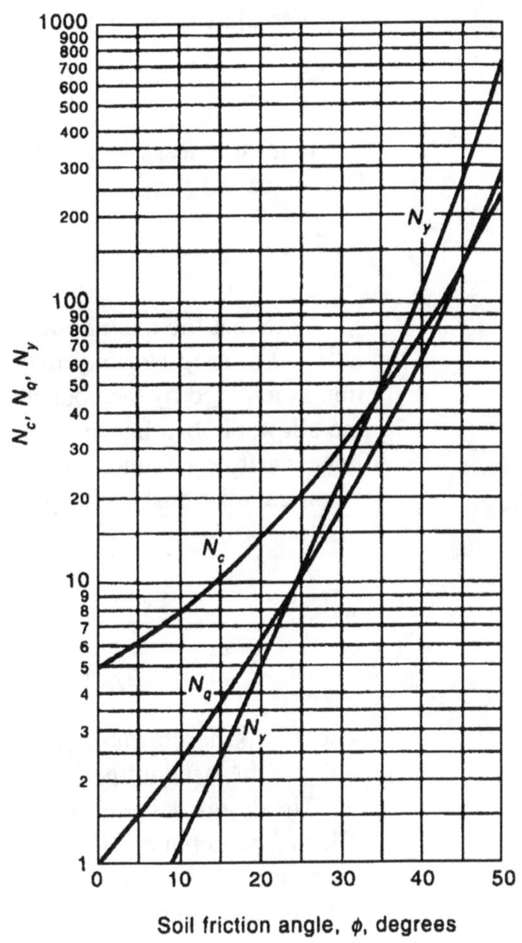

Figure 15.18

The variations of N_q, N_c, and N_γ with soil friction angle ϕ are shown in Figure 15.18.

The shape factors for rectangular foundations are as follows (DeBeer, 1970):

$$F_{cs} = 1 + \left(\frac{B}{L}\right)\left(\frac{N_q}{N_c}\right) \tag{15.54}$$

$$F_{qs} = 1 + \left(\frac{B}{L}\right)\tan\phi \tag{15.55}$$

$$F_{\gamma s} = 1 - 0.4\left(\frac{B}{L}\right) \tag{15.56}$$

The depth factors ($D_f \leq B$) are as follows (Hansen, 1970):

$$F_{qd} = 1 + 2\tan\phi(1 - \sin\phi)^2\left(\frac{D_f}{B}\right) \tag{15.57}$$

$$F_{cd} = F_{qd} - \frac{1 - F_{qd}}{N_q \tan\phi} \tag{15.58}$$

$$F_{\gamma d} = 1 \tag{15.59}$$

For the $\phi = 0$ condition,

$$F_{\gamma d} = 1 + 0.4\left(\frac{D_f}{B}\right) \tag{15.60}$$

A shallow foundation is generally accepted as one in which

$$\frac{D_f}{B} \leq 1 \tag{15.61}$$

In most cases, a factor of safety (F_s) of 3 to 5 is used to estimate the allowable bearing capacity, q_{all},

$$q_{\text{all}} = \frac{q_u}{F_s} \tag{15.62}$$

Note that the magnitude of q_{all} obtained by applying a factor of safety to the ultimate bearing capacity may be too high to control the settlement of a foundation as required by the code and designer. For that reason, it is important to check the allowable bearing pressure based upon settlement considerations. This bearing capacity for foundations supported on sand can be obtained by the following empirical relationships, as proposed by Meyerhof (1956):

$$q_{\text{all}}\left(\text{kip/ft}^2\right) = \frac{N}{4} \quad \left(\text{for } B \leq 4 \text{ ft}\right) \tag{15.63}$$

and

$$q_{\text{all}}\left(\text{kip/ft}^2\right) = \frac{N}{4}\left(\frac{B+1}{B}\right)^2 \quad \left(\text{for } B > 4 \text{ ft}\right) \tag{15.64}$$

where q_{all} is the allowable bearing capacity for a 1-inch settlement of the foundation, and N is the standard penetration number obtained during field soil exploration.

The design value of N should be estimated by taking into consideration the N values for a depth of $2B$ to $3B$, measured from the bottom of the foundation. Cone penetration resistance values can also be used to estimate q_{all} (for a 1-inch settlement of the foundation). This was also proposed by Meyerhof (1956).

$$q_{\text{all}}\left(\text{lb/ft}^2\right) = \frac{q_c\left(\text{lb/ft}^2\right)}{15} \quad \left(\text{for } B \leq 4 \text{ ft}\right) \tag{15.65}$$

$$q_{\text{all}}\left(\text{lb/ft}\right)^2 = \frac{q_c\left(\text{lb/ft}^2\right)}{25}\left(\frac{B+1}{B}\right)^2 \quad \left(\text{for } B > 4 \text{ ft}\right) \tag{15.66}$$

where q_c = cone penetration resistance.

Example 15.10

For a shallow foundation, $D_f = 3$ ft, $B = 3$ ft, $L = 6$ ft; for the soil, $\phi = 28°$, $c = 600$ lb/ft^2, $\gamma = 115$ lb/ft^3. Determine the allowable bearing capacity using a factor of safety of 5.

Solution

For $\phi = 28°$, $N_c = 25.8$, $N_q = 14.72$, and $N_\gamma = 16.72$.

$F_{cs} = 1 + (B/L)(N_q/N_c) = 1 + (3/6)(14.72/25.8) = 1.285$

$F_{qs} = 1 + (B/L) \tan \phi = 1 + (3/6) \tan 28 = 1.266$

$F_{\gamma s} = 1 - 0.4(B/L) = 1 - 0.4(3/6) = 0.8$

$F_{qd} = 1 + 2 \tan \phi (1 - \sin \phi)^2 (D_f/B) = 1.3$

$$F_{cd} = F_{qd} - \left(\frac{1 - F_{qd}}{N_q \tan\phi}\right) = 1.3 - \left(\frac{1-13}{14.72 \tan 28}\right) = 1.338$$

$F_{\gamma d} = 1$

$$q_u = (600)(1.285)(1.338)(25.8) + (115)(3)(1.266)(1.3)(14.72)$$
$$+ \left(\frac{1}{2}\right)(115)(3)(0.8)(1)(16.72)$$
$$= 26{,}615 + 8358 + 2307 = 37{,}280 \text{ lb/ft}^2$$
$$q_{\text{all}} = q_u/5 = 37{,}280/5 = 7456 \text{ lb/ft}^2$$

DEEP (PILE) FOUNDATIONS

Piles and caissons are generally classified as deep foundations. For these, the failure surface in the soil below the tip does not extend to the ground surface. Several methods are now available to estimate the ultimate and allowable bearing capacities of a pile foundation. In any case, the estimated values obtained by different methods vary widely. Hence, extreme caution and judgment should be used in arriving at the design value. Figure 15.19 shows a pile foundation in a saturated clay ($\phi = 0$ condition). The ultimate load-carrying capacity (Q_u) of the pile can be estimated as follows:

$$Q_u = Q_p + Q_s \quad \textbf{(15.67)}$$

where Q_p is the point resistance, and Q_s is the resistance along the pile surface due to adhesion at the pile-clay interface (skin friction).

$$Q_p = 9c_u A_p \quad \textbf{(15.68)}$$

where

c_u = the undrained shear strength of the soil under the pile tip
A_p = cross-sectional area of the pile tip

$$Q_s = \alpha c_u pL \quad \textbf{(15.69)}$$

where

α = empirical adhesion factor (suggested values given in Figure 15.20)
p = perimeter of the pile cross section
L = length of the pile

Thus, combining Equations (15.67), (15.68), and (15.69),

$$Q_u = 9c_u A_p + \alpha c_u pL \quad \textbf{(15.70)}$$

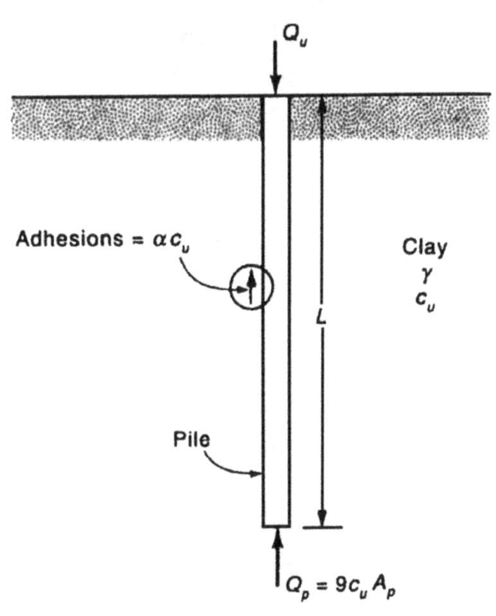

Figure 15.19

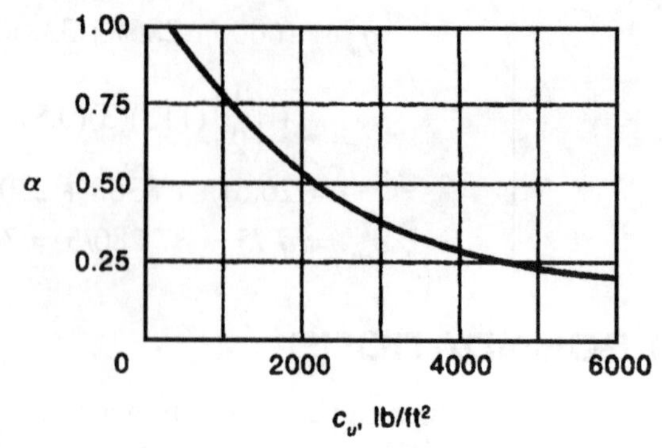

Figure 15.20

Figure 15.21(a) shows a pile in a sandy soil ($c = 0$). For this case,

$$Q_u = Q_p + Q_s \tag{15.71}$$

where

Q_p = point resistance
Q_s = skin friction

For this case (Meyerhof, 1976),

$$Q_p \text{ (lb/ft}^2\text{)} = A_p q N_q^* \leq A_p 1000 N_q^* \tan \phi \tag{15.72}$$

where

A_p = cross-sectional area of the pile tip
q' = effective overburden pressure at the level of the pile tip
N_q^* = bearing capacity factor [Figure 15.21(b)]
ϕ = soil friction angle

The skin friction, Q_s, can be estimated from the following relationships:

$$Q_s = \Sigma p(\Delta L) f \tag{15.73}$$

where

p = perimeter of the pile cross section
f = unit frictional resistance at a depth z

$$f = K \sigma_v' \tan \delta \tag{15.74}$$

where

K = earth pressure coefficient
σ_v' = effective vertical stress at a depth z
δ = soil-pile interface friction angle (≈1/3 to 2/3 ϕ)

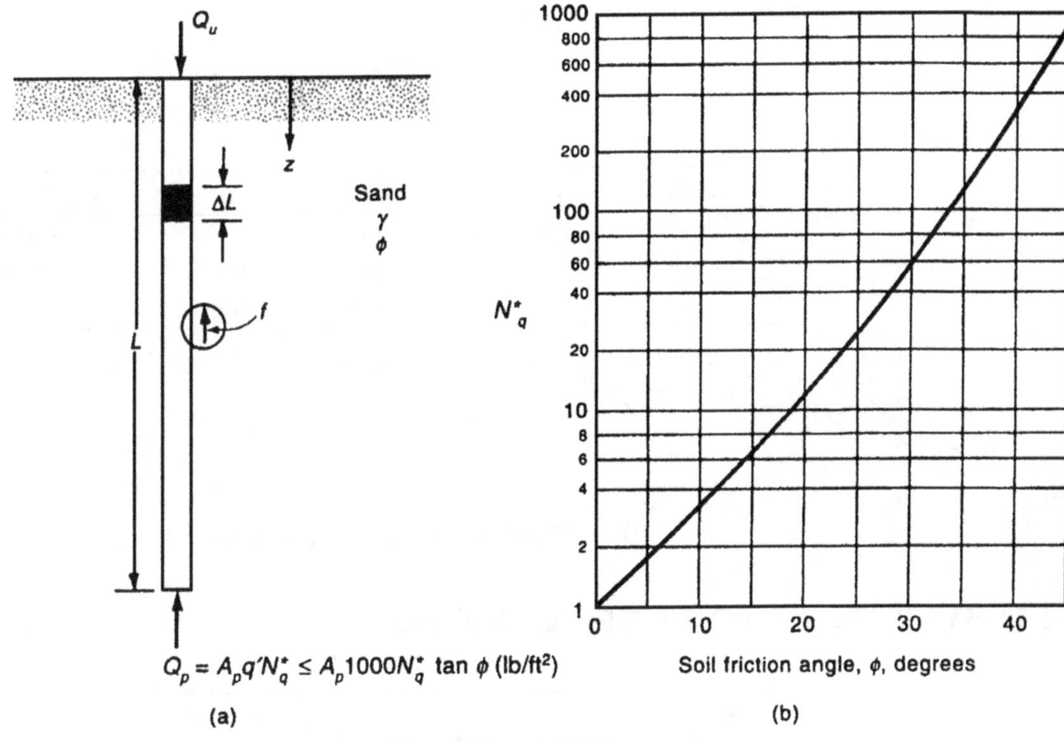

Figure 15.21

For bored piles

$$K \approx K_o \approx 1 - \sin\phi \tag{15.75}$$

and for driven piles

$$K = K_o \text{ to } 1.5K_o \tag{15.76}$$

Hence, for pile in homogeneous dry sand [Equations (15.73) and (15.74)],

$$Q_s = \frac{1}{2}\gamma L^2 Kp \tan\delta \tag{15.77}$$

A large factor of safety (6 to 8) is usually recommended to obtain the allowable load-carrying capacity of a pile.

Example 15.11

A concrete pile having a cross section of 1 ft × 1 ft is shown in Exhibit 4. Determine the load-carrying capacity of the pile.

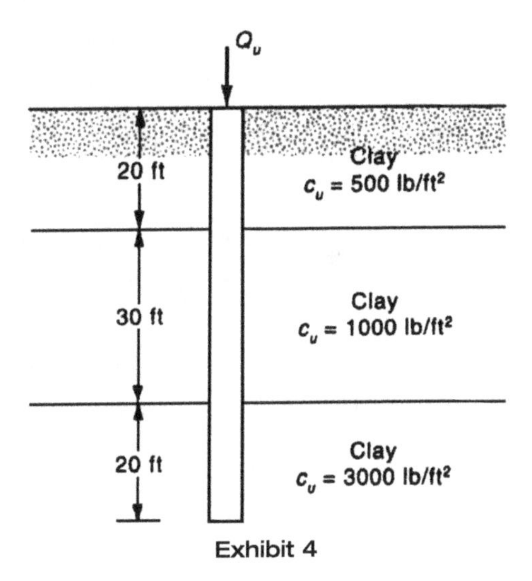

Exhibit 4

Solution

$$Q_p = 9c_uA_p = (9)(3000)(1 \times 1) = 27{,}000 \text{ lb}$$
$$Q_s = \Sigma \alpha c_u p \Delta L$$

ΔL (ft)	p (ft)	c_u (lb/ft^2)	α*	$\alpha c_u p \Delta L$ (lb)
20	4	500	0.925	37,000
30	4	1000	0.775	93,000
20	4	3000	0.375	90,000
				Σ220,000

*From Figure 15.20.

So

$$Q_u = 27{,}000 + 220{,}000 = 247{,}000 \text{ lb} = 247 \text{ kips}$$

SELECTED SYMBOLS AND ABBREVIATIONS

Symbol or Abbreviation	Description
D_{10}	grain size in mm
C_z	coefficient of gradation
G_s	specific gravity of soil solids
W	weight of soil sample
W_s	weight of soil solids
W_w	weight of water in soil sample
V	volume of soil sample
V_s	volume of soil solids
V_v	volume of voids
V_w	volume of water
V_a	volume of air
e	void ratio
n	porosity
S	degree of saturation
w	moisture content
γ	moist unit weight
γ_d	dry unit weight
γ_{sat}	saturated unit weight
γ_w	unit weight of water
D_r	relative density
e_{max}	void ratio in the loosest state
e_{min}	void ratio in the densest state
$\gamma_{d(max)}$	maximum dry unit weight
$\gamma_{d(min)}$	minimum dry unit weight
$\gamma_{d(field)}$	dry unit weight of field sample
LL	liquid limit
PL	plastic limit
PI	plasticity index
SL	shrinkage limit

Symbol or Abbreviation	Description
v	discharge velocity
k	coefficient of permeability
i	hydraulic gradient
L	length of soil sample
h	loss of head
q	flow through sample in unit time
A	area of sample cross section at right angle to direction of flow
N_f	number of flow channels
N_d	number of drops
H	head difference, upstream to downstream
σ	total stress at a point in a soil mass
u	pore water pressure
σ'	effective stress
γ'	effective unit weight
$i_{c\gamma}$	hydraulic gradient for quicksand
Δp	increase of vertical stress
Q	point load on surface
r	$\sqrt{x^2 + y^2}$
L, B	length and width of a rectangular foundation
q	distributed load of foundation
z	vertical distance below foundation
Δp_{av}	average increase of vertical stress
OCR	overconsolidation ratio
p_c	preconsolidation pressure
p	effective overburden pressure
C_c	compression index
C_s	swell index
S	settlement
U	average degree of consolidation
T_v	time factor
c_v	coefficient of consolidation
H_d	length of drainage path
s	shear strength of soil
c	cohesion
ϕ	drained friction angle
c_u	undrained shear strength
q_u	unconfined compression strength
σ'_h	effective horizontal pressure
σ'_v	effective vertical pressure
K_O	at-rest earth pressure coefficient
P_h	total lateral force per unit length of wall
K_a	Rankine active earth pressure coefficient
σ'_a	Rankine active pressure
α	angle between backfill and horizontal
σ'_p	passive earth pressure
K_p	Rankine passive earth pressure coefficient

Symbol or Abbreviation	Description
P_p	Rankine passive force per unit length of wall
P_a	Rankine active force per unit length of wall
σ_w	hydrostatic pressure
F_{cs}, F_{qs}, $F_{\gamma s}$	shape factors
F_{cd}, F_{qd}, $F_{\gamma d}$	depth factors
N_c, N_q, N_γ	bearing capacity factors
D_f	depth of a shallow foundation
F_s	safety factor
q_{all}	allowable bearing capacity
N	standard penetration number
Q_u	ultimate load capacity of pile
Q_p	point resistance of pile
Q_s	pile surface resistance
A_p	cross-sectional area of the pile tip
α	empirical adhesion factor
p	perimeter of pile cross section
L	length of pile
N_q^*	Meyerhof's bearing capacity factor (deep foundation)
f	unit frictional resistance on pile
δ	soil-pile interface friction angle
K	earth pressure coefficient

REFERENCES

Boussinesq, J. *Application des Potentials à L'Etude de L'Equilibre et du Mouvement des Solides Elastiques*. Gauthier-Villars, Paris, 1883.

DeBeer, E. E. Experimental Determination of the Shape Factors and Bearing Capacity Factors of Sand. *Geotechnique*. Vol. 20, No. 4, pp. 387–411, 1970.

Hansen, J. B. A Revised and Extended Formula for Bearing Capacity. Danish Geotechnical Institute, *Bulletin 28*, Copenhagen, 1970.

Meyerhof, G. G. Penetration Tests and Bearing Capacity of Cohesionless Soils. *Journal of the Soil Mechanics and Foundations Division*, American Society of Civil Engineers. Vol. 82, No. SM1, pp. 1–19, 1956.

Meyerhof, G. G. Bearing Capacity and Settlement of Pile Foundations. *Journal of the Geotechnical Engineering Division*, American Society of Civil Engineers. Vol. 102, No. GT3, pp. 197–228, 1976.

Terzaghi, K. *Theoretical Soil Mechanics*. Wiley, New York, 1943.

Vesic, A. S. Analysis of Ultimate Loads of Shallow Foundations. *Journal of the Soil Mechanics and Foundations Division*, American Society of Civil Engineers. Vol. 99, No. SM1, pp. 45–73, 1973.

Recommendations for Further Study

Das, B. M. *Principles of Geotechnical Engineering*, 5th ed. Brooks/Cole, Pacific Grove, CA, 2002.

Das, B. M. *Principles of Foundation Engineering*, 5th ed. Brooks/Cole, Pacific Grove, CA, 2004.

PROBLEMS

15.1 A moist soil specimen has a volume of 0.15 m^3 and weighs 2.83 kN. The water content is 12%, and the specific gravity of soil solids is 2.69. Determine

a. Moist unit weight, γ
b. Dry unit weight, γ_d
c. Void ratio, *e*
d. Degree of saturation, *S*

15.2 For a soil deposit in the field, the dry unit weight is 14.9 kN/m^3. From the laboratory, the following were determined: $G = 2.66$, $e_{max} = 0.89$, $e_{min} = 0.48$. Find the relative density in the field.

15.3 For a sandy soil, the maximum and minimum void ratios are 0.85 and 0.48, respectively. In the field, the relative density of compaction of the soil is 29.3 percent. Given $G = 2.65$, determine the moist unit weight of the soil at $w = 10\%$.

15.4 Refer to the flow net shown in Figure 15.6. Given $k = 0.03$ cm/min, $H_1 = 10$ m, and $H_2 = 1.8$ m, determine the seepage loss per day per meter under the sheet pile construction.

15.5 The results of a sieve analysis of a granular soil are as follows:

U.S. Sieve No.	Sieve Opening (mm)	Percent Retained on Each Sieve
4	4.75	0
10	2.00	20
40	0.425	20
60	0.25	30
100	0.15	20
200	0.075	5

Determine the uniformity coefficient and coefficient of gradation of the soil.

15.6 For a normally consolidated clay of 3.0 m thickness, the following are given:

Average effective pressure = 98 kN/m^2
Initial void ratio = 1.1
Average increase of pressure in the clay layer = 42 kN/m^2
Compression index = 0.27

Estimate the consolidation settlement.

15.7 An oedometer test in a normally consolidated clay gave the following results.

Average Effective Pressure (kN/m²)	Void Ratio
100	0.9
200	0.82

Calculate the compression index.

SOLUTIONS

15.1

a. $$\gamma = \frac{W}{V} = \frac{2.83}{0.15} = 18.87 \text{ kN/m}^3$$

b. $$\gamma_d = \frac{\gamma}{1+w} = \frac{1887}{1+\left(\frac{12}{100}\right)} = 16.85 \text{ kN/m}^3$$

c. $$\gamma_d = \frac{G\gamma_w}{1+e}; \quad e = \frac{G\gamma_w}{\gamma_d} - 1 = \frac{(2.69)(9.81)}{16.85} - 1 = 0.566$$

d. $$S = \frac{wG}{e} \times 100 = \frac{(0.12)(2.69)}{0.566} \times 100 = 57.03\%$$

15.2 In the field

$$\gamma_d = \frac{G\gamma_w}{1+e}; \quad e = \frac{G\gamma_w}{\gamma_d} - 1 = \frac{(2.66)(9.81)}{14.9} - 1 = 0.75$$

$$D_d = \frac{e_{max} - e}{e_{max} - e_{min}} = \frac{0.85 - 0.75}{0.89 - 0.48} = 34\%$$

15.3

$$D_d = 0.293 = \frac{e_{max} - e}{e_{max} - e_{min}} = \frac{0.85 - e}{0.85 - 0.45}; \quad e = 0.733$$

$$\gamma = \frac{G\gamma_w(1+w)}{1+e} = \frac{(2.65)(9.81)(1+0.1)}{1+0.733} = 16.5 \text{ kN/m}^3$$

15.4

$$Q = k\frac{N_f}{N_d}H = \frac{(0.03 \times 60 \times 24 \text{cm/day})}{100}\left(\frac{4}{6}\right)(10 - 1.8) = 2.36 \text{m}^3\text{/day/m}$$

15.5

Sieve Opening (mm)	Cumulative Percent Passing
4.75	100
2.00	80
0.425	60
0.25	30
0.15	10
0.075	5

So $D_{60} = 0.425$ mm; $D_{30} = 0.25$ mm; $D_{10} = 0.15$ mm

$$c_u = \frac{D_{60}}{D_{10}} = \frac{0.425}{0.15} = 2.83$$

$$c_c = \frac{D_{30}^2}{D_{60} \times D_{10}} = 0.98$$

15.6

$$\Delta H = \frac{C_c H}{1+e_i} \log\left(\frac{p_i + \Delta p}{p_i}\right) = \frac{(0.27)(3)}{1+1.1} \log\left(\frac{98+42}{98}\right) = 0.0597 \text{ m} = 59.7 \text{ mm}$$

15.7

$$C_c = \frac{e_1 - e_2}{\log\left(\frac{p_2}{p_1}\right)} = \frac{0.9 - 0.82}{\log\left(\frac{200}{100}\right)} = 0.266$$

CHAPTER 16

Transportation Engineering

Robert W. Stokes

OUTLINE

HIGHWAY CURVES

Simple (Circular) Horizontal Curves

The location of highway centerlines is initially laid out as a series of straight lines (tangent sections). These tangents are then joined by circular curves to allow for smooth vehicle operations at the design speed selected for the highway. Figure 16.1 shows the basic geometry of a simple circular curve. If the two tangents intersecting at PI are laid out, and the angle Δ between them is measured, only one other element of the curve need be known to calculate the remaining elements. The radius of the curve (R) is the other element most commonly used.

Circular Curve Formulas

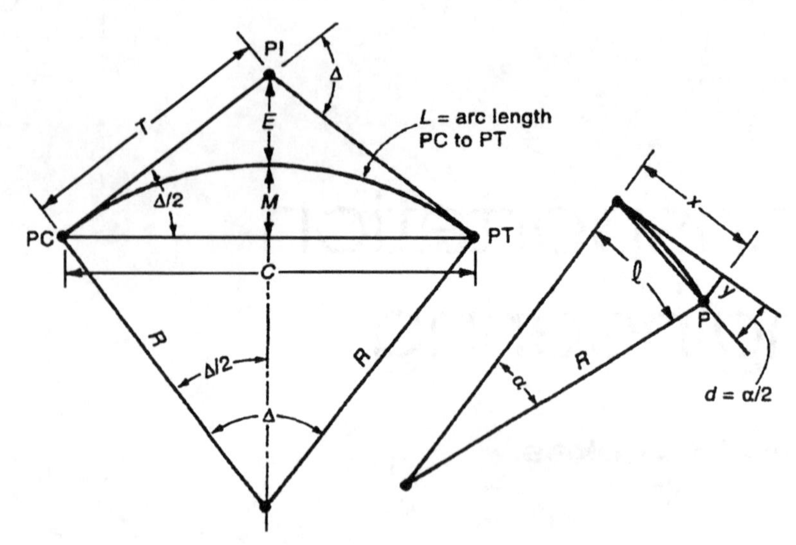

Figure 16.1

The following symbols are defined as shown in Figure 16.1:

PC = Point of curvature (beginning of curve)
PI = Point of intersection
PT = Point of tangency (end of curve)
Δ = Intersection or central angle, degrees
R = Radius of curve, m
L = Length of curve, m
E = External distance, m
M = Middle ordinate, m
C = Long chord, m
P = Length of arc between any two points on curve, m
α = Central angle subtended by arc P, degrees
d = Deflection angle for any arc length P, degrees
x = Distance along tangent from PC or PT to set any point P on curve, m
y = Offset (normal) from tangent at distance x to set any point P on curve, m

Then

$L = \Delta R/57.2958$
$T = R \tan \Delta/2$
$E = R(\sec \Delta/2 - 1) = R \text{ exsec } \Delta/2$
$M = R(1 - \cos \Delta/2) = R \text{ vers } \Delta/2$
$C = 2R \sin \Delta/2$
$R = (C^2 + 4M^2)/8M$ (formula for estimating R from field measurements)
$P = (\alpha \times R)/57.2958$
$d = \alpha/2 = 1718.873\ P/R$ (in minutes)
$= 28.64789\ P/R$ (in degrees)

For any length x,

$$y = R - (R^2 - x^2)^{1/2}$$

For any length P,

$$X = R \sin \alpha$$
$$Y = R(1 - \cos \alpha) = R \text{ vers } \alpha$$

Example **16.1**

Given the following horizontal curve data, determine the curve radius, R; length of curve, L; stationing (sta) of the PC; stationing of the PT; the long chord, C; and deflection angle, d, of a point on the curve 30 m ahead of the PC.

$$\text{PI} = \text{sta } 2 + 170.00 \text{ (km)}$$
$$\Delta = 41°10'$$
$$T = 115.00 \text{ m}$$

Solution

$$\text{PC sta} = \text{PI sta} - T = 2170.00 - 115.00 = 2 + 055.00$$
$$R = T/\tan(\Delta/2) = 115.00/0.3755 = 306.22 \text{ m}$$
$$L = (\Delta R)/57.2958 = 220.02 \text{ m}$$
$$\text{PT sta} = \text{PC sta} + L = 2055.00 + 220.02 = 2275.02 \text{ m}$$
$$C = 2R \sin(\Delta/2) = 2(306.22)(0.3516) = 215.32 \text{ m}$$
$$d = (28.64789\text{P})/R = [(28.64689)(30)]/306.22 = 2.81°$$

Horizontal Curve Design

The minimum radius of horizontal curvature is determined by the dynamics of vehicle operation and sight distance requirements. The minimum radius necessary for the vehicle to remain in equilibrium with respect to the incline of a horizontal curve is given by the following formula:

$$R = V^2/[127\,(e + f)]$$

where V = vehicle speed in km/h, e = rate of superelevation ("banking") in m/m of width, and f = allowable side friction factor. The equation can be used to solve for the minimum radius for a given speed, or the maximum safe speed for a given radius.

Example **16.2**

An existing horizontal curve has a radius of 235 m. What is the maximum safe speed on the curve? Assume $e = 0.08$ and $f = 0.14$.

Solution

$$R = V^2/[127\,(e + f)]$$
$$V = [R[127\,(e + f)]]^{0.5}$$
$$V = [235\,[127\,(0.08 + 0.14]]^{0.5} = 81 \text{ km/h}$$

Vertical Curves

The vertical-axis parabola (see Figure 16.2) is the geometric curve most commonly used in the design of vertical highway curves. The general equation of the parabola is

$$Y = aX^2 + bX + c$$

where the constant a is an indication of the rate of change of slope, b is the slope of the back tangent (g_1), and c is the elevation of the PVC.

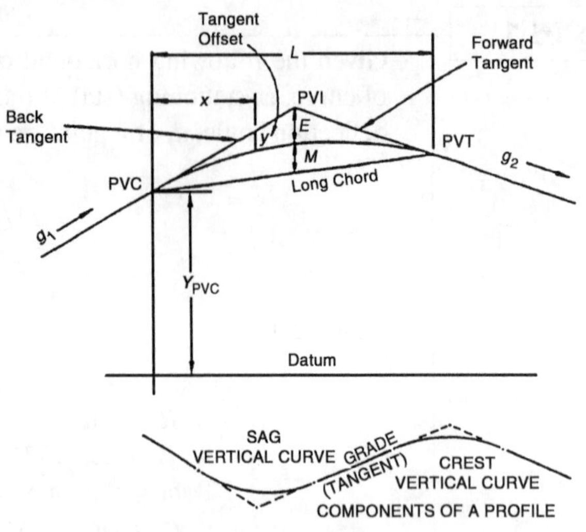

Figure 16.2

Vertical Curve Formulas

The following notation is defined as shown in Figure 16.2:

L = Length of curve (horizontal)
PVC = Point of vertical curvature
PVI = Point of vertical intersection
PVT = Point of vertical tangency
g_1 = Grade of back tangent (decimal)
g_2 = Grade of forward tangent (decimal)
a = Parabola constant
y = Tangent offset
E = Tangent offset at PVI
M = Middle ordinate
r = Rate of change of grade
K = Rate of curvature
A = Algebraic difference in grades
x = Horizontal distance from PVC to point on curve
x_m = Horizontal distance to min/max elevation on curve

Then

$A = g_2 - g_1$
$a = (g_2 - g_1)/2L = A/2L$
$r = (g_2 - g_1)/L = A/L$
$K = L/A$
$E = a(L/2)^2 = (AL)/8 = M$
$y = ax^2$
$xm = -g_1/(2a) = -g_1K$
Tangent elevation = $Y_{PVC} + g_1x$
Curve elevation = $Y_{PVC} + g_1x \pm ax^2$ (add for sag curves, subtract for crest curves)

Example 16.3

Determine the following for the sag vertical curve shown in Exhibit 1

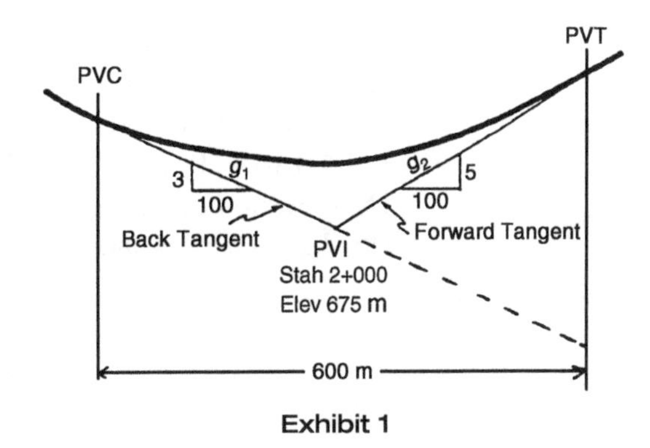

Exhibit 1

a). Stationing and elevation of the PVC

b). Stationing and elevation of the PVT

c). Elevation of the curve 100 m from the PVC

d). Elevation of the curve 400 m from the PVC

e). Stationing and elevation of the low point of the curve

Solution

a). The curve is symmetric about the PVI, therefore:

$$\text{Sta PVC} = \text{Sta PVI} - L/2 = 2000 - 600/2 = 1 + 700.0$$
$$\text{Elev of the PVC} = \text{Elev PVI} + g_1(L/2) = 675 + 0.03(300) = 684.0 \text{ m}$$

b).
$$\text{Sta PVT} = \text{Sta PVI} + L/2 = 2000 + 600/3 = 2 + 300.0$$
$$\text{Elev of the PVT} = \text{Elev PVI} + g_2(L/2) = 675 + 0.05(300) = 690.0 \text{ m}$$

c). The elevation of the curve at any distance x from the PVC is equal to the tangent offset + the tangent elevation. To calculate the tangent offset, it is first necessary to evaluate the constant a.

$$a = (g_2 - g_1)/2L = [0.05 - (-0.03)]/[2(600)] = 0.000067$$

Tangent offset at $x = 100$ m from the PVC $= ax^2 = 0.000067\ (100)^2 = 0.67$ m
Tangent elevation 100 m from the PVC = PVC elev + g_1x
$= 684.0 \text{ m} + (-0.03)100 = 681.0$ m

Curve elevation = tangent offset + tangent elev = 0.67 + 681.0 = 681.67 m

d). The elevation of the curve 400 m from the PVC can be calculated relative to the forward tangent or relative to the extension of the back tangent. Relative to the extension of the back tangent:

Tangent offset $= ax^2 = 0.000067\ (400)^2 = 10.67$ m
Elev of the extended back tangent = PVC elev + g_1x
$= 684.0 + (-0.03)400 = 672.0$ m
Curve elev = tangent offset + tangent elev = 10.67 + 672.0 = 682.67 m

e). The low point of the curve occurs at a distance x_m from the PVC.

$$x_m = -g_1/(2a) = -(-0.03)/[2(0.000067)] = 225 \text{ m (from the PVC)}$$

Sta of the low point = Sta PVC + 225 = 1700 + 225 = 1 + 925.0
The tangent offset at 225 m from the PVC = $ax^2 = 0.000067(225)^2 = 3.38$ m
The tangent elevation at this distance = PVC elev + g_1x
= 684.0 + (–0.03)225 = 677.25 m
The elevation of the low point = tangent offset + tangent elevation
= 677.25 + 3.38 = 680.63 m

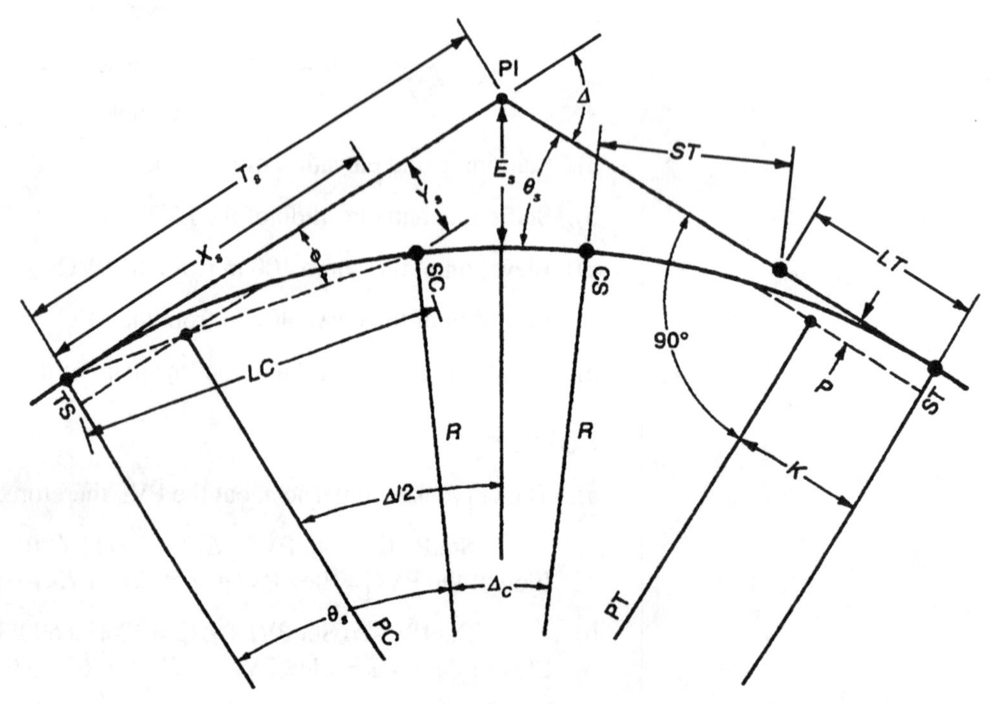

Figure 16.3

Transition (Spiral) Curves

The spiral curve is used to allow for a transitional path from tangent to circular curve, from circular curve to tangent, or from one curve to another that have substantially different radii. The minimum length of spiral curve needed to achieve this transition can be computed from the following formula.

$$P_s = V^3/(46.7RC)$$

where P_s = minimum length of spiral (m), V = design speed (km/hr), R = circular curve radius (m), and C = rate of increase of centripetal acceleration (m/s³). The value $C = 0.6$ is generally used for highway curves.

Transition (Spiral) Curve Formulas

The basic geometry of the spiral curve is shown in Figure 16.3. The spiral is defined by its parameter A and the radius of the simple curve it joins. The product of the radius (r) at any point on the spiral and the corresponding spiral length (P) from the beginning of the spiral to that point is equal to the product of the radius (R) of the simple curve it joins and the total length (P_s) of the spiral as shown in the following equation:

$$rR = RR_s = \text{constant} = A^2$$

The following notation is defined as shown in Figure 16.3:

ℓ_s = total length of spiral from TS to SC
ℓ = spiral length from TS to any point on spiral
L_c = total length of circular curve
R = radius of circular curve
T_s = total tangent distance from the PI to the TS or ST
Δ = deflection angle between the tangents
θ_s = central angle of the entire spiral (spiral angle)
θ = central angle of any point on the spiral
Δ_c = central angle of simple curve
E_s = external distance
LC = long chord
LT = long tangent
ST = short tangent
X_s = tangent distance from TS to SC
x_s = tangent distance from TS to any point on the spiral
Y_s = tangent offset at the SC
y_s = tangent offset at any point on the spiral
k = simple curve coordinate (abscissa)
p = simple curve coordinate (ordinate)
ϕ = deflection angle at TS from the initial tangent to any point on the spiral
TS = tangent to spiral
SC = spiral to circular curve
CS = circular curve to spiral
ST = spiral to tangent
C_s = spiral deflection angle correction factor

Then

$$\theta_s = 28.648\,(P_s/R)$$
$$\theta = (P/P_s)^2\theta_s$$
$$\phi = (\theta/3) - C_s$$
$$C_s\,(\text{seconds}) = 0.0031\theta^3 + 0.0023\theta^5 \times 10^{-5}$$
$$Y_s = (\ell_s^2/6R) - (\ell_s^4/133R^3)$$
$$y_s = (\ell^2/6R) - (\ell^4/336R^3)$$
$$X_s = \ell_s - (\ell_s^3/40R^2)$$
$$p = Y_s - R(1 - \cos\theta_s)$$
$$k = X_s - R\sin\theta_s$$
$$T_s = (R + p)\tan(\Delta/2) + k$$
$$\text{ST} = Y_s/\sin\theta_s$$
$$\text{LT} = X_s - (Y_s/\tan\theta_s)$$

All distances are in meters and all angles are in degrees unless noted otherwise.

Example 16.4

A transition spiral is being designed to provide a gradual transition into a circular curve with a radius of 435 m and a design speed of 90 km/h. The PI of the curve is at Station 0 + 625.00, and the deflection angle between the tangents is 33.67 degrees. Determine the minimum length of spiral required and the stationing of the TS and the ST.

Solution

Assuming a value of $C = 0.6$, the minimum length of the spiral can be determined as follows.

$$\ell_s = V^3/(46.7RC) = (90)^3/(46.7 \times 435 \times 0.6) = 60 \text{ m}$$

To determine the stationing of the TS and the ST, proceed as follows:
Compute the spiral angle.

$$\theta_s = 28.648\ (P_s/R) = 28.648\ (60/435) = 4 \text{ degrees}$$

Compute Y_s and X_s.

$$Y_s\ (\ell_s^2/6R) - (\ell_s^4/336R^3) - = [(60)^2/(6)(435)] - [(60)^4/(336)(435)^3] = 1.379 \text{ m}$$
$$X_s = \ell_s - (\ell_s^3/40R^2) = 60 - [(60)^3/(40)(435)^2] = 59.972 \text{ m}$$

Compute p and k.

$$p = Y_s - R(1 - \cos\theta_s) = 1.379 - 435(1 - \cos 4°) = 0.319 \text{ m}$$
$$k = X_s - R\sin\theta_s = 59.972 - (435)\sin 4° = 29.628 \text{ m}$$

Compute the spiral tangent length.

$$T_s = (R + p)\tan(\text{D}/2) + k$$
$$= (435 + 0.319)\tan(33.67/2) + 29.628 = 161.349 \text{ m}$$

Compute the central angle of the circular curve.

$$\Delta_c = \Delta - 2\theta_s = 33.67 - 2(4) = 25.67 \text{ degrees}$$

Compute the length of the circular curve.

$$L = (\Delta_c R)/57.2958 = [25.67(435)]57.2958 = 194.891 \text{ m}$$

Compute the stationing of the TS.

$$\text{TS Sta} = \text{PI Sta} - T_s = 0 + 625 - 0 + 161.349 = 0 + 463.651$$

Compute the stationing of the ST.

$$\text{ST Sta} = \text{TS Sta} + L + 2y\,\ell_s$$
$$= 0 + 463.651 + 0 + 194.891 + 2(60) = 0 + 778.542$$

SIGHT DISTANCE

The ability of drivers to see the road ahead is of utmost importance in the design of highways. This ability to see is referred to as **sight distance** and is defined as the length of highway ahead that is visible to the driver.

Stopping sight distance is the sum of the distance traveled during the perception-reaction time and the distance traveled while braking to a stop. The stopping sight distance can be determined from the following equation.

$$S = 0.278Vt + V^2/[254\ (f \pm g)]$$

where S = stopping sight distance (m), V = vehicle speed (km/hr), t = perception-reaction time (assumed to be 2.5 sec), f = coefficient of friction, and g = grade (*plus* for uphill, *minus* for downhill), expressed as a decimal.

Sight Distance on Simple Horizontal Curves

The required middle ordinates (M) for clear sight areas to satisfy stopping sight distance (S) requirements as a function of the radii (R) of simple horizontal curves can be determined using the following equation. The stopping sight distance is measured along the centerline of the inside lane of the curve.

$$M = R[1 - \cos(28.65\, S/R)]$$

Example 16.5

A large outcropping of rock is located at M (middle ordinate) = 10 m from the centerline of the inside lane of a proposed highway curve. The radius of the proposed curve is 120 m. What speed limit would you recommend for this curve? Explain your answer. Assume $f = 0.28$, $e = 0.10$, perception-reaction time = 2.5 sec, and level grade.

Solution

To solve this problem you must determine whether sight distance or the radius of the curve controls the speed.

Determine the available sight distance.

$$M = R[1 - \cos(28.65\, S/R)]$$
$$S = (R/28.65)\cos^{-1}[(R - M)/R]$$
$$= (120/28.65)\cos^{-1}[(120 - 10)/120] = 98.67 \text{ m}$$

Determine the maximum safe speed on the curve as a function of stopping sight distance.

$$S = 0.278Vt + V^2/[254(f \pm g)]$$
$$98.67 = 0.278\text{V}(2.5) + V^2/[254(0.28)]$$
$$V^2 + 49.43\text{V} - 7017.41 = 0$$

and from the quadratic equation

$$V = 62.6 \text{ km/hr}$$

Determine the maximum safe speed on the curve as a function of the radius.

$$R = V^2/[127(e + f)]$$
$$V = [127R(e + f)]^{0.5} = [127(120)(0.10 + 0.28)]^{0.5} = 76.1 \text{ km/hr}$$

Because the sight distance is sufficient only for a speed of about 62.6 km/hr, set speed limit at 60 km/hr.

Sight Distance on Vertical Curves

Formulas for stopping and passing sight distances for *vertical curves* are summarized following. The formulas are based on an assumed driver eye height (h_1) of 1070 mm. The height of object (h_2) above the roadway surface is assumed to be 150 mm for stopping sight distance and 1300 mm for passing sight distance.

Formulas for Sight Distance on Vertical Curves

The following terms are defined as shown in Figure 16.4:

L = Length of vertical curve, m
A = Algebraic difference grades, %
S = Sight distance, m
K = Vertical curvature, L/A

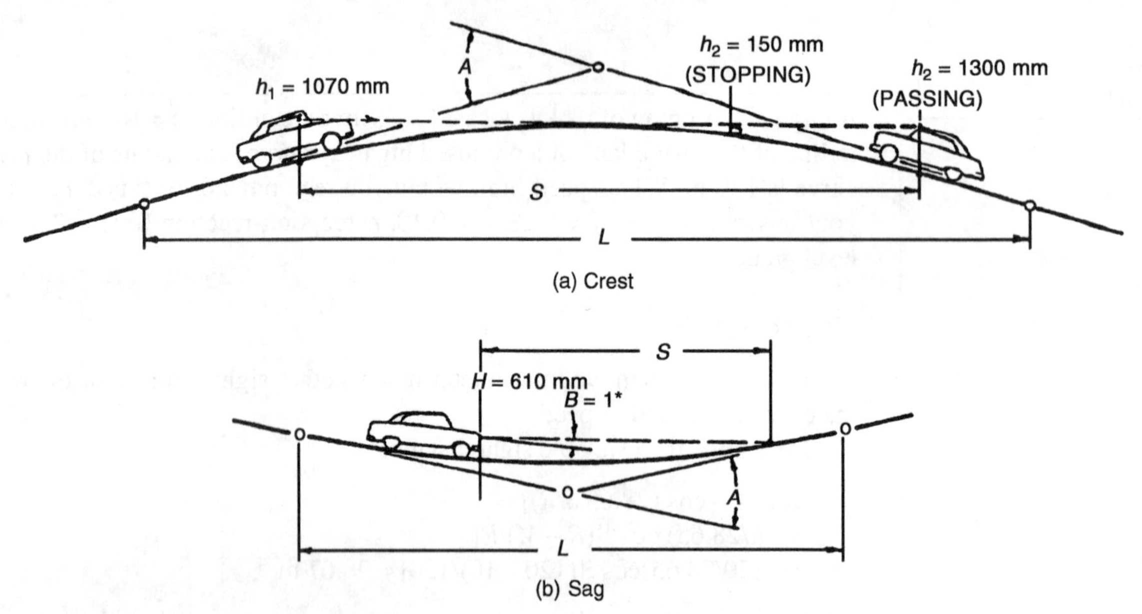

Figure 16.4

H = Headlight height, m
B = Upward divergence of light beam, degrees

Crest:

$$L = AS^2/100\left(\sqrt{2h_1} + \sqrt{2h_2}\right)^2 \quad \text{when } S < L$$

The formula is different for when $S > L$, but does not apply to geometric design criteria.

$$\text{Stopping sight distance: } L = AS^2/404 \text{ or } K = S^2/404$$
$$\text{Passing sight distance: } L = AS^2/946 \text{ or } K = S^2/946$$

Sag:

$$L = AS^2/200\,(H + S \tan B) \text{ when } S < L$$

For passenger cars:

$$L = AS^2/(122 + 3.5\,S) \text{ or } K = S^2/(122 + 3.5\,S)$$

Example 16.6

A crest vertical curve joins a +2% grade with a –2% grade. If the design speed of the highway is 95 km/hr, determine the minimum length of curve required. Assume $f = 0.29$ and the perception-reaction time = 2.5 sec. Also assume $S < L$.

Solution

Determine the stopping sight distance required for the design conditions.

$$S = 0.278Vt + V^2/[254\ (f \pm g)] = 0.278\ (95)(2.5) + (95)^2/[254(0.29 - 0.02)]$$
$$= 197.62\ \text{m}$$

(Note that the worst-case value for g is used.)

Determine the minimum length of vertical curve to satisfy the required sight distance.

$$L = AS^2/404 = [4(197.62)^2]/404 = 386.67\ \text{m}$$

TRAFFIC CHARACTERISTICS

The traffic on a roadway may be described by three general parameters: volume or rate of flow, speed, and density.

Traffic **volume** is the number of vehicles that pass a point on a highway during a specified time interval. Volumes can be expressed in daily or hourly volumes or in terms of subhourly rates of flow. There are four commonly used measures of daily volume.

1. **Average annual daily traffic** (AADT) is the average 24-hour traffic volume at a specific location over a full year (365 days).
2. **Average annual weekday traffic** (AAWT) is the average 24-hour traffic volume occurring on weekdays at a specific location over a full year.
3. **Average daily traffic** (ADT) is basically an estimate of AADT based on a time period less than a full year.
4. **Average weekday traffic** (AWT) is an estimate of AAWT based on a time period less than a full year.

Highways are generally designed on the basis of the **directional design hourly volume (DDHV)**.

$$\text{DDHV} = \text{AADT} \times K \times D$$

where AADT = average annual daily traffic (veh/day), K = proportion of daily traffic occurring in the design hour, D = proportion of design hour traffic traveling in the peak direction of travel.

For design purposes, K often represents the proportion of AADT occurring during the **thirtieth highest hour** of the year on rural highways, and the fiftieth highest hour of the year on urban highways.

Traffic volumes can also exhibit considerable variation within a given hour. The relationship between hourly volume and the maximum 15-minute rate of flow within the hour is defined as the **peak-hour factor (PHF)**.

$$\text{PHF} = V_h/(4V_{15})$$

where V_h = hourly volume (vph), V_{15} = maximum 15-minute rate of flow within the hour (veh).

The second major traffic stream parameter is **speed**. Two different measures of average speed are commonly used. **Time mean speed** is the average speed of all vehicles passing a point on the highway over a given time period. **Space mean speed** is the average speed obtained by measuring the instantaneous speeds of all vehicles on a section of roadway. Both of these measures can be calculated from a series of measured travel times over a measured distance from the following equations.

$$\mu_t = (\Sigma(d/t_i))/n$$
$$\mu_s = (nd)/\Sigma t_i$$

where μ_t = time mean speed (m/sec or km/hr), μ_s = space mean speed (m/sec or km/hr), d = distance traversed (m or km), n = number of travel times observed, t_i = travel time of ith vehicle (sec or hr).

Because the space mean speed weights slower vehicles more heavily than time mean speed does, it results in a lower average speed. The relationship between these two mean speeds is

$$\mu_t = \mu_s + \sigma/\mu_s$$

where σ = the variance of the space speed distribution.

Example 16.7

The following travel times were observed for four vehicles traversing a 1-km segment of highway.

Vehicle	Time (minutes)
1	1.6
2	1.2
3	1.5
4	1.7

Calculate the space and time mean speeds of these vehicles.

Solution

The space mean speed is

$$\mu_s = (nd)/\Sigma t_i = 4(1)/(1.6 + 1.2 + 1.5 + 1.7) = 0.67 \text{ km/min} = 40.0 \text{ km/hr}$$

The time mean speed is

$$\mu_t = \Sigma(d/t_i)/n = [(1/1.6) + (1/1.2) + (1/1.5) + (1/1.7)]/4$$
$$= 0.68 \text{ km/min} = 40.8 \text{ km/hr}$$

The third measure of traffic stream conditions, **density** (sometimes referred to as **concentration**), is the number of vehicles traveling over a unit length of highway at a given instant in time.

The general equation relating flow, density, and speed is

$$q = k\mu_s$$

where q = rate of flow (vph), μ_s = space mean speed (km/hr), and k = density (veh/km).

The relationship between density and flow shown in Figure 16.5 is commonly referred to as the "fundamental diagram of traffic flow." Note in the figure that (1) when density is zero, the flow is also zero; (2) as density increases, the flow also increases; and (3) when density reaches its maximum, referred to as jam density (k_j), the flow is zero.

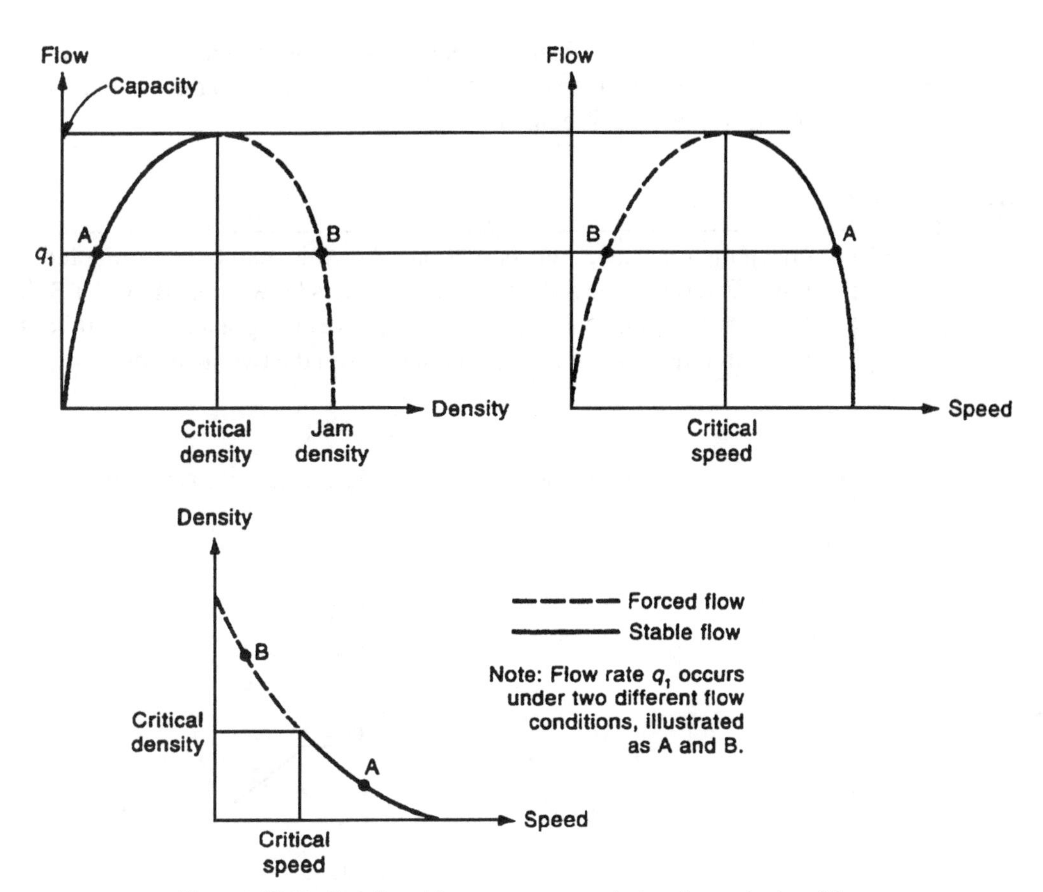

Figure 16.5 Relationships among speed, density, and rate of flow

EARTHWORK

One of the major objectives in evaluating alternative route locations is to minimize the amount of cut and fill. A common method of determining the volume of earthwork is the **average end area** method. This method is based on the assumption that the volume between two consecutive cross sections is the average of their areas multiplied by the distance between them.

$$V = L(A_1 + A_2)/2$$

where V = volume (meters³), A_1 and A_2 = end areas (m²)

L = distance between cross sections (m)

In situations where there is a significant difference between A_1 and A_2, it may be advisable to calculate the volume as a pyramid:

$$V = (1/3)(\text{area of base})(\text{length})$$

where V is in cubic meters, the area of the base is in square meters, and the length is in meters.

The average end area and the pyramid methods provide "reasonably accurate" estimates of volumes of earthwork. When a more precise estimate of volume is desired, the **prismoidal formula** is frequently used:

$$V = L(A_1 + 4A_m + A_2)/6$$

where V = volume (m³), A_1 and A_2 = end areas (m²), and A_m = middle area determined by averaging corresponding *linear* dimensions (*not* the end areas) of the end sections (m²).

When materials from cut sections are moved to fill sections, shrinkage factors (generally in the range of 1.10 to 1.25) are applied to the fill volumes to determine the quantities of fill required.

Example 16.8

Given the trapezoidal cross sections shown in Exhibit 2 from a temporary access ramp, determine the volume in cubic meters between stations 4 + 320 and 4 + 400 by (1) the average end area method, and (2) the prismoidal method. Comment on any differences in the results obtained from the two methods.

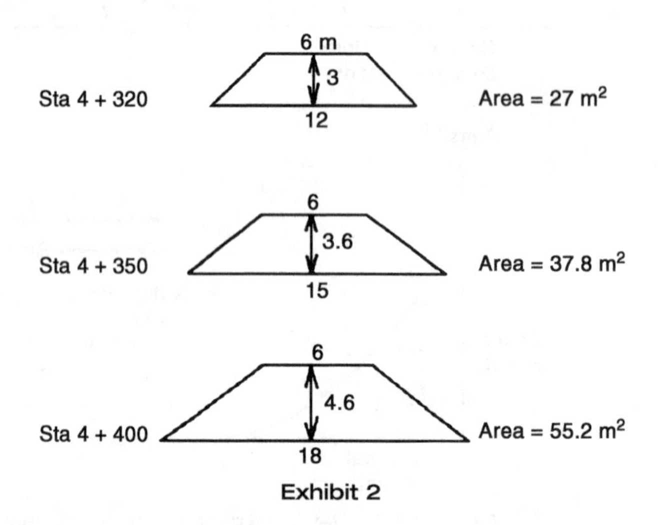

Exhibit 2

Solution

1. For the average end area method:

$$V = L/2\,(A_1 + A_2)$$

The total volume is the sum of the volumes between stations 4 + 320 and 4 + 350 and between 4 + 350 and 4 + 400.

$$V = (30/2)(27.0 + 37.8) + (50/2)(37.8 + 55.2) = 3{,}297.9\text{ m}^3$$

2. For the prismoidal method:

$$V = (L/6)(A_1 + 4A_m + A_2)$$

The middle areas (A_m) are calculated from the averages of the base, top, and height dimensions of the trapezoids.

The middle area between stations 4 + 320 and 4 + 350 is $[(6.0 + 13.5)/2] \times 3.3 = 32.2\text{ m}^2$. The middle area between stations 4 + 350 and 4 + 400 is $[(6.0 + 16.5)/2] \times 4.1 = 46.1\text{ m}^2$.

The total volume is the sum of the volumes between stations 4 + 320 and 4 + 350 and between 4 + 350 and 4 + 400.

$$V = (30/6)[27.0 + 4(32.2) + 37.8] + (50/6)[37.8 + 4(46.1) + 55.2] = 3{,}279.7\text{ m}^3$$

Comment: The prismoidal volumes are theoretically more precise than the average end area volumes. The average end area method tends to overestimate volumes, as in this case.

REFERENCES

American Association of State Highway and Transportation Officials (AASHTO). *A Policy on Geometric Design of Highways and Streets*. Washington, DC, 1994.

Garber, N. J. and Hoel, L. A. *Traffic and Highway Engineering*. West Publishing Co., St. Paul, MN, 1988.

Hickerson, T. F. *Route Location and Design*, 5th ed. McGraw-Hill, New York, 1967.

Institute of Transportation Engineers (ITE). *Traffic Engineering Handbook*, 4th ed. Washington, DC, 1992.

Newnan, D. G., ed. *Civil Engineering License Review*, 12th ed. Engineering Press, Austin, TX, 1995.

Transportation Research Board (TRB). *Highway Capacity Manual*, Special Report 209, 3rd ed. TRB, Washington, DC, 1994.

PROBLEMS

16.1 For a simple horizontal curve with D = 12 degrees, R = 400 m, and PI at Station 0 + 241.782, the stationing of the PC and PT, respectively, would be most nearly

a. 0 + 197.7, 0 + 283.8
b. 0 + 197.7, 0 + 283.5
c. 0 + 158.0, 0 + 283.8
d. 0 + 195.6, 0 + 279.4

16.2 The minimum radius (m) for a simple horizontal curve with a design speed of 110 km/hr is most nearly (assume e = 0.08 and f = 0.14)

a. 425
b. 395
c. 550
d. 435

16.3 A vehicle hits a bridge abutment at a speed estimated by investigators as 25 km/hr. Skid marks of 30 m on the pavement (coefficient of friction, f = 0.35) followed by skid marks of 60 m on the gravel shoulder (f = 0.50) approaching the abutment are observed at the accident site. The grade is level. The initial speed (km/hr) of the vehicle was at least

a. 105
b. 90
c. 110
d. 95

For Questions 16.4 and 16.5, assume a + 3.9% grade intersects a + 1.1% grade at station 0 + 625.0 and elevation 305.0 m.

16.4 The minimum length (m) of the vertical curve for a design speed of 80 km/hr is most nearly (assume perception-reaction time = 2.5 sec, f = 0.30, and $S < L$)

a. 85
b. 130
c. 55
d. 110

16.5 The elevation (m) of the middle point of the curve is most nearly

a. 305.00
b. 305.45
c. 304.45
d. 304.55

The areas (in m²) of the cut sections of a proposed highway are shown below.

Stations	Area, m^2
5 + 000	27.9
5 + 050	326.1
5 + 100	965.3
5 + 200	1651.8
5 + 300	2126.6

16.6 The total volume of cut (to the nearest cubic meter) between stations 5 + 000 and 5 + 100 is most nearly (use the average end area method):

a. 35,615
b. 34,865
c. 44,420
d. 41,135

A 11-m-wide vertical wall roadway tunnel is to be constructed using the cut and cover method. The cross sections to be excavated are shown below.

Stations	Area, m^2
1 + 000	20
1 + 100	30
1 + 130	35
1 + 200	40
1 + 260	45
1 + 300	40
1 + 400	35

16.7 The volume of earth in cubic meters to be excavated between stations 1 + 130 and 1 + 260 is most nearly (use the prismoidal method):

a. 5165
b. 5585
c. 7860
d. 2340

SOLUTIONS

16.1 **b.** Calculate the tangent length:

$$T = R \tan(\Delta/2) = 400 \tan(12/2) = 42.042 \text{ m}$$

Calculate the length of the curve:

$$L = (R\Delta)/57.2958 = 83.776 \text{ m}$$
$$\text{PC sta} = \text{PI sta} - T = 0 + 241.782 - 0 + 042.042 = 0 + 199.740$$
$$\text{PT sta} = \text{PC sta} + L = 0 + 199.740 + 0 + 083.776 = 0 + 283.516$$

16.2 **d.** R = V 2/[127(e + f)] = (110)2/[127(0.08 + 0.14)] = 433 m

16.3 **a.** The only known speed is the final collision speed of 25 km/hr. Therefore, consider the braking distance on the gravel first.

$$\text{Braking distance} = (V_0^2 - V^2)/[254(f \pm g)]$$

where V_0 and V represent the initial and final speeds, respectively.

$$\text{Braking distance (gravel)} = 60 \text{ m} = (V_0^2 - 25^2)/(254)(0.5)$$

Solving for V_0, V_0 = 90.8 km/hr = speed at the beginning of the gravel and at the end of the pavement skid. Therefore, for the pavement skid,

$$\text{Braking distance (pavement)} = 30 \text{ m} = (V_0^2 - 90.8^2)\,(254)(0.35)$$
$$\text{Solving for } V_0,\ V_0 = 104.5 \text{ km/hr}$$

Comment: Note that this solution does not account for any speed reduction prior to the beginning of the skid marks on the pavement. Therefore, we conclude that the vehicle speed was *at least* 104.5 km/hr.

16.4 **b.** The required stopping sight distance is

$$S = 0.278\,Vt + V^2/[254(f \pm g)]$$
$$= 0.278(80)(2.5) + (80)^2/[254(0.30 + 0.011)] = 136.62 \text{ m}$$

The minimum length of crest vertical curve to satisfy the required stopping sight distance is

$$L = (|A|S^2)/404 = [|(1.1 - 3.9)|(136.62)^2]/404 = 129.36 \text{ m}$$

16.5 **d.** The midpoint offset is

$$E = (AL)/8 = [(0.011 - 0.039)(129.36)]/8 = -0.45 \text{ m}$$

Therefore,

Elevation of the midpoint of the curve
= elevation of the PVI (305 m) – 0.45 m = 304.55 m

Note: The elevation of the midpoint could also be found using the basic properties of the parabola. The offset at the midpoint of the curve ($x = L/2$) is

$$y = ax^2 = [(g_2 - g_1)/2L](L/2)^2 = [(0.011 - 0.039)/2(129.36)]/(129.36/2)^2$$
$$= -0.45 \text{ m}$$

16.6 d.

Station	Area, m^2	Total Area	Average Area	Distance, m	Volume[a], m^3
5 + 000	27.9				
		354	177.0	50	8850.0
5 + 050	326.1				
		1291.4	645.7	50	32,285.0
5 + 100	965.3				
					Total Volume = 41,135.0

[a]Volume = distance × average area

16.7 a. The bases of the cross sections are constant (11 m). The heights of the cross sections vary between stations but can be determined by dividing the cross section areas by 11 m. The prismoidal formula is

$$V = (L/6)(A_1 + 4A_m + A_2)$$

Determine the volume between stations 1 + 130 and 1 + 200.

$$A_m = [(3.2 + 3.6)/2]11 = 37.4 \text{ m}^2$$
$$V = (70/6)[35 + 4(37.4) + 40] = 2620.3 \text{ m}^3$$

Determine the volume between stations 1 + 200 and 1 + 260.

$$A_m = [(3.6 + 4.1)/2]11 = 42.4 \text{ m}^2.$$
$$V = (60/6)[40 + 4(42.4) + 45 = 2546.0 \text{ m}^3]$$

Total volume = 2620.3 + 2546.0 = 5166.3 m^3

CHAPTER 17

Environmental Engineering

OUTLINE

WASTEWATER FLOWS

Wastewater flows comprise domestic and industrial wastewaters, infiltration, inflow, and storm water. Most modern sanitary sewers are separated from storm water systems, so these flows are treated separately. Modern sewers are constructed so that inflow rates are assumed to be negligible.

Domestic flows are determined from water use rates, with typical values being 0.57 m^3 per day per capita. Industrial and infiltration flows vary widely. Specific data are required based upon industry and production rates.

SEWER DESIGN

Hydraulics of Sewers

Sewers are designed as open channels, usually with a circular cross section. Flows in sewers are modeled using Manning's equation (see Chapter 3) as

$$V = \frac{1}{n} R^{2/3} S^{1/2} \tag{17.1}$$

where

V = velocity (m/s)
n = Manning coefficient, 0.013
R = hydraulic radius (m)
S = slope of hydraulic grade line

The relation between the hydraulic radius and the flow depth for a circular cross section is a complex relationship; as a result, flows in partially full sewers are calculated using nomographs like that in Figure 17.1.

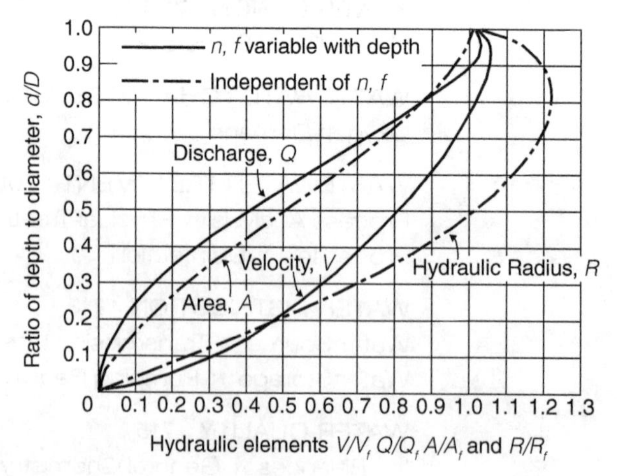

Figure 17.1

Design calculations usually involve knowing the slope of the sewer (S) and the partially full flow rate, Q. The nomograph is solved by assuming a pipe diameter (D), solving for the full flow rate, Q_f. From the ratio of Q/Q_f, the ratio of partial depth to pipe diameter (d/D) is obtained from the nomograph, from which the depth of flow is calculated. Sanitary sewers are designed to carry the peak flow with a depth of flow from one-half to full.

WASTEWATER CHARACTERISTICS

Wastewater can be characterized by a large number of parameters. Typical concentrations for design are given in Table 17.1. Per capita loading rates for 5-day biochemical oxygen demand and total suspended solids are 100 and 120 g/capita-d, respectively.

Table 17.1 Typical composition of domestic wastewater

Constituent	Concentration (mg/L)
Dissolved solids	700
Volatile dissolved solids	300
Total suspended solids	220
Volatile suspended solids	135
5-day biochemical oxygen demand	200
Organic nitrogen	15
Ammonia nitrogen	25
Total phosphorus	8

Oxygen Demand

Theoretical Oxygen Demand

Theoretical oxygen demand (ThOD) is a value calculated as the oxygen required to convert organic compounds in the wastewater to carbon dioxide and water. To determine the ThOD, the chemical formula of the waste must be known.

Chemical Oxygen Demand

Chemical oxygen demand (COD) is an empirical parameter representing the equivalent amount of oxygen that would be used to oxidize organic compounds under strong chemical oxidizing conditions. The COD and the ThOD are approximately equal for most wastes.

Biochemical Oxygen Demand

Biochemical oxygen demand (BOD) is an empirical parameter representing the amount of oxygen that would be used to oxidize organic compounds by aerobic bacteria. The test involves seeding of diluted wastewater and measurement of the oxygen depletion after five days in 300-mL glass bottles. The 5-day BOD (BOD_5) is related to the ultimate BOD as

$$BOD_5 = BOD_L(1 - 10^{-kt}) \quad \textbf{(17.2)}$$

where

BOD_L = the maximum BOD exerted after a long time of incubation (mg/ L);
k = the base-10 BOD decay coefficient (d^{-1})

The BOD test is accomplished at 20°C; the decay coefficient can be converted to other temperatures as

$$k_T = k_{20}\theta^{(T-20)} \quad \textbf{(17.3)}$$

where

k_T, k_{20} = BOD decay coefficient at temperature T and 20°C, respectively
θ = temperature correction coefficient, 1.056 for 20–30°C and 1.135 for 4–20°C

Nitrogenous Oxygen Demand

Nitrogenous oxygen demand results from the oxidation of ammonia to nitrate by nitrifying bacteria. The overall equation is

$$NH_4^+ + 2O_2 = NO_3^- + 2H^+ + 2H_2O \quad \textbf{(17.4)}$$

Nitrogenous oxygen demand is not included in either the BOD_5 or the COD value.

WASTEWATER TREATMENT

Wastewater treatment in the United States is managed under the Federal Water Pollution Control Act Amendments of 1972. Design of wastewater treatment facilities focuses on the two parameters BOD and total suspended solids (TSS). The general approach is to link together a series of unit processes to remove BOD and TSS sequentially to meet the discharge requirements. The unit processes are typically chosen to minimize treatment costs.

Process Analysis

Types of Reactions

Most reactions in wastewater treatment are considered to be homogeneous. In homogeneous reactions, the reaction occurs throughout the liquid, and mass transfer effects can be ignored. Chlorination is an example of a homogeneous reaction. In a heterogeneous reaction, the reactants must be transferred to a reactive site and the products must be transferred away from the reactive site. Biological reactions in bacterial films are an example of heterogeneous reactions.

Reaction Rates

Chemical reactions have many different forms, although the most common form is the conversion of a single product to a single reactant:

$$A \rightarrow B \quad \textbf{(17.5)}$$

If the rate of the reaction is zero-order, then the rate of conversion of A is

$$\frac{dA}{dt} = -k_0 \quad \textbf{(17.6)}$$

where k_0 = the zero-order reaction coefficient (mg/L-d).

If the rate of the reaction is first-order, then the rate of conversion of A is

$$\frac{dA}{dt} = -k_1[A] \quad \textbf{(17.7)}$$

where k_1 = first-order reaction coefficient (d^{-1}).

Reactor Types

Three important reactor types are batch, complete-mix, and plug flow, as shown in Figure 17.2. Batch reactors have no influent or effluent and have a hydraulic detention time of t_b, which is the time between filling and emptying. Complete-mix reactors have constant influent and effluent flow rates and are mixed so that spatial gradients of concentration are near zero. They have a hydraulic detention time as

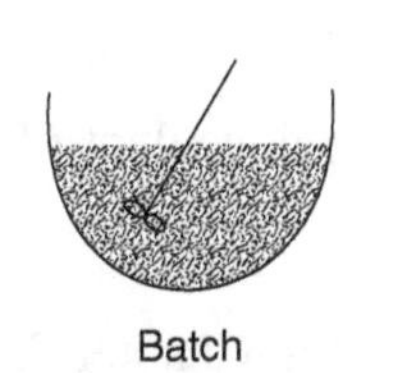

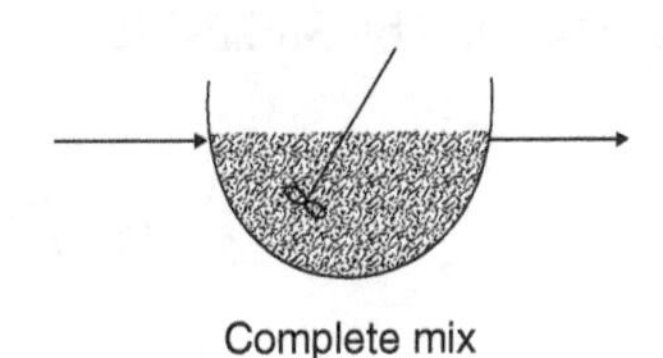

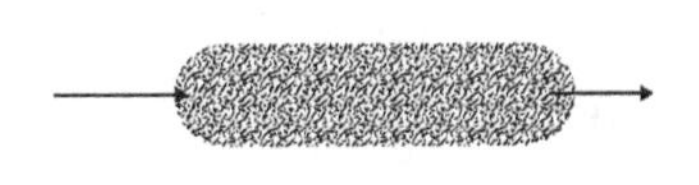

Figure 17.2

$$\theta_h = \frac{V}{Q} \quad (17.8)$$

where

θ = hydraulic detention time (d)
V = reactor volume (L)
Q = influent and effluent flow rate (L/d)

Plug flow reactors have constant influent and effluent flow rates but lack internal mixing. The reactors tend to be long and narrow, and the transport of water occurs from one end to the other. Their hydraulic detention time is

$$t_r = \frac{L}{v_e} \quad (17.9)$$

where

t_r = hydraulic detention time (d)
L = length of reaction (m)
v_e = liquid flow velocity (m/d)

The application of mass balances to the three types of reactors, assuming either zero- or first-order reaction rates, results in the following equations for the effluent concentration of A at steady state:

Batch Reactor, Zero-Order

$$A = A_0 - k_0 t \quad (17.10)$$

Complete-Mix, Zero-Order

$$A = A_0 - k_0 \theta_h \quad (17.11)$$

Plug Flow, Zero-Order

$$A = A_0 - k_0 t_r \quad (17.12)$$

Batch Reactor, First-Order

$$A = A_0 e^{(-k_1 t_b)} \quad (17.13)$$

Complete-Mix, First-Order

$$A = \frac{A_0}{1 + k_1 \theta_k} \quad (17.14)$$

Plug Flow, First-Order

$$A = A_0 e^{-k_1 t_r} \quad (17.15)$$

Plug flow reactors will consistently give lower effluent concentrations as compared to complete-mix reactors for all reaction rates greater than zero.

Physical Treatment Processes

Sedimentation

Sedimentation is divided into four types: discrete (Type 1); flocculent (Type 2); zone (Type 3); and compression (Type 4).

Discrete Sedimentation Discrete sedimentation is described in Chapter 7. In discrete sedimentation, the removal is determined by the terminal settling velocity. Settling velocities for small Reynolds numbers can be estimated by Stokes' law as

$$V_p = \frac{g(\rho_s - \rho_w)d^2}{18\mu} \tag{17.16}$$

where

V_p = particle settling velocity (m/s)
g = gravitational constant (9.8 m/s^2)
ρ_p, ρ_w = density of particle and water, respectively (kg/m^3)
μ = water viscosity (N-s/m^2)

For particles with V_p less than the overflow rate of the sedimentation basin, V_c, the removal R is

$$R = \frac{V_p}{V_c} \tag{17.17}$$

R is 1.0 for all particles with $V_p > V_s$.

Flocculent Sedimentation Flocculent sedimentation is similar to Type 1 settling except that the particle size increases as the particle settles in the sedimentation reactor. Type 2 or flocculent settling occurs in coagulation/flocculation treatment for water and in the top of secondary sedimentation basins.

Removal under flocculent sedimentation is determined empirically. Data must be collected to give isolines for percent removal as shown in Figure 17.3. Given such data, the total removal is

$$R = \sum \left(\frac{\Delta h_n}{h_s}\right)\left(\frac{R_n + R_{n+1}}{2}\right) \tag{17.18}$$

where the symbols are as illustrated in Figure 17.3.

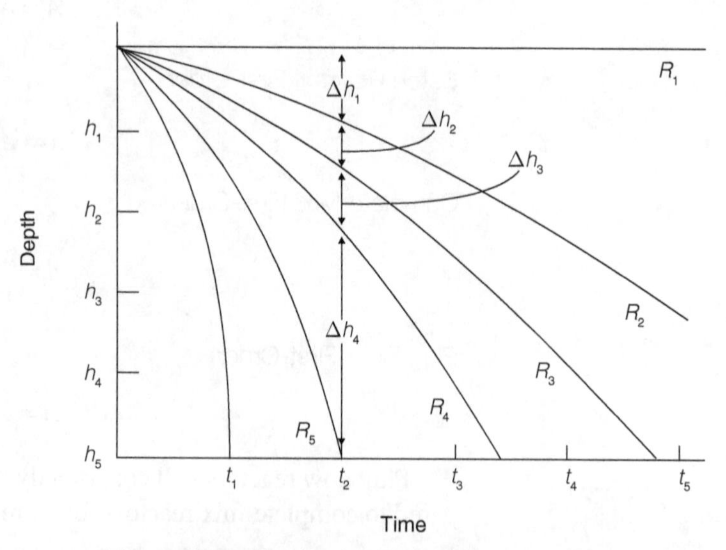

Figure 17.3 Empirical analysis of Type 3 settling

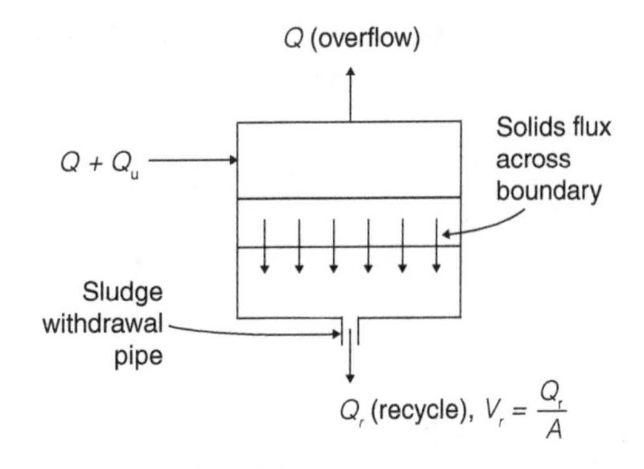

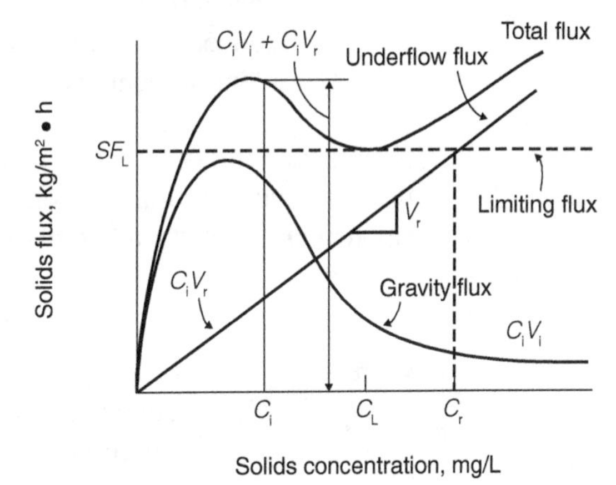

Figure 17.4 Schematic of secondary clarifier and empirical analysis of Type 3 settling

Zone Settling Zone settling occurs in the bottom of a secondary sedimentation basin, where the concentration of the solids increases to over 5000 mg/L. A schematic of a sedimentation basin coupled to a biological treatment reactor with cell recycle is shown in Figure 17.4. Under such conditions, the modeling of the zone sedimentation process is based upon the solids flux through the bottom section of the basin. The solids flux results from two components: gravity sedimentation and liquid velocity from the underflow.

$$SF = SF_g + SF_r \tag{17.19}$$

where

SF, SF_g, SF_r = solids flux, solids flux due to gravity, and solids flux due to recycle, respectively (kg/m²-d)

The resulting relationship is

$$SF = C_iV_i + C_iV_r \tag{17.20}$$

where

C_i = concentration of solids at a specified height in the basin (mg/L)
V_i = gravity settling velocity of sludge with concentration C_i (m/d)
V_r = downward fluid velocity from recycle (m/d)
= Q_r/A_H
A_H = horizontal area of sedimentation basin

Equation (17.19) is graphed in Figure 17.4. Under a constant recycle flow and a defined sludge, the transport of solids through the bottom of the clarifier becomes limited by the limiting solids flux rate (SF_L) shown in Figure 17.4. Estimates of SF_L

are obtained by graphical analysis as illustrated in Figure 17.4. The horizontal area of the sedimentation basin is then calculated as

$$A = \frac{(Q+Q_r)(C_{\text{infl}})}{\text{SF}_L} \tag{17.21}$$

or

$$A = \frac{Q_r C_r}{\text{SF}_L} \tag{17.22}$$

where

C_{infl} = concentration of TSS in influent to sedimentation basin
C_r = concentration of TSS in recycle from sedimentation basin
Q, Q_r = plant influent and recycle flow rates

Compression Sedimentation Compression sedimentation involves the slow movement of water through the pores in a sludge cake. The process is modeled as an exponential decrease in height of the sludge cake as

$$H_t - H_\infty = (H_0 - H_\infty)e^{-i(t-t_0)} \tag{17.23}$$

where

H_t, H_∞, H_0 = sludge height after time t, a long period, and initially
i = compression coefficient.

Biological Treatment Processes

Biological treatment involves the conversion of organic and inorganic compounds by bacteria with subsequent growth of the organisms. The common biological treatment processes in wastewater treatment involve the removal of various substrates including BOD, ammonia, nitrate, and organic sludges.

Description of Homogeneous Processes

The rate of removal of a substrate is given as

$$r_{\text{su}} = -\frac{k \times S}{(K_s + S)} \tag{17.24}$$

where

r_{su} = rate of substrate removal (mg/L-d)
k = maximum substrate removal rate (mg/mg-d)
S, K_s = substrate and half-velocity coefficient, respectively (mg/L)

The growth of organisms is given as

$$r_X = Yr_{\text{su}} - k_d X \tag{17.25}$$

where

r_X = rate of growth of bacteria (mg/L-d)
Y = substrate yield rate (mg/mg)
k_d = bacterial decay coefficient (d^{-1})
X = bacterial concentration, usually expressed as TVSS (mg/L)

Complete-Mix Reactor Without Cell Recycle For the complete-mix reactor without cell recycle shown in Figure 17.5(a), the governing equations are

$$X = \frac{Y(S_0 - S)}{(1 + k_d\theta_h)} \quad \textbf{(17.26)}$$

$$S = \frac{K_s(1 + \theta_h k_d)}{\theta_h(YK - k_d) - 1} \quad \textbf{(17.27)}$$

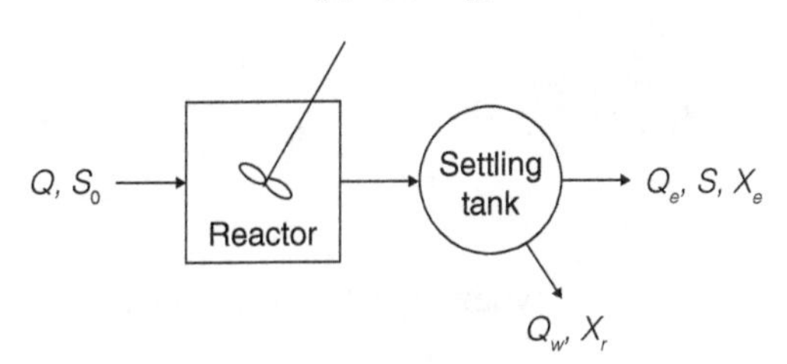

(a) Complete-Mix Without Cell Recycle

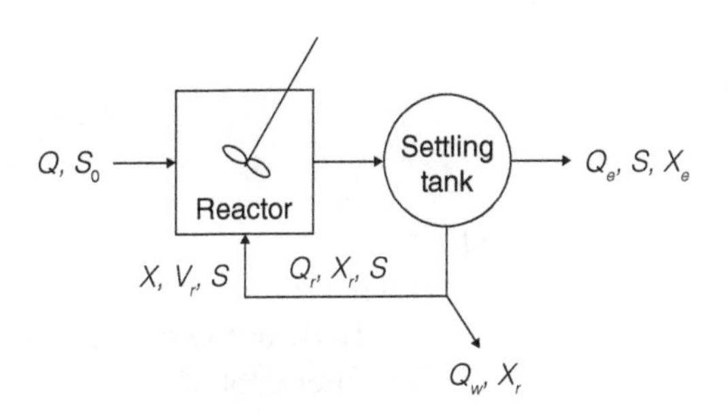

(b) Complete-Mix With Cell Recycle

Figure 17.5 Schematic of complete-mix reactors

Complete-Mix Reactor with Cell Recycle For the complete-mix reactor with cell recycle shown in Figure 17.5(b), the governing equations are

$$\frac{1}{\theta_c} = Y\left(\frac{kSX}{S + K_s}\right) - k_d \quad \textbf{(17.28)}$$

where θ_c = solids retention time (d);

$$\frac{F}{M} = \frac{S_0}{\theta_h X} \quad \textbf{(17.29)}$$

where F/M = food-to-microorganism ratio (mg/mg-d); and

$$X = \frac{\theta_c Y(S_0 - S)}{\theta_h(1 + k_d\theta_c)} \quad \textbf{(17.30)}$$

$$S = \frac{K_s(1 + \theta_c k_d)}{\theta_c(Yk - k_d) - 1} \quad \textbf{(17.31)}$$

Process design proceeds from selection of a solids retention time (θ_c) as the controlling parameter. The volume of the reactor is determined from Equation (17.30) with $\theta_h = V/Q$. The waste sludge flow rate is computed from

$$\theta_c = \frac{VX}{Q_w X_w + Q_e X_e} \quad \textbf{(17.32)}$$

where

Q_w, Q_e = sludge waste and effluent flow rates (L/d)

X_w = sludge waste concentration (X if from reaction, X_r if from the clarifier) (mg/L)

X_e = effluent TSS concentration (mg/L), usually 20 mg/L

Plug Flow Reactor with Cell Recycle The cell concentration in a plug flow reactor can be estimated using Equation (17.30). The effluent substrate concentration can be estimated from

$$\frac{1}{\theta_c} = \frac{Yk(S_0 - S)}{(S_0 - S) + \left(1 + \frac{Q_r}{Q}\right)K_s \ln(S_i / S)} - k_d \qquad \textbf{(17.33)}$$

where S_i = the diluted substrate concentration entering the reactor (mg/L).

Description of Heterogeneous Processes

Trickling filters are designed using an empirical equation as

$$A = Q\left(\frac{-\ln(S_{\text{eff}}/S_{\text{infl}})}{k_1 D}\right)^2 \qquad \textbf{(17.34)}$$

where

A = horizontal area

k_1 = BOD decay coefficient

D = filter depth

Disinfection

Disinfection in wastewater treatment is almost universally accomplished using chlorine gas. The objective of disinfection is to kill bacteria, viruses, and amoebic cysts. Effluent standards for secondary treatment require fecal coliform levels of less than 200 and 400 per 100 mL for 30-day and 7-day averages, respectively.

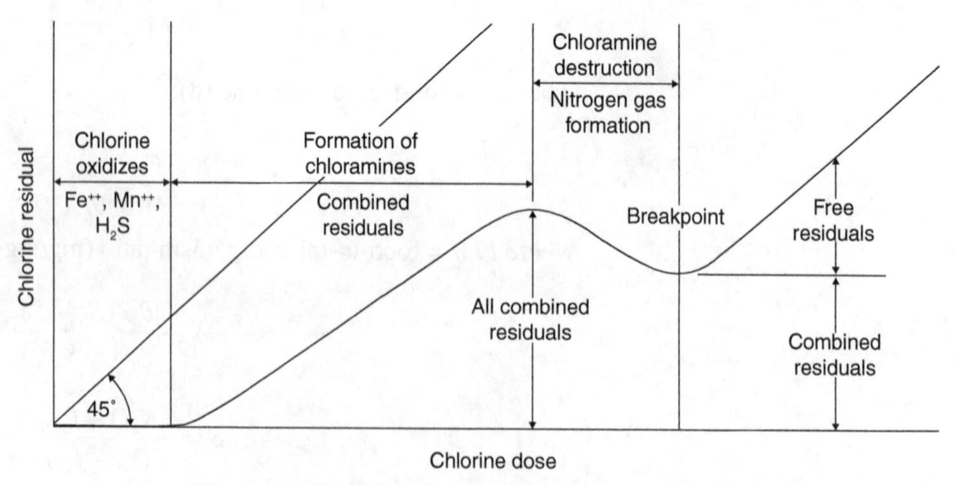

Figure 17.6 Chlorine dosage versus chlorine residual

Modeling of disinfection is described in Chapter 7. The major difference between chlorination of water supplies and that of wastewater is the reaction of chlorine with ammonia to produce chloramines as

$$NH_3 + HOCl \rightarrow NH_2Cl + H_2O$$

Chloramines can undergo further oxidation to nitrogen gas with their subsequent removal.

The stepwise oxidation of various compounds is shown in Figure 17.6. As chlorine is initially added, it reacts with easily oxidized compounds such as reduced iron, sulfur, and manganese. Further addition results in the formation of chloramines, collectively termed **combined chlorine residual**. Further addition of chlorine results in destruction of the chloramines and ultimately to the formation of free chlorine residuals as HOCl and OCl^-. The development of free chlorine residual is termed **breakpoint chlorination**.

WATER DISTRIBUTION

Water distribution systems involve a water source, a transmission line, a distribution network, pumping, and storage.

Water Source

The importance of the water source is primarily related to water elevation and water quality. The water elevation will determine the pumping required to maintain adequate flows in the network.

Water quality is largely determined by the source. Typical water sources include reservoirs, lakes, rivers, and underground aquifers. Surface water sources typically have low levels of dissolved solids but high levels of suspended solids. As a result, treatment of such sources requires coagulation, followed by sedimentation and filtration, to remove suspended solids. Groundwater sources are low in suspended solids but may be high in dissolved solids and reduced compounds. Groundwater, if high in dissolved solids, requires chemical precipitation or ion exchange for removal of dissolved ions.

Transmission Line

A transmission line is required to convey water from the sources and storage to the distribution system. Arterial main lines supply water to the various loops in the distribution system. The main lines are arranged in loops or in parallel to allow for repairs. Such lines are designed using the Hazen-Williams formula for single pipes (Equation 17.35) and the equivalent-pipe method for single loops.

$$v = kCr^{0.63}s^{0.54} \quad \textbf{(17.35)}$$

where

v = pipe velocity (m/s)
k = constant (2.79)
C = roughness coefficient (100 for cast iron pipe)
r = hydraulic radius (m)
s = slope of hydraulic grade line (m/m)

The equivalent-pipe method involves two procedures: conversion of pipes of unequal diameters in series into a single length of pipe, and conversion of pipes in parallel flow from the same into a single pipe with a single diameter. The headloss is maintained through each conversion. Multiple loops are designed using Hardy Cross methodology.

The Distribution Network

The distribution network comprises the arterial mains, distribution mains, and smaller distribution piping. The controlling design variable for water distribution networks is the pressure under maximum flow, with a minimum value of 140 to 280 kPa and a maximum value of 690 kPa.

Flow for the water distribution network is calculated as the sum of domestic, irrigation, industrial and commercial, and fire requirements. The various flows for a given community are typically obtained from historical data of water use. Without such data, estimates can be made from average values listed in Table 17.2.

Table 17.2 Water flow rates

Flow	Method of Estimation
Average domestic flow	Population served, 0.4 m^3/cap-d, 3,000 to 10,000 persons/km^2
Irrigation flow	Maximum of 0.75 times average daily flow for arid climates
Industrial/commercial flow	Computed from known industries and area of commercial districts

Water Storage

Water storage is required to maintain pressure in the system, to minimize pumping costs, and to meet emergency demands. Typically the storage is placed so that the load center or distribution network is between the water source and the storage, as shown in Figure 17.7.

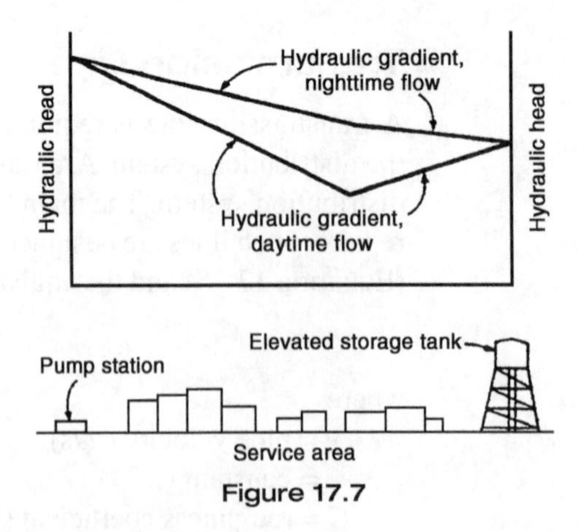

Figure 17.7

Storage requirements are calculated based upon variations in hourly flow and fire requirements. If the pumping capacity is set at some value less than the maximum hourly flow, then all flow requirements above that value will have to be provided by storage.

Pumping Requirements

The pumping system needs adequate capacity for design flows, taking into account the supply from storage. The required head must be adequate to maintain 140 kPa at the load center during maximum flows.

WATER QUALITY

The design of water treatment facilities requires calculations of concentrations and masses of various constituents in water and chemical additions. Common elements and ions of various constituents are listed in Table 17.3, and common water treatment chemicals are listed in Table 17.4. The equivalent weight is equal to the molecular weight divided by the absolute value of the valence.

Ion Balances

Electroneutrality requires that water have an equal number of equivalents of cations and anions. Often, bar graphs are used to show this relationship, as shown in Figure 17.8. From such diagrams, concentrations of alkalinity, carbonate (calcium and magnesium associated with alkalinity), and noncarbonate hardness (calcium and magnesium in other forms) can be easily identified.

General Chemistry Concepts

A number of concepts from general chemistry are required for engineering calculations related to water treatment.

Oxidation-Reduction Reactions

Oxidation-reduction (redox) reactions involve the transfer of electron from an electron donor to an electron acceptor. The easiest method to construct balanced redox reactions is through the use of half-reactions. A list of common half-reactions related to water treatment are in Table 17.5.

Table 17.3 Common elements and radicals

Name	Symbol	Atomic Weight	Valence	Equivalent Weight
Aluminum	Al	27.0	+3	9.0
Calcium	Ca	40.1	+2	20.0
Carbon	C	12.0	–4	
Chlorine	Cl	35.5	–1	35.5
Fluorine	F	19.0	–1	19.0
Hydrogen	H	1.0	+1	1.0
Iodine	I	126.9	–1	126.9
Iron	Fe	55.8	+2	27.9
			+3	
Magnesium	Mg	24.3	+2	12.15
			+4	
			+7	

(continued)

Nitrogen	N	14.0	−3	
			+5	
Oxygen	O	6.0	−2	8.0
Potassium	K	39.1	+1	39.1
Sodium	Na	23.0	+1	23.0
Ammonium	NH_4^+	18	+1	18.0
Hydroxyl	OH^-	17.0	−1	17.0
Bicarbonate	HCO_3^-	61.0	−1	61.0
Carbonate	CO_3^{2-}	60.0	−2	30.0
Nitrate	NO_3^-	46.0	−1	46.0
Hypochlorite	OCl^-	51.5	−1	51.5

Henry's Law

Henry's Law states that the weight of any dissolved gas is proportional to the pressure of the gas.

$$C_{\text{equil}} = \alpha\, p_{\text{gas}} \tag{17.36}$$

where

C_{equil} = equilibrium dissolved gas concentration
P_{gas} = partial pressure of gas above liquid
α = Henry's Law constant

Equilibrium Relationships

For an equilibrium chemical equation expressed as

$$A + B \rightleftharpoons C + D \tag{17.37}$$

the relationship of concentrations at equilibrium can be approximated as

$$K_{\text{eq}} = \frac{[C][D]}{[A][B]} \tag{17.38}$$

where

K_{eq} = equilibrium constant
[ ] = molar concentrations

Table 17.4 Common inorganic chemicals for water treatment

Name	Formula	Usage	Molecular Weight	Equivalent Weight
Activated carbon	C	Taste and odor	12.0	
Aluminum sulfate	$Al_2(SO_4)_3 \cdot 14.3H_2O$	Coagulation	600	100
Ammonia	NH_3	Chloramines, disinf.	17.0	
Ammonium fluosilicate	$(NH_4)_2SiF_6$	Fluoridation	178	
Calcium carbonate	$CaCO_3$	Corrosion control	132	66.1
Calcium fluoride	CaF_2	Fluoridation	78.1	
Calcium hydroxide	$Ca(OH)_2$	Softening	74.1	37.0

(continued)

Calcium hypochlorite	$Ca(ClO)_2 \cdot 2H_2O$	Disinfection	179	
Calcium oxide	CaO	Softening	56.1	28.0
Carbon dioxide	CO_2	Recarbonation	44.0	22.0
Chlorine	Cl_2	Disinfection	71.0	
Chlorine dioxide	ClO_2	Taste and odor	67.0	
Ferric chloride	$FeCl_3$	Coagulation	162	54.1
Ferric hydroxide	$Fe(OH)_3$		107	35.6
Fluorosilicic acid	H_2SiF_6	Fluoridation	144	16.0
Oxygen	O_2	Aeration	32.0	
Sodium bicarbonate	$NaHCO_3$	pH adjustment	84.0	84.0
Sodium carbonate	Na_2HCO_3	Softening	106	53.0
Sodium hydroxide	NaOH	pH adjustment	40.0	40.0
Sodium hypochlorite	NaClO	Disinfection	74.4	
Sodium fluosilicate	Na_2SiF_6	Fluoridation	188	

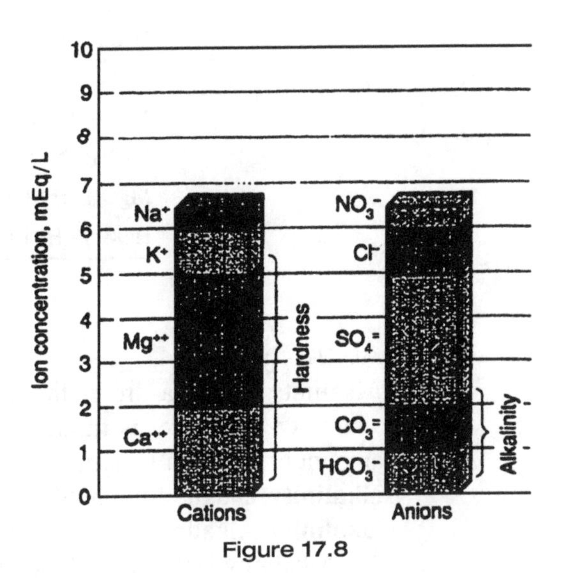

Figure 17.8

Some commonly used equilibrium constants are listed in Table 17.6.

A common equilibrium relationship is the disassociation of water, which is given as

$$H_2O \rightleftharpoons H^+ + OH^- \quad \textbf{(17.39)}$$

which has a K_{eq} value of 10^{-7}. The hydrogen ion concentration is typically represented by the pH value, or the negative log of the hydrogen ion concentration:

$$pH = -\log[H^+] \quad \textbf{(17.40)}$$

Table 17.5 Common half-reactions for water treatment

Reduced Element	Half-Reaction
Cl	$\frac{1}{2}Cl_2 + e^- \rightarrow Cl^-$
Cl	$\frac{1}{2}ClO^- + H^+ + e^- \rightarrow \frac{1}{2}Cl^- + \frac{1}{2}H_2O$
Cl	$\frac{1}{8}ClO_4^- + H^+ + e^- \rightarrow \frac{1}{8}Cl^- + \frac{1}{2}H_2O$
Fe	$\frac{1}{2}Fe^{2+} + e^- \rightarrow \frac{1}{2}Fe$
Fe	$Fe^{3+} + e^- \rightarrow Fe^{2+}$
Fe	$\frac{1}{3}Fe^{3+} + e^- \rightarrow \frac{1}{3}Fe$
I	$\frac{1}{2}I_2 + e^- \rightarrow I^-$
N	$\frac{1}{8}NO_3^- + \frac{5}{4}H^+ + e^- \rightarrow \frac{1}{8}NH_4^+ + \frac{3}{8}H_2O$
N	$\frac{1}{5}NO_3^- + \frac{6}{5}H^+ + e^- \rightarrow \frac{1}{10}N_2 + \frac{3}{5}H_2O$
O	$\frac{1}{4}O_2 + H^+ + e^- \rightarrow \frac{1}{2}H_2O$

Table 17.6 Common equilibrium constants

Equation	K_{eq}
$H_2CO_3 \rightleftharpoons H^+ + HCO_3^-$	$10^{-6.4}$
$HCO_3^- \rightleftharpoons H^+ + CO_3^{2-}$	$10^{-10.3}$
$NH_3 + H_2O \rightleftharpoons NH_4^+ + OH^-$	$10^{-4.7}$
$CaOH^+ \rightleftharpoons Ca^{2+} + OH^-$	$10^{-1.5}$
$MgOH^+ \rightleftharpoons Mg^{2+} + OH^-$	$10^{-2.6}$
$HOCl \rightleftharpoons H^+ + OCl^-$	$10^{-7.5}$

Alkalinity

Alkalinity is a measure of the ability of water to consume a strong acid. In the alkalinity test, water is titrated with a strong acid to pH 6.4 and to pH 4.5. The amount of acid that is used to reach the first pH endpoint is termed the **carbonate alkalinity**, and the amount of acid needed to reach the second end point is the **total alkalinity**. Alkalinity is expressed as calcium carbonate.

Alkalinity is the sum of bicarbonate, carbonate, and hydroxide concentrations minus hydrogen ion concentration. For most sources to be used for drinking water, alkalinity can be approximated as carbonate and bicarbonate. Carbonate alkalinity is the predominant species when the pH is above 10.8, and bicarbonate is the predominant species when the pH is between 6.9 and 9.8.

Solubility Relationships

The equilibrium between a compound in its solid crystalline state and its ionic form in solution, where

$$X_aY_b = aX^{b+} + bY^{a-} \quad \textbf{(17.41)}$$

is given by

$$K_{sp} = [X^{b+}]^a[Y^{a-}]^b \quad \textbf{(17.42)}$$

where K_{sp} = solubility product. Solubility products for common precipitation reactions in water treatment are given in Table 17.7.

Table 17.7 Common solubility products

Compound	K_{sp}
Magnesium carbonate	4×10^{-5}
Magnesium hydroxide	9×10^{-12}
Calcium carbonate	5×10^{-9}
Calcium hydroxide	8×10^{-6}
Aluminum hydroxide	1×10^{-32}
Ferric hydroxide	6×10^{-38}
Ferrous hydroxide	5×10^{-15}
Calcium fluoride	3×10^{-11}

Dissolved Oxygen Relationships

The dissolved oxygen concentration in a stream is determined by the rate of oxygen consumption, commonly caused by the discharge of oxygen-demanding substances, and the rate of reoxygenation from the atmosphere.

Biochemical Oxygen Demand

The concentration of oxygen-demanding substances in a river is commonly expressed as biochemical oxygen demand or BOD as described in Chapter 5. BOD is measured as the oxygen that is removed in a 20°C, five-day test and is expressed as BOD_5. This value has to be converted to an ultimate BOD or L value at the temperature of interest as

$$L = \frac{BOD_5}{1 - e^{-5k_{20}}} \tag{17.43}$$

where k_{20} = BOD decay constant at 20°C. Typical k_{20} values for organic wastes range from about 0.20 to 0.30.

The rate of oxygen consumption by BOD is related to the ultimate BOD concentration as

$$r_{BOD} = k_T L \tag{17.44}$$

Oxygen Deficit

The dissolved oxygen in water is determined by Henry's Law, based upon a 20% partial pressure in the atmosphere. Saturation values as a function of temperature can be extrapolated from known values of 14.6, 11.3, 9.1, and 7.5 mg/L at 0, 10, 20, and 30°C, respectively.

At a given temperature the dissolved oxygen concentration can be expressed also as a deficit or

$$D = (C_{sat} - C) \tag{17.45}$$

where

C_{sat}, C = saturation and actual dissolved oxygen concentration (mg/L)

D = dissolved oxygen deficit at given temperature

Mixing

Assuming complete mixing in a river, the concentration of dissolved oxygen and ultimate BOD in a receiving stream is related as

$$C_0 = \frac{Q_r C_r + Q_w C_w}{Q_r + Q_w} \tag{17.46}$$

where

C_r, C_w = concentration of constituents in river and waste, respectively
Q_r, Q_w = flow rate of river and waste, respectively

Reoxygenation

Reoxygenation occurs by diffusion of oxygen from the atmosphere to the river. The rate of reoxygenation or reaeration is given as

$$r_{\text{reoxy}} = k_2 D \tag{17.47}$$

where k_2 = reaeration coefficient (1/d).

Oxygen Sag Model

The dissolved oxygen concentration in a stream after the addition of oxygen-demanding wastes can be expressed in differential form as

$$\frac{dD}{dt_r} = r_{\text{BOD}} - r_{\text{reoxy}} \tag{17.48}$$

where t_r = travel time in the river from the point of waste addition (d).

The resulting dissolved oxygen profile is shown in Figure 17.9. After addition of the waste, the dissolved oxygen decreases to a maximum deficit (D_c) or minimum dissolved oxygen level at a distance x_c or travel t_c. Past this point, the deficit decreases as the river recovers.

Equation (17.48) can be integrated to

$$D = \frac{kL_0}{k_2 - k}\left(e^{-kt_r} - e^{-k_2 t_r}\right) + D_0 e^{-k_2 t_r} \tag{17.49}$$

where D_0, L_0 = oxygen deficit and ultimate BOD concentration at point of waste discharge in the river, respectively.

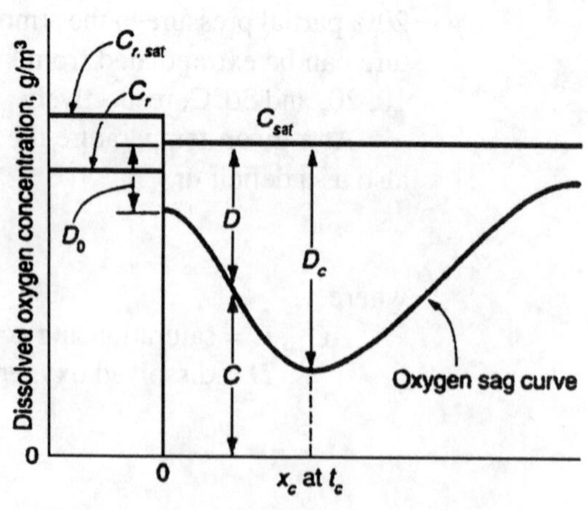

Figure 17.9

WATER TREATMENT

Water treatment is used to alter the quality of water to make it chemically and bacteriologically safe for human consumption. Common sources are groundwater and surface waters. Groundwater treatment may involve removal of pathogens, removal of iron and manganese, and removal of hardness (Figure 17.10). Surface water treatment typically involves simultaneous removal of pathogens, suspended solids, and taste- and odor-causing compounds (Figure 17.11). All of these treatment processes are composed of unit processes.

Sedimentation

After flocculation, the larger particles are removed by sedimentation. Settling is also used in water treatment after oxidation of iron or manganese. Flocculated particles have densities of about 1400 to 2000 kg/m.

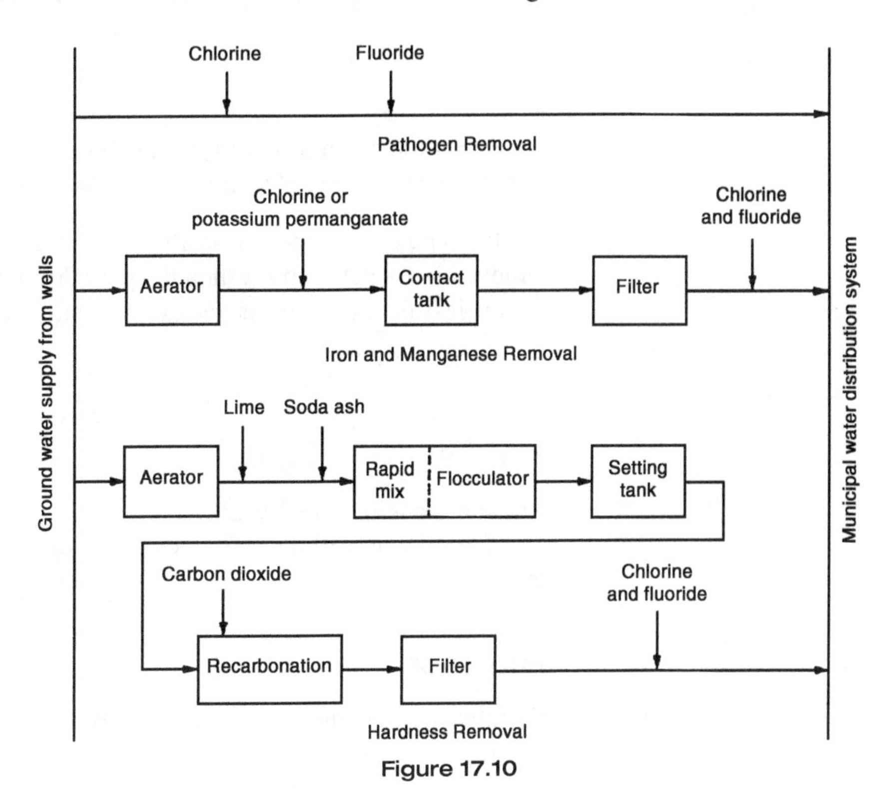

Figure 17.10

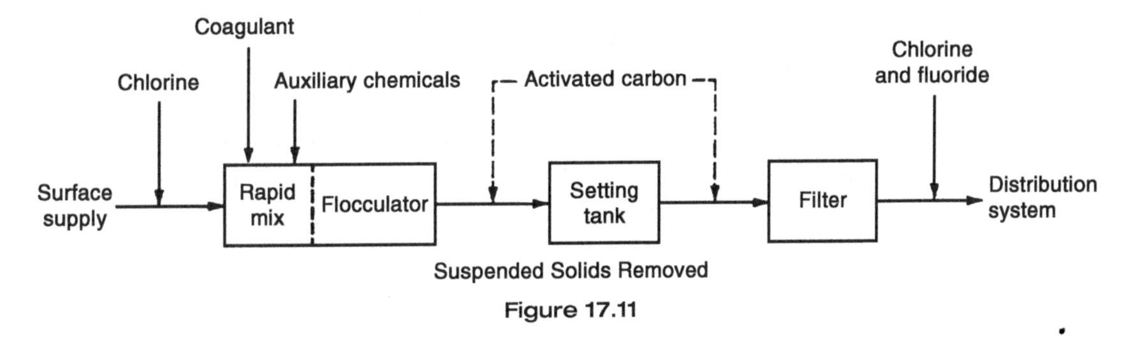

Figure 17.11

Settling velocities of particles are determined by Stokes' law for Reynolds numbers less than 0.3:

$$v_s = \frac{g(\rho_p - \rho_w)d_p^2}{18\mu} \tag{17.50}$$

where
v_s = terminal settling velocity (m/s)
g = gravitational constant (9.8 m/s^2)
d_p = particle diameter (m)
μ = water viscosity (0.001 kg/m-s)

For particles coming into a sedimentation basin, the particles enter at all depths. A critical settling velocity related to the sedimentation basin can be calculated as

$$v_{sc} = \frac{Q}{A_s} \tag{17.51}$$

where
Q = flow rate into sedimentation basin (m^3/s)
A_s = surface area of sedimentation basin (m^2)

In standard sedimentation theory, if the settling rate of the particles to be removed is greater than v_{sc}, then the particles will be 100% removed. For particles with v_s less than v_{sc}, the particles will be only partially removed, with the removal given as

$$R = \frac{v_s}{v_{sc}} \tag{17.52}$$

where R = decimal removal.

The critical settling velocities used for design are typically expressed as an overflow rate.

Filtration

Filtration in water treatment is commonly accomplished with granular filters. Media for such filters may include sand, charcoal, and garnet; the filters can involve one, two, or several different filter media. The water is applied to the top of the filter and is collected through underdrains.

Filtration is a complex process involving entrapment, straining, and absorption. As the filtration process proceeds, the headloss associated with the water flow through the porous media increases as the filtered material accumulates in the pores. In addition, the number of particles passing through the filter increases, with a subsequent increase in effluent turbidity. When either the allowable headloss or allowable effluent quality is exceeded, then the filter has to be backwashed to remove the accumulated material.

Softening

The removal of hardness is accomplished using line/soda ash softening or ion exchange.

Lime/Soda Ash Softening

Lime/soda ash softening occurs by the addition of calcium hydroxide and sodium bicarbonate to form a chemical precipitate, which is removed by sedimentation and filtration.

For waters containing excess, the calcium hydroxide reacts with carbon dioxide and carbonate hardness as

$$CO_2 + Ca(OH)_2 \rightarrow CaCO_3 + H_2O$$
$$Ca(HCO_3)_2 + Ca(OH)_2 \rightarrow CaCO_3 + 2H_2O$$
$$Mg(HCO_3)_2 + 2Ca(OH)_2 \rightarrow 2CaCO_3 + Mg(OH)_2 + 2H_2O$$

The sodium bicarbonate and calcium hydroxide react with the noncarbonate hardness as

$$Ca^{2+} + Na_2CO_3 \rightarrow CaCO_3 + H_2O$$
$$Mg^{2+} + Ca(OH)_2 \rightarrow Mg(OH)_2 + Ca^{2+}$$

Recarbonation is required after treatment to remove the excess lime, magnesium hydroxide, and carbonate, reducing the pH to about 8.5 to 9.5.

Based upon the foregoing equations, the requirements of lime (L) and soda ash (SA) in mEq/L are

$$L = CO_2 + HCO_3^- + Mg^{2+} \quad \textbf{(17.53)}$$

$$SA = Ca^{2+} + Mg^{2+} - Alk \quad \textbf{(17.54)}$$

The lime requirement needs to be increased about 1 Eq/m^3 to raise the pH to allow precipitation of the magnesium hydroxide.

Ion Exchange

Ion exchange processes are used to remove ions of calcium, magnesium, iron, and ammonium. The exchange material is a solid that has functional groups that replace ions in solution for ions on the exchange material. Ion exchange materials can be made to remove cations or anions. Most materials are regenerated by flushing with solutions with high concentrations of either H^+ or Na^+.

Ion exchange process is controlled by the exchange capacity and the selectivity. The **exchange capacity** expresses the equivalents of cations or anions that can be exchanged per unit mass. The **selectivity** for ion B as compared to ion A is expressed as

$$K_{B/A} = \frac{\chi_{R-A.}[A^+]}{\chi_{R-B}[B^+]} \quad \textbf{(17.55)}$$

where

$K_{B/A}$ = selectivity coefficient for replacing A on resin with B;
χ_{R-A}, χ_{R-B} = mole fractions of A and B for the absorbed species.

Chlorination

Chlorination is the most common form of disinfection for treatment of water. Chlorine is a strong oxidation agent that destroys organisms by chemical attack. Chlorine is usually added in the form of chlorine gas at a level of 2 to 5 mg/L. Other common forms are chlorine dioxide, sodium hypochlorite, and calcium hypochlorite.

Chlorine in water undergoes an equilibrium reaction to form HOCl and OCl^-. The hypochlorite is a weak acid as listed in Table 17.6. HOCl is a more effective disinfectant than OCl^-; as such, the effectiveness is strongly dependent upon pH.

The effectiveness of chlorination depends upon the time of contact, the chlorine concentration, and the concentration of organisms. The effect of time is modeled as

$$\frac{N_t}{N_0} = e^{-kt^m} \tag{17.56}$$

where

N_t, N_0 = number of organisms at time t and time zero
k = decay coefficient (1/d)
t = time (d)
m = empirical coefficient (usually 1)

Fluoridation

Fluoridation of water supplies has been shown to reduce dental caries dramatically. Optimum concentrations are about 1 mg F/L. The most commonly used compounds are sodium fluoride, sodium fluosilicate, and fluorosilicic acid. Fluoride is usually added after coagulation or lime/soda ash softening, because high calcium concentration can result in the fluorides precipitating.

Activated Carbon

Activated carbon is added in water treatment to adsorb taste- and odor-causing compounds and to remove color. The process usually involves the direct addition of powdered activated carbon, with removal in the sedimentation and filtration units. The required dosage is usually determined empirically using laboratory tests.

Sludge Treatment

The sources of sludge in water treatment include sand, silt, chemical sludges, and backwash solids. These sludge are typically concentrated in settling basins or lagoons. The sludges can be further concentrated by drying on sand drying beds or by centrifugation.

PROBLEMS

17.1 What is the required slope for a 0.3-m diameter circular sewer as shown in Exhibit 17.1, with a flow of 0.014 m^3/s and a minimum velocity of 0.6 m/s?

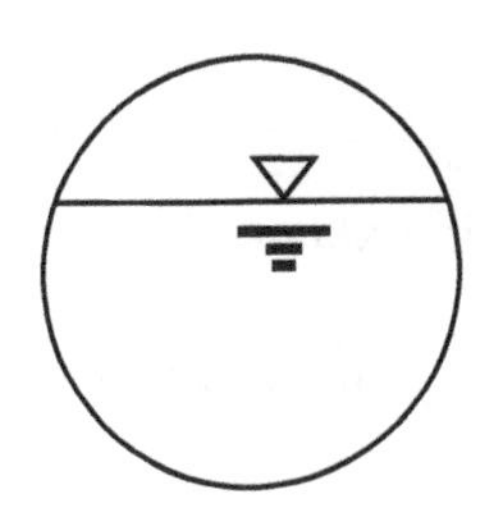

Exhibit 17.1

17.2 Compute the carbonaceous and nitrogenous oxygen demands for a waste with a chemical formula of $C_5H_7NO_2$.

17.3 A lake with a volume of 5×10^6 m^3 has a freshwater flow of 20 m^3/s. A waste is dumped into the lake at a rate of 50 g/s with a decay rate of 0.2/d. What is the steady-state concentration? Assume that the lake is completely mixed.

17.4 A waste with a flow of 2.8 L/s is discharged to a small stream with a flow of 141 L/s. The waste has a 5-d BOD of 200 mg/L ($k = 0.2$/d). What is the BOD after one day's travel in the stream?

17.5 A sedimentation basin has an overflow of 3 ft/hr. The influent wastewater has a particle distribution as follows:

Percent of Particles	Settling Velocity (m/hr)
20	0.30 to 0.61
30	0.61 to 0.91
50	0.91 to 1.22

Determine the total removal in the sedimentation basin.

17.6 Disinfection with chlorine is known to be a first-order reaction. The first-order decay rate under a given concentration of chlorine is measured as 0.35/hr. The flow rate is 12,000 L per hour, and the desired removal of organisms is from 10×10^6/100mL to less than 1/100mL. Determine

a. Volume of reactor required assuming the use of a complete-mix reactor
b. Volume of reactor required assuming the use of plug flow reactor

17.7 A Type 3 settling test was conducted on a waste activated sludge. The results were as follows:

MLSS (mg/L)	Settling Velocity (m/hr)
4000	2.4
6000	1.2
8000	0.55
10,000	0.31
20,000	0.06

Determine the limiting solids flux if the concentration of the recycled solids is 10,000 mg/L.

17.8 Find the effluent soluble BOD and reactor cell concentration for an aerobic complete-mix reactor with no recycle.

$$k = 10 \text{ mg/mg-d}$$
$$K_s = 50 \text{ mg/L}$$
$$k_d = 0.10\text{/d}$$
$$Y = 0.6 \text{ mg/mg}$$
$$S_0 = 200 \text{ mg BOD/L}$$
$$\theta_h = \theta_c = 2\text{d}$$

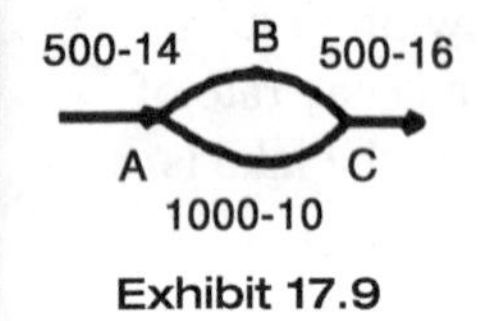

Exhibit 17.9

17.9 Find a single pipe to replace the pipe loop shown in Exhibit 17.9.

17.10 How many grams of oxygen are required to burn 1 gram of methane?

$$CH_4 + O_2 \rightarrow CO_2 + H_2O$$

17.11 A liter of water is at equilibrium with an atmosphere containing a partial pressure of 0.1 atm of CO_2. How many grams are dissolved in the water? (α = 2.0 g/L-atm)

17.12 How much HOCl is present in a solution containing 0.1M chlorine at pH 8?

17.13 How many grams of fluoride would be present in a solution saturated with calcium fluoride?

17.14 A 5-d BOD and ultimate BOD are measured at 180 mg/L and 200 mg/L, respectively. What is the decay coefficient?

17.15 A wastewater with a dissolved oxygen concentration of 1 mg/L is discharged to a river. The river is at 20°C and saturated with dissolved oxygen. If the flows are 2.8×10^{-2} m/s and 2.8 m/s for the wastewater and the river, respectively, what is the oxygen deficit after mixing?

17.16 A water treatment plant is required to produce an average treated water flow 14.4×10^4 m^3/d from a water source of the following characteristics:

Chlorines	92 mg/L
Potassium	31 mg/L
Sodium	14 mg/L
Sulfates	134 mg/L
Calcium	94 mg/L
Magnesium	28 mg/L
Alkalinity	135 mg/L as $CaCO_3$
pH	7.8
Temperature	21°C
Total solids	720 mg/L

Determine an ion balance for the water.

17.17 A water contains silt particles with a uniform diameter of 0.02 mm and a specific gravity of 2.6. What removal is expected in a clarifier with an overflow rate of 12 m/d (300 gal/ft^2-d)?

17.18 The carbonaceous oxygen demand of 200 mg/L of glycine (H_2NCH_2COOH) is:

a. less than the nitrogenous oxygen demand
b. depends on the reaction rate
c. greater than 120 mg/L
d. greater than 100 mg/L

17.19 A completely mixed lagoon has a waste flow of 1 m^3/s of a BOD waste with a decay coefficient of 0.3/d and a concentration of 100 mg/L. If the effluent BOD must be 20 mg/L or less, then:

a. the volume required is about 1×10^6 m^3
b. the volume required is about 2×10^6 m^3
c. the volume required is about 0.5×10^6 m^3
d. an increase in temperature would increase the volume required

17.20 A completely mixed reactor with cell recycle is designed to treat a municipal waste. Assuming removal kinetics follow the equation:

$$r_{su} = -\frac{k \times S}{(K_s + S)}$$

Which of the following statements is correct?

a. The effluent substrate concentration decreases with an increase in θ_c.
b. The food:microorganism level is independent of θ_c.
c. Microbe concentrations will be smaller than in the no-recycle case.
d. θ_c is independent of effluent quality.

17.21 Which of these statements are correct?

1. Chlorination of wastewater effluents requires more chlorine than chlorination of drinking water.
2. Chlorination of wastewater effluents requires three moles of chlorine for each mole of ammonia.
3. Chlorination of wastewater effluents is used to improve effluent quality.
4. Chlorination of wastewater effluents oxidizes other chemicals such as ferrous iron.

a. All of the above statements are correct.
b. None of the above statements is correct.
c. Only statements 1 and 3 are correct.
d. Only statements 1, 3, and 4 are correct.

17.22 A chemical is removed in a completely mixed reactor at a zero-order rate of 0.25 mg/L-d. The influent concentration is 10 mg/L and the hydraulic detention time is 10 days. For this reactor, which of these statements are correct?

1. The effluent concentration would be the same as that for a batch reactor with a 2-day retention time.
2. The effluent concentration will be about 7.5 mg/L.
3. The effluent quality would improve with longer detention times.
4. The effluent quality would improve with increased mixing.

a. All of the above statements are correct.
b. None of the above statements is correct.
c. Only statements 1 and 2 are correct.
d. Only statements 1, 3, and 4 are correct.

SOLUTIONS

17.1

$$D_f = 0.30 \text{ m}$$
$$Q = 0.014 \text{ m}^3\text{/s}$$
$$V = 0.6 \text{ m/s}$$

Need V/V_f or Q/Q_p or A/A_f or R/R_f

$$A_f = \frac{\pi D_f^2}{4} = 0.071 \text{ m}^2$$
$$A = \frac{Q}{V} = 0.023 \text{ m}^2$$
$$\frac{A}{A_f} = 0.32$$

From nomograph (Exhibit 17.1a):

$$d/D_f = 0.35 \; d = 0.105 \text{ m}$$
$$R/R_f = 0.75$$
$$R_f = \frac{\pi D_f^2/4}{\pi D} = \frac{D_f}{4} = 0.075 \text{ m}$$
$$R = (0.75)(0.075) = 0.056 \text{ m}$$

From Manning's equation:

$$V = \frac{1}{n} R^{2/3} S^{1/2}$$
$$S = \left(\frac{nV}{1.R^{2/3}} \right)^2$$
$$= \left(\frac{(0.013)(0.6 \text{ m/s})}{(1.)(0.056)^{2/3}} \right)^2$$
$$= 0.0028 \text{ m/m}$$

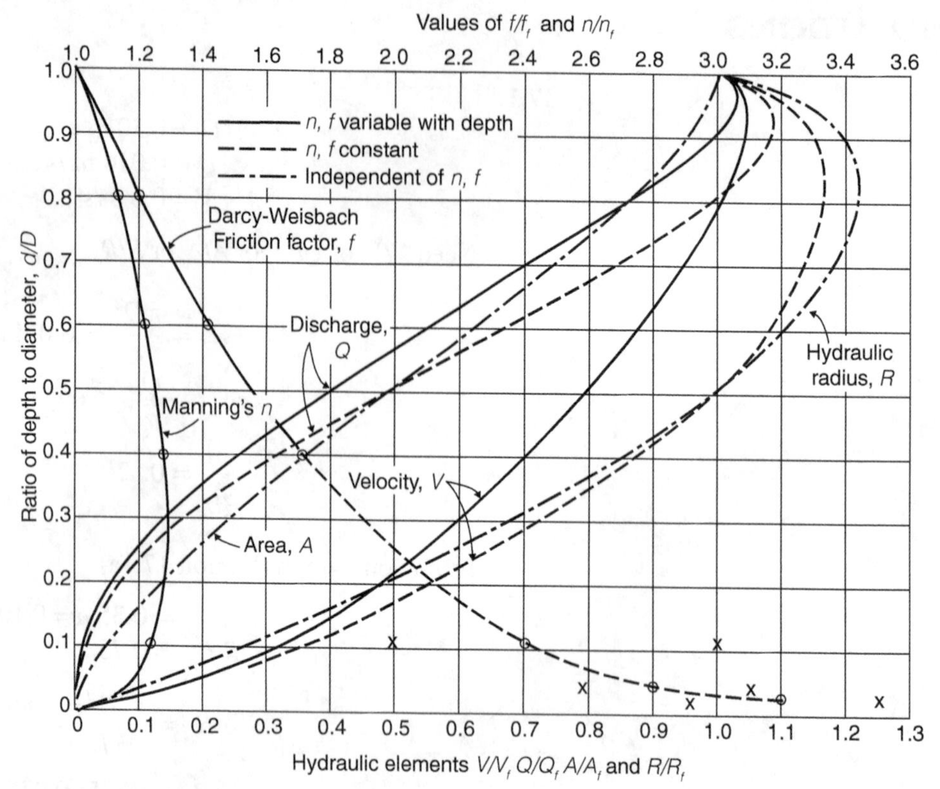

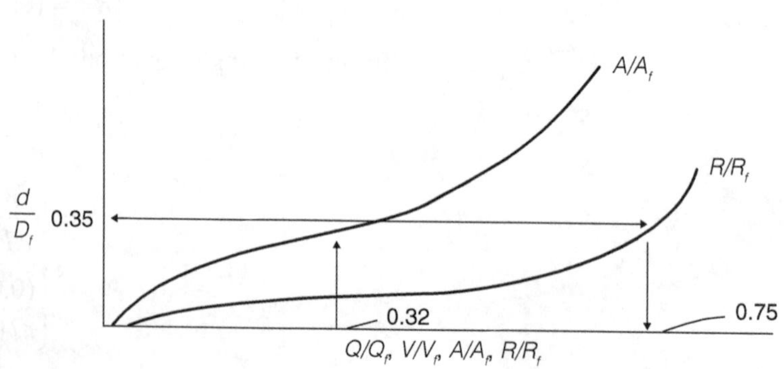

Exhibit 17.1a

17.2 For ThOD:

$$C_5H_7NO_2 + 5O_2 \rightarrow 5CO_2 + NH_3 + 2H_2O$$

For NOD:

$$NH_3 + 2O_2 \rightarrow HNO_3 + H_2O$$

For carbonaceous oxygen demand:

$$\frac{5 \text{ moles } O_2}{1 \text{ mole } C_5H_7NO_2} \times \frac{32 \text{ g } O_2}{\text{mole } O_2} \times \frac{1 \text{ mole } C_5H_7NO_2}{113 \text{ g}} = \frac{1.4 \text{ g } O_2}{\text{g } C_5H_7NO_2}$$

For nitrogenous oxygen demand:

$$\frac{2 \text{ moles } O_2}{1 \text{ mole } C_5H_7NO_2} \times \frac{32 \text{ g } O_2}{\text{mole } O_2} \times \frac{1 \text{ mole } C_5H_7NO_2}{113 \text{ g}} = \frac{0.57 \text{ g } O_2}{\text{g } C_5H_7NO_2}$$

17.3

$$C = \frac{S}{Q + kV}$$

$$= \frac{50 \text{ g/s}}{20 \text{ m}^3\text{/s} + (0.2\text{/d})(5 \times 10^6 \text{ m}^3)\left(\frac{1 \text{ d}}{8.64 \times 10^4 \text{s}}\right)}$$

$$= 1.58 \text{ g/m}^3$$

17.4

$$\text{BOD}_L = \frac{\text{BOD}_5}{(1 + e^{-5k})}$$

$$= \frac{20 \text{ mg/L}}{\left(1 - e^{-\left(5\text{d} \times \frac{2}{\text{d}}\right)}\right)}$$

$$= 294 \text{ mg/L}$$

After mixing,

$$\text{BOD}_i = \frac{(294 \text{ mg/L})(2.8 \text{ L/s})}{141.6 \text{ L/s} + 2.8 \text{ L/s}}$$

$$= 5.76 \text{ mg/L}$$

Assume the stream is plug flow.

$$\text{BOD} = \text{BOD}_i e^{-kt}$$
$$= (5.76 \text{ mg/L})e^{-(0.21\text{d})(1\text{d})}$$
$$= 4.7 \text{ mg/L}$$

17.5

Particles	Avg V_p (cm/hr)	V_c (cm/hr)	R (%)
2.54–6.45	3.81	7.62	50
6.45–7.62	6.35	7.62	83
7.62–10.16	8.89	7.62	100

$$R_{\text{total}} = (0.2)(0.5) + (0.3)(0.83) + (0.5)(1.0)$$
$$= 0.85 \text{ or } 85\%$$

17.6

a.

$$\frac{A}{A_0} = \frac{1}{1 + k_1 \frac{v}{Q}} = \frac{1}{10 \times 10^6}$$

$$1 + k_1 \frac{v}{Q} = 1 \times 10^7$$

$$V = \frac{(1 \times 10^7)(12{,}000 \text{ L/hr})}{0.35\text{/hr}}$$

$$= 3.4 \times 10^{11} \text{ L}$$

b.

$$\frac{A}{A_0} = e^{-kv/Q} = \frac{1}{10\times10^6}$$

$$-k\frac{v}{Q} = -16.1$$

$$V = \frac{-16.1(12{,}000\text{ L/hr})}{0.35\text{/hr}}$$

$$= 5.5\times10^5\text{ L}$$

17.7

MLSS (mg/L)	v_i (ft/hr)	x_iv_i (mg-ft/L-hr)	v_i (cm/hr)	x_iv_i (kg/m²-d)
4000	7.8	3.12 × 10⁴	19.81	4.63
6000	3.8	2.28 × 10⁴	9.65	3.11
8000	1.8	1.44 × 10⁴	4.57	1.97
10,000	1.0	1.00 × 10⁴	2.54	1.37
20,000	0.2	0.80 × 10⁴	0.51	1.09

From graphical solution (see Exhibit 17.7)

Limiting solids flux = 7 kg/m²-d

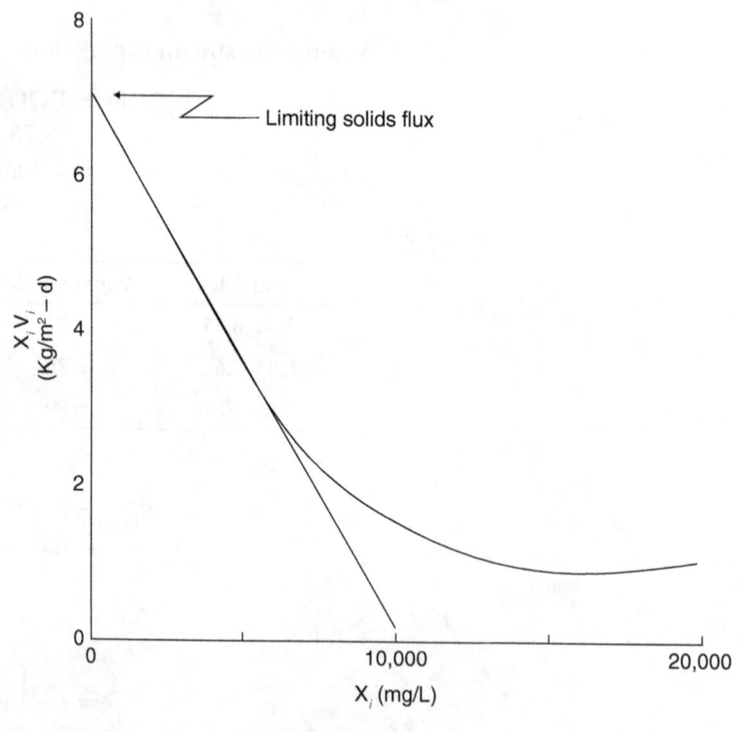

Exhibit 17.7

17.8

$$s=\frac{K_s\left(1+\theta_h k_d\right)}{\theta_h\left(Y_k-k_d\right)-1}$$

$$=\frac{\left(50\text{ mg}/L\right)\left(1+\left(2\text{ d}\right)\left(0.10/\text{d}\right)\right)}{\left(2\text{ d}\right)\left(\frac{0.6\text{ mg}}{\text{mg}}\times\frac{10\text{ mg}}{\text{mg}-\text{d}}-\frac{0.10}{\text{d}}\right)^{-1}}$$

$$=5.7\text{ mg/L}$$

$$X=\frac{4\left(S_0-S\right)}{1+k_d\theta_h}=\frac{\left(\frac{0.6\text{ mg}}{\text{mg}}\right)\left(200\text{ mg/L}-5.7\text{ mg/L}\right)}{1+\left(0.10/\text{d}\right)\left(2\text{ d}\right)}$$

$$=97\text{ mg/L}$$

17.9 Assume a flow in A-B-C of 4 cfs. Use the nomograph in Exhibit 17.9a.

h_L in AB = 5.8'/1000'
= 2.9 ft

h_L in BC = 3.0'/1000'
= 1.5'

Total h_L = 4.5'
= 4.5'/1000'

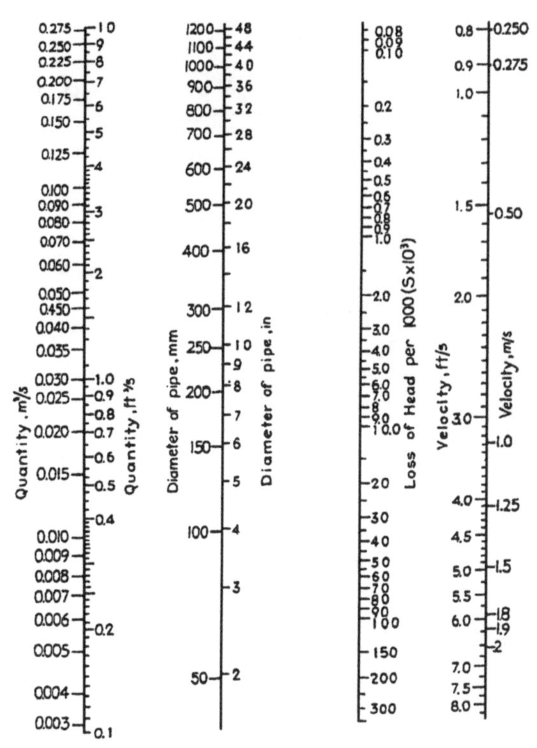

Exhibit 17.9a Flow in old cast iron pipes. (Hazen-Williams C = 100)

Use ϕ = 14.5", L = 1000' for a h_L of 4.5' for A-C.
then Q = 1.4 cfs.

$$Q_{\text{Total}} = 4 + 1.4 = 5.4 \text{ cfs}$$
Equivalent diameter for loop = 16.2"
Equivalent length for loop = 1000'

17.10 Balance the equation as

$$CH_4 + 2O_2 \rightarrow CO_2 + 2H_2O$$

$$gO_2 = 1 \text{ g } CH_4 \times \frac{1 \text{ mol } CH_4}{16 \text{ g } CH_4} \times \frac{2 \text{ moles } O_2}{1 \text{ mole } CH_4} \times \frac{32 \text{ g } O_2}{1 \text{ mole } O_2}$$
$$= 4 \text{ g } O_2$$

17.11

$$C_{\text{equil}} = \alpha P \text{ gas}$$
$$= 2.0 \text{ g} / 1 - \text{atm} \times 0.1 \text{ atm}$$
$$= 2.0 \text{ g / liter}$$

$$M_{CO_2} = C \times V$$
$$= 0.2 \text{ g} / 1 \times 1 \text{ liter}$$
$$= 0.2 \text{ g}$$

17.12

$$[H^+] = 10^{-pH} = 10^{-8}$$
$$HOCl \Leftrightarrow H^+ + OCl^-$$
$$\frac{[H^+]\,[OCl^-]}{[HOCl]} = 10^{-7.5}$$
$$\frac{[H^+](0.1 - HOCl)}{[HOCl]} = 10^{-7.5}$$
$$\frac{(0.1 - [HOCl])}{[HOCl]} = 3.16$$
$$[HOCl] = 0.024 \text{ M}$$

17.13

$$CaF_2 \rightarrow Ca^{2+} + 2F^-$$
$$[Ca^{2+}][F^-]^2 = 3 \times 10^{-11}$$
$$\left[\frac{1}{2}F^-][F^-\right]^2 = 3 \times 10^{-11}$$
$$[F^-] = [2(2 \times 10^{-11})]^{1/3}$$
$$= 4.2 \times 10^{-4} \text{ M}$$
$$= 8.0 \times 10^{-3} \text{ g/liter}$$

17.14

$$1 - e^{-k(5d)} = BOD_5/L$$
$$= 180/300 = 0.6$$
$$k = 0.18/d$$

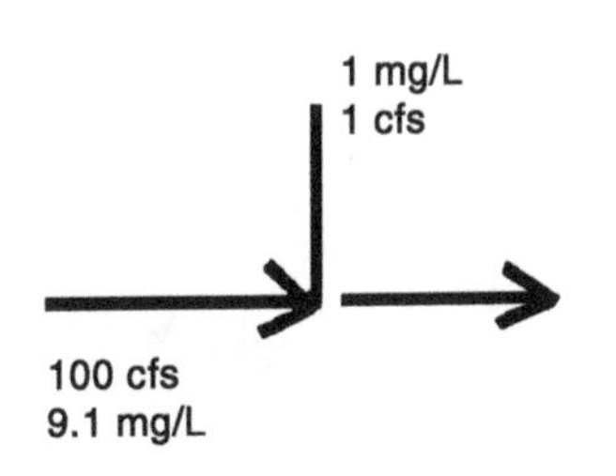

Exhibit 17.15

17.15

$$C = \frac{(100\text{ cfs})(9.1\text{ mg/L}) + (1\text{ cfs})(1\text{ mg/L})}{101\text{ cfs}}$$
$$= 9.0\text{ mg/L}$$
$$D = C_{sat} - C$$
$$= 0.1\text{ mg/L}$$

17.16 The milliequivalents per liter for the ions are:

Ion	Conc (mg/L)	Equiv. wt.	Conc (mEq/L)
Na^+	14	23	0.61
K^+	31	39	0.79
Mg^{2+}	28	12.2	2.30
Ca^{2+}	94	20	4.7
		Total	8.4
SO_4^{2-}	134	48	2.79
Cl^-	92	35.5	2.59
$HCO_3^- + CO_3^{2-}$	135	50	2.70
		Total	8.1

When the concentration of the alkalinity (expressed as $CaCO_3$) is converted to mEq/L, it will be equal to the sum of the concentrations of bicarbonate and carbonate.

17.17

$$v_s = \frac{g(\rho_p - \rho_w)d_p^2}{18\mu}$$

$$= \frac{(9.8\text{ m/s}^2)(1600\text{ kg/m}^3)(2\times10^{-5}\text{ m})^2}{(18)(1\times10^{-3}\text{ N-s/m}^2)}$$

$$= 3.5\times10^{-5}\text{ m/s}$$

$$v_{sc} = \frac{300\text{ gal}}{\text{ft}^2\text{-d}} \times \frac{0.0017\text{ m/hr}}{1\text{ gal/ft}^2\text{-d}} \times \frac{1\text{ hr}}{3600\text{ s}}$$

$$v_{sc} = 12\text{ m/d} \times \frac{\text{d}}{86,400\text{ s}} = 1.4\times10^{-4}\text{ m/s}$$

$$= 1.4\times10^{-4}\text{ m/s}$$

$$k = \frac{3.5\times10^{-5}\text{ m/s}}{14\times10^{-5}\text{ m/s}} = 0.25\text{ or }25\%$$

17.18 c.

17.19 a.

$$\frac{A}{A_0} = \frac{1}{1 + K_1 \dfrac{v}{Q}}$$

$$0.2 = \frac{1}{1 + K_1 \dfrac{v}{Q}}$$

$$1 + K_1 \frac{v}{Q} = 5$$

$$v = \frac{4Q}{K_1} = \frac{3.46 \times 10^5}{0.3} = 1.1 \times 10^6 \text{ m}^3$$

17.20 a.

17.21 d.

17.22 c.

CHAPTER 18

Construction

OUTLINE

Approximately 10% of the civil FE/EIT exam concerns construction management topics; candidates can expect to see six construction management questions. According to NCEES, the following specific topics may be covered:

- Procurement methods
- Allocation of resources
- Contracts and contract law
- Project scheduling
- Engineering economics
- Project management
- Construction estimating

This chapter covers key terms, concepts, and techniques of construction management that you may encounter on the exam. Coverage of engineering economics, which is also tested in the morning portion of the FE/EIT exam, can be found in Chapter 6.

PROCUREMENT METHODS

Three important procurement methods in construction management are as follows:

1. Design/bid/build
2. Design/build
3. Construction manager

Design/Bid/Build

Design/bid/build is the conventional procurement method that solicits bids from contractors for construction, following completion of the design by an architecture/ engineering (AE) firm. (See Figure 18.1.) It is a sequential process, with contractors bidding on a complete set of plans. Most often the lowest bid is sought. The client establishes separate contracts with the AE and the contractor for their work. There is no contractual relationship between the AE and contractor, although the client may hire the AE to perform inspection functions.

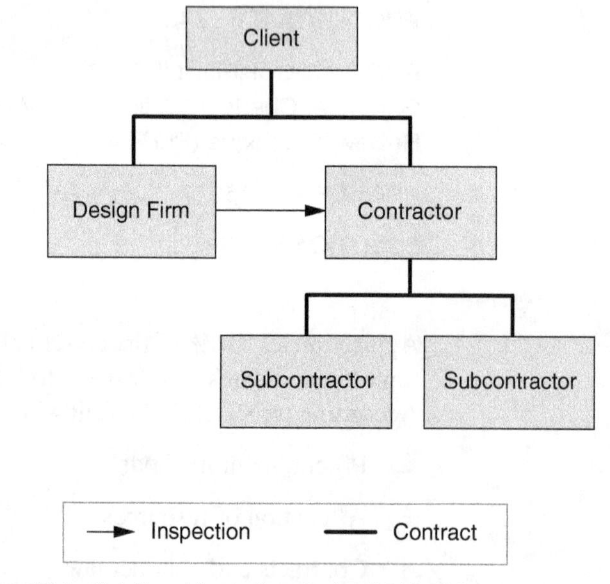

Figure 18.1 Relationships in the design/bid/build procurement method

Design/Build

The design/build method selects one firm to both design and build the project. (See Figure 18.2.) It is usually advantageous for the client to deal with one firm for both design and construction, often termed "turn-key." Disputes between designer and contractor are handled within the one firm. Coordination is also better. One major advantage is that design and construction can be done concurrently. This is often called "fast track" or "phased construction," meaning that construction on some aspects of the project can begin while other aspects are still in design. Design/build is often used on large, complex projects with tight time schedules or for those that are not completely defined initially. The disadvantage to the client is that he or she does not know the total cost until the project is complete.

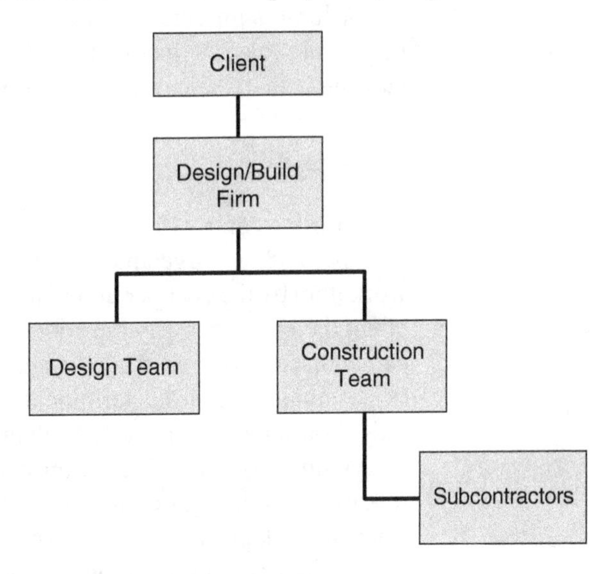

Figure 18.2 Relationships in the design/build procurement method

Construction Manager

The construction manager method uses a construction manager (CM), hired by the client to manage the work of the AE and the contractor, who each have contracts with the client. (See Figure 18.3.) This usually results in time and cost savings due to the close supervision of the design and construction efforts.

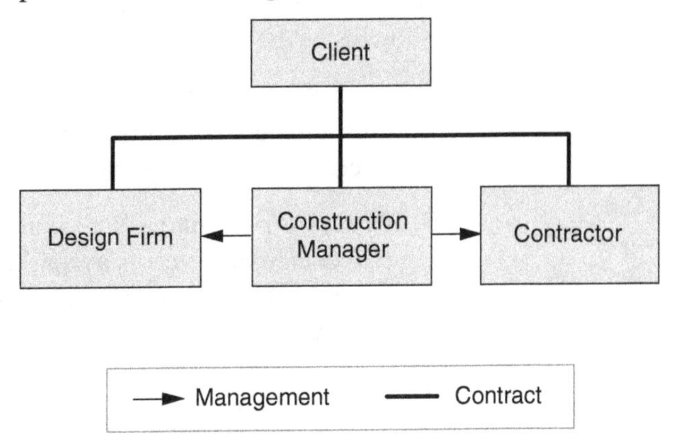

Figure 18.3 Relationships in the construction manager procurement method

CONTRACT TYPES

In this section, we will review three main contract types for construction management: lump sum, unit price, and cost plus.

Lump Sum

A lump sum contract stipulates a total price for all work to be performed, including all material, labor, equipment, overhead, and profit. This contract requires a detailed set of plans and quantity estimates. Any difference in costs requires a change order, or a formal change in the contract, agreed on by both parties. Lump sum contracts are usually used with conventional procurement.

A lump sum contract is advantageous to the client, who knows exactly how much the project will cost. In addition to requiring a detailed set of plans, a disadvantage is that changes are more difficult to make once construction has started.

Unit Price

A unit price contract is used when quantities may not be known, often with projects that include large amounts of excavation. The contractor bids a price for each work item by the cost per unit (for example, cost per cubic yard of excavation). The client then agrees to pay the contractor for each unit of work performed. A low bid can be used, based on the contractor's bidding on an engineer's estimated quantities. Actual quantities determine the total price paid to the contractor. All unit price bids should include any overhead and profit.

A unit price contract is an excellent one to use when exact quantities are unknown, but it requires agreement between contractor and client as to actual quantities of work performed. Adjustments are usually made to the unit price whenever large differences in quantities are encountered.

Cost Plus

A *cost plus* contract is used when the total scope of work is not known. The client agrees to pay the contractor for the actual work performed and for overhead. The contractor's profit is the *plus* portion of the contract. This contract type is useful for emergency construction or for projects that must be started before design is complete. The *plus* portion of the terms can have several variations, including the following:

- *Cost plus fixed fee*. This contract pays the contractor for costs and overhead plus a fixed amount for markup/profit, which gives the contractor incentive to complete the project as quickly and economically as possible.
- *Cost plus percentage*. This contract pays the contractor for costs plus a percentage of the costs as markup/profit. Under this type contract, there is no incentive for the contractor to be efficient. On the contrary, the more expensive the project is, the more the contractor makes.

CONTRACTS AND CONTRACT LAW

The morning portion of the FE exam may contain some general questions on agreements and contracts. Here, we will focus on legal issues specific to construction management.

Bidding

The bid process begins with a Notice to Bidders, indicating the scope and location of the project, client, availability of plans and specifications to be used, date and location of submission, and bond requirements. Bidders may be required to attend a prebid meeting/site visit to ask questions, clarify scope, and see the actual project site.

Interested contractors will estimate material and labor costs, time and equipment requirements, and overhead and profit desired. The completed bid must be submitted at the proper location no later than the time required. Late submissions will be rejected.

Bonding

On all public construction projects, and most private projects, three types of bonds are obtained by contractors prior to submitting a bid and are obtained from a surety company:

1. *Bid bond.* Typically 5% to 20% of estimated project cost. The bid bond guarantees that the bidder will enter into a contract with the client if the bidder is the low bidder. If the low bidder does not sign a contract, the amount of the bid bond is used by the client to offset either the cost of the next lowest bid or the cost of rebidding the project.
2. *Performance bond.* Typically 100% of project cost. This bond guarantees that the contractor will perform the specified work in accordance with the contract. If the contractor defaults on the contract, the bond is the upper amount that the surety company will incur to arrange completion of the project.
3. *Payment (labor and material payment) bond:* Typically 100% of project cost. The payment bond guarantees that the contractor will pay for all materials and labor used on the project, protecting the client from liens against the project by third parties.

CONSTRUCTION ESTIMATING

In preparing a bid, one of the most commonly estimated items is earthwork excavation. The following sections present equations used in estimating earthwork quantities.

Soil Volume Changes

In earthwork operations, material will change in volume depending on its position in the construction operation. A given weight of soil will occupy different volumes depending on whether it is in its natural (bank) condition, loose, or compacted

(Table 18.1). Generally, soils will swell 30% to 40% between natural and loose conditions and will shrink 10% to 20% from natural to compacted condition. The

Table 18.1 Soil volume differences

Bank yards	Density = 2000 lb/yd^3 = 2000 lb/yd^3
Loose yards (30% swell)	Density = 2000 lb/1.3yd^3 = 1538 lb/yd^3
Compacted yards (0.75 shrinkage factor)	Density = 2000 lb/0.75 yd^3 = 2667 lb/yd^3

Table 18.2 Soil unit weights and change factors

	Unit Weight (lb/cu yd)						
	Bank	**Loose**	**Compacted**	**Swell (%)**	**Shrinkage (%)**	**Swell Factor**	**Shrinkage Factor**
Clay	3000	2310	3750	30	20	0.77	0.80
Earth	3100	2480	3450	25	10	0.80	0.90
Rock, loose	4600	3060	3550	50	–30*	0.67	1.30*
Sand/gravel	3200	2860	3650	12	12	0.89	0.88

* Compacted rock is not as dense as bank rock.

quantities of earth in the process of excavating from in-place to loose to compacted condition are measured in bank cubic yards (BCY), loose cubic yards (LCY), and compacted cubic yards (CCY).

Swell factors and shrinkage factors are used to determine the volume changes. Swell factor measures the increase in volume from natural (bank) to loose condition (e.g., from natural to dump truck). Swell factor is often termed the load factor and is defined as:

Swell factor = Loose unit weight ÷ Bank unit weight
Swell factor = 1 ÷ (1 + swell)

Loose volume is multiplied by the swell factor to obtain the bank volume.

The shrinkage factor measures the decrease in volume from bank to compacted condition. It is defined as:

Shrinkage factor = Bank unit weight ÷ Compacted unit weight
Shrinkage factor = 1 – Shrinkage

Bank volume is multiplied by the shrinkage factor to obtain the compacted volume.

Typical soil unit weights and change factors are summarized in Table 18.2.

Spoil Banks/Piles

Excavated (loose) material needs to be stored in either banks (triangular cross section) or piles (conical shape). The volume of the spoil bank/pile is measured in loose cubic yards (LCY). Dimensions can be calculated if the angle of repose (angle between the slope and horizontal plane) of the soils is known. Table 18.3 lists some typical angles of repose.

Table 18.3 Angles of repose for common excavation materials

Material	Angle of Repose (R) in Degrees
Clay	35
Common earth, dry	32
Common earth, moist	37
Gravel	35
Sand, dry	25
Sand, wet	37

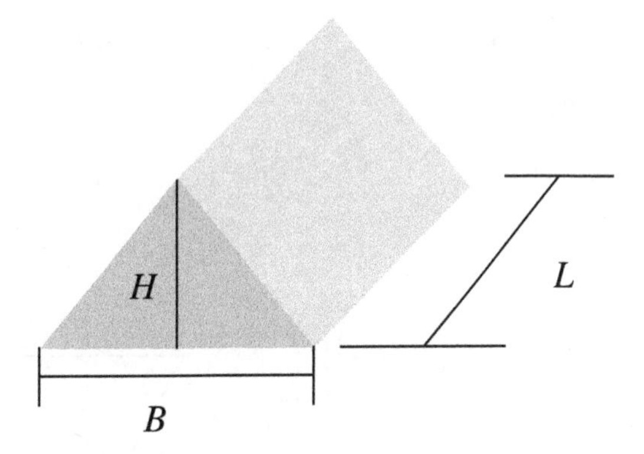

Figure 18.4 A triangular spoil bank

For a triangular spoil bank, as shown in Figure 18.4, relevant equations are as follows:

$$\text{Volume} = \text{Section area} \times \text{Length}$$
$$\text{Volume} = (\tfrac{1}{2}BH)\,L$$
$$B = [(4V \div (L \times \tan R)]^{1/2}$$
$$H = (B \times \tan R) \div 2$$

Or,

$$\text{Volume} = (B^2\,L\tan R)/4$$

where
- B = base width (ft)
- H = pile height (ft)
- L = pile length (ft)
- R = angle of repose (degrees)
- V = volume (cu ft)

For a conical spoil pile, as shown in Figure 18.5, the relevant equations are:

$$\text{Volume} = 1/3 \times \text{Base area} \times \text{Height}$$
$$D = (24V/\pi \tan R)^{1/3}$$
$$H = D/2 \times \tan R$$

where
- D = diameter of pile base (ft)
- H = height of pile (ft)

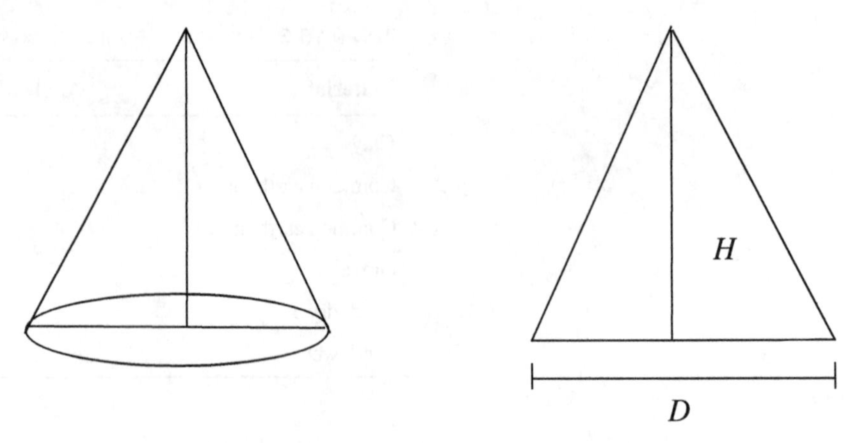

Figure 18.5 Conical spoil pile

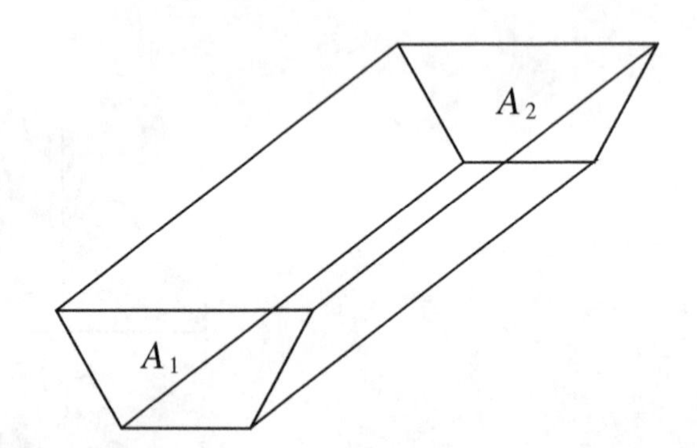

Figure 18.6 Geometry for the average end area method

Pit Excavation

In order to calculate the volume of excavation, the surface area is divided into regular-sized grids. The depth of excavation is determined for each corner of each grid square by subtracting the cut elevation from the surface elevation. The volume of excavation is found by:

$$\text{Volume (BCY)} = \text{Area of surface (sq ft)} \times \text{Average depth of cut (ft)}$$

Linear Cut/Fill

Road excavation (cut) or material added (fill) are calculated by 100–foot station. The volume of cut or fill is calculated by the average end area or the trapezoidal methods.

Average End Area Method

Figure 18.6 shows the geometry for the average end area method of calculating cut/fill volume.

$$\text{Volume (cu yd)} = \frac{(A_1 + A_2)}{2} \, (L) \, \frac{1 \text{ yd}^3}{27 \text{ ft}^3}$$

where

A_1 and A_2 = area in sq ft for respective end areas
L = length between end areas (ft)

Prismoidal Method

Figure 18.7 shows the geometry for the prismoidal method of calculating cut/fill volume.

$$\text{Volume (cu yd)} = \frac{(A_1 + 4A_m + A_2)}{6}\ (L)\ \frac{1\ \text{yd}^3}{27\ \text{ft}^3}$$

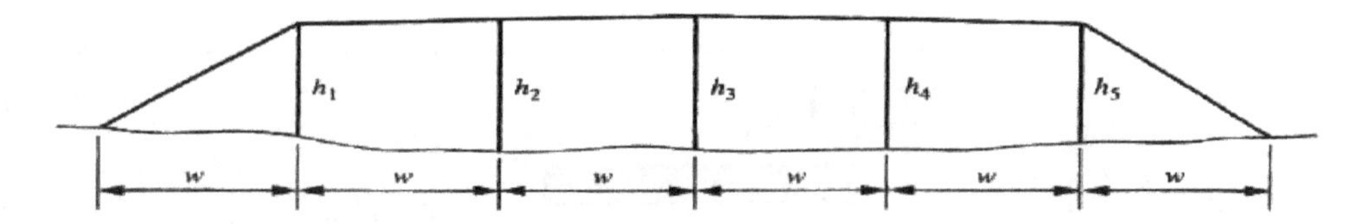

Figure 18.7 Geometry for the prismoidal method

where

A_1 and A_2 = end areas

A_m = area at the midpoint of the length

The end areas can be determined by breaking the cross section into triangles and trapezoids, determining the area of each and summing the area:

$$\text{Area of triangle} = hw/2$$

where

h = height of triangle

w = width of triangle

$$\text{Area of trapezoid} = (h_1 + h_2)/2 \times w$$

where

h_1 and h_2 = lengths of parallel sides

w = distance between parallel sides

The general rule for calculation of multiple trapezoidal areas is:

$$\text{Area} = w(h_0/2 + h_1 + h_2 + \ldots + h_{(n-1)} + h_n/2)$$

In general, the prismoidal method will produce a more accurate volume than the average end area method. It is common that excavation be expressed in BCY, hauling in LCY, and final volume in CCY. It is important to work in one measure and then convert to the required condition.

PRODUCTIVITY

Equipment productivity can be calculated by:

$$\text{LCY/hr} = \text{Cycles per hour} \times \text{Bucket payload (LCY) per cycle}$$

The payload is calculated by multiplying the bucket size by a fill factor, which ranges from 0.40 to 1.10, depending on the type of material being loaded.

PROJECT SCHEDULING

Several scheduling techniques are commonly used on construction projects. In this review, we will focus on Gantt/bar charts, CPM, and PERT.

Gantt/Bar Charts

Gantt/bar charts are the simplest of scheduling techniques. They indicate each task or work by start and end date. The chart may indicate milestones and percent completion of each task. It does not indicate relationships between activities. Figure 18.8 shows a sample bar chart.

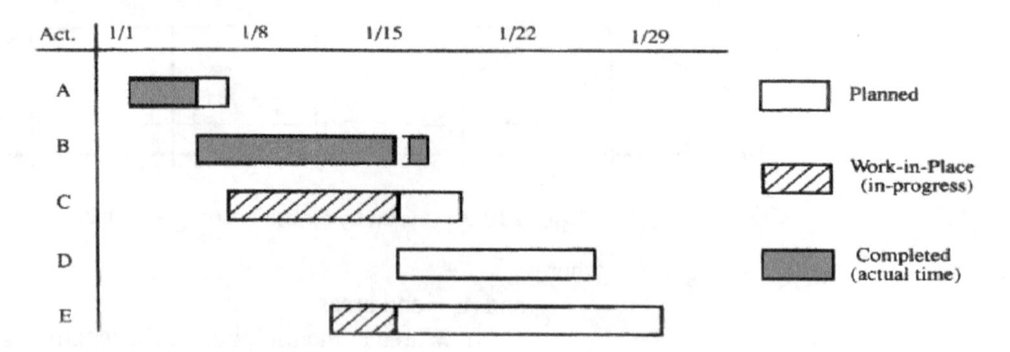

Figure 18.8 A generic Gantt/bar chart

Critical Path Method (CPM)

CPM diagrams may be either activity-on-arrow (AOA) or activity-on-node (AON). Both methods indicate the logical relationship between activities and can be used to indicate the shortest completion time and the activities that must be completed on time to ensure the timely completion of the project, the critical path. Calculations will also determine how much time activities may be delayed without delaying either the entire project (total float) or the succeeding activities (free float).

Activity-on-Arrow (AOA)

In AOA notation, each activity is represented by an arrow. Activities may be labeled with a name or with the numbers of the nodes at the start and end of the activity. This is known as *ij* notation. Each activity has a duration measured in days and begins and ends at a node.

The notation shown in Figure 18.9 is as follows:

- ES = early start, the earliest the activity can start (all immediate preceding activities completed)
- EF = early finish, the earliest the activity can finish = ES + duration
- LS = late start, the latest the activity can start without delaying project completion
- LF = late finish, the latest the activity can finish without delaying project completion
- TF = total float, the amount the activity can be delayed without delaying completion of project
- FF = free float, the amount the activity can be delayed without delaying succeeding activities

No activity leaving a node can be started until all activities entering the node have been completed. The CPM diagram in Figure 18.10 shows a project with five activities. The duration for each activity is shown below the activity letter. Activity d1 is a *dummy* activity, which has a duration of zero days but shows a logical dependence of activity E not being able to start until activity B is completed (as well as activity C).

In order to determine the critical path, a forward pass through the CPM diagram is performed to calculate the ES and EF:

ES = latest EF of all immediate preceding activities
EF = ES + Duration

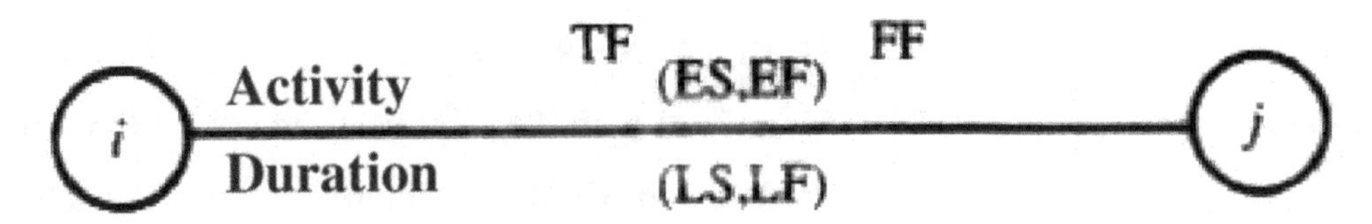

Figure 18.9 *ij* notation from a CPM diagram

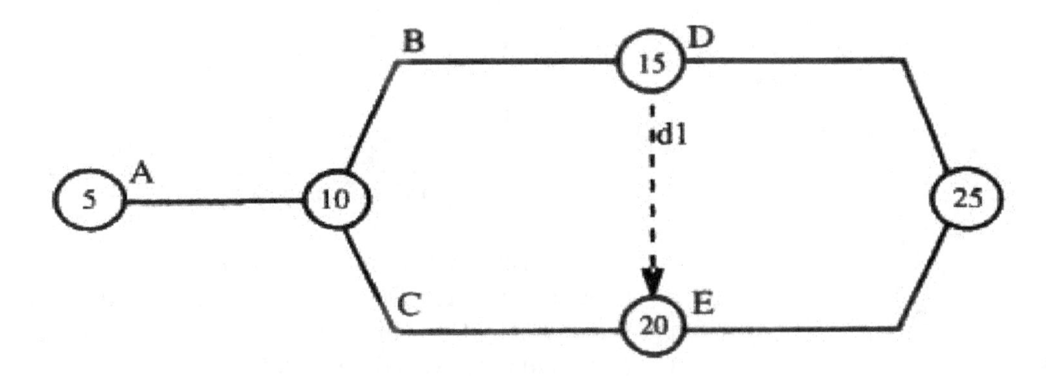

Figure 18.10 CPM diagram—activity-on-arrow method

This is followed by a backward pass to calculate the LS and LF:

LF = earliest LS of all succeeding activities
LS = LF – Duration

Total float is then calculated as TF = LS – ES. Finally, free float is calculated as the difference between the ES of each succeeding activity minus the EF of the activity (smallest value is selected).

Example 18.1

Using the CPM diagram shown in Figure 18.10 and the following activity durations, determine the project duration, critical path, and float.

Activity	Duration	Activity	Duration
A	5	D	2
B	4	E	4
C	3		

Solution

Calculations for activity E are shown as an example.

ES = 9 (Activity C finishes at end of day 8, but activity B, with dummy, ends at end of day 9)

EF = ES + Duration = 9 + 4 = 13

LF = 13 (the ES of succeeding activity, which is project end)

LS = LF – Duration = 13 – 4 = 9

TF = LS – ES = 9 – 9 = 0

FF = 13 – 13 = 0 (ES of project end is 13; EF of E is 13)

Using the CPM diagram in Figure 18.10, a table is set up to record all values:

Activity	Duration	ES	EF	LS	LF	TF	FF	Critical?
A	5	0	5	0	5	0	0	*
B	4	5	9	5	9	0	0	*
C	3	5	8	6	9	1	1	
D	2	9	11	11	13	2	2	
E	4	9	13	9	13	0	0	*

* Indicates activities on the critical path

Any activity with a total float of zero is on the critical path. Activities with a total not equal to zero can be delayed that number of days without delaying the project. Activities with free float not equal to zero can be delayed that number of days until the next activity is delayed. Free float does not have to equal total float.

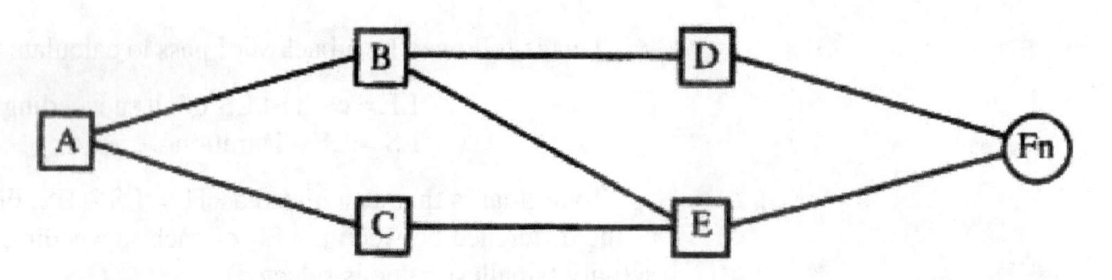

Figure 18.11 CPM diagram—activity on node methodŽ

Activity-on-Node (AON)

(ES,EF)

TF	FF
Activity	
Duration	

(LS,LF)

Figure 18.12 Typical node labeling

An alternate method of drawing the CPM diagram is to use activity on node. In this method, activities are represented by the node, not by the arrow. Nodes may be a circle (bubble diagram) or a rectangle (often termed precedence). AON is popular because most scheduling software uses this method. The AON diagram in Figure 18.11 shows the same project as the AOA diagram in Figure 18.10.

Calculations are performed in the same manner as with the AOA diagram. Instead of placing time calculation information above and below the arrow, information is usually placed inside the node, as shown in Figure 18.12.

One advantage of precedence diagrams is that the location of the tail and head of the line connecting two activity nodes can indicate more detailed dependencies between the two activities. This is shown in Table 18.4.

Table 18.4 Node configurations show activity dependencies

CPM Node Diagram	Relationship	Description
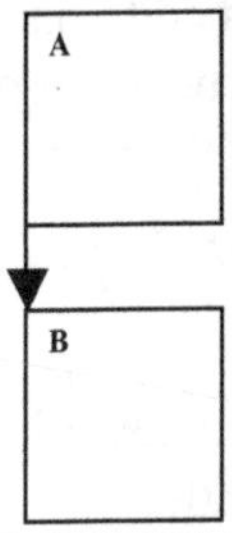 A B	Start-to-start	The start of B depends on the start of A.
Wb 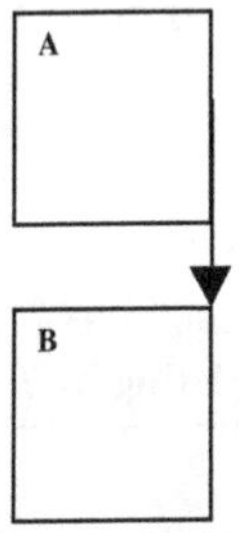 A B	Finish-to-finish	The finish of B depends on the finish of A.
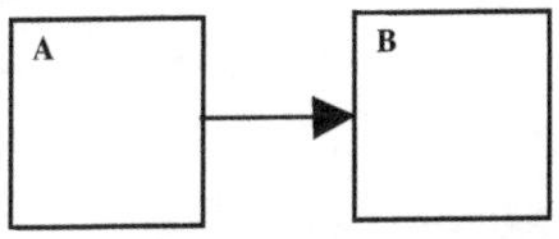 A B	Finish-to-start	The start of B depends on the finish of A.
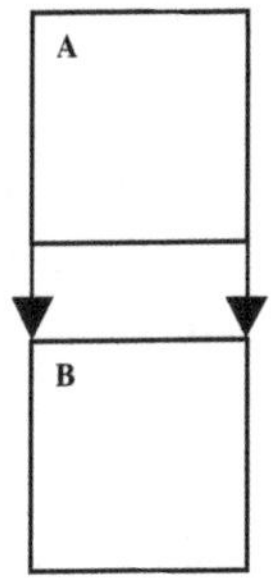 A B	Combination	The start and finish of B depend on the start and finish of A.

Program Evaluation and Review Technique (PERT)

PERT uses the CPM logic and statistics to determine the most likely time of completion for a project. Each activity has an optimistic completion time, pessimistic completion time, and a most likely completion time. For each activity, an expected completion time (t_e) is calculated, along with a standard deviation (σ) and a variance (ν). Using normal distribution tables, a probable completion date is calculated as follows:

$$t_e = (a + 4m + b)/6$$
$$\sigma_{te} = (b - a)/6$$
$$\nu = \sigma_{te}^{2}$$

where

a = optimistic activity duration
m = most likely duration
b = pessimistic activity duration

The variance (V) for the critical path is the sum of the variances for the activities on the critical path, and $\sigma_{TE} = V^{1/2}$.

The critical path is assumed to have a normal distribution. Therefore, the probability of completion of a project can be determined using a Z table as follows (see Table 18.5):

$$Z = (T_s - T_E)/\sigma_{TE}$$

where

Z = number of standard deviations from the mean
T_E = mean of critical path
σ_{TE} = standard deviation of critical path = $V^{1/2}$
T_s = date of interest

Table 18.5 A Z table

Z	P, probability of completing by T_s	Z	P, probability of completing by T_s
–3.0	0	+0.1	.54
–2.5	.01	+0.2	.58
–2.0	.02	+0.3	.62
–1.5	.07	+0.4	.66
–1.4	.08	+0.5	.69
–1.3	.10	+0.6	.73
–1.2	.12	+0.7	.76
–1.1	.14	+0.8	.79
–1.0	.16	+0.9	.82
–0.9	.18	+1.0	.84
–0.8	.21	+1.1	.86
–0.7	.24	+1.2	.88
–0.6	.27	+1.3	.90
–0.5	.31	+1.4	.92
–0.4	.34	+1.5	.93
–0.3	.38	+2.0	.98
–0.2	.42	+2.5	.99
–0.1	.46	+3.0	1.00
0	.50		

The CPM diagram shown in Figure 18.13 shows three durations for each activity. The first is the optimistic time (a), the second is the most likely (m), and the third is the pessimistic time (b).

The expected duration (t_e) is found as:

$$t_e = (2 + 4 \times 5 + 14)/6 = 36/6 = 6$$

The standard deviation (σ_{te}) is found as:

$$\sigma_{te} = (14 - 2)/6 = 2$$

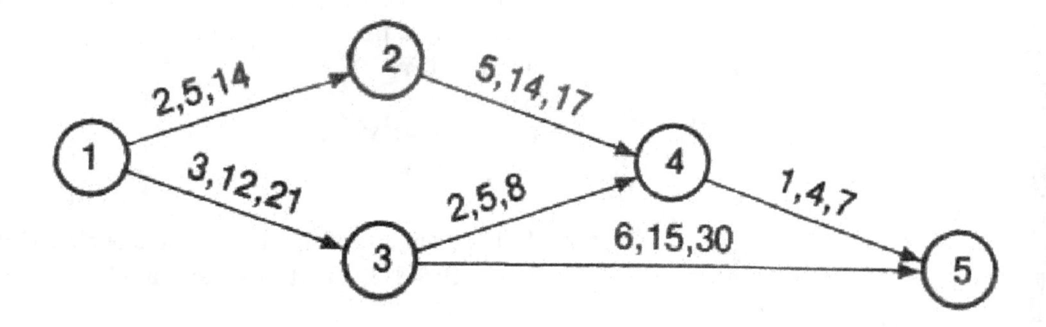

Figure 18.13 An example CPM

The variance is:

$$\nu = \sigma_{te}^{2} = 2^2 = 4$$

Performing the same calculations for all the activities yields the following:

Activity	t_e	s_{te}	n
1,2	6	2	4
1,3	12	3	9
2,4	13	2	4
3,4	5	1	1
4,5	4	1	1
3,5	16	4	16

Using the expected durations for each activity (t_e) and the CPM, the critical path is 1 – 3, 3 – 5, with a duration of 12 + 16 = 28 days. The variance (V) = 9 + 16 = 25.

$$\sigma_{TE} = 25^{1/2} = 5.$$

The probability of completing this project by the end of day 23 is found by:

$$Z = (23 - 28)/5 = -1.0$$

The Z table shows a probability of 0.16 for $Z = -1.0$. Therefore, there is a 16% probability of completing the project by the end of day 23.

PROBLEMS

18.1 A dump truck can haul 15 loose cubic yards of common earth. If the material has a swell factor of 0.80, how many bank cubic yards must be excavated to fill the dump truck?

a. 15 BCY
b. 12 BCY
c. 18.75 BCY
d. 20 BCY

18.2 If the shrinkage of the material in question 18.1 is 10%, how much volume will 100 BCY of this material fill when compacted?

a. 90 CCY
b. 100 CCY
c. 111 CCY
d. 99 CCY

18.3 What is the volume of excavation in a gravel pit, if the depths of cut at the corners are 6.0′, 5.8′, 7.6′, and 8.2′? The area is 20 ft by 20 ft.

a. 102.2 BCY
b. 306.6 BCY
c. 102.2 LCY
d. 306.6 LCY

18.4 An excavator will be used to load trucks. The contractor has the choice of four excavators, each having a different cycle time and bucket size. Which excavator is the most productive?

	Cycle Time (min)	Bucket Size (LCY)
Excavator A	0.32	1.00
Excavator B	0.25	0.75
Excavator C	0.50	1.50
Excavator D	0.60	2.00

a. Excavator A
b. Excavator B
c. Excavator C
d. Excavator D

18.5 Using PERT, if a project has an expected duration of 50 days, with a project variance of 16 days, what is the probability of the project being completed by the end of day 48?

a. 31%
b. 69%
c. 50%
d. 55%

18.6 Using the information in question 18.5, on which day will the project be completed with 90% probability?

a. 45
b. 46
c. 55
d. 56

Use the CPM diagram in Exhibit 18.6 to answer questions 18.7, 18.8, and 18.9.

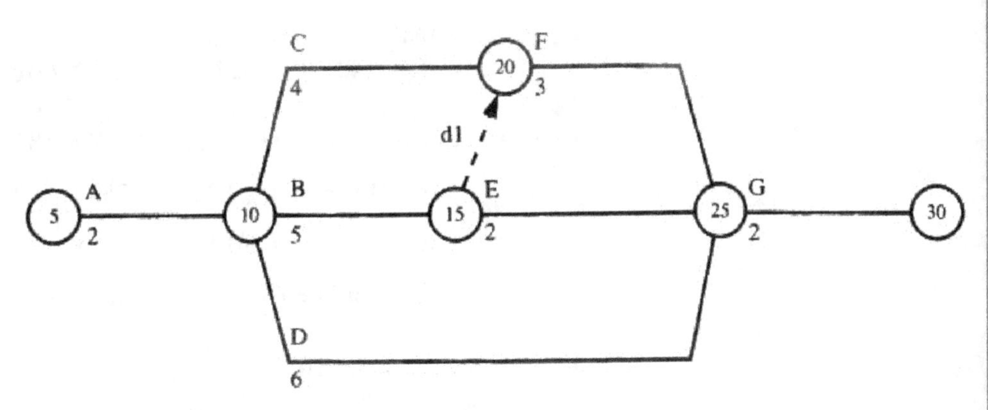

Exhibit 18.6 CPM diagram

18.7 What is the duration of the project?

a. 11 days
b. 12 days
c. 13 days
d. 10 days

18.8 Which activities are on the critical path?

a. A-B-E-G
b. A-D-G
c. A-C-F-G
d. A-B-F-G

18.9 How long can activity D be delayed without delaying the project completion?

a. 1 day
b. 2 days
c. 3 days
d. 0 days

SOLUTIONS

18.1 **b.** BCY = LCY × Swell factor = 15 × 0.8 = 12 BCY

18.2 **a.** Shrinkage factor = 1 – Shrinkage = 1 – 0.1 = 0.90

CCY = BCY × Shrinkage factor = 100 × 0.90 = 90 CCY

18.3 **a.** Average depth of cut = (6.0 + 5.8 + 7.6 + 8.2)/4 = 6.9 ft

Area = 20′ × 20′ = 400 sq ft

Volume = (400 × 6.9)/27 = 102.2 BCY (excavation is measured in BCY)

18.4 **d.** Excavator A = (60/0.32) × 1 CY = 187.5 LCY

Excavator B = (60/0.25) × 0.75 CY = 180 LCY

Excavator C = (60/0.50) × 1.5 CY = 180 LCY

Excavator D = (60/0.60) × 2.0 CY = 200 LCY

18.5 **a.** $\sigma_{TE} = V^{1/2} = 16^{1/2} = 4$

$Z = (48 - 50)/4 = -0.50$

From Z table, $p = 0.31 = 31\%$

18.6 **d.** From Z table, working backwards, $p = 0.90$. This gives $Z = 1.3$.

$Z = 1.3 = (x - 50)/4$

$x = 4\,(1.3) + 50 = 55.2$ during 56th day

18.7–18.9 Sample calculations for activity F are shown.

ES = 7 (Activity C finishes at end of day 6, but activity B ends at end of day 7)

EF = ES + Duration = 7 + 3 = 10

LF = 10 (the ES of succeeding activity, G)

LS = LF – duration = 10 – 3 = 7

TF = LS – ES = 7 – 7 = 0

FF = 10 – 10 (ES of G is 10; EF of F is 10)

Using the CPM diagram shown in Exhibit 18.6, a table is set up to record values for all activities:

Activity	Duration	ES	EF	LS	LF	TF	FF	Critical?
A	2	0	2	0	2	0	0	*
B	5	2	7	2	7	0	0	*
C	4	2	6	3	7	1	1	
D	6	2	8	4	10	2	2	
E	2	7	9	8	10	1	1	
F	3	7	10	7	10	0	0	*
G	2	10	12	10	12	0	0	*

18.7 **b.** See the preceding calculations.

18.8 **d.** Activities A-B-F-G have 0 total float, so they are on the critical path.

18.9 **b.** See the preceding table.

CHAPTER 19

Surveying

OUTLINE

Approximately 11% of the civil FE/EIT exam concerns surveying topics; candidates can expect to see roughly seven surveying questions. According to NCEES, the following specific topics may be covered:

- Angles, distances, and trigonometry
- Area computations
- Closure
- Coordinate systems (e.g., GPS, state plane)
- Curves (vertical and horizontal)
- Earthwork and volume computations
- Leveling (e.g., differential, elevations, and percent grades)

This chapter covers key terms, concepts, and analytical techniques of surveying that you may encounter on the exam. We begin with a glossary of terms and a brief summary of basic trigonometry; this information will be useful throughout the rest of the chapter. Coverage of vertical and horizontal curves can be found in Chapter 16, "Transportation Engineering." That chapter also contains some additional discussion of earthwork and volume computations.

GLOSSARY OF SURVEYING TERMS

Backsight (BS) Elevation (rod reading) obtained by sighting back station. Also called Plus Sight (+S).

Benchmark A fixed reference point or object, the elevation of which is known.

Contour An imaginary line of constant elevation on the ground surface.

Deflection angle An angle measured to a line from the extension of the preceding line.

Departure The orthographic projection of a line on the east-west axis of the survey. East departures are considered positive.

Foresight (FS) Elevation (rod reading) obtained by sighting forward station. Also called Minus Sight (– S).

Height of instrument (HI) The elevation of the line of sight of the telescope above the survey station or control point.

Height of target The elevation of the target or prism above the survey station or control point.

Least count The smallest graduation shown on a vernier. The least count of an instrument is the smallest possible measurement that can be made without interpolation.

Level surface A curved surface, every element of which is normal to a plumb line.

Latitude (traverse) The orthographic projection of a line on the north-south axis of the survey. North latitudes are considered positive.

Latitude (astronomical) Angle measured along a meridian north (positive) and south (negative) from the equator. Latitude varies from 0° to 90°.

Longitude Angle measured at the pole, east or west from the prime meridian. Longitude varies from 0° to 180° east or 180° west.

Meridian (astronomical) An imaginary line on the earth's surface having the same astronomical longitude at every point.

Traverse A succession of lines for which distance and horizontal angles are measured. A traverse may be closed, where the traverse ends at the point from which it started, or open.

Vernier An auxiliary scale placed alongside the main scale of an instrument, by means of which the fractional parts of the least division of the main scale can be measured precisely.

Zenith The zenith is the point on the celestial sphere where the gravity vector, extended upward, intersects it.

Zenith angle An angle formed between two intersecting lines in a vertical plane where one of these lines is directed toward the zenith.

BASIC TRIGONOMETRY

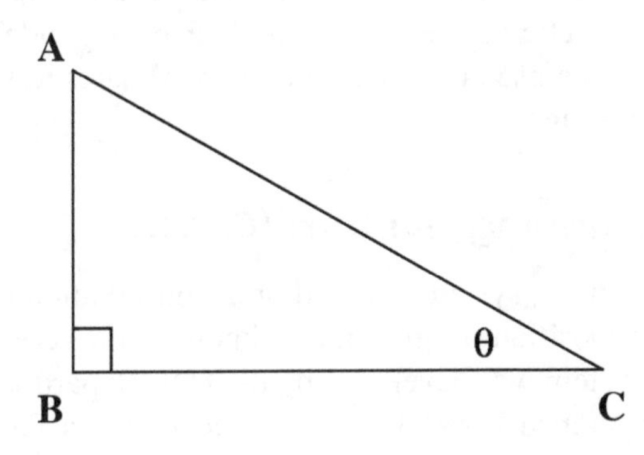

For a Right Triangle

$$\sin\theta = \frac{AB}{AC}$$

$$\cos\theta = \frac{BC}{AC}$$

$$\tan\theta = \frac{AB}{BC}$$

Pythagorean theorem $AB^2 + BC^2 = AC^2$

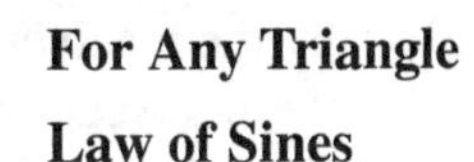

For Any Triangle

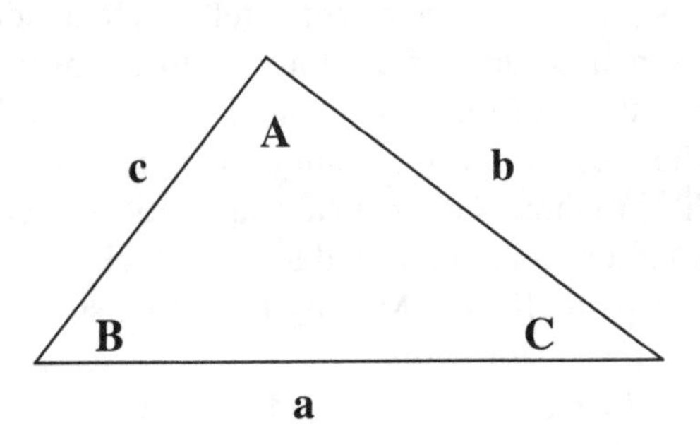

Law of Sines

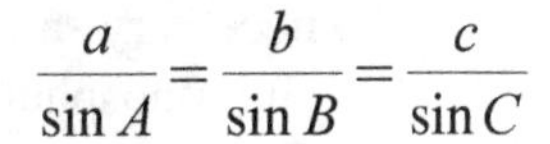

$$\frac{a}{\sin A} = \frac{b}{\sin B} = \frac{c}{\sin C}$$

Law of Cosines

$$a^2 = b^2 + c^2 - 2bc\cos A$$
$$b^2 = a^2 + c^2 - 2ac\cos B$$
$$c^2 = a^2 + b^2 - 2ab\cos C$$

TYPES OF SURVEYS

Plane surveying methods consider the surface of the earth as a plane. Curvature of the earth surface is neglected. This type of survey is appropriate for small geographical areas.

Geodetic surveying takes into account the true (near-spherical or spheroidal) shape of the earth. When the survey covers a large geographical area, neglecting the curvature of the earth introduces significant errors.

COORDINATE SYSTEMS

State Plane Coordinate System (SPCS)

Standardized in 1983 (*NOAA Manual NOS NGS 5*), the state plane coordinate system is a map projection system based on the North American Datum of 1983 (NAD 83). Given the geodetic coordinates (latitude and longitude) of a point, the equations in SPCS can be used to convert these coordinates to state plane coordinates (northings and eastings).

Global Positioning System (GPS)

The global positioning system is based on a constellation of nongeosynchronous satellites, whose position at any time is known to a very high degree of precision. The GPS system was developed by the U.S. Department of Defense in 1978 and is officially named NAVSTAR. GPS receivers on the ground calculate the distance to these known reference points by measuring a time lag (phase shift) between a coded signal generated by the satellite and an identical signal generated by the receiver. When distances from four satellites are measured simultaneously, the intersection of the four imaginary spheres reveals the location of the receiver. Knowledge of the elevation of the point on the ground can reduce the number of necessary satellites to three. GPS data are susceptible to errors due to atmospheric and ionospheric effects, ephemeris and atomic clock errors, numerical errors, and errors due to multipath effects. Moving receivers, such as those in vehicles, are less susceptible to multipath errors.

Differential GPS may be used to achieve subcentimeter accuracy. The sum total of all errors inherent in measurements is estimated using a second receiver located at a known position. This error estimate is then used to adjust the computed position of other receivers in the same general locale.

STATIONING

The concept of stations is commonly used for linear projects, such as in transportation engineering. In the United States, a common convention is to utilize 100-foot stations. Thus, a distance of 320.76 feet may be expressed as 3 + 20.76 stations. Earthwork (haulage) costs are often expressed in typical units of $/yd^3 - sta. This means the cost of hauling a soil volume of 1 yd^3 a distance of 1 station (100 ft).

CHAINING TECHNIQUES

Distances used in surveying are generally horizontal distances, not sloped distances (e.g., property lines on deeds, stationing, layout of building corners, etc.). Even though chaining as a distance measurement method has been rendered obsolete by modern optical instruments, the history of surveying has involved the use of chains and tapes (typically steel) for measuring distance. The results obtained by chaining need to be corrected for extension errors (created by the application of a tension on the tape to prevent excessive sag), thermal elongation errors (created by the difference between the average ambient temperature and the tape's standardization temperature), and sag errors (created by the self-weight induced sag of a tape that is held between two elevated supports).

Tension Correction

$$e_P = \frac{(P - P_0)L}{AE} \tag{19.1}$$

where

P = applied tension
P_0 = tension at which tape or chain was standardized
L = length of tape or chain
A = cross-sectional area of tape or chain
E = modulus of elasticity of tape or chain (steel has E = 29,000 kips/in^2 or 200 GPa)

Note: Applied tension greater than the standard tension makes the tape longer than its standard length, thereby recording a lesser tape reading (negative error), and the correction should be added (positive correction).

Temperature Correction

$$e_T = L\alpha(T - T_0) \tag{19.2}$$

where

α = coefficient of thermal expansion of chain or tape (steel has α = 6.5 × 10^{-6}/°F = 11.6 × 10^{-6}/°C)
L = measured length
T = ambient temperature
T_0 = standardization temperature

Note: Ambient temperature greater than the standard temperature makes the tape longer than its standard length, thereby recording a lesser tape reading (negative error), and the correction should be added (positive correction).

Sag Correction

$$e_S = \frac{w^2 L^3}{24P^2} \tag{19.3}$$

where

w = weight of tape per unit length
L = horizontal distance between level supports
P = applied tension

Note: Tape sag makes recorded length greater than actual (positive error), and the correction should be subtracted (negative correction).

DIFFERENTIAL LEVELING

Differential leveling is the process of measuring the difference in elevation between points as illustrated in Figure 19.1. A transit level is used in combination with a rod or staff to determine elevations of points by obtaining differential vertical measurements. All elevations are referenced to a known elevation or **benchmark**. Table 19.1 shows typical field data for a series of staff readings obtained at successive stations A, B, C, . . . in a traverse. In this example, the traverse is started at station A (known elevation 1000 ft) and ended at station D (elevation 1010 ft). In each case, the height of instrument (HI) is equal to the elevation of the back station plus the backsight (BS) (e.g., 1000.00 + 6.35 = 1006.35) or the elevation of the forward station plus the foresight (FS) (e.g., 1003.90 + 2.45 = 1006.35).

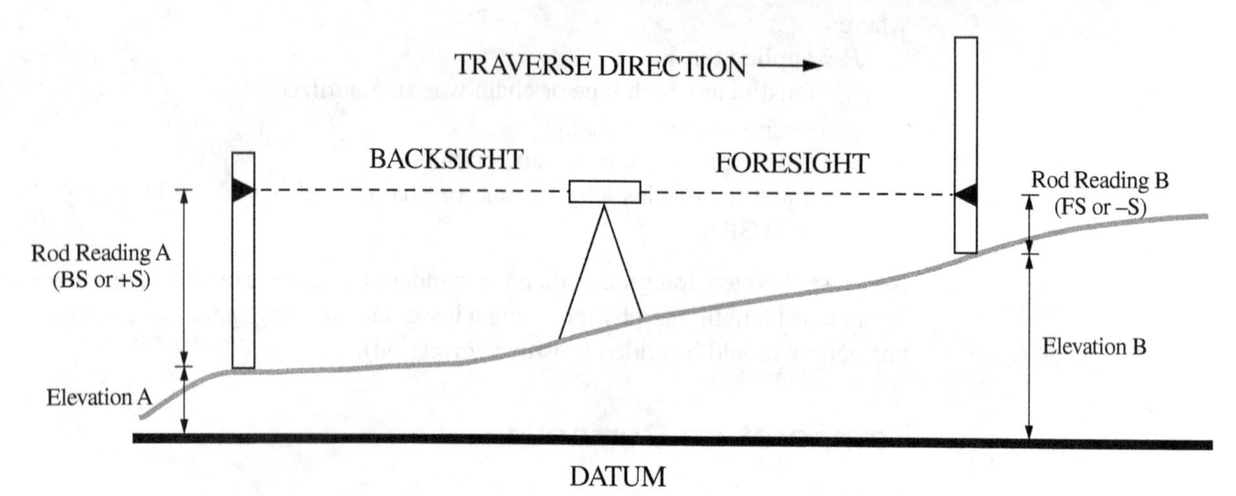

Figure 19.1 Differential leveling

Table 19.1 Differential leveling example

Station	BS	HI	FS	Elevation	Notes
BM-A	6.35			1000.00	Benchmark – marker no. ES-12
		1006.35			
B	9.29		2.45	1003.90	
		1013.19			
C	6.15		3.78	1009.41	
		1015.56			
BM-D			5.56	1010.00	Benchmark – marker no. ES-13
Σ=	**21.79**	Σ=	**11.79**		**Check**
					ΣBS – ΣFS = 21.79 – 11.79 = +10.00
					Ending elevation = 1000.00 + 10.00 = 1010.00 OK

As the telescope is leveled, the line of sight is horizontal (or level). A graduated rod is held vertically at a point of known elevation. The elevation of the benchmark *plus* the rod reading at the benchmark gives the HI, which is the elevation of the line of sight. If the rod is now held vertically at some other location, its unknown elevation can be calculated as the elevation of the line of sight *minus* the rod reading. Thus, with successive locations (1, 2, . . .) of the level, we may write as follows:

$$HI_1 = Elevation_A + BS_A$$
$$Elevation_B = HI_1 - FS_B$$
$$HI_2 = Elevation_B + BS_B$$
$$Elevation_C = HI_2 - FS_C$$

etc. which can also be expressed as follows:

$$Elevation_A + BS_A = Elevation_B + FS_B$$

ANGLES AND DISTANCES

Azimuth

The azimuth of a line is the horizontal angle measured to the line from a specific meridian (usually north) in a particular direction (usually clockwise). Figure 19.2 shows the azimuths of lines AB and BA. The back azimuth of a line is the azimuth of the line running in the reverse direction. When the azimuth is less than 180°, the back azimuth equals the azimuth plus 180°, and when the azimuth is greater than 180°, the back azimuth equals the azimuth minus 180°.

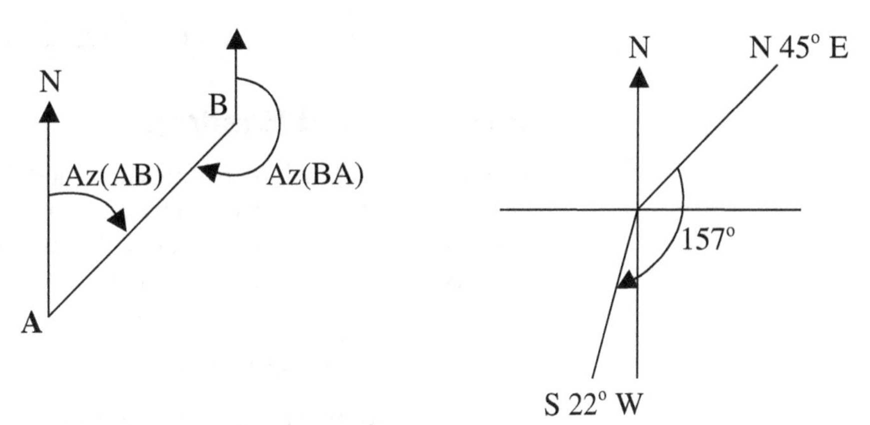

Figure 19.2 Azimuth and back-azimuth of line bearings

Figure 19.3 Bearings of lines and angle between lines

Bearings of lines are directional (horizontal) angles with respect to a meridian (north or south), measured at the originating point on the line. For example, the bearing of a line headed in the north-east direction can be written as N45°E.

The angle between two lines whose bearings are given may be calculated as the difference between their azimuths. For example, if the two lines are N45°E and S22°W, then the first step will be to describe them as azimuths. It is helpful to visualize the angles: S22°W is in the third quadrant; 22° west of south, as shown in Figure 19.3. The azimuth of N45°E is 45°, and the azimuth of S22°W is 22° + 180° = 202°. Therefore, the angle between the two lines is 202° – 45° = 157°.

Latitude and Departure

In the rectangular coordinate system, where the north-south meridian serves as the y-axis and the east-west line serves as the x-axis, the projections of a line are termed **latitude** (y-projection) and **departure** (x-projection), as shown in Figure 19.4. The departure is considered positive to the east (i.e., the line shows an *increase in easting*), and the latitude is considered positive to the north (i.e., the line shows an *increase in northing*).

If the bearing of the line shown in Figure 19.4 is NθE, where θ is the angle between the north meridian and the line (also known as the **azimuthal angle**), then the latitude is given by $L\cos\theta$ and the departure by $L\sin\theta$, where L is the length of line AB.

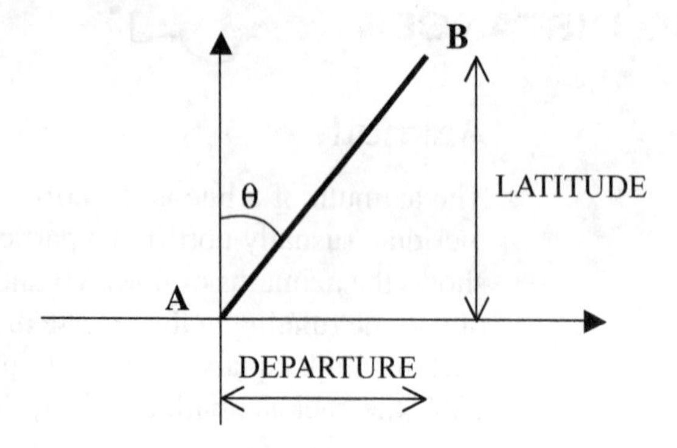

Figure 19.4 Latitude and departure

Northings and Eastings

Whereas latitudes and departures are north-south and east-west projections, respectively, of a line segment, **northings** and **eastings** are coordinates of points in a traverse. Thus, for line segment AB, the latitude is the difference of the northings of points A and B, and the departure is the difference of the eastings.

Interior and Exterior Angles

For a closed traverse (polygon with n sides), the following relations must hold:

$$\text{Sum of all exterior angles} = 180° \ (n + 2)$$

$$\text{Sum of all interior angles} = 180° \ (n - 2)$$

For successive lines in a traverse, interior angles may be calculated from azimuths:

$$\text{Azimuth}_1 + 180° - \text{Interior angle} = \text{Azimuth}_2$$

This may also be stated as:

$$\text{Interior angle} = \text{Azimuth}_1 + 180° - \text{Azimuth}_2 = \text{Back azimuth}_1 - \text{Forward azimuth}_2$$

The bearings of lines AB, BC, CD, and DA in Figure 19.5 may be written as follows:

AB	S 60° E
BC	S 17° E
CD	S 75° W
DA	N 23° W

In each of these cases, either the north or south direction is used as the reference meridian, utilizing the acute angle. This notation is traditional. In order to calculate interior angles, one should first convert each of these bearings to an equivalent azimuth (from north), with the angle measured clockwise, that is, toward the east. Thus, the azimuths are:

$$Az_{AB} = 120°, Az_{BC} = 163°, Az_{CD} = 255°, Az_{DA} = 337°$$

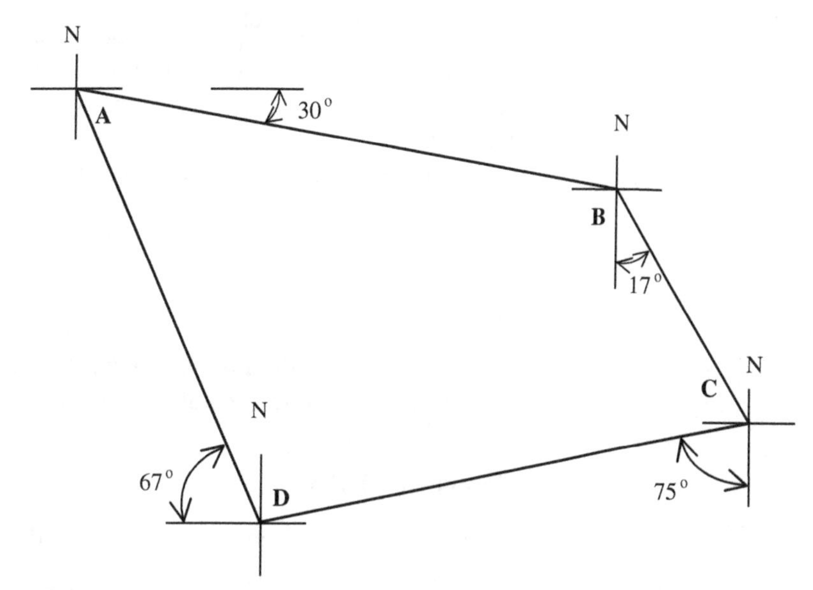

Figure 19.5 Illustration of relation between bearings and interior angles

Using these azimuths, we can calculate the interior angles at A, B, C, and D as:

$$\text{Angle A} = \text{Back azimuth DA} - \text{Forward azimuth AB}$$
$$= 337° + 180° - 120° = 397° = 37° \text{ (modulo 360°)}$$

$$\text{Angle B} = \text{Back azimuth AB} - \text{Forward azimuth BC}$$
$$= 120° + 180° - 163° = 137°$$

$$\text{Angle C} = \text{Back azimuth BC} - \text{Forward azimuth CD}$$
$$= 163° + 180° - 255° + 180° = 88°$$

$$\text{Angle D} = \text{Back azimuth CD} - \text{Forward azimuth DA}$$
$$= 255° + 180° - 337° + 180° = 98°$$

Note that the sum of the interior angles is 360° (as appropriate for a quadrilateral).

TRAVERSE CLOSURE

A closed traverse must, in a perfect world, be closed; that is, it must have no closure error. However, in starting a traverse from a particular station and following several paths to finally return to the originating station, there are many length and angle measurements, each of which could have some degree of error. Errors can be instrument error or human error. The purpose of traverse closure methods is to distribute the closure error to all parts of the traverse, thereby attempting to approximate the true orientation of the lines and angles in the traverse.

Whereas greater accuracy in this correction may be achieved by more involved methods, such as least-squares techniques, these are likely to be beyond the scope of the FE/EIT exam. Some less computationally demanding techniques using coordinate adjustment are described next.

Compass Rule

This method for traverse closure is appropriate when accuracy of angular measurements is about the same as accuracy of distance measurements. The closure error has two components in Cartesian coordinates—the northing error (δy) and the easting error (δx). In this method, the coordinate error is distributed in proportion to the length of traverse lines. The assumption is that the greatest error will come from the longest shots.

Thus, every line in the traverse has its northing and easting adjusted according to

$$\text{Northing adjustment} = \frac{L_i}{\sum L_i} \times \delta y$$

where L_i is the length of the line being adjusted and δy is the northing closure error.

Easting and elevation values may also be adjusted using the same concept.

Transit Rule

In this method, the coordinate error is distributed in proportion to the amount that various coordinates change between points. Thus, latitudes and departures of lines are adjusted, rather than their coordinates. This method is appropriate when accuracy of angular measurements is much better than accuracy of distance measurements:

$$\text{Latitude adjustment} = \frac{|LAT_i|}{\sum |LAT_i|} \times \delta y$$

$$\text{Departure adjustment} = \frac{|DEP_i|}{\sum |DEP_i|} \times \delta x$$

where LAT_i is the latitude (y-projection) of the line being adjusted and δy is the northing closure error of the traverse, and DEP_i is the departure (x-projection) of the line being adjusted and δx is the easting closure error of the traverse.

Application of the Transit Rule

Let us consider the steps in applying the transit rule, using the data shown in Table 19.2 for a traverse, ABCD. The second and third columns of the table show lengths and azimuthal angles for lines in the traverse. Latitude and departure calculated values are given in the fourth and fifth columns.

First, we need to calculate the closure error for latitude and departure. This is found by summing the latitude and departure values of all the lines. In a perfect closed traverse, these values would sum to zero, but as you can see, the latitude and departure closure errors are 0.054 feet and 0.492 feet, respectively.

The negative of the closure error is the **closure correction**, and it must be distributed in proportion to each segment. So, the next step is to determine the appropriate proportions. The proportion to use is the ratio of the absolute latitude (or departure) of each segment to the sum of these values for all lines in the traverse. The results of these calculations are shown in the sixth and seventh columns of Table 19.2.

Next, the closure correction is distributed proportionally for each line. These calculations are shown in the final two columns of Table 19.2.

Table 19.2 Data for transit rule example

Line	L Length (ft)	u Azimuth Angle	Latitude (ft)	Departure (ft)	$\frac{\lvert LAT_i \rvert}{\sum \lvert LAT_i \rvert}$	$\frac{\lvert DEP_i \rvert}{\sum \lvert DEP_i \rvert}$	ΔLAT	ΔDEP
AB	890.32	51° 50′ 40″	+550.039	+700.091	0.1774	0.2916	–0.010	–0.144
BC	1392.85	158° 56′ 40″	–1299.853	+500.413	0.4193	0.2085	–0.023	–0.103
CD	1079.35	256° 36′ 00″	–250.137	–1049.966	0.0807	0.4374	–0.004	–0.215
DA	1011.20	351° 28′ 00″	+1000.006	–150.047	0.3226	0.0625	–0.017	–0.031
			+0.054	**+0.492**			–0.054	–0.492

Finally, the corrections are added to the original latitudes and departures to obtain corrected values, as shown in Table 19.3.

Table 19.3 Adjusted latitude and departure values after applying the transit rule

Line	Latitude (ft)	Departure (ft)
AB	+ 550.029	+ 699.947
BC	– 1299.876	+ 500.311
CD	– 250.141	– 1050.181
DA	+ 999.988	– 150.077
	Σ = 0.000	Σ = 0.000

Solving the same problem using the compass rule, calculations are shown in Table 19.4. The only difference between the two methods is the way in which the ratio is calculated for distributing the correction. For the compass rule, this ratio is calculated from the lengths of the lines AB, BC, CD, and so on, rather than in terms of the latitudes and departures as in the transit rule. Table 19.5 shows the corrected values.

Table 19.4 Data for compass rule example

Line	L Length (ft)	u Azimuth Angle	Latitude (ft)	Departure (ft)	$\frac{L_i}{\sum L_i}$	ΔLAT	ΔDEP
AB	890.32	51° 50′ 40″	+550.039	+700.091	0.2036	–0.011	–0.100
BC	1392.85	158° 56′ 40″	–1299.853	+500.413	0.3184	–0.017	–0.157
CD	1079.35	256° 36′ 00″	–250.137	–1049.966	0.2468	–0.013	–0.121
DA	1011.20	351° 28′ 00″	+1000.006	–150.047	0.2312	–0.013	–0.114
	Σ = 4373.72		**+0.054**	**+0.492**		–0.054	–0.492

Table 19.5

Line	Latitude (ft)	Departure (ft)
AB	+ 550.027	+ 699.991
BC	– 1299.870	+ 500.257
CD	– 250.151	– 1050.087
DA	+ 999.983	– 150.160
	$\Sigma = 0.000$	$\Sigma = 0.000$

AREA OF A TRAVERSE

In this section, we will review two methods of calculating the area of a traverse: by coordinates and by double meridian distance.

By Coordinates

To calculate the area of a closed traverse ABCD by coordinates, the formula is

$$A = \left| \sum \frac{1}{2} y_i (x_{i-1} - x_{i+1}) \right|$$

Coordinates x_i and y_i of points A, B, C, and D are often given as eastings and northings, respectively.

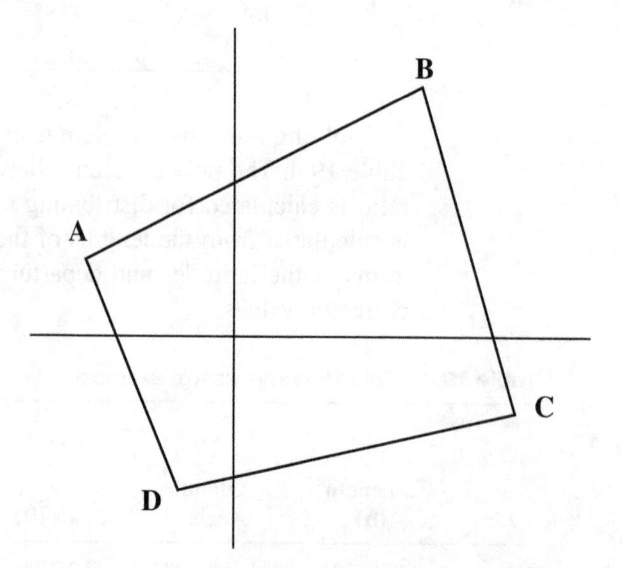

For example, for a closed traverse ABCD, columns 2 and 3 of Table 19.6 specify the coordinates of vertices A, B, C, and D.

Therefore, A = 2,272,500 ÷ 2 = 1,136,250 sq ft = 26.085 acres

Table 19.6 Data for area of traverse by coordinates

Station	*Y* Northing (ft)	*X* Easting (ft)	X_{i-1} 2 X_{i11}	Y_i (X_{i21} 2 X_{i11})
A	+500	–250	–100 – 450	–275,000
B	+1050	+450	–250 – 950	–1,260,000
C	–250	+950	–450 – (–100)	–137,500
D	–500	–100	+950 – (–250)	–600,000
				–2,272,500

By Double Meridian Distance

To calculate the area of a closed traverse ABCD by double meridian distance, the formula is

$$A = \left| \sum \frac{1}{2} LAT_i DMD_i \right|$$

where the double meridian distance (DMD) is given by

$$DMD_i = DMD_{i-1} + D_{i-1} + D_i$$

For example, for a closed traverse ABCD, Table 19.7 specifies the latitudes and departures of lines AB, BC, CD, and DA.

Therefore, A = 2,272,500 ÷ 2 = 1,136,250 sq ft = 26.085 acres

Table 19.7 Data for area of traverse by double meridian distance

Segment	Δ*Y* Latitude (ft)	Δ*X* Departure (ft)	$DMD_i = DMD_{i-1}$ 1 D_{i-1} 1 D_i	$LAT_i \times DMD_i$
AB	+550	+700	0 + (–150) + 700 = 550	550 × 550 = 302,500
BC	–1300	+500	550 + 700 + 500 = 1750	–1300 × 1750 = –2,275,000
CD	–250	–1050	1750 + 500 – 1050 = 1200	–250 × 1200 = –300,000
DA	+1000	–150	1200 – 1050 – 150 = 0	1000 × 0 = 0
				–2,272,500

Note that for the very first line in the traverse, the preceding DMD value has to be assumed. The assumed value is arbitrary and immaterial. The final DMD value, once calculated, must equal the initially assumed value. Obviously, an assumed value of zero reduces the number of computed terms by one, as the last of the LAT × DMD products becomes zero.

In the preceding example, the first line in the traverse is AB. The preceding line is DA. However, for the purpose of calculating the DMD for AB, the DMD for DA has been assumed to be zero. Upon calculating the DMD for DA, it must therefore also be zero.

AREA UNDER AN IRREGULAR CURVE

When calculating area under a curve (with known vertical ordinates), or volume of earthwork (with known end areas), several numerical schemes are available. Of these, the trapezoidal rule and Simpson's rule are commonly used. A requirement of both formulas is that the stations (marked as $i = 0, 1, 2$, etc., in Figure 19.6) be spaced equal distance apart (shown as Δ in the following equations). Also, for Simpson's rule, the number of intervals (n) must be even.

Note that in Figure 19.6 the first station is marked as $i = 0$ (rather than $i = 1$). This is significant for Simpson's rule, because it involves different weighting factors for odd and even i values (4 and 2, respectively).

Using **linear** approximation between (regularly spaced) nodes, we have the **trapezoidal rule**:

$$A = \frac{\Delta}{2}\left[y_0 + y_n + 2\sum_{i=1}^{n-1} y_i\right]$$

Using quadratic approximation between (regularly spaced) nodes, we have **Simpson's rule** (n must be even):

$$A = \frac{\Delta}{3}\left[y_0 + y_n + 4\sum_{\substack{\text{odd}\\ i}} y_i + 2\sum_{\substack{\text{even}\\ i}} y_i\right]$$

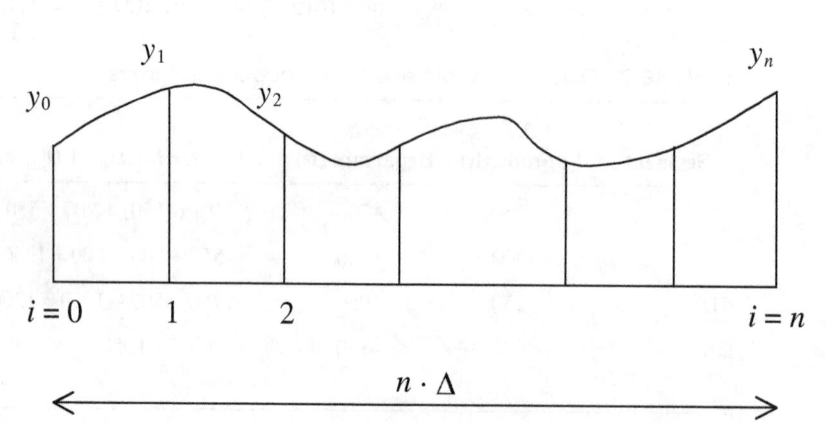

Figure 19.6 Calculating area under an irregular curve

Table 19.8 Sample data for calculating area under a curve

i	x_i (m)	y_i (m)
0	0.0	2.6
1	0.5	3.5
2	1.0	2.4
3	1.5	4.2
4	2.0	4.1
5	2.5	2.0
6	3.0	3.0

For the data shown in Table 19.8, we apply the trapezoidal rule as follows:

$$A = \frac{0.5}{2}[2.6 + 3.0 + 2(3.5 + 2.4 + 4.2 + 4.1 + 2.0)]$$
$$= 9.5 \text{ m}^2$$

For the same data, applying Simpson's rule, we have:

$$A = \frac{0.5}{3}[2.6 + 3.0 + 4(3.5 + 4.2 + 2.0) + 2(2.4 + 4.1)]$$
$$= 9.57 \text{ m}^2$$

The same algorithms may also be used to compute earthwork volumes, given end areas at regularly spaced stations (Figure 19.7). In such a case, the end areas serve as ordinates (y), whereas the station coordinates serve as x values. Using the average end area method, the volume between two stations may be expressed as:

$$V = L\left(\frac{A_1 + A_2}{2}\right)$$

When calculating volumes at a location where a fill transitions to a cut (or vice versa), we might encounter regions where one of the end areas is negligible (Figure 19.8). For better accuracy, such volumes should be calculated using the pyramid formula, rather than the average end area method. The formula is as follows:

$$V = L\left(\frac{A}{3}\right)$$

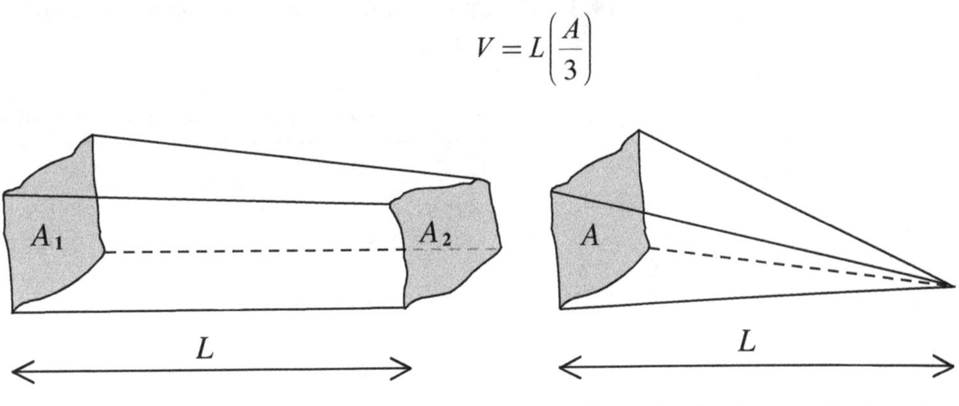

Figure 19.7 Volume using average end area

Figure 19.8 Volume of pyramid

PROBLEMS

19.1 The back-tangent to a horizontal circular curve has a bearing of N 34° 44′ 35″W. The deflection angle between tangents is 67° to the right. If the PC is located at coordinates 4453.51 m N, 643.29 m W, and the tangent length is 850.32 m, what are the coordinates of the PT?

a. 5657.19 N, 888.18 W
b. 5850.95 N, 643.29 W
c. 5871.32 N, 674.06 W
d. 4184.70 N, 147.33 E

19.2 The following data were obtained as foresight and backsight angles at a station in a traverse. What is the horizontal angle at this station?

		Plate Reading
Backsight	Direct reading	00° 00′ 00″
	Reverse reading	180° 00′ 05″
Foresight	Direct reading	51° 40′ 22″
	Reverse reading	231° 40′ 16″

a. 51°40′ 22″
b. 51°40′ 16.5″
c. 51°40′ 11″
d. 51°40′ 15″

19.3 The table below describes a traverse ABCDE. Determine the interior angle at C.

Line	Azimuth Angle	Bearing	Length (m)	Deflection
AB	132	–	237.12	–
BC	–	–	156.18	47° 30′ left
CD	158° 02′ 30″	–	349.65	–
DE	–	–	165.76	105° 16′ 45″ left

a. 73° 32′ 30″
b. 106° 27′ 30″
c. 74° 43′ 15″
d. 84° 30′ 00″

19.4 For the traverse shown below, calculate the coordinates of station D if the coordinates of A are (562.34 m N, 760.27 m W).

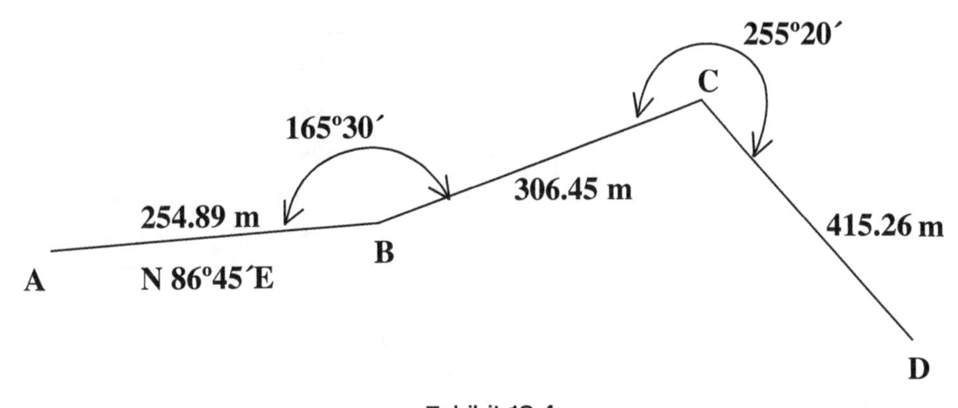

Exhibit 19.4

a. 1331.29 N, 1002.94 W
b. 319.67 N, 1002.94 E
c. 1331.29 N, 8.68 W
d. 319.67 N, 8.68 E

19.5 Two tangents intersecting at an angle of 35°45′ are to be joined by a horizontal circular curve. If the degree of curve (based on a 100-ft arc) is 9° 30′, compute the tangent distance and the radius of the curve.
a. 185.12 ft
b. 194.51 ft
c. 574.00 ft
d. 94.85 ft

19.6 The table below shows differential leveling data using a transit level. The starting station is of known elevation. Find the elevation of station D.

Station	B.S (m)	F.S. (m)	Elevation (m)	Notes
A	3.95	–	500.00	Benchmark
B	2.47	6.34		
C	3.81	5.51		
D	–	6.78		

a. 508.40 m
b. 489.77 m
c. 510.23 m
d. 491.60 m

19.7 The table below shows length and azimuthal angles for lines in a closed traverse, ABCD. What is the correction to the departure of CD, using the transit rule?

Line	Length (ft)	Azimuth Angle
AB	850.00	80° 30′
BC	1250.00	136° 15'
CD	1000.00	220° 30'
DA	1850.00	325° 20'

a. + 0.192 ft
b. – 0.192 ft
c. – 0.343 ft
d. + 0.343 ft

19.8 Assuming the earth to be a perfect sphere of radius 6370 km, what is the curved distance (measured on the surface) between two points 1° of longitude apart on the 32nd parallel (32° N or 32° S latitude)?
a. 94,284 m
b. 42,371 m
c. 84,732 m
d. 47,142 m

SOLUTIONS

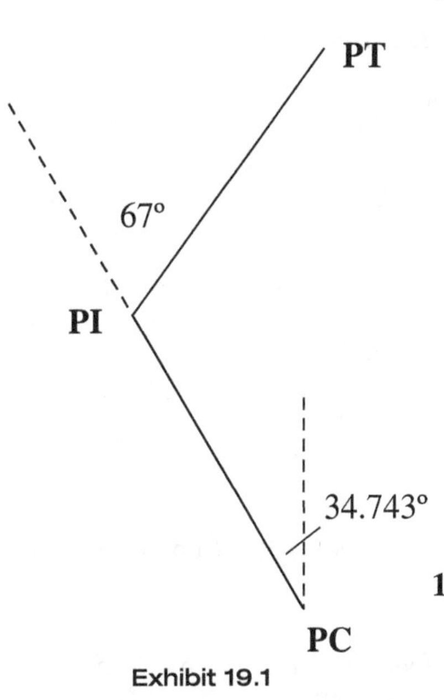

Exhibit 19.1

19.1 **c.** Bearing angle for forward tangent = 67° – 34° 44′ 35″ = N32° 15′ 25" E (32.257 deg)

Coordinates of PI:

Northing = 4453.51 + (850.32)cos(34.743°) = + 5152.23

Easting = – 643.29 – (850.32)sin(34.743°) = – 1127.88

Coordinates of PI: 5152.23 N, 1127.88 W

Coordinates of PT:

Northing = 5152.23 + (850.32)cos(32.357°) = + 5871.32

Easting = – 1127.88 + (850.32)sin(32.357°) = – 674.06

Coordinates of PT: 5871.032 N, 674.06 W

19.2 **b.** Backsight readings differ by 180° 00′ 05″

Error 00° 00′ 05″

Adjusted backsight readings:

Direct: 00° 00′ 02.5″

Reverse: 180° 00′ 02.5″

Foresight readings differ by 179° 59′ 54″

Error 00° 00′ 06″

Adjusted foresight readings:

Direct: 51° 40′ 19″

Reverse: 231° 40′ 19″

Corrected Angle 51° 40′ 19 – 00° 00′ 02.5″ = 51° 40′ 16.5″

19.3 **b.** The azimuthal angles for all lines may be calculated starting with AB:

Az(AB) = 132° (given)

Az(BC) = 132° – 47° 30′ = 84° 30′ (Bearing BC = N 84° 30′ 00″ E)

Az(CD) = 158° 02′ 30″ (given) (Bearing BC = S 21° 57′ 30″ E)

Az(DE) = 158° 02′ 30″ – 105° 16′ 45″ = 52° 45′ 45″

Interior angle at C = Azimuth(BC) – Azimuth(CD) + 180° = 84° 30′ – 158° 02′ 30″ +180° = 106° 27′ 30″

19.4 **d.**

Line	Length (m)	Azimuth (degrees)	Lat =Lcos(Az) (m)	Dep =Lsin(Az) (m)
AB	254.89	86.75	+ 14.45	+ 254.48
BC	306.45	72.25	+ 93.43	+ 291.86
CD	415.26	147.58	– 350.55	+ 222.61

Starting with coordinates of A (+ 562.34, – 760.27), we obtain the coordinates of D as follows:

$$562.34 + 14.45 + 93.43 - 350.55 = +319.67$$
$$-760.27 + 254.48 + 291.86 + 222.61 = +8.68$$

Coordinates of D are (319.67 N, 8.68 E).

19.5 b. $I = 35° \, 45' = 35.75°$

$D = 9° \, 30' = 9.5°$

$$R = \frac{5729.578}{9.5} = 603.11 \text{ ft}$$

$$T = R \tan \frac{I}{2} = 603.11 \times \tan 17.875° = 194.51 \text{ ft}$$

19.6 d. Sum of all backsight values = 10.23

Sum of all foresight values = 18.63

Elevation of D = Elevation of A + Sum of backsight – Sum of foresight = 491.60 m

19.7 b. Sum of departures = $\Sigma(L \sin \theta)$ = 850sin80.5° + 1250sin136.25° + 1000sin220.5° + 1850sin325.33° = 838.343 + 864.391 – 649.448 – 1052.282 = + 1.004

$$\text{Departure correction for CD} = \frac{649.448}{838.343 + 864.391 + 649.448 + 1052.282} \times -1.004 = -0.192 \text{ ft}$$

19.8 a. If the equatorial radius is R, the radius at a latitude θ is $R\cos\theta$.

The circumference of this small circle is $2\pi R\cos\theta$, which is divided into 360 degrees of longitude.

Thus, at this latitude, each degree of longitude is equivalent to a distance of $2\pi R\cos\theta \div 360$.

The answer is 94.284 km.